Unification of the Fundamental Particle Interactions

ETTORE MAJORANA INTERNATIONAL SCIENCE SERIES

Series Editor:
Antonino Zichichi
European Physical Society
Geneva, Switzerland

(PHYSICAL SCIENCES)

Volume 1 **INTERACTING BOSONS IN NUCLEAR PHYSICS**
Edited by F. Iachello

Volume 2 **HADRONIC MATTER AT EXTREME ENERGY DENSITY**
Edited by Nicola Cabibbo and Luigi Sertorio

Volume 3 **COMPUTER TECHNIQUES IN RADIATION TRANSPORT AND DOSIMETRY**
Edited by Walter R. Nelson and T. M. Jenkins

Volume 4 **EXOTIC ATOMS '79: Fundamental Interactions and Structure of Matter**
Edited by Kenneth Crowe, Jean Duclos, Giovanni Fiorentini, and Gabriele Torelli

Volume 5 **PROBING HADRONS WITH LEPTONS**
Edited by Giuliano Preparata and Jean-Jacques Aubert

Volume 6 **ENERGY FOR THE YEAR 2000**
Edited by Richard Wilson

Volume 7 **UNIFICATION OF THE FUNDAMENTAL PARTICLE INTERACTIONS**
Edited by Sergio Ferrara, John Ellis, and Peter van Nieuwenhuizen

Unification of the Fundamental Particle Interactions

Edited by

Sergio Ferrara
CERN
Geneva, Switzerland
and
INFN
Frascati, Italy

John Ellis
CERN
Geneva, Switzerland

and

Peter van Nieuwenhuizen
State University of New York at Stony Brook
Stony Brook, New York

Plenum Press · New York and London

Library of Congress Cataloging in Publication Data

Europhysics Study Conference on Unification of the Fundamental Particle Interactions, Erice, Italy, 1980.
Unification of the fundamental particle interactions.

(Ettore Majorana international science series: Physical sciences; v. 7)
Includes index.
1. Nuclear reactions—Congresses. 2. Unified field theories—Congresses. 3. Supergravity—Congresses. 4. Scherk, Joel. I. Ferrara, S. II. Ellis, John, 1946- III. Van Nieuwenhuizen, P. IV. Title. V. Series.
QC793.9.E92 1980 539.7'54 80-24447
ISBN 0-306-40575-X

Proceedings of the Europhysics Study Conference on Unification of the Fundamental Particle Interactions, held in Erice, Sicily, Italy, March 17–24, 1980.

Printed in the United States of America

This volume is dedicated to Joel Scherk

PREFACE

This volume constitutes the Proceedings of a Europhysics Study Conference held in Erice, Sicily from March 17 to 24, 1980. The objective of the meeting was to bring together practitioners of two different approaches to the unification of the fundamental particle interactions: supersymmetry and supergravity on the one hand, and grand unified gauge theories on the other hand. The hope was that exposure to each others' ideas and problems would at least aid mutual comprehension, and might start people thinking how to develop a synthesis of the two approaches which could avoid their individual shortcomings.

It is not clear to us how successful the conference was in achieving these objectives. On the one hand many important advances in supersymmetric theories were reported which were primarily of a technical nature, while some interesting attempts to probe the phenomenological consequences of supersymmetry and supergravity were also presented. On the other hand there was considerable interest in phenomenological aspects of grand unified theories such as proton decay, neutrino masses and oscillations, and links with cosmology. There was also some work on model-building but relatively few purely technical advances. A few speakers tried to build bridges between the formalism of supersymmetry or supergravity and the phenomenologically successful gauge theories of elementary particle interactions. Perhaps some of these attempts at cross-fertilization may bear some interesting fruit, and we hope and believe that such efforts will prove more numerous and successful in the future. If participating in the Conference or reading these Proceedings gives anyone an idea for furthering the unification of the fundamental interactions, then we would feel that our objective was achieved.

Our future efforts will be greatly weakened by the tragic absence of Joel Scherk, one of the most valued contributors to our field and a physicist of great creativity and breadth. He died shortly after participating in this Conference and we have dedicated these Proceedings to his memory. He left behind a manuscript for part of his talk which we have supplemented by the remainder of the transparencies he showed. We would like to thank Bruno Zumino for editing Joel's talk and providing a short introduction to his work.

We are grateful to many people and organizations for all the help and support necessary for holding the Conference and producing these Proceedings. In particular we thank the European Physical Society, the Italian Ministry of Public Education, the Italian Ministry of Scientific and Technological Research, the Italian National Research Council and the Sicilian Regional Government for their sponsorship, and the staff of the Ettore Majorana Centre, especially S.A. Gabriele, for their support and organizational assistance.

We also thank Tatiana Fabergé, Kathie Hardy and Nanie Perin for their invaluable help in preparing the final form of the typescripts. Finally, we are immensely grateful to Sheila Navach for her major aid throughout the preparation of the Conference and its Proceedings. She organized us both as organizers and editors, gallantly thwarting all our lapses into hapless chaos while never losing her bluff good humour.

Sergio Ferrara
John Ellis
Peter van Nieuwenhuizen

CONTENTS

SUPERGRAVITIES: SUCCESSES AND PROBLEMS*

S. Deser

Department of Physics
Brandeis University
Waltham, Massachusetts 02254

We review some of the properties and difficulties of supergravity models and their higher spin extensions - hypergravities. To emphasize the resemblance of supergravity to other non-abelian theories, we first derive it from its free field constituents. Next, the renormalizability problem is discussed, and contrasted with that of ordinary gravity. Some speculations about possible renormalizability are presented. Finally, as one possible generalization, we consider hypergravities, in which the graviton is replaced by spin $\geq 5/2$ particles as the highest component of the multiplet.

INTRODUCTION

The two main topics of this Conference, Grand Unified Theories (GUTs) and graded, gravitationally unified, theories (guts) represent very different approaches to the possible synthesis of the fundamental gauge interactions. I will be dealing (in old-fashioned component language) with aspects of supergravities which may be useful to practitioners of GUTs in assessing the formal structure and possible generalizations of guts as alternate gauge models.

Consider first some of the differences between GUTs and guts. The former aim at unification of the physically observed interactions excluding gravitation, through conventional internal symmetries; yet the mass scale necessary to achieve this turns out to be amazingly close (to within a mere factor of 10^{-4} or less) to the Planck mass. This is a very surprising numerical coincidence when we realize that previously this quantum gravity scale (10^{-32} cm) seemed far too remote from the 10^{-15} cm of hadronic physics to warrant serious consideration. On the other hand, supergravity attempts to unify quantum gravity with lower spin particles already at the kinematical level, through symmetries involving different spins. The usual no-go theorems forbidding fraternization across spins are avoided because the (super) symmetries involved are graded (fermionic) ones, and as a bonus internal symmetries are incorporated as well. Indeed, there is a very close connection between the maximal spin of a supermultiplet and the maximal number of its constituents. Of course, there may be additional internal symmetries as a result of the self-interactions and this is an important possibility for realistic applications of supergravity, since it is well-known that the largest system there (with spin ≤ 2) has, on the fact of it, the multiplicities associated with O(8). So, unlike GUTs, supergravity does bring space-time into particle systematics but may have trouble being realistic enough vis a vis the other interactions.

Next, consider renormalizability. GUTs have no problem here, precisely because they only involve vector gauge fields with a dimensionless coupling constant. In contrast, supergravity inherits the power-counting renormalizability difficulties of general relativity along with the dimensional Newton constant. It is therefore miraculous that these models of gravity-matter interaction are nevertheless finite at both one- and two-loop levels precisely because of their additional local supersymmetry. It is only at three loops that possible invariant counterterms begin to arise. It should be recalled, however, that there are as yet no explicit calculations beyond one loop and that we still do not have a regularization scheme guaranteed to preserve supersymmetry to all orders. Nevertheless, supergravity _is_ better than gravity, both in its finiteness for the first two loops, and in having far fewer possible counterterms at all succeeding orders. I will present later some very speculative remarks about how the latter fact may be significant. So, GUTs are renormalizable because gravity-free, while guts face (albeit in a milder form) the gravitationally inherent difficulties of dimensional coupling.

Finally, as regards generalizations, it is clear that GUTs can choose from various internal invariance groups, while guts are very tight models: only by involving higher spins ($\geq 5/2$) could larger elementary multiplets be obtained. We call such models hypergravities.

In the following, we will be dealing with three topics closely related to the above considerations. First, we will give a physical approach to supergravity itself. Here the aim will be to show that simple O(1) supergravity, at least, is a perfectly normal gauge theory, which is unique and can be completely derived in a couple of steps from its free particle content, just like the Einstein or Yang-Mills fields. We shall start with a free massless helicity (± 2, $\pm 3/2$) system in flat space. Masslessness of spin ≥ 1 particles is well-known to pose strong conservation restrictions on their sources. As in conventional gauge theories, the system (if it is to be interacting at all) must then have as its sources precisely the Noether currents corresponding to the global symmetries of the free fields. A bootstrap on these currents leads (in two easy steps) to full supergravity, without imposing any a priori local gauge requirements, and with consistency built in. Despite possible ambiguities in the procedure, it is in fact unique.

The second topic will deal with the renormalization problems of supergravity and their contrast to those of gravity. Absence of one and two loop invariants and the reason the (in)famous three-loop invariant exists will be sketched, as will some speculations on the general n-loop problem.

Finally, we will consider the coupling of higher spin fields, especially s = 5/2, to gravity in order to see some of the problems associated with hypergravities. The mechanisms which insured a consistent spin 3/2-gravity coupling are no longer available, and we exhibit the type of difficulties which must be overcome if these models are to be viable.

II. SUPERGRAVITY FROM FIRST PRINCIPLES

The work described in this section was performed in collaboration with D. Boulware and J.H. Kay; details will be found in Ref. 1.

To fix ideas and notation, we begin with pure gravity, whose starting-point is the free massless spin 2 field. As is well-known[2] qualitative phenomenological reasoning on the one-graviton exchange responsible for the long-range aspects of gravity shows that the free graviton necessarily has s = 2, m = 0. Its free, quadratic, action may be written as

$$I_2^E = \int d^4x \left[\eta_{\mu a} h_{\nu b} \, {}^*R_L^{*\mu\nu ab}(\omega) + \tfrac{1}{2} \eta_{\mu a} \eta_{\nu b} \, {}^*R_Q^{*\mu\nu ab}(\omega) \right] \tag{2.1}$$

where R_L, R_Q are the linear and quadratic parts of the usual Einstein curvature formed from the affinities $\omega_{\mu ab} \equiv -\omega_{\mu ba}$:

$$R^L_{\mu\nu ab} \equiv \partial_\mu \omega_{\nu ab} - \partial_\nu \omega_{\mu ab}$$

$$R^Q_{\mu\nu ab} \equiv -\omega_{\mu ac} \omega_\nu{}^c{}_b + \omega_{\nu ac} \omega_\mu{}^c{}_b \, . \tag{2.2}$$

The double-dual is defined by

$${}^*R^{*\mu\nu ab} \equiv \tfrac{1}{4} \varepsilon^{\mu\nu\alpha\beta} \varepsilon^{abcd} R_{\alpha\beta cd} \, . \tag{2.3}$$

Here $\eta_{\mu a}$ is the Minkowski metric diag(+++-) and both Greek and Latin indices range from 0 to 3. The potential $h_{\mu a}$ is non-symmetric, and in this first-order formulation (h,ω) are to be varied independently. The free-field possesses a local abelian invariance to ensure that only helicity ±2 is present. This invariance,

$$\delta h_{\mu a} = \partial_\mu \xi_a(x) \quad , \; \delta\omega = 0 \tag{2.4}$$

implies the Bianchi identity

$$\partial_\nu \left({}^*R_L^{*\mu\nu ab} \eta_{\mu a} \right) \equiv 0 \tag{2.5}$$

on the fied equation. There is also the global abelian invariance (corresponding to constant translations) under

$$\begin{aligned} \delta h_{\mu a} &= -\omega_{\mu a b}\rho^b \\ \delta\omega_{\rho c d} &= (\eta_{\rho b}\eta_{\mu c}\eta_{\nu d} - \tfrac{1}{2}\eta_{\rho d}\eta_{\mu c}\eta_{\nu b} + \tfrac{1}{2}\eta_{\rho c}\eta_{\mu d}\eta_{\nu b})\, {}^*R_L^{*\mu\nu a b} \rho_a \, , \end{aligned} \tag{2.6}$$

as well as one corresponding to constant Lorentz rotations which we do not consider further here as it is automatically taken care of and gives nothing new. Clearly (2.5) requires that any sources $T^{\nu b}$ of the linear equation be conserved (on the ν index). This is possible, as is also physically clear, only if the $T^{\nu b}$ includes contributions from the gravitational field itself. For on the one hand, a conserved $T^{\nu b}$ is provided precisely by the Noether current associated with the global invariance (2.6), and on the other we know that this current is essentially unique. Indeed, the stress tensor for any quantum system is unique to the desired order in momentum.[3] [This is just the equivalence principle.] Thus, coupling to matter requires a self-coupling to the Noether current whose generic form is $T_2^{\nu b}(h,\omega) \sim (h\,\partial\omega + \omega\omega)$. The dimensions of $T^{\nu b}$ require that the coupling constant κ in ${}^*R^*_L = \kappa\, T_2$ be dimensional. Furthermore, the original linear action (2.1) must acquire a cubic term to yield this nonlinear field equation. Its form is

$$I_3^E = \kappa \int d^4x \left[\tfrac{1}{2} h_{\mu a} h_{\nu b}\, {}^*R_L^{*\mu\nu ab} + h_{\mu a}\eta_{\nu b}\, \varepsilon^{\mu\nu\lambda\rho}\varepsilon^{abcd}\omega_{\lambda c f}\omega_{\rho d}{}^f \right] . \tag{2.7}$$

But we must now also include in the field equations the contribution T_3 from (2.7) to the Noether current; its form is $T_3 \sim \kappa\, h\omega\omega$. The field equations now being cubic, they require a quartic addition to the action,

$$I_4^E = -\kappa^2/4 \int d^4x\, h_{\mu a} h_{\nu b}\, \varepsilon^{\mu\nu\lambda\rho}\varepsilon^{abcd}\omega_{\lambda c f}\omega_{\rho d}{}^f . \tag{2.8}$$

The process stops here, since I_4^E has no explicit derivatives to generate a further Noether contribution and $T_4^{\nu b} \equiv 0$. The final (and necessarily consistent) action then reads

$$I^E \equiv I_2^E + I_3^E + I_4^E = \frac{\kappa^{-2}}{2}\int d^4x\,(\eta_{\mu a} + \kappa h_{\mu a})(\eta_{\nu b} + \kappa h_{\nu b})\,{}^{**}R_{L+Q}^{\mu\nu ab}(\omega) \tag{2.9}$$

where we have added a total divergence $\int \eta\eta\, {}^{*}R_L$ and R_{L+Q} is the full curvature, namely the sum $R_L + R_Q$ of (2.2). This is of course the usual Einstein action in terms of the vierbeins $e_{\mu a} \equiv (\eta_{\mu a} + \kappa h_{\mu a})$. It is automatically invariant under the non-abelian local generalization of (2.4) plus (2.6) in which the parameter ρ^a is identified with $\xi^a(x)$, namely the usual coordinate invariance

$$\begin{aligned} \delta e_{\mu a} &= D_\mu \xi_a(x) \\ \delta \omega_{\rho cd} &= |e|^{-1}(e_{\rho b} e_{\mu c} e_{\nu d} - \tfrac{1}{2} e_{\mu c} e_{\rho d} e_{\nu b} + \tfrac{1}{2} e_{\mu d} e_{\rho c} e_{\nu b})\,{}^{**}R^{\mu\nu ab}\,\xi_a(x) \end{aligned} \tag{2.10}$$

where the covariant derivative is $D_\mu \xi_a = (\partial_\mu \xi_a - \omega_{\mu a}{}^c \xi_c)$ and the tangent space indices (a,b,...) are moved by the Minkowski metric η_{ab}.

One can show, using the Ward identities, that this action is unique (up to higher derivative nonminimal terms such as R^2) despite possible ambiguities arising from different field definitions or addition of identically conserved "superpotentials" to the $T^{\nu b}$. The theory is also automatically invariant under local Lorentz transformations, but the latter would not have given rise to an independent self-coupling constant κ'.

We now come to supergravity, where we augment the free action (2.1) by a free spin 3/2, m=0 partner described by a Majorana vector-spinor gauge field $\psi_\mu(x)$. We do this because we know that free massless fields of adjoining spins (s,s-1/2) enjoy a global supersymmetry (the other possible choice, s=5/2, will be dealt with later). The Rarita-Schwinger action describing a free helicity $\pm 3/2$ particle is given by

$$I_2^{RS} = -i/2 \int d^4x\, \bar{\psi}_\mu \gamma_5 \gamma^a \eta_{\nu a}\, {}^{*}f_L^{\mu\nu}(\psi) \tag{2.11}$$

where

$$f^L_{\alpha\beta}(\psi) \equiv \partial_\alpha \psi_\beta - \partial_\beta \psi_\alpha \quad , \; {}^{*}f^{\mu\nu} \equiv \tfrac{1}{2} f_{\alpha\beta}\, \varepsilon^{\mu\nu\alpha\beta} . \tag{2.12}$$

Note the resemblance of (2.12) to (2.1); in both cases the action is a product of a potential times a (dualized) gauge invariant field strength. The local abelian invariance which ensures pure helicity 5/2 content is just the analog of the Maxwell one,

$$\frac{1}{2}\delta\psi_\mu = \partial_\mu \alpha(x) \quad , \quad \delta f^L_{\mu\nu} = 0 \tag{2.13}$$

and the corresponding Bianchi identity tells us that the free Rarita-Schwinger operator,

$$R^\mu_L \equiv \eta_{\nu a}\gamma_5\gamma^a\, {}^*f^{\mu\nu}_L \tag{2.14}$$

is conserved :

$$\partial_\mu R^\mu_L \equiv 0 \tag{2.15}$$

Note that the gauge function $\alpha(x)$ is a spinor, and of course the local spin 2 and spin 3/2 invariances (2.4, 2.13) are independent.

The free system has two global invariances, namely the generalization of the original translation transformation (2.6) to include

$$\delta\psi_\mu = f^L_{\mu\nu}\,\rho^\nu \tag{2.16}$$

and a new supersymmetry which mixes the fields:

$$\delta h_{\mu a} = i\bar{\beta}\gamma_a\psi_\mu \quad , \quad \delta\psi_\mu = -\omega_{\mu ab}\sigma^{ab}\beta \quad ; \sigma^{ab} \equiv \tfrac{1}{4}[\gamma^a, \gamma^b] . \tag{2.17}$$

Here β is a <u>constant</u> Majorana spinor parameter (constant spinors are well-defined in flat space), and there is also a $\delta\omega \sim \bar{\beta}(\partial\psi)$.

There are now two a priori independent Noether currents, namely the conserved $T^{\nu\sigma}$ associated with (2.6, 2.16) and a spinorial current J^μ generated by (2.17), which has the form $J_2 \sim \omega\psi + h\partial\psi$:

$$J^\mu_2 = i/2\, \epsilon^{\lambda\nu\mu\rho}\bar{\psi}_\lambda\gamma_5\eta_{\nu a}\{\gamma^a, \sigma^{cd}\}\omega_{\rho cd} + i\, {}^*\bar{f}^{\mu\nu}_L h_{\nu a}\gamma_5\gamma^a \tag{2.18}$$

when all the γ matrices are made explicit. Both bootstraps must now be performed for consistency, and a priori we expect two separate self-coupling constants as a result (or even three, if we also count Lorentz rotations). That is, the cubic addition to the action which will yield the field equations ${}^*R^{*\nu a}_L = \kappa\, T^{\nu a}_2$ and $R^\mu_L = \kappa' J^\mu_2$ would have the form $I_3 \sim \int \kappa\, h T_2 + \int \kappa' J_2\psi \sim \kappa A + \kappa' B$. However there is a common term between the two parts of I_3 which forces the identification $\kappa = \kappa'$, a feature which persists in I_4 . We will not trace the now familiar steps leading to I_4, but simply give the final result (again both T_4 and J_4 vanish identically). The total action, which is the sum of $(I_2^E+I_2^{RS})$ with the nonlinear (I_3+I_4) terms is of course the supergravity action in first order form,

$$I^{SG} = \frac{\kappa^{-2}}{2}\int d^4x\, e_{\mu a} e_{\nu b} {}^{**}R^{\mu\nu ab}(\omega) - i/2 \int \bar{\psi}_\mu \gamma_5 \gamma^a e_{\nu a} {}^*f^{\mu\nu} d^4x \tag{2.19}$$

where ${}^*f^{\mu\nu} \equiv \frac{1}{2}\varepsilon^{\mu\nu\alpha\beta} f_{\alpha\beta}$ as before. However the full field strength

$$f_{\alpha\beta} \equiv D_\alpha(\omega)\psi_\beta - D_\beta(\omega)\psi_\alpha \tag{2.20}$$

is now a covariant spinorial curl, since $D_\alpha(\omega) \equiv (\partial_\alpha - \frac{1}{2}\omega_{\alpha ab}\sigma^{ab})$ is the covariant derivative acting on a spinor.

It can be shown (using the Ward identities) that I^{SG} is unique; there can be no non-minimal contact terms $\sim \kappa^2\psi^4$ for example. Consistency of the two gauge field equations is also guaranteed: for the divergence of the Einstein equations $G^{\mu a} = T^{\mu a}(\psi)$, by the usual (covariant) conservation of $T^{\mu a}(\psi)$ which is a consequence of the Rarita-Schwinger equation; for the divergence of the latter by the fact that it is proportional to the Einstein equations,

$$D_\mu(\gamma_5\gamma_\nu {}^*f^{\mu\nu}) \sim [G^{\mu a} - T^{\mu a}(\psi)]\gamma_a\psi_\mu \tag{2.21}$$

and so vanishes on shell.

The supergravity action (2.19) is manifestly invariant under the local coordinate group given by (2.10) together with the generalization $\delta\psi_\mu = e^{\nu a}\xi_a(x) f_{\mu\nu}$ of (2.16), and under local lorentz rotations. More significantly, the fermionic gauge invariances also become local, as they had better since the constant spinor β is no longer well-defined in curved space. They are just the generalization of (2.13) and (2.17) with β identified with α (x), namely

$$\delta\psi_\mu = 2D_\mu\alpha(x) \equiv (2\partial_\mu\alpha - \omega_{\mu ab}\sigma^{ab}\alpha), \quad \delta e_{\mu a} = i\bar{\alpha}\gamma_a\psi_\mu, \quad \delta\omega \sim \bar{\alpha}D\psi. \tag{2.22}$$

Thus the graviton and the spinor form a gauge doublet; a profound consequence is that space-time is no longer an invariant concept. Note also that the O(1) simple supergravity action has the special property that it is polynomial in the fundamental variables (e, ω, ψ); for example, the inverse vierbein $e^{\mu a}$ does not enter, which it does when spin 1 is involved. This would make it difficult to derive the extended models in a finite number of steps in the same way, although it may still be possible to do so in the higher dimensional formulations mentioned later.

III. ULTRAVIOLET BEHAVIOR OF SUPERGRAVITY

Having analyzed the tree structure of supergravity, let us now turn to its quantum loop behavior. It is clear that the presence of a coupling constant κ with dimension of length, corresponding to momentum dependence in the sources $T^{\mu\alpha}$ and J^{ν}, leads to naive power-counting difficulties. Consider for example pure gravity, expanded about flat space in powers of $\kappa h_{\mu\alpha}$. The one-loop graviton bubble diverges quartically because there are two propagators $\sim(p^{-2})^2$ but two vertices $\sim(p^{+2})^2$. At two loops, there are three more propagators and two more vertices, together with a second (d^4p') integration, leading to a behavior $\sim\Lambda^4(\kappa^2\Lambda^2)$, etc. The same power counting applies to gravity-matter coupling, whether to bosons or to spinors (whose propagator and vertex go as p^{-1} and p^{+1} respectively).

If we use dimensional regularization, we can identify the possible counterterms ΔI_n in pure gravity, for example, with the coordinate invariants of dimension (4+2n) since $\kappa^{2n}\Delta I_n \sim L^{-4}$. All geometric invariants can be constructed from the curvature and its covariant derivatives, each curvature having dimension 2. Thus, the possible counterterms for the first three loops have the generic form (to be added to $I^E \sim \kappa^{-2}\int d^4x\, R$):

$$\Delta_{1-3} I^E \sim \int R^2_{\mu\nu\alpha\beta} + \kappa^2\int R^3_{\mu\nu\alpha\beta} + \kappa^4\int R^4_{\mu\nu\alpha\beta} \tag{3.1}$$

For physical processes, we are really only interested in ΔI with the external gravitons on-shell ($R_{\mu\nu} = 0$). In that case, one may use a well-known geometric result that there are only four independent scalars which can be constructed algebraically from the Weyl tensor $C_{\mu\nu\alpha\beta}$. [The latter is that part of the curvature independent of $R_{\mu\nu}$: $C_{\mu\nu\alpha\beta} \sim R_{\mu\nu\alpha\beta} - (R_{\mu\alpha}g_{\nu\beta} \cdots) + R(g_{\mu\alpha}g_{\nu\beta} + \cdots)$.] These are

$$A \equiv tr\, C^2 \qquad B \equiv tr\, CCC$$
$$A^* \equiv tr\, C\,{}^*C \qquad B^* \equiv tr\, CC\,{}^*C \tag{3.2}$$

where ${}^*C$ is the single dual (on either pair of indices) of the Weyl tensor; they correspond to the two independent scalars F^2, F^*F) for the vector field. The one-loop term is therefore uniquely A and only B contributes at 2 loops, by parity invariance. At 3 loops there are two invariants A^2 and $(A^*)^2$. Up to this order at least there are no additional derivative-dependent terms, because e.g. $\int CDDC \sim \int C^3$ and $\int CDCDC \sim \int C^4$ by using the cyclic and Ricci identities. The one-loop term $\int C^2$ is in fact present (by explicit calculation), but harmless because (like $\int F^* F$) it is a topological invariant which does not contribute to usual scattering processes (both C^2 and CC^* are total divergences). If topology is also of interest, then those terms do renormalize the Euler and Pontriagin numbers. Thus, at one loop, pure gravity is "accidentally" finite on shell whatever the actual coefficients of the invariant may be. Since no 2 loop calculations have been performed - the combinations of elementary 3- and 4-point vertices involved here is astronomical - we do not yet know whether $\int C^3$ is actually present. [In the one-loop calculations[4,5] at least, all allowed invariants do appear with non-zero coefficients explicitly.] So pure gravity is, in principle at least, nonrenormalizable from 2 loops on, where there are counterterms which do not vanish on shell. The general statement would be that to determine n-point scattering one has to specify the coefficients of all new invariants with n powers of the curvature.

When coupling to matter is included, the same gravitational counterterms are possible, but in addition there are also terms involving external matter particles. At one-loop order, gravity + spin 0, 1/2 or 1 matter invariants have been calculated explicitly[4,5] and in each case there is already one-loop nonrenormalizability. The one-loop pattern includes terms such as

$$\Delta_1 I^{E+M} \sim \frac{1}{\epsilon}\int [a R^2_{....} + b\kappa R_{..}T^{..} + c\kappa^2 T^2_{..}]\, d^4x \tag{3.3}$$

where T is the stress-tensor. Even on shell, where $\kappa T^{\mu\nu}$ can be replaced by the Einstein tensor $G^{\mu\nu}$, these terms fail to vanish.

A priori, gravity - spin 3/2 coupling should have the same difficulties. It is here, however, that miracles occur due to the additional gauge symmetry linking gravity and spin 3/2 (separate matter gauge invariances such as that of spin 1 do not help because all they say is that the matter-dependent parts of the counterterms must reflect these invariances). The first miracle is that at one loop (3.3) does vanish on combined mass shell,[6-8] as we should expect: Since the purely gravitational contributions are necessarily total divergences up to terms which vanish on Einstein shell, their supersymmetric completions should also be total divergences up to terms vanishing on the full supergravity mass shell. The terms of $\Delta_1 I^{SG}$ group themselves into squares of the Einstein $(G - T)^2$ or Rarita-Schwinger $(R_\mu)^2$ equations plus the topological invariant $\int C^2$ plus its supersymmetric companion, a total divergence $\sim \kappa^2 \int \bar{f}\partial f$. At 2 loops, the counterterm must be the supersymmetric completion of $\int C^3$. But there is no such supersymmetric quantity,[6-7] and hence not even a candidate for a 2 loop counterterm. This can be understood as follows: under a supersymmetry transformation, the field strengths C and f rotate into each other. So $\delta C^3 \sim C^2 f$ and this must be cancelled by varying a Cff term. However, the latter also transforms into an fff part which cannot be cancelled by anything else (and does not vanish by itself).

The above miracles hold also for the extended models, basically because the chain leading to the other multiplet members must start from the original 0(1) pieces, which we have just seen to be harmless. We may conclude, then, that naive power counting nonrenormalizability due to κ is defeated by supersymmetry even though the system can include all spins from 2 down to zero interacting with gravity. Note the importance of the full supersymmetry: coupling a global matter supermultiplet, say (1/2, 0) or (1, 1/2) or (1, 1/2, 1/2, 0) to gravity alone does <u>not</u> help - and the spin 3/2 component's contributions are essential to cancel dangerous internal graviton radiative corrections. On the other hand, this suggests that supergravity plus (necessarily supersymmetric) sources (which is possible for N = 4) is no better than gravity plus sources, since one can form one-loop supersymmetric invariants from the matter system alone which do not vanish on shell. The simplest example is [6]

$$\Delta_1 I^M \sim \int d^4x \left[T^{\mu\nu} T_{\mu\nu} - i \bar{J}^\mu \not{\partial} J_\mu + \tfrac{3}{2} C^\mu \Box C_\mu \right] \tag{3.4}$$

where $T^{\mu\nu}$, J^μ, C^α are respectively the stress tensor, supercurrent and axial vector of some 0(1) global multiplet; this follows from the known behavior under supersymmetry of these quantities.[9]

The invariant (3.4) provides the hint for the existence of a three-loop invariant for 0(1) supergravity.[6] Consider for simplicity the free (2, 3/2) multiplet. As is well known, the stress-tensor of the free spin 2 field is not gauge-invariant under (2.4), and neither is the supercurrent abelian invariant. Consequently, one has to find acceptable (gauge invariant) generalizations of (T,J,C). These are provided by the Bel-Robinson tensor $B_{\mu\nu\alpha\beta}$ and corresponding generalizations $J^{\mu\nu\beta}$, and $C^{\mu\nu\alpha}$. The conserved tensor B is quadratic in the invariant curvatures rather than in first derivatives of $h_{\mu a}$, and has long been known in gravitation as the analog $\sim(RR + R^*R^*)$ of the Maxwell stress tensor $(FF + {}^*F^*F)$. Because the dimension of B^2 is L^{-8}, it can only appear at 3 loops when multiplied by κ^4. This simple linear picture of the origin of the three-loop invariant has since been verified by its explicit form in the full nonlinear theory,[10] so there is no doubt as to the invariance of

$$\Delta_3 I^{SG} = \int d^4x \left[T^{\mu\nu\alpha\beta} T_{\mu\nu\alpha\beta} - i \bar{J}^{\mu\nu\alpha} \not{\partial} J_{\mu\nu\alpha} + \tfrac{3}{2} C^{\mu\nu\alpha} \Box C_{\mu\nu\alpha} \right] \tag{3.5}$$
$$T^{\mu\nu\alpha\beta} \sim B^{\mu\nu\alpha\beta} + (\bar{f} \gamma \partial f)^{\mu\nu\alpha\beta}$$

for 0(1); whether or not it actually has a non-vanishing coefficient is of course not known. Note that there is only one invariant, that whose gravitational part is $B^2 \sim (C^2)^2 + (CC^*)^2$, whereas pure gravity could have an arbitrary combination of A^2 and $(A^*)^2$. Indeed, it is now known[13] that (as a consequence of supersymmetry) all supergravity invariants must have a form which vanishes when C is self- (or antiself-) dual and the spin 3/2 field has a single helicity, $f_{\mu\nu} = \pm\gamma_5 f_{\mu\nu}$ (although this would only lead to a finite theory if all solutions were also of this type).

Is there any hope that extended models will lead to improvement, for example because the additional internal invariance will forbid the equivalent of (3.5)? At the linearized level it is known explicitly that this is not the case for O(2) where a generalized version of (3.5) has been constructed.[12] In the absence of simple technology for writing arbitrary invariants in these models, there are at present two ways to proceed. The first (which was used for O(2)) is a more or less empirical addition of likely terms to embody internal invariance, which would rapidly become a prohibitive task, especially for the most interesting maximal O(8) case. Instead, there is a much more elegant route involving the dimensional reduction technique, which was in fact the way the full O(8) action was first constructed.[13] There is a direct relation between a simple supersymmetry action in higher dimension and the corresponding O(N) form in four dimensions; one simple drops all dependence on the higher coordinates, while keeping all the new field variables (for example a 5-D vector A_α corresponds to a 4-vector plus a scalar, and a metric $g_{\alpha\beta}$ becomes the sum of a four-tensor $g_{\mu\nu}$, vector $g_{5\mu}$ and scalar g_{55}). The usefulness of this technique for constructing higher invariants has first been tested in the simpler context of matter multiplets,[14] in which a 6D vector multiplet's quadratic invariant reduces to the O(2) invariant of the type discussed earlier, and the method is very simple in principle (except that γ-algebra becomes messier as D increases!). [We are of course not talking about radiative corrections in D dimensions, but only use the higher D as a method of constructing invariants at four dimensions.]

The vector multiplet with action $I_6 = \int d^6x \, [-\frac{1}{4}F_{\alpha\beta}^2 + i\bar{\lambda}\gamma\lambda]$ has not only T, J and C but additional conserved "currents" such as $\epsilon^{\mu\nu\alpha\beta\lambda\sigma} F_{\alpha\beta} F_{\lambda\sigma}$ which contribute to the invariant. The explicit calculation of the equivalent of (3.5) for the key O(8) case is almost complete.[15] Here, one wishes to construct an invariant in 11-dimensions corresponding to the tree action with leading terms

$$I_{11} = \int (d^{11}x) \, [\frac{\kappa^{-2}}{2} R - i/2 \, \bar{\Psi}_\alpha \Gamma^{[\alpha\beta\delta]} D_\beta \Psi_\delta - \frac{1}{3} F_{\alpha\beta\gamma\delta}^2 + \cdots] \qquad (3.6)$$

whose 4-D reduction is O(8). Here $A_{[\alpha\beta\gamma]}$ is the totally skew generalization of a vector field, $F_{\dots} = \partial_{[\cdot} A_{\dots]}$ and Ψ_α is an 11-D spinor. The parts already obtained do reduce, as they must, to 4D pieces such as the Bel-Robinson term, which must be the starting point in any 3-loop invariant. So the higher models apparently do not avoid the 3-loop problem, at least if we trust linearized approximation arguments. [It is conceivable that the "accidental" higher symmetries possessed by the 4-D action could be relevant in ruling out some otherwise dangerous invariants]. We complete this brief summary of the renormalization difficulties of supergravity with a very tentative and speculative suggestion that perhaps some (if not all) of the higher loop invariants might be removed by field redefinition, as were the one-loop terms, which vanished on shell. Let us first recall how field redefinition works at one loop.[4] There, one needed to absorb infinite terms of the form $\Delta_1 I^{\epsilon} \sim 1/\epsilon \int (\alpha R_{\mu\nu}^2 + \beta R^2) d^4x \sqrt{-g}$. Noting that redefining the metric, $g_{\mu\nu} \to g_{\mu\nu} + \delta g_{\mu\nu}$ alters the tree action by $\kappa^{-2} \int G^{\mu\nu} \delta g_{\mu\nu}$, it is clear that $\delta g_{\mu\nu} \sim \kappa^2/\epsilon\,(a R_{\mu\nu} + b g_{\mu\nu} R)$ can cancel these terms. It is also clear that terms depending on the Weyl tensor only, which do not vanish on shell, e.g., the two loop C^3 of pure gravity or the 3-loop $[(C^2)^2 + (C\tilde{C})^2]$ cannot be so removed. Clearly, one would like to get rid of these by varying terms $\sim \int d^4x\, C^2 \sqrt{-g}$ or $\sim \int d^4x\, C\tilde{C}$. But we have seen that they are topological invariants, which means that they do not change if $g_{\mu\nu}$ does. On the other hand, if we continue them to dimension n, they are no longer inert. For example,

$$\delta \int d^n x\, C^2 \sqrt{-g} = \int [(CC)_{\mu\nu} - \tfrac{1}{4} g_{\mu\nu} C^2]\, \delta g^{\mu\nu} = \int B^{\mu\nu} \delta g_{\mu\nu} \qquad (3.7)$$

where the Bach tensor $B^{\mu\nu}$ vanishes identically only at n=4, $B^{\mu\nu} \sim \epsilon$. Then, by setting $\delta g_{\mu\nu} \sim C^2 g_{\mu\nu}$ or $\sim (CC)_{\mu\nu}$, one can remove the $\int (C^2)^2$ part of the 3-loop term (but could not get rid of $\int C^3$ in 2 loop gravity). To handle $\int (CC^*)^2$ is a bit trickier, because in the corresponding invariant $\int d^n x (CC^*)$, one has to dimensionally continue the Levi-Civita symbol $\epsilon^{\mu\nu\alpha\beta}$. According to the prescription recently given by Siegel[16] and discussed also by Townsend, one should let it retain four indices, but make them range over n. In that case, $\epsilon^{\mu\nu\alpha\beta}$ is no longer a proper density and one needs an extra factor of $(\sqrt{-g})^{n-4}$ which again permits a variation, such that $\delta\, 1/\epsilon \int d^n x (C\tilde{C}) \sim \delta\, 1/\epsilon \int (d^n x) C \epsilon C (\sqrt{-g})^{\epsilon} \sim \int (C\tilde{C}) g^{\mu\nu} \delta g_{\mu\nu}$ and clearly $\delta g_{\mu\nu} \sim (CC^*) g_{\mu\nu}$ now removes $\int (CC^*)^2$. The only terms this technique would not remove among the generic set $A^n A^{*2m} B^{p} B^{*2q}$ are those in which n = m = 0. But it may well be that such terms are excluded by supersymmetry, for the same reason that B itself is: there is no invariant of which B is a component. There are of course gaps in this preliminary argument. Some of them are: (1) the method seems critically dependent on a particular (dimensional) regularization method. (2) One must face the role of possible derivative-dependent invariants $C.D^n\, C\, D^m\, C$ ... (3) the parts involving spin 3/2 must also work in the same way as the pure graviton terms we have considered. (4) Once one intends to go to all orders, $\delta g_{\mu\nu}$ is no longer infinitesimal and the contribution to order κ^{4n} of the κ^{2n} part of $\delta g_{\mu\nu}$, etc, must be included as well.

IV. HYPERGRAVITIES

We have seen that, although supergravities provide the only known unification mechanism for the Einstein field, the restrictions on the maximal number of component fields may be too strong for realistic application. Since raising the maximum spin automatically extends the maximum size of a multiplet, we discuss here the properties of such models, in particular that of the simplest extension to spin (5/2,2). The action of the free fields does form a global invariant, but this is no guarantee that a satisfactory interacting theory exists. One way to see why there might be problems is that the supersymmetric Noether current is still a spin 3/2 object j^μ, unlike the gauge field to which it should couple, so the bootstrap we discussed in Sec. II would run into difficulties. It is also known that there are no realizations[17-19] of a "spin 3/2" supersymmetry algebra with vector-spinor parameters (and therefore no useful spin 5/2 Noether current).

The theory of a free spin 5/2 field can be developed in two alternate formulations and I will be discussing here approaches developed in collaboration with C. Aragone.[20] There is by now a growing literature on this subject and on higher spin field systematics in general.[21] One may use either a symmetric tensor-spinor $\psi_{\mu\nu} = \psi_{\nu\mu}$ or a nonsymmetric vierbein $\psi_{\mu\tilde{a}}$, with $\gamma^a \psi_{\mu\tilde{a}} \equiv 0$ to describe the field. The respective abelian invariances are then $\delta\psi_{\mu\nu} = \partial_\mu \xi_{\tilde{\nu}} + \partial_\nu \xi_{\tilde{\mu}}$ or $\delta\psi_{\mu\tilde{a}} = \partial_\mu \xi_{\tilde{a}}$, with $\gamma \cdot \xi \equiv 0$ in both cases. The minimal coupling to gravity leads to the same problems in both formulations. For example in the tensor formulation, the free field equations state that the Christoffel symbol formed from $\psi_{\mu\nu}$ is γ-traceless,

$$\gamma^\alpha \Omega_{\mu\nu\alpha} = 0 \qquad , \Omega_{\mu\nu\alpha} \equiv (\partial_\nu \psi_{\mu\alpha} + \partial_\mu \psi_{\nu\alpha} - \partial_\alpha \psi_{\mu\nu}) \tag{4.1}$$

corresponding to $\gamma_\alpha f^{\mu\alpha}_{\psi} = 0$ for spin 3/2. When gravitation is included, the divergence of the field equation involves the Weyl tensor explicitly, and not just the Einstein tensor as in spin 3/2, so that even on Einstein mass shell there are Weyl-dependent terms remaining. These have a special form, however. If we denote by $S^{\mu\nu}(\psi, e) = 0$ the spin 5/2 field equation in presence of gravity, we find schematically

$$D_\mu S^{\mu\nu} \sim C^+ \psi^R + C^- \psi^L + G\psi \tag{4.2}$$

where $C^\pm$ are the self- and anti-self dual parts of the Weyl tensor and $\psi^{R,L}$ are its $\frac{1}{2}(1 \pm i\gamma_5)$ helicity projections. Thus, if the curvature is half-flat and ψ has only one helicity, these new terms vanish; otherwise the whole Weyl tensor or ψ field must be zero, which is clearly unacceptable. However, these conditions are too strong in Minkowski (but not necessarily in Euclidean) signature, since there are no Weyl-Majorana spinors ($\psi^R \equiv 0 \Rightarrow \psi = 0$ for example). The same pattern holds for coupling of higher half-integer spin fields to gravity. We are not really interested in the above O(1) hypergravity but rather in extended ones. Here further problems arise because O(2) will involve in addition a second graviton as well as a spin 3/2 field. Now the coupling of two "gravitons" (or of a linear spin 2 field with gravity) is itself fraught with difficulties[22] quite apart from the question of how symmetry breaking could leave only one true graviton. These difficulties are also traceable to consistency requirements and it is not clear how they can be overcome here, let alone for "O(9)" models with 1 spin 5/2, 9 "gravitons" etc. Of course, "no-go" theorems are often misleading, as evidenced by the success of spin (2,3/2) coupling. Even spin (3/2,1) coupling, which is inconsistent by itself (it very much resembles (5/2,2) in having consistency terms of the form $F^+\psi^R + F^-\psi^L$), can be restored when gravity is included.[23] This suggests that one should perhaps start with spin 3 (rather than 5/2) as the top particle. Of course massless fields with $s > 2$ are well-known[24] to require sources which vanish at zero momentum, but if such a theory does exist, its interest would lie in its $s \leq 2$ sector anyhow. The higher spins would just be there to permit bigger multiplets. One way to study this question more systematically would be to consider actions in $D > 11$, which are known[25] to involve spins > 2; the consistency problem for a given maximum helicity might be more transparent there. It is of course quite possible that only an infinite tower of spins is at all consistent, as in string models.

Whatever their outcome, the recent investigations of higher spins have at least furnished us with elegant unified formulations for all free half integral or integral massless fields and given us an idea of the problems involved in constructing interacting models.

V. SUMMARY

We have tried to survey three aspects of super-models. First, we saw that supergravity is a unique self-interacting (doubly) local gauge theory. All its properties flow uniquely from the pure helicity (2,3/2) content of its free field ancestors, together with the global invariances of the free doublet. Both consistency and the local non-abelian gauge invariances emerge as automatic consequences of the (necessary) self-coupling procedure. Purely special relativistic quantum theoretical requirements have led to a geometric theory in which spacetime is curved, but not invariantly defined; only the unified multiplet of gravity plus lower spins is. Second, we noted that the new gauge invariances also improved ultraviolet behavior vis a vis quantum gravity: gravity plus matter, when forming a gauge multiplet, is a finite theory through 2 loops at least, and even the possible counterterms which exist from 3 loops onward are of a very special structure. Finally, we mentioned hypergravities, which involve s $\geqslant$ 5/2, as one way to generalize supergravities. With maximal s = 5/2 they seem to suffer both from (albeit rather special) consistency constraints as well as from the problems arising when more than one "graviton" is present. However, it is not excluded that more ambitious models, involving higher spins, can be constructed.

We may conclude with one certain effect of GUTs: with it mass scales approaching the Planck value but no natural place for gravitation, it is bound to spur new guts attempts to reach that remaining unification.

REFERENCES

* Work supported in part by NSF Grant #78-09644-A01.

1. D. Boulware, S. Deser, and J.H. Kay, Physica 96A, 141 (1979).
2. D. Boulware and S. Deser, Ann. Phys. 89, 193 (1975).
3. C. Orzalezi, J. Sucher, and C.H. Woo, Phys. Rev. Lett. 21, 1550 (1968); L.S. Brown (unpublished).
4. G. 't Hooft and M. Veltman, Ann. Inst. Henri Poincare 20, 69 (1974).
5. S. Deser and P. van Nieuwenhuizen, Phys. Rev. Lett. 32, 245 (1974); Phys. Rev. D10, 401, 411 (1974); S. Deser, P. van Nieuwenhuizen and H.-S. Tsao, ibid. 337 (1974).
6. S. Deser, J.H. Kay, and K.S. Stelle, Phys. Rev. Lett. 38, 527 (1977).
7. M.T. Grisaru, Phys. Lett. 66B, 75 (1977); E. Tomboulis, ibid 67B, 417 (1977).
8. M.T. Grisaru, P. van Nieuwenhuizen, and J.A.M. Vermaseren, Phys. Rev. Lett. 37, 1662 (1976); P. van Nieuwenhuizen and J.A.M. Vermaseren, Phys. Lett. 65B, 263 (1976); Phys. Rev. D16, 248 (1977).
9. S. Ferrara and B. Zumino, Nucl. Phys. B87, 207 (1975).
10. M. Kaku, P.K. Townsend, and P. van Nieuwenhuizen, Phys. Rev. Lett. 39, 1109 (1977); S. Ferrara and B. Zumino, Nucl. Phys. B134, 301 (1978).
11. R.E. Kallosh, Zh. ETF Pisma 29, 192, 493 (1979); Nucl. Phys. B165, 119 (1980); S.M. Christensen, S. Deser, M.J. Duff and M.T. Grisaru, Phys. Lett. 84B, 411 (1979).
12. S. Deser and J.H. Kay, Phys. Lett. 76B, 400 (1978).
13. E. Cremmer, B. Julia and J. Scherk, Phys. Lett. 76B, 409 (1978); J. Scherk and J.H. Schwarz, ibid 57B 463 (1975); E. Cremmer and J. Scherk, Nucl. Phys. B103, 399 (1976). L. Brink, J.H. Schwarz and J. Scherk, ibid B121, 77 (1977).
14. S. Deser and U. Lindstrom, Phys. Lett. 90B, 68 (1980).
15. A. Karlhede and U. Lindstrom (private communication).
16. W. Siegel, Princeton Inst. for Adv. Study preprints (1980).
17. R. Haag, J.T. Lopuszanski and M. Sohnius, Nucl. Phys. B88, 513 (1975).

18. F.A. Berends, J.W. Van Holten, P. van Nieuwenhuizen and B. de Wit, Leiden University preprint to be published in J. Phys. A (1980).
19. C. Aragone and S. Deser, Proc. of the Supergravity Workshop (North Holland, 1979) eds P. van Nieuwenhuizen and D.Z. Freedman, p. 257.
20. C. Aragone and S. Deser, Phys. Lett. 86B, 161 (1979); Phys. Rev. D21, 352 (1980); Brandeis University preprint to be published in Nuclear Phys. B (1980).
21. J. Schwinger, Particles, Sources and Fields (Addison-Wesley, 1970); J. Fang and C. Fronsdal, Phys. Rev. D18, 3630 (1978); T. Curtright, Phys. Lett. 85B, 219 (1979); F.A. Berends, J.W. Van Holten, P. van Nieuwenhuizen and B. de Wit, Nucl. Phys. B154, 261 (1979), Phys. Lett. 83B, 188 (1979); B. deWit and D.Z. Freedman, Phys. Rev. D21, 358 (1980).
22. C. Aragone and S. Deser, Brandeis University preprint to be published in Nuovo Cimento B (1980); ibid 3A, 709 (1971).
23. D.Z. Freedman and A. Das, Nucl. Phys. B120, 221 (1977); S. Deser and B. Zumino, Phys. Rev. Lett. 38, 1433 (1977).
24. S. Weinberg, Phys. Lett. 9, 245 (1974); Phys. Rev. B135, 1049 (1964), 138, 988 (1965).
25. W. Nahm, Nucl. Phys. B135, 149 (1978).

NEUTRINO MASSES, LEPTON NUMBER VIOLATION

AND UNIFICATION

R. Barbieri

CERN

1211 Geneva 23, Switzerland

1. INTRODUCTION

Among the energy scales of elementary particle physics, the following quantities are compatible with being zero: the graviton, the gluon and photon masses, the electron's decay width Γ_e, the proton's decay width Γ_p, and, finally, the neutrino masses m_ν. Whereas we have a dynamical understanding of the vanishing of the first four in terms of suitable gauge principles, $\Gamma_e = 0$ being guaranteed by electric charge conservation, the same is apparently not true for the vanishing (or the smallness) of Γ_p and m_ν. In fact, the supposedly conserved relevant quantities, baryon and lepton numbers, could not possibly correspond to unbroken local symmetries*. For a believer in the currently fashionable Gauge Symmetry Dogma, i.e., that exact symmetries in Nature are local ones, this is enough to want Γ_p and m_ν to be different from zero. Now, of course, the very important recent work on baryon creation in the Universe gives also, for the first time, a good "experimental" reason for the proton to decay. It seems therefore worth while to give a discussion of the possible generation of neutrino masses. This is what I shall do in this talk, trying to keep the various arguments on general grounds, rather than going into model dependent details.

---

*) It has been known for a long time[1] that a massless boson coupled to a baryon or a lepton number would introduce discrepancies in the Eötvös experiment unless it had an ultraweak coupling.

It is well known that the standard SU(2) × U(1) theory with the "minimal" Higgs structure, a scalar SU(2)-doublet, describes a strictly massless neutrino. The omission of the right-handed neutrino prevents a Dirac mass term $\bar{\nu}_R \nu_L$, whereas the Majorana mass term $\nu_L \nu_L$ is forbidden by exact lepton number (L) conservation.

However, L is not gauged and, as such, not sacred. Leaving aside the uninteresting possibility of breaking L and getting a zeroth order Majorana mass from the vacuum expected value of a scalar SU(2)-triplet -- which would also explicitly break the $I = \frac{1}{2}$ rule for the strength of neutral currents -- one may think quite generally of some new physics emerging at a mass M greater or much greater than M_W, and containing an inherent L violation. Approximate SU(2) × U(1) invariance at this large mass scale M, together with dimensional counting, leads to the following dominant effective L-violating interaction

$$\frac{f}{M} \psi \phi^2 \psi \ , \tag{1}$$

where f is a dimensionless coupling, $\psi = (\nu_L, e_L)$ and ϕ is the "minimal" Higgs doublet*. This interaction would produce a Majorana neutrino mass through the vacuum expected value of ϕ

$$m_\nu \simeq \frac{b}{\alpha M} M_W^2 \ . \tag{2}$$

Depending on the explicit model, f can vary over a wide range of values: from a Yukawa coupling squared $f \simeq \lambda^2$ to the gauge coupling squared $f \simeq \alpha$. Taking for λ the average value $\lambda^2 \simeq 10^{-4}\,\alpha$ one gets

$$m_\nu \simeq (10^{-2} - 10^{+2}) \frac{M_W}{M} \,\text{GeV} \ .$$

Turned round the other way, the cosmological limit of a few eV for m_ν requires M to be postponed to $10^9 - 10^{13}$ GeV. On the other hand, if M is as high as the Planck mass, one may still have a neutrino mass of order 10^{-5} eV, which is not without consequences for solar neutrino physics.

---

*) The omitted SU(2) indices are meant to be saturated to give a scalar quantity. A possible family structure is also not explicitly indicated.

2. L-VIOLATION IN GRAND UNIFIED THEORIES

The central question then becomes: Do we have any good reason for expecting this L-violating physics to come in at some superhigh energy? How can we be more explicit about neutrino masses and, vice versa what would a neutrino mass signal teach us about the underlying physics?

Herein lies the connection with grand unified theories. In this theoretical framework, there is a basic distinction between theories where a C- and a P-operator can be defined, e.g., the schemes based on SO(10) or E_6, and theories where P-violation is intrinsic, like the smallest scheme based on SU(5). In the first case the right-handed neutrino ν_R participates in the gauge interactions and B-L (baryon minus lepton number), which takes the place of L in the following discussion, is a gauged charge[3],

$$B - L = 2(Q - T_{3L} - T_{3R})$$

which has to be broken. On the other hand, in the case of SU(5) with its reducible representation for fermions, $\underline{\bar{5}} + \underline{10}$, ν_R is absent and (B-L) is allowed to obey an exact global conservation law, through a mechanism which is interesting in itself[4]. B-L can be written as a linear combination

$$B - L = X + \frac{4}{5}(Q - T_{3L})$$

of a gauged charge plus an ungauged quantum number X, that is -3/5 for $\underline{\bar{5}}$ and 1/5 for $\underline{10}$, thus commuting with SU(5). Then the spontaneous violation of both X and $(Q-T_{3L})$ with the linear combination B-L exactly conserved gets rid of the Goldstone boson associated with X-violation, since it is "eaten" by the gauge boson carrying $(Q-T_{3L})$. True enough, this picture is tied to the "minimal" Higgs structure for SU(5), but we consider very ugly the possible complication of the Higgs sector, e.g. through a $\underline{10}$ or $\underline{15}$ of scalars, which leads to an explicit violation of X in the Lagrangian, and *a fortiori* of B-L *. Therefore, within the SU(5) model, very much like in the low-energy SU(2) × U(1) theory, one is led to exclude a neutrino mass, either of Dirac or of Majorana type, or at least to postpone it to some post-SU(5) interaction, e.g., gravity, with only a very tiny effect expected, may be of order $m_\nu \sim 10^{-4} - 10^{-5}$ eV [6].

---

*) A number of papers exist on this subject, probably initiated by S. Glashow[5].

In the alternative scenario, provided by basically left-right symmetric theories, on the one hand a Dirac neutrino mass is expected, since ν_R is present, and on the other Majorana masses for ν_L and ν_R, since B-L is violated*. Here a clue to understanding the smallness of the observed left-handed neutrino mass is offered by the different transformation properties under weak SU(2) of the various mass terms: $\nu_R\nu_R$ is a singlet, $\bar{\nu}_R\nu_L$ transforms as a doublet, whereas $\nu_L\nu_L$ transforms as a triplet. One is then led to associate the right-handed neutrino Majorana mass, M, with the huge mass characteristic of the breaking of the unifying group down to $SU(3) \times SU(2) \times U(1)$, and the Dirac neutrino mass with a generic quark mass, m, perhaps with its typical generational structure. Barring for the moment a direct $\nu_L\nu_L$ mass term, the observed left-handed neutrino would then acquire, through the diagonalization of the appropriate 2×2 matrix, an effective Majorana mass of roughly m^2/M [8]. This is, of course, a special case of Eq. (2), ($m = \lambda\langle\phi_2\rangle = (\lambda/e) M_W$), but now we have physical reasons for it.

Several comments are in order here. The first point concerns the size of a possible direct $\nu_L\nu_L$ mass term which could come from the vacuum expected value of a Higgs ϕ_3 which is a triplet under SU(2). This question is tied to the "gauge hierarchy" problem, which requires a "miniaturization" with respect to M of the mass for the Higgs SU(2) doublet ϕ_2 needed to break SU(2) and to give masses to quarks and leptons. The relevant SU(2) invariant effective potential for ϕ_3, after the first breaking at M has taken place, will look like

$$V = \mu^2\phi_3^2 + \lambda\phi_3\phi_2^2$$

with μ and λ both of order M, if no other unnatural "miniaturization" occurs. But then the minimization of V implies a vacuum expected value for ϕ_3 of order $\langle\phi_2\rangle^2/M$, giving in turn a direct Majorana mass to ν_L of the same order of the induced term already discussed.

A second comment concerns the possibility that no Higgs field couples to $\nu_R\nu_R$, so that the right-handed neutrino Majorana mass cannot come from a direct vacuum expected value. A general rule is, however, that it is in any case hard to prevent a huge mass being

*) The possibility[7] of having (B-L) violated locally but conserved globally, which would still forbid Majorana masses, seems unnatural.

induced by radiative corrections with the previous picture substantially unchanged*. In fact, in models with a large fermionic representation, like E_6, the right-handed neutrino is likely to be only one member of a numerous family of superheavy fermions[10].

A final comment concerns (B-L)-violation, which could also show up in proton decay. A general argument[11] based on dimensional counting and SU(3) × SU(2) × U(1) invariance at the superheavy scale M, very similar to the one that has led to Eq. (1), gives however

$$\frac{\Delta(B-L) \neq 0}{\Delta(B-L) = 0} \simeq \frac{M_W}{M}$$

for the ratio of amplitudes of (B-L)-violating versus (B-L)-conserving processes. As a consequence, e.g., $p \to e^+\pi^0$ is allowed, but $p \to e^-\pi^+\pi^+$ is not. In the simplest picture, with two basic scales only, M_W and M, the (B-L)-violating forces at M could show up through a neutrino mass signal but not in proton decays. The detection of modes like $p \to e^-\pi^+\pi^+$ would indicate the presence of an intermediate scale between M_W and M, which invalidates the dimensional arguments. In constructing models of this kind, with a medium-heavy mass, an eye has to be kept however to the B-$\bar{B}$ asymmetry in the Universe, which could get washed out.

3. CONCLUSIONS

Theories with parity as a short-distance symmetry lead rather naturally to a small but non-vanishing ν_L mass. A reference formula for the size of the effect is $m_\nu \simeq m^2/M$ with M a huge Majorana mass of the ν_R field, associated with the breaking of the group down to SU(3) × SU(2) × U(1) and m a typical quark mass, most likely that of charge 2/3. This is because of the Pati-Salam SU(4) [12] which relates neutrinos with charge 2/3 quarks, and is contained in the prototypes of these theories, SO(10) or E_6. Ten GeV for m requires $M \simeq 10^{11}$ GeV in order to saturate the cosmological bound (m_ν of a few eV). This value is not too far from the currently preferred mass $\simeq 10^{14}$ GeV of the superheavy gauge bosons. Fermions have a tendency to be substantially lighter than the corresponding vectors; this would be especially true for the Majorana mass M if it is generated by radiative corrections[9].

---

*) An example of such a theory, based on SO(10), has been proposed by E. Witten[9].

In view of these concepts, the search for neutrino oscillations appears to be of overwhelming importance. A combined effort in all different kinds of possible experiments (reactors, accelerators, deep mines, and solar neutrino observations) may indeed lead to a positive result since the range of values indicated, $m_\nu \sim 10^{-5}$-1 eV, overlaps considerably with the expected sensitivities. Even a particularly low value $m_\nu \simeq 10^{-4}$-10^{-5} eV could, perhaps, be detected by the observation of a time dependence in the solar neutrino flux[6].

Acknowledgements

I would like to thank J. Ellis, M.K. Gaillard, D.V. Nanopoulos, F. Strocchi, A. Masiero and G. Morchio for interesting discussions. I would also like to thank the organizers, J. Ellis, S. Ferrara and P. Van Nieuwenhuizen for organizing such a successful meeting.

REFERENCES

1. T.D. Lee and C.N. Yang, Phys. Rev. 98 (1955) 101.
2. S. Weinberg, Phys. Rev. Letters 43 (1979) 1566.
3. R. Barbieri, D.V. Nanopoulos, G. Morchio and F. Strocchi, Phys. Letters 90B (1980) 91.
4. P. Ramond, Preprint CALT-68-709 (1979).
5. S. Glashow, Cargèse Lectures (1979), to be published.
6. R. Barbieri, J. Ellis and M.K. Gaillard, Phys. Letters 90B (1980) 249.
7. R. Barbieri, G. Morchio and F. Strocchi, Nuovo Cimento Letters 26 (1979) 445; and Ref. 3.
8. M. Gell-Mann, P. Ramond and R. Slansky, to be published; P. Ramond, Ref. 4; H. Georgi and D.V. Nanopoulos, Nucl. Phys. B155 (1979) 52.
9. E. Witten, Preprint HUTP-79-A076 (1979).
10. R. Barbieri, D.V. Nanopoulos, CERN preprint TH.2810, to appear in Phys. Letters B.
11. D.V. Nanopoulos, D. Sutherland and A. Yildiz, Preprint HUTP-79/A038 (1979). S. Weinberg, Ref. 2; F. Wilczek and A. Zee, Phys. Rev. Letters 43 (1979) 1571.
12. J.C. Pati and A. Salam, Phys. Rev. D10 (1974) 275.

EXCEPTIONAL GROUPS FOR GRAND UNIFICATION

Berthold Stech

Institut für Theoretische Physik der
Universität Heidelberg
Philosophenweg 16, D-6900 Heidelberg

ABSTRACT

A general discussion of exceptional groups as candidates for the grand unification of fundamental interactions is given. Two E_6 models are discussed in some detail. It is demonstrated that in particular the "top model of E_6" has a number of advantages over SU(5) and SO(10) models. The spontaneous symmetry breaking of E_6 via 27 and 351 dimensional Higgs fields can give the correct pattern of vector boson and fermion masses and provides for interesting relations.

1. SOME PROPERTIES OF EXCEPTIONAL GROUPS

The use of exceptional groups in grand unified theories[1,2] was suggested for several reasons:

1) Exceptional groups always contain an SU(3) subgroup - to be identified with the colour group of Quantum Chromodynamics - as an essential ingredient[3,4,5]. In the defining representations only SU(3) singlets (leptons) and triplets (quarks) appear (the group E_8, however, has also an SU(3) octet in its lowest representation).

2) Exceptional groups bigger than G_2 contain additional SU(3) subgroups which can be used as flavour groups. They provide in a natural way the desired weak and electromagnetic charges for quarks and leptons[4,6,7].

3) Exceptional groups are free of anomalies.

4) The exceptional group E_6 has a perfect cyclic lepton-quark-antiquark symmetry[6].

Typical properties of exceptional groups can already be seen by considering the simplest group G_2. G_2 is a minimal group which extends SU(3): $G_2 \supset SU(3) = SU(3)_{colour}$. The seven-dimensional defining representation of G_2 decomposes with respect to SU(3) as follows*):

$$\underset{\sim}{7} = \underset{\sim}{1} + \underset{\sim}{3}^* + \underset{\sim}{3} \qquad \ell(x) \quad q^{\gamma}(x) \quad \hat{q}_{\gamma}(x) \tag{1}$$

In (1) I have attached a lepton field $\ell(x)$ to the singlet of $SU_C(3)$ and a quark and an antiquark field to the triplets. ℓ, q, $\hat{q}$ denote two-component Weyl spinor fields which are chosen to be left-handed.

The adjoint representation has dimension 14

$$\underset{\sim}{14} = \underset{\sim}{8} + \underset{\sim}{3}^* + \underset{\sim}{3} \tag{2}$$

In terms of the above lepton and quark fields the 14 generators read

$$\begin{aligned} F_{\gamma'}^{\gamma} &= (\hat{q}_{\gamma})^* \hat{q}_{\gamma'} - (q^{\gamma'})^* q^{\gamma} - \tfrac{1}{3}\delta_{\gamma'}^{\gamma}(\hat{q}^*\hat{q} - q^* q) \\ F^{\gamma} &= \ell^* q^{\gamma} - (\hat{q}_{\gamma})^* \ell + \tfrac{1}{\sqrt{2}} \varepsilon^{\gamma\gamma'\gamma''}(q^{\gamma'})^* \hat{q}_{\gamma''} \\ F_{\gamma} &= (F^{\gamma})^+ \end{aligned} \tag{3}$$

The step operators**) F^i perform a cyclic change of quark into lepton, lepton into antiquark, and antiquark into quark. This establishes an interesting, but not yet perfect, lepton-quark-antiquark symmetry. From (3) follows also another important property: it is possible to define a parity operator π and a charge conjugation operator C which leave the commutation relation unchanged:

---

*) Mathematical properties of exceptional groups useful for physicists are contained in papers by P. Ramond[8] and B. Wybourne[9].

**) For convenience, the integral signs have been left out in (3).

$$\begin{aligned}
&\Pi\, \ell(x)\, \Pi^{-1} = \sigma_2\, \ell^*(x_0 - \vec{x}) \;, \quad C\, \ell(x)\, C^{-1} = \ell(x) \\
&\Pi\, q^\gamma(x)\, \Pi^{-1} = \sigma_2\, (\hat{q}_\gamma(x_0, -\vec{x}))^* \;, \quad C\, q^\gamma(x)\, C^{-1} = \hat{q}_\gamma(x) \\
&\Pi\, \hat{q}_\gamma(x)\, \Pi^{-1} = \sigma_2\, (q^\gamma(x_0, -\vec{x}))^* \;, \quad C\, \hat{q}_\gamma(x)\, C^{-1} = q^\gamma(x) \\
&\{\Pi^2, \psi(x)\}_+ = 0 \;, \quad [C^2, \psi(x)]_- = 0
\end{aligned} \tag{4}$$

In (4) σ_2 denotes the second Pauli spin matrix. As a result each representation of G_2 transforms into itself by a parity or charge conjugation operation and the $SU_C(3)$ current is automatically a <u>pure vector</u> current as ought to be the case for a useful model. These properties are also valid for the remaining exceptional groups F_4, E_6, E_7 and E_8. The two-component fields can be put together to form four-component spinor fields: the $\underset{\sim}{7}$-representation of G_2 consists of a Dirac quark and a Majorana lepton

$$\underset{\sim}{q}^\gamma(x) = \begin{pmatrix} q^\gamma(x) \\ \sigma_2 (\hat{q}_\gamma(x))^* \end{pmatrix} , \quad \underset{\sim}{\ell}(x) = \begin{pmatrix} \ell(x) \\ \sigma_2\, \ell^*(x) \end{pmatrix} \tag{5}$$

2. WHICH EXCEPTIONAL GROUP ?

In Table 1 the exceptional groups, with some of their maximal subgroups and their defining and adjoint representations[8,9], are listed.

<u>Table 1</u>

The exceptional groups

	defining representation	adjoint representation
$G_2 \supset SU_C(3)$	$\underset{\sim}{7}$	$\underset{\sim}{14}$
$F_4 \supset SU(3) \otimes SU_C(3)$	$\underset{\sim}{26}$	$\underset{\sim}{52}$
$E_6 \supset SU(3) \otimes SU(3) \otimes SU_C(3)$	$\underset{\sim}{27}$	$\underset{\sim}{78}$
$E_7 \supset SU(6) \otimes SU_C(3)$	$\underset{\sim}{56}$	$\underset{\sim}{133}$
$E_8 \supset E(6) \otimes SU(3)$	$\underset{\sim}{248}$	$\underset{\sim}{248}$

The group F_4 could in principle be used as a gauge group for grand unification because it contains an SU(3) flavour group which gives fractional charges to quarks and integer charges to leptons in a natural way. However, the antiquark fields as well as the quark fields and the antilepton fields as well as the lepton fields transform non-trivially with respect to this flavour group: F_4 gives rise to a vector-like theory with an unrenormalized weak angle $\sin^2 \theta_W^0 = 3/4$ incompatible with experimental findings. Also, from a theoretical point of view, an F_4 model is hardly tenable: this group allows an invariant fermion mass term to appear in the Lagrangian. Thus, in an F_4 grand unified theory, all fermions could easily acquire very large masses.

The groups E_7 and E_8 also give rise to vector-like models with $\sin^2 \theta_W^0 = 3/4$. The mathematical reason is that these groups have, like G_2 and F_4, only real (pseudoreal) representations. The only exceptional group with complex representations is E_6. In particular the lowest representation of E_6, the $\underset{\sim}{27}$, which should describe the fermions, is a complex representation. Among the exceptional groups E_6 is, therefore, uniquely singled out as the best candidate for grand unification . Models based on E_6 are <u>not</u> vector-like and give the unrenormalized weak angle $\sin^2 \theta_W^0 = 3/8$ which is an acceptable value. Moreover, as will be seen, E_6 models possess the highest possible lepton-quark-antiquark symmetry which is - at least aesthetically - very appealing. E_6 contains SO(10) and thereby also SU(5) as subgroups. With respect to SO(10) the 27 fermions decompose as follows:

$$\underset{\sim}{27} = \underset{\sim}{1} + \underset{\sim}{10} + \underset{\sim}{16} \tag{6}$$

As is well known, the $\underset{\sim}{16}$ of SO(10) describes a fermion family (generation)[10]. E_6 is the <u>minimal</u> group with a complete fermion family in the defining representation !

I will describe in detail models based on E_6 in the next chapter. Before, let me return shortly to the groups E_7 and E_8. These groups have E_6 as subgroups:

$$E_7 \supset E_6 \otimes U(1) \qquad E_8 \supset E_6 \otimes SU(3) \tag{7}$$

Can E_7 and E_8 be used as a generalization of E_6 models if such models turn out to be successful ? Since E_7 and E_8 give rise to vector-like theories, as was mentioned above, at least half of the corresponding states must be removed or shifted to very high energies by some unknown mechanism. So far, no satisfactory suggestion has been put forward. Nevertheless it is perhaps worth pointing out that models based on E_8 as a generalization of E_6 would be interesting. The reason is that the additional SU(3) in E_8 could be used as a family (generation) group[11]. One obtains three families with 27 fermions in each family, i.e., three $\underset{\sim}{16}$ multiplets of SO(10):

$$\begin{array}{l} \underset{\sim}{248} = (\underset{\sim}{27}, \underset{\sim}{3}) + (\underset{\sim}{27}^*, \underset{\sim}{3}^*) + (\underset{\sim}{78}, \underset{\sim}{1}) + (\underset{\sim}{1}, \underset{\sim}{8}) \\ SO(10) \downarrow \\ \quad (\underset{\sim}{16}, \underset{\sim}{3}) \end{array} \tag{8}$$

However, additional unsolved problems with new states arise: colour octet flavour singlet quarks, isotriplet leptons, etc.

3. E_6 FOR GRAND UNIFICATION

Representations and generators of the exceptional group E_6 can be classified by a maximal subgroup decomposition

$$E_6 \supset \underbrace{SU_L(3) \otimes SU_R(3)}_{\text{flavor group}} \otimes \underbrace{SU_C(3)}_{\text{color group}} \tag{9}$$

The flavour group is, therefore, a direct product of two SU(3) groups which are denoted by left (L) and right (R) for reasons which will become obvious in a moment. This group must contain the Glashow-Salam-Weinberg group as subgroup:

$$\begin{aligned} SU_L(3) \otimes SU_R(3) &\supset SU_L(2) \otimes U_L(1) \otimes U_R(1) \\ &\supset SU_L(2) \otimes U(1) \end{aligned} \tag{10}$$

The observed left-handed charged weak currents are $SU_L(2)$ currents and the particle charges are obtained from the direct sum of the two SU(3) charge operators

$$\begin{aligned} Q &= Q_L + Q_R \\ Q_{L,R} &= I^3_{L,R} + \frac{1}{2} Y_{L,R} \end{aligned} \tag{11}$$

The fermion representation decomposes with respect to $SU_L(3) \otimes SU_R(3) \otimes SU_C(3)$ as follows:

$$\underset{\sim}{27} = \underset{q_\alpha^\gamma}{(\underset{\sim}{3}, \underset{\sim}{1}, \underset{\sim}{3}^*)} + \underset{\hat{q}_\gamma^\beta}{(\underset{\sim}{1}, \underset{\sim}{3}^*, \underset{\sim}{3})} + \underset{\ell_\beta^\alpha}{(\underset{\sim}{3}^*, \underset{\sim}{3}, \underset{\sim}{1})} \tag{12}$$

Thus, each left-handed, two-component fermion field carries two SU(3) indices: a quark field carries a left index (α) and an upper colour index (γ), an antiquark field carries a right index (β) and a lower colour index, and a lepton field carries a left and a right index. Note that (12) has been written in a permutation symmetric form which immediately demonstrates a complete lepton-quark-antiquark symmetry[6].

For later reference, and for illustration, particle names - from the electron family - may be introduced

$$q_\alpha \qquad \hat{q}^\beta \qquad \ell^\alpha{}_\beta$$

$$\begin{pmatrix} u \\ d \\ h \end{pmatrix} \qquad (\hat{u}\ \hat{d}\ \hat{h}) \qquad \begin{pmatrix} N_1 & E^- & e^- \\ E^+ & N_2 & \nu_e \\ e^+ & \hat{\nu}_e & N_3 \end{pmatrix} \tag{13}$$

The group $SU_L(3)$ acts vertically, the group $SU_R(3)$ acts horizontally on (13). Colour indices on the quark fields have been omitted, h stands for a new (heavy) quark, E for a new (heavy) lepton, and N_1, N_2, N_3 denote neutral lepton fields. (One has to keep in mind, however, that the symmetry breaking to be discussed later on will introduce particle mixing.) We see that E_6 does not lead to a vector-like theory; in particular we note that $\hat{u}$, $\hat{d}$, e^+, $\hat{\nu}_e$ are singlets with respect to $SU_L(2)$.

The adjoint representation $\underset{\sim}{78}$ of E_6 decomposes with respect to $SU_L(3) \otimes SU_R(3) \otimes SU_C(3)$ in the following way:

$$\begin{aligned} \underset{\sim}{78} = {} & \underbrace{(\underset{\sim}{8},\underset{\sim}{1},\underset{\sim}{1})}_{F_L} + \underbrace{(\underset{\sim}{1},\underset{\sim}{8},\underset{\sim}{1})}_{F_R} + \underbrace{(\underset{\sim}{1},\underset{\sim}{1},\underset{\sim}{8})}_{F_C} \\ & + \underbrace{(\underset{\sim}{3},\underset{\sim}{3},\underset{\sim}{3})}_{F} + \underbrace{(\underset{\sim}{3}^*,\underset{\sim}{3}^*,\underset{\sim}{3}^*)}_{F^+} \end{aligned} \tag{14}$$

Equations (12) and (14) differ in form from those given in the mathematical literature where they are written in a less symmetric way. F_L, F_R, F_C denote the generators of the SU(3) groups. The generators F perform the cyclic change of quarks into leptons, leptons into antiquarks, antiquarks into leptons:

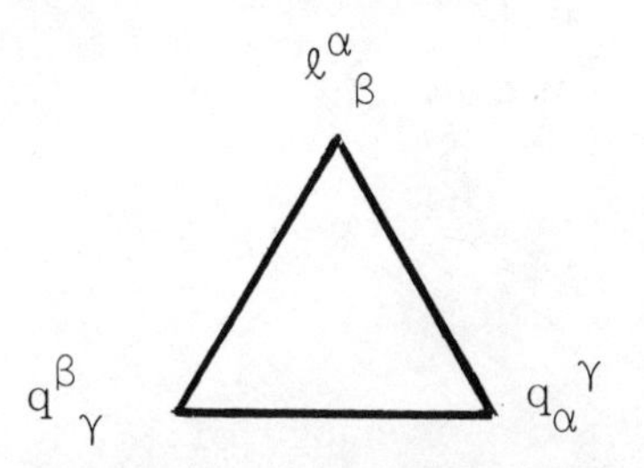

Fig. 1 The lepton-quark-antiquark triangle symmetry of E_6.

$$F^{\alpha\beta\gamma} = \left(\ell^{\alpha'}{}_{\beta}\right)^{*} q^{\gamma}{}_{\alpha''}\, \varepsilon^{\alpha\alpha'\alpha''} + \left(\hat{q}^{\beta'}{}_{\gamma}\right)^{*} \ell^{\alpha}{}_{\beta''}\, \varepsilon^{\beta\beta'\beta''} + \left(q_{\alpha}{}^{\gamma'}\right) \hat{q}^{\beta}{}_{\gamma''}\, \varepsilon^{\gamma\gamma'\gamma''} \tag{15}$$

$$F_{\alpha\beta\gamma} = \left(F^{\alpha\beta\gamma}\right)^{+}$$

As in the case of G_2, a parity operator π and a charge conjugation operator C can be defined. These operators interchange L and R indices, for instance

$$C\, \ell^{\alpha}{}_{\beta}\, C^{-1} = \ell^{\beta}{}_{\alpha} \tag{16}$$

The E_6 Lagrangian is π and C-invariant. An additional global SU(N) symmetry arises if the Lagrangian contains N fermion families, i.e., if it contains N $\underset{\sim}{27}$-multiplets.

The limit of unbroken E_6 symmetry is easy to discuss: the 78 vector bosons are all massless. Also, all fermion masses vanish because the product of two $\underset{\sim}{27}$ representations does not contain a singlet:

$$\underset{\sim}{27} \otimes \underset{\sim}{27} = \left(\underset{\sim}{27}^{*} + \underset{\sim}{351}\right)_{symm} + \left(351'\right)_{antisymm} \tag{17}$$

Because there is only one gauge coupling constant one has the relation

$$\alpha_{strong}(SU_C(3)) = \alpha_{weak}(SU_L(2)) = \frac{8}{3}\alpha_{el.magn.}$$

$$\sin^2\theta_W^0 = \frac{3}{8} \tag{18}$$

Proton decay occurs via the exchange of $(\underset{\sim}{3}, \underset{\sim}{3}, \underset{\sim}{3})$ vector bosons which transform quarks into leptons and antiquarks. Obviously, as in other grand unified theories, a superstrong symmetry breaking is necessary to modify the above relations and to describe the physical situation.

Before discussing the symmetry breaking I will describe in a phenomenological way two different E_6 models which suggest themselves[6,12]. Which one of these two, if any, is realized in Nature depends on details of the symmetry breaking.

A. The Topless Model - A specific version of E_6

In this model one adds to E_6 a new parity-like operator $P \neq \pi$ with $[P^2,\psi] = 0$ [12].

The lowest fermion representation of $E_6 \otimes P$ (with $P \neq 1$) is $\underset{\sim}{54} = \underset{\sim}{27} + \underset{\sim}{27}$, i.e., it consists of two fermion families ($N = 2$). The first family should contain the up and the down quark, and the electron and its neutrino, the second one the charmed and the strange quark, and the μ meson and its neutrino. In this minimal case there is no room for a top quark besides u and c. The bottom quark b ($\simeq$ 4.8 GeV) and the heavy lepton τ ($\simeq$ 1.8 GeV) are not in a new family but must be identified with states already present in (13):

$$\begin{aligned} \underset{\sim}{q}_3 &= \underset{\sim}{h} = b\,(4.8\ \text{GeV}) \\ \underset{\sim}{\ell}'_2 &= \underset{\sim}{E}^- = \tau^-(1.8\ \text{GeV}) \\ \ell^2_2 &= N_2 = \nu_\tau \end{aligned} \tag{19}$$

This identification is only possible if the symmetry breaking mechanism is such that ν_e and N_2 remain almost massless while $E = \tau$ and N_1 obtain sizeable masses. No simple mechanism will do that[13]. Nevertheless, by introducing an additional E_6 singlet neutrino S, it is possible to write down a mass matrix which has the desired properties*)

	N_2	N_1	S
N_2	0	m	0
N_1	m	0	m'
S	0	m'	m_s

	ν_e	$\hat{\nu}_e$
ν_e	0	m
$\hat{\nu}_e$	m	M

(20)

If m_S is very small or zero while m, m' are of order m_τ, an eigenstate of small mass mixes with N_2. Because of this neutral lepton mixing the τ lifetime is reduced by the factor $1/(1 + m^2/m'^2)$ compared to the usual expectation. The ν_e neutrino mass is very small for M sufficiently large (I will come to this point again in the section on symmetry breaking).

A second problem of the topless version of E_6 concerns the strangeness changing neutral decays which would occur by family mixing but which are not observed experimentally. Only the following mixing of quarks can be allowed[16]:

$$\begin{array}{ccc} u & \longleftrightarrow & c \\ d \updownarrow & & s \updownarrow \\ b & & \tilde{b} \end{array} \qquad (21)$$

In other words, Cabibbo mixing is only admissible in the charge 2/3 sector. This requirement complicates to some extent the allowed Higgs structure.

However, if one is willing to accept all this, the topless model[12,11] makes striking predictions:

i) No top quark exists besides u and c.

ii) A second b-quark ($\tilde{b}$) exists with $m_b \simeq m_{\tilde{b}} \cdot m_s/m_d$

*) Mass matrices with an additional singlet neutrino have first been suggested by Georgi and Nanopoulos[14]. The second mass matrix in (20) has been used independently by Gell-Mann, Ramond and Slansky[15], H. Ruegg and T. Schücker[13], and the author[14].

iii) A second heavy lepton ($\tilde{\tau}$) exists with $m_{\tilde{\tau}} \simeq m_\tau \cdot m_s/m_d$.

iv) Neutral heavy leptons N, $\tilde{N}$ exist with $m_{\tilde{N}} \simeq m_\tau$, $m_{\tilde{N}} = m_{\tilde{\tau}}$.

v) The b-quark decays preferably into non-charmed channels $b \to ue^-\bar{\nu}_e$ etc., $B \to \pi$'s by $b \leftrightarrow d$ mixing

vi) Flavour-changing neutral current processes occur: $b \to de^+e^-$, $B^- = b\hat{u} \to \pi^-e^+e^-$ etc. with a branching ratio of about 1-2%.

The last two points are rather easy to check once the B's are found. Thus, this version of E_6 should be a target for experimentalists who like to shoot down models. Since the assumptions necessary to construct this model are somewhat artificial, I give the second version of E_6, to which we come now, a bigger chance to survive.

B. The top E_6 Model - A second version of E_6

A superstrong breaking of E_6 which involves a Higgs field transforming as a $\underset{\sim}{27}$ representation will in general break E_6 down to SO(10) or a smaller group (see below). If this happens the fermions in the $\underset{\sim}{10}$ of SO(10) which appear in the decomposition (6) obtain very high masses[6]. These particles which are in the $\underset{\sim}{10}$ (q_3, $\ell^1{}_2$, $\ell^1{}_1$, $\ell^2{}_2$ of (13)) cannot be identified with known quarks and leptons any more. The same applies to the corresponding particles in other families. Thus one has to place the b-quark and the heavy lepton τ in the $\underset{\sim}{16}$ representation of a third family as one does in the SO(10) model[10,14]. In this case a top quark as a partner of the b is required. This "top model" of E_6 [6,11] with three (or more) families is certainly not easy to distinguish from an SO(10) model. But it is more elegant, more consistent, and more predictive than this model. From the above discussion one sees already that the E_6 model provides a reason why the $\underset{\sim}{16}$ of SO(10) and not the $\underset{\sim}{10}$ describes a fermion family.

Let us now discuss the symmetry breaking is some detail. First we note that Higgs fields transforming as a $\underset{\sim}{27}$ and as a $\underset{\sim}{351}$ of E_6 are needed. With a $\underset{\sim}{27}$ alone one would obtain the neutrino mass beginning equal to the up-quark mass of the same family. Furthermore, E_6 could at most break down to SO(8). On the other hand, including a $\underset{\sim}{351}$ Higgs field and choosing the vacuum expectation values of its components appropriately, a breaking of the local E_6 symmetry down to $U_Q(1) \otimes SU_C(3)$ can be accomplished. The scalar fields in the $\underset{\sim}{27}$ and $\underset{\sim}{351}$ representation can formally be constructed from two-fermion operators according to (17). Scalar fields in the adjoint representation $\underset{\sim}{78}$, which have also been considered, could, on the other hand, only be formed using at least four fermion fields. Therefore, if some kind of dynamical symmetry breaking is at work, the $\underset{\sim}{27}$ and $\underset{\sim}{351}$ breaking will be favoured. This point has recently been emphasized by Barbieri and Nanopoulos[17].

The Higgs field in the reducible $\underset{\sim}{27} + \underset{\sim}{351}^*$ representation may be written as a symmetric 27 × 27 matrix ϕ^*_{ij} with 756 independent real components*). In Table 2 examples of "stability groups" are listed, i.e., subgroups of E_6 which leave certain combinations of vacuum expectation values ϕ_{ij} invariant. The corresponding combinations are exhibited in column 2. They are characterized in this table by their E_6 transformation properties using the symmetric product of two $\underset{\sim}{27}$-fermion representations (12). To each invariant combination of the ϕ's corresponds also an invariant fermion mass-matrix relation of the same form. For example, when E_6 is broken down to SO(10) one finds from Table 2

$$m(\ell^3{}_3) \neq 0, \quad m(q_3) = m(\ell^1{}_2) = m(\ell^1{}_1, \ell^2{}_2) \neq 0 \tag{22}$$

The fermion $\ell^3{}_3$ has become a massive Majorana lepton and the quark q_3, the lepton $\underset{\sim}{\ell}{}^1_2$, and the four-component field formed from ℓ^1_1 and $\tilde{\ell}^2{}_2$ are very massive Dirac particles. In the breaking to SU(5) in addition non-diagonal terms appear which give rise to the mixings

$$\hat{q}^3 \leftrightarrow \hat{q}^2, \quad \ell^1{}_2 \leftrightarrow \ell^1{}_3, \quad \ell^2{}_2 \leftrightarrow \ell^2{}_3 \tag{23}$$

In the third column of Table 2 the intersection of the stability group with the vector-like group F_4 is considered. The generators of the intersection group are those generators which also leave the Higgs field combination defining F_4 invariant.

Table 2 shows that breaking E_6 down to the vector-like group F_4 gives masses to all fermions. In particular, the 26 states in the decomposition.

$$\underset{\sim}{27} \xrightarrow{F_4} \underset{\sim}{1} + \underset{\sim}{26}$$

pair off to a fermion-antifermion condensate. I will assume that this pairing will occur in all three families as a consequence of the gauge interaction. The F_4 mass relation among the Dirac particles is simply

$$m^\circ(\underset{\sim}{q}_1) = m^\circ(\underset{\sim}{q}_2) = m^\circ(\underset{\sim}{q}_3) = m^\circ(\underset{\sim}{\ell}{}'_2) = m^\circ(\underset{\sim}{\ell}{}'_3) = m^\circ(\underset{\sim}{\ell}{}^2_3) \tag{24}$$

---

*) Antisymmetric matrices can occur for Higgs fields which connect different fermion familes.

Table 2

E_6 broken by the reducible representation $\phi_{ij} = (27 + 351^*)_6$

$SU_Q(4)$: contains the generator of electric charge Q.

$SU_Y(4)$: contains the generator $Y = Y_L + Y_R$.

Stability group	Invariant elements ϕ_{ij} and invariant fermion mass relations	Intersection with F_4
F_4	351 $\frac{15}{6}(\ell^1{}_1 + \ell^2{}_2 + \ell^3{}_3)^2 + K$ 27* $\begin{cases} K = q_1\hat{q}^1 + q_2\hat{q}^2 + q_3\hat{q}^3 + \ell^1{}_2\ell^2{}_1 + \\ \ell^1{}_3\ell^3{}_1 + \ell^2{}_3\ell^3{}_2 - \ell^1{}_1\ell^2{}_2 - \ell^1{}_1\ell^3{}_3 - \ell^2{}_2\ell^3{}_3 \end{cases}$	F_4
SO(10)	351 $\ell^3{}_3\ell^3{}_3$ 27* $q_3\hat{q}_3 + \ell^1{}_2\ell^2{}_1 \quad \ell^1{}_1\ell^2{}_2$	SO(9)
SU(5)	351 $\ell^3{}_3\ell^3{}_3,\ \ell^3{}_2\ell^3{}_3,\ \ell^3{}_2\ell^3{}_2$ 27* $\begin{cases} q_3\hat{q}^3 + \ell^1{}_2\ell^2{}_1 - \ell^1{}_1\ell^2{}_1, \\ q_3\hat{q}^2 - \ell^1{}_3\ell^2{}_1 \quad \ell^1{}_1\ell^2{}_3 \end{cases}$	$SU_Q(4)$
$SU_L(2) \otimes$ $SU_R(2) \otimes$ $SU(4)$	351 $\begin{cases} \ell^3{}_3\ell^3{}_3, \\ q_3\hat{q}^3 - \frac{3}{2}(\ell^1{}_2\ell^2{}_1 - \ell^1{}_1\ell^2{}_2) \end{cases}$ 27* $q_3\hat{q}^3 + \ell^1{}_2\ell^2{}_1 - \ell^1{}_1\ell^2{}_2$	$SU_{L+R}(2) \otimes$ $SU_Y(4)$
$U(1) \otimes$ $SU_L(2) \otimes$ $SU_C(3)$	351 $\begin{cases} \ell^3{}_3\ell^3{}_3,\ \ell^3{}_2\ell^3{}_3,\ \ell^3{}_2\ell^3{}_2, \\ q_3\hat{q}^3 - \frac{3}{2}(\ell^1{}_2\ell^2{}_1 - \ell^1{}_1\ell^2{}_2), \\ q_3\hat{q}^2 + \frac{3}{2}(\ell^1{}_3\ell^2{}_1 - \ell^1{}_1\ell^2{}_3) \end{cases}$ 27* $\begin{cases} q_3\hat{q}^3 + \ell^1{}_2\ell^2{}_1 - \ell^1{}_1\ell^2{}_2, \\ q_3\hat{q}^2 - \ell^1{}_3\ell^2{}_1 + \ell^1{}_1\ell^2{}_3 \end{cases}$	$U_Q(1) \otimes$ $SU_C(3)$

The corresponding mass scale should, however, be extremely small compared to the mass scales involved in the superstrong breaking of E_6.

As seen from Table 2, E_6 can be broken superstrongly down to the Glashow-Salam-Weinberg group $U(1) \times SU_L(2) \times SU_C(3)$ if the Higgs potential gives suitable vacuum expectation values to ϕ. The very weak fermion-antifermion pairing which occurs in addition will then reduce the local symmetry to $U_Q(1) \otimes SU_C(3)$. According to Table 2 the superstrong breaking of E_6 down to $U(1)$ $SU_L(2) \otimes SU_C(3)$ gives high masses to q_3, $\ell^3{}_3$, $\ell^3{}_2$, $\ell^1{}_2$ and $\ell^1{}_1$, and also causes particle mixings. It is of interest to note that the masses of the quarks $q_1(u, c, t)$ are not affected, they retain their values given by (24). The leptons $\ell^1{}_3(e^-,\mu^-,\tau^-)$, on the other hand, mix with their heavy partners $\ell^1{}_2(E^-,..)$ according to the mass matrix

	ℓ^1_3	ℓ^1_2
ℓ^3_1	m^0	0
ℓ^2_1	M_2	M_1

$$M_2, M_1 >>> m^0(q_1) \tag{25}$$

One finds for the final "light" lepton masses (e^-, μ^-, τ^-)

$$m(\text{charged lepton}) \simeq m^0(q_1)\frac{1}{\sqrt{1 + M_2^2/M_1^2}} \tag{26}$$

The "light" lepton masses are, therefore, smaller than the up-quark masses even at the scale of the grand unification mass. Similarly, the down-quark masses turn out to be smaller than the up-quark masses because of the $\hat{q}^2 \leftrightarrow \hat{q}^3$ mixing. Experimentally this latter result is not true in the case of the lightest family: one has $m_d > m_u$. Small corrections, which are unessential for the higher families, may be the origin for this discrepancy[18,17].

The superstrong breaking of E_6 down to $U(1) \otimes SU_L(2) \otimes SU_C(3)$ involves a spontaneous breakdown of parity conservation. According to the Table 2 this part of the breaking necessarily produces large Majorana masses for the antineutrinos $\hat{\nu} = \ell^3{}_2$. This fact is encouraging because it gives us a reason for the smallness of the neutrino masses: according to (24) and because of $\ell^2{}_3 \leftrightarrow \ell^2{}_2$ mixing the masses of the Dirac neutrinos should be equal to the masses of charged leptons($\ell^1{}_3$). A high Majorana mass of the antineutrino, however, changes this situation drastically. From the mass matrix (20) one obtains for the neutrino masses

$$m_\nu \simeq \frac{(m(\text{charged lepton}))^2}{M_{\hat{\nu}}} \tag{27}$$

which will be very small for a super-high value of $M_{\hat{\nu}}$[11,13,15].

The qualitative results

$$\begin{aligned} m(\text{charged lepton}) &< m(q_1) \\ m(q_2) &< m(q_1) \\ m(\nu) &\ll m(q_1) \end{aligned} \tag{28}$$

are encouraging. Comparing now different families one expects -- neglecting family mixing --

$$\begin{aligned} \frac{m_e}{m_u} \approx \frac{m_\mu}{m_c} &\simeq \frac{m_\tau}{m_t} \\ \left(\frac{m_e}{m_d} \approx\right) \frac{m_\mu}{m_s} &\simeq \frac{m_\tau}{m_b} \end{aligned} \tag{29}$$

i.e., a top quark mass $m_t \simeq 22$ GeV *). While the relations (29) should be independent of the energy scale at which one measures (in the region below the scale of the main breaking), these ratios themselves are scale-dependent according to the renormalization group equations[20].

Let us finally ask how the vacuum expectation values, the boson and fermion masses and mixing angles could be determined. A general renormalizable E_6-invariant potential depending on ϕ_{ij} contains a large number of parameters. It is not very sensible to look for the minimum of such a very general potential. The desired breaking pattern could most probably be obtained but one would not learn much. Since in an E_6 model the Higgs mechanism is likely to be dynamically generated, it appears to be more appropriate and at least worth trying to use in the tree approximation the highest possible symmetry, i.e., the symmetry of the free Lagrangian without

*) This mass value for the top quark has also been suggested in Refs 19 and 17.

mass parameter, and then obtain an effective dynamical potential from the one loop calculation according to S. Coleman and E. Weinberg[21]. My motivation for this choice is the appealing self-consistent and almost parameter-free connection between all the particles which arise in this way.

$$V_{eff} = \frac{3}{64\pi^2} \mathrm{Tr}\, M_V^4 \ln \frac{M_V^2}{\mu^2} - \frac{4}{64\pi^2} \mathrm{Tr}\, m_F^4 \ln \frac{m_F^2}{\mu^2} \tag{30}$$

In this equation μ is a scale parameter and M_V and m_F stand for vector meson and fermion mass matrices:

$$M_V^{2\,ab} = g^2 (t^a\phi \cdot t^b\phi) \qquad a,b = 1 \ldots 78$$

$$m^2_{F\,ij} = \lambda^2 g^2 \phi_{ik} \phi^*_{kj} \qquad i,j = 1 \ldots 27 \tag{31}$$

t^a : E_6 generator

g : gauge coupling constant

λg : Yukawa coupling constant

(Actually, one can use a Yukawa coupling constant for the $\underset{\sim}{27}$, and a different one for the $\underset{\sim}{351}$ representation.)

The task is now to find in terms of μ and λ the values for which V_{eff} takes its smallest value.

To see the effect of such a potential let me consider for a moment the simpler case where E_6 is broken by a single Higgs field in the adjoint representation $\underset{\sim}{78}$. In this particular case the general renormalizable Higgs potential is O(78)-invariant in tree approximation, i.e., it automatically possesses the symmetry of the free Lagrangian. The potential (30) now selects the stability group. One finds that the U(1) ⊗ SO(10) orbit has the lowest energy, i.e., E_6 breaks down to U(1) ⊗ SO(10). This result was first obtained by F. Harvey[22].

The program of finding the orbit of lowest energy of the potential (30) with the reducible representation ϕ^*_{ij} (i,j = 1 ... 27) has been recently started by Norbert Dragon and myself. It is a very involved task. So far we have the following results (for $\lambda = 0$):

i) The SO(10) orbit has a lower energy than the F_4 orbit. The components of ϕ_{351} remain zero, only ϕ_{27} breaking occurs on the SO(10) orbit.

ii) With a pure ϕ_{27} breaking the potential (30) has its absolute minimum on the SO(10) orbit. Vacuum expectation values which would break E_6 to SO(9) or SO(8) give a higher value of the potential energy.

iii) If we admit the ϕ_{351} Higgs field, less symmetric orbits are favoured: the SU(5) orbit has a lower energy than the SO(10) orbit. Components of ϕ_{351} which break parity conservation appear. They are small compared to the ϕ_{27} components.

iv) The orbit $U(1) \otimes SU_L(2) \otimes SU_C(3)$ has still a lower energy than the SU(5) orbit.

The completion of this study appears to be valuable. By calculating the vacuum expectation values which give the lowest value of the potential (30) the physical relevance of this potential can be checked*). One could see to what extent stability groups like SO(10), SO(9), SU(5) etc. play an intermediate role between E_6 and the Glashow-Salam-Weinberg group $U(1) \otimes SU_L(2) \otimes SU_C(3)$. If for instance the breaking of parity conservation occurs at a scale very small compared to the original scale of the main breaking, say at 10 GeV, the stability group $SU_L(2) \otimes SU_R(2) \otimes SU(4)$ which is also listed in Table 2, is relevant. It provides for right-handed currents and currents transforming quarks into leptons at this energy. This is not excluded in view of the observed stability of the proton since the currents of this group do not contribute to proton decay contrary to SU(5) and SO(10) currents. According to (27) the detection of a small neutrino mass is an indication of a "low" mass scale for the restoration of parity conservation.

In conclusion one can say that E_6, especially in the form of the top model, is a very interesting candidate for a grand unifying gauge theory. It appears to be superior to SU(5) and SO(10) models

a) because the fermion families are in the lowest representation;
b) because down-quark and up-quark masses are connected with lepton masses;

---

*) Of course, the result could be that the potential (30) prefers an orbit with a symmetry being too low. If this is the case one has to introduce a phenomenological mass parameter in order to prevent this collapse of symmetry.

c) because of a remarkable lepton-quark-antiquark symmetry, and

d) because the desired breaking pattern is obtainable from scalar fields which transform as the product of two fermion fields.

As for a precise determination of $\sin^2 \theta_W$ and the lifetime estimate for the proton, it has, however, not yet reached the quantitative status of the SU(5) model. The number of families remains an open problem but the model strongly suggests an identical number of observable particle states in each family, identical chiral current couplings, tiny neutrino masses and a strong similarity between the fermion mass spectra.

It is a pleasure to acknowledge helpful discussions with Yoav Achiman and Norbert Dragon. Part of this work was done during the author's stay at CERN in the first few months of 1979. I thank Jacques Prentki for making this visit possible.

REFERENCES

1. J.C. Pati and A. Salam, Phys. Rev. D8 (1973) 1240.
2. H. Georgi and S.L. Glashow, Phys. Rev. Letters 32 (1974) 438.
3. M. Günyadin and F. Gürsey, Phys. Rev. D9 (1974) 3387.
4. F. Gürsey, P. Ramond and P. Sikivie, Phys. Letters 60B (1976) 177.
5. M. Gell-Mann, P. Ramond and R. Slansky, Rev. Mod. Phys. 50 (1978) 721.
6. Y. Achiman and B. Stech, Phys. Letters 77B (1978) 389.
7. Q. Shafi, Phys. Letters 79B (1978) 301.
8. P. Ramond, Caltech report 68-577 (1976).
9. B.G. Wybourne, Classical Groups for Physicists, Wiley, New York (1974);
   B.G. Wybourne and M.J. Bowick, Aust. J. Phys. 30 (1977) 259-286.
10. H. Fritzsch and P. Minkowski, Ann. Phys. 93 (1975) 193;
    M. Chanowitz, J. Ellis and M.K. Gaillard, Nucl. Phys. B128 (1977) 506.
11. B. Stech, invited talk given at the Rutherford Lab. Study Week-end on Unification, May 1979, unpublished.
12. Y. Achiman and B. Stech, in: New Phenomena in Lepton-Hadron Physics, ed. by D. Fries and J. Wess, Plenum Publishing Corporation, p. 303 (1979).
13. H. Ruegg and T. Schücker, Nucl. Phys. B161 (1979) 388.
14. H. Georgi and D. Nanopoulos, Nucl. Phys. B155 (1979) 52.
15. M. Gell-Mann, P. Ramond and R. Slansky, unpublished.
16. H. Georgi and A. Pais, Phys. Rev. D19 (1979) 2746.
17. R. Barbieri and D.V. Nanopoulos, Phys. Letters 91B (1980) 369.

18. J. Ellis, M.K. Gaillard, Phys. Letters 88B (1979) 315.
19. S. Pakvasa, H. Sugawara, Phys. Letters 82B (1979) 105.
20. A.J. Buras, J. Ellis, M.K. Gaillard and D.V. Nanopoulos, Nucl. Phys. B135 (1978) 66.
21. S. Coleman and E. Weinberg, Phys. Rev. D7 (1973) 1888.
22. F. Harvey, Nucl. Phys. B163 (1980) 254.

A SIMPLE INTRODUCTION TO COMPLEX MANIFOLDS

Luis Alvarez-Gaumé and Daniel Z. Freedman

Institute for Theoretical Physics
State University of New York
Stony Brook, L.I., New York 11794

Dedicated to the memory of Joel Scherk, who has made many creative contributions to supersymmetry and supergravity.

The material discussed at the conference consisted of two applications of Kahler differential geometry to supersymmetric nonlinear σ-models [1,2]. Since this material will be published in the standard literature, there is little point in presenting it again here. However, in the course of our work we found that Kahler geometry is a beautiful subject not well-known to most theoretical physicists and the essential part not as accessible as one might hope from the many mathematical textbooks. Kahler geometry not only has applications in supersymmetric σ-models, as was shown first by Zumino [3] and then in our work, but also in quantum gravity where every self-dual gravitational instanton is a Kahler manifold [4].

For reasons stated above, we thought that an introductory treatment of complex manifolds and Kahler geometry prepared by and for theoretical physicists would be a useful contribution to these proceedings. We shall be very concrete and will emphasize local considerations such as the concrete formulae for the connection and curvature which were useful in our work. Mathematicians may legitimately object that our approach misses key points and merely skims the surface of a deep subject. On the other hand, as physicists hungry for information of direct use, we were not entirely happy with the treatment of the mathematics texts [5-8]. Therefore we simply say that the viewpoint we develop here has been useful to us, and we hope it will be useful for readers. If our view changes upon further digestion of the mathematical literature, then so be it.

The prerequisites for this course are few. We assume familiarity with the notion of an n-dimensional manifold M as a set of points in patchwise 1:1 correspondence with open sets of Euclidean n-space R^n such that n real parameters $x^1, x^2, .. x^n$ are coordinates on M. A Riemannian structure will be imposed from the outset so that we assume that there is a smooth symmetric positive definite metric tensor $g_{ij}(x^k)$ and line element, Levi-Civita connection and curvature tensor as follows:

$$ds^2 = g_{ij}\, dx^i dx^j$$

$$\Gamma^i{}_{jk} = \frac{1}{2} g^{im} (\partial_j g_{km} + \partial_k g_{jm} - \partial_m g_{jk})$$

$$R^i{}_{jkl} = \partial_k \Gamma^i{}_{jl} + \Gamma^i{}_{km} \Gamma^m{}_{lj} - (l \leftrightarrow k) \tag{1}$$

The Ricci tensor is $R_j = R^i{}_{jik}$. A contravariant vector field $V^i(x)$ and a covariant vector field $V_i(x)$ are defined as sets of n-functions on M which transform under a change of coordinates from $x^i \to x'^i(x^k)$ as

$$V'^i(x') = \frac{\partial x'^i}{\partial x^j} V^j(x)$$

$$V'_i(x') = \frac{\partial x^j}{\partial x'^i} V_j(x) \tag{2}$$

Tensors are defined by the obvious extension of (2) so that the transformation property of the metric tensor in particular is

$$g'_{ij}(x') = \frac{\partial x^k}{\partial x'^i} \frac{\partial x^l}{\partial x'^j} g_{kl}(x) \tag{3}$$

The covariant derivative

$$D_i V_j = \partial_i V_j - \Gamma^k_{ij} V_k$$

$$D_i V^j = \partial_i V^j + \Gamma^j{}_{ik} V^k \tag{4}$$

and its extension for tensors operates on a tensor and generates a new tensor with one more covariant index. For the curvature tensor we have the cyclicity, Bianchi, and Ricci identities

$$R^i{}_{jkl} + R^i{}_{klj} + R^i{}_{ljk} = 0$$

$$D_m R^i{}_{jkl} + D_k R^i{}_{jlm} + D_l R^i{}_{jmk} = 0$$

$$[D_i, D_j] T^k{}_l = R^k{}_{mij} T^m{}_l - R^m{}_{lij} T^k{}_m \tag{5}$$

Aside from these familiar relations, some knowledge of differential forms will be helpful but not essential. The reason we belabor such familiar material is that (1-5) are all applicable in complex manifolds and simplify in Kahler manifolds.

The viewpoint we will take is that an n-dimensional complex manifold is simply a 2n-dimensional real manifold parameterized by n-complex coordinates. This is mathematically correct only for local considerations which is sufficient for us. We will outline later the more rigorous view of the textbooks. To obtain complex coordinates one simply takes a real coordinate system $x^1 x^2 \ldots,$ $x^n, x^{n+1}, \ldots x^{2n}$ on M and defines

$$\begin{aligned} z^\alpha &= x^\alpha + i\, x^{\alpha+n} \\ \bar{z}^\alpha &= x^\alpha - i\, x^{\alpha+n} = z^{\bar{\alpha}} \end{aligned} \tag{6}$$

We then take Z^A to be a set of 2n complex coordinates where the index A runs first through the n unbarred or "holomorphic" indices α and then through the n barred or "antiholomorphic" indices $\bar{\alpha}$. We consider that (6) is a coordinate transformation of the type ordinarily considered in Riemannian geometry except that the new coordinates are complex. Complexity causes no difficulty and all formulae of Riemannian geometry, such as (1-5) are valid for complex manifolds when expressed in terms of 2n valued indices A,B,C.... In particular there is a real line element given by $ds^2 = g_{AB} dZ^A dZ^B$ with metric $g_{AB} = g_{BA}$ obtained from the Riemannian form (1), using the coordinate transformations (3) and (6). Reality is insured by the conditions $g_{\alpha\beta} = \bar{g}_{\bar{\alpha}\bar{\beta}}$ and $g_{\alpha\bar{\beta}} = \bar{g}_{\beta\bar{\alpha}}$.

A vector field $V_A(Z,\bar{Z})$ on a complex manifold has 2n components, V_α and $V_{\bar{\alpha}}$. We say that the vector field is real if $V_\alpha = \bar{V}_{\bar{\alpha}}$. A vector field obtained by the coordinate transformation (6) from a vector field on a 2n dimensional real manifold is real. Contravariant vectors V^A and tensors of arbitrary rank $T_{AB}{}^C$ can be viewed in a similar way. The covariant derivative is

$$D_A V_B = \frac{\partial}{\partial Z^A} V_B - \Gamma^C_{AB} V_C = \partial_A V_B - \Gamma^C_{AB} V_C$$
$$D_A V^B = \partial_A V^B + \Gamma^B_{AC} V^C \tag{7}$$

where the connection is defined just as in (1), i.e.

$$\Gamma^A_{BC} = \frac{1}{2} g^{AD} (\partial_B g_{CD} + \partial_C g_{BD} - \partial_D g_{BC}) \tag{8}$$

where g^{AB} is the inverse of g_{AB}, i.e. $g^{AC} g_{CB} = \delta^A{}_B$. One should note that the splitting of an index A into α and $\bar{\alpha}$ is not preserved by a general transformation of complex coordinates $Z'^\alpha = f^\alpha(Z,\bar{Z})$. However, the splitting is preserved by the special class of "holomorphic" coordinate transformation defined by $Z'^\alpha = f^\alpha(Z)$. Under the subset of coordinate changes, holomorphic indices α of any tensor transform into holomorphic indices α', and antiholomorphic indices $\bar{\alpha}$ transform into antiholomorphic indices $\bar{\alpha}'$.

We now define a Hermitean manifold as a complex manifold where there is a preferred class of coordinate systems in which unmixed components of the metric tensor vanish ($g_{\alpha\beta} = g_{\bar{\alpha}\bar{\beta}} = 0$). The line element then takes the Hermitean form $ds^2 = 2g_{\alpha\bar{\beta}} dZ^\alpha d\bar{Z}^\beta$. Coordinate systems for which this form holds are said to be adapted to the Hermitian structure. Hermitean metrics are definitely a restriction. Readers can show this by formulating the equations for a coordinate transform from a general line element $ds^2 = g_{AB} dZ^A dZ^B$ to a Hermitean one. There are in general too many equations to possess a solution. One can easily show that connection coefficients of the form $\Gamma^\alpha_{\bar{\beta}\bar{\gamma}}$ and $\Gamma^{\bar{\alpha}}_{\beta\gamma}$ vanish in a Hermitean manifold [6]. The geometrical meaning of this can be seen if we formally allow Z^α and $Z^{\bar{\alpha}}$ to be independent (i.e. $\bar{Z}^\alpha \neq Z^{\bar{\alpha}}$). Then we can show that a vector of type (1,0), i.e. of the form $(V^\alpha, 0)$, remains of type (1,0) under parallel transport along curves of the form $(Z^\alpha(t), \bar{Z}^\alpha_0)$ where the $Z^\alpha(t)$ are arbitrary functions of an affine parameter and the $\bar{Z}$ are constants [6]. An analogous property holds for vectors of type (0,1).

We adopt the custom of always writing the antiholomorphic index of the metric tensor $g_{\alpha\bar{\beta}}$ or its inverse $g^{\alpha\bar{\beta}}$ on the right, which is permissible since $g_{\alpha\bar{\beta}} = g_{\bar{\beta}\alpha}$. Note that the inverse satisfies $g^{\alpha\bar{\beta}} g_{\gamma\bar{\beta}} = \delta^{\alpha}{}_{\gamma}$. On any Hermitean manifold we can define the fundamental 2-form

$$\Phi = -2i\, g_{\alpha\bar{\beta}}\, dz^{\alpha} \wedge d\bar{z}^{\beta} \tag{9}$$

Its exterior derivative is given by

$$\begin{aligned} d\Phi = & -i(\partial_{\gamma} g_{\alpha\bar{\beta}} - \partial_{\alpha} g_{\gamma\bar{\beta}})\, dz^{\gamma} \wedge dz^{\alpha} \wedge d\bar{z}^{\beta} \\ & - i(\partial_{\bar{\gamma}} g_{\alpha\bar{\beta}} - \partial_{\bar{\beta}} g_{\alpha\bar{\gamma}})\, d\bar{z}^{\gamma} \wedge dz^{\alpha} \wedge d\bar{z}^{\beta} \end{aligned} \tag{10}$$

A Kahler manifold is a Hermitian manifold whose fundamental 2-form is closed, i.e. $d\Phi = 0$. The necessary and sufficient condition for this is that the metric tensor is curl free, i.e.

$$\begin{aligned} \partial_{\gamma} g_{\alpha\bar{\beta}} - \partial_{\alpha} g_{\gamma\bar{\beta}} &= 0 \\ \partial_{\bar{\gamma}} g_{\alpha\bar{\beta}} - \partial_{\bar{\beta}} g_{\alpha\bar{\gamma}} &= 0 \end{aligned} \tag{11}$$

These conditions imply that locally (i.e. in each coordinate patch), $g_{\alpha\bar{\beta}}$ can be represented by

$$g_{\alpha\bar{\beta}} = \frac{\partial}{\partial z^{\alpha}} \frac{\partial}{\partial \bar{z}^{\beta}} F(z,\bar{z}) \tag{12}$$

where $F(Z,\bar{Z})$ is called the Kahler potential. Note that the metric is invariant under changes of the potential of the form $F(Z,\bar{Z}) \to F(Z,\bar{Z}) + g(Z) + h(\bar{Z})$.

An immediate consequence of (11) is that many connection components vanish, specifically $\Gamma^{\alpha}_{\bar{\beta}\gamma} = \Gamma^{\alpha}_{\beta\bar{\alpha}} = 0$ in addition to $\Gamma^{\alpha}_{\bar{\beta}\bar{\gamma}} = \Gamma^{\bar{\alpha}}_{\beta\gamma} = 0$, which required only the Hermitean structure. Thus the only non-vanishing connection components are those of the form [5]

$$\begin{aligned} \Gamma^{\alpha}{}_{\beta\gamma} &= g^{\alpha\bar{\varepsilon}}\, \partial_{\beta} g_{\gamma\bar{\varepsilon}} \\ \Gamma^{\bar{\alpha}}{}_{\bar{\beta}\bar{\gamma}} &= g^{\varepsilon\bar{\alpha}}\, \partial_{\bar{\beta}} g_{\varepsilon\bar{\gamma}} \end{aligned} \tag{13}$$

Notice the simplification of the standard formula (1) or (8). Upon tracing one finds

$$\Gamma_{\alpha}{}^{\gamma}{}_{\gamma} = \partial_{\alpha} \log \det g_{\gamma\bar{\delta}} \tag{14}$$

which is also true for Riemannian manifolds.

There are equally stunning simplifications in the curvature tensor R_{ABCD} which is defined by the standard formula (1), with 2n-valued complex indices, and therefore has all the usual symmetries. One finds, using the simplifications of Γ^{A}_{BC} that only the following component types are non-vanishing in general:

$$R^{\alpha}{}_{\beta\gamma\bar{\delta}} = -R^{\alpha}{}_{\beta\bar{\delta}\gamma} \qquad R^{\bar{\alpha}}{}_{\bar{\beta}\gamma\bar{\delta}} = -R^{\bar{\alpha}}{}_{\bar{\beta}\bar{\delta}\gamma}$$
$$R_{\alpha\bar{\beta}\gamma\bar{\delta}} = -R_{\bar{\beta}\alpha\gamma\bar{\delta}} = -R_{\alpha\bar{\beta}\bar{\delta}\gamma} = R_{\bar{\beta}\alpha\bar{\delta}\gamma} \tag{15}$$

and we have the simple formulae [5].

$$R^{\alpha}{}_{\beta\gamma\bar{\delta}} = -\partial_{\bar{\delta}}\Gamma^{\alpha}{}_{\beta\gamma}$$
$$R_{\alpha\bar{\beta}\gamma\bar{\delta}} = \partial_{\gamma}\partial_{\bar{\delta}}g_{\alpha\bar{\beta}} - g^{\epsilon\bar{\eta}}\partial_{\gamma}g_{\alpha\bar{\eta}}\partial_{\bar{\delta}}g_{\epsilon\bar{\beta}} \tag{16}$$

Most important is the simplification in the Ricci tensor $R_{\alpha\bar{\beta}} = R^{\gamma}{}_{\alpha\gamma\bar{\beta}}$ which takes the form

$$R_{\alpha\bar{\beta}} = -\partial_{\alpha}\partial_{\bar{\beta}}\log\det g_{\gamma\bar{\delta}} \tag{17}$$

We also note that the cyclicity and Bianchi identities simplify in a Kahler manifold and become

$$R_{\alpha\bar{\beta}\gamma\bar{\delta}} = R_{\gamma\bar{\beta}\alpha\bar{\delta}}$$
$$D_{\epsilon}R_{\alpha\bar{\beta}\gamma\bar{\delta}} = D_{\gamma}R_{\alpha\bar{\beta}\epsilon\bar{\delta}} \tag{18}$$

while the Ricci identity implies that $[D_{\alpha}, D_{\beta}] = 0$ on any tensor (but $[D_{\alpha}, D_{\bar{\beta}}] \neq 0$ in general.)

Readers are now well equipped to shuffle indices on Kahler manifolds. One can observe that all formulae can be written directly in terms of the Kahler potential by substituting (12). For example, it follows from (17) that in a Ricci flat manifold

$$\log\det g_{\gamma\bar{\delta}}(Z,\bar{Z}) = f(Z) + f(\bar{Z}) \tag{19}$$

By a coordinate transformation one can usually eliminate the right side. Then by substituting (12) one finds

$$\det(\partial_{\alpha}\partial_{\bar{\beta}}F(z,\bar{z})) = 1 \tag{20}$$

This is the Monge-Ampere equation, which can be viewed as a non-linear partial differential equation for an unknown scalar $F(Z,\bar{Z})$. The solution then determines a Ricci flat Kahler manifold. This equation has been solved to determine one Kahler description [9] of the Eguchi-Hanson metric (which is distinct from the form found in our work [2]).

A very interesting consequence of Kahler geometry is the result that a Kahler manifold of complex dimension 2 is automatically a self-dual gravitational instanton, i.e. a self-dual solution of the Euclidean signature Einstein equations. The converse is also true [4], but we prove here only the direct statement which depends only on the change from complex to real coordinates defined by $(Z^1, Z^2, \bar{Z}^1, \bar{Z}^2) \to (x^1, x^2, x^3, x^4)$ with $Z^i = x^i + i\,x^{i+2}$ and $\bar{Z}^i = x^i - i x^{i+2}$. The curvature tensor in complex components then becomes

$$R_{ijkl} = \frac{\partial Z^A}{\partial x^i}\frac{\partial Z^B}{\partial x^j}\frac{\partial Z^C}{\partial x^k}\frac{\partial Z^D}{\partial x^l} R_{ABCD} \tag{21}$$

We now use the two facts that the only non-vanishing curvature tensor components are of the form $R_{\alpha\bar{\beta}\gamma\bar{\delta}}$, while self-duality implies

$$\begin{aligned} R_{ij12} &= R_{ij34} \\ R_{ij13} &= R_{ij42} \\ R_{ij14} &= R_{ij23} \end{aligned} \tag{22}$$

Since self-duality affects only the last two indices, we need only check the real and complex components of a 2nd rank antisymmetric tensor $T_{\alpha\bar{\beta}} = -T_{\bar{\beta}\alpha}$ with vanishing unmixed components. By direct computation we find

$$\begin{aligned} T_{12}\big|_{real} &= T_{1\bar{2}} + T_{\bar{1}2} \\ T_{34}\big|_{real} &= T_{1\bar{2}} + T_{\bar{1}2} \end{aligned} \tag{23}$$

so that $T_{12} = T_{34}$. The other two relations of (22) are proved similarly, so that self-duality holds.

One of the simplest sets of Kahler manifolds are the complex projective spaces CP^n defined as the homogeneous spaces $U(n+1)/$

$U(n) \times U(1)$ when $U(p)$ is the unitary group. Since $U(n+1)/U(n)$ can be identified with the sphere $\sum_{J=1}^{n+1} \bar{u}^j u^j = 1$ in n+1 complex dimensions we see that a point of CP^n is the equivalence class defined by a point $(u^1 \cdots u^{n+1})$ on the sphere together with all points which differ from this one representative by a phase, i.e. $(w^1 \cdots w^{n+1}) = e^{i\alpha}(u^1 \cdots u^{n+1})$. We shall obtain an explicit parameterization of CP^n as done in the physics literature of the CP^n non-linear σ-models [10]. We consider a constrained Lagrangian for n+1 complex scalar fields $u^i(x)$.

$$\mathcal{L} = \partial_\mu \bar{u}^i \partial^\mu u^i \qquad \sum_{i=1}^{n+1} \bar{u}^i u^i = 1 \tag{24}$$

The invariance group of this Lagrangian is $U(n+1)$. We must now incorporate the fact that a common $U(1)$ phase of the $u^i(x)$ is irrelevant, and we do this by gauging a $U(1)$ subgroup of $U(n+1)$ by using the Lagrangian

$$\mathcal{L} = (\partial_\mu + iA_\mu)\bar{u}^i (\partial^\mu - iA^\mu) u^i \tag{25}$$

Since A_μ is now an auxiliary field we solve for it and resubstitute obtaining

$$\mathcal{L} = \partial_\mu \bar{u}^i \partial^\mu u^i + \frac{1}{4}(\bar{u}^i \overleftrightarrow{\partial}_\mu u^i)^2$$
$$\sum_{i=1}^{n+1} \bar{u}^i u^i = 1 \tag{26}$$

We can now simultaneously fix the $U(1)$ gauge and solve the constraint by expressing the $u^i(x)$ in terms of n-independent complex fields $Z^i(x)$ by

$$(u^1, \ldots, u^{n+1}) = \frac{1}{\sqrt{1 + z\cdot\bar{z}}}(z^1, \ldots, z^n, 1)$$
$$z\cdot\bar{z} = \sum_{\alpha=1}^{n} \bar{z}^\alpha z^\alpha \tag{27}$$

We substitute (27) in (26) and find the Lagrangian

$$\mathcal{L} = \frac{1}{1+z\cdot\bar{z}}\left(\delta_{\alpha\beta} - \frac{\bar{z}_\alpha z_\beta}{1+z\cdot\bar{z}}\right)\partial_\mu z^\alpha \partial^\mu \bar{z}^\beta \tag{28}$$

Then a metric of CP^n is defined by (28) and takes the form

$$g_{\alpha\bar{\beta}} = \frac{1}{1+Z\cdot\bar{Z}}\left(\delta_{\alpha\beta} - \frac{\bar{Z}_\alpha Z_\beta}{1+Z\cdot\bar{Z}}\right) \tag{29}$$

which is curl free and has Kahler potential $F(Z,\bar{Z}) = \ln(1+Z\cdot\bar{Z})$.

In the final part of this note we will attempt to give a simple outline of the rigorous approach to complex manifolds. Up to now we have been technically incorrect in supposing that a n-dimensional complex manifold is simply a real 2n-dimensional manifold viewed from complex coordinates. The correct definition of a complex manifold is a set of points which can be covered by a set of complex coordinate charts, i.e. mappings from open sets of C^n. In the intersection of two charts the transition functions are required to be holomorphic i.e. the coordinate transformations $Z'^\alpha = f^\alpha(Z)$ i.e. $\frac{\partial Z'^\alpha}{\partial \bar{Z}^\beta} = 0$. The mathematics textbooks develop considerable machinery relevant to the question of when a 2n dimensional real manifold is actually a complex manifold. We shall discuss some of this apparatus here in order to familiarize readers with some of the concepts in the mathematics literature.

A key element in the rigorous treatment is the concept of a complex structure which is nothing else but the formalization of "multiplication by i" smoothly over the manifold, i.e. an operation on geometrical objects whose square is minus the identity. For example on flat R^{2n}, the canonical complex structure is given by the matrix

$$(J_i{}^j) = \begin{pmatrix} 0 & \mathbb{1} \\ -\mathbb{1} & 0 \end{pmatrix} \tag{30}$$

Since any 2n dimensional manifold is locally equivalent to R^{2n}, the natural definition of a complex structure is a (1,1) tensor field $J_i^j(x)$ with the property

$$J_i{}^j(x) J_j{}^k(x) = -\delta_i{}^k .$$

A 2n-dimensional real manifold which has a smooth complex structure is called an almost complex manifold. Not all 2n-dimensional manifolds admit a complex structure, a famous counterexample is the 4-dimensional sphere S^4 (indeed, only the spheres S^2 and S^6 admit almost complex structures [5]).

An almost complex space is called almost Hermitean if there is a Riemannian metric g_{ij} which is invariant under the action of the complex structure

$$J_i{}^k J_j{}^l g_{kl} = g_{ij} \tag{31}$$

In this case $J_{ij} = J_i^{\ k} g_{kj}$ is antisymmetric. One can show that this is equivalent to the existence of complex coordinates $Z^A = (Z^\alpha, \bar{Z}^\alpha)$ such that $g_{\alpha\beta} = g_{\bar{\alpha}\bar{\beta}} = 0$. An almost Hermitean space is called an almost Kahler space if the fundamental 2-form (9) is closed.

One can show [5,6] that an almost complex structure over a 2n dimensional real manifold gives a complex manifold if and only if its Nijenhuis tensor

$$N_{ij}^{\ \ k} = J_i^{\ l}(\partial_l J_j^{\ k} - \partial_j J_l^{\ k}) - J_j^{\ l}(\partial_l J_i^{\ k} - \partial_i J_l^{\ k}) \tag{32}$$

vanishes. If $N_{i,j}^{\ \ k} = 0$ and (31) holds, we have a complex Hermitean manifold, and if in addition (9) is closed, we have a Kahler manifold. In any Kahler manifold the complex structure has vanishing covariant derivatives, $D_k J_{ij} = 0$, so that the complex structure is parallel. Conversely, if $D_k J_{ij} = 0$ then the manifold is Kahler.

Finally we discuss the holonomy group of a Kahler manifold (which was the subject of a running debate between one of the authors and B. Zumino at the Conference, and which could not be settled because Erice lacked a copy of Kobayashi and Nomizu or Lichnerowicz.) The simplest way to define the holonomy group of a general n-dimensional real manifold is to consider a point p and all closed loops passing through it. If we take a tangent vector V^i at p and parallel transport it around the loop, we find another tangent vector V'^i which is related to V^i by an element of GL(n,R). The subset of GL(n,R) determined in this way from all vectors and all paths is a group which is the holonomy group. The restricted holonomy group is the subgroup defined by paths which can be shrunk to a point. If the manifold is connected, the holonomy groups at any two points p and p' are the same. Hence the holomony group is always a subgroup of GL(n,R) and always a subgroup of O(n) if parallel transport is performed with the Levi-Civita connection. Thus in a complex n-dimensional manifold we would expect that the holomony group is a subgroup of O(2n)(for the Levi-Civita connection). For a Kahler manifold the holonomy group is typically smaller. If the Kahler space is Ricci flat, then the restricted holonomy group is a subgroup of SU(n) [7]. If the Kahler space has non-zero Ricci tensor, then the holonomy group has a non-discrete center, typically a U(1) factor [7].

Readers should now be able to follow the recent applications of Kahler geometry in supersymmetric σ-models [1-3] and to appreciate the power of these geometrical ideas. We hope that they will be stimulated to study the subject further and find further physical applications.

REFERENCES

1. L. Alvarez-Gaumé and D.Z. Freedman, Stony Brook preprint ITP-SB-80-13 (1980, to be published in Physical Review D.
2. L. Alvarez-Gaumé and D.Z. Freedman, Stony Brook preprint ITP-SB-80-14 (1980), to be published in Physics Letters B.
3. B. Zumino, Phys. Letters 87B (1979) 203.
4. M.F. Atiyah, N. Hitchin and I.M. Singer, Proc.Roy.Soc. A362 (1978) 425.
5. S. Kobayashi and K. Nomizu, "Foundations of Differential Geometry", Vol. II (Wiley Interscience, New York, 1963).
6. K. Yano, "Differential Geometry on Complex and Almost Complex Manifolds" (Macmillan Ed. Co., 1965).
7. A. Lichnerowicz, "Théorie Globale des connections et des groupes d'holonomie", Consiglio Nazionale delle Ricerche, Roma (1962).
8. S. Goldberg, "Curvature and Homology" (Academic Press, New York, 1962).
9. G.W. Gibbons and C.N. Pope, Comm. Math. Phys. 66 (1979) 267.
10. H. Eichenherr, Nuclear Phys. B146 (1978) 215.

PRECISE PREDICTIONS OF GUTs AND OF THE HIGGS MESON MASS[×]

K. T. Mahanthappa and Marc A. Sher[*]

Department of Physics
University of Colorado, Boulder
Boulder, Colorado 80309

At this meeting, we will hear a great deal about various grand unified theories (GUTs). We would, of course, eventually like to know which, if any, of these GUTs provides an accurate description of nature. While certain experimental results, such as neutrino oscillations, could eliminate some GUTs, distinguishing between them all necessitates precise theoretical predictions of certain quantities and precise experimental determination of these quantities. This is somewhat difficult since the symmetry of GUTs is, in most cases, only manifest at momentum transfers above 10^{14} GeV (note that colliding e^+e^- rings with such energies would have a radius slightly larger than that of the observed universe, although by the time they are built, the universe may have expanded sufficiently). Predictions can be made[1], however, using the renormalization group equations to extrapolate the current values of coupling constants and masses to high momentum transfers.

×) Talk presented by M.A. Sher

*) Present address: University of California, Santa Cruz 94064.

In this talk, I will first discuss the main predictions of GUTs, starting with the simplest GUT, SU(5)[2]. These predictions[1-3] include the proton lifetime, τ_p, the b-quark mass, m_b, and the weak mixing angle, $\sin^2\theta$. The most precise method of determining the mixing angle, and thereby precisely testing SU(5), will then be discussed. It will then be shown that the simple form of SU(5), i.e. with only a $\underline{5}$ of Higgs breaking the SU(3) x SU(2) x U(1) symmetry, is unfortunately the only GUT in which precise predictions are possible. I will then note that the measurement of the Higgs meson mass will provide another accurate determination of the mixing angle, if the symmetry breakdown is caused by radiative corrections.[4] This determination will necessitate calculation of the $O(\alpha^2)$ radiative corrections to the effective potential in the Weinberg-Salam model. First, let us consider the predictions of the simple form of SU(5).

The details of the calculation of τ_p and $\sin^2\theta$ are discussed in these Proceedings by D. Ross[5], and the details of the b-quark mass calculations can be found in Ref. 6. The effects of two-loop and scala meson contributions to the beta-functions, as well as fermion and vector boson threshold effects are included. We have checked the results of Refs. 5 and 6 and modified the initial conditions to ensure that the same definitions of the coupling constants are employed. The results are listed in Table I. We list m_b, τ_p and $\sin^2\theta$ as a function of Λ and $\alpha_s(Q^2 = 4m_w^2)$, defined in the symmetric-momentum subtraction scheme.

The uncertainty in the b-quark mass is approximately 300 MeV, primarily due to three-loop effects. There is a somewhat greater uncertainty due to the differing definitions of the quark mass. We have drawn a line corresponding to $\Lambda = 0.4$ GeV. If Λ is greater than this, then the b-quark mass prediction will be too high to be consistent with the observed upsilon mass. As will be discussed later, the uncertainty[5] in τ_p is, given Λ, a factor of 10. We have drawn a line corresponding to $\Lambda = 0.15$ GeV, or $\alpha_s(2m_w) = 0.13$.

Table I

Predictions of m_b, τ_p and $\sin^2\theta(m_W)$ in SU(5) as a function of Λ

Λ(GeV)	$\alpha_s(2\ m_W)$	m_b(GeV)	τ_p(years)	$\sin^2\theta(m_W)$
0.10	0.127	4.7	0.8×10^{29}	0.208
0.17	0.132	4.9	2.0×10^{29}	0.207
0.24	0.137	5.1	4.4×10^{29}	0.206
0.30	0.142	5.3	1.0×10^{30}	0.205
0.36	0.147	5.5	2.1×10^{30}	0.204
0.42	0.152	5.7	4.0×10^{30}	0.202

Uncertainties in these values are discussed in the text.

The value of m_x for Λ = .30 GeV is 4.3×10^{14} GeV.

If Λ is smaller than this, then the predicted proton lifetime will be too short to be consistent with the observed[7] lower limit of 10^{30} years. Note that an order of magnitude increase in this limit will bring the two lines dangerously close together, and an increase of two orders of magnitude will rule out the simple SU(5) model.

The most precise prediction of the model is $\sin^2\theta$. The uncertainty, primarily due to threshold effects, is approximately 1%. It is not correct to compare the results with the "experimental" value[8] of $\sin^2\theta = 0.230 \pm 0.015$. This value is determined from tree-level calculations, thus there is no distinction between, say, $\sin^2\theta \equiv e^2(q^2)/g^2(q^2)$ for any q^2 and $\sin^2\theta \equiv 1 - m_W{}^2/m_Z{}^2$. Since radiative corrections can change the results by several percent, an additional uncertainty must be added. Once radiative corrections to experiments have been calculated and the definition used for $\sin^2\theta$ matched with our definition, then this uncertainty will disappear. Our definition is $\sin^2\theta = e^2(m_W{}^2)/g^2(m_W{}^2)$, where the coupling constants are defined in the symmetric-momentum subtraction scheme, with fermion (but not vector boson) thresholds included. (Our results for $\sin^2\theta$ agree to within 0.001 with those of Refs. 5 and 9).

Present neutral current experiments give $\sin^2\theta$ with uncertainties of 5 - 10%. The most accurate determination of $\sin^2\theta$ in the near future will most likely come from measurement of the W and Z vector boson masses. Using the full expression (including radiative corrections[10] and all fermions) for muon decay, for our definition of $\sin^2\theta$ we find*)

$$m_W = 38.0/\sin\theta \text{ GeV} \qquad (1$$

*) Marciano[11] has $m_W = 38.5/\sin\theta$. His definition of $\sin^2\theta$ does include radiative corrections to muon decay (~1%), but not fermion thresholds (~1%). Using our definition changes his 38.5 to 38.1. Either definition is acceptable, as long as it agrees with that calculated in SU(5).

Radiative corrections to the m_Z to m_W mass ratio have also been calculated.[12] Again, using our definition of $\sin^2\theta$, m_Z/m_W can be determined.

The results[13] for m_W and m_Z are listed in Table II. The uncertainties in the calculations of m_W and m_Z are approximately 500 and 600 MeV, respectively, for given Λ.

These quantities are the most precise predictions of the simple SU(5) model. In order to compare the predictions of SU(5) with those of other GUTs, similar calculations in other GUTs must be performed. A measurement of τ_p, $\sin^2\theta$, m_W and/or m_Z will then enable us experimentally to distinguish between them. Unfortunately, the simple form of SU(5) is the only GUT in which such precise predictions are possible. The reason is as follows.

Table II

Predictions of SU(5) with three fermion families and one Higgs doublet

Λ(GeV)	$\alpha_s(2m_W)$	$\sin^2\theta(m_W)$	m_W(GeV)	m_Z(GeV)
0.10	0.127	0.208	83.2	93.7
0.17	0.132	0.207	83.4	93.9
0.24	0.137	0.206	83.6	94.1
0.30	0.142	0.205	83.8	94.3
0.36	0.147	0.204	84.0	94.4
0.42	0.152	0.203	84.2	94.6

In SU(5), the second stage of symmetry breaking is due to either a 5 of Higgs, a 45 of Higgs, or both. For both multiplets, one color-singlet SU(2) doublet acquires a mass of $0(m_W)$, while the remaining Higgs mesons, all colored, acquire masses of $0(m_X)$. One assumption that has been made in the above calculations is that the masses of the colored Higgs mesons are sufficiently close to m_X that their effects on the renormalization group equations are negligible. We now critically examine this assumption,[14] and show that, except in the case of the simplest Higgs structure, it is not justified.

Let us first consider the case in which only a 5 breaks the symmetry.[3] In addition to the light Higgs doublet, the 5 contains a colored triplet (SU(2) singlet) of Higgs with $(\text{mass})^2$ of $0(\beta m_X^2)$, where β is a parameter in the Higgs potential related to the mixing of the 5 and 24 of Higgs. We have no knowledge of the value of β. However, it has been shown that β must be smaller than unity in order that the SU(3) x SU(2) x U(1) minimum be the absolute minimum. If β is extremely small, however, renormalization effects will produce an effective β of $0(\alpha^2)$, where $\alpha = g_G^2/4\pi \approx 0.024$. Thus we conclude that $0(\alpha^2) < \beta < 0(1)$, and thus

$$\alpha m_X \lesssim m_H \lesssim m_X$$

where m_H is the mass of the Higgs triplet. Since $m_H \lesssim m_X$, there will be an effect on the renormalization group equations which, neglecting thresholds, occurs for $q^2 > 4m_H^2$. Since the closer m_H is to m_X, the smaller the effect on the renormalization group equations, we will set $m_H = \alpha m_X$ in order to get the largest possible effect. The difference in the effects between $m_H = m_X$ and $m_H = \alpha m_X$ will yield the uncertainties in the calculations of $\sin^2\theta$, m_X (and thus τ_p) and m_b caused by our lack of knowledge of β, and thus of m_H.

The calculation is rather simple. We simply modify the beta-functions to include an additional scalar color triplet once $q^2 > 4\alpha^2 m_X^2$. For definiteness, we set $\alpha_s(2m_W) = 0.137$. The results are in Table III. It is clear that m_X is uncertain by about 5%,

$\sin^2\theta$ by 0.5% and m_b is unchanged. τ_p is thus uncertain by about 20%. These uncertainties are much smaller than the previous uncertainties; thus the neglect of the effects of the triplet is justified. If a 45 breaks the symmetry, however, one expects a larger effect, simply because there are so many more colored scalars in the multiplet. We now turn to these alternate Higgs structures involving a 45.

The 45 decomposes[15] into seven SU(3) x SU(2) multiplets of Higgs given by (8,2) $(6^*,1)$, (3,3), $(3^*,2)$, $(3^*,1)$, (3,1) and (1,2). The (1,2) is the usual Higgs doublet and acquires a light mass, while the other six multiplets acquire heavy masses. Call these masses m_i (i = 1-6) in the order they occur above. Their effects on the renormalization group equations can be easily found from Ref. 16. The full potential is given in Ref. 15. In this case, there are 5 β_i's which mix the 45 and 24 of Higgs. As before, the m_i^2 are proportional to $O(\beta_i m_x^2)$ and cannot be determined precisely. However, as before, we can conclude that

$$\alpha m_x \lesssim m_i \lesssim m_x$$

In this instance, the largest change in $\sin^2\theta$ and m_x occurs when some of the Higgs multiplets have masses αm_x and others have masses m_x. We have examined all possible combinations of masses consistent with the Higgs potential.

Table III

Values of $\sin^2\theta$, m_x and m_b for different values of the mass of the colored triplet of the 5 of Higgs, m_H

m_H	$\sin^2\theta(m_w)$	m_x(GeV)	m_b(GeV)
$m_H = m_x$ (no effect)	0.206	3.5×10^{14}	5.2
$m_H = \alpha m_x$	0.205	3.7×10^{14}	5.2

The results for some choices of the m_i are given in Table IV, which includes extreme values. Since we have no knowledge of the actual masses, any of the results for $\sin^2\theta$ and m_X are possible. It can be seen that the uncertainty in $\sin^2\theta$ is 4%, and in m_X is a factor of 2.8. Thus, if a 45 is involved in the Higgs structure of SU(5), then the uncertainty due to our lack of knowledge of the colored scalar masses is 4% in $\sin^2\theta$ and a factor of 60 (or $(2.8)^4$) in the proton lifetime.

Table IV

Values of $\sin^2\theta$ and m_X for various values of the m_i (i = 1-6)*

m_i/m_X	$\sin^2\theta(m_W)$	m_X(GeV)
α(i = 1,2,3,4,5,6)	0.205	3.7×10^{14}
α(i = 3,4,5,6) 1(i = 1,2)	0.209	1.3×10^{14}
α(i = 1,2,6) 1(i = 3,4,5)	0.201	10.3×10^{14}
α(i = 3) 1(i = 1,2,4,5,6)	0.214	2.8×10^{14}
α(i = 2,5,6) 1(i = 1,3,4)	0.199	4.3×10^{14}

which are the masses of the (8,2), (6,1), (3,3), (3*,2), (3*,1) and (3,1) of the 45 of Higgs, respectively. We have assumed three fermion families for definiteness.

In the simplest form of SU(5) in which only a 5 breaks the SU(3) x SU(2) x U(1) symmetry, the uncertainty, as discussed earlier, in $\sin^2\theta$ is 1.0% and in τ_p is a factor of 5. In any form of SU(5) which contains a 45, the respective uncertainties, given Λ, are at least 5% and a factor of 600. The uncertainty in m_b, in SU(5) models which contain a 45, is, of course, very model-dependent, but turns out to be several hundred MeV.

In any other GUT, there are a large number of Higgs mesons which require heavy masses. Thus, calculations in these models will have similarly large uncertainties. We therefore see that the only GUT which can make precise predictions is the simplest form of SU(5). The other GUTS could be eliminated (if, say $\tau_p > 10^{35}$ years), but cannot be precisely tested.

Let us now return to the simple form of SU(5). There possibly exists one additional method of determining $\sin^2\theta$ precisely -- by determining the mass of the Higgs meson. In the standard model, the mass of the Higgs meson is arbitrary, yet it can be calculated precisely using an increasingly popular assumption made by S. Coleman and E. Weinberg.[4] They assumed that the bare, or classical, mass of the Higgs, μ^2, is identically zero, and thus the symmetry breakdown is entirely due to quantum corrections. The expression for the Higgs mass, m_s, is

$$\frac{m_s^2}{m_W^2} = \frac{3e^2}{16\pi^2 \sin^2\theta} (1 + 1/2 \sec^4\theta) \qquad (2$$

This assumption is totally ad hoc, yet does have some strong motivating factors. It was shown by Weinberg[17] that a large gauge hierarchy can be obtained if, after the first stage of symmetry breaking, one or more Higgs mesons has either zero or very small bare mass. Thus, μ^2 must be very small. In fact, if one believes that the grand unification scale is the true scale in nature, then the requirement that m_s be less than 1000 GeV or so (a perfectly reasonable requirement) forces μ^2 to be less than 10^{-24} (in units of m_X). So, μ^2 must be very, very small. The hope of theoretical physics is to understand all of the arbitrary parameters in nature. Presumably, zero will be a simpler number to explain than 10^{-24}. Also, if there are no bare masses in the Lagrangian, then the "hierarchy of hierarchies" can be explained.[18] The physical basis of this assumption may come from superunified theories unifying GUTs with supersymmetry.[19] We thus assume that μ^2 vanishes and the Higgs mass is therefore calculable.

In this case, the Higgs mass is approximately 10 GeV. The phenomenology of such a Higgs is enormously complicated due to the presence of the bottomonium states (upsilon) in the same mass region. Ellis et al.[20] have studied this phenomenology, including the possibility of strong mixing between the Higgs meson and bottomonium states. They also pointed out that the value of the mass is very sensitive to sin θ ($m_s \sim 1/\sin^2\theta$) and thus a measurement of the mass would allow a precise determination of $\sin^2\theta$ under the above assumption. This would constitute a precise test of SU(5).

A higher order calculation of Equation (2) to $O(e^4)$ would be useful for several reasons: (i) a measurement of the mass would then allow a determination of the precise value of $\sin^2\theta$, the comparison of which, with the values determined in neutral current experiments (once radiative corrections to these experiments are calculated), leads to the test of the assumption $\mu^2 = 0$; (ii) as discussed above, assuming $\mu^2 = 0$, the precise value of $\sin^2\theta$ obtained from the measured value of m_s provides a test of grand unified models; (iii) detection techniques differ very much according to whether the mass of the Higgs scalar is below or above that of the various upsilon states, and some clue as to its relative position could be obtained by using the precisely determined value of $\sin^2\theta$ in SU(5) in the higher order formula for m_s.

With this in view, we calculate the $O(e^4)$ terms of the scalar-vector mass ratio.[21] In deriving Equation (2), the one-loop effective potential and the tree-level masses for m_s and m_W are needed. To calculate the next order corrections, we need the two-loop effective-potential, the one-loop vector propagator, the one-loop scalar propagator and the renormalization constants associated with g and g' of SU(2) and U(1) respectively.

The most general form of the effective potential through two loops is:

$$V(\phi) = \frac{1}{4}\lambda\phi^4 + \frac{1}{16\pi^2}\frac{1}{4}[a_1\lambda^2 + a_2\lambda g^2 + a_3\lambda g'^2 + a_4 g^4$$
$$+ a_5 g^2 g'^2 + a_6 g'^4]\phi^4 \ln(g^2\phi^2/4M^2)$$
$$+ \frac{1}{(16\pi^2)^2}\frac{1}{4}\{[b_1\lambda^3 + b_{21}\lambda^2 g^2 + b_{22}\lambda^2 g'^2 + b_{31}\lambda g^4$$
$$+ b_{32}\lambda g^2 g'^2 + b_{33}\lambda g'^4 + b_4 g^6 + b_5 g^4 g'^2 + b_6 g^2 g'^4$$
$$+ b_7 g'^6]\phi^4 + [c_1\lambda^3 + c_{21}\lambda^2 g^2 + c_{22}\lambda^2 g'^2 + c_{31}\lambda g^4$$
$$+ c_{32}\lambda g^2 g'^2 + c_{33}\lambda g'^4 + c_4 g^6 + c_5 g^4 g'^2 + c_6 g^2 g'^4$$
$$+ c_7 g'^6]\phi^4 \ln(g^2\phi^2/4M^2) + [d_1\lambda^3 + d_{21}\lambda g^2$$
$$+ d_{22}\lambda^2 g'^2 + d_{31}\lambda g^4 + d_{32}\lambda g^2 g'^2 + d_{33}\lambda g'^4 + d_4 g^6$$
$$+ d_5 g^4 g'^2 + d_6 g^2 g'^4 + d_7 g'^6]\phi^4 \ln^2(g^2\phi^2/4M^2)\}$$

where a_i, b_i, c_i and d_i are coefficients to be determined. Unlike the case of scalar QED, there are two gauge coupling constants and hence the logarithms of coupling constants cannot be absorbed by choosing the renormalization point judiciously. For simplicity, we choose our renormalization point to be m_W. The choice of M is arbitrary, and cannot affect the physical results, but it must be noted that the coupling constants are defined at m_W. The coefficients a_i, b_i, c_i and d_i may thus contain $\ln(m_z{}^2/m_W{}^2)$ terms. Most of the two-loop integrals, as a result, must be done numerically.

The vector-meson and scalar-meson propagators can be written as

$$G_{\mu\nu}^{-1}(p^2) = -ig_{\mu\nu}\{p^2 - \frac{1}{4} g^2<\phi>^2 - \frac{1}{16\pi^2} g^2(yp^2 - z<\phi>^2) + O(\lambda p^2, g^2p^4)\}$$

and

$$G^{-1} = i \{ p^2 - \frac{1}{16\pi^2} g^2p^2x - \frac{d^2V}{d\phi^2}\Big|_{\phi=<\phi>} + O(\lambda p^2, g^2p^4)\}$$

where x, y and z are coefficients to be determined. The terms of $O(\lambda p^2, g^2p^4)$ will not enter into our calculation to $O(e^4)$. The particle masses are given by the zeroes of the inverse propagators. Finally, we must express the results in terms of renormalized coupling constants. We define

$$g^2 = g_R^2 (1 - \frac{1}{16\pi^2} g_R'^2 s)$$

$$g'^2 = g_R'^2 (1 - \frac{1}{16\pi^2} g_R'^2 r)$$

where s and r are to be determined. We consider here only the finite parts of the renormalization; the infinite parts (which do affect the two-loop potential) are incorporated into the potential.

Minimizing the potential and finding the poles of the propagators yields, dropping the subscript R,

$$\frac{m_s^2}{m_W^2} = \frac{1}{16\pi^2} \{8a_4g^2 + 8a_5g'^2 + 8a_6g'^4/g^2\}$$

$$+ \frac{1}{(16\pi^2)^2} \{(8a_4x - 8a_4s + 8a_2a_4 + 4c_4 + 8d_4 - 8a_4y - 32a_4z)g^4 + (8a_5x + 8a_3a_4 + 8a_2a_5 + 4c_5 + 8d_5 - 8a_5y - 32a_5z)g^2g'^2 + (8a_6x + 8a_3a_5 + 8a_2a_6 - 8a_5r + 8c_6 + 8d_6 + 8a_6s - 8a_6y - 32a_6z)g'^4 + (8a_3a_6 - 16a_6r + 4c_7 + 8d_7)g'^6/g^2\}$$

As we see, there is no need to calculate the terms in the effective potential proportional to ϕ^4 nor the terms with a λ in one of the vertices.

The relevant quantities have been calculated: see Ref. 21 for details. The two-loop potential was calculated in the Feynman gauge using the method of Ref. 22. A strong check on the calculation is the cancellation of divergences in the potential and renormalized propagators.

The major complication arises due to the fact that one cannot just subtract the vector self-energies on shell to determine the renormalized propagator. This is because the photon renormalization constant is not independent of the W and Z renormalization constants, and thus, if we subtract on shell the self-energies of the W and Z; then we can't subtract on shell the self-energy of the photon without violating Ward identities. We must use the renormalization procedure in the Feynman gauge of Refs. 23 and 24.

The result is given in Table V. Previously,[20] a value of 10.0 GeV for m_s was quoted for $\sin^2\theta = 0.208$ (the definition of $\sin^2\theta$ is irrelevant at the $O(\alpha)$ level). If the t-quark mass is larger than 15 GeV, the result is lowered by $6(m_t/15\ \text{GeV})^4$MeV. We estimated our uncertainties at 50 - 100 MeV.

Thus assuming both dynamical symmetry breaking and the simple form of SU(5), we find that the Higgs meson mass is 10.8 ± 0.3 GeV. This will be above the states of bottomonium, thus the best method of detection[20] will be in toponium decays.

Of course, if the Higgs mass is not close to 10.8 GeV (say, either less than 9 GeV or greater than 12 GeV), then the assumption that $\mu^2 = 0$ will not be correct, and the value of the mass will not enable a determination of $\sin^2\theta$, and thereby be a test of SU(5). If the mass of the Higgs is outside the above range, there are three possibilities: either SU(5) is wrong (or has a more complicated Higgs structure), $\mu^2 \neq 0$ or there are heavy fermions. In this case, it will probably be necessary to find m_w or m_z to determine the reason for the discrepancy.

Table V

Production of Higgs meson mass in SU(5)
Assuming dynamical symmetry breaking

Λ(GeV)	$\alpha_s(2m_w)$	$\sin^2\theta(m_w)$	m_s(GeV)
0.10	0.127	0.208	10.7
0.17	0.132	0.207	10.7
0.24	0.137	0.206	10.8
0.30	0.142	0.205	10.8
0.36	0.147	0.204	10.9
0.42	0.152	0.203	10.9

In conclusion, we have examined the precise predictions that are made in the GUT based on SU(5). The simple form of the model allows a very restricted range of values for $\sin^2\theta$, m_w and m_z. Assuming symmetry breaking by radiative corrections, it also allows a restricted range of values for the Higgs mass. Higher order calculations of m_w and m_z were previously done; we calculated higher order corrections to m_s. In any other group, or in a more complicated form of SU(5), lack of knowledge of parameters in the Higgs potential preclude any precise test.

REFERENCES

1. H. Georgi, H. Quinn and S. Weinberg, Phys. Rev. Letters 33 (1974) 451.
2. H. Georgi and S. Glashow, Phys. Rev. Letters 32 (1975) 438.
3. A.J. Buras, J. Ellis, M.K. Gaillard and D.V. Nanopoulos, Nuclear Phys. B135 (1978) 66.
4. S. Coleman and E. Weinberg, Phys. Rev. D7 (1973) 1888.
5. D.A. Ross, these proceedings p. 623 , see also T. Goldman and D.A. Ross, Caltech preprint CALT-68-759 (1980).
6. D.V. Nanopoulos and D.A. Ross, Nuclear Phys. B157 (1979) 273.
7. J. Learned, F. Reines and A. Soni, Phys. Rev. Letters 43 (1979) 907.
8. C.Y. Prescott et al., Phys. Letters 77B (1978).
9. A. Din, G. Girardi and P. Sorba, Phys. Letters 91B (1980) 77.
10. T. Appelquist, J. Primack and H. Quinn, Phys. Rev. D6 (1972) 2998 and Phys. Rev. D7 (1973) 2998.
    A. Lahanas, University of Athens preprint (1980).
11. W. Marciano, Phys. Rev. D20 (1979) 274.
12. W. Marciano, Nuclear Phys. B84 (1975) 132.
13. K.T. Mahanthappa, and M.A. Sher, Colorado preprint COLO-HEP 13 (to be published in Phys. Rev.) (1980).
14. G.P. Cook, K.T. Mahanthappa and M.A. Sher, Phys. Letters 90B (1980) 298.
15. P.H. Frampton, S. Nandi and J.J.G. Scanio, Phys. Letters 85B (1979) 255.
16. D. Gross and F. Wilczek, Phys. Rev. 38 (1973) 3633.
17. S. Weinberg, Phys. Letters 82B (1979) 387.
18. J. Ellis, M.K. Gaillard, A. Peterman and C.T. Sachrajda, Nuclear Phys. B164 (1980) 253.
19. J. Ellis, M.K. Gaillard and B. Zumino, CERN preprint TH.2842/ LAPP TH-16 (1980).
20. J. Ellis, M.K. Gaillard, D.V. Nanopoulos and C.T. Sachrajda, Phys. Letters 83B (1979) 339.
21. K.T. Mahanthappa and M.A. Sher, Colorado preprint COLO-HEP-18 (1980). See also Ref. 24.
22. S.Y. Lee and A.M. Sciaccaluga, Nuclear Phys. B96 (1975) 435.
23. D.A. Ross and J.C. Taylor, Nuclear Phys. B151 (1973) 125.
24. S. Sakakibara, Aachen preprint 79/17 (1979).

ATTEMPTS AT SUPERUNIFICATION*)

John Ellis
CERN, Geneva, Switzerland

Mary K. Gaillard
LAPP, Annecy-le-Vieux, France

Luciano Maiani
CERN, Geneva, Switzerland and University of Rome

Bruno Zumino
CERN, Geneva, Switzerland

1. INTRODUCTION

At present most particle physicists believe that the known interactions are described by a renormalizable gauge theory, namely

$$SU(3)_{colour} \otimes SU(2)_{left} \otimes U(1) \tag{1}$$

and many theorists believe in addition that these interactions become unified[1] at a mass scale of about 10^{15} GeV. The minimal grand unified theory (GUT) which can incorporate the product group (1) is the SU(5) of Georgi and Glashow[2]. This picture has met with some phenomenological success. The assumptions of a common unification mass scale for the three low energy gauge groups and of essentially no new physics below this scale determine[3] the weak angle which parameterizes neutral current couplings at the value[4],[5]

$$\sin^2\theta_W = 0.206 \pm 0.006, \tag{2}$$

compatible with values obtained from analyses[6],[7] of the present experimental data

---

*) Presented by Mary K. Gaillard

$$\left.\begin{aligned} \sin^2\theta_W &= 0.229 \pm 0.008 \text{ (exp.)} \pm 0.006 \text{ (theor.)} \quad &\text{(Ref. 6)} \\ \sin^2\theta_W &= 0.238 \pm 0.011 &\text{(Ref. 7)} \end{aligned}\right\} \tag{3}$$

In addition, the minimal GUT relates quark and lepton masses[2] at 10^{15} GeV. Evaluation of symmetry breaking effects[8] gives, for example the ratio[4),9)]

$$m_b/m_\tau \simeq 3, \tag{4}$$

again compatible with experiment. A further crucial test of at least the minimal GUT will soon be forthcoming with searches for nucleon decay with the predicted lifetime[10]

$$\tau_{proton} \simeq 10^{30\pm2} \text{ years.} \tag{5}$$

However, the existing models for grand unification also have conspicuous failures which strongly suggest that they are incomplete. They contain far too many free parameters: most of the fermion masses and weak mixing angles are undetermined, as well as the couplings and even the field content of the scalar sector. There is a notorious "gauge hierarchy" problem which is the difference of about 12 orders of magnitude in the mass scales at which the initial breaking

$$\text{GUT} \to SU(3)_c \otimes SU(2)_L \otimes U(1) \tag{6}$$

and the final breaking

$$SU(2)_L \otimes U(1) \to U(1)_{e.m.} \tag{7}$$

are found to occur. This is in fact the same problem of parameter adjustment, but it appears particularly unpalatable as it requires a readjustment of parameters in the scalar potential to many orders in perturbation theory. Finally, what is perhaps the most glaring defect of grand unification is that it excludes gravity which we know must become important at a mass scale $m_P \simeq 10^{19}$ GeV, and another hierarchy problem[11] is to understand why the GUT unification scale is about four orders of magnitude lower.

On the other hand, while there is no sign of the phenomenological relevance of supersymmetry, let alone supergravity, the only known technology which might allow a marriage of renormalizable gauge theories with gravitational interactions is extended supergravity[12]. The largest such theory known is $N = 8$ supergravity in which the elementary fields include 28 vector bosons which form the adjoint representation of a global SO(8) symmetry. Even if SO(8) is gaugeable, it is too small[12] to contain the observed gauge group of Eq. (1). Therefore this approach appeared to be a dead end until Cremmer and Julia discovered[13] that SO(N) supergravity theories have a concealed, nonlinear, local unitary symmetry: U(N) for N = 4, 5, 6 and SU(8) for N = 7 or 8. They conjectured that the local gauge symmetry may become dynamically realized.

A supporting example for this conjecture comes from CP^{n-1} models in two dimensions[14]. This model contains n complex scalar fields with the normalization condition

$$\sum_{i=1}^{n} |A_i|^2 = 1 \tag{8}$$

so that it in fact contains 2n - 1 real scalar fields. The Lagrangian is

$$\mathcal{L} = - \sum_{i=1}^{n} (\partial_\mu - iV_\mu) A_i^* (\partial_\mu + iV_\mu) A_i \tag{9}$$

Since the vector field V_μ has no kinetic energy term it is not a propagating field and is determined by the equations of motion in terms of the scalar fields:

$$\frac{\delta \mathcal{L}}{\delta V_\mu} = 0 = -2V_\mu + i \sum A_i^* \overleftrightarrow{\partial}_\mu A_i \tag{10}$$

The Lagrangian (9) satisfies a formal U(1) gauge invariance:

$$A_i \rightarrow e^{i\Lambda} A_i, \; V_\mu \rightarrow V_\mu - \partial_\mu \Lambda \tag{11}$$

which can be used to reduce the number of scalar degrees of freedom to 2n - 2 by a particular choice of gauge. While the U(1) gauge invariance of the Lagrangian (9) is apparently a formal artifact, investigation of the two dimensional theory in the 1/N expansion shows[14] that the propagator for the composite vector field operator, which is just the conserved U(1) current, develops a pole. So also does the partner fermion in a supersymmetric version of the two dimensional CP^{n-1} model[15),12)].

In SO(8) supergravity, the supermultiplet of elementary fields is self conjugate and contains spins 0 to 2 with multiplicities determined by the binomial expansion coefficients[16]: 1 spin 2, 8 spin 3/2, 28 vectors, 56 spin 1/2 and 70 scalars. The scalars transform according to the $\underline{70}$ representation of SO(8) which is an explicit global invariance of the theory. Making use of the implicit SU(8) local gauge invariance[13] one can introduce 63 additional scalar fields which correspond to the parameters of the gauge transformations. This gives a total of 133 scalar fields which form a representation of a non-compact E_7 which is also a global invariance of the theory. The original SO(8) global invariance group is a subgroup of both E_7 and SU(8).

The 133 scalar fields can be represented by a 56 × 56 matrix

$$S = \begin{pmatrix} u_{[AB]}{}^{[MN]} & v_{[AB][MN]} \\ \bar{v}^{[AB][MN]} & \bar{u}^{[AB]}{}_{[MN]} \end{pmatrix} \qquad (12)$$

where square brackets imply antisymmetrization of indices, A,B = 1,...,8 transform under SU(8), and M,N = 1,...,8 transform under E_7. (The elements of (12) satisfy additional constraints so that the number of independant components is indeed 133). One may construct an E_7 invariant composite vector operator from the matrix product

$$(\partial_\mu S)\, S^{-1} = \begin{pmatrix} 2Q_\mu{}_{[A}^{[C}\,\delta_{B]}^{D]} & P_{\mu[A\,B\,C\,D]} \\ \bar{P}_\mu^{[A\,B\,C\,D]} & 2\bar{Q}_\mu{}^{[C}_{[A}\,\delta^{D]}_{B]} \end{pmatrix} \qquad (13)$$

which is a nonlinear function of the elementary scalar fields; the P_μ contains a linear term which is the gradient of a scalar, while $Q_\mu{}^A{}_B$ is a traceless 8 × 8 matrix which transforms according to the adjoint representation of SU(8) and is in fact the part of the conserved SU(8) current matrix constructed from the scalar fields. Cremmer and Julia constructed explicitly only this part of the Q_μ, but their generalization to supercovariant operators should correspond to the full SU(8) conserved currents, containing fermion field bilinears, etc.

We conjecture that the composite operators so constructed become propagating fields which correspond to the standard gauge fields of low energy phenomenology, the photon, colored gluons, the W,Z of weak interactions, as well as the X and Y of SU(5) and additional (super heavy) gauge fields associated with the generation group. We further conjecture that the additional fields of the low energy theory, the observed quarks and leptons and the scalars needed for symmetry breaking, are dynamical realizations of composite field operators which belong to the same N = 8 supersymmetric multiplet as the conserved current. An objection to this conjecture might be that, contrary to the case for the CP^{n-1} models, the N = 8 supergravity theory is infrared finite in perturbation theory, invalidating the previous analogy. We would then appeal to some as yet undiscovered non-perturbative phenomenon as being responsible; the point is really that such a dynamical realization of composite operators appears at present as the only hope for reconciling the particle spectrum of the low energy gauge theory with supergravity.

Once this viewpoint is adopted, we are led to a unique candidate theory for the embedding of a GUT in a supergravity theory, namely SO(8) supergravity with an SU(5) GUT. The argument is based on three points: a) the largest supergravity theory is SO(8), so the largest dynamically realized gauge group is SU(8), b) the minimal GUT is SU(5), and c) since all observed particles are contained in a common supermultiplet, its algebra must contain a non-trivial generation group for which the minimal candidate is SU(2). Since U(5) and U(6) do not contain $SU(5) \otimes SU(2)$, let alone anything larger, we are immediately led to SO(7) or SO(8) supergravity with

$$SU(8) \ni \text{GUT} \otimes \text{Generation group}.$$

Since SU(8) has rank 7, and the minimal GUT has rank 4, inclusion of a generation group restricts us to GUTs with rank 4, 5 and 6. Georgi and Glashow[2] showed some time ago that the only acceptable rank 4 GUT is SU(5). Similarly, the only viable[8] rank 5 GUT is SO(10), but this is not contained in SU(8). It is then easy to see that SU(8) does not contain the product of SU(2) with a suitable*) rank 6 group. We are thus uniquely restricted to

$$SU(8) \to SU(5)_{GUT} \otimes SU(3)_{Gen} \otimes U(1).$$

The philosophy which we tentatively adopt is the following. We plead ignorance of the dynamics of the underlying fully supersymmetric and SU(8) gauge invariant theory, but assume that composite states are generated dynamically, and that at, say, the Planck mass the full symmetry breaks by some unspecified dynamical mechanism down to SU(5), possibly, but not necessarily, with an SU(8) or $SU(5) \otimes SU(3)$ invariance at an intermediate stage above the mass scale of 10^{15} GeV where we believe SU(5) to break through a conventional Higgs mechanism. We use the initial symmetries only to study the particle spectrum and constrain couplings; our strategy consists of two steps. First we look for the supermultiplet which contains the conserved currents of SU(8), which

---

*) We are implicitly assuming that the GUT by definition unifies only the presently observed interactions and does not account for generation replication; the assumption that GUT commutes with the generation group, which will be relaxed in section 4, allows it to be made fully natural[8] in the Glashow-Weinberg[17] sense.

are assumed to be realized on mass shell as the corresponding gauge bosons. We further assume that this multiplet contains all the particles relevant to the "observed" low energy gauge theory (fermions and Higgs scalars), as well as many unobserved states including some of high spin. We ignore the fields contained in the original supergravity multiplet (preons) as being superconfined, with the exception of the graviton which is an overall singlet. Finally we adopt a conventional, purely group theoretical, symmetry breaking scenario according to which whatever states which can acquire invariant masses at each stage of symmetry breaking do so, and ask what states remain massless at each level. That is, do we end up with an acceptable low mass particle spectrum? We shall see in section 3 that the answer is negative, and in section 4 that assuming we misidentified the supermultiplet and/or the overall pattern of symmetry hierarchies does not alleviate the problem. In section 5 we consider an alternative conjecture[18], according to which the effective theory of the self-interacting composite fields breaks dynamically at the Planck mass to the maximal renormalizable sub theory, and conclude that this approach appears somewhat more promising.

2. THE SUPERMULTIPLET CONTAINING THE SU(8) VECTOR FIELDS

We take the point of view that the vector fields in the adjoint representation of SU(8) are described by composite fields which are at least bilinear in the fundamental fields of the theory. The expressions given by Cremmer and Julia[13] are functions of the scalar fields alone, but it is clear that in a supersymmetric description there will be additional terms containing the other fundamental fields (which have spins up to 2), and that the vector fields are just the fields describing the conserved SU(8) currents. Therefore, the problem of finding the other members of the supermultiplet is the same as that of finding the supermultiplet of the "supercurrent", to which belong, in particular, the spinor currents (spin 3/2) and the (chiral) vector currents (spin 1). The corresponding particle states can then be obtained by putting on the mass shell of the composite particles the fields of the supercurrent multiplet.

The supercurrent multiplet is not known for extended supersymmetry for general N. However, it is known in great detail[19] for N = 1. For N = 2 it is known for a simple special case[20]; from the experience with N = 1, one can assume that the fields and the transformation formulas obtained in the special case are of general validity for N = 2. In both the N = 1 and N = 2 cases we take the supercurrent for a massless supermultiplet of basic fields. It is not difficult to see which is the on-mass shell supermultiplet for the composite fields. Since these two cases follow exactly the same pattern, one can extrapolate and make a "calculated" guess for higher N.

Consider for instance $N = 1$. For massless fundamental fields the supercurrent multiplet consists of the axial vector current $J_\mu(5)$, of the spinor current J_μ, and the symmetric energy momentum tensor $\Theta_{\mu\nu}$. They are all three conserved and

$$\gamma^\mu J_\mu = \Theta_\mu{}^\mu = 0.$$

As it is shown in the Appendix of Ref. 19), these fields can be used to describe a massive supermultiplet of spins (2, 3/2, 3/2, 1). It suffices to put them on the mass shell, by requiring all fields to satisfy in addition the Klein-Gordon equation for mass m. As $m \to 0$, this massive supermultiplet decomposes into massless supermultiplets having the spins indicated: (2, 3/2) + (3/2, 1) + + (1, 1/2) + (1/2, 0, 0). There are two spin 1 fields, but it is easy to see that the axial current gives rise to the spin 1 field of the massless supermultiplet (3/2, 1) (the other spin 1 field comes from the $m \to 0$ limit of the original massive spin 2).

Similarly, for $N = 2$, the fields of the supercurrent multiplet given in Ref. 3) can be used to describe a massive supermultiplet having spins 2, 4 × (3/2), 6 × (1), 4 × (1/2), 0. (We take the reduced multiplet corresponding to vanishing central change). As $m \to 0$ this massive supermultiplet decomposes. Among the resulting massless supermultiplets one finds an $N = 2$ supergravity multiplet with spins 2, 2 × (3/2), 1, and a supermultiplet with spins 2 × (3/2), 4 × 1, 2 × (1/2). The latter is the only one which has 3 spin 1 fields in the adjoint representation of SU(2), plus a spin 1 singlet. The additional massive supermultiplets have spin 1 or lower, but the spin 1 are not in the adjoint representation.

The cases $N = 1$ and $N = 2$ indicate that, for general N, the supermultiplet of states with the vectors in the adjoint representation can be constructed as follows. One starts with states $(+3/2)^A$, with $A = 1,\ldots N$, of positive helicity 3/2, belonging to the fundamental representation N of SU(N). Applying to them the N spinor charges*), one obtains states $(+1)^A{}_B$ belonging to the $N \otimes \bar{N}$. This decomposes into the adjoint representation plus a singlet. Continuing this operation one obtains states $(+1/2)^A{}_{[BC]}$, $(0)^A{}_{[BCD]}$, $(-1/2)^A{}_{[BCDE]}$ etc. These tensors are totally antisymmetrized in the lower indices, which all take values 1,...N. The procedure stops when one reaches helicity 3/2 − N/2. The states of this last helicity are in the same representation $(3/2 - N/2)^A$ as the original helicity 3/2 states, since a totally antisymmetric tensor with N indices is an SU(N) singlet. To all the above states one must add the TCP conjugate states $(-3/2)_A$, $(-1)_A{}^B$ etc.

*) This procedure for constructing multiplets of states is standard. See, for instance Ref. 21).

We have constructed the supermultiplet for general N by extrapolating from the known cases for N = 1 and 2. We can further argue that the general result must be correct on the grounds that the supergravity current is unique since it is the source of supergravity. It is also worth pointing out that the massless multiplets so obtained are precisely the particle states of superconformal gravity (but, hopefully, in the present case, without the accompanying ghosts). We may believe the algebraic structure of the supermultiplet which we are studying is more general than the specific SO(8) model which was our original motivation. Therefore some of our subsequent analysis may have a wider validity than this specific model.

3. THE PARTICLE CONTENT OF THE COMPOSITE MULTIPLET

Applying the construction obtained in the preceding section for the multiplet of the supercurrent on zero mass shell, we obtain the states of Table 1 and their TCP conjugates. (We start with negative rather than positive helicity because we want to embed left-handed 10-plets and $\bar{5}$-plets in the usual SU(5) gauge theory; this simply means that we define the spinor operator of supersymmetry transformations which is an 8 of SU(8) to be left-handed). Except for the first and last entries, each helicity state transforms according to a reducible multiplet of SU(8), $(h)^a{}_{[bc\ldots]}$, which decomposes into its trace, $(h)^a{}_{[ab\ldots]}$, and a traceless part which is the larger irreducible representation in each case. In addition to the spin 1 63-plet which we identify with the SU(8) gauge bosons, we are interested in fermions and scalars. The left-handed spin 1/2 traceless multiplets decompose under $SU(5) \otimes SU(3)_{Gen.}$ as

$$
\begin{aligned}
216 &= (10,\bar{3}) + (\bar{5},\bar{3}) + (\overline{45},1) + (1,3) + (5,1) \\
&\quad + (1,\bar{6}) + (24,3) + (5,8). \\
\overline{504} &= (45,3) + (40,\bar{3}) + (24,1) + (15,1) + (10,1) \\
&\quad + (10,8) + (\overline{10},6) + (\overline{10},\bar{3}) + (5,3) + (\bar{5},\bar{3})
\end{aligned}
\tag{14}
$$

The first two terms represent obvious candidates for the three generations of light fermions. The scalar 420-plet decomposes according to

$$
\begin{aligned}
420 &= (24,\bar{3}) + (5,3) + (5,\bar{6}) + (45,3) + (\overline{40},1) \\
&\quad + (\overline{10},\bar{3}) + (\bar{5},1) + (10,1) + (10,8) + (1,\bar{3})
\end{aligned}
\tag{15}
$$

Table 1. Helicity states of the composite multiplet of SU(8) gauge vectors.

Helicity	-3/2	-1	-1/2	0	+1/2	+1	+3/2	+2	+5/2
SU(8) content	$\overline{8}$	1, 63	8, 216	28, 420	56, 504	70, $\overline{378}$	$\overline{56}$, $\overline{168}$	$\overline{28}$, $\overline{36}$	$\overline{8}$

If the generation triplet of 24's acquires a vacuum expectation value at a scale of 10^{15} GeV, it will serve to break[2),4)] SU(5) down to $SU(3)_c \otimes SU(2)_L \otimes U(1)$, and any combination of 5's and 45's can be used[2),4)] to break $SU(2)_L \otimes U(1)$ down to $U(1)_{e.m.}$ at the second stage of the usual GUT hierarchy. One would like the same objects to generate fermion masses. The (5,3), $(5,\overline{6})$ and (45,3) all have $SU(5) \otimes SU(3)_{Gen}$ invariant couplings to the fermion bilinear

$$(10,\overline{3})_L \; (\overline{5},\overline{3})_L$$

which is the conventional source of masses for charge -1/3 quarks and charged leptons. However, in the fermion bilinear which can give masses to charge 2/3 quarks

$$(10,\overline{3})_L \; (10,\overline{3})_L = (\overline{5} + \overline{45} + 50, \; 3 + \overline{6})$$

only the symmetric part, $(\overline{5},\overline{6}) + (\overline{45},3) + (50,\overline{6})$, can contribute, and none of these representations is contained in the multiplet (15) nor in the scalar 28-plet. Retaining the aesthetic and simple identification of the observed fermions with the $(10,\overline{3}) + (\overline{5},\overline{3})$ would then imply that the top quark mass arises from radiative corrections, which seems somewhat hard to swallow even if generation SU(3) is already broken at the Planck mass. However, there is also a $(10,1)_L + (10,8)_L$ in the conjugate of the helicity +1/2 504, some of which might also be identified with the three 10's of SU(5), if one is willing to abandon a simple interpretation of the generation group. We will see in section 5 that this complication may indeed be what happens.

There is also, of course, a lot of unwanted garbage in the states of Table 1. We should remark at this point that we checked that the SU(8) gauge currents are not anomaly free at the preon level, nor at the composite level either, and presumably the spinor currents coupled to higher spin states also have anomalies. The singularities arising from anomalies could induce large masses in some states and effectively remove them from the theory below the Planck mass. However at this stage we simply ignore the question of what mechanisms induce masses (and also the question of what keeps some scalars effectively massless so that the usual SU(5) gauge hierarchy can go through), and take it as an empirical "fact" that SU(5) is a good gauge symmetry above 10^{15} GeV and $SU(3)_C \otimes SU(2)_L \otimes U(1)$ is good above 10^2 GeV. Then, as in conventional Higgs and super-Higgs symmetry breaking mechanisms, those states which can acquire SU(5) gauge invariant masses will have $m \simeq m_P$, those masses which are $SU(3) \otimes SU(2) \otimes U(1)$ invariant will be $m \simeq 10^{15}$ GeV, and those which break $SU(2)_L$ will be $m \simeq 10^2$ GeV (with the very light fermion masses possibly corresponding to zeros in the original mass matrix). A spin 1/2 fermion, for example, in a representation r of SU(5) acquires a mass $O(m_P)$ if the reducible representation $(8)_L + (216)_L + (\overline{56})_L + (\overline{504})_L$ contains the conjugate representation $\overline{r}$ as well as r. For a massive higher spin fermion, rotational invariance in its rest frame demands that all helicity states n, n-1,..., -n occur. Therefore elimination of a spin 3/2 object in representation r requires that $r + \overline{r}$ be contained in the -3/2 representation 8 + 56 + 168 and also in the -1/2 representation given above. One sees immediately that the charged members of the SU(5) $\overline{5}$-plet contained in the $\overline{8}$-plet of -5/2 states will never acquire a mass if conventional SU(5) charge assignments are used and if electromagnetism remains unbroken. One might try to argue that the confined color triplet of spin 5/2 has such highly non-renormalizable interactions that its bound states acquire masses beyond the reach of present experiment, but we certainly cannot tolerate a massless charged lepton, which must remain forever massless as it has no helicity partner.

Since the program described up to here has obviously failed, we tried several alternatives which we outline in the next section.

4. ALTERNATIVE SCENARIOS

A possibility might be that we misidentified the multiplet containing the conserved vector current. We also tried a multiplet with the structure

$$(-2)\,[AB],\ (-3/2)\,\genfrac{}{}{0pt}{}{[AB]}{[C]},\ (-1)\,\genfrac{}{}{0pt}{}{[AB]}{[CD]},\ (-1/2)\,\genfrac{}{}{0pt}{}{[AB]}{[CDE]},\ldots,\ (+2)\,[AB]\,. \qquad (16)$$

The supermultiplet (16) contains 3/2 as its highest half-integral spin, and this falls in a complicated reducible multiplet where we might hope to find at least opposite charge partners for everyone. This turns out not to be the case.

A second possibility is that we have been taking our conventional view of GUT too seriously, and that we should bypass SU(5) as an intermediate, physically relevant symmetry. Our previous difficulty was that the electromagnetic theory was not vector-like. We can cure this by imposing charge symmetry on the basic supermultiplet. The only charge assignment which does that and which preserves the factorization and observed charge spectrum of the observed group (1) is

$$8 = [3_{\text{colour}}(Q = 0),\ 2_{SU(2)_L}(Q,\ Q - 1),\ -Q,\ -Q + 1,\ 0].$$

We found that in order to give masses to all spin 3/2 states we had to impose an additional and amusing charge quantization condition, $Q = n/3$, which gave us quarks and leptons with the conventional charges as well as integrally charged coloured quarks and fractionally charged leptons (which might possibly be found[22] on Niobium balls). However, even in this scenario we found it necessary to abandon partially the principle of attributing all possible invariant masses at each stage of breaking; some spin 1/2 self-conjugate SU(2) triplets had to remain massless until after SU(2) breaking to feed left over 3/2 states which would otherwise have eaten desirable fermions.

A much worse result was that we were left with many massless unpaired helicity states which were non-singlet and non-self-conjugate under colour SU(3). The previous argument wishing away bound states of high spin quarks would not seem applicable to spin 1/2 states. In addition, even if one could wish away all the bound states of such quarks, it would seem difficult to wish away their loop contributions to gluon self-couplings at low energies, and we would expect them to generate parity violation in strong interactions. We could try to impose that the basic supermultiplet be self conjugate, and thus vector-like, with respect to both charge and colour. For $N \leq 8$, the only such possibility is SU(8) with

$$8 = [3_c(Q),\ \bar{3}_c(-Q),\ 2_L(Q',\ -Q')]. \tag{17}$$

The observed SU(2) charge structure dictates $-Q' = Q' - 1$, so $Q' = 1/2$, and we found that there is no choice of Q which reproduces the observed particle charge spectrum. (Another possibility, not a priori vector-like, for embedding $SU(3)_c \otimes SU(2)_L$ in SU(8) would be with a fundamental 6-plet transforming like a colour triplet and an SU(2) singlet; this does not work either.)

The major difficulty is simply that the composite multiplets of SO(8) supergravity appear as chiral multiplets of dynamical SU(8). This chirality property is a priori appealing as it may be relevant to the apparent masslessness of the neutrino and the observed chirality of weak interactions. Our problem is one of two much chirality; we are unable to make any interactions vector-like. Is there any way to salvage the marriage of low energy gauge theories with supergravity? One might argue that any composite supermultiplet containing vectors which transform according to the adjoint representation of the unitary group might be the one containing dynamically realized physical states, on the grounds that the vector interactions have to be those of a Yang-Mills theory if they are to result in an effective renormalizable theory. One could further argue that the condition $N = 8$ is not necessary. Since the supermultiplets contain large irreducible representations of helicity 1/2 states, containing repeated representations of subgroups, it may be that an explicit non trivial generation group is not a necessary condition for a realistic theory. Therefore we asked the question: is there any suitable irreducible supermultiplet of $SO(N)$ supersymmetry, with the basic N-plet transforming under $SU(N) \to SU(3)_{colour}$ according to

$$N = [3_{colour}, (N-3)_{colour\ singlets}],$$

and which contains spin 1 states which transform according to the adjoint representation of $SU(N)$? A general irreducible multiplet is constructed[21] by assigning helicity h to some irreducible representation R_o and applying the (automatically totally antisymmetrized) N spinor charges to obtain states of helicity decreasing (or increasing) successively by one half unit. Thus we consider a multiplet of the form

$$S = [R_o(h), R_{oA}(h-1/2), \ldots, R_{o[A, \ldots A_k]}(h-k/2\equiv 1), \ldots, R_o(h-N/2)] \tag{18}$$

and impose the condition that states of given half-integer helicity in $S + S^{\dagger}$ (the TCP conjugate) be vector-like under $SU(3)$, that is, that each set of helicity states contain equal numbers of r and $\bar{r}$ where r is any non-self-conjugate representation of $SU(3)$. Our analysis of this general case led to the conclusion that no such irreducible multiplet can be found, at least for $5 \leq N \leq 8$, which are the limiting values imposed by the minimal rank incorporating the observed gauge interactions and the maximal supergravity theory.

Perhaps even the restriction to an irreducible composite supermultiplet is too strong, and it might be possible to construct a viable theory using some superposition of multiplets. There is no obvious way to see that this is possible, and it would require a better group theoretical technology than we have found to either construct an example or prove that it is not possible. Another escape route is to conjecture that a supergravity theory with $N > 8$ might be constructed by means of a non-linear realization of the supersymmetry. For $N \geq 10$ it is trivial to impose the required vector-like conditions by choosing, for example, a decomposition of SU(10) $\to$ SU(5) according to

$$10 = (5, \bar{5}).$$

In this case even SU(5) is initially vector-like, and the observed chirality would have to be attributed to symmetry breaking effects.

There is yet an alternative possibility, which seems to us at present the most plausible one, which is that the initial philosophy was misguided. We have assumed that composite states of high spin must necessarily acquire group invariant masses in the way we conventionally attribute masses to elementary fields described by an elementary field theory. In fact we have no notion of how dynamical symmetry breaking may occur in supergravity theories, nor of the dynamics of a self-coupled composite supermultiplet, and in particular no indication that its effective interactions could ever be made renormalizable even if the underlying supergravity theory were shown to be renormalizable on the preon level. In the following section, we outline an alternative program[18] based largely on intuition and experimental evidence for the reality of renormalizable gauge theories.

5. AN EFFECTIVE RENORMALIZABLE SUB-THEORY

Present experimental data strongly support the belief that low energy physics is described by renormalizable field theory, which as far as we know could be valid up to energies not much below the Planck mass; this is precisely the assumption made in obtaining the GUTs predictions (2), (4) and (5). The renormalizable theories that we know consist solely of Yang-Mills vector fields, spin 1/2 fermions, and scalar fields.

There is an unpublished argument of Veltman[23], which we understand as follows: if composite states with masses that are much smaller than their inverse size have effective interactions which can be described by perturbation theory at energy scales much lower than the inverse size, then these interactions must be renormalizable. Otherwise singularities would arise in the computation of vertex functions for which the only available cut-off is the inverse size.

The above two points encourage us to exploit the following conjecture[18]: the would-be supersymmetry of the effective Lagrangian describing the self-interaction of the dynamically realized composite supermultiplet in the SO(8) supergravity theory is dynamically broken down to a sub-theory containing only renormalizable interactions. Those states which do not have renormalizable interactions may acquire masses $\sim (\text{size})^{-1} \sim m_P$ and decouple from the effective "low" ($\lesssim m_P$) energy theory. Within this philosophy we are simply ignoring the previous question as to where all the states in the original supermultiplet find helicity partners with the same $SU(3)_c \otimes U(1)_{e.m.}$ transformation properties to form a rotationally invariant massive state. Since we are dealing with composite states, they could simply become unbound at the scale m_P or possibly marry helicity components of some other composite supermultiplet in a non-supercovariant way.

We thus arbitrarily reduce the original composite supermultiplet to its renormalizable sector which to our knowledge can contain only the Yang-Mills vectors, spin 1/2 fermions and scalars of a renormalizable gauge theory which could be as large as SU(8). Aside from restrictions on the multiplet structure, are we left with any predictive power at all? An initial working hypothesis[18)] might be that a remnant of the underlying supersymmetry of the interactions survives in the coupling constants of the residual renormalizable theory as specified at the Planck mass. Since perturbation theory is by assumption applicable below the Planck mass, the low energy theory becomes in principle completely determined.

Returning to our original arguments that the only viable embedding of a GUT in a supergravity theory requires SO(8) supergravity, and that the dynamically realized supermultiplet is the one of Table 1, we reduce the theory to that part which contains only the 63 gauge bosons of SU(8), the 216 (+ 8?) left-handed spinors, the 504 (+ 56?) right-handed spinors, the 420 (+ 28?) scalars, and the TCP conjugates of these states. At this stage there is a certain amount of ambiguity as to how much of the fermion and scalar sector we should actually retain. Our criterion of renormalizability requires us to make the theory anomaly free which implies the elimination of some fermions. It is a priori unclear as to at what stage and in what way this should be done. Also, in addition to the non-Yang-Mills $\overline{378}$ + 70 of vector bosons, we have discarded the SU(8) singlet accompanying the 63 as it does not couple to a conserved current (this is because of the self-conjugate nature of the fundamental supermultiplet which is unique to the SO(8) supergravity theory). It could be that the trace parts (8 + $\overline{56}$, 28) of the fermion and scalar reducible SU(8) multiplets go out in the wash with the vector singlet, but this cannot be inferred from the renormalizability criterion alone. In either case, the group theoretical structure of our renormalizable sub-theory is much more general than our starting point which was SO(8) supergravity; starting with vector fields in the adjoint SU(N) representation, application of N supersymmetry transformation operators will generate the same fermion and scalar states although they can presumably not be made to form a closed multiplet.

The next step in our program is to extract a candidate effective Lagrangian for the surviving fields. To this effect we assume that there exists an effective Lagrangian describing the self-coupling of the entire composite supermultiplet which is superficially invariant under global SO(8) supersymmetry and local SU(8) gauge symmetry. We then extract that part which corresponds to a renormalizable gauge theory describing the interactions of the above subset of fields. There are several methods which we can appeal to. Relations have been derived[24] for S matrix elements in an unbroken extended supersymmetric theory. By imposing these constraints on the amplitudes calculated in the tree approximation in perturbation theory, one can, in principle, reconstruct an effective Lagrangian which has the required invariance properties. In practice this approach is lengthy, and since we are only interested in a subset of couplings, we can make use of supersymmetric Lagrangians in the existing literature by noticing that the full symmetry contains many smaller symmetries. The eight indices of the basic SO(8) [SU(8)] vector (spinor) play a double role: they label the eight supersymmetries of the theory and the eight fundamental gauge transformation properties. We may consider a sub-symmetry which is the product of $N < 8$ supersymmetries with $8 - N$ variations under unitary gauge transformations:

$$SO(N)_{\text{supersymmetry}} \otimes SU(8 - N)_{\text{Yang-Mills}}. \tag{19}$$

Lagrangians for theories invariant under the product group (19) are known[25] for $N \leq 4$. The $N = 4$ case in fact describes a real supermultiplet, but is equivalent to a complex $N = 3$ theory which is relevant for us since the right-handed fermions in Table 1 are not the conjugates of the left-handed fermions. The use of known Lagrangians symmetric under (19), with the additional constraint of a full local SU(8) symmetry allows us to partially constraint some couplings, namely:

$$\mathcal{L}\left[(420)^4\right] + \mathcal{L}\left[(216)^2 \times 420\right] + \mathcal{L}(216 \times \overline{504} \times 420) \tag{20}$$

together with the SU(8) gauge couplings. Even this limited sector of the effective Lagrangian displays some interesting features[18], which we summarize.

a) The quadratic coupling of the 420 vanishes for certain non trivial scalar field configurations, so the local SU(8) gauge symmetry can be spontaneously broken at the tree level. If we may neglect non-perturbative effects, the minimum of the potential is then determined by radiative corrections, which in the one-loop approximation take the form[26]

$$V(\phi) = \sum_i \eta_i \, M_i^4 \, (\phi) \ln \left[M_i^2(\phi)/\mu^2\right], \tag{21}$$

where μ is the field strength for which the tree Lagrangian couplings are specified, here assumed to be the Planck mass, $M_i(\emptyset)$ is the mass acquired by the ith particle in the tree approximation when $\emptyset$ is the scalar vacuum expectation value, and $\eta_i = +3, +1$ and -4 for vectors, scalars and fermions, respectively. At this stage it is essential that the high spin states have decoupled from the effective Lagrangian; if the full supermultiplet were retained, the vacuum degeneracy would persist to all orders in perturbation theory (if it could be defined!).

b) By inspection of the $SO(2) \otimes SU(6)$ and $SO(3) \otimes SU(5)$ sub-Lagrangians which were used to construct the fully $SU(8)$ invariant quartic coupling, it is easy to show that the tree potential vanishes for certain field configurations where the only non-vanishing components of the 420-plet $\psi^i{}_{[jkl]}$ have one lower index equal to the upper index:

$$\langle \psi^i{}_{[ikl]} \rangle \neq 0 \qquad (22)$$

One can see immediately (excluding the unlikely possibility that the true minimum occurs for zero field strength), that the largest simple unitary group which can remain unbroken at the Planck mass is $SU(5)$. This is because the tensor $\psi^i{}_{[jkl]}$ contains no singlets of $SU(6)$ or of $SU(7)$, but the diagonal form (22) does contain an $SU(5)$ singlet.

To pursue our preliminary analysis further we make the simplifying assumption that the true vacuum expectation value has non-vanishing elements only for fixed values of (k,l) in (22), i.e. that it transforms according to the adjoint representation of an $SU(6)$. Then we can make the following further observations.

c) The surviving group will be a rank 5 subgroup of $SU(6)$. From the study[11] of adjoint breaking in $SU(5)$, we can make some guesses as to how the radiative corrections will determine the preferred direction of symmetry breaking in $SU(6)$ space. For positive η, the lowest vacuum is generally that with the greatest degree of (non-vanishing) mass degeneracy. Thus the contribution of gauge boson loops alone would choose a symmetry breaking to $SU(3) \otimes SU(3) \otimes U(1)$. The scalar loop contribution is more complicated to analyse, but is likely to go in the same direction. However, since the fermion loop contribution has a negative coefficient, it will prefer the solution with maximum mass degeneracy, namely $SU(5) \otimes U(1)$. Since we have more fermions (with weight factor 4) than bosons, they may turn out to play a determining role. In the subsequent remarks we assume that the surviving symmetry at the Planck mass is $SU(5) \otimes U(1)$.

d) Keeping only the 420-plet of scalars, at least two adjoints and one 5-plet of SU(5) remain massless at the tree level. Such massless scalar multiplets are precisely what is needed to implement further symmetry breakdown at lower mass scales through radiative corrections[27),11)]. In conventional perturbation theory, in fact, the massless 5-plet would have been eaten by a 5-plet of ultra-heavy gauge bosons of SU(8). However, the analysis of anomalies discussed below shows that this vector multiplet should already have acquired a mass through its anomalous current divergence. It remains to be seen whether the masslessness of the SU(2) doublet of the 5-plet will survive SU(5) breaking. (There are of course other 5-plets in the 28, some of which may remain massless if their couplings are included in the effective Lagrangian.)

e) If we examine the Yukawa couplings contained in the partial Lagrangian (20), we find that the $(\overline{5}, \overline{3})_L$ of 216 [see Eq. (14)] remains massless, but the $(10, \overline{3})_L$ acquires masses through its Yukawa couplings to combinations of $(\overline{10}, 6)_L$ and $(\overline{10}, \overline{3})_L$ contained in $\overline{504}$. It therefore cannot be identified with the three generations of "massless" 10's in the conventional SU(5) theory. Since the set $(216)_L + (\overline{504})_L$ contain altogether 12 10's and 9 $\overline{10}$'s, including the self-couplings of $\overline{504}$ to 420 will necessarily leave at least three 10's massless. These surviving states can have SU(8) allowed tree-level couplings to the $(\overline{5}, 1)$ of the scalar 420, providing a zeroth order t quark mass upon SU(2) breaking. However, the set $(216)_L + (\overline{504})_L$ also contains 6 $\overline{5}$'s and 12 5's, which would leave us with at least 6 massless 5's (and not necessarily any $\overline{5}$'s). What actually survives in the low energy effective theory depends on how the initial fermion content is reduced to an anomaly free subset.

From an analysis of SU(N) anomalies we learn the following. The effective SU(8) gauge theory cannot be made anomaly free. Sub-theories based on SU(7) or SU(6) which are anomaly free cannot be made vector-like under $SU(3)_{colour} \otimes U(1)_{e.m.}$ unless they are completely vector-like. On the other hand there is a set of fermions contained in $(216)_L + (\overline{504})_L$ which has no SU(5) anomalies, is chiral, and is vector-like under $SU(3)_{colour} \otimes U(1)_{e.m.}$, namely:

$$(45 + \overline{45})_L + 4(24)_L + 9(10 + \overline{10})_L + \\ + 3(5 + \overline{5})_L + 9(1)_L + 3(\overline{5} + 10)_L . \tag{23}$$

This choice is maximal in the sense that it contains the largest number of anomaly free helicity states which can be found in $(216)_L + (\overline{504})_L$. (There is another subset with the same number of states which is excluded phenomenologically because it is not vector-like under $SU(3)_{colour}$.) We note with satisfaction that it contains precisely three generations of $(\overline{5} + 10)_L$ which is the minimal number compatible with observation, and the preferred number for giving a satisfactory calculation of the b quark mass[4),9)]. The remaining content of (23) is vector-like under SU(5) and can acquire SU(5) invariant (Dirac or Majorana) masses $O(m_P)$. We note also that the content of (23) is too large to fit into one of the vector-like anomaly free SU(6) or SU(7) sub-theories of SU(8).

A final feature of our effective Lagrangian is that the β function is extremely positive when the full fermion and scalar multiplets are included with the Yang-Mills fields of SU(8) ($\beta = (147.5)\frac{g^3}{48\pi^2}$ for the traceless parts). If there are a sufficient number of superheavy fermions which acquire masses only at the SU(5) breaking scale of 10^{15} GeV where $\alpha_{GUT} \simeq 1/40$, the coupling may become strong[28)] at m_P, which is desirable if we wish to attribute the primary dynamical breakdown of supersymmetry to a non-perturbative strong coupling effect.

In summary, arguments based on the form of the Higgs potential as determined from the conjectured underlying symmetry, and on the requirement of an anomaly free but non vector-like fermion sector, both point to SU(5) as the largest gauge group which can remain unbroken after the conjectured dynamical breakdown of the SO(8) supersymmetry at the Planck mass. Superficially, at least, the content of the residual gauge theory contains the elements necessary to populate the low energy particle spectrum and to permit a perturbative SU(5) theory to emerge from a strong coupling theory at the Planck mass, with the further breakdown of SU(5) at lower energies arising from radiative corrections. It remains to be seen whether such a program can be made to work in detail.

ACKNOWLEDGEMENTS

We would like to thank D.V. Nanopoulos, H. Römer, P. Sikivie and M. Veltman for encouragement and instructive discussions.

REFERENCES

1. J.C. Pati and A. Salam, Phys. Rev. D8:1240 (1973).
2. H. Georgi and S.L. Glashow, Phys. Rev. Letters 32:438 (1974).
3. H. Georgi, H.R. Quinn and S. Weinberg, Phys. Rev. Letters 33:451 (1974).
4. A.J. Buras, J. Ellis, M.K. Gaillard and D.V. Nanopoulos, Nucl. Phys. B135:66 (1978).
5. K.T. Mahanthappa and M.A. Sher, Univ. of Colorado preprint COLO-HEP 13 (1979).

6. P. Langacker et al., U. of Penn. preprint COO-3071-243 (1979).
7. M. Roos and I. Liede, Helsinki Univ. preprint HU-TFT 79-27 (1979).
8. M.S. Chanowitz, J. Ellis and M.K. Gaillard, Nucl. Phys. B128: 506 (1977).
9. D. Ross and D.V. Nanopoulos, Nucl. Phys. B157:273 (1979).
P. Binétruy and T. Schücker, CERN preprint TH-2802 (1980), submitted to Nucl. Phys. B.
10. See, e.g.:
W.J. Marciano, Rockefeller Univ. preprint COO-2232B-195 (1980);
T. Goldman and D.A. Ross, Caltech preprint CALT-68-759 (1980);
J. Ellis, M.K. Gaillard, D.V. Nanopoulos and S. Rudaz, preprint LAPP-TH-14/CERN-TH-2833 (1980);
and references contained in the above.
11. J. Ellis, M.K. Gaillard, A. Peterman and C.T. Sachrajda, Nucl. Phys. B164:253 (1980).
12. See e.g.:
B. Zumino in Proc. Einstein Symposion Berlin, Lecture Notes in Physics 100:114, Ed. H. Nelkowski, A. Herman, H. Poser, R. Schrader and R. Seiler (Springer-Verlag, Berlin 1979) and references therein. The first attempt to embed low energy gauge theories in SO(8) supergravity using the basic supermultiplet was reported by M. Gell-Mann, Talk at the 1977 Washington Meeting of the American Physical Society.
13. E. Cremmer and B. Julia, Phys. Letters 80B:48 (1978) and Nucl. Phys. B 159:141 (1979). A previous attempt to relate this work to phenomenology has been made by T. Curtright and P. Freund, in "Super Gravity", Proc. of the Supergravity workshop at Stony Brook, Sept. 1979, Ed. P. Van Nieuwenhuizen and D.Z. Freedman, (North Holland, Amsterdam, 1979), p. 197. Telegraphic accounts of our work on embedding GUTs in supergravity have been reported in:
J. Ellis, Proc. E.P.S. Int. Conf. on High Energy Physics, Geneva June 1979 (CERN, 1979), p. 940,
M.K. Gaillard and L. Maiani, 1979 Cargèse Summer Institute Lectures, LAPP preprint TH-09 (1979).
14. A. d'Adda, P. Di Vecchia and M. Lüscher, Nucl. Phys. B146:63 (1978);
E. Witten, Nucl. Phys. B149:285 (1979).
15. A. d'Adda, P. Di Vecchia and M. Lüscher, Nucl. Phys. B152:125 (1979).
16. B. De Wit and D.Z. Freedman, Nucl. Phys. B130:150 (1977).
17. S.L. Glashow and S. Weinberg, Phys. Rev. D15:1958 (1977).
18. J. Ellis, M.K. Gaillard and B. Zumino, CERN preprint TH-2842/ LAPP preprint TH-16 (1980).
19. S. Ferrara and B. Zumino, Nucl. Phys. B87:207 (1975).
20. M. Sohnius, Phys. Letters 81B:8 (1979).
21. D.Z. Freedman, in "Recent Results in Gravitation, Cargèse 1978", Ed. M. Levy and S. Deser (Plenum Press, New York, 1978), p. 549.

22. G.S. La Rue, W. M. Fairbank and A.F. Hebard, Phys. Rev. Letters 38:1011 (1977);
G.S. La Rue, W. M. Fairbank and J.D. Phillips, Phys. Rev. Letters 42:142 (1979).
23. M. Veltman, private communication; arguments similar in spirit have probably occured to many people, for example, G. t'Hooft, Cargèse Lecture Notes (1979);
K. Wilson, G. Parisi and L. Susskind, private communications.
24. A. Salam and J. Strathdee, Nucl. Phys. B80:499 (1974);
M.T. Grisaru, H.N. Pendleton and P. Van Nieuwenhuizen, Phys. Rev. D15:996 (1976);
M.T. Grisaru and H.N. Pendleton, Nucl. Phys. B124:81 (1977).
25. S. Ferrara and B. Zumino, Nucl. Phys. B79:413 (1974)
P. Fayet, Nucl. Phys. B13:135 (1976);
L. Brink, J.H. Schwarz and J. Scherk, Nucl. Phys. B121:77 (1977);
F. Gliozzi, J. Scherk and D. Olive, Nucl. Phys. B122:253 (1977).
26. S. Coleman and E. Weinberg, Phys. Rev. D7:1888 (1977).
27. S. Weinberg, Phys. Letters 82B:387 (1979).
28. L. Maiani, G. Parisi, R. Petronzio, Nucl. Phys. B 36:115 (1978) and references therein.

UNIFICATION THROUGH A SUPERGROUP

Yuval Ne'eman*

Department of Physics and Astronomy
Tel Aviv University, Tel Aviv, Israel
and
Center for Particle Theory
University of Texas, Austin, Texas, U.S.A.

ABSTRACT

The representations of supergroups provide the most precise fit to the known set of fundamental physical matter fields. SU(2/1) describes the unified weak electromagnetic interactions and SU(5 + k/1) contains $SU(3)_{colour}$ and 2^k exact "generations". Moreover, SU(2/1) predicts that the fourth states of the lepton multiplets (i.e., ν_R^0) decouple, whereas they do not in quark multiplets, and SU(n/1) predicts that the colour degree of freedom is SU(3). We provide all relevant constructions.

---

* Wolfson Chair Extraordinary in Theoretical Physics; supported in part by the U.S. - Israel Binational Science Foundation and in part by the U.S. Dept. of Energy Contract EY-76-S-05-3992.

1. SU(2/1) AND THE QUANTUM NUMBERS OF THE SPINOR FIELDS

It has been suggested[1,2] that the supergroup SU(2/1), when gauged in a way appropriate to an internal supergauge, generates the unified electroweak (or asthenodynamic) interaction. The resulting Lagrangian roughly corresponds to a highly restricted Salam-Weinberg $SU(2)_L \otimes U(1)$ model.

We have applied a theorem stating that the assignment of different gradings[3] between v_L (left) and v_R (right) chiralities within the spinor field vector space v, ensures that all matrices of $SU(2)_L \otimes U(1)$ be supertraceless.

$$\mathrm{str}\, M = \mathrm{tr}\, M_L - \mathrm{tr}\, M_R = 0 \tag{1.1}$$

Proof: (a) The electric charges of different chiral components are identical.

$$Q(v_L^m) = Q(v_R^m)$$

$$\mathrm{str}\, M\{Q\} = 0$$

(b) $$\mathrm{tr}\, M_L\{SU(2)_L\} = 0 \, , \quad M_R\{SU(2)_L\} \equiv 0 \tag{1.2}$$

(c) $$Q = I_3^L + \tfrac{1}{2} U$$

(I_3^L is the third component of the weak left-handed isospin, U is the weak hypercharge.) Thus, since (b) implies $\mathrm{str}\, M\{I_3^L\} = 0$, we have

$$\mathrm{str}\, M\{U\} = 0$$

Postulating that supertracelessness reflects the presence of a supergroup, we have assumed that this is SU(2/1). The defining (irreducible) representation $\mathcal{D}(\frac{1}{2},\frac{1}{2})$ in the notation of Scheunert et al.[4] (which we denote as SNR in this paper) carries the eigenvalues

$$I_3^L = \mathrm{diag}\,(\tfrac{1}{2}, -\tfrac{1}{2} / 0) \, , \quad U = \mathrm{diag}\,(1, \tfrac{1}{2}) \tag{1.3}$$

which fits the antileptons $\left((\overline{e_L^-})_R^+ , (\overline{\nu_L^0})_R \,/\, (\overline{e_R^-})_L^+ \right)$

Note that in Ref. 1 we had picked the lepton eigenvalues for our definitions. Our present choice (1.3) provides for positive eigenvalues. It has the additional advantage of fitting the SNR notation $(\frac{1}{2}U_{mid}, I^{L}_{3max})$ characterizing the eigenvalues of the state with highest U within the set with highest I^{L}_{3}. The SU(2/1) hypothesis scores highly in providing us with a (star-Hermitean) irreducible representation $\mathcal{D}(1/6,1/2)$ fitting the quarks

$$\left(u_L^{2/3}, d_L^{-1/3} / d_R^{-1/3}, u_R^{2/3}\right)$$

precisely[5]

$$\left.\begin{aligned} \mathcal{D}(1/6, 1/2; I_3^L) &= \mathrm{diag}\,(1/2, -1/2 / 0, 0) \\ \mathcal{D}(1/6, 1/2; U) &= \mathrm{diag}\,(1/3, 1/3 / -2/3, 4/3) \end{aligned}\right\} \quad (1.4)$$

Note that the finite representations of SU(2/1) are those of SL(2/1), and had been listed[4] by SNR prior to the formulation of our theory. The definition of unitarity for SU(2/1) is based on (id denotes the identity matrix)

$$M^{+} = (M^{T})^{x}, \quad M^{+}M = \mathrm{id}.$$

where M^T is ordinary transposition, and the superscript x denotes complex conjugation with inversion of the order of Grassmann generating elements within each matrix element. As long as we do not have a complete quantum theory of internal supergauges, we do not know how relevant unitarity and Hermiticity are in our representations. Condition (1.1) ensures super unimodularity.

$$\mathrm{sdet}\, M := \det(M_R) \cdot \det(M^{-1})_L = 1 \quad (1.5)$$

Note that the four-dimensional representation in (1.4) depends on a parameter[6] fixing the $1/2\ U_{mid} = 1/6$ SNR eigenvalue. This parameter b translates the eigenvalues of $\mathcal{D}(U)$, without affecting the dimension. Taking, for reasons that we shall touch upon later, b = 0 for the leptons (and more generally, $\frac{1}{2}U_{mid} = b - \frac{1}{2}$)

$$\begin{aligned} \mathcal{D}(-1/2, 1/2; I_3^L) &= \mathrm{diag}\,(1/2, -1/2 / 0, 0) \\ \mathcal{D}(-1/2, 1/2; U) &= \mathrm{diag}\,(-1, -1 / -2, 0) \end{aligned} \quad (1.6)$$

as befits

$$(\nu^0_L, e^-_L / e^-_R, \nu^0_R)$$

We get for b = 1 the four antileptons

$$\left((\overline{e^-_L})^+_R, (\overline{\nu^0_L})_R / (\overline{\nu^0_R})_L, (\overline{e^-_R})_L \right)$$

given by $\mathcal{D}(\frac{1}{2},\frac{1}{2})$.

The quarks (1.4) correspond to b = 2/3, and the antiquarks

$$\left((\overline{d^{-1/3}_L})^{1/3}_R, (\overline{u^{2/3}_L})^{-2/3} (\overline{u_R^{2/3}})^{-2/3}_L, (\overline{d^{-1/3}_R})^{1/3}_L \right)$$

to b = 1/3 or $\mathcal{D}(-1/6,1/2)$.

Note that the representations in which the absolute values of the two SNR quantum numbers coincide, or alternatively, in which b is an integer, are reducible. They are not completely reducible, but b = 1 has the three-dimensional set (1.3) as an invariant subspace. Integer b implies integer U (and even U for ν_R), i.e., integer electric charges Q, as fixed by (1.2). SU(2/1) may thus be said to predict that for integer charges, one state

$$(\nu^0_R \text{ or } (\overline{\nu^0_R})_L)$$

disconnects.

2. PHYSICAL INTERPRETATION OF THE VECTOR SPACE STATISTICS

Our introduction of SU(2/1) was motivated by the finding[7] that the ghosts χ^a, introduced in the quantization procedure of a Yang-Mills theory, correspond to proper geometric "classical" objects, natural to the Principal Fibre Bundle description[8]. We refer the reader to the re-derivation of the Becchi-Rouet-Stora equations[9],

(W^a_μ is the Yang-Mills field potential, ψ^m a matter field)

$$s W_\mu^a = (D_\mu \chi)^a \tag{2.1}$$

$$s \chi^a = -\tfrac{1}{2}[\chi,\chi]^a \tag{2.2}$$

$$s v^m = [\chi, v]^m = \psi^m \tag{2.3}$$

$$s^2 W_\mu^a = 0, \quad s^2 \chi^a = 0, \quad s^2 v^m = 0 \tag{2.4}$$

in which the ghosts χ^a are seen to represent the "vertical" pieces of the complete connection on the principal bundle P(M,G). We use equations (2.3) to define an "effective" matter-ghost ψ^m, with statistics opposed to those of v^m. Note that in the original derivation [10]

$$s = -\frac{\partial}{\partial\lambda}\delta_B \tag{2.5a}$$

in its action on W^a_μ, χ^a and ψ^m. Here δ_B represents an ordinary gauge transformation, in which the parameter $\alpha^a(x)$ has been factorized into two fermionic factors,

$$\alpha^a(x) = \chi^a(x)\lambda \tag{2.5b}$$

where λ is a constant Grassmann anticommuting element. In our classical treatment[1,5] we extended Eqs. (2.1) - (2.4) to the case of a supergroup (with a detailed presentation in Ref. 11). All multiplets take on new components with opposite grading (assuming that G in (2.1) -(2.4) is the maximal even subgroup of the supergroup $\tilde{G}$). We now write

$$\begin{pmatrix} W_\mu^a \\ \chi_\mu^i \end{pmatrix} = \varepsilon_\mu \begin{pmatrix} \chi^a \\ w^i \end{pmatrix} \tag{2.6}$$

$$s \begin{pmatrix} v^m \\ \psi^2 \end{pmatrix} = \begin{pmatrix} \psi^m \\ v^2 \end{pmatrix} \tag{2.7}$$

where $\varepsilon_\mu = s^{-1}D$, a formal solution of (2.1) χ^i_μ is a new vector-ghost (fermionic, like $D_\mu\chi^i$ in (2.1)) and w^i is the Goldstone-Higgs multiplet. In SU(2/1) it has the correct phenomenological assignment!

We use Latin letters for physical fields and Greek for ghosts. The indices a,i and m,r represent even-odd components in the algebra and in the matter representations, respectively. The action of s may be inverted, as can be seen in (2.5a). Equation (2.7) and its inverse allow us to interpret the representations of SL(2/1). Each of b = 0,1 and b = 2/3,1/3 have to appear twice, with inverted statistics, as on both sides of (2.7). In the first, the field v^m corresponding to the highest weight is a physical chiral spinor and the components with opposite gradings are ghost-fields ψ^r related by s to the physical v^r in the second representation. Similarly, s acting on v^m generates the ghosts ψ^m which fill out the highest weight chirality in that second representation. This is known as our "ghost interpretation" of the supergroup and its representation. We shall not deal here with the gauge fields (2.6), but note that our classical treatment[1,5] yielded 4/3 g^2 for the w^4 coupling and 250 GeV for the mass of the Higgs field. We also had $\sin^2\theta_W = \frac{1}{4}$ as the unrenormalized Weinberg angle, with a conjecture that it is not renormalized except for the SU(2/1) breakdown itself (i.e., $\sim$ 100 GeV rather than the 10^{15} GeV of GUTs).

Without a quantum treatment, these results should be considered as tentative. We encountered some difficulties in dealing with the gauge multiplet. The Killing metric

$$g_{AB} = \mathrm{str}\,(M_A M_B) \tag{2.8}$$

is not positive-definite in the g_{uu} even direction (that of the U(1) in the even subgroup $SU(2)_L \times U(1)$. It is also antisymmetric in the odd directions (whereas w^4 requires a symmetric metric). We deal with these issues elsewhere and prefer to concentrate on the kinematical aspects which are as spectacular in SU(2/1) as they were in the original unitary symmetry, hadronic SU(3).

However, our ghost interpretation is not the only one which has been suggested. Other interpretations have been proposed, based on extending space-time by additional Bose[2,12] or Fermi dimensions[13,14,15,16], etc. The results presented in the rest of this paper are independent of the ghost conjecture and can be adopted in either interpretation.

3. THE FUNDAMENTAL REPRESENTATIONS OF SL(n/m)

In a recent study[6], we have constructed a class of "fundamental" representations of SL(n/m). These resemble the representations we used[17] to describe the supergravity gauge multiplet. For $m \neq 1$ they are given by (k is an integer)

$$\Lambda^0(V) \otimes S^k(X) \oplus \Lambda^1(V) \otimes S^{k-1} \oplus \ldots\ldots \oplus \Lambda^n(V) \otimes S^{k-n}(X) \qquad (3.1)$$

where $\Lambda(V)$ is the Grassmann manifold constructed on the Fermi subspace $V = \Lambda^1$ of the defining vector space $W = V + X$. The S^{k-i} is the space of homogeneous polynomials of degree $k - i$ on X^* and $S^j = 0$ for $j < 0$. In the special case $m = \dim X = 1$, (3.1) can be replaced by

$$\Lambda^0(V) \otimes F^b \oplus \Lambda^1(V) \otimes F^{b-1} \oplus \ldots . \oplus \Lambda^n(V) \otimes F^{b-n} \qquad (3.2)$$

where $b \in \mathbb{C}$ and F^{b-i} is the one-dimensional space of possibly multivalued functions defined on X and homogeneous of degree $b - i$. The dimensionality of (3.2) is 2^n, the same as that of $\Lambda(V)$. For (3.1) we recover the defining representation by taking $k = 1$ since we are left with

$$\Lambda^0(V) \otimes X \oplus \Lambda^1(V) \otimes \mathbb{C}$$

In extended supergravity the odd generators behave like an irreducible n-dimensional representation of SO(n) and have helicity $j_z = \frac{1}{2}$. One starts with the up-graviton with $j_z = 2$ and acts with the odd annihilation operators once, getting n $j_z = 3/2$ states forming an irreducible n-dimensional representation.

Here, working with the simpler $SL(n) \supset SU(2/1)$, we have altogether three parameters: two in picking the eigenvalues of U in the defining representation, and the quantum number b. We are given the highest weight, i.e., U_{mid}. Assuming for SL(n/1)

$$I_3^L = \mathrm{diag}(\tfrac{1}{2}, -\tfrac{1}{2}, \underbrace{0, \ldots\ldots 0}_{n-2 \text{ zeros}}/0) \qquad (3.3)$$

and

$$U = \mathrm{diag}\,(\alpha, \alpha, \underbrace{(\beta) \ldots\ldots}_{n-5\ \text{times}}, \underbrace{(\gamma)}_{3\ \text{times}} \ /T) \tag{3.4}$$

$$T = 2\alpha + (n-5)\beta + 3\gamma$$

Note that if the U_{max} is the highest value of U in the representation b and if the lowest value is $-U_{max}$, in a self-conjugate representation we have

$$U_{max} = \frac{n-1}{2} T \tag{3.5}$$

$$b = \frac{n-1}{2} \tag{3.6}$$

Equation (3.6) reduces our freedom of choice to two parameters (and U_{max} = 2 from phenomenology). To construct the states of the representation and denoting by E the set of diagonal operators I^L, U, etc., we have for their eigenvalues

$$\Lambda^0 \quad : e_{max} \text{ (given)}$$

$$\Lambda^1 \quad : (i^{th} \text{ state}) \ e_{max} - T + E_i$$

$$\Lambda^2 \quad : (i, j^{th} \text{ state}) \ e_{max} - 2T + E_i + E_j$$

etc.

E_i are the eigenvalues of E in the defining representation.

We have sketched here the mechanism for the construction of the representation's vector space. We refer the reader to Ref. 6 for proofs and for the actual construction of the representation.

4. UNIFICATION: PRECISE NUMBERS OF GENERATIONS AND THE PREDICTION OF SU(3) COLOUR

It is remarkable that with such a restricted system, the results should again be so spectacular. Using the choice

$$SL(n/1)\ ,\ n-S=K\geq 0;\ SL(n/1)\supset SU(2/1)\otimes SU(3)_c$$

$$b=\frac{n-1}{2} \tag{4.1}$$

$$U=\mathrm{diag}\left(\frac{-K}{4+K},\ \frac{-K}{4+K},\ \left(\frac{4}{4+K}\right)_{K\ \text{times}},\ \left(\frac{4-2K}{3(4+K)}\right)_{3\ \text{times}}\ \Bigg|\ \frac{4}{4+K}\right)$$

$$I_3^L=\mathrm{diag}\left(\tfrac{1}{2},-\tfrac{1}{2},(0)_{K+3\ \text{times}}\qquad \big|\ 0\ \right)$$

we get a 2^n-dimensional representation containing precisely 2^k "generations" of the basic 16 fermion fields v^m and their 16 matter ghosts ψ^m. The discrete s operator in (2.7) is thus embedded in the superalgebra for $n \geq 5$. This set of 32 fields makes up the $b = 2$ representation of SL(5/1) which is, itself, reducible (though not completely reducible, a common case in superalgebras). This feature of reducibility and an invariant 2^{n-1} subspace occurs for all SL(n/1) if b is an integer. In the case of SL(5/1), in approaches which do not distinguish between fields and ghosts, the irreducible 16 dimensional invariant subspace already contains a full set of fields. In either case, the fields sometimes appear through their charge-conjugates, as in the Georgi-Glashow SU(5). However, if we take SL(6/1) as the basic physical set, its irreducible 64 dimensional $b = 5/2$ representation is charge-conjugation symmetric and contains all 16 fields, their ghosts and the charge-conjugate states of both. Depending on the treatment, this could be regarded as one, two or four generations. C-symmetry might be important for the removal of Adler-Bell-Jackiw anomalies.

We have introduced colour SU(3) through the step SL(2/1) → SL(5/1). Our quantum numbers (4.1) produce an SU(5) which does not coincide with the Georgi-Glashow SU(5) assignments, since our $U \not\subset SU(5)$ of $SL(5/1) \supset SU(5) \times U(1)$ as can be seen in (4.1). However, it is also possible in SL(5/1) to identify U with an SU(5) quantum number as in the Georgi-Glashow model, as was suggested by A. Salam (private communication) and J.G. Taylor[16,6].

Beyond this spectacularly unique fit with the generation issue, it has also been shown[18] that the supergroup structure requires colour to be represented by SU(3). Indeed, these fundamental representations impose a unique subdivision in the eigenvalues of U in (4.1), generating the $SU(3)_{colour}$ group commuting with SL(2/1), if we require that the colour group SU(r) appear in representation r for all quarks.

One important result of such a unification is the fact that all the matter fields (and their ghosts, in that interpretation) can be considered as composite. The fundamental constituent is the fractionally charged n + 1 dimensional primitive field with n fermions and one ghost. The charges will depend on the final selection for k.

In the multigeneration systems beyond SL(5/1), the correlation between chirality and the grading vanishes.

REFERENCES

1. Y. Ne'eman, Phys. Lett. 81B (1979) 190.
2. D.B. Fairlie, Phys. Lett. 82B (1979) 97.
3. L. Corwin, Y. Ne'eman and S. Sternberg, Rev. Mod. Phys. 47 (1975) 573.
4. M. Scheunert, W. Nahm and V. Rittenberg, Jour. Math. Phys. 18 (1977) 155.
5. Y. Ne'eman and J. Thierry-Mieg, Proc. Salamanca Int. Conf. on Differential Geometry Methods in Mathematical Physics, to be published by Springer Verlag, Lecture Notes in Mathematics (1979).
6. Y. Ne'eman and S. Sternberg, Proc. Natl. Acad. Sci. U.S.A., in press.
7. J. Thierry-Mieg, Thèse de Doctorat d'Etat, Orsay, France (1978); J. Thierry-Mieg, Jour. Math. Phys,, in press (1980); J. Thierry-Mieg, Nuovo Cim. A., in press (1980).
8. Y. Ne'eman, T. Regge and J. Thierry-Mieg, Proc. XIX Int. Conf. on High Energy Physics (Tokyo 1978) Eds S. Homma, M. Kawaguchi and H. Miyazawa (Phys. Soc. Jap. Pub. 1979) pps 552-554; J. Thierry-Mieg and Y. Ne'eman, Annals of Phys. vol. 123, 2 (1979) 247-273.
9. Y. Ne'eman and J. Thierry-Mieg, Proc. VIII Int. Conf. on the Applications of Group Theory to Physics (Kiryat-Anavim 1979) Eds L. Horwitz and Y. Ne'eman, (Annals of the Israeli Physical Society, 1980).
10. C. Becchi, A. Rouet and R. Stora, Com. Math. Phys. 42 (1975) 127.
11. Y. Ne'eman and J. Thierry-Mieg, Proc. Nat'l Acad. Sci. U.S.A. 77 (1980) 2353.
12. E.J. Squires, Phys. Lett. 92B (1979) 395.
13. J.G. Taylor, Phys. Lett. 83B (1979) 331.
14. P.H. Dondi and P.D. Jarvis, Phys. Lett. 84B (1979) 75.
15. J.G. Taylor, Phys. Lett. 84B (1979) 79.
16. J.G. Taylor, Phys. Rev. Lett. 43 (1979) 826.

17. M. Gell-Mann and Y. Ne'eman, (1976) unpublished; first quoted in D.Z. Freedman, Phys. Rev. Lett. 38 (1977) 105.
18. Y. Ne'eman and S. Sternberg, to be published (1980).

SUPERSPACE

B. Zumino

CERN

1211 Geneva 23, Switzerland

1. INTRODUCTION

Superspace is an extension of ordinary space-time. Its points are labelled not only by commuting bosonic (vectorial) co-ordinates but also in addition by anticommuting fermionic (spinorial) co-ordinates. Superspace gives a geometric picture of supersymmetry and provides a technique for finding representations of the supersymmetry algebra by fields. In this lecture I shall give first a brief discussion of rigid (flat) superspace[1-4] which has rigid* supersymmetry as a group of motions. Then I shall discuss curved flexible superspace[5-7]. The latter is relevant to a geometric description of supergravity. My aim is to provide an elementary introduction for non-experts, but I shall also make a few remarks and present some new results which may be of interest to the experts.

2. RIGID SUPERSPACE

Written in two component spinor notation the algebra of extended rigid supersymmetry is

$$\{Q_{\alpha i}, \bar{Q}^{j}_{\dot\beta}\} = -2\,\delta^{j}_{i}(\sigma^{m})_{\alpha\dot\beta}P_{m} \;;\quad \{Q_{\alpha i}, Q_{\beta j}\} = \{\bar{Q}^{i}_{\dot\alpha}, \bar{Q}^{j}_{\dot\beta}\} = 0$$

$$[Q_{\alpha i}, P_{m}] = [\bar{Q}^{i}_{\dot\alpha}, P_{m}] = 0 \;;\quad [P_{m}, P_{n}] = 0\,, \tag{1}$$

---

*) I prefer the word rigid to the word global, to denote a symmetry with x independent parameters, because global has a very different topological connotation.

where P_m is the momentum operator which generates translations and $Q_{\alpha i}$ (i = 1, ... N) the supersymmetry spinorial generators*. Clearly the algebra (1) admits as automorphism the Lorentz group, under which P_m transforms as a vector and $Q_{\alpha i}$, $\bar{Q}^i_{\dot\alpha}$ as (Majorana) spinors. It also admits as automorphism SU(N) (actually U(N)), for which $Q_{\alpha i}$ are in the N and $\bar{Q}^i_{\dot\alpha}$ in the $\bar{N}$ representation. We are interested in finding representations of the algebra (1) in terms of fields**. For this we employ the technique of superspace.

The points of superspace are labelled by bosonic co-ordinates x^m and fermionic co-ordinates $\theta_{\alpha i}$, $\bar\theta^i_{\dot\alpha}$

$$x^m x^n - x^n x^m = 0 \quad ; \qquad \theta_{\alpha i}\theta_{\beta j} + \theta_{\beta j}\theta_{\alpha i} = \bar\theta^i_{\dot\alpha}\bar\theta^j_{\dot\beta} + \bar\theta^j_{\dot\beta}\bar\theta^i_{\dot\alpha} = 0$$

$$x^m\theta_{\alpha i} - \theta_{\alpha i}x^m = x^m\bar\theta^i_{\dot\alpha} - \bar\theta^i_{\dot\alpha}x^m = 0 \ ; \ \theta_{\alpha i}\bar\theta^j_{\dot\beta} + \bar\theta^j_{\dot\beta}\theta_{\alpha i} = 0 \ . \tag{2}$$

The algebra (1) can be realized as a group of rigid motions in superspace, in which the spinorial generators correspond to the infinitesimal transformations

$$\begin{cases} \delta x^m = i(\theta\sigma^m\bar\zeta - \zeta\sigma^m\bar\theta) \\ \delta\theta = \zeta \ , \quad \delta\bar\theta = \bar\zeta \end{cases} \tag{3}$$

of infinitesimal totally anticommuting spinorial parameters $\zeta^{\alpha i}, \bar\zeta^{\dot\alpha}_i$. The commutator of two transformations like (3) is given by a translation

$$\begin{cases} [\delta_2,\delta_1]\, x^m = 2i(\zeta_2\sigma^m\bar\zeta_1 - \zeta_1\sigma^m\bar\zeta_2) \\ [\delta_2,\delta_1]\,\theta = [\delta_2,\delta_1]\,\bar\theta = 0 \ . \end{cases} \tag{4}$$

corresponding to (1). The anticommutation relation becomes a commutation relation because the parameters $\zeta,\bar\zeta$ anticommute (with each other and with $Q,\bar{Q}$).

---

*) In this lecture I do not consider the generalization of (1) involving central charges[8,9]. For some recent developments making use of the idea of central charges, see the lectures by Sohnius, Stelle and West in these proceedings.

**) Representations in terms of on-mass shell states can be obtained very easily by observing that the momentum is diagonal and therefore (1) can be reduced to a Clifford algebra. For details see, for example, Ref. 10.

A superfield is a function $V(x,\theta;\bar\theta)$ in superspace. It can represent the algebra (1) if it is taken to transform like a scalar under (3)

$$\delta V(x,\theta,\bar\theta) = \left(\delta x^m \partial_m + \delta\theta \frac{\partial}{\partial\theta} + \delta\bar\theta \frac{\partial}{\partial\bar\theta}\right) V(x,\theta,\bar\theta) \ . \tag{5}$$

If one expands $V(x,\theta,\bar\theta)$ in a power series in θ and $\bar\theta$, the series stops after a finite number of terms, since the square of each component of θ or $\bar\theta$ vanishes by (2). The coefficients of the expansion are ordinary fields, functions of the bosonic variables x^m. Each additional θ or $\bar\theta$ in the expansion corresponds to a change in statistics and in spin (by $\frac{1}{2}$). By the transformation (5) the component fields transform into each other, they form a supermultiplet of fields.

The field representations given by (5) are in general reducible. If one defines the spinorial "covariant" derivatives

$$D_{\alpha i} = \frac{\partial}{\partial\theta^{\alpha i}} + i(\sigma^m\bar\theta)_{\alpha i}\,\partial_m$$

$$\bar D^{i}_{\dot\alpha} = -\frac{\partial}{\partial\bar\theta^{\dot\alpha}_{i}} - i(\theta\sigma^m)^{i}_{\dot\alpha}\,\partial_m \tag{6}$$

$$\left(\partial_m \equiv \frac{\partial}{\partial x^m}\right)$$

one easily verifies that they commute with the transformations (3),(5). Therefore, one can use them to impose differential constraints on a superfield or a set of superfields, which are preserved by a supersymmetry transformation. In extended supersymmetry, where the indices i,j take several values, θ and $\bar\theta$ have many independent components and a superfield has many independent component fields, some of high spin (and mass dimension, see below). To obtain sensible Lagrangian theories, it is essential to find the minimal irreducible pieces. Although this is done in principle by imposing differential constraints constructed with the covariant derivatives (6), in practice this program has been worked out only in a few relatively simple cases. The difficulty comes mainly from the fact that the covariant derivatives do not anticommute, like simple fermionic derivatives. They satisfy the commutation relations

$$\{D_{\alpha i}, \bar D^{j}_{\dot\beta}\} = -2i\,\delta^{j}_{i}(\sigma^m)_{\alpha\dot\beta}\,\partial_m$$
$$\{D_{\alpha i}, D_{\beta j}\} = \{\bar D^{i}_{\dot\alpha}, \bar D^{j}_{\dot\beta}\} = 0 \ . \tag{7}$$

Still, the covariant derivatives are relatively simpler to use. Observe that the supersymmetry transformation (5) can be written in terms of covariant derivatives

$$\delta V(x,\theta,\bar{\theta}) = \left[\zeta D + \bar{\zeta}\bar{D} + 2i(\theta\sigma^m\bar{\zeta} - \zeta\sigma^m\bar{\theta})\partial_m\right] V \ . \tag{8}$$

Instead of expanding in $\theta,\bar{\theta}$, one can obtain the various component fields of the superfield V by applying to V the above spinorial covariant derivatives and then setting $\theta = \bar{\theta} = 0$. The order of differentiation is important, of course, since the covariant derivatives do not anticommute completely, but the difference is seen by (7) to involve extra terms which are just ordinary x derivatives of lower component fields. The various definitions are, therefore, equivalent and one knows how to connect them. The transformation properties of the component fields can be obtained from (8) by applying the appropriate covariant derivatives and setting $\theta = \bar{\theta} = 0$. On the right-hand side, the covariant derivatives commute with the differential operator in the square bracket and apply directly to V.

The last component in the $\theta,\bar{\theta}$ expansion (highest power) is obtained (up to x derivatives) by applying to V all spinorial covariant derivatives with different indices. If one applies to it one more spinorial covariant derivative, one obtains zero or an x derivative. This observation, together with (8), shows that the last component of a superfield transforms under supersymmetry by an x derivative. It is, therefore, a candidate for a Lagrangian, since its integral over x (the action) is invariant under supersymmetry. From this argument we see also that it is not necessary to take the last component of a superfield for a Lagrangian. Any superfield having the property that a spinorial covariant derivative applied to it equals an x derivative, will give a candidate for a Lagrangian (setting $\theta = \bar{\theta} = 0$ in it). In the following I shall mention examples of this, but first I must recall some basic facts about integration over anticommuting variables.

3. INTEGRATION IN SUPERSPACE AND SUPERDETERMINANTS

Integration over anticommuting variables has been used in quantum field theory for a long time. Its properties have been codified by Berezin[11]. We have in mind here the analogue of the definite integral (from $-\infty$ to $+\infty$) for ordinary functions, not the analogue of the primitive of a function. Consider a function of one variable. When it exists, the definite integral is translationally invariant

$$\int_{-\infty}^{+\infty} dx\ f(x + a) = \int_{-\infty}^{+\infty} dx\ f(x) \tag{9}$$

For functions of one anticommuting variable θ, which means that

$$\theta^2 = 0 \tag{10}$$

we shall postulate the analogous property

$$\int_\theta f(\theta + \alpha) = \int_\theta f(\theta) \ . \tag{11}$$

Now, because of (10), the most general function of one anticommuting variable θ is linear in θ

$$f(\theta) = a + \theta\beta \ . \tag{12}$$

Therefore (11) means

$$\int_\theta (a + \theta\beta + \alpha\beta) = \int_\theta (a + \theta\beta) \tag{13}$$

or, assuming linearity,

$$\int_\theta \alpha\beta = 0 \ . \tag{14}$$

This equation is generalized by postulating that the integral of any constant vanishes. In order to obtain a non-trivial result in (13) one must then assume that the integral of θ itself does not vanish, and it is usual to normalize it to 1. In conclusion, one assumes the basic relations

$$\int_\theta \text{const} = 0 \qquad \int_\theta \theta = 1 \ , \tag{15}$$

together with linearity. Observe that this corresponds to taking

$$\int_\theta f(\theta) \equiv \frac{\partial}{\partial\theta} f(\theta) \ . \tag{16}$$

In spite of this identity, it is often convenient to use the integral notation, especially when there are accompanying bosonic integrations. In many cases the integral notation is more inspiring, for instance there exists a theory of Fourier transforms (with completeness relations etc.) for functions of anticommuting variables. The delta function for anticommuting variables is

$$\delta(\theta - \theta') = \theta - \theta' \ . \tag{17}$$

Indeed, one sees from (15) that

$$\int_\theta (\theta - \theta')(a + \theta\beta) = a + \theta'\beta \ . \tag{18}$$

The integral of a derivative vanishes

$$\int_\theta \frac{\partial}{\partial\theta} f(\theta) = 0 . \tag{19}$$

This follows, e.g., from (16) since the square of $\partial/\partial\theta$ vanishes. Therefore, integration by parts can be used

$$\int_\theta f(\theta) \frac{\partial}{\partial\theta} g(\theta) = \mp \int_\theta \left[\frac{\partial}{\partial\theta} f(\theta)\right] g(\theta) , \tag{20}$$

where the plus sign has to be used if $f(\theta)$ is itself fermionic.

For a function of several variables one just applies the definition (15) successively for each variable. Observe, however, that the order counts. From (16) one sees that

$$\int_{\theta_1} \int_{\theta_2} f = - \int_{\theta_1} \int_{\theta_2} f . \tag{21}$$

When the integrand depends also on ordinary commuting variables, the integrations over them are defined in the usual way. Changes of variables can be performed in an integral. The integrand must be multiplied by the Jacobian of the transformation, which is defined in the usual way but by using the concept of superdeterminant discussed below. Finally, we observe that, while integration over a commuting variable changes the dimensions of a quantity in the same way as multiplication by x, on the contrary integration over an anticommuting variable θ changes it as division by θ. This is obvious from (16).

Let M be a matrix of the form

$$M = \begin{pmatrix} A & \Gamma \\ \Delta & B \end{pmatrix} , \tag{22}$$

where the submatrices A and B have bosonic matrix elements while Γ and Δ have fermionic elements (even and odd elements of a Grassman algebra, respectively). Matrices like M can be combined linearly and multiplied with each other and the generic matrix of this kind has an inverse. We define the trace by

$$\mathrm{Tr}\, M = \mathrm{Tr}\, A - \mathrm{Tr}\, B . \tag{23}$$

The reason for the minus sign in this definition is that we wish the basic property of the trace

$$\mathrm{Tr}\, M_1M_2 = \mathrm{Tr}\, M_2M_1 \tag{24}$$

to hold. Indeed we have

$$\begin{aligned}\mathrm{Tr}\, M_1M_2 &= \mathrm{Tr}(A_1A_2 + \Gamma_1\Delta_2) - \mathrm{Tr}(\Delta_1\Gamma_2 + B_1B_2)\\ &= \mathrm{Tr}(A_2A_1 - \Delta_2\Gamma_1) - \mathrm{Tr}(-\Gamma_2\Delta_1 + B_2B_1)\\ &= \mathrm{Tr}\, M_2M_1 .\end{aligned} \tag{25}$$

The superdeterminant can be defined from

$$\det M = \exp \mathrm{Tr} \ln M . \tag{26}$$

Because of (24), it satisfies the product property

$$\det (M_1M_2) = (\det M_1)(\det M_2) . \tag{27}$$

Indeed, writing

$$\ln M = N , \tag{28}$$

one has

$$\ln (M_1M_2) = N_1 + N_2 + \tfrac{1}{2} \left[N_1,N_2\right] + \ldots , \tag{29}$$

where the dots denote multiple commutators. Taking the trace, all commutator terms vanish by (24), and one has

$$\mathrm{Tr} \ln (M_1M_2) = \mathrm{Tr} \ln M_1 + \mathrm{Tr} \ln M_2 \tag{30}$$

which establishes (27). It follows from (26) that, if

$$M = 1 + X , \tag{31}$$

where 1 is the unit matrix and X is infinitesimal, then

$$\det M = 1 + \mathrm{Tr}\, X . \tag{32}$$

More generally, one finds in the standard way, combining (27) and (32), that

$$\delta \det M = (\det M)\, \mathrm{Tr}\, M^{-1}\, \delta M \tag{33}$$

for any infinitesimal variation δM. An explicit form for the determinant can be obtained by writing M in the standard form

$$M = \begin{pmatrix} C & \Sigma \\ 0 & 1 \end{pmatrix} \begin{pmatrix} 1 & 0 \\ \emptyset & D \end{pmatrix} . \tag{34}$$

Comparing with (22) one finds

$$C = A - \Gamma B^{-1}\Delta, \qquad \Sigma = \Gamma B^{-1}$$
$$\emptyset = \Delta \qquad , \qquad D = B .$$

On the other hand (34) gives, using (23), (26) and (27),

$$\det M = \det C / \det D . \tag{36}$$

Therefore,

$$\det M = \frac{\det(A - \Gamma B^{-1}\Delta)}{\det B} . \tag{37}$$

An equivalent form is

$$\det M = \frac{\det A}{\det(B - \Delta A^{-1}\Gamma)} \tag{38}$$

In (37) and (38) the determinants occurring on the right-hand side are ordinary determinants of matrices with bosonic elements.

4. EXAMPLES OF DYNAMICS IN SUPERSPACE

The first is the well-known example of the "chiral" superfield. We are in N = 1 (simple) supersymmetry, so we have no indices i,j. We take a complex superfield ϕ, subject to the invariant constraints

$$\bar{D}_{\dot{\alpha}}\phi = 0 , \quad D_\alpha\bar{\phi} = 0 . \tag{39}$$

The component fields are the values for $\theta = \bar{\theta} = 0$ of

$$A = \phi , \quad \chi_\alpha = D_\alpha\phi , \quad F = -\tfrac{1}{4} D^\alpha D_\alpha\phi \tag{40}$$

together with their complex conjugates. All other spinorial covariant derivatives can be reduced to x derivatives of these fields by using (7) (with no δ_i^j for N = 1) and therefore do not give independent component fields. As a free action in superspace one can take

$$\int d^4x \int_\theta \int_{\bar\theta} \; (\phi\bar\phi + \Lambda^\alpha D_\alpha \bar\phi + \bar\Lambda_{\dot\alpha} \bar D^{\dot\alpha} \phi) \;, \tag{41}$$

where we have introduced Lagrange multipliers Λ^α and $\bar\Lambda_{\dot\alpha}$ for the constraints (39). The integral over θ is over both components, same for $\bar\theta$. Variation of $\bar\phi$ gives

$$\phi = D_\alpha \Lambda^\alpha \;. \tag{42}$$

The multiplier Λ^α can be eliminated by applying the operator $D^\beta D_\beta$, since the product of three anticommuting D_α derivatives vanishes, and one obtains the equation of motion

$$D^\beta D_\beta \phi = 0 \;, \tag{43}$$

which contains compactly the equations of motion for the component fields

$$\Box A = 0 \;, \quad \sigma^m \partial_m \bar\chi = 0 \;, \quad F = 0 \;. \tag{44}$$

The action (41) is an integral over all of superspace. It can be transformed as follows

$$\int d^4x \int_\theta \int_{\bar\theta} \; \phi\bar\phi = \int d^4x \int_\theta \bar D \bar D(\phi\bar\phi) = \int d^4x \int_\theta \phi \; \bar D \bar D \bar\phi \;. \tag{45}$$

In the first step we have used the equivalence of integration and differentiation for anticommuting variables (the additional terms in the covariant derivatives vanish by integration over x). In the second step we have used the constraints (39). In (45) the integration is over a submanifold of superspace, the Lagrangian density $\phi\bar D\bar D\bar\phi$ is chiral

$$\bar D_{\dot\alpha}(\phi \bar D\bar D\bar\phi) = 0 \tag{46}$$

which shows that the integral (45) is independent of $\bar\theta$. Finally, one can replace also the θ integrations by covariant derivatives and write the action as an integral over x alone

$$\int d^4x \; DD\bar D\bar D(\phi\bar\phi) \;. \tag{47}$$

This simple example shows that there are various possibilities for writing an action in superspace. It can be given as an integral over all of superspace or over submanifolds. This is especially important in the case of extended supersymmetry. Since the dimensions of θ are $\dim \theta = \dim \bar\theta = (\dim x)^{\frac{1}{2}}$, the various component fields in a given superfield differ not only in spin and statistics but also in dimension. In the case of extended supersymmetry, when there are many independent components of θ and $\bar{\theta}$, the difference in dimension between the first and the last component can be very high. Dimensional arguments can often be used to exclude the possibility that the Lagrangian density in x space, which must have dimensions equal to $(\dim x)^{-4}$, be the last component of a superfield. One of the intermediate components is sometimes the correct choice, corresponding to an integral over a submanifold in superspace. In this case, however, as in the last form of (45), one must have relations like (46) which show that the integral is independent of the particular choice of submanifold.

While the above example of the chiral multiplet is relatively trivial, because the Lagrangian can also be written as the last component of a supermultiplet, there are some interesting examples in extended supersymmetry where this is not possible, and the Lagrangian can only be obtained as a sum of terms with two covariant derivatives applied to a superfield. The first is the N = 2 so-called "hypermultiplet" with central charge[12] as described in superspace by Sohnius[9]. The second is the six-dimensional simply supersymmetric Yang-Mills theory described in superspace by Siegel[13] (which, by dimensional reduction, gives the N = 2 Yang-Mills theory in four dimensions). Wess and I[14] have shown that the Lagrangian of this theory can be written as a sum of terms with two covariant derivatives. One expects an analogous situation for higher extended supersymmetries. It is of considerable interest to work out in detail the dynamics of theories with an action which is given by an integral over a submanifold in superspace: how one derives the equations of motion, quantization, perturbation theory, etc.

5. CURVED SUPERSPACE

Curved, flexible, superspace is to flat, rigid, superspace like the Riemannian space of general relativity is to Minkowski space. There is no *a priori* group of rigid motions, instead the geometry is determined dynamically. There are two versions of this geometry of superspace, which differ in the choice of the structure group (tangent space group), the super-Riemannian geometry[5] and the non-Riemannian geometry[6,7] appropriate to supergravity. I do not have the time here to go into a detailed description, for which I refer to the literature cited*. In this lecture I shall discuss only some basic notions which are common to the two approaches.

---

*) See also the lectures by Brink, Ferrara, Grimm and Nath in these proceedings.

The points in curved (differentiable) superspace are labelled by a finite number of curvilinear co-ordinates $z^M = (x^m, \theta^\mu)$, the x^m are bosonic (commuting), the θ^μ fermionic (anticommuting). More precisely, they are respectively even and odd elements of a (very large if not infinite) Grassmann algebra. Co-ordinate transformations in superspace mix x and θ but in such a way that the new x's are still even and the new θ's odd. An infinitesimal co-ordinate transformation is specified by infinitesimal parameters $\xi^M(z)$. For instance, a scalar V(z) transforms as

$$\delta V = \xi^M \partial_M V \ , \quad \partial_M \equiv \frac{\partial}{\partial z^M} \ , \tag{48}$$

a tensor $T_{LM}{}^N(z)$ transforms as

$$\begin{aligned} \delta T_{LM}{}^N = \xi^S \partial_S T_{LM}{}^N + U_L{}^{L'} T_{L'M}{}^N \\ + U_M{}^{M'} T_{LM'}{}^N (-1)^{(m+m')(\ell+1)} - T_{LM}{}^{N'} U_{N'}{}^N \end{aligned} \tag{49}$$

where

$$U_M{}^N \equiv \partial_M \xi^N \ . \tag{50}$$

These transformation laws are the same as in an ordinary bosonic space except for some sign factors due to the anticommuting nature of some quantities. In (49) we have used the customary convention that the small indices in the exponent are 0 or 1 according to whether the corresponding capital index is bosonic or fermionic. In addition to scalars, covariant and contravariant vectors and tensors, in a differentiable superspace one can define densities $\mathfrak{D}(z)$ which transform as

$$\delta \mathfrak{D} = \partial_S (\xi^S \mathfrak{D}) (-1)^s \ . \tag{51}$$

Since a density changes by a sum of terms each of which is a derivative, the integral of a density over all of superspace is invariant under (51). Therefore densities can be used as Lagrangians.

Differential forms can be defined in a differentiable superspace. One takes the differentials dx^m to be odd and $d\theta^\mu$ to be even, the opposite of x^m and θ^μ. This simple grading has the advantage of simplicity, one is still working with a Grassmann algebra.

All quantities are assigned a parity which for x^m, θ^μ, dx^m, $d\theta^\mu$ is $g = 0, 1, 1, 0$, respectively, and the commutation relations for any two quantities are

$$A_1A_2 = (-1)^{g_1g_2} A_2A_1 \ . \tag{52}$$

Following Ref. 6, in the physics literature one has often used a double grading (g,h), which for x^m, θ^μ, dx^m, $d\theta^\mu$ is (0,0), (1,0), (0.1), (1,1), respectively, and the commutation relations

$$A_1A_2 = (-1)^{g_1g_2+h_1h_2} A_2A_1 \ . \tag{53}$$

It is easy to see that the simple and the double grading are equivalent. Let x^m, θ^μ, dx^m, $d\theta^\mu$ have simple grading and satisfy (52). Define

$$\begin{aligned} X^m &= x^m \otimes 1 \\ \Theta^\mu &= \theta^\mu \otimes \sigma_3 \\ DX^m &= dx^m \otimes \sigma_1 \\ D\Theta^\mu &= d\theta^\mu \otimes (-i\sigma_2) \\ D &= d \otimes \sigma_1 \end{aligned} \tag{54}$$

with the usual Pauli matrices. It is easy to check that the commutation relations of the capital quantities are those of the table

	X^m	Θ^μ	DX^m	$D\Theta^\mu$
X^m	1	1	1	1
Θ^μ	1	-1	1	-1
DX^m	1	1	-1	-1
$D\Theta^\mu$	1	-1	-1	1

namely those appropriate to the double grading (53). This equivalence applies in particular to the different quantization rules which have been given by various authors[15] for bosonic ghosts (which behave like dx^m) and for fermionic ghosts (which behave like $d\theta^\mu$).

The σ matrices drop out in all physical amplitudes. Finally, we observe that the differentiation operator

$$d = dx^m \frac{\partial}{\partial x^m} + d\theta^\mu \frac{\partial}{\partial \theta^\mu} \tag{55}$$

satisfies

$$d^2 = 0 \ . \tag{56}$$

This is almost obvious for simple grading. By the last of (54) it is also true for double grading. From now on we shall use simple grading, which is slightly easier to use and mnemonically more convenient.

6. DENSITIES AND TENSORS

In ordinary bosonic space there is a relation between densities and certain tensors. A totally antisymmetric tensor with as many lower indices as the dimension of the space has only one independent component, which transforms as a density. Because covariant antisymmetric tensors correspond to differential forms, this gives a relation between certain differential forms and densities. For some time no such connection was known in superspace. More recently, due to some work by Russian authors[16], which I have somewhat developed[17], such a relation has been found.

Consider a superspace with p bosonic and q fermionic coordinates $z^M = (x^m, \theta^\mu)$, where $m = 1,2 \ldots p$ and $\mu = p+1, \ldots p+q$. Let a tensor

$$T_{L_1 L_2 .. L_p}{}^{M_1 M_2 .. M_q} \tag{57}$$

be graded antisymmetric in its p lower indices and graded symmetric in its q upper indices. Graded symmetric (for instance) means that it is symmetric for the exchange of two Bose or one Bose and one Fermi index, antisymmetric for the exchange of two Fermi indices. We assume for definiteness that the component

$$T_{\ell_1 \ell_2 .. \ell_p}{}^{\mu_1 \mu_2 .. \mu_q} \tag{58}$$

of the tensor (57) is bosonic. The parity of the other components is then fixed. Such a tensor is partially reducible. This is easy to see in the very simple case of a superspace with one x and one θ. Let the components of the tensor be denoted by

$$T_L{}^M = \begin{pmatrix} \Gamma & \mathcal{D} \\ \mathcal{A} & \Delta \end{pmatrix} \tag{59}$$

and let

$$\partial_L \xi^M = \begin{pmatrix} a & \beta \\ \gamma & d \end{pmatrix} \ . \tag{60}$$

We use Latin letters for bosonic, Greek letters for fermionic quantities. The transformation law analogous to (49) gives

$$\begin{aligned} \delta'\mathcal{D} &= a\mathcal{D} - \mathcal{D}d + \beta\Delta - \Gamma\beta \\ \delta'\Gamma &= \underline{a\Gamma} + \beta\mathcal{A} - \underline{\Gamma a} - \mathcal{D}\gamma \\ \delta'\Delta &= \gamma\mathcal{D} + \underline{d\Delta} - \mathcal{A}\beta - \underline{\Delta d} \\ \delta'\mathcal{A} &= \gamma\Gamma + d\mathcal{A} - \mathcal{A}a - \Delta\gamma \end{aligned} \tag{61}$$

where we have abbreviated

$$\delta' \equiv \delta - \xi^M \partial_M \ . \tag{62}$$

The terms underlined in (61) cancel. It is clear that the transformation (61) admits the invariant subspace

$$\Gamma + \Delta = 0 \ , \quad \mathcal{A} = 0 \ . \tag{63}$$

If we substitute these into (61) we obtain

$$\delta'\mathcal{D} = \mathcal{D}(a - d) \tag{64}$$

which is the transformation law of a density. In the general case the same situation occurs. A tensor like (57) can be reduced (maximally) by imposing certain relations among its components. After the reduction, its component (58) (it is totally antisymmetric in the lower indices and also in the upper indices and therefore it reduces to one independent component) transforms like a density, it does not mix with the other components. One can write

$$T_{\ell_1\ell_2..\ell_p}{}^{\mu_1\mu_2..\mu_q} = \varepsilon_{\ell_1\ell_2..\ell_p}\,\varepsilon^{\mu_1\mu_2..\mu_q}\,\mathcal{D} \quad . \tag{65}$$

where $\mathcal{D}$ is a density. Furthermore, one can show that there exists a co-ordinate system where all components of the tensor vanish except for (65).

Just as in an ordinary bosonic space, one can introduce in superspace linear frames and the associated supervielbein matrix $E_M{}^A(z)$ and its inverse $E_A{}^M(z)$. These transform like a covariant respectively contravariant vector with respect to the index M. The index $A = (a,\alpha)$ takes as many values as M, that is $a = 1, \ldots p$ bosonic and $\alpha = p + 1, \ldots p + q$ fermionic. With respect to the index A, $E_M{}^A$ transforms linearly like a contravariant tangent space vector, $E_A{}^M$ like a covariant tangent space vector. It is now easy to see that the superdeterminant of the supervielbein matrix

$$E = \det E_M{}^A \tag{66}$$

transforms like a density under general co-ordinate transformations. Infinitesimally the supervielbein transforms as

$$\delta E_M{}^A = \xi^L \partial_L E_M{}^A + \partial_M \xi^L E_L{}^A \tag{67}$$

Using (33) one finds

$$\begin{aligned} \delta E &= \xi^L \partial_L E + E \quad E_A{}^M \partial_M \xi^L E_L{}^A (-1)^a \\ &= \xi^L \partial_L E + E \cdot \partial_L \xi^L (-1)^\ell \\ &= \partial_L (\xi^L E)(-1)^\ell \end{aligned} \tag{68}$$

which is the transformation property of a superspace density. Under linear transformations on A, the superdeterminant E is invariant.

One can define the differential forms

$$dz^M E_M{}^A = dx^m E_m{}^A + d\theta^\mu E_\mu{}^A \tag{69}$$

and their "duals"

$$E_A{}^M \, \overline{dz_M} = E_A{}^m \, \overline{dx_m} + E_A{}^\mu \, \overline{d\theta_\mu} \; . \tag{70}$$

The quantities $\overline{dx_m}$ are taken to be commuting (even) while $\overline{d\theta_\mu}$ are anticommuting (odd), and they are to transform covariantly. So they are the "opposite" of the dx^m, $d\theta^\mu$ (observe that they are not differentials, there is no operation $\bar{d}$). The volume element in superspace is

$$\prod_{a=1}^{p} dz^L E_L{}^a \prod_{\alpha=p+1}^{p+q} E_\alpha{}^M \overline{dz_M} =$$

$$= dz^{L_p} \dots dz^{L_1} T_{L_1 \dots L_p}{}^{M_1 \dots M_q} \overline{dz_{M_q}} \dots \overline{dz_{M_p}} \ . \qquad (71)$$

The tensor T defined by (71) is of the type (57) and one can verify that it satisfies the maximal reduction relations mentioned after that equation. For instance,

$$T_{\lambda_1 \dots \lambda_p}{}^{m_1 \dots m_q} = 0 \qquad (72)$$

Finally, using (38) or

$$\det E_M{}^A = (\det E_m{}^a) \cdot (\det E_\alpha{}^\mu) \qquad (73)$$

it is easy to verify that (65) is satisfied in this case with

$$\mathcal{D} \propto \det E_M{}^A \ . \qquad (74)$$

Acknowledgement

The content of Section 4 was inspired by conversations with Julius Wess.

REFERENCES

1. D.V. Volkov and V.P. Akulov, Phys. Letters 46B (1973) 109.
2. A. Salam and S. Strathdee, Nuclear Phys. B76 (1974) 477.
3. S. Ferrara, J. Wess and B. Zumino, Phys. Letters 51B (1974) 239.
4. For a review of rigid supersymmetry including superspace methods see P. Fayet and S. Ferrara, Phys. Reports 32 (1977) 251.
5. R. Arnowitt, P. Nath and B. Zumino, Phys. Letters 56B (1975) 81; R. Arnowitt and P. Nath, Phys. Letters 56B (1975) 117; 65B (1976) 73.

6. B. Zumino, Proc. of the Conf. on Gauge Theories and Modern Field Theory, Northeastern University, September (1975), eds R. Arnowitt and P. Nath (MIT Press); J. Wess and B. Zumino, Phys. Letters 73B (1977) 361.
7. For a review of curved superspace with other references, see B. Zumino, Cargèse Lectures, in Recent Developments in Gravitation, eds M. Levy and S. Deser (Plenum Press 1979).
8. R. Haag, J.T. Lopuszanski and M. Sohnius, Nuclear Phys. B88 (1975) 257.
9. M. Sohnius, Nuclear Phys. B138 (1978) 109.
10. D.Z. Freedman, Cargèse Lecture, in Recent Developments in Gravitation, eds. M. Levy and S. Deser (Plenum Press 1979).
11. F.A. Berezin, The Method of Second Quantization, New York, Academic Press (1966).
12. P. Fayet, Nuclear Phys. B113 (1976) 135.
13. W. Siegel, Phys. Letters 80B (1979) 220.
14. J. Wess and B. Zumino, in preparation.
15. See P. van Nieuwenhuizen, Stony Brook preprint ITP-SP-80-7 (1980).
16. F.A. Berezin, Preprint ITEP-71, Moscow (1979) and references therein.
17. B. Zumino, in preparation.

ELEVEN-DIMENSIONAL SUPERGRAVITY IN SUPERSPACE

Sergio Ferrara

Laboratoire de Physique Théorique de l'Ecole Normale
Superieure - CNRS, Paris, France
and Laboratori Nazionali di Frascati, INFN, Frascati, Italy

INTRODUCTION

In the present scenario of our theoretical understanding of fundamental particle interactions, extended supergravity appears to be the only viable gauge quantum field theory which may succeed in the attempts at superunification of electroweak, strong and gravitational interactions.

The way supergravity should be used in such a framework is not yet clear[1,2], but investigations of the gauge structure of extended supergravity theories and attempts[3,4] to construct a phenomenological viable model which embeds the minimal GUT consistent with QCD and the electroweak theory seem to indicate that (N = 8 extended) supergravity may be viewed as a preconstituent theory for quarks, leptons and vector bosons. The Yang-Mills Lagrangian of a GUT should then emerge as a dynamically generated (through radiative corrections) low energy (compared to the Planck scale 10^{19} GeV) Lagrangian for composite constituents (quarks, leptons...) which look point-like at scales well below the Planck mass.

Such a scenario has been recently proposed by Ellis, Gaillard, Maiani and Zumino[3] following a previous investigation by Cremmer and Julia[5]. The latter constructed the maximally extended N = 8 supergravity theory (N refers to the number of gauged spinorial generators) and discovered the remarkable fact that such a theory intrinsically contains a hidden local SU(8) gauge symmetry without propagating gauge vector bosons. They further conjectured, in analogy with two-dimensional CP^N models,[6] that these gauge fields may acquire a kinetic term through (probably non-perturbative) quantum effects. Ellis et al.[7] explored the consequences of this hypothesis and argued that under suitable dynamical breaking of the original symmetries of the composite supermultiplet containing the vector particles which gauge SU(8), one is led to an effective anomaly-free subtheory which uniquely contains SU(5) as GUT and with three (flavor) generations of particle species.

These attempts at superunification are using supergravity as an underlying fundamental Lagrangian field theory from which everything else should follow. Why ? The reason is twofold.

Supergravity is the gauge theory of supersymmetry,[8] the only known symmetry which is capable of unifying internal and space-time symmetries[9] in a consistent way overcoming previous no-go theorems.

Supergravity has a much better ultraviolet behaviour than Einstein theory. It provides the first example of a gravitational theory giving finite S-matrix elements for the scattering of gravitons with matter fields at the first two-loop orders of perturbation theory.[10] In extended supergravities with higher N one has examples in which particle states with helicity range from zero up to two give rise to interactions with finite S matrix-elements in perturbation theory and there are further indications[11,12] that the maximally extended N = 8 model may have unexpected convergence properties which may improve the situation of quantum divergences beyond the first two loops.

In this connection, a challenging problem is a better understanding of the as yet unknown ultraviolet behaviour of N = 8 extended supergravity and a major goal would be the construction of the basic multiplet which may be relevant for the analysis of possible counterterms that one may encounter in perturbation theory.

As a preliminary investigation towards these problems we have considered a manifestly supersymmetric formulation of simple supergravity in D = 11 dimensions.[13] This theory, through dimensional reduction, reproduces the N = 8 maximally extended supergravity in four dimensions and is responsible for the hidden local SU(8) symmetry present in the dimensionally reduced four-dimensional theory. It is important to stress that the hidden local SU(8) and global E_7 symmetries present in N = 8 supergravity were discovered[5] by using the technique of dimensional reduction. However the enlargement of the naively expected local SO(7) and global GL(7,R) symmetries to the bigger local SU(8) and global E_7 symmetries came as a surprise and they are really a consequence of the supersymmetric structure of the model.[14] This is the reason that enables us to think that the superspace formulation of eleven-dimensional supergravity may tell us something about the origin of these unexpected higher symmetries. The main output of our investigation is that all the gauge covariant quantities of eleven-dimensional supergravity are embedded in a single irreducible supermultiplet described by a superfield $W_{rstu}(x,\theta)$ which is a totally antisymmetric tensor with respect to its flat Lorentz (11-dimensional) vector indices. Proper use of this superfield should allow us to understand the origin of these hidden symmetries and possibly the analysis of counterterms of N = 8 extended supergravity in D = 4 space-time dimensions.

SUPERGRAVITY IN 11-DIMENSIONAL MINKOWSKI SPACE

Before discussing simple supergravity in 11 dimensions in the formulation of ref.15 let us point out the importance of considering field theories in higher dimensional space-time :

1. Dimensional reduction of field theories from higher dimensional space-time might explain the origin of (some) internal symmetries (Kaluza-Klein). At the same time it gives a natural separation between gauged (local) and rigid (global) symmetries.

2. Non trivial dependence of fields upon the extra coordinates gives rise to mass generation in lower dimensions. This is particularly interesting in the broken version of N = 8 supergravity (N refers to the number of spinorial generators) in four dimensions.[16] The symplectic symmetry of the unbroken theory in five dimensions implies mass relations for the particle spectrum of the broken four-dimensional theory[17]. These mass formulae have important consequences at the quantum level[16] (finiteness of the one-loop induced cosmological constant). This constant is badly (quartically) divergent in theories without local supersymmetry.

3. Simple supersymmetry in higher dimensions gives rise to extended supersymmetry in lower dimensions. Dimensionally reduced Yang-Mills theory in D = 10 dimensions reproduces maximally extended N = 4 supersymmetric Yang-Mills theory in four dimensions.[18] Dimensional reduction of simple supergravity in D = 11 dimensions reproduces maximally extended N = 8 supergravity in four dimensions.[5,14,15]

4. On shell supersymmetry in higher dimensions can be used in order to give an off-shell realization of the supersymmetry algebra in lower dimensions.[19] This trick can be used only for even N and transforms the O(N) on-shell symmetry into an Sp(N) off-shell symmetry.

5. Eleven-dimensional supergravity in superspace can be formulated (on the mass-shell) in terms of a single superfield $W_{rstu}(x,\theta)$.[13,20] The associated supermultiplet (prefield

strength) should play a major role in the understanding of the hidden local SU(8) and global E_7 symmetries present in the dimensionally reduced theory. Moreover, the analogy with N = 1 supergravity in four dimensions[21] may suggest that possible counterterms in perturbation theory should be polynomial in the superfield W_{rstu} and its covariant derivatives.

After these considerations, let us come back to simple gravity in D = 11 dimensions.[15] We give some preliminaries : we use a space-time metric η_{rs}: $\eta_{rr} = 1, -1 \ldots -1$; $\eta_{rs} = 0$ $r \neq s$ $r, s = 1, \ldots 11$. We use a Majorana representation for the 32 x 32 γ matrices. They are all purely imaginary, γ^0 antisymmetric and γ^r symmetric. Spinors are real and $\bar{\chi} = \chi\gamma^0$. We denote by $\gamma^{\mu_1 \ldots \mu_m}$ the antisymmetrized product $\gamma^{[\mu_1} \ldots \gamma^{\mu_m]}$ (with a weight factor $1/n!$). We note that $\gamma^0\gamma^{\mu_1}$, $\gamma^0\gamma^{\mu_1\mu_2}$, $\gamma^0\gamma^{\mu_1 \ldots \mu_5}$ are symmetric while γ^0 , $\gamma^0\gamma^{\mu_1\mu_2\mu_3}$, $\gamma^0\gamma^{\mu_1 \ldots \mu_4}$ are antisymmetric. The matrices $\gamma^{\mu_1\mu_2}$ span the algebra of SO(1,10) while the matrices γ^{μ_1} , $\gamma^{\mu_1\mu_2}$, $\gamma^{\mu_1 \ldots \mu_5}$ span the algebra of S p (32). The field content of (N = 1) eleven dimensional supergravity is given by an 11-bein field $e_{\mu r}(x) (\mu, r = 1 \ldots 11)$ a Rarita-Schwinger Majorana spinor $\psi_{\mu a}(x) (\mu = 1 \ldots 11, a = 1 \ldots 32)$ and a totally antisymmetric world tensor $A_{\mu\nu\rho}(x)$ $(\mu, \nu, \rho = 1 \ldots 11)$ An important fact to be contrasted with the extended four-dimensional theories, is that all these fields are gauge fields. This was of great help in restricting the form of possible interaction terms appearing in the final Lagrangian.

The supergravity Lagrangian is[15] : (we put the gravitational constant K = 1)

$$\mathcal{L}_{SG}^{D=11} = -\frac{1}{4} e R(\omega) - \frac{i}{2} e \bar{\psi}_\mu \gamma^{\mu\nu\rho} \mathcal{D}_\nu [(\omega + \hat{\omega})/2] \psi_\rho$$
$$- \frac{1}{48} e F_{\mu\nu\rho\sigma} F^{\mu\nu\rho\sigma} - \frac{1}{192} e \left(\bar{\psi}_\mu \gamma^{\mu\nu\rho\sigma\lambda\tau} \psi_\nu + 12 \bar{\psi}^\rho \gamma^{\lambda\tau} \psi^\sigma \right) \left(F_{\rho\sigma\lambda\tau} + \right.$$
$$\left. + \hat{F}_{\rho\sigma\lambda\tau} \right) + \frac{2}{(12)^4} \varepsilon^{\mu_1 \ldots \mu_{11}} A_{\mu_1\mu_2\mu_3} F_{\mu_4 \ldots \mu_7} F_{\mu_8 \ldots \mu_{11}} \tag{1}$$

and is left invariant (up to a four-divergence) by the following local supersymmetry variations :

$$\delta e_{\mu r} = -i\bar{\epsilon}\gamma_r\psi_\mu$$

$$\delta\psi_\mu = D_\mu(\hat{\omega})\epsilon + T_\mu{}^{\rho\sigma\tau\lambda}\epsilon\hat{F}_{\rho\sigma\tau\lambda}$$

$$\delta A_{\mu\nu\rho} = \frac{3}{2}\bar{\epsilon}\gamma_{[\mu\nu}\psi_{\rho]} \tag{2}$$

where

$$\hat{\omega}_{\mu rs} = \omega^{0}_{\mu rs}(e) + \frac{i}{2}(\bar{\psi}_\mu\gamma_s\psi_r - \bar{\psi}_\mu\gamma_r\psi_s + \bar{\psi}_s\gamma_\mu\psi_r)$$

$$\omega_{\mu rs} = \hat{\omega}_{\mu rs} - \frac{i}{4}\bar{\psi}_\rho\gamma_{\mu rs}{}^{\rho\sigma}\psi_\sigma$$

$$\hat{F}_{\rho\sigma\tau\lambda} = F_{\rho\sigma\tau\lambda} - 3\bar{\psi}_{[\rho}\gamma_{\sigma\tau}\psi_{\lambda]}$$

$$F_{\rho\sigma\tau\lambda} = 4\partial_{[\rho}A_{\sigma\tau\lambda]}$$

$$T^{rstuv} = \frac{1}{(12)^2}(\gamma^{rstuv} - 8\gamma^{[stu}\eta^{v]r}) \tag{3}$$

$\mathcal{L}^{D=11}_{SG}$ is also invariant under general coordinate transformations, local Lorentz rotations and the following (Abelian) gauge transformation :

$$\delta A_{\mu\nu\rho} = 3\partial_{[\mu}\xi_{\nu\rho]} \tag{4}$$

The Euler-Lagrange equations which come from $\delta\mathcal{L}^{D=11}_{SG} = 0$ can be written in a supercovariant form

$$\gamma^{rst}\hat{\mathcal{D}}_s\psi_t = 0$$

$$\hat{\mathcal{D}}_t\hat{F}^{turs} + \frac{1}{(24)^2}\varepsilon^{mnpqvwxyurs}\hat{F}_{mnpq}\hat{F}_{vwxy} = 0$$

$$\hat{R}_t{}^n{}_{,sn} - \frac{1}{2}\eta_{ts}\hat{R}_{mn,}{}^{mn} = \frac{1}{24}\left[\eta_{ts}(\hat{F}_{mnpq})^2 - 8\hat{F}_{mnpt}\hat{F}^{mnp}{}_s\right] \tag{5}$$

(the "hat" denotes supercovariant differentiation).

We observe that in (5), contrary to what happens in (1) the gauge transformation given in (4) has become superfluous because $A_{\mu\nu\rho}$ appears in (5) only through its field strength $F_{\mu\nu\rho\sigma}$ which is obviously invariant under (4).

We now come back to the particle content of $\mathcal{L}_{SG}^{D=11}$. In order to describe the free-particle spectrum it is sufficient to consider the quadratic part of $\mathcal{L}_{SG}^{D=11}$. Due to the residual gauge-invariances of the kinetic terms, the free particle states are described by irreducible representations of SO(9), the little group of a light-like momentum in eleven dimensions. For each field we have the following corresponding set of degrees of freedom :

$$e_{\mu r}(x) \rightarrow \frac{(D-1)(D-2)}{2} - 1 = 44 \quad \text{degrees of freedom}$$

$$\psi_{\mu a}(x) \rightarrow \frac{32}{2}(D-3) = 128 \quad \text{degrees of freedom}$$

$$A_{\mu\nu\rho}(x) \rightarrow \frac{(D-1)(D-2)(D-3)}{3!} = 84 \quad \text{degrees of freedom} \tag{6}$$

In D = 4 dimensions these fields correspond to fields classified by representations of SO(7) as follows

$$g_{\mu\nu}: \quad \underset{(1)}{g_{\mu\nu}}, \quad \underset{(7)}{g_{\mu i}}, \quad \underset{(27+1)}{g_{ij}}$$

$$A_{\mu\nu\rho}: \quad \underset{(7)}{A_{\mu\nu i}}, \quad \underset{(35)}{A_{ijk}}, \quad \underset{(21)}{A_{\mu ij}}$$

$$\psi_{\mu a}: \quad \underset{(8)}{\psi_{\mu a\alpha}}, \quad \underset{(56)}{\psi_{ia\alpha}} \tag{7}$$

In terms of SO(7) representations one has a 2^+ singlet, a 1^+ 7-plet, a 1^- 21-plet, 27+1 scalars and 35 pseudoscalars.

Moreover one has a 7-plet of antisymmetric pseudotensor fields which are equivalent to a 7-plet of scalars.

As far as the fermions are concerned one has an 8-plet of spin-3/2 fields and 56 spin 1/2 fields which lie in the product of the spinor x vector representation of SO(7). It was shown in ref. 5 (see also (14)) that one can perform suitable dual transformations (on the 21 pseudovector fields and on the 7 antisymmetric pseudotensors) in such a way that the SO(7) symmetry can be enlarged to SO(8) and one gets particle fields in conventional antisymmetric tensor representations of SO(8) :

$$e_{\mu r},\ \psi_{\mu a}^{i},\ A_{\mu}^{ij},\ \chi_a^{ijk},\ A^{ijkl},\ B^{ijkl};\quad \left(A(B)^{ijkl} = \pm\eta \frac{\varepsilon^{ijklmnpq}}{24} A(B)_{mnpq} \quad \eta=\pm 1\right)$$

We now consider the group structure that emerges from the gauge algebra defined through the transformation laws in (2).

This group structure plays a crucial role in the connection of the component formulation of supergravity given in this section and the formulation of the same theory over superspace.

The group composition of two arbitrary elements of the gauge group of eleven-dimensional supergravity is given by

$$(\xi_1^\mu, \epsilon_1^a, \Omega_1^{rs}, \xi_{1\mu\nu}) \otimes (\xi_2^\mu, \epsilon_2^a, \Omega_2^{rs}, \xi_{2\mu\nu}) =$$

$$= (\xi_{12}^\mu, \epsilon_{12}^a, \Omega_{12}^{rs}, \xi_{12\mu\nu}) \tag{8}$$

where the symbol $\otimes$ means the infinitesimal group composition defined by the commutator of two infinitesimal variations $[\delta_1, \delta_2] = \delta_{12}$ and the resulting parameter gives the general composition rule of any two elements of the local gauge algebra.

The relevant part of (8) is the following:

$$(0, \epsilon_2^a, 0, 0) \otimes (0, \epsilon_1^a, 0, 0) = (\xi_{12}^\mu, \epsilon_{12}^c, \Omega_{12}^{rs}, \xi_{12\mu\nu}) \tag{9}$$

where

$$\xi_{12}^\mu = -i\,\bar{\epsilon}_1 \gamma^\nu \epsilon_2$$

$$\epsilon_{12}^c = -\xi_{12}^\nu \psi_\nu^c$$

$$\Omega_{12}^{rs} = \frac{1}{72}\bar{\epsilon}_1 (\gamma^{rs\mu\nu\rho\sigma} - 24\, e^{s\mu} e^{r\nu} \gamma^{\rho\sigma}) \epsilon_2 \hat{F}_{\mu\nu\rho\sigma} + \xi_{12}^\nu \hat{\omega}_\nu^{rs}$$

$$\xi_{12\mu\nu} = \frac{3}{2}\bar{\epsilon}_1 \gamma_{\mu\nu} \epsilon_2 - 3\, \xi_{12}^\rho A_{\rho\mu\nu} \tag{10}$$

We observe that ξ_{12}^μ as well as $\xi_{12\,\mu\nu}$ contain terms which are field independent. In the superspace approach this last fact will imply a non trivial θ dependence of the supervierbein E_Λ^A and of the supergauge field $A_{\Lambda\Sigma\Delta}$ in the absence of gravitational disturbance (flat limit).

Before closing this section we would like to make some additional comments.

The gauge algebra defined in (8), (9) through the transformation laws (2) closes only where the spin 3/2 field equations are used. If one performs multiple commutators the group structure requires the complete set of equations of motion given by (5).

The reason for this lack of closure of the gauge algebra is due to the fact that the fields $e_{\mu r}$, $\psi_{\mu a}$ and $A_{\mu\nu\rho}$ do not provide a representation of the supersymmetry algebra off the mass-shell. It is easy to see that there are 145 fermions more than bosons when the equations of motion are not fulfilled. This fact has the immediate consequence that any superfield built only in terms of $e_{\mu r}$, $\psi_{\mu a}$,$A_{\mu\nu\rho}$ will be automatically put on the mass-shell.

Another important problem is to understand the geometric origin of the field $A_{\mu\nu\rho}$. An interesting possibility would be that of deriving eleven-dimensional supergravity as the contraction of a gauge theory of a simple graded Lie group whose even part is Sp(32) [22].

An appealing possibility could be the identification of $A_{\mu_1\mu_2\mu_3}$ with the dual of the extra connection for the 462 generators $\gamma^{\mu_1 \cdots \mu_5}$ of the Sp(32) Lie algebra spanned by γ^{μ_1}, $\gamma^{\mu_1\mu_2}$ and $\gamma^{\mu_1 \cdots \mu_5}$.

SUPERSPACE FORMULATION OF ELEVEN-DIMENSIONAL SUPERGRAVITY

In the previous sections we have seen that the maximally extended N = 8 four-dimensional supergravity theory corresponds to simple supergravity in eleven dimensions. The latter theory has a very simple field content and this fact was very helpful to guess[15] the possible interaction terms which may have occurred in $\mathcal{L}_{SG}^{D=11}$. In a recent investigation with E. Cremmer[13] it seemed natural to further exploit the supersymmetric structure of the theory by considering eleven-dimensional supergravity in a 43-dimensional superspace[23] parametrized by Minkowski coordinates x^μ $(\mu=1 \cdots 11)$ and spinorial (anticommuting) Majorana coordinates $\theta^\alpha (\alpha=1 \cdots 32)$. A given point in superspace is therefore labelled by a couple of bosonic and fermionic coordinates $Z^\Lambda=(x^\mu, \theta^\alpha)$ and the index Λ runs from 1 to 43.

We introduce moving frames into superspace by defining a 43-bein $E_\Lambda^{\hat{A}}(z)$ with its inverse $E_{\hat{A}}^{\Lambda}(z)$ $(E_{\hat{A}}^{\Lambda} E_\Lambda^{B}= \delta_A^B,\ E_\Lambda^{A} E_A^{\Sigma}= \delta_\Lambda^\Sigma)$. We use Latin capital letters A = (a, r) for flat indices and Greek capital letters $\Lambda=(\alpha, \mu)$ for world indices. Letters from early in the alphabet refer to Fermi labels while later letters refer to Bose labels.

We introduce an affine connection $\Omega_\Lambda(z)$ which is Lie-algebra valued over SO(1,10) and, in terms of E_Λ^A and Ω_Λ, we define the supertorsion tensor

$$T_{AB}^{\ \ C} = (-)^{\Lambda(B+\Pi)} E_A^{\ \Lambda} E_B^{\ \Pi} \left(\mathcal{D}_\Lambda E_\Pi^{\ C} - (-)^{\Lambda\Pi} \mathcal{D}_\Pi E_\Lambda^{\ C} \right) \tag{11}$$

and the Lie-algebra valued supercurvature

$$R_{AB} = (-)^{\Lambda(B+\Pi)} E_A^{\ \Lambda} E_B^{\ \Pi} \left(\partial_\Lambda \Omega_\Pi - (-)^{\Lambda\Pi} \partial_\Pi \Omega_\Lambda + [\Omega_\Lambda, \Omega_\Pi\} \right) \tag{12}$$

The symbol $[X, Y\}$ denotes the graded Poisson bracket $XY - (-)^{XY} YX$.

The components of the supercurvature read :

$$R_{AB}^{\ \ rs} = (-)^{\Lambda(B+\Pi)} E_A^{\ \Lambda} E_B^{\ \Pi} \Big(\partial_\Lambda \Omega_\Pi^{\ rs} - (-)^{\Lambda\Pi} \partial_\Pi \Omega_\Lambda^{\ rs} + \\ + [\Omega_\Lambda^{\ rt} \Omega_\Pi^{\ t's} - (-)^{\Lambda\Pi} \Omega_\Pi^{\ rt} \Omega_\Lambda^{\ t's}] \eta_{tt'} \Big)$$

$$R_{AB}{}^{c}{}_{d} = \frac{1}{4} R_{AB}^{\ \ rs} (\gamma_{rs})^c{}_d \ , \quad R_{AB}^{\ \ cs} = R_{AB}^{\ \ rb} = 0 \tag{13}$$

Covariant derivatives are given by

$$\mathcal{D}_\Lambda E_\Pi^{\ r} = \partial_\Lambda E_\Pi^{\ r} + \Omega_\Lambda{}^{r}{}_{s} E_\Pi^{\ s}$$

$$\mathcal{D}_\Lambda E_\Pi^{\ a} = \partial_\Lambda E_\Pi^{\ a} + \frac{1}{4} \Omega_\Lambda^{\ rs} (\gamma_{rs})^a{}_b E_\Pi^{\ b} \ , \quad \mathcal{D}_A = E_A^{\ \Lambda} \mathcal{D}_\Lambda \tag{14}$$

flat indices are raised and lowered with the metrics $\eta^{rs} = \eta_{rs}$, $(\gamma^0)^{ab} = (\gamma^0)_{ab}$ and the usual $\gamma^{\mu_1 \cdots \mu_n}$ matrices have indices $(\gamma^{\mu_1 \cdots \mu_n})^a{}_b$.

Torsion, curvature and covariant derivatives are related by

$$[\mathcal{D}_A, \mathcal{D}_B\} = -2 T_{AB}^{\ \ C} \mathcal{D}_C + R_{AB} \tag{15}$$

The superspace geometry which reproduces the component formulation of the previous section is specified by the following constraints

$$T_{rs}^{\ \ t} = T_{ab}^{\ \ c} = T_{as}^{\ \ t} = T_{ab}^{\ \ r} - \frac{i}{2}(\gamma^0\gamma^r)_{ab} = 0 \tag{16}$$

From the Jacobi identities $[[\mathcal{D}_A, \mathcal{D}_B\}, \mathcal{D}_C\} = 0$ it follows that $T_{AB}^{\ \ C}$ and R_{AB} satisfy the Bianchi identities

$$B_{ABC}^{\ \ \ \ D} = 2\mathcal{D}_{[A} T_{BC\}}^{\ \ \ D} + 4 T_{[AB}^{\ \ \ C'} T_{C'C\}}^{\ \ \ D} + R_{[AB,C\}}^{\ \ \ \ \ D} = 0 \tag{17}$$

$$\mathcal{D}_{[A} R_{BC\}} + 2\, T_{[AB}^{\ \ \ D} R_{DC\}} = 0 \tag{18}$$

The Lie-algebra valued identity (18) is actually a consequence of (17).[24] The use of (17) and (16) enables us to solve all components of $T_{AB}^{\ \ C}$ and R_{AB} in terms of a single irreducible superfield W_{rstu} totally antisymmetric in its flat-vector indices and satisfying the differential constraints

$$(\gamma^{rst}\mathcal{D})_a W_{rstu}(x,\theta) = 0 \tag{19}$$

The non vanishing components of $T_{AB}^{\ \ C}$ and R_{AB} are

$$T_{rs}^{\ \ a} = -\frac{1}{42}(\gamma^{tu}\gamma^0)^{ba}\mathcal{D}_b W_{rstu}$$

$$T_{ar}^{\ \ c} = \frac{1}{2} W_{pstu}(T_r^{\ pstu})^c{}_a$$

$$R_{ab}^{\ \ mn} = (\gamma^0 S)_{ab}^{\ \ mnuvzw} W_{uvzw}$$

$$R_{as}^{mn} = -\frac{i}{42}\left[(\gamma^0\gamma^m\gamma^{rt}\gamma^0\mathcal{D})_a W_{rt}{}^m{}_s - (\gamma^0\gamma^m\gamma^{rt}\gamma^0\mathcal{D})_a W_{rt}{}^n{}_s + (\gamma^0\gamma^s\gamma^{rt}\gamma^0\mathcal{D})_a W_{rt}{}^{mn}\right]$$

$$R_{mn}{}^b{}_c = -\frac{1}{21}(\gamma^{rs}\gamma^0)^{ba}\mathcal{D}_c\mathcal{D}_a W_{rsmn} - (T_{[m}{}^{xypq}\mathcal{D}_{n]})_c^b W_{xypq} - [T_m{}^{xypq}, T_n{}^{uvzw}]_c^b W_{xypq} W_{uvzw} \tag{20}$$

where $S_{mn}{}^{uvzw} = \frac{1}{72}\left(\gamma_{mn}{}^{uvzw} + 24\,\delta_m^{[u}\delta_n^{v}\gamma^{zw]}\right)$

The superfield eq. (19) is nothing but the supercovariant form of the eleven-dimensional supergravity equations of motion given in (5). The $\psi_{\mu a}$ field equation is given by the $\theta = 0$ component while the $A_{\mu\nu\rho}$ and $e_{\mu r}$ field equations are given by the θ component.

In order to describe the $A_{\mu\nu\rho}$ field as a component of some superfield the superspace quantities introduced so far are not sufficient. We must then introduce a tensor superfield with supergauge transformations

$$\delta A_{\Lambda\Gamma\Delta} = 3\,\partial_{[\Lambda}\Sigma_{\Gamma\Delta\}} \qquad \Sigma_{\Gamma\Delta} = -(-)^{\Gamma\Delta}\Sigma_{\Delta\Gamma} \tag{21}$$

and its field strength

$$F_{ABCD} = (-)^{\Sigma(\Lambda+B+\Gamma+C+\Delta+D)+\Lambda(\Gamma+C+\Delta+D)+\Gamma(\Delta+D)} E_A^{\Sigma} E_B^{\Lambda} E_C^{\Gamma} E_D^{\Delta} F_{\Sigma\Lambda\Gamma\Delta}$$

$$F_{\Sigma\Lambda\Gamma\Delta} = 4\,\partial_{[\Sigma}A_{\Lambda\Gamma\Delta\}} \tag{22}$$

The field strength is assumed to satisfy the following constraints

$$F_{astu} = F_{abcu} = F_{abcd} = F_{rsab} - \frac{1}{2}(\gamma^0 \gamma_{rs})_{ab} = 0 \tag{23}$$

Then the Bianchi identities imply

$$F_{rstu}(x,\theta) = W_{rstu}(x,\theta) \tag{24}$$

Therefore it follows that all components of $T_{AB}^{\ \ C}$, R_{AB} and F_{ABCD} are in fact functions of W_{rstu}. It is easy to see that this implies that the gauge superfields $E_\Lambda^{\ a}$, $E_\Lambda^{\ r}$, $\Omega_\Lambda^{\ rs}$ and $A_{\Lambda\Gamma\Delta}$ are not independent and a major problem is to solve them in terms of some prepotential superfield. It is not difficult to realize for instance that the constraints (16) already imply that $\Omega_\Lambda^{\ rs}$ and $E_\Lambda^{\ A}$ are only functions of $E_a^{\ \Lambda}$. Another problem is that the constraints (16) are too stringent to specify the superspace geometry. In fact they also imply the field equations as seen through the differential constraint (19) on the prefield strength W_{rstu}.

We now comment on the derivation of these results. We have used a procedure called gauge completion and already used in refs[25-28]. We have demanded a gauge choice for the potentials $E_\Lambda^{\ A}$, $\Omega_\Lambda^{\ rs}$, $A_{\Lambda\Sigma\Delta}$ and the superparameters Ξ^Λ, Λ^{rs}, $\Sigma_{\Gamma\Delta}$ such that :

$$\begin{aligned}
&E_\mu^{\ r}(x,\theta=0) = e_\mu^{\ r}(x), \quad E_\mu^{\ a}(x,\theta=0) = \psi_\mu^{\ a}(x)\\
&A_{\mu\nu\rho}(x,\theta=0) = A_{\mu\nu\rho}(x)\\
&\Omega_\mu^{\ rs}(x,\theta=0) = \hat{\omega}_\mu^{\ rs}(e,\psi)\\
&\Xi^\mu(x,\theta=0) = \xi^\mu(x), \quad \Xi^\alpha(x,\theta=0) = \epsilon^\alpha(x)\\
&\Sigma_{\mu\nu}(x,\theta=0) = \xi_{\mu\nu}(x)\\
&\Lambda^{rs}(x,\theta=0) = \lambda^{rs}(x)
\end{aligned} \tag{25}$$

The superparameters Ξ^{Λ} , Λ^{rs} and $\Sigma_{\Gamma\Delta}$ refer to general coordinate, Lorentz and supergauge transformations in superspace. Then the requirement that the gauge algebra (8)-(10) in Minkowski space be compatible with the group-composition of the superparameters in superspace is sufficient to give all higher Θ components of the superparameters once the $\Theta = 0$ components are identified as dictated by (25). In the same way this method applies to any superfield by comparing its transformation rules in Minkowski space with the corresponding one in superspace. In order to compute (16), (20), (23) and (24) it is sufficient to compute these relations at $\theta = 0$ because they can be extended by covariance to all orders in θ . Thus we have to know the linear θ terms in $E_{\Lambda}^{A}(x,\theta)$, $\Omega_{\Lambda}^{rs}(x,\theta)$ and $A_{\Lambda\Gamma\Delta}(x,\theta)$. This also requires the knowledge of the ϵ^{α} dependence of the terms linear in Θ of the superparameters Ξ^{Λ}, Λ^{rs} , $\Sigma_{\Lambda\Delta}$.

The Θ dependent part of these quantities can be computed from the superspace variations

$$\delta E_{\Lambda}^{A} = \Xi^{\Pi}\partial_{\Pi} E_{A}^{\Lambda} + \partial_{\Lambda}\Xi^{\Pi} E_{\Pi}^{A} + (\Lambda^{rs} X_{rs})^{A}{}_{B} E_{\Lambda}^{B}$$

$$\delta\Omega_{\Lambda}^{rs} = \Xi^{\Pi}\partial_{\Pi}\Omega_{\Lambda}^{rs} + \partial_{\Lambda}\Xi^{\Pi}\Omega_{\Pi}^{rs} - (\Omega_{\Lambda}^{rt}\Lambda_{t}{}^{s} - r\leftrightarrow s) - \partial_{\Lambda}\Lambda^{rs}$$

$$\delta A_{\Lambda\Gamma\Delta} = \Xi^{\Pi}\partial_{\Pi} A_{\Lambda\Gamma\Delta} + \partial_{[\Lambda}\Xi^{\Pi} A_{\Pi\Gamma\Delta\}} + 3\,\partial_{[\Lambda}\Sigma_{\Gamma\Delta\}}$$

and group-composition (26)

$$\Xi_{12}^{\Lambda} = \Xi_{2}^{\Pi}\partial_{\Pi}\Xi_{1}^{\Lambda} - (1\leftrightarrow 2)$$

$$\Lambda_{12}^{rs} = \Xi_{2}^{\Pi}\partial_{\Pi}\Lambda_{1}^{rs} + \Lambda_{2}^{rt}\Lambda_{1t}{}^{s} - \Lambda_{2}^{st}\Lambda_{1t}{}^{r} - (1\leftrightarrow 2)$$

$$\Sigma_{12\,\Gamma\Delta} = \Xi_{2}^{\Pi}\partial_{\Pi}\Sigma_{1\,\Gamma\Delta} + \partial_{[\Gamma}\Xi_{2}^{\Pi}\Sigma_{1\,\Pi\Delta\}} - (1\leftrightarrow 2) \tag{27}$$

and we have omitted in (27) the variation from the field dependent part of the parameters because it does not contribute to the linear θ terms we are interested to compute. These quantities up to θ terms were explicitly given in ref.(13).

To check the $(\theta=0)$ Bianchi identities (17) $B_{ABC}{}^{D}=0$ we needed $T_{AB}{}^{C}$ to order θ. This required the knowledge of the supersymmetry variations of $\hat{F}_{rstu}$ and $\hat{\mathcal{D}}_{[r}\psi_{s]}$ which are

$$\delta \hat{F}_{rstu} = 6\, \bar{\epsilon}\gamma_{[rs}\hat{\mathcal{D}}_{t}\psi_{u]}$$

$$\delta \hat{\mathcal{D}}_{[r}\psi_{s]} = \Big\{ \tfrac{1}{8} \hat{R}_{rsmn}\gamma^{mn} + \tfrac{1}{2}\big[T_{r}{}^{tuvw}, T_{s}{}^{xyzp}\big]\hat{F}_{tuvw}\hat{F}_{xyzp} + T_{[s}{}^{tuvw}\hat{\mathcal{D}}_{r]}\hat{F}_{tuvw}\Big\}\epsilon \tag{28}$$

and $\hat{R}_{rs}{}^{mn}(x) = 2\left(\mathcal{D}_{[r}\omega_{s]}{}^{mn} + \hat{\omega}_{[r}{}^{mt}\hat{\omega}_{s]}{}^{n}{}_{t}\right) = R_{rs}{}^{mn}(x,\theta=0)$

We also used the following trick which is valid for any world scalar superfield

$$\delta_{\epsilon}\phi(x,\theta=0) = \epsilon^{\alpha}(\partial_{\alpha}\phi)(x,\theta=0) = \epsilon^{a}(\mathcal{D}_{a}\phi)(x,\theta=0) \tag{29}$$

as a consequence of the fact that

$$\mathcal{D}_{a}\phi(x,\theta=0) = E_{a}^{\Lambda}\mathcal{D}_{\Lambda}\phi(x,\theta=0) = \partial_{a}\phi(x,\theta=0)$$

$$\left(E_{a}^{\alpha}(x,\theta=0) = \delta_{a}^{\alpha}\,,\; E_{a}^{\mu}(x,\theta=0)=0\right)$$

The Bianchi identities (17) are all identically fulfilled with the exception of $B_{abr}{}^{c}=0$ which implies the $\psi_{\mu a}$ field equation and by covariance the differential constraint (19). To prove the Bianchi identities, knowledge of different forms of the $\psi_{\mu a}$ field equations was required. They are

$$\gamma^{\mu_1 \ldots \mu_N \rho} \hat{\mathcal{D}}_{[\rho} \psi_{\sigma]} = -\frac{1}{2} N \gamma^{[\mu_1 \ldots \mu_{N-1}} (\hat{\mathcal{D}}^{\mu_N]} \psi_\sigma - \mathcal{D}_\sigma \psi^{\mu_N]})$$

$$\gamma^{\mu_1 \ldots \mu_N \rho \sigma} \hat{\mathcal{D}}_{[\rho} \psi_{\sigma]} = -N(N-1) \gamma^{[\mu_1 \ldots \mu_{N-2}} \hat{\mathcal{D}}^{\mu_{N-1}} \psi^{\mu_N]}$$

(30)

The latter relation is a consequence of the former.

The above discussion completes the description of the superspace formulation of 11-dimensional supergravity on the mass-shell. The main result is the fact that all supercovariant quantities can be expressed in terms of a single superfield $W_{rstu}(x,\theta)$ satisfying a very simple differential constraint. Eq. (19) gives all the field equations of the theory. The next step further would be to solve the identities (17), (18) without the constraint (19). This would provide the off-shell structure of the theory and could give some insight in the fundamental problem of the ultraviolet divergences of the dimensionally reduced N = 8 extended supergravity theory in four-dimensions.

It is interesting to observe finally that the constraints (16) and (23) imply that the potentials E_Λ^A and $A_{\Lambda\Sigma\Delta}$, in the flat limit, have a non trivial θ dependence. It is in fact easy to see that the vacuum solutions, in absence of gravitational disturbance, are given by :

$$E^{o\,r}_{\mu} = \delta^r_\mu \,,\quad E^{o\,a}_{\mu} = 0 \,,\quad E^{o\,a}_{\alpha} = \delta^a_\alpha \,,\quad E^{o\,r}_{\alpha} = \frac{i}{2} (\bar{\theta}\gamma^r)_\alpha$$

$$A^o_{\mu\nu\rho} = A^o_{\mu\nu\alpha} = A^o_{\alpha\beta\gamma} = 0 \qquad A^o_{\mu\nu\alpha} = -\frac{1}{4} (\bar{\theta}\gamma_{\mu\nu})_\alpha$$

$$\Omega^{o\,rs}_{\Lambda} = 0$$

(31)

Eqs. (31) show that 11-dimensional flat superspace has a vanishing supercurvature but non-vanishing supertorsion and $A_{\Lambda\Gamma\Delta}$ curvature. The potential $A_{\Lambda\Gamma\Delta}$ therefore plays a crucial role also in the structure of rigid supertransformations.

REFERENCES

1. M. Gell-Mann, Talk given at the 1977 Washington meeting of the American Physical Society. See also,
   M. Gell-Mann, P. Ramond and R. Slansky, "Supergravity", Proc. of the Supergravity Workshop at Stony Brook, eds P. van Nieuwenhuizen and D.Z. Freedman (North Holland Publ. Co., Amsterdam, 1979), p. 315.
2. P. Fayet, these proceedings, p. 589.
3. J. Ellis, M.K. Gaillard, L. Maiani and B. Zumino, these proceedings, p.69.
4. For an earlier attempt, see, T. Curtright and P. Freund in "Supergravity", Proceedings of the Supergravity Workshop at Stony Brook, eds P. van Nieuwenhuizen and D.Z. Freedman (North Holland Publ. Co., Amsterdam, 1979), p. 197.
5. E. Cremmer and B. Julia, Phys. Letters 80B (1978) 48; Nuclear Phys. B159 (1979) 141.
6. A. D'Adda, P. Di Vecchia and M. Lüscher, Nuclear Phys. B146 (1978) 63; Nuclear Phys. B152 (1979) 125.
7. J. Ellis, M.K. Gaillard and B. Zumino, CERN preprint TH.2842/ LAPP-TH 16 (1980).
8. D.Z. Freedman, P. van Nieuwenhuizen and S. Ferrara, Phys. Rev. D13 (1976) 3214;
   S. Deser and B. Zumino, Phys. Letters 62B (1976) 335.
9. The first extended supergravity model has been constructed by S. Ferrara and P. van Nieuwenhuizen, Phys. Rev. Letters 36 (1976) 1669.
10. For a review, see P. van Nieuwenhuizen and M.T. Grisaru in "Deeper Pathways in High-Energy Physics", Proceedings of Orbis Scientiae, Coral Gables 1977, eds B. Kursonoglu, A. Perlmutter and L.F. Scott (Plenum Press, N.Y., 1977),p.233.
11. B. de Wit and S. Ferrara, Phys. Letters 81B (1979) 317.
12. S.M. Christensen, M.J. Duff, G.W. Gibbons and M. Roček, NSF-ITP-80-14 (1980).
13. E. Cremmer and S. Ferrara, Phys. Letters 91B (1980) 61.
14. E. Cremmer, these proceedings, p.
15. E. Cremmer, B. Julia and J. Scherk, Phys. Letters 84B (1979) 83.
16. J. Scherk and J.H. Schwarz, Phys. Letters 82B (1979) 60; Nuclear Phys. B153 (1979) 61;
    E. Cremmer, J. Scherk and J.H. Schwarz, Phys. Letters 84B (1979) 83.
17. F. Gliozzi, J. Scherk and D. Olive, Nuclear Phys. B122 (1977) 253.
    L. Brink, J.H. Schwarz and J. Scherk, Nuclear Phys. B121 (1977) 77.
19. M.F. Sohnius, K.S. Stelle and P.C. West, ICTP/79-80, ICTP/79-80/22 and these proceedings, p. 187.

20. See also L. Brink and P. Howe, Phys. Letters 91B (1980) 384.
21. S. Ferrara and B. Zumino, Nuclear Phys. B134 (1978) 301;
S. Ferrara and P. van Nieuwenhuizen, Phys. Letters 78B (1978) 573.
22. A. D.Adda, R. D'Auria, P. Fré and T. Regge, Turin preprint (1979) to appear in Rivista del Nuovo Cimento;
R. D'Auria, P. Fré and T. Regge, these proceedings, p. 171.
23. J. Wess and B. Zumino, Phys. Letters 66B (1977) 361.
24. N. Dragon, Z. Phys. C2 (1979) 29.
25. P. Nath and R. Arnowitt, Phys. Letters 65B (1976) 63.
26. L. Brink, M. Gell-Mann, P. Ramond and J.H. Schwarz, Phys. Letters 74B (1978) 336.
27. J. Wess and B. Zumino, Phys. Letters 79B (1978) 394.
28. S. Ferrara and P. van Nieuwenhuizen, in "Supergravity", Proceedings of the Supergravity Workshop at Stony Brook, eds P. van Nieuwenhuizen and D.Z. Freedman (North Holland Publ. Co., Amsterdam, 1979) p. 221.

N = 8 SUPERGRAVITY

Eugène Cremmer

Laboratoire de Physique Théorique
de l'Ecole Normale Supérieure
Paris (FRANCE)

INTRODUCTION

After the trip through superspace in 11 dimensions under the direction of S. Ferrara,[1] I am supposed to bring you back to ordinary 4 dimensional space-time by presenting to you the N = 8 supergravity, a work done in collaboration with B. Julia[2].

I shall first sketch the dimensional reduction from 11 to 4. Then I shall describe the extension of symmetries in 4 dimensions SO(7) local x GL(7,R) global $\longrightarrow$ SO(8) local x SL(8,R) global $\longrightarrow$ SU(8) local x $E_{7(+7)}$ global
explaining briefly the G/H non linear σ-model especially for G non compact ; the structure of E_7 and its subgroup SU(8) and the chiral dual invariance of the theory on shell. Next, all these notions will be used to present the N = 8 supergravity and parametrization for the scalar fields.

Then the content and the symmetries of extended supergravities will be given. Using simple counting arguments, I shall give the guessed symmetries for maximal extended supergravities in D dimensions and describe in particular the D = 5 N = 8 supergravity.

DIMENSIONAL REDUCTION FROM 11 TO 4

To derive the N = 8 supergravity in 4 dimensions from the N = 1 supergravity in 11 dimensions, we shall use the simplest dimensional reduction[3] i.e. starting from $\Phi(x_M)$ M = 0,1...10 we shall make the restriction that Φ does not depend on the extra coordinates x_i (i = 4...10). Each 11 dimensional index will be split into a 4 dimensional-part and an external part. We shall use the SO(10,1) local invariance to choose a particular gauge for the elfbein $e_M{}^A$: $e_m{}^\alpha = 0$ (m = 4...10 ; α = 0, ...3), then in 4 dimensions we shall be left with the local invariance SO(3,1) x SO(7). The reparametrization invariance in 11 dimensions will become the reparametrization invariance in 4 dimensions x GL(7,R) global x 7 abelian gauge invariances (with parameter ξ^i).

The Abelian gauge invariance in 11 dimensions associated with A_{MNP} turns into three Abelian gauge invariances with parameters $\xi_{ij}, \xi_{\mu i}, \xi_{\mu\nu}$. In order to define in 4 dimensions fields which do not mix the various gauge transformations $(\xi^i, \xi_{ij}, \xi_{\mu i}, \xi_{\mu\nu})$, it is useful to perform the dimensional reduction on fields with flat indices in 11 dimensions and redefine fields with curved indices in 4 dimensions by using the vierbein $e_\mu{}^\alpha$. We must also perform Weyl rescaling in order to have a canonical Einstein term.

In order to agree with the content in fields and parity of N = 8 supergravity, we have to perform duality transformations. This eliminates the auxiliary fields $A_{\mu\nu\rho}$, transforms the tensor field $A_{\mu\nu i}$ into scalar fields ϕ^i and the pseudo vector fields $A_{\mu ij}$ into vector fields B_μ^{ij} . This is possible because $A_{\mu\nu\rho}$, $A_{\mu\nu i}$ and $A_{\mu ij}$ appear in the Lagrangian in 4 dimensions only through their field strengths $F_{\mu\nu\rho\sigma}$, $F_{\mu\nu\rho i}$, $F_{\mu\nu ij}$ which can be defined by the following constraints :

$$F_{\mu\nu\rho\sigma} = 4\, \partial_{[\mu} A_{\nu\rho\sigma]} \Longleftrightarrow \text{no constraints}$$

$$F_{\mu\nu\rho i} = 3\, \partial_{[\mu} A_{\nu\rho]i} \Longleftrightarrow \varepsilon^{\mu\nu\rho\sigma} \partial_\sigma F_{\mu\nu\rho i} = 0$$

$$F_{\mu\nu ij} = 2\, \partial_{[\mu} A_{\nu]ij} \Longleftrightarrow \varepsilon^{\mu\nu\rho\sigma} \partial_\sigma F_{\mu\nu ij} = 0$$

These constraints are introduced via Lagrange multipliers ϕ^i and B_σ^{ij} into the Lagrangian which becomes

$$\mathcal{L}(F_{\mu\nu\rho\sigma}, F_{\mu\nu\rho i}, F_{\mu\nu ij}) + \phi^i \varepsilon^{\mu\nu\rho\sigma} \partial_\sigma F_{\mu\nu\rho i} + B_\sigma^{ij} \varepsilon^{\mu\nu\rho\sigma} \partial_\rho F_{\mu\nu ij}$$

The duality transformations are performed by integrating over $F_{\mu\nu\rho\sigma}$, $F_{\mu\nu\rho i}$ and $F_{\mu\nu ij}$. This leads to the new Lagrangian

$$\mathcal{L}'(\partial_\sigma \phi^i, G_{\mu\nu}^{ij}) \qquad \text{with} \qquad G_{\mu\nu}^{ij} = 2\, \partial_{[\mu} B_{\nu]}^{ij}$$

For the fermionic fields, a 32 component Majorana spinor in 11 dimensions turns into 8 four component Majorana spinors in 4 dimensions corresponding to the tensor product decomposition of 32 x 32 matrices into $\gamma^\alpha \otimes 1_A{}^B$ and

$$\gamma_5 \otimes (\Gamma^a)_A{}^B$$

(α = 0,...3 ; a = 4...10 ; A,B = 1...8).

To obtain fermionic fields with canonical kinetic terms we need to perform some scaling and rediagonalization. In particular the spin 1/2 field λ_{ABC} is related to $\psi_i{}^A$ by $\lambda_{ABC} \sim (\Gamma^i)_{[AB}\psi_{iC]}$

The final content is summarized in the following scheme

$e_M^A \; (g_{MN}) \longrightarrow e_\mu^\alpha \; (g_{\mu\nu})$ 1
$B_\mu^i \; (g_{\mu i})$ (7)
$e_i^r \; (g_{ij})$ [28]
(28)
[35]
$A_{MNP} \longrightarrow A_{\mu\nu\rho}$ AUXILIARY
$A_{\mu\nu i} \sim \phi^i$ [7]
[70]
$A_{\mu ij} \sim B_\mu^{ij}$ (21)
A_{ijk} [35]

$\psi_M \longrightarrow \psi_\mu^A$ 8
$\psi_i^A \sim \lambda_{ABC}$ 56

The residual symmetries are now

- reparametrization in 4 dimensions
- local Lorentz invariance SO(3,1).
- Abelian gauge invariance for B_μ^i and B_μ^{ij}
- global GL(7,R)
- local SO(7).

Let us note that because of duality transformations, the supersymmetry as it is deduced directly from 11 dimensions is only valid on shell.

HIDDEN SYMMETRIES

From what happens in other extended supergravities N $\leq$ 4, we expect to have at least SO(8) symmetry off shell and SU(8) symmetry on shell.

From SO(7) Local x GL(7,R) Global to SO(8) Local x SL(8,R) Global

It is easy to feel the SO(8) symmetry through the following facts :

(1) $(\Gamma^a)_A{}^B$, and $(\Gamma^{ab})_A{}^B$ $a = 1\ldots 7$ form a representation of the Lie Algebra of SO(8).
(2) It is tempting to recombine B_μ^{ij} and B_μ^i by defining $B_\mu^{i8} \simeq B_\mu^i$
(3) as well as to recombine g^{ij} and ϕ^i to form a 35 representation of SO(8).

Neglecting the pseudoscalars A_{ijk} it is easy to see that the bosonic part of the Lagrangian (except the R term) can be rewritten as

$$\mathcal{L}_B = -\frac{e}{16}\partial_\mu S^{i'j'}\partial^\mu S_{i'j'} - \frac{e}{8} S_{i'k'}S_{j'\ell'}G_{\mu\nu}^{i'j'}G^{\mu\nu k'\ell'}$$

with $i', j' \ldots = 1, \ldots 8$

$$G_{\mu\nu}^{i'j'} = (G_{\mu\nu}^{ij}, -\tfrac{1}{2}G_{\mu\nu}^{i})$$

$$S^{i'j'} = \Delta^{-3/4}\begin{pmatrix} \Delta g^{ij} - \phi^i\phi^j, & \phi^j \\ \phi^i, & -1 \end{pmatrix}$$

$$\Delta = -\det(g_{ij}) \quad ; \quad \det S = 1$$

The matrix $S_{i'j'}$ being the inverse of $S^{i'j'}$.

At this stage we discover an SL(8,R) global symmetry which extends the global GL(7,R) derived from the dimensional reduction. We can then extend the local SO(7) to local SO(8) by defining a square root of S^{ij} defined up to a local SO(8) transformation. If this is trivial for the pure bosonic part, it is necessary for the coupling to fermions. The situation is therefore analogous to that of the pure gravity theory when we use the vierbein formalism of Weyl. The vierbein e_μ^α is an element of GL(4,R) (16 parameters) defined up to a local Lorentz transformation of SO(3,1) $\Lambda_\alpha{}^\beta$ (6 parameters) which therefore corresponds to 10 parameters. It is an element of the coset GL(4,R)/SO(3,4) and is related to the metric (invariant under the local Lorentz group) by $g_{\mu\nu} = e_\mu^\alpha \eta_{\alpha\beta} e_\nu^\beta$. In the same way we can define an element of SL(8,R) v_i^a (63 parameters) defined up to a local SO(8) transformation (28 parameters) which corresponds to 35 parameters. It is an element of the coset SL(8,R)/SO(8) and is related to the metric S_{ij} by $S_{ij} = v_i^a \eta_{ab} v_j^b$. The scalar fields will then be described by a generalized non linear model associated with the coset SL(8,R)/SO(8). Let us make a digression on general σ- models associated with a coset G/H and explain how the problems due to the eventual non compactness of G (as is the case here) are solved.

Generalized Non Linear σ -Models

The non linear σ- models associated with a group G are described by the Lagrangian

$$\mathcal{L} = -\mathrm{Tr}(\partial_\mu v\, \partial^\mu v^{-1}) = +\mathrm{Tr}[(v^{-1}\partial_\mu v)^2]$$

where v is an element of G and therefore $v^{-1}\partial_\mu v$ is an element of the Lie algebra of G . If G is non compact, the Killing metric contains some - signs and the Hamiltonian of such a theory theory is not positive definite.

As usual this problem is solved by introducing an extra gauge invariance which eliminates the ghost fields. Let H be the maximal compact subgroup of G and let us ask for an invariance under global G x local H

$$v \rightarrow g\, v\, h(x)$$

Introducing a connexion Ω_μ for the local group H we can construct the Lagrangian

$$\mathcal{L} = \mathrm{Tr}[(v^{-1}D_\mu v)^2]$$

where $D_\mu \mathcal{V} = \partial_\mu \mathcal{V} - \mathcal{V}\Omega_\mu$ is the covariant derivative with respect to H, Ω_μ having no kinetic terms is not propagating and can be eliminated by the use of its equations of motion

$$\Omega_\mu = (\mathcal{V}^{-1}\partial_\mu \mathcal{V})_{//}$$

and the Lagrangian becomes

$$\mathcal{L} = \mathrm{Tr}\left[(\mathcal{V}^{-1}\partial_\mu \mathcal{V})_\perp^2\right]$$

where $//$ and $\perp$ mean parallel and perpendicular to H with respect to the Killing metric. This Lagrangian now leads to a positive definite Hamiltonian and describes dim G - dim H scalar fields.

In a special gauge $\mathcal{V}$ can be generated by the elements of the Lie algebra of G $\perp$ to the elements of Lie algebra of H (denoted by W). G acts on W in a non linear way. The maximum group acting linearly on W is isomorphic to H and corresponds to global transformations

$$\mathcal{V} \to h^{-1}\mathcal{V}h \qquad h \in H$$

This applies in particular to SL(8,R)/SO(8).

From SO(8) Local x SL(8,R) Global to SU(8) Local x E_7 Global

Now we should like to describe all scalars and pseudoscalars together as a coset G/H. We expect the maximal linear group acting on physical fields to be SU(8). Therefore, postulating H = SU(8) (63 generators), the 70 scalars and pseudoscalars should be described by a coset G/H where the dimension of G is 70 + 63 = 133. This is the number of generators of E_7. In fact there exists a non-compact form of E_7 (denoted $E_{7(+7)}$) which has SU(8) as maximum compact subgroup (+7 = 70 - 63 is the signature of the non-compact group).

Before going farther in the description of the theory, let us describe E_7 in a pedestrian way by the infinitesimal transformations acting on the vectors of the fundamental representation of dimension 56 : X = (x^{ij}, y_{ij}) where

$$x^{ij} = -x^{ji} \quad ; \quad y_{ij} = -y_{ji} \quad ; \quad i,j = 1 \cdots 8$$

$$\delta x^{ij} = \Lambda^i{}_k x^{kj} + \Lambda^j{}_k x^{ik} + \frac{1}{24}\varepsilon^{ijk\ell mnop}\Sigma_{mnop}\, y_{k\ell}$$

$$\delta y_{ij} = \Lambda_i{}^k y_{kj} + \Lambda_j{}^k y_{ik} + \Sigma_{ijk\ell}\, x^{k\ell}$$

with $\Lambda^i{}_i = 0 \;,\; \Lambda^i{}_k = -\Lambda_k{}^i$:

defining the subgroup SL(8,R) and where the Σ_{ijk} are totally antisymmetric in i, j, k, ℓ. There exists a bilinear simplectic invariant

$$X_1^t \Omega X_2 \equiv x_1^{ij} y_{2ij} - x_2^{ij} y_{1ij}$$

and a quartic invariant which defines E_7, while the maximal compact subgroup SU(8) is defined by

$$\Lambda^i{}_j = -\Lambda^j{}_i \;(\text{subgroup } SO(8))\;;\; \Sigma_{ijk\ell} = \frac{1}{24}\varepsilon^{ijk\ell mnop}\Sigma_{mnop}$$

$$\delta(x^{ij} \pm i y_{ij}) = (2\Lambda^{[i}{}_{[k}\delta^{j]}{}_{\ell]} \pm i\Sigma_{ijk\ell})(x^{k\ell} \pm i y_{k\ell})$$

Since E_7 and SU(8) are extensions of GL(7,R) and SO(8) we expect that E_7 will not act on fermion fields (if the gauge of SU(8) is unfixed) and that only the local SU(8) will act on them. Therefore the matrix $\mathcal{V}$ of E_7 will play the role of a vielbein connecting fields transforming under the global E_7 and fields transforming under local SU(8). Generalizing the SL(8,R) invariance previously discussed, we expect that the vector fields will be transformed by E_7, but since there are only 28 of them and the smallest representation of E_7 is of dimension 56, this has to be done in a very peculiar way : the chiral dual transformations.

Vector Fields and Chiral Dual Transformations

For extended supergravities N = 1...3 and 4, the invariances U(N) ($N \leq 3$) [4] and SU(4) x SU(1,1) for N = 4 [5] are realized only on shell because they make use of chiral dual transformations which interchange the Bianchi identity for the vector fields with their equations of motion. It is the case for E_7 which interchanges

$$\partial_\mu (e\tilde{G}^{\mu\nu ij}) = 0$$

with
$$\partial_\mu (e\tilde{H}^{\mu\nu}{}_{ij}) = 0$$

where
$$\tilde{H}^{\mu\nu}_{ij} = \frac{4}{e}\frac{\delta\mathcal{L}}{\delta G^{ij}_{\mu\nu}}$$

$G^{\mu\nu ij}$ and $H^{\mu\nu}{}_{ij}$ can be indentified with the 56 representations of E_7 x^{ij} and y_{ij} respectively, we shall denote by $\mathcal{F}$ this 56 vector of E_7 :

$$\mathcal{F}_{\mu\nu} = \begin{pmatrix} G_{\mu\nu}{}^{ij} \\ H_{\mu\nu ij} \end{pmatrix}$$

Since H and G are not independent $\mathcal{F}$ must satisfy a constraint. This comes from the coupling to fermion fields which occurs only through $\mathcal{V}\mathcal{F}_{\mu\nu}$ invariant under E_7 and through $\gamma_{\mu\nu}$ which satisfies $\gamma^{\mu\nu} = i\gamma_5\tilde{\gamma}^{\mu\nu}$. It therefore suggests the constraint $\mathcal{V}\mathcal{F}_{\mu\nu} = \Omega\mathcal{V}\tilde{\mathcal{F}}_{\mu\nu}$ (in absence of fermions in H).

This can be checked directly on the Lagrangian and defines $\mathcal{V}$ in a triangular gauge which is then shown to be an E_7 matrix. Finally all couplings in the Lagrangian are shown to be consistent with the required symmetry SU(8) local and E_7 global using this particular $\mathcal{V}$.

The supersymmetry transformations are partially deduced from the one in 11 dimensions and completed by requiring the covariance with respect to these symmetries.

In order to put the Lagrangian in a simpler form, it is useful to make a change of fields which will have now simple transformation properties under SU(8). This change makes all $(\Gamma^a)_A{}^B$ matrices disappear and requires many properties of these matrices. Now all the fields (fermions and bosons) will be labelled by SU(8) indices A, B or E_7 indices M,N.

Let us now describe the theory.

N = 8, E_7/SU(8) Supergravity

The content of the N = 8 supergravity is 1 graviton $e_\mu^{\ \alpha}$; 8 spin 3/2 particles $\psi_{\mu A}$; 28 vector particles $B_\mu^{\ MN} = -B_\mu^{\ NM}$ (N, M = 1...8), 56 spin 1/2 particles λ_{ABC} (antisymmetric in ABC) and 70 scalar (pseudoscalar) particles described by an element of the coset E_7/SU(8) or W_{ABCD} [6]

The invariances of the theory are

- reparametrization invariance in 4 dimensions
- local Lorentz invariance SO(3,1)
- local N = 8 supersymmetry
- 28 Abelian gauge invariances
- local SU(8)
- global non compact $E_{7(+7)}$ on shell (A subgroup SL(8,R) is valid off shell).

The theory could be built directly in a rather simple way by imposing the last two symmetries at the same time as the others.

The SO(8) symmetry which classifies the physical states is obtained as a diagonal global subgroup SO(8) $\subseteq$ SO(8) local $\otimes$ SO(8) global $\subset$ SU(8) local $\otimes$ E_7 global.

It is useful to rewrite the infinitesimal transformations of E_7 in the fundamental representation of dimension 56 in a way which features the SU(8) subgroup. The 56 dimensional vector space is described by $(z_{AB}, \bar{z}^{AB})$
with z_{AB} complex $z_{AB} = -z_{BA}$; $\bar{z}^{AB} = (z_{AB})^*$; $A, B = 1 \dots 8$

$$\delta z_{AB} = \Lambda_A{}^C z_{CB} + \Lambda_B{}^C z_{AC} + \Sigma_{ABCD} \bar{z}^{CD}$$
$$\delta \bar{z}^{AB} = \bar{\Lambda}^A{}_C \bar{z}^{CB} + \bar{\Lambda}^B{}_C \bar{z}^{AC} + \bar{\Sigma}^{ABCD} z_{CD}$$

$\Lambda_A{}^C$ is a traceless antihermitian 8 x 8 matrix (belonging to the Lie Algebra of SU(8)).

Σ_{ABCD} is totally antisymmetric, $\bar{\Sigma}^{ABCD} = (\Sigma_{ABCD})^*$ and we have the constraint

$$\Sigma_{ABCD} = \frac{\eta}{24} \varepsilon_{ABCDEFGH} \bar{\Sigma}^{EFGH} \qquad (\eta = \pm 1)$$

This gives the structure of the Lie algebra of E_7 necessary to study the scalar fields. Let $\mathcal{V}$ be the 56 x 56 matrix $\in$ E_7 describing the scalar fields

$$\mathcal{V} = \begin{pmatrix} u_{AB}{}^{MN} ; & V_{ABMN} \\ \bar{V}^{ABMN} ; & \bar{U}^{AB}{}_{MN} \end{pmatrix}$$

E_7 will act on the right of $\mathcal{V}$ (indices M N) and SU(8) local on the left (indices A, B). It is now easy to construct the Lagrangian for the scalar fields as a σ- model on the coset E_7/SU(8). $\partial_\mu \mathcal{V} \mathcal{V}^{-1}$ is an element of the Lie algebra of E_7 and therefore takes the form

$$\partial_\mu \mathcal{V} \mathcal{V}^{-1} = \begin{pmatrix} 2\, Q_{\mu[A}{}^{[C} \delta_{B]}{}^{D]} & ; P_{\mu ABCD} \\ \bar{P}_\mu{}^{ABCD} & ; 2\, \bar{Q}_\mu{}^{[A}{}_{[C} \delta^{B]}{}_{D]} \end{pmatrix}$$

$Q_{\mu A}{}^B$ and $P_{\mu ABCD}$ are now invariant under E_7. $P_{\mu ABCD}$ is a covariant tensor for SU(8) which satisfies the same constraints as Σ_{ABCD} . $Q_{\mu A}{}^B$ is the connexion for the local SU(8) transformations so that

$$D_\mu \mathcal{V} \mathcal{V}^{-1} = \begin{pmatrix} 0 & ; & P_{\mu ABCD} \\ \bar{P}_\mu{}^{ABCD} & ; & 0 \end{pmatrix}$$

and the Lagrangian for the scalar field is simply

$$\mathcal{L} \sim \mathrm{Tr}\left[(D_\mu \mathcal{V} \mathcal{V}^{-1})^2\right] \sim P_{\mu ABCD} \bar{P}^{\mu ABCD}$$

Let us now concentrate on the vector fields B_μ^{MN}. Defining

$$G_{\mu\nu}^{MN} = \partial_\mu B_\nu^{MN} - \partial_\nu B_\mu^{MN} \quad \text{and} \quad H_{\mu\nu MN} = -\frac{4}{e}\frac{\delta \tilde{\mathcal{L}}}{\delta G_{\mu\nu}^{MN}}$$

$$\mathcal{F}_{\mu\nu} = \frac{1}{\sqrt{2}} \begin{pmatrix} G_{\mu\nu}^{MN} + i H_{\mu\nu MN} \\ G_{\mu\nu}^{MN} - i H_{\mu\nu MN} \end{pmatrix}$$

$\mathcal{F}$ is a 56 vector for E_7 and a scalar for SU(8). Such transformations are true only for B_μ^{MN} on shell

$$\partial_\mu \tilde{\mathcal{F}}_{\mu\nu} = 0 \quad \Rightarrow \quad \mathcal{F}_{\mu\nu} = \sqrt{2} \begin{pmatrix} \partial_{[\mu} B_{\nu]}^{MN} + i\, \partial_{[\mu} C_{\nu]MN} \\ \partial_{[\mu} B_{\nu]}^{MN} - i\, \partial_{[\mu} C_{\nu]MN} \end{pmatrix}$$

Using the 56-bein $\mathcal{V}$ we can form a scalar for E_7 since under E_7

$$\mathcal{V} \to \mathcal{V} E^{-1}; \quad \mathcal{F} \to E\, \mathcal{F} \qquad \text{then}$$

$$\mathcal{V} \mathcal{F}_{\mu\nu} \equiv \begin{pmatrix} \mathcal{F}_{\mu\nu AB} \\ \bar{\mathcal{F}}_{\mu\nu}{}^{AB} \end{pmatrix} \quad \text{is a scalar for } E_7.$$

Since $H_{\mu\nu}$ and $G_{\mu\nu}$ are not independent, $\mathcal{F}$ must satisfy a constraint which is

$$\hat{\mathcal{F}}_{\mu\nu AB} = i\, \tilde{\hat{\mathcal{F}}}_{\mu\nu AB}$$

where $\hat{\mathcal{F}}_{\mu\nu AB} = \mathcal{F}_{\mu\nu AB} +$ quadratic fermionic terms such that $\hat{\mathcal{F}}$ is covariant under N = 8 supersymmetry.

The fermionic fields are directly coupled to $\gamma^{\mu\nu} \hat{\mathcal{F}}_{\mu\nu AB}$ which is consistent with the constraint on $\mathcal{F}_{\mu\nu AB}$.

Since $\psi_{\mu A}$ and λ_{ABC} are Majorana spinors, SU(8) transformations $\Lambda'_A{}^B + i\Lambda''_A{}^B$ are to be understood as $\Lambda'_A{}^B + i\gamma_5 \Lambda''_A{}^B$. We can also use Weyl spinors $\psi_{R \atop L} = \frac{1 \pm \gamma_5}{2}\psi$, then $\psi_{\mu A(R)}$ transforms as 8 for SU(8) and $\psi^A_{\mu(L)}$ as $\underline{8}$. All derivatives of spinors must be covariant with respect to SU(8).

We can now write the complete Lagrangian for the N = 8 supergravity

$$
\begin{aligned}
\mathcal{L} = & -\frac{e}{4} R(\omega, e) + \frac{1}{2} \varepsilon^{\mu\nu\rho\sigma} \bar{\psi}_{\mu A} \gamma_\sigma \gamma_5 (\delta_A^B D_\nu(\omega) - \varphi_{\nu A}{}^B) \psi_{\rho B} \\
& + \frac{e}{8} G_{\mu\nu}^{MN}(B) \tilde{H}^{\mu\nu}_{MN}(B, \mathcal{V}, \psi, \lambda) + \frac{ie}{12} \bar{\lambda}_{ABC} \gamma^\mu (\delta_A^D D_\mu(\omega) - 3\varphi_{\mu A}{}^D) \lambda_{BCD} \\
& + \frac{e}{24} P_{\mu ABCD} \bar{P}^{\mu ABCD} + \frac{e}{6\sqrt{2}} \bar{\psi}_{\mu A} \gamma^\nu \gamma^\mu (\bar{P}_\nu{}^{ABCD} + \hat{\bar{P}}_\nu{}^{ABCD}) \lambda_{BCD} \\
& + \frac{e}{8\sqrt{2}} \Big\{ \bar{\psi}_{\mu A} \gamma^{[\nu} \hat{\mathcal{F}}_{AB} \gamma^{\mu]} \psi_{\nu B} - \frac{i}{\sqrt{2}} \bar{\psi}_{\mu C} \hat{\mathcal{F}}_{AB} \gamma^\mu \lambda_{ABC} \\
& \qquad - \frac{\eta}{72} \varepsilon^{ABCDEFGH} \bar{\lambda}_{ABC} \hat{\mathcal{F}}_{DE} \lambda_{FGH} \Big\}
\end{aligned}
$$

where $\hat{\mathcal{F}}_{AB} = \gamma^{\mu\nu} \hat{\mathcal{F}}_{\mu\nu AB}$

$\hat{P}_{\mu ABCD} = P_{\mu ABCD} +$ bilinear fermionic fields (covariant under supersymmetry)

$\omega =$ supercovariant Lorentz connexion.

The supersymmetry transformation laws are given by

$$
\delta_S e_\mu{}^\alpha = -i \bar{\varepsilon}_A \gamma^\alpha \psi_{\mu A}
$$

$$
\delta_S \mathcal{V} \mathcal{V}^{-1} = -2\sqrt{2} \begin{pmatrix} 0 & ; \bar{\varepsilon}^{(L)}_{[A} \lambda^{(R)}_{BCD]} + \frac{\eta}{24} \varepsilon_{ABCDEFGH} \bar{\varepsilon}^E_{(R)} \lambda^{FGH}_{(L)} \\ \bar{\varepsilon}^{[A}_{(R)} \lambda^{BCD]}_{(L)} + \frac{\eta}{24} \varepsilon^{ABCDEFGH} \bar{\varepsilon}^{(L)}_E \lambda^{(R)}_{FGH} ; & 0 \end{pmatrix}
$$

The zeros in $\delta_S \mathcal{V} \mathcal{V}^{-1}$ show that $\varphi_{\mu A}{}^B$ is supercovariant but $P_{\mu ABCD}$ is not. The transformation laws for $B_\mu{}^{MN}$ (off shell) and $C_{\mu MN}$ (on shell) can be put together in a E_7 covariant form

$$\delta_S \begin{pmatrix} B_\mu^{MN} + i C_{\mu MN} \\ B_\mu^{MN} - i C_{\mu MN} \end{pmatrix} = -2\sqrt{2}\, \mathcal{V}^{-1} \begin{pmatrix} \bar{\epsilon}^{(L)}_{[A} \psi^{(R)}_{\mu B]} - \frac{i\sqrt{2}}{4} \bar{\epsilon}^{(R)C} \gamma_\mu \lambda^{(R)}_{ABC} \\ \bar{\epsilon}^{[A}_{(R)} \psi^{B]}_{\mu(L)} - \frac{i\sqrt{2}}{4} \bar{\epsilon}^{(L)}{}_C \gamma_\mu \lambda^{ABC}_{(L)} \end{pmatrix}$$

$$\delta_s \lambda^{(R)}_{ABC} = -i\sqrt{2}\, \hat{P}_{\mu ABCD} \gamma^\mu \epsilon^D_{(L)} + \frac{3}{4} \hat{\mathcal{F}}_{[AB}\, \epsilon^{(R)}_{C]}$$

$$\delta_s \psi^{(R)}_{\mu A} = \left(D_\mu(\omega) \delta_A^B - \mathcal{Q}_{\mu A}{}^B \right) \epsilon^{(R)}_B - \frac{i}{4\sqrt{2}} \hat{\mathcal{F}}_{AB} \gamma_\mu \epsilon^B_{(L)}$$
$$+ i\, \bar{\lambda}^{(L)}_{ABC} \gamma^\alpha \lambda^{DBC}_{(L)} \gamma_\alpha \gamma_\mu \epsilon^{(R)}_D - \frac{1}{\sqrt{2}} \bar{\psi}^B_{\mu(R)} \gamma^\alpha \lambda^{(R)}_{ABC} \gamma_\alpha \epsilon^C_{(L)}$$

We can now fix a gauge for the local SU(8). This will give some specific parametrization for the matrix $\mathcal{V}$ (scalar fields) but will not change the Lagrangian. This however will modify the supersymmetry transformation laws because this gauge fixing requires field dependent SU(8) gauge transformations. A particular useful gauge is the Hermitian gauge defined by $\mathcal{V} = \mathcal{V}^+$. $\mathcal{V}$ is then generated by the elements of the Lie algebra of $E_7 \perp$ to SU(8) and therefore can be written

$$\mathcal{V} = \exp X \qquad \text{with} \qquad X = \begin{pmatrix} 0 & W_{ABCD} \\ \bar{W}^{ABCD} & 0 \end{pmatrix}$$

where W satisfies $\bar{W}^{ABCD} = \frac{\eta}{24} \varepsilon^{ABCDEFGH} W_{EFGH}$

Following a general method for parametrizing coset spaces we can then introduce the variables

$$y_{AB,CD} = \left(W \frac{\mathrm{th}\sqrt{\bar{W}W}}{\sqrt{\bar{W}W}} \right)_{AB,CD}$$

(all y's are not independent because of the constraints on W) then $\mathcal{V}$ becomes

$$\mathcal{V} = \begin{pmatrix} \frac{1}{\sqrt{1-y\bar{y}}} \, , & y \frac{1}{\sqrt{1-\bar{y}y}} \\ \bar{y} \frac{1}{\sqrt{1-y\bar{y}}} \, , & \frac{1}{\sqrt{1-\bar{y}y}} \end{pmatrix}$$

The action of E_7 is then very simple on y. It is defined by

$$\mathcal{V}(y)\begin{pmatrix} A & B \\ \bar{B} & \bar{A} \end{pmatrix} = \begin{pmatrix} \mathcal{U} & 0 \\ 0 & \mathcal{U}^* \end{pmatrix} \mathcal{V}(y')$$

where $\mathcal{U}$ is an irrelevant SU(8) matrix. Then

$$y' = (A + y\bar{B})^{-1}(B + y\bar{A})$$

The y's are called the inhomogeneous coordinates of the coset.

Using these variables we can compute $P_{\mu ABCD}$ and $Q_{\mu A}{}^{B}$. We get

$$P_{\mu ABCD} = \left(\frac{1}{\sqrt{1 - y\bar{y}}} \partial_\mu y \frac{1}{\sqrt{1 - \bar{y}y}} \right)_{ABCD}$$

(note that $P_{\mu ABCD}$ automatically satisfies the right constraints) and

$$|P_{\mu ABCD}|^2 = \mathrm{Tr}\left(\frac{1}{1 - y\bar{y}} \partial_\mu y \frac{1}{1 - \bar{y}y} \partial_\mu \bar{y} \right)$$

can be rewritten treating the variables $y, \bar{y}$ as independent

$$|P_{\mu ABCD}|^2 = -\frac{\partial}{\partial y_{ABCD}} \frac{\partial}{\partial \bar{y}^{A'B'C'D'}} \left(\mathrm{Tr}\,\mathrm{Log}(1 - y\bar{y}) \right) \partial_\mu y_{ABCD}\, \partial^\mu \bar{y}^{A'B'C'D'}$$

(In order to prove that it is a Kählerian manifold[7] we should need the same structure with independent variables). It should be noted that $Q_{\mu A}{}^{B}$ is more complicated.

We can also simply write $F_{\mu\nu AB}$ and $H_{\mu\nu AB}$ in terms of $G_{\mu\nu AB}$

$$F_{\mu\nu AB} = \sqrt{2}\left(\sqrt{1 - y\bar{y}}\, \frac{1}{1 - \bar{y}} \right)_{AB}{}^{CD} G_{\mu\nu\, CD} \quad + \text{ fermionic terms}$$

$$H_{\mu\nu\,AB} = \eta_{AB\,CD}\, \mathcal{G}_{\mu\nu}^{\;CD} + \text{fermionic terms}$$

(where i is replaced by $\frac{1}{2}\varepsilon^{\mu\nu\rho\sigma}$)

$$\eta_{AB,CD} = \left(\frac{\mathbb{1}+y}{\mathbb{1}-y}\right)_{AB,CD} \quad \text{and} \quad \mathbb{1}_{AB,CD} = \frac{1}{2}\left(\delta_{AC}\delta_{BD} - \delta_{AD}\delta_{BC}\right)$$

(Note that $G_{\mu\nu}^{\;CD}$ and $H_{\mu\nu\,AB}$ are not SU(8) tensors).

In fact when we reduce the theory from N = 8 to N = 4 we can see that the variable y was used[5] $(y_{ABCD} \rightarrow \frac{1}{2}\varepsilon_{ABCD}(A-iB)$ and not W .

EXTENDED SUPERGRAVITIES IN 4 DIMENSIONS

We can reduce the theory form N = 8 to N = 7, 6, ... 1 by consistent truncation. We find that the scalar fields are always described as a coset G/H. We give below the content and the symmetries of these extended supergravities.

N	8	7	6	5	4
Spin 2	1	1	1	1	1
Spin 3/2	8	7+1	6	5	4
Spin 1	28	21+7	15+1	10	6
Spin 1/2	56	35+21	20+6	10+1	4
Spin 0	70	35+35	15+15	5+5	1+1
Global group rank	$E_{7(+7)}$ 7	$E_{7(+7)}$ 7	SO*(12) 6	SU(5,1) 5	SU(4) x SU(1,1) 4
Local group rank	SU(8) 7	SU(8) 7	U(6) 6	U(5) 5	U(4) 4

We recover in particular the curious symmetry SU(1,1) found for the supergravity N = 4.

MAXIMAL SUPERGRAVITIES IN D DIMENSIONS

We can conjecture that all maximal supergravities (obtained by dimensional reduction of N = 1 supergravity in 11 dimensions) have the same property : namely the scalar fields are always described by a non linear model on a coset G/H. Using counting arguments we can then guess what the local and global symmetries of these supergravities are.

For D odd the global symmetry will be a symmetry of the Lagrangian but for D even (D = 4, 6, 8) the global symmetry will only be true on shell because it will require duality type transformation.

We list these symmetries below

	Global symmetry	Local symmetry
D = 9	GL(2,R)	SO(2)
D = 8	$E_{3(+3)} \equiv SL(3,R) \times SL(2,R)$	SO(3) x SO(2)
D = 7	$E_{4(+4)} \equiv SL(5,R)$	SO(5)
D = 6	$E_{5(+5)} \equiv SO(5,5)$	SO(5) x SO(5)
D = 5	$E_{6(+6)}$	$US_p(8)$
D = 4	$E_{7(+7)}$	SU(8)
D = 3	$E_{8(+8)}$	SO(16)

We note the appearance of the E group series.* Up to now it has not been understood. The structure for D odd is simpler since there is no duality and the group which classifies the physical states is the group H (obtained as a diagonal subgroup of H global x H local or G global x H local).

In particular it is quite easy to construct the supergravity N = 8 in 5 dimensions by conjecturing the global symmetry $E_{6(+6)}$ and the local symmetry $US_p(8)$. This has been done in collaboration with J. Scherk and J. Schwarz[8]. The physical states are classified by $US_p(8)$ (Note that $US_p(2N)$ classify the massive physical states in 4 dimensions) characterized by the symplectic metric Ω_{MN} (M,N = 1... 8).

It contains 1 graviton $e_\mu{}^\alpha$; 8 gravitinos $\psi_\mu{}^A$, 27 vector particles $A_\mu{}^{MN} = - A_\mu{}^{NM}$ with $\Omega_{MN} A_\mu{}^{MN} = 0$; 48 spin 1/2 particles χ^{ABC} (antisymmetric in ABC and $\Omega_{AB}\chi^{ABC} = 0$), 42 scalar particles described by the coset $E_{6+(6)}/US_p(8)$ (78-36=42)

* We thank W. Nahm for pointing out to us the equivalences of the groups SL(3,R) x SL(2,R), SL(5.R) and SO(5,5) with E_3, E_4 and E_5.

or in a linearized approximation by ϕ^{ABCD} antisymmetric and traceless. All these fields are pseudo-real i.e. complex with the condition :

$$\psi_\mu{}^A = \gamma_5 \psi^*_\mu{}_A \; ; \; \chi^{ABC} = \gamma_5 \chi^*_{ABC}$$

$$A_\mu{}^{MN} = A^*_\mu{}_{MN} \; ; \; \phi_{ABCD} = \phi^{*ABCD}$$

where the indices are raised with the symplectic metric Ω

$E_{6(+6)}$ has a fundamental representation of dimension 27 described by z^{MN} ($z^{MN} = -z^{NM}$, $z^{MN}\Omega_{MN} = 0$; $z^{MN} = z^*{}_{MN}$) . The infinitesimal transformations of z^{MN} are given by

$$\delta z^{MN} = \Lambda^M{}_P z^{PN} + \Lambda^N{}_P z^{MP} + \Sigma^{MN}{}_{PQ} z^{PQ}$$

where $\Lambda^M{}_P$ is an antihermitian matrix such that $\Lambda^M{}_P \Omega_{QM} \equiv \Lambda_{QP} = \Lambda_{PQ}$ (it generates the subgroup $US_p(8)$) and Σ_{MNPQ} is antisymmetric, pseudoreal and traceless for Ω .

There is no quadratic invariant constructed from z^{MN} (27 x 27 does not contain 1) but there exists a trilinear invariant

$$J = z^{MN}\Omega_{NP} z^{PQ} \Omega_{QR} z^{RS} \Omega_{SM}$$

The vector fields $A_\mu{}^{MN}$ will be in the 27 representation of E_6. The scalar fields will be in the $\overline{27}$ representation for E_6 and 27 for $US_p(8)$ and described by a 27 x 27 matrix $\mathcal{V}_{MN}{}^{AB}$ The spinor fields will be scalar as usual for E_6.

As for $E_7/SU(8)$ it is easy to construct a connexion for $US_p(8)$ from the scalar fields

$$\mathcal{V}^{-1}\partial_\mu \mathcal{V} \equiv \mathcal{V}^{MN}{}_{CD}\, \partial_\mu \mathcal{V}_{MN}{}^{AB} = 2\,\varphi_\mu{}^{[A}{}_{[C}\,\delta^{B]}{}_{D]} + P_\mu{}^{AB}{}_{CD}$$

so that

$$\mathcal{V}^{MN}{}_{CD}\, D_\mu \mathcal{V}_{MN}{}^{AB} = P_\mu{}^{AB}{}_{CD}$$

It is also useful to define the metric on the coset space $E_6/US_p(8)$

$$\mathcal{G}_{MN,PQ} = \Omega_{AC}\Omega_{BD}\,\mathcal{V}_{MN}{}^{AB}\,\mathcal{V}_{PQ}{}^{CD}$$

Using the supersymmetry to fix all the coefficients of the expression invariant under E_6 and $US_p(8)$ we can now write the Lagrangian for N = 8 supergravity in 5 dimensions . We note that the cubic invariant J enters explicitly into the Lagrangian (this shows that the maximum symmetry is really E_6) through a pure self interacting term for the vector fields of the same type as the one for the tensor field A_{MNP} in 11 dimensions.

$$
\begin{aligned}
e^{-1}\mathcal{L} = & -\frac{1}{4}R - \frac{i}{2}\bar{\Psi}_\mu^{A}\gamma^{\mu\nu\rho}D_\nu\psi_{\rho A} - \frac{1}{8}g^{\mu\rho}g^{\nu\sigma}\mathcal{J}_{MN,PQ}F_{\mu\nu}^{MN}F_{\rho\sigma}^{PQ} \\
& + \frac{i}{12}\bar{\chi}^{ABC}\gamma^\mu D_\mu\chi_{ABC} - \frac{1}{24}g^{\mu\nu}D_\mu \mathcal{V}^{AB}{}_{MN}D_\nu\mathcal{V}^{MN}{}_{AB} \\
& - \frac{e^{-1}}{12}\varepsilon^{\mu\nu\rho\sigma\lambda}F_{\mu\nu}^{MN}\Omega_{NP}F_{\rho\sigma}^{PQ}\Omega_{QR}A_\lambda^{RS}\Omega_{SM} \\
& + \frac{i}{3\sqrt{2}}P_\rho^{ABCD}\bar{\Psi}_{\mu A}\gamma^\rho\gamma^\mu\chi_{BCD} \\
& + \frac{i}{4}\mathcal{V}^{AB}{}_{MN}F_{\mu\nu}^{MN}\Big\{\bar{\Psi}_A^{\rho}\gamma_{[\rho}\gamma^{\mu\nu}\gamma_{\sigma]}\psi_B^{\sigma} - \frac{1}{\sqrt{2}}\bar{\Psi}_\rho^{C}\gamma^{\mu\nu}\gamma^\rho\chi_{ABC} \\
& \qquad + \frac{1}{2}\bar{\chi}_{ACD}\gamma^{\mu\nu}\chi_B{}^{CD}\Big\}
\end{aligned}
$$

+ quartic fermionic terms.

CONCLUDING REMARKS

The N = 8 supergravity in four dimensions has been shown to have a local symmetry SU(8) whose gauge fields are composite fields. Moreover, it has a global non-compact symmetry E_7 on shell and the scalar fields are described in a geometrical way as the coset $E_7/SU(8)$. No such interpretation has been found up to now for the spin-1/2 fields.

Before the discovery of this local SU(8) symmetry all attempts to make a connection between this maximal supergravity and low energy phenomenology were hopeless[9]. We could introduce gauge coupling for the SO(8) symmetry but it is well known that SO(8) does not contain the minimal SU(3) x SU(2) x U(1) required to describe strong, electromagnetic and weak interactions ; moreover the potential for scalar fields which is generated by the introduction of the gauge coupling constant is unbounded from below.

Now by analogy with the CP^{N-1} models in two dimensions, we conjecture that this local SU(8) symmetry becomes dynamical at the quantum level, in particular the composite gauge fields become propagating fields as well as all the fields in the same supermultiplets. In this spirit, supergravity at the classical level is considered as a theory of preons, and the fundamental particles appearing in grand unified theories are bound states. It has been shown by J. Ellis et al[11] that, conjecturing some presently unknown dynamical symmetry breaking, we can extract an SU(5) grand unified theory.

Of course, we implicitly assumed that this theory is meaningful at the quantum level i.e. the theory is finite : in this case the gauge coupling constant should be computable from "geometry" since we have only one mass scale (namely the Planck mass). In fact this theory has probably more chances to be finite than other supergravities ($N \leqslant 6$) ; two intuitive arguments are in favour of such a property : (i) the maximal supersymmetric Yang Mills theory (N = 4) is much more convergent than for $N \leqslant 2$ and could even be finite[12].(ii) There is no quantum correction to the cosmological constant of the SO(N) gauge supergravities (or de Sitter supergravities) for $N \geqslant 5$[13].

In order to study the quantum theory and try to prove these conjectures which could lead via a dynamical symmetry breaking to a predictive grand unified theory, we need the knowledge of auxiliary fields or a superspace formulation (which has been shown to be consistent on shell with E_7 global and SU(8) local)[14].

REFERENCES

1. E. Cremmer, B. Julia, and J. Scherk, Phys. Letters 76B (1978) 409;
   E. Cremmer and S. Ferrara, Phys. Letters 91B (1980) 61;
   S. Ferrara, these proceedings, p. 119.
2. E. Cremmer and B. Julia, Phys. Letters 80B (1978) 48; Nuclear Phys. B159 (1979) 141.
3. J. Scherk and J.H. Schwarz, Phys. Letters 57B (1975) 463.
   E. Cremmer and J. Scherk, Nuclear Phys. B103 (1976) 399.
4. S. Ferrara, J. Scherk and B. Zumino, Nuclear Phys. B121 (1977) 393.
5. E. Cremmer, J. Scherk and S. Ferrara, Phys. Letters 68B (1977) 234; Phys. Letters 74B (1978) 61;
   E. Cremmer and J. Scherk, Nuclear Phys. B127 (1977) 259.
6. Partial results using W_{ABCD} have been obtained by B. De Wit and D.Z. Freedman, Nuclear Phys. B130 (1977) 105.
   B. De Wit, Nuclear Phys. B155 (1979) 530.
7. B. Zumino, Phys. Letters 87B (1979) 203.
8. E. Cremmer, J. Scherk and J.H. Schwarz, Phys. Letters 84B (1979) 83.
9. M. Gell-Mann, Talk at the 1977 Washington meeting of the American Physical Society (unpublished).

10. A. D'Adda, P. Di Vecchia and M. Lüscher, Nuclear Phys. B146 (1978) 63.
E. Witten, Nuclear Phys. B149 (1979) 285.
11. J. Ellis, M.K. Gaillard, L. Maiani and B. Zumino, these proceedings, p. 69.
J. Ellis, M.K. Gaillard and B. Zumino, CERN preprint TH.2842/ LAPP TH-16 (1980).
12. M. Grisaru, these proceedings, p. 545.
13. M. Duff, these proceedings, p. 369.
S.M. Christensen, M.J. Duff, G.W. Gibbons and M. Roček, Santa Barbara preprint (1980).
14. L. Brink, these proceedings, p. 157.
L. Brink and P. Howe, Phys. Letters 88B (1979) 268.

THE ON-SHELL N=8 SUPERGRAVITY IN SUPERSPACE

Lars Brink

Institute of Theoretical Physics
S-412 96 Göteborg
Sweden

ABSTRACT

The N=8 supergravity theory is presented in superspace. Only the physical fields are kept so that the theory is on-shell. The global E7 and local SU(8) invariances arise naturally.

INTRODUCTION

There is great hope that extended supergravity is, in some sense, the correct description of a unified picture of all the interactions. It is thus important to get an understanding of these models which is as deep and thorough as possible. In the construction of extended supersymmetric theories, two methods have been very useful. The first one, which, so far, has been the most successful, is to use the dimensional reduction technique originally suggested by Kałuza and Klein. One starts in a dimension of space-time which is higher than four and then lets the extra dimensions become cyclic and infinitesimal such that they can be integrated away in the action, leaving a four-dimensional theory. The second method is to describe the supersymmetric theory in a superspace, which is (in the global case) a homogeneous space of the supersymmetry algebra. In this space, which consists of the Minkowski space together with anticommuting coordinates, the supersymmetry is manifest and a field on this space, a superfield, represents in a compact way a lot of ordinary Minkowski space fields.

The largest possible extended supergravity theory with a linearly realized spectrum is the N=8 theory, where N refers to the number of fermionic generators in the supersymmetry algebra. This model has been constructed and presented in great detail by Cremmer and Julia[1]. They obtained the theory by dimensionally reducing the 11-dimensional N=1 supergravity[2] and found, as a result, that the corresponding 4-dimensional theory has a local SU(8) invariance as well as a global (non-compact) E_7 invariance. Extended supergravity for arbitrary N has been studied in superspace by MacDowell[3], but his approach is based on the "natural" extended supergroup Osp(N ;4) and should therefore lead to an N=8 theory with a local SO(8) gauge invariance (where the corresponding gauge fields are the 28 vector fields of the N=8 supersymmetry representation, as opposed to the SU(8) case, where the gauge fields are composite). This report, which is based on work by Paul Howe and myself[4], will describe the superspace formulation of the SU(8) model.

The approach that we have used is pragmatic in the sense that we anticipate the result of Cremmer and Julia by building in the SU(8) invariance as one of the starting assumptions. The technique then used is essentially dimensional analysis and the equations of the theory take the form of constraints on certain components of the supertorsion tensors. We will keep only components that correspond to dynamical fields or space derivatives thereof. In this way we force the theory on to the mass shell. We do this only because it simplifies the calculations. Our present understanding of the theory is therefore incomplete and the off-shell description of this theory remains to be constructed.

The order of the presentation is as follows: general formalism; solution to the Bianchi identities; the problem of the scalar and vector fields and conclusions. For convenience, notational conventions are summarized in Table 1.

GENERAL FORMALISM

The superspace to be used has coordinates $z^{m}=(x^{\mu},\theta^{\mathcal{A}},\bar{\theta}^{\mathcal{A}'})$ where x^{μ} is the normal bosonic coordinate and $\theta^{\mathcal{A}}$ a set of 8 2-component Grassmann spinor coordinates. We shall use the formalism of differential forms and a detailed discussion of this method has been given by Wess[5]. The basic fields of the theory are the moving frame which is related to the coordinate one-forms by the vielbein matrix:

$$E^{M}=dz^{m}E_{m}{}^{M} \tag{1}$$

The vielbein fields are essentially the gravity fields on this space and the generalized equivalence principle that is underlying our theory demands a local Lorentz invariance in the tangent space together with an internal invariance, which is chosen to be SU(8). To implement these invariances we introduce the connection (gauge field)

$$\Omega_M{}^N = dz^m \Omega_{m,M}{}^N \tag{2}$$

A prototype tangent space vector transforms as

$$\delta X^M = X^N L_N{}^M \tag{3}$$

where

$$\begin{aligned}
&L_{uv} = -L_{vu} \\
&L_A{}^B = \delta_a{}^b L^i{}_j + \delta^i{}_j L_a{}^b \\
&L_{A'}{}^{B'} = -\bar{L}_{A'}{}^{B'} \\
&L_{ab} = \tfrac{1}{2}(\sigma^{uv})_{ab} L_{uv} \\
&\bar{L}_j{}^i = -L^i{}_j \,; \quad L^i{}_i = 0
\end{aligned} \tag{4}$$

and all others are cyclic. $\Omega_M{}^N$ has the same symmetries as $L_M{}^N$ although it transforms inhomogeneously

$$\delta\Omega_M{}^N = -L_M{}^Q \Omega_Q{}^N + \Omega_M{}^Q L_Q{}^N - dL_M{}^N \tag{5}$$

Using Ω, we may define a covariant exterior derivative, for example

$$DX^M = dX^M + X^N \Omega_N{}^M \tag{6}$$

There are two basic field strengths, the torsion

$$T^M = DE^M = \tfrac{1}{2} E^P \wedge E^Q T_{QP}{}^M \tag{7}$$

and the curvature

$$R_M{}^N = d\Omega_M{}^N + \Omega_M{}^Q \wedge \Omega_Q{}^N = \tfrac{1}{2} E^P \wedge E^Q R_{QP,M}{}^N \tag{8}$$

the latter having the same symmetries as $L_M{}^N$.

The torsion and the curvature are not independent of each other and from their definitions and the properties of exterior differentiation one derives the Bianchi identities

$$DT^M = E^N \wedge R_N{}^M \tag{9}$$

$$DR_M{}^N = 0 \tag{10}$$

the first of which reads in components

$$\sum_{(MNP)} R_{MN,P}{}^{Q} - D_M T_{NP}{}^{Q} - T_{MN}{}^{S} T_{SP}{}^{Q} = 0 \tag{11}$$

where $\sum_{(MNP)}$ means the graded cyclic sum.

A useful theorem assures us that if (9) is satisfied so is (10), if one uses the commutation properties of two D's [6]. Furthermore, it has been shown that the torsion completely determines the curvature via (9) because of the latter's symmetry properties (4)[6].

The problem to be solved now is to find torsions which are consistent with (11), and which only contain the physical degrees of freedom. This means that we do not allow for any auxiliary fields at this stage and hence the resulting theory will be locally supersymmetric only on-shell. As was said before this is done only to simplify the complicated calculations we have to do and should be remedied later.

The physical spectrum of the N=8 model that we shall find is the following:

1 graviton,	spin-2,	$e_{\mu}{}^{u}$
8 gravitinos,	spin-3/2,	$\psi_{\mu i}{}^{a}$
28 vector particles,	spin-1,	$A_{\mu ij}$
56 spinors,	spin-1/2,	W_{ijka}
70 scalars,	spin-0,	$V_{ijk\ell}$

SOLUTION TO THE BIANCHI IDENTITIES

In order to generate the N=8 spectrum, we require a starting point and it is interesting to observe that for N=4, the spectrum of the supersymmetric Yang-Mills, which is the maximally supersymmetric such theory, may be generated from a set of scalar superfields[7]. This is a useful property, since these fields have dimension -1 (length). Expanding it in θ we find at order θ the canonical spinor fields and at the order θ^2, the field strength for the vector field. Unfortunately this cannot happen for the N=8 theory. We require 70 scalars and this corresponds to a 4-index (totally antisymmetric) field. There is no such torsion and also no such curvature since we are using SU(8) and not SO(8). We shall therefore try to obtain the spectrum from spinor superfields. However, since these fields have dimension -3/2, we can only hope to get the bosonic fields (including the scalars) occuring with space derivatives. There is one possible place for the spinor fields to occur in the torsions, namely

$$T_{AB}{}^{C'} = \epsilon_{ab}\,\overline{W}^{ijkc'} \tag{12}$$

W is totally antisymmetric on ijk and hence corresponds to a $\underset{\sim}{56}$ of Weyl spinor superfields.

We now want to choose a set of conditions on torsions such that we derive (12) from the Bianchi identities, i.e. we want conditions which force all other components of $T_{AB}{}^{C'}$ which are not of the form $\epsilon_{ab}\,\overline{W}^{ijk\ell'}$ to be zero. In analogy with the N=1 theory[8] we try to choose as many of the torsions as possible to be zero or constants. Our choice is

$$\begin{aligned} &T_{AB}{}^{u} = T_{AB}{}^{C} = T_{AB'}{}^{C'} = T_{Au}{}^{v} = T_{uv}{}^{w} = 0 \\ &T_{AB'}{}^{u} = -i(\sigma^u)_{ab'}\,\delta^i_j \end{aligned} \tag{13}$$

This only leaves a few torsions undetermined, namely

$$T_{uA}{}^{B},\; T_{uA}{}^{B'},\; T_{uv}{}^{C}$$

These torsions may be decomposed into irreducible representations of SL(2,C)xSU(8) and many of these parts are also determined via the Bianchi identities (9). In this process we find some parts which are not determined. These we fix in terms of W on general dimensional and symmetry grounds, such that they do not impose extra unwanted constraints on W. In this way we throw away some of the auxiliary fields. It is, however, hard to judge which of these parts are genuine auxiliary fields and which are parts that have to be fixed also in the presence of such fields. To include all the auxiliaries one really has to start all over again and change the set (13) appropriately. Hence we make these fixings of the undetermined parts. Since we now keep only the physical degrees of freedom, the Bianchi identities must be powerful enough also to provide us with the equations of motion, since it is only with those imposed that the physical system can be locally supersymmetric, and our formalism is manifestly locally supersymmetric.

Proceeding in this fashion we arrive at expressions for all the torsions (and hence the curvatures) completely in terms of W, its derivatives and multilinear combinations. In addition W is constrained to contain only the physical degrees of freedom in its expansion in θ. The components of any arbitrary superfield S may be expressed as the leading (θ=0) components of the superfields formed from S by taking spinorial covariant derivatives[9], i.e. S's components are

$$S|_{\theta=0}\ ;\ D_A S|_{\theta=0}\ ;\ \bar{D}_{A'} S|_{\theta=0}\ ;\ D_A D_B S|_{\theta=0}\ \text{etc.}$$

Using this technique we can find all independent components of W, the remainder being expressible as vector derivatives and multilinear combinations of this basic set. We find W's components to be

$$\begin{aligned}
&W_{ijka}|_{\theta=0}\\
&M_{ab\,ij} = \tfrac{1}{6} D^k_{(a} W_{b)ijk}|_{\theta=0}\\
&P_{ab'\,ijkl} = -\tfrac{i}{2}\bar{D}_{a'i} W_{jklb}|_{\theta=0}\\
&\qquad = \tfrac{1}{4!}\epsilon_{ijklmnpq}\,\bar{P}_{a'b}{}^{mnpq}\\
&U_{(abc)i} = -\tfrac{1}{7} D^j_{(a} M_{bc)ij} + \text{bilinears}|_{\theta=0}\\
&A_{(abcd)} = \tfrac{1}{8} D^i_{(a} U_{bcd)i} + \text{bilinears}|_{\theta=0}
\end{aligned} \tag{14}$$

We will show that M and P are essentially the field strengths for the vector and scalar fields respectively. U is an $\underset{\sim}{8}$ of spin-3/2 'Weyl' spinors, by which we mean that part of the spin-3/2 field strength which does not enter into the Rarita-Schwinger equation, and A is the spin-2 Weyl spinor.

Besides these fields we have the vierbein and gravitino fields themselves, which may be identified as the θ=0 components of parts of the vielbein,

$$E_\mu{}^u|_{\theta=0} = e_\mu{}^u\ ;\ E_\mu{}^A|_{\theta=0} = \Psi_\mu{}^A \tag{15}$$

of the physical states. In Tables 2 and 3 the resulting torsions and curvatures are given.

THE SCALAR AND VECTOR FIELDS

In the process of solving the Bianchi identities we did not obtain the vector and scalar fields explicitly. We only got objects which looked like field strengths in the correct representations for the vectors and the scalars. However, since the theory is manifestly locally supersymmetric, the system of equations derived from the Bianchi identities must provide us with conditions on these field strengths such that they can be written in terms of the fundamental vector and scalar fields. The spin-1 field strength form a $\underset{\sim}{28}$ under SU(8). However, there can be no SU(8)-covariant spin-1

Bianchi identity from which we could deduce the existence of the spin-1 potentials. This strongly suggests that the theory must have an additional global symmetry. In order to find this symmetry we introduce the following objects

$$R^{ij}{}_{kl} = -2\,\delta^{[i}{}_{[k} R^{j]}{}_{l]} \tag{16}$$

$$\Omega^{ij}{}_{kl} = -2\,\delta^{[i}_{[k} \Omega^{j]}{}_{l]} \tag{17}$$

$$P_{Jklm} = E^M P_{MJklm} \tag{18}$$

$$F_{kl} = \tfrac{1}{2} E^N \wedge E^M F_{MN,kl} \tag{19}$$

where $\Omega^i{}_j (R^i{}_j)$ is the SU(8) part of the connection (curvature), and

$$\begin{aligned} P_{ujklm} &= \tfrac{1}{2} (\sigma_u)^{ab'} P_{ab'jklm} \\ P_{Ajklm} &= 2\,\delta^{i}_{[j} W_{klm]a} \\ P_{A'jklm} &= \tfrac{2}{4!} \epsilon_{ijklmnpq} \bar{W}_{a'}{}^{npq} \end{aligned} \tag{20}$$

$$\begin{aligned} F_{AB',kl} &= F_{A'B',kl} = F_{uA,kl} = 0 \\ F_{AB,kl} &= -2i\,\epsilon_{ab}\,\delta^{[ij]}_{[kl]} \\ (\sigma^u)_{aa'} F_{uB',kl} &= \epsilon_{a'b'} W_{jkl} \\ (\sigma^u)_{aa'} (\sigma^v)_{bb'} F_{uv,kl} &= -i\epsilon_{ab} \bar{N}_{a'b'kl} - i\epsilon_{a'b'} M_{abkl} \end{aligned} \tag{21}$$

with M as in (14) and

$$\bar{N}_{a'b',ij} = \tfrac{1}{72} \epsilon_{ijklmnpq} \bar{W}_{a'}{}^{klm} \bar{W}_{b'}{}^{npq} \tag{22}$$

Then as a consequence of the Bianchi identities and our solution, we find the following identities:

$$D P_{ijkl} = 0 \tag{23}$$

$$R^{ij}{}_{kl} = \bar{P}^{ijkl} \wedge P_{mnkl} \,. \tag{24}$$

$$DF_{ij} = \bar{F}^{kl} \wedge P_{ijkl} \tag{25}$$

We can write these equations in a compact form by introducing a block matrix notation. We put

$$\hat{\Omega} = \begin{pmatrix} Q & \bar{P} \\ P & \bar{Q} \end{pmatrix} \tag{26}$$

Equations (23) and (24) may then be written

$$d\hat{\Omega} - \hat{\Omega} \wedge \hat{\Omega} = 0 \tag{27}$$

This shows that $\hat{\Omega}$ is a pure gauge field and a closer investigation shows that the corresponding group is E_7. This is the global symmetry group found by Cremmer and Julia. We can now write

$$\hat{\Omega} = V^{-1} dV \tag{28}$$

where V is an element of E_7 and may be written as a 56x56 matrix

$$V = \begin{pmatrix} U^{IJ}{}_{ij} & \bar{V}^{IJ}{}_{ij} \\ V_{IJ\,ij} & \bar{U}_{IJ}{}^{ij} \end{pmatrix} \tag{29}$$

The global E_7 acts only on the capital indices and $\hat{\Omega}$ is invariant under transformations of this type. As has been explained by Cremmer and Julia, V may be used to convert local SU(8) indices to global E_7 indices. Hence we can use this matrix to convert the SU(8) covariant spin-1 field strengths to field strengths covariant under the global group. Then we can have Bianchi identities for those objects, which can imply the existence of spin-1 potentials. To do this we define a 56 of spin-1 2-forms under E_7 by

$$(F_{IJ}, \bar{F}^{IJ}) = (F_{ij}, \bar{F}^{ij})\, V^{-1} \tag{30}$$

In terms of F_{IJ}, the identity (25) is simply

$$dF_{IJ} = 0 \tag{31}$$

This is a correct Bianchi identity proving that F_{IJ} can be decomposed into spin-1 potentials. However, F_{IJ} is complex

$$F_{IJ} = G_{IJ} + i H^{IJ} \tag{32}$$

but the spin-1 part of F_{ij} is M_{abij} which is complex self-dual, and hence G and H are essentially dual versions of one another and so we really do have 28 spin-1 fields and not 56.

According to (29) there are 133 scalar fields in V. However, not all of them are dynamical, since we still have the SU(8) invariance, where the gauge fields are written in terms of components of V. Fixing the gauge completely fixes 63 scalar fields and leaves us with 70 dynamical scalars.

We have hence proven that our set of equations includes 28 dynamical spin-1 fields and 70 dynamical spin-0 ones, which are the correct numbers for N=8 supergravity.

CONCLUDING REMARKS

We have here exhibited the construction of the N=8 supergravity on-shell in superspace. The construction aimed at finding the theory corresponding to the one found in x-space by Cremmer and Julia and was much guided by that work. The next problem is to find the corresponding construction with the auxiliary fields included. The technique to use will be exactly the same as the one used here. However, this time we do not know a priori the total field content, if we do not know the auxiliary field content from some other source. This presents quite a hard problem. We also do not know the exact symmetries of the off-shell theory. This is another problem since in our construction the choice of local symmetry group in the tangent space was an important ingredient.

Finally, although we have concentrated exclusively on the N=8 theory, our approach is also applicable for smaller N, although for N<4, there may be additional complications due to the absence of scalar fields from which one could construct the composite SU(N) gauge potentials. Indeed, in these cases it is probably sufficient to use a global SU(N) group, as has been shown for the N=2 off-shell theory[10].

TABLE 1: CONVENTIONS

A. Indices

	Tangent space	Curved space
All	MNPQ	$\mathcal{m}\ \mathcal{n}$
Spinor & internal	ABCDEF	$\mathcal{A}\ \mathcal{B}$
Spinor	abcdef	$\alpha\ \beta$
Internal (SU(8))	ij..... t	$\xi\ \eta$
Vector	uvw.... z	$\mu\ \nu$
SL(8,R), E_7	IJKL	—

B. Summation convention & spinors

$$X^M Y_M = X^\mu Y_\mu + X^A X_A - X^{A'} Y_{A'}$$

$$X^A = X^a_i \quad ; \quad X^{A'} = X^{a'i}$$

$$Y_A = Y_a^i \quad ; \quad Y_{A'} = Y_{a'i}$$

$$X_{ab'} = (\sigma^\mu)_{ab'} X_\mu$$

$$(\sigma^\mu)_{ab'} = (1, \vec{\sigma}) \; ; (\sigma^{\mu\nu})_{ab} = \frac{1}{4}\{(\sigma^\mu)_{ac'}(\sigma^\nu)^{c'}{}_b - \mu \leftrightarrow \nu\}$$

$$X^a_i = \epsilon^{ab} X_{bi} \; ; \; \epsilon^{ab} = \epsilon_{ab} = \begin{pmatrix} 0 & 1 \\ -1 & 0 \end{pmatrix}$$

Metric $\eta_{\mu\nu} = (1,-1,-1,-1)$

$\epsilon_{\mu\nu xy} : \epsilon^{0123} = 1$

Complex conjugation replaces spinor indices by primed ones and raises or lowers internal ones, e.g. $(x^a_i)^+ = \bar{x}^{a'i}$.

TABLE 2: TORSIONS

$$T_{AB}{}^\mu = T_{AB}{}^C = T_{AB'}{}^{C'} = T_{A\mu}{}^\nu = T_{\mu\nu}{}^w = 0$$

$$T_{AB'}{}^\mu = -i(\sigma^\mu)_{ab'}\delta^i_j$$

$$T_{AB}{}^{C'} = \epsilon_{ab}\bar{W}^{ijkc'}$$

$$(\sigma^\mu)_{aa'} T_{\mu B}{}^{C'} = i\epsilon_{ab}\bar{M}_{a'}{}^{c'jk} + i\delta_{a'}{}^{c'} N_{ab}{}^{jk}$$

$$M_{ab\,ij} = \frac{1}{6} D^k{}_{(a} W_{b)ijk}$$

$$N_{ab}{}^{jk} = -\frac{1}{72}\epsilon^{jklmnpqr} W_{lmna} W_{pqrb}$$

$$(\sigma^\mu)_{aa'} T_{\mu B}{}^C = -\frac{i}{2}\epsilon_{ab} J^c{}_{a'}{}^j{}_k + \frac{i}{24}\delta^j_k[\epsilon_{ab} J^c{}_{a'} + \delta_a{}^c J_{ba'}]$$

$$J_{ab'}{}^i{}_j = \bar{W}_{b'}^{ikl} W_{jkla} ; \; J_{aa'} = J_{aa'}{}^k{}_k$$

$$(\sigma^\mu)_{aa'}(\sigma^\nu)_{bb'} T_{\mu\nu}{}^C = \epsilon_{ab}\bar{V}_{a'b'}{}^c{}_k + \epsilon_{a'b'}\{U_{ab}{}^c{}_k + \delta_a{}^c U_{bk} + \delta_b{}^c U_{ak}\}$$

TORSIONS: cont.

$$V^{i}_{abc'} = \frac{1}{2}\bar{W}_{c'}{}^{ilm} M_{lmab} + \frac{i}{3}\bar{P}^{ilmn}_{c'(a} W_{b)lmn}$$

$$\bar{U}^{i}_{c'} = -\frac{i}{18}\bar{P}^{ilmn}_{c'c} W_{lmn}{}^{c} - \frac{1}{54} A^{i}{}_{lmn}\bar{W}^{lmn}_{c'}$$

$$\bar{U}^{i}_{a'b'c'} = -\frac{1}{7}\bar{W}^{ilm}_{(a'}\bar{N}_{b'c')lm} - \frac{1}{7}\bar{D}_{(a'k}\bar{M}^{ik}_{b'c')}$$

$$P_{ba'ijkl} = -\frac{i}{2}\bar{D}_{a'i} W_{jklb}$$

$$A^{i}{}_{jkl} = -\frac{1}{2} D_{a}{}^{i} W_{jkl}{}^{a} = -\frac{1}{4!}\epsilon_{jklmnpqr}\bar{W}_{a'}{}^{imn}\bar{W}^{pqra'}$$

TABLE 3: CURVATURES

$$R_{AB,cd} = R^{ij}_{ab,cd} = \epsilon_{ab} N_{cd}{}^{ij}$$

$$R_{AB,c'd'} = \epsilon_{ab}\bar{M}_{c'd'}{}^{ij}$$

$$R_{AB',cd} = \frac{1}{2}\epsilon_{ac}\{J_{db'}{}^{i}{}_{j} - \frac{1}{6}\delta^{i}_{j} J_{db'}\} + \frac{1}{2}\epsilon_{ab}\{J_{cb'}{}^{i}{}_{j} - \frac{1}{6}\delta^{i}_{j} J_{cb'}\}$$

$$(\sigma^{v})^{bb'} R^{i}_{av,cd} = \frac{3i}{4}(\epsilon_{ac}\epsilon_{bd} + \epsilon_{ad}\epsilon_{bc})\bar{U}^{i}{}_{b'}$$
$$+\frac{i}{4}(\epsilon_{cd} V^{i}_{abb'} + 2\epsilon_{bd} V^{i}_{acb'} - 4\epsilon_{ad} V^{i}_{bcb'})$$

$$(\sigma^{v})^{bb'} R^{i}_{av,c'd'} = i\epsilon_{ab}\bar{U}^{i}_{b'c'd'} + \epsilon_{b'd'} R^{i}_{abc'} + \epsilon_{b'c'} R^{i}_{abd'}$$

$$R^{i}_{abc'} = \frac{i}{4} V^{i}_{abc'} - \frac{i}{4}\epsilon_{ab}\bar{U}^{i}{}_{c'}$$

$$(\sigma^{u})_{aa'}(\sigma^{v})_{bb'} R_{uv,cd} = \epsilon_{ab} B_{a'b'cd} + \epsilon_{a'b'} A_{abcd}$$

$$B_{a'b'cd} = \frac{1}{16}\Big[4 M_{cdkm}\bar{M}_{a'b'}{}^{km} + 4 N_{cd}{}^{km}\bar{N}_{a'b'km}$$
$$-\frac{2i}{3} D_{a'(c}\bar{W}_{b'}{}^{ijk} W_{d)ijk} + \frac{2i}{3}\bar{W}_{b'}{}^{ijk} D_{a'(c} W_{d)ijk}$$
$$-\frac{5}{4} J_{(ca'}{}^{k}{}_{m} J_{d)b'}{}^{m}{}_{k} + \frac{7}{144} J_{(ca'} J_{d)b'}$$
$$+\frac{2}{3} P_{da'ijkl}\bar{P}_{cb'}{}^{ijkl} + (a' \leftrightarrow b')\Big]$$

$$A_{abcd} = \frac{1}{16}\Big[2 D^{k}_{(c} U_{abd)k} + 4 N_{(ac}{}^{km} M_{bd)km}$$
$$+\epsilon_{ac} A_{bd} + \epsilon_{bc} A_{ad} + \epsilon_{ad} A_{bc} + \epsilon_{bd} A_{ac}\Big]$$

CURVATURES: cont.

$$A_{bd} = \frac{i}{12} D_{ba'} \bar{W}^{a'ijk} W_{dijk} - \frac{i}{12} \bar{W}^{a'ijk} D_{ba'} W_{dijk}$$
$$- \frac{i}{36} \bar{W}^{a'ijk} D_{da'} W_{bijk} + \frac{5}{288} \epsilon_{bd} J_{aa'}{}^{i}{}_{i} J^{aa'j}{}_{i}$$
$$- \frac{1}{1728} \epsilon_{bd} J_{aa'} J^{aa'} + \frac{1}{6} \epsilon_{bd} \bar{M}_{a'b'}{}^{ij} \bar{N}^{a'b'}{}_{ij}$$
$$- \frac{17}{32} \epsilon_{bd} M_{acij} N^{acij} + \frac{9}{16} M_{a(dij} N_{b)}{}^{aij}$$
$$- \frac{1}{9} \bar{P}_{d}{}^{c'ijkl} P_{c'bijkl}$$

$$R_{AB}{}^{k}{}_{l} = \epsilon_{ab} \bar{A}^{ijk}{}_{l} + \delta^{i}_{l} N_{ab}{}^{jk} + \delta^{j}_{l} N_{ab}{}^{ik}$$

$$R_{AB'}{}^{k}{}_{l} = - J_{ab'}{}^{ik}{}_{jl} + \frac{1}{2} \delta^{i}_{j} J_{ab'}{}^{k}{}_{l} - \frac{1}{2} \delta_{j}{}^{k} J_{ab'}{}^{i}{}_{l}$$
$$- \frac{1}{2} \delta^{i}_{l} J_{ab'}{}^{k}{}_{j} + \frac{1}{4} \delta^{k}_{l} J_{ab'}{}^{i}{}_{j} - \frac{1}{12} \delta^{i}_{j} \delta^{k}_{l} J_{ab'}$$
$$+ \frac{1}{6} \delta_{j}{}^{k} \delta^{i}_{l} J_{ab'}$$

$$J_{ab'}{}^{ik}{}_{jl} = \bar{W}_{b'}{}^{ikm} W_{jlma}$$

$$R_{Au}{}^{k}{}_{l} = - \bar{P}_{u}{}^{ikmn} W_{lmna} - \frac{1}{3} \delta^{i}_{l} \bar{P}_{u}{}^{kmnr} W_{mnra}$$
$$+ \frac{1}{6} \delta^{k}_{l} \bar{P}_{u}{}^{imnr} W_{mnr}$$

$$R_{uv}{}^{i}{}_{j} = \frac{2}{3} \bar{P}_{[u}{}^{ilmn} P_{v]lmn}$$

References

1. E. Cremmer and B. Julia, Phys. Letters 80B, 48 (1978) and Nucl. Phys. B159, 141 (1979).
2. E. Cremmer, B. Julia and J. Scherk, Phys. Letters 76B, 409 (1978).
3. S. MacDowell, Phys. Letters 80B, 212 (1979).
4. L. Brink and P. Howe, Phys. Letters 88B, 268 (1979).
5. J. Wess, Topics in Quantum Field Theory and Gauge Theories, Salamanca 1977, ed. J.A. de Azcárraga.
6. N. Dragon, Z. Phys. C2, 29 (1979).
7. M.F. Sohnius, Nucl. Phys. B136, 461 (1978).

8. J. Wess and B. Zumino, Phys. Letters 66B, 361 (1977);
R. Grimm, J. Wess and B. Zumino, Phys. Letters 73B, 415 (1978);
L. Brink, M. Gell-Mann, P. Ramond and J.H. Schwarz, Phys. Letters 74B, 336 (1978);
L. Brink, M. Gell-Mann, P. Ramond and J.H. Schwarz, Nucl. Phys. B145, 93 (1968);
W. Siegel and S.J. Gates Jr., Nucl. Phys. B147, 77 (1979).
9. J. Wess and B. Zumino, Phys. Letters 79B, 394 (1978).
10. B. de Wit and J.W. van Holten, Nucl. Phys. B155, 530 (1979);
E.S. Fradkin and M.A. Vassiliev, Lett. Nuovo Cimento 25, 79 (1979);
P. Breitenlohner and M. Sohnius, Nucl. Phys. to be published;
L. Castellani, P. van Nieuwenhuizen and S.J. Gates Jr., Stony Brook preprint ITP-SB-80-9 (1980).

SUPERGRAVITY AND COHOMOLOGY THEORY:

PROGRESS AND PROBLEMS IN D = 5

R. D'Auria, P. Fré and T. Regge

Istituto di Fisica Teorica, Università di Torino, Italy
and
Istituto di Fisica Nucleare, Sezione di Torino, Italy

In this contribution, emphasizing new developments, we plan to review the group manifold-rheonomic symmetry approach to supergravity[1,2] which has already been presented to other conferences[3]. In particular, we want to emphasize the central role of the mathematical concept of the graded Lie algebra cohomology class, which gives a constructive criterion for Lagrangians and which was not discussed in previous papers. The cohomological foundations of geometrical theories will be fully explained in a forthcoming paper[4].

The machinery of a geometrical field theory on a (super) group manifold is introduced as follows. Let G be a (super) group and $\mathbb{G}$ its (graded) Lie algebra. A basis of $\mathbb{G}$ is given by the generators T_A (A = 1, ..., n) which satisfy:

$$[T_A, T_B\} = C^L_{AB} T_L \tag{1}$$

On the manifold G, whose co-ordinates y^M are the group parameters themselves, we consider a $\mathbb{G}$ -valued 1-form (the <u>pseudo-connection</u>):

$$\mu = \mu^A T_A = dy^M \mu_M^{.A} T_A \tag{2}$$

which will be the fundamental field of the theory.

In order to write down an action for μ^A we have to consider two kinds of objects which can be constructed out of it. One is the curvature 2-form R:

$$R = d\mu + \mu\wedge\mu \Longrightarrow R^A \cdot T_A = R$$
$$R^A = d\mu^A + \frac{1}{2} C^A_{\cdot BC}\, \mu^B \wedge \mu^C \tag{3}$$

(when R = 0 we say that μ is a left-invariant 1-form); the other is the cochain ν^i. A cochain ν^i is a p-form with an index i in some finite-dimensional representation $D(T_A)$ of $\mathbb{G}$ such that it admits the following expansion

$$\nu^i = \nu^i_{A_1 \ldots A_p}\, \mu^{A_1} \wedge \ldots \wedge \mu^{A_p} \tag{4}$$

where $\nu^i_{A_1 \ldots A_p}$ are constant numbers. With respect to this definition we remark that any p-form can be expanded in the basis of the μ^A (which is complete) but in general its components will be functions of the y^M co-ordinates and not constants, as we have assumed to be the case for the cochain. We can perform two operations on the cochains: one is the covariant derivative which maps a p cochain into a p + 1 cochain

$$\nabla : \nu^i \mapsto \nabla\nu^i = d\nu^i + \mu^A \wedge D(T_A)^i_{\cdot j}\, \nu^j \tag{5}$$

where $D(T_A)^i_{\cdot j}$ is the matrix representation of the generator T_A; the other operation is the contraction $A\rfloor$ which maps p cochains into (p - 1) cochains. For every tangent vector $\vec{T}_A$ such that

$$\mu^A(\vec{T}_B) = \delta^A_B \tag{6}$$

and for every cochain (4) we define:

$$\underline{A}\rfloor \nu^i = p\, \nu^i_{A A_2 \ldots A_p}\, \mu^{A_2} \wedge \ldots \wedge \mu^{A_p} \tag{7}$$

Combining the contraction and the covariant derivative we also obtain a third operation which does not change the degree of the cochain and which is called the Lie derivative:

$$L_A \nu^i = \underline{A}\rfloor \nabla \nu^i + \nabla(\underline{A}\rfloor \nu^i) \tag{8}$$

These operations have several important formal properties and they all have a deep geometrical meaning which is basic to the discussion of Chevalley cohomology theory[5]; more will be said about them in a forthcoming paper[4]; here we just note that for any ν^i we get

$$\nabla\nabla\nu^i = R^A \wedge D(T_A)^i_{\cdot j}\, \nu^j \tag{9}$$

With these ingredients the action of our typical field theory will be the following:

$$\mathcal{A}[\mu] = \int_{M_{p+2}} R^A[\mu] \wedge \nu_A[\mu] \tag{10}$$

where the integration domain M_{p+2} is an arbitrary (p + 2) dimensional hypersurface of the manifold G, and the variational principle requires $\mathcal{A}[\mu]$ to be an extremum independently of the particular choice of M_{p+2}. From (10) we get the following equation of motion

$$\nabla\nu_A + (-)^{AB} R^B \wedge \underline{A}\rfloor \nu_B = 0 \tag{11}$$

which is an equation for (p + 1)-forms holding on the whole G-manifold. This latter statement means that the projection of (10) on any combination of (p + 1) tangent vectors $\vec{T}_{C_1}, \vec{T}_{C_2}, \ldots, \vec{T}_{p+1}$ is an equation of motion

$$\left(\nabla\nu_A + R^B \wedge \underline{A}\rfloor \nu_B (-)^{AB}\right)\left(\vec{T}_{C_1}, \ldots, \vec{T}_{C_{p+1}}\right) = 0 \tag{12}$$

Up to this point ν_A is fully arbitrary and therefore there is no criterion for selecting a particular Lagrangian. The criterion comes from the physical interpretation of the group G. In a theory which aims to be an extension of general relativity there must be a vacuum solution which corresponds to a flat space (= space without curvature) admitting symmetry under a group of motions. Excited states are the deformations of this flat space and are no longer symmetrical under the original group. The idea of geometrical theories on group manifolds is that G is indeed the group of motions of the vacuum which therefore corresponds to a left-invariant μ^A ($R^A = 0$). Such a physical requirement has the far-reaching consequence that $R^A = 0$ must be a solution of (11). This means

$$\nabla\nu_A = 0 \quad \text{if} \quad R^A = 0 \tag{13}$$

Now it is remarkable that (13) is preciyely the definition of a cocycle in Chevalley cohomology theory.

A cocycle is, in fact, a cochain which is covariantly closed, where closed means that its covariant derivative is zero modulo curvature. This immediately leads us into the realm of cohomology. In fact, calling coboundary an ω_A cochain which is covariantly exact, namely, is the covariant derivative of some other cochain:

$$\omega_A = \nabla\varphi_A \tag{14}$$

because of (9), we find that a coboundary is always a cocycle

$$\nabla\omega_A = \nabla\nabla\varphi_A = -C^B_{DA} R^D \wedge \varphi_B = 0 \text{ if } R^D = 0 \tag{15}$$

but the reverse is not always true. The equivalence classes of cocycles of degree p modulo the coboundaries of the same degree are called $H^p(\mathfrak{G},D)$, the pth cohomology group of $\mathfrak{G}$ in the D representation (in our case D is the coadjoint representation). These cohomology classes are in a finite and small number, and depend entirely on the structure of the (super) group. For instance, a fundamental theorem[5] states that for any faithful representation of G there are no non-trivial cohomology classes if G is semi-simple. Now it is of the utmost importance that the action (10) depends only on the cohomology class and not on the particular cocycle representing it. In fact, if to ν_A we add a coboundary

$$\nu_A' = \nu_A + \nabla \varphi_A \tag{16}$$

the new action is

$$\mathcal{A}'[\mu] = \mathcal{A}[\mu] + \int_{M_{p+2}} R^A \wedge \nabla \varphi_A \tag{17}$$

where the second term on the right-hand side of (17) is, due to the Bianchi identity $\nabla R^A \equiv 0$, a pure divergence:

$$\int R^A \wedge \nabla \varphi_A = \int d\left(R^A \wedge \varphi_A\right) \tag{18}$$

and therefore does not contribute to the equations of motion. Hence we conclude that the possible geometrical theories which can be constructed with a supergroup G are in one-to-one correspondence with its cohomology classes. In particular, the already quoted theorem on semi-simple groups would rule out theories based on them. This difficulty can be overcome with the introduction of a larger cohomology theory which we shall discuss elsewhere[4]. For the purpose of this talk, we limit our discussion to the case of a non-semi-simple G which already includes the examples of gravity, ordinary D = 4 supergravity, and of the five-dimensional theory we shall discuss in the second part of this paper.

Once the correspondence between Lagrangians and cohomology has been established, a further restriction on the domain of possible theories comes from the observed fact that reasonable theories, although not invariant under the full G. are, however, exactly gauge invariant under some subgroup $H \subset G$. In most cases, H is the Lorentz group. Therefore, besides being a representative of a cohomology class, ν_A must be such that the action (10) is invariant under gauge transformations of H:

$$\mu^A \longmapsto \mu^A + \nabla\epsilon^A \quad \begin{cases} \epsilon^K = 0 \\ \epsilon^H \neq 0 \end{cases} \tag{19}$$

(In (19) we have called H an index belonging to the subalgebra $\mathbb{H}$ of $\mathbb{G}$ and K an index belonging to the complement $\mathbb{K}$ of $\mathbb{H}$ in $\mathbb{G}$: $\mathbb{G} = \mathbb{H} \oplus \mathbb{K}$). Again the Chevalley theory comes to rescue us by supplying the concept of $\mathbb{H}$-orthogonal $\mathbb{G}$ cohomology classes. By definition, a $\mathbb{G}$ cohomology class is orthogonal to the subalgebra $\mathbb{H}$ if, for all its representatives ν_A, we have

$$\left.\begin{aligned} {}_{H}\rfloor \nu_A &= 0 \\ L_H \nu_A &= 0 \end{aligned}\right\} \text{ if } T_H \in \mathbb{H} \tag{20}$$

In a forthcoming paper it will be shown that the orthogonality of ν_A is sufficient to guarantee the gauge invariance of the action under H so that we can conclude by stating the following: "*For any pair* (G,H) *of a (super) group* G *and one of its sub-groups the possible geometrical theories are in one-to-one correspondence with the* $\mathbb{H}$*-orthogonal cohomology classes of* $\mathbb{G}$".

When ν_A is a cocycle its covariant derivative must be, by definition, proportional to the curvature and indeed we can show that :

$$\nabla\nu_A = R^B \wedge {}_B\rfloor \nu_A \tag{21}$$

Therefore, Eq. (11) becomes

$$R^B \wedge \left({}_{\underline{B}}\rfloor \nu_A + (-)^{AB} {}_{\underline{A}}\rfloor \nu_B \right) = 0 \tag{22}$$

which, when projected on all possible combinations of tangent vectors, becomes an algebraic equation for the intrinsic curvature components:

$$R^A_{\cdot BC} = R^A(\vec{T}_B, \vec{T}_C) \tag{23}$$

The essential features of the field theory described by our action (10) are determined by what sort of relations among $R^A_{\cdot BC}$ we get from (22). In order to discuss the various cases let us write the (graded) Lie algebra $\mathbb{G}$ in the following way:

$$\mathbb{G} = \mathbb{H} \oplus \mathbb{K} = \mathbb{H} \oplus \mathbb{I} \oplus \mathbb{O} \tag{24}$$

where $\mathbb{H}$ is the subalgebra and the complement $\mathbb{K}$ has been further decomposed into two subspaces which we shall call Inner and Outer, respectively.

First possibility

The only solution of (22) is $R^A_{\cdot BC} = 0$ for all values of A, B,C. In this case, the theory contains only the vacuum $R^A = 0$. It is a trivial theory.

Second possibility

Equation (22) admits solutions with some $R^A_{\cdot BC} \neq 0$. In all cases the theory is non-trivial but its properties are critically dependent on how many and which ones are the independent intrinsic component $R^A_{\cdot BC}$ parametrizing the most general solution of (22). The reason is that any theory on a group manifold, described by an action principle of type (10), being the action of a topological invariant, is symmetrical under an infinitesimal general co-ordinate transformation $y^M \mapsto y^M + \xi^M$. Such a transformation can be re-written as the following shift of the pseudo-connection $\mu^A \mapsto$ $\mapsto \mu^A + \delta\mu^A$, where (see (2)):

$$\delta\mu^A = \nabla\epsilon^A - 2\mu^F\epsilon^G R^A_{\cdot FG} \tag{25}$$

and

$$\epsilon^A = \xi^M \mu_M^{\cdot A} \tag{26}$$

It is apparent from (25) that, if some components $R^A_{\cdot F_1F_2}$ are determined to be zero, the co-ordinate transformation in the corresponding direction actually becomes a gauge transformation (19). Because of this, a theory based on an H-orthogonal cohomology class, which is automatically H-gauge invariant, must be H-factorized, namely (22) must imply:

$$H\text{-factorization} \Longleftrightarrow {}_{H}\rfloor R^A = 0 \quad \text{if } T_H \in \mathbb{H} \tag{27}$$

Therefore, for a geometrical theory based on ν_A of type (20), the only non-vanishing components are those in the directions of $\mathbb{K}$. However, not all these components are independent; we call II (= inner), the space spanned by those directions of $\mathbb{K}$ such that the components of the curvature along them are independent. The complementary space ◎ (= outer) is such that all the curvature components in such directions are just linear combinations of the inner components:

$$R^A_{OX} = \mathcal{C}^{A,I_1I_2}_{OX,B} R^B_{I_1I_2} \tag{28}$$

where χ is any index, 0 belongs to ◎, I_1, I_2 belong to II and $\mathcal{C}^{A,I_1I_2}_{0,x,B}$ are some constant coefficients.

Rheonomic symmetry

We say that a theory is rheonomic symmetrical when the subspace $\mathbb{II}$ can be identified with space-time and $\circledcirc$ is non-empty. Identification with space-time means that $\mathbb{II}$ is spanned by the translation generators whose number d matches the number of dimensions of the Lorentz group SO(1,D - 1) contained in H. For example, in ordinary D = 4 supergravity $\mathbb{G}$ is the graded Poincaré algebra, $\mathbb{H}$ the SO(1,3) Lorentz algebra and $\mathbb{K} = \mathbb{P} \oplus \mathbb{Q}$ the direct sum of the translations $\mathbb{P}$ and of the supersymmetries $\mathbb{Q}$. The theory is rheonomic symmetrical because $\mathbb{II} = \mathbb{P}$ which contains exactly four translations, and $\circledcirc = \mathbb{Q}$.

When a theory is rheonomic symmetrical Eq. (25), supplemented with Eq. (28), tells us that the restriction of theory to space-time $\mathbb{II}$ contains as many extra symmetries besides $\mathbb{H}$ as there are generators in $\circledcirc$. In fact, every $\circledcirc$ general co-ordinate transformation is, due to (28), a transformation which involves only the space-time fields and their derivatives. In the case of supergravity, as discussed in Refs 2) and 3), the extra symmetry (rheonomic symmetry) is supersymmetry: however, the fermionic character of the transformation is accidental. In the D = 5 supergravity, which we shall presently discuss, we encounter an example of bosonic rheonomic symmetry.

D = 5 supergravity

It is well known that a supersymmetric theory of gravitation is 5-space-time dimensions besides the graviton must contain a complex spin 3/2 gravitino and also a spin-1 field. This is in order to match the number of physical degrees of freedom in the bosonic and in the fermionic sector[6]. In fact, in D = 5 the graviton has five and the vector field three degrees of freedom which together make eight; on the other hand, eight is precisely the number of polarizations of the complex gravitino. In view of this, the minimal supergroup G apt to describe D = 5 supergravity must have the following 24 generators:

$$\begin{array}{lllcl} \text{Lorentz} & SO(1,4): & J_{ab} & = & 10 \; + \\ \text{Translations} & T_5: & P_a & = & 5 \; + \\ \text{Internal} & U(1): & Z & = & 1 \; + \\ \text{Supersymmetry} & \left\{\begin{matrix} Q \\ \bar{Q} \end{matrix}\right. : & & = & \underline{8} \; = \\ & & & & 24 \end{array} \tag{29}$$

It turns out that with these generators we can span the graded Lie algebra of SU(2,2|1) [7,8] or of one of its contractions. If we call ω^{ab}, V^a, B, ξ, $\bar{\xi}$ the components of the pseudoconnection μ, respectively, along J_{ab}, P_a, Z, Q, and $\bar{Q}$ the structure of SU(2,2|1) is given by the Maurer-Cartan equations which we obtain when the curvature is equal to zero in the following definitions:

$$
\begin{aligned}
R^{ab} &= d\omega^{ab} + \omega^{ac} \wedge \omega^{bd} \eta_{cd} + V^a \wedge V^b - i \bar{\xi} \wedge \Sigma^{ab} \xi \\
R^a &= dV^a - \omega^{ab} \wedge V^c \eta_{bc} - \frac{i}{2} \bar{\xi} \wedge \Gamma^a \xi \\
R^{\otimes} &= dB - i \bar{\xi} \wedge \xi \\
\rho &= d\xi + \frac{i}{2} \omega^{ab} \wedge \Sigma_{ab} \xi + \frac{i}{2} V^a \wedge \Gamma_a \xi - \frac{3i}{4} B \wedge \xi \\
\bar{\rho} &= \rho^{+} \Gamma_0
\end{aligned}
\qquad (30)
$$

where η_{ab} is the flat metric of D = 5 Minkowski space and Γ_a the D = 5 gamma matrices ($\Sigma_{ab} = i/4\, [\Gamma_a, \Gamma_b]$. SU(2,2 1) is a semi-simple supergroup and therefore we do not expect non-trivial cohomology classes. We can, however, obtain a non-semi-simple supergroup with the same number of generators performing the contraction. This is done by redefining

$$
\omega^{ab\prime} = \omega^{ab} ; \quad V^{a\prime} = e V^a ;
$$

$$
B' = eB ; \quad \xi' = \sqrt{e}\, \xi ; \quad R^{ab\prime} = R^{ab} ; \quad R^{a\prime} = e R^a ;
$$

$$
R^{\otimes\prime} = e R^{\otimes} ; \quad \rho' = \sqrt{e}\, \rho
$$

and performing the limit $e \to 0$ in Eq. (30). In this way we obtain the structural equations of the contracted non-semi-simple SU(2,2|1)

$$\begin{aligned}
R^{ab} &= \mathcal{R}^{ab} \\
R^{a} &= \mathcal{D}V^{a} - \frac{i}{2}\bar{\xi}\wedge\Gamma^{a}\xi \\
R^{\otimes} &= dB - i\,\bar{\xi}\wedge\xi \\
\rho &= \mathcal{D}\xi
\end{aligned} \tag{31}$$

where R^{ab} is the curvature of the SO(1,4) Lorentz subgroup $R^{ab} = d\omega^{ab} + \omega^{ac}\,\omega^{bd}\eta_{cd}$ and $\mathcal{D}V^a$ and $\mathcal{D}\xi$ are the SO(1,4) covariant derivatives of V^a and ξ, respectively:

$$\mathcal{D}V^{a} = dV^{a} - \omega^{ac}\wedge V^{d}\eta_{cd} \quad ; \quad \mathcal{D}\xi = d\xi + \frac{i}{2}\omega^{ab}\wedge\Sigma_{ab}\xi$$

Given the group and its curvature, in order to write an action we have to find the cohomology classes in the co-adjoint representation. To do this we have to know the form of the covariant derivative ∇ in such a representation. Let then, the action written as:

$$\mathcal{A} = \int\left\{-\frac{1}{2}R^{ab}\wedge\nu_{ab} - R^{a}\wedge\nu_{a} + \frac{3}{4}R^{\otimes}\wedge\nu_{\otimes} - \bar{m}\wedge\rho - \bar{\rho}\wedge m\right\} \tag{32}$$

It follows that the covariant derivative of the adjoint cochain $(\nu_{ab}, \nu_a, \nu_\otimes, m)$ of degree p is the following one:

$$\begin{aligned}
\nabla\nu_{ab} &= \mathcal{D}\nu_{ab} + V_a\wedge\nu_b - V_b\wedge\nu_a - i\left(\bar{\xi}\Sigma_{ab}m - (-)^{p}\bar{m}\Sigma_{ab}\xi\right) \\
\nabla\nu_{a} &= \mathcal{D}\nu_{a} \\
\nabla\nu_{\otimes} &= d\nu_{\otimes} \\
\nabla m &= \mathcal{D}m - \frac{i}{2}\Gamma^{a}\xi\wedge\nu_{a} + \frac{3}{4}i\,\xi\wedge\nu_{\otimes}
\end{aligned} \tag{33}$$

Now as the theory we want to construct has got to include five-dimensional gravitation, ν_{ab} must have, in addition, the Einstein term $\varepsilon_{abijk}V^i\wedge V^j\wedge V^k$ (in fact the component of the pseudoconnection along the translation generator P_a is to be identified with the fünfbein). This means that ν should be the most general cohomology class of order three containing the Einstein term. By explicit computations we have determined the complete cohomology group of order three of G = $\overline{SU(2,2|1)}$ orthogonal to the Lorentz group H = SO(1,4) [4]. It turns out to be composed of four elements so that the most general ν, which is a linear combination of these elements, contains three arbitrary parameters (in fact the over-all constant in front of the ν is irrelevant). Explicitly we find

$$\nu_{ab} = -\epsilon_{abijk}V^i\wedge V^j\wedge V^k + (\alpha_1-3)V_a\wedge V_b\wedge B + 2\alpha_2\,\bar{\xi}\wedge\Sigma_{ab}\xi\wedge B + 2\alpha_3\,\bar{\xi}\wedge\Gamma_{[a}\xi\wedge V_{b]}$$

$$\nu_a = \alpha_1\left(\frac{i}{4}\bar{\xi}\wedge\Gamma_a\xi\wedge B - \frac{i}{2}\bar{\xi}\wedge\xi\wedge B\right)$$

$$\nu_\otimes = i\,\bar{\xi}\wedge\xi\wedge B - 2i\,\bar{\xi}\wedge\Gamma_a\xi\wedge V^a$$

$$\mathcal{M} = -3\Sigma_{ab}\xi\wedge V^a\wedge V^b + \frac{i}{2}\left(3-\frac{\alpha_1}{2}\right)\Gamma_m\xi\wedge V^m\wedge B + i\alpha_2\Sigma_{ab}\xi\wedge V^a\wedge V^b \tag{34}$$

In a recent paper[9], two of us have studied the bosonic limit of (34) as an independent theory. In this limit one sets $\xi = 0$ and disregards the corresponding Q generators. In this way G becomes $ISO(1,4) \otimes U(1)$, namely, the direct product of the Poincaré group in D = 5 times a U(1) internal group. The multiplet ν reduces to $\nu_{ab} = -\varepsilon_{abijk} V^i \wedge V^j \wedge V^k + \text{cost}\, x V_a \wedge V_b \wedge B$ which is the most general SO(1,4) orthogonal cohomology class of order three for $ISO(1,4) \otimes U(1)$. The corresponding action:

$$\mathcal{A} = \int \left\{ R^{ab} \wedge V^i \wedge V^j \wedge V^k \epsilon_{abijk} + \text{cost}\, R^{ab} \wedge V_a \wedge V_b \wedge B \right\} \tag{35}$$

is, as it should be, gauge invariant under SO(1,4) but not under U(1). From the equations of motion, however, it follows that the theory is factorized and rheonomic symmetrical. In fact, the components of the curvature along SO(1,4) are all zero while on the other hand, we have:

$$\begin{aligned} R^{a}_{\cdot bc} &= \text{cost}\, \eta_{bm} \eta_{cn} \epsilon^{amnpq} R^{\otimes}_{\cdot pq} \\ R^{a}_{\cdot \otimes m} &= \text{cost}\, \eta^{ab} R^{\otimes}_{bm} \\ R^{\otimes}_{\cdot \otimes m} &= 0 \\ R^{rs}_{\cdot \otimes k} &= \text{cost}\, \epsilon^{rsabc} R^{it}_{\cdot bc} \eta_{ik} \eta_{at} \end{aligned} \tag{36}$$

These equations tell us that the independent curvature components are $R^{\otimes}$ pq and $R^{ab}_{\cdot}$ pq. All the other components can be expressed in terms of these. Such an occurrence is indeed what we named rheonomic symmetry and it guarantees that the original theory (35) restricted to the inner subspace II (spanned by the $5 V^a$) admits an extra U(1) symmetry whose infinitesimal form is:

$$\begin{aligned}
\delta V^a &= \mathrm{cost} \times \epsilon^{\otimes} V^c R^{\otimes}{}_{\cdot bc}\, \eta^{ab} \\
\delta B &= d\epsilon^{\otimes} \\
\delta \omega^{ab} &= \mathrm{cost} \times \epsilon^{\otimes} V^k \epsilon^{abrst} R^{ij}{}_{|st}\, \eta_{ik} \eta_{rj}
\end{aligned} \tag{37}$$

The meaning of this symmetry transformation becomes apparent when one turns from the first to the second order formalism for the space-time restriction of the theory (35). The transition to the second order description is obtained by the feed-back into (35) of Eqs (36) which can be solved for the spinor connection ω_μ^{ab} in terms of V_μ^a, B_μ and their derivatives. Once this is done the resulting second order action is

$$\begin{aligned}
A^{(2nd\ order)} = \int \Big\{ (\det V)\, R^{\mu\nu}_{\mu\nu} + \mathrm{cost}\ F_{\mu\nu} F^{\mu\nu} + \\
+ \mathrm{cost}'\, \epsilon^{\lambda\mu\nu\rho\sigma} F_{\lambda\mu} F_{\nu\rho} B_\sigma \Big\}\, d^5x
\end{aligned} \tag{38}$$

where $R^{\mu\nu}_{\mu\nu}$ is the usual curvature scalar and $F_{\mu\nu} = 1/2(\partial_\mu B_\nu - \partial_\nu B_\mu)$. This theory is obviously invariant under the transformation

$$\delta V^a_\mu = \mathrm{cost}\ \epsilon^{\otimes}\, V^{a|\nu} F_{\nu\mu}\,; \quad \delta B_\mu = \partial_\mu \epsilon^{\otimes}$$

which is the component transcription of (37) and quite remarkably exhibits the trilinear coupling of the spin 1 field which is a well-established feature of D = 5 supergravity[10].

The conclusion is that the cohomology argument has reproduced the correct bosonic sector of D = 5 supergravity in the same way as it has reproduced D = 4 supergravity. It is therefore very surprising that the complete theory based on the most general cohomology class (34) admits only the vacuum solution (all components of the curvature equal to zero) for all values of the parameters.

This result, which will be fully discussed in Ref. 4), seems to suggest that some of the existing second order supersymmetric theories have no first order parents on the group manifold and this might be the explanation why no action in superspace has been found for them.

REFERENCES

1. Y. Ne'eman and T. Regge, Rivista del Nuovo Cimento 1, number 5, p.1 (1978).
2. A. D'Adda, R. D' Auria, P. Fré and T. Regge, "Geometrical Formulation of Supergravity Theories on Orthosymplectic Supergroup Manifolds", to be published in Rivista del Nuovo Cimento (1980).
3. A. D' Adda, R. D'Auria, P. Fré and T. Regge, "Gauge Theories of Supergravity on Orthosymplectic Supergroup Manifolds", Proceedings Firenze Workshop (June 1979), to be published by J. Hopkins University;
   A. D'Adda, R. D'Auria, P. Fré and T. Regge, "Geometrical Formulation of Supergravity as a Theory on a Supergroup Manifold", in "Supergravity", Proc. of the Stony Brook Workshop, eds P. van Nieuwenhuizen and D.Z. Freedman, North-Holland Publ. Co., Amsterdam,(1979), p. 85
4. R. D'Auria, P. Fré and T. Regge,"Graded Lie Algebra Cohomology and Supergravity", forthcoming paper.
5. C. Chevalley and S. Eilenberg, Trans. Amer. Math. Soc. 63 (1948) 85.
6. E. Cremmer, B. Julia and J. Scherk, Phys. Letters 76B (1978) 409.
7. P. Freund and I. Kaplanski, J. Math. Phys. 17 (1976) 228.
8. J. Lukierski, Wroclaw preprint 436 (1978)
9. R. D'Auria and P. Fré, Torino preprint IFTT 323, Nucl. Phys. B, in press (1980).
10. E. Cremmer, J. Scherk and J.H. Schwarz, Phys. Letters 84B (1979) 83.

SUPERSYMMETRIC YANG-MILLS THEORIES

Martin F. Sohnius and Kellogg S. Stelle

The Blackett Laboratory, Imperial College,
London, SW7 2BZ

and

Peter C. West

Department of Mathematics, King's College
London WC2R 2LS

A comprehensive discussion of the off-shell structure of supersymmetric gauge theories is presented. We review various methods for finding the auxiliary fields which are necessary to permit a manifest realization of supersymmetry in a field theory.

This article covers the topics treated in separate presentations by the three authors, based mainly on the joint work of references [1, 2].

Chapter 1 begins with an introduction that explains the importance of manifestly supersymmetric formalisms for the study of supersymmetric quantum theories and outlines the two equivalent manifestly supersymmetric formalisms. Two methods of obtaining these off-shell formalisms are described. One gives an algebraic derivation of the superspace constraints necessary for the formulation of supersymmetric theories, the other uses a knowledge of on-shell states and an analysis of ghost excitations in higher derivative Lagrangians to find the auxiliary fields necessary for off-shell and manifestly supersymmetric theories in Minkowski space. These techniques are used to find the off-shell formulation of the N=4 Abelian supersymmetric gauge theory.

Chapter 2 describes a method for generating off-shell formulations of any extended supersymmetric theory. The method is based upon the canonical formalism of mechanics and constructs the action of a four-dimensional theory by Legendre transformation: it is the

Hamiltonian of a five-dimensional theory. Examples given in detail are the scalar hypermultiplet of N=2 supersymmetry and the N=4 supersymmetric Yang-Mills theory.

Chapter 3 describes how the various supersymmetric Yang-Mills theories are generated from the minimal theory in ten space-time dimensions. The effects of dimensional reduction by compactification and Legendre transformation are studied with particular emphasis on the structure of the generated internal symmetry. A diagrammatic overview of the relationships between all on- and off-shell supersymmetric Yang-Mills theories is presented, and the superspace formulations for the N=2 and N=4 theories in four dimensions are given.

In the Conclusion, we summarize open problems and indicate possible applications of the presented methods in supergravity.

1. THE ROLE OF CENTRAL CHARGES IN EXTENDED SUPERSYMMETRIC THEORIES*)

1.1 Introduction

In this and the following chapters we present several methods of finding representations of the supersymmetry group. With the discovery of these representations we are able to formulate supersymmetric theories in a way that makes the supersymmetry manifest. To stress the importance of these formulations, we shall first consider their role in answering some of the interesting questions of supersymmetry at the present time.

(i) Does supersymmetry lead to the unification of gravity and quantum mechanics ? It is well known that there is a considerable reduction in the number of divergences of supersymmetric theories. Part of this reduction is due to the fact that renormalization respects the supersymmetry, and so only supersymmetric counter-terms are allowed. However, there is also the phenomenon of "miraculous" cancellations. These are situations where, although there exist supersymmetric counter-terms, their coefficients are zero. An example of this restriction of counter-terms by supersymmetry is the case of two-loop simple supergravity where there exists no supersymmetric counter-term. However, at three loops there does exist a supersymmetric counter-term and it would require a "miraculous" cancellation for its coefficient to be zero. The analogous situation for the larger extended theories is unknown and it is hoped that the additional symmetry could lead to more promising results. What is required in these theories is a

* This chapter corresponds to the lecture given by P.C.W.

tensor calculus to construct the possible supersymmetric counter-terms as well as a technique for calculating the Feynman graphs necessary for evaluation of the coefficients of these counter-terms.

(ii) Does supersymmetry lead to a finite four dimensional field theory ? The maximally extended supersymmetric Yang-Mills theory has a β-function which vanishes up to the two-loop order. This is an example of the "miraculous" cancellations referred to above and it will be extremely interesting to know whether the β- function is zero to all orders.

(iii) Does supersymmetry lead to a unification of all the forces of nature ? The maximally extended supergravity theory contains particles of all spins from spin 0 to spin 2.

It is unknown how the fundamental particles that occur in this theory could correspond to those in Nature, although it is clear that there are not a sufficient number of them to correspond to all of the quarks, leptons and gauge bosons now seen. It has been suggested that the particles now observed could be accounted for by bound states of this theory. This hope has been encouraged by the discovery that the theory possesses an SU(8) symmetry and by the further speculation that the possible bound states fall into representations of this SU(8) symmetry.

None of these three questions was voiced in the early investigations of supersymmetry and it is possible that the aspirations for supersymmetry will change further as the subject develops. However, in order to answer these questions we need to be able to calculate effectively, particularly at the quantum level.

To perform all but the simplest of such calculations requires a formulation of these theories in which the supersymmetry is manifest. There are two reasons for this. First, from a practical viewpoint, calculations performed in a non-supersymmetric way are very lengthy because one must take into account every particle of the theory separately. Calculating in a manifestly supersymmetric fashion, on the other hand, treats many particles as one entity. For example, the maximally extended supergravity theory contains particles of many different spins, but a manifestly supercovariant formulation would treat them as just one entity. Second, if one does not work in a manifestly supersymmetric formalism one cannot guarantee that the quantum theory possesses the same symmetries as the classical theory. This is particularly true in the cases where the fields of the theory carry a representation of the supersymmetry group only if the classical field equations are imposed. Unfortunately, the maximally extended supergravity is formulated in just such a way and, until recently, so was the maximally extended Yang-Mills theory.

The above considerations are somewhat similar to the need for manifestly Lorentz covariant theories. One can imagine the difficulties encountered in Yang-Mills theories if we regarded each component of the vector potential as distinct rather than as a four-vector. At the quantum level the development of Lorentz-covariant Feynman rules or path integrals was necessary to be able to calculate with speed and confidence.

There are two ways in which supersymmetric theories can be formulated in manifestly supersymmetric fashion. The first method relies upon finding sets of component fields whose transformations form a representation of the supersymmetry group. Given such sets of fields (supermultiplets) one can find, as with any other symmetry, rules for combining these supersymmetric multiplets together to form new supermultiplets, and rules for constructing invariants. This method, called tensor calculus, does not make use of extra coordinates beyond those of ordinary space-time. The difficulty in this construction is the initial step, that of finding the supermultiplets. This difficulty stems from a phenomenon unique to supersymmetry. Namely, given a particular theory in which the fields form a representation of the supersymmetry group on-shell (that is, when the classical field equations of the theory hold) it is necessary to add additional fields to the on-shell fields in order to find a representation of the supersymmetry group without the use of any field equations. These additional fields are called auxiliary fields and will not propagate in the particular theory being considered, although they may propagate in other theories constructed from the same supermultiplet. The auxiliary fields are often very difficult to find and their absence represents the main stumbling block to the further development of supersymmetric theories.

The second method is to work in a space having additional fermionic dimensions, called superspace. This construction will be explained later. Although in a superspace formulation one always has a representation of the algebra, there is a difficulty corresponding to that of finding the auxiliary fields, namely the necessity of finding superspace constraints.

These two formulations are equivalent and we wish to stress that there is now a well known path from one formulation to the other. Given a theory formulated in superspace we can find the auxiliary fields in ordinary space and then the associated tensor calculus. Conversely, given a theory with its auxiliary fields and hence a tensor calculus formulation, we can, in a straight-forward way, use the technique called gauge completion to find its superspace formulation. It is worth noting that although almost all theories were first found in their auxiliary field formulation, it is the superspace formulation which allows the easy calculation of quantum processes.

Although we know manifestly supersymmetric formulations for all interesting theories which have only one or two supersymmetry generators, up until recently we have not known such formulationss for any of the more interesting supersymmetric theories with more than two spinorial generators. However, all these theories have been given in terms of fields which do not carry a representation of the algebra and so can clearly not be expressed in a manifestly supersymmetric way. As a measure of the difficulty of the problem it is worth noting that it was two years after the discovery of supersymmetry that a formulation of simple supergravity without auxiliary fields was found. It required another two years to find a formulation with a simple set of auxiliary fields.

It is the purpose of this article to describe ways of finding manifestly supersymmetric formulations. In particular, we will give such a formulation of the theory which may be finite - the maximally extended Yang-Mills theory. These methods are worth studying in their own right as they give considerable insight into the structure of supersymmetric theories. This chapter is divided into two parts, the first part is concerned with a method of finding the superspace formulation of supersymmetric theories. The second part works in ordinary space and uses an examination of ghost excitations to predict auxiliary fields. The auxiliary fields of the maximally extended Yang-Mills theory are found in this way. Since these formulations are equivalent, we will encounter similar ideas in the two parts and will see the considerable importance of central changes.

Before moving on to the first part we will describe the supersymmetry group on which all discussions are based. This group occupies a unique place in theoretical physics; it is the only known group that combines space-time and internal symmetries in a Lorentz-covariant way. That is, it is the only known non-trivial extension of the Poincaré group to include internal symmetries. It has been shown that no Lie group could achieve this result [3]. The algebra associated with the supersymmetry group has anti-commutators as well as commutators[4]. It is a graded Lie algebra.

The generators of this algebra are those of the translations P_μ, and of the Lorentz rotations, $J_{\mu\nu}$ of the Poincaré group, N spinorial charges, Q_α^i, $Q_{\dot\beta j}$ $(i, j = 1 \rightarrow N)$, the central charges Z^{ℓ} and the internal symmetry generators, T_a. Here we use two-component spinor notation. The graded Lie algebra is

$$\{Q_\alpha{}^i, \bar{Q}_{\dot\beta j}\} = 2\delta^i_j (\sigma^\mu)_{\alpha\dot\beta} P_\mu, \{Q_\alpha{}^i, Q_\beta{}^j\} = 2\epsilon_{\alpha\beta}(a^\ell)^{ij} Z_\ell$$

$$\{\bar{Q}_{\dot\alpha i}, \bar{Q}_{\dot\beta j}\} = 2\epsilon_{\dot\alpha\dot\beta}(a^{\ell *})_{ij} Z_\ell, [Q_\alpha{}^i, P_\mu] = 0 = [\bar{Q}_{\dot\beta j}, P_\mu]$$

$$[Q_{\alpha i}, J_{\mu\nu}] = \frac{i}{2}(\sigma_{\mu\nu})_\alpha{}^\beta Q_{\beta i}, [T_a, Q_\alpha{}^i] = f_a{}^i{}_j \quad Q_\alpha{}^j$$

$$[T_a, T_b] = i f_{ab}{}^c T_c \quad , [Z^\ell, Q_\alpha{}^i] = 0 = [Z^\ell, \bar{Q}_{\dot\beta j}] \tag{1.1}$$

where

$$(a^\ell)^{ij} = -(a^\ell)^{ji}, (a^{\ell *})_{ij}$$

is the complex conjugate of $(a^e)^{ij}$ and

$$(\sigma^{\mu\nu})_\alpha{}^\beta = -\frac{1}{2}((\sigma^\mu)_{\alpha\dot\delta}(\sigma^\nu)^{\beta\dot\delta} - (\mu \leftrightarrow \nu))$$

plus those for the Poincaré group.

The central charges Z^ℓ commute with all the generators of the graded Lie algebra and only occur in the anticommutator of two undotted (or dotted) spinorial charges[5]. It is quite consistent to set them to zero without affecting the rest of the algebra, (a Wigner-Inönü contraction). This result together with the historical accident that most experience with supersymmetry is based on the N = 1 case, where there are no central charges, lead to the feeling that central charges were bizarre objects not encountered in the usual run of events. We wish to show that central charges are essential in most situations and setting them to zero in the algebra at the outset will, more often than not, make vital constructions impossible. It is interesting to note that if one wishes to combine space-time and internal symmetries one has to accept not only the spinorial generators of supersymmetry but also a bosonic central charge generator. We will later show that the non-vanishing of this generator is equivalent to a dependence of the fields on a fifth dimension. The central charge generates translations in this fifth dimension. This is in contrast to the mechanism of dimensional reduction which uses extra dimensions to generate internal symmetries. The internal symmetry generators act on the spinorial generators. What the internal group is depends upon the particular system under study and whether or not that system has a central charge.

1.2 Superspace, Constraints and the Role of Central Charges

1.2.1 Superspace

The N = 1 superspace [6] is an eight-dimensional space labelled by coordinates $x^\mu, \theta^\alpha, \theta^{\dot\beta}$. The x^μ are the commuting coordinates of Minkowski space and the $\theta^\alpha, \theta^{\dot\beta}$ are additional anticommuting coordinates. We refer the reader to the lecture by B. Zumino for the details of this construction. The supersymmetry is realized by the following transformations

$$x^\mu \rightarrow x^\mu + i\theta^\alpha(\sigma^\mu)_{\alpha\dot\beta}\zeta^{\dot\beta} - i\zeta^\alpha(\sigma^\mu)_{\alpha\dot\beta}\theta^{\dot\beta},$$

$$\theta^\alpha \rightarrow \theta^\alpha + \zeta^\alpha \quad ; \quad \theta^{\dot\beta} \rightarrow \theta^{\dot\beta} + \zeta^{\dot\beta}, \tag{1.2}$$

which form a representation of the supersymmetry algebra. Superspace is in fact the coset space of the super-Poincaré group divided by the Lorentz group; just as Minkowski space is the coset space of the Poincaré group divided by the Lorentz group. The transformations of equation (1.2) are the natural action of the group on its coset.

The ordinary derivatives on this space are not covariant under supersymmetry; the covariant derivatives are

$$D_\mu = \partial_\mu \,,\; D_\alpha = \frac{\partial}{\partial\theta^\alpha} + i(\not\partial)_{\alpha\dot\beta}\theta^{\dot\beta} \,,\; D_{\dot\beta} = -\frac{\partial}{\partial\theta^{\dot\beta}} - i\,\theta^\alpha(\not\partial)_{\alpha\dot\beta}, \tag{1.3}$$

and are denoted collectively by D_A. This structure reflects the fact that superspace is a space with zero curvature but non-zero torsion.

The N = 2 superspace is a thirteen dimensional superspace labelled by coordinates $x^\mu, z, \theta^\alpha{}_i, \theta^{\dot\beta j}$. The x^μ are the bosonic coordinates of Minkowski space and the θ's are the fermionic coordinates. The z is a commuting (bosonic) coordinate and corresponds to the presence of a central charge in the supersymmetry algebra. This space is the coset space of the N = 2 super-Poincaré group divided by the Lorentz group. A realization of the N = 2 supersymmetry algebra is provided by the following set of transformations on this space.

$$x^\mu \rightarrow x^\mu + \omega^\mu{}_\nu x^\nu + a^\mu + i\theta^\alpha{}_j(\sigma^\mu)_{\alpha\dot\beta}\zeta^{\dot\beta j} - i\zeta^\alpha{}_j(\sigma^\mu)_{\alpha\dot\beta}\theta^{\dot\beta j}$$

$$z \rightarrow z + \omega + i a\,\zeta^\alpha_j\theta_\alpha{}^j + i a^*\theta^{\dot\beta j}\zeta_{\dot\beta j}$$

$$\theta^\alpha{}_i \rightarrow \theta^\alpha{}_i + \frac{i}{4}\omega^{\mu\nu}(\sigma_{\mu\nu})^\alpha{}_\beta\theta^\beta{}_i + \zeta^\alpha{}_i$$

$$\theta^{\dot\beta j} \rightarrow \theta^{\dot\beta j} + \frac{i}{4}\omega^{\mu\nu}(\bar\sigma_{\mu\nu})^{\dot\alpha}{}_{\dot\beta}\theta^{\dot\beta j} + \zeta^{\dot\beta j} \tag{1.4}$$

where a is an arbitrary constant.

The corresponding covariant derivatives are denoted collectively by D_A and are given by

$$D_\alpha{}^i = \frac{\partial}{\partial\theta^\alpha{}_i} + i(\not\partial)_{\alpha\dot\beta}\theta^{\dot\beta i} - ia\theta_\alpha{}^i\frac{\partial}{\partial z}, \quad D_{\dot\beta j} = -\frac{\partial}{\partial\theta^{\dot\beta j}} - i\theta^\alpha{}_j(\not\partial)_{\alpha\dot\beta} - ia^*\theta_{\dot\beta j}\frac{\partial}{\partial z}.$$

$$D_\mu = \partial_\mu \quad , \quad D_z = \frac{\partial}{\partial z} \equiv \partial_z \quad . \tag{1.5}$$

These covariant derivatives obey the relations

$$\{D_\alpha{}^i, D_{\dot\beta j}\} = -2i(\sigma^\mu)_{\alpha\dot\beta}\delta_j{}^i\partial_\mu, \quad \{D_\alpha{}^i, D_\beta{}^j\} = -2ia\epsilon_{\alpha\beta}\epsilon^{ij}\partial_z$$

$$\{D_{\dot\alpha i}, D_{\dot\beta j}\} = +2ia^*\epsilon_{\dot\alpha\dot\beta}\epsilon_{ij}\partial_z, \quad [D_\alpha{}^i, \partial_\mu] = 0 = [D_{\dot\beta j}, \partial_\mu]. \tag{1.6}$$

The presence of the additional bosonic coordinate, z, is somewhat surprising in view of the fact that we must be able to extract from the superspace theory a theory in ordinary Minkowski space with an invariant action. The fact that the Taylor expansion in the fermionic coordinates of the superfield terminates eventually allows us to work with just the coefficients of this expansion. The same argument cannot be applied to the bosonic coordinate z. In particular, the four dimensional action of Minkowski space will have to be invariant under the central charge transformations, which are translations in the z coordinate. How to achieve this will be explained in Chapter 2.

1.2.2 Superspace formulation of N = 1 Yang-Mills theory

Before moving to the general method of constructing supersymmetric theories we will illustrate the difficulties encountered and their subsequent resolution in the context of a particular model, the N = 1 Yang-Mills theory [7]. The theory's spectrum contains one spin-1 and one spin-$\frac{1}{2}$ massless particle.

The Lagrangian in ordinary space was found to be

$$\mathcal{L} = -\frac{1}{4}F_{\mu\nu}{}^2 - \frac{i}{2}\bar\lambda\not\partial\lambda + \frac{1}{2}D^2 \tag{1.7}$$

where $F_{\mu\nu} = \partial_\mu A_\nu - \partial_\nu A_\mu$.

It is invariant under the supersymmety transformations

$$\delta A_\mu = i\bar\xi\gamma_\mu\lambda$$

$$\delta\lambda = -\sigma^{\mu\nu}F_{\mu\nu} + \gamma_5 D\xi$$

$$\delta D = i\bar\xi\gamma_5\not\partial\lambda \quad . \tag{1.8}$$

These transformations form a representation of the algebra of equation (1.1) (when augmented by gauge transformations). Although there are three fields A_μ, λ and D in the Lagrangian, only

A_μ and λ give rise to on-shell states. The field D is of dimension two, does not propagate and vanishes on-shell. Such fields are called auxiliary fields. If we set D to zero, then the transformations become

$$\delta A_\mu = i \bar{\xi} \gamma_\mu \lambda \quad , \quad \delta \lambda = - \sigma^{\mu\nu} F_{\mu\nu} \xi . \tag{1.9}$$

These transformations still form an invariance of the action of equation (1.7) if D is set to zero. However, the transformations of equation (1.9) only form a representation of the algebra when the equations of motion hold, that is on-shell. The need for auxiliary fields to provide a representation of the group algebra on fields is a phenomenon unique to supersymmetry and, except in very rare cases, necessary in all supersymmetric theories. Without these fields we do not possess a representation of the group and consequently cannot make use of the usual group theoretic concepts to construct actions and make the symmetry manifest.

There is a simple argument which demonstrates the need for auxiliary fields in general supersymmetry theories. It follows from the supersymmetry algebra that in a supersymmetric theory where the fields do form a representation, the number of fermionic fields equals the number of bosonic fields. Since imposing the equations of motion is a supersymmetric procedure, there are the same number of fermionic and bosonic on-shell states. For example, in the theory above we have on-shell two bosonic degrees of freedom for A_μ and two fermonic degrees for λ . Off-shell, we have three for A_μ (4 minus 1 gauge component), one for D and four for λ_α . The auxiliary field D is necessary to make up the four bosonic fields required to balance the four fermionic ones. In general, we can see that the auxiliary fields are necessary to account for the fact that Bose and Fermi fields yield different numbers of degrees of freedom when one goes on-shell.

Knowing the auxiliary fields we could use the method of "gauge completion" [8] to find the superspace formulation of the N = 1 Yang-Mills theory. However, we now illustrate, in the context of this theory, a general method of finding the superspace formulations which does not require a knowledge of auxiliary fields. We work entirely in superspace, so all our fields are functions of x^μ, θ^α and $\theta^{\dot\beta}$. In analogy with ordinary space we covariantize the supersymmetric covariant derivatives, with respect to the Yang-Mills group, [9]; namely

$$\mathcal{D}_A = D_A - A_A \cdot Y , \tag{1.10}$$

where Y are the Yang-Mills generators. The field strengths, F_{AB} are defined by

$$\{ \mathcal{D}_A , \mathcal{D}_B \}_\pm = - T_{AB}{}^C \mathcal{D}_C - F_{AB} \cdot Y , \tag{1.11}$$

where $T_{AB}{}^{C}$ are zero except for $T_{\alpha\dot{\beta}}{}^{\mu} = T_{\dot{\beta}\alpha}{}^{\mu} = +2i\,(\sigma^{\mu})_{\alpha\dot{\beta}}$

It follows that

$$F_{BC}\cdot Y = +\mathcal{D}_B A_C\cdot Y - \mathcal{D}_C A_B\cdot Y(-1)^{BC} - \{A_B\cdot Y, A_C\cdot Y\}_{\pm} + T_{BC}{}^{P}A_P \tag{1.12}$$

The final term in this expression has no analogue in ordinary space and corresponds to the fact that superspace has non-zero torsion. The object $T_{AB}{}^{C}$ is in fact the torsion tensor.

Unfortunately, at this point the Minkowski space analogue is no longer useful. The superfield $A_B(x^{\mu}, \theta^{\alpha}, \bar{\theta}^{\dot{\beta}})$ contains many component fields; in particular it contains fields of spin greater than one.

The appearance of high spins in the formalism is a symptom of a much more serious disease. This disease is the inability of some of the irreducible representations of rigid supersymmetry to survive in the presence of the local coupling. In fact, the ability to reduce A_B into separate irreducible representations is just what is required for the elimination of high spin fields. Whereas it is not clear whether the presence of high spins in A will lead to physical particles with spin greater than one, it is clear that the absence of an irreducible representation means that the particles in that representation cannot be coupled to the Yang-Mills fields.

To illustrate these points we take the case of a scalar superfield $\phi(x^{\mu}, \theta^{\alpha}, \bar{\theta}^{\dot{\beta}})$. The fact that superspace carries a reducible representation of the Lorentz group allows us to write the supersymmetric equation

$$D_{\alpha}\phi = 0 \quad . \tag{1.13}$$

This equation defines a non-constant field, ϕ_{+} and shows that ϕ forms a reducible representation of supersymmetry. The field $\phi_{+}(x^{\mu}, \theta^{\alpha}, \bar{\theta}^{\dot{\beta}})$ is in fact irreducible and describes a scalar, a pseudoscalar, a spinor and two dimension-two auxiliary fields; it is the simplest possible representation of supersymmetry. Let us now consider this representation in the presence of Yang-Mills coupling. Its defining condition must be Yang-Mills covariant as well as supersymmetrically covariant, i.e.

$$\mathcal{D}_{\alpha}\phi = 0 \quad . \tag{1.14}$$

This implies that $\{ \mathcal{D}_\alpha , \mathcal{D}_\beta \} \phi = 0 = - F_{\alpha\beta} \cdot Y \phi$

$$\leadsto \qquad F_{\alpha\beta} = 0 . \tag{1.15}$$

Hence the preservation of this representation requires the covariant constraint of equation (1.15). The preservation of the representation defined by $D_{\dot\beta} \phi = 0$ implies that

$$F_{\dot\alpha\dot\beta} = 0 . \tag{1.16}$$

At this stage we have only ensured the preservation of these two particular representations, but it turns out that this is sufficient to ensure the existence of all irreducible representations in the presence of the Yang-Mills coupling.

There is another type of constraint which it is desirable to impose. Consider $F_{\alpha\dot\beta}$, which is given by

$$F_{\alpha\dot\beta} \cdot Y = + D_\alpha A_{\dot\beta} \cdot Y + D_{\dot\beta} A_\alpha \cdot Y - \{ A_\alpha \cdot Y , A_{\dot\beta} \cdot Y \} + 2i (\sigma^c)_{\alpha\dot\beta} A_c \cdot Y \tag{1.17}$$

Since $F_{\alpha\dot\beta}$ is a covariant object it follows that $A_c - \frac{1}{4} (\sigma_c)^{\alpha\dot\beta} F_{\alpha\dot\beta}$ has exactly the same transformation property as A_c. Consequently it can be used as a gauge potential, making the presence of A_c redundant. We can achieve this by setting $F_{\alpha\dot\beta} = 0$, which then can be solved algebraically for A_c in terms of A_α and $A_{\dot\beta}$. This situation is analogous to that in general relativity where the connection, $\omega_{\mu a}{}^b$ can be expressed in terms of the vierbein $e_\mu{}^a$ by setting the torsion tensor to zero. We call this type of constraint a conventional constraint. This constraint, unlike the constraints $F_{\dot\alpha\dot\beta} = F_{\alpha\beta} = 0$, is not demanded by any principle, it is simply a matter of convenience : a wish to formulate the theory without unnecessary fields.

We have adopted the constraints

$$F_{\dot\alpha\dot\beta} = F_{\alpha\beta} = 0 = F_{\alpha\dot\beta} \tag{1.18}$$

to cure the diseases of the absence of representations and the appearance of surplus fields in the theory. The symptom of these diseases, the appearance of high spin fields, is now no longer present. It can be shown by use of the Bianchi identities of the theory that the only fields present are A_μ, λ_α and D.

The action [9] constructed out of the remaining field strengths is

$$I = \int d^4x \{ \mathcal{D}^\alpha \mathcal{D}_\alpha (W_\beta W^\beta) + \text{h.c.} \} , \tag{1.19}$$

where $W_\beta = (\sigma^a)_{\beta\dot\gamma} F_a{}^{\dot\gamma}$.

It agrees with the theory given in equation (1.7).

1.2.3 Algebraic Origin of Constraints

The difficulties encountered in the superspace formulation of the N = 1 Yang-Mills theory occur in all superspace formulations which involve a local invariance. In the theories that do not have a local symmetry, superspace constraints such as that given in equation (1.13) are necessary to define irreducible representations. In these cases it is the definition of the irreducible representations itself which leads to the elimination of high spin fields. The cure presented for the N = 1 Yang-Mills theory is part of a much more general programme which can be formulated by the following requirements [10]:

(i) Find the conditions necessary to ensure that the generic representations of rigid supersymmetry are preserved in the presence of the local symmetry.

(ii) Solve algebraically for as many of the gauge fields as possible by choosing conventional constraints.

(iii) In the case of supergravity, a further type of constraint must be imposed, which is necessary to make the superconformal invariance of the constraints depend upon an irreducible superfield parameter [10].

Application of this programme leads us to set to zero some of the covariant tensors of the theory, which in turn places constraints on the fields of the theory. In the case of Yang-Mills invariance, these are the gauge fields A_B and in the case of supergravity, the supervierbeins, $E_A{}^M$ and the super-connections, $\Omega_{MA}{}^B$. The problems created by the occurrence of the high spin fields are solved by these constraints. The programme given above accounts successfully for all N = 1 and 2 superspace theories having a local invariance and we refer the reader to references [10], [11] for its detailed exposition. We explain its application to N = 2 supersymmetric Yang-Mills theory in the next section.

1.2.4 N = 2 Supersymmetric Yang-Mills Theory

To apply the programme given above we must first discover what are the irreducible representations of rigid N = 2 supersymmetry [10]. The simplest such representation is called the hypermultiplet [12] and contains 2 Majorana spinors, 4 scalars and 4 auxiliary fields. The superfield, $\phi_i(x^\mu, z, \theta^\alpha{}_i, \theta^{\dot\beta}{}_j)$, which has an internal symmetry index i, is a reducible representation of supersymmetry. The irreducible hypermultiplet is described by this superfield when subject to the covariant conditions [13]

$$\sum_{(ij)} D_{\alpha i}\phi_j = 0 = \sum_{(ij)} D_{\dot\beta i}\phi_j . \tag{1.20}$$

The symbol $\sum_{(ij)}$ means symmetrize in i and j.

From these conditions we can by successive application of further covariant derivatives and a knowledge of their algebra derive the equations

$$D_{\beta j} D_{\alpha}{}^{k} \phi_k = -4i\, a\, \epsilon_{\alpha\beta} D_z \phi_j \,, \tag{1.21}$$

$$\partial^2 \phi_k = + |a|^2 \partial_z^2 \phi_k \,. \tag{1.22}$$

Consequently, if ϕ_k possesses no central charge then $D_z \phi \equiv \partial_z \phi = 0$ which implies that $\partial^2 \phi = 0$. This is an equation of motion rather than an equation which could result from a representation defining condition. We may conclude that the defining conditions of equation (1.20) imply that the hypermultiplet has a non-zero central charge. This is an example of an important phenomenon: the central charge transformations of an irreducible representation prevent its defining conditions from implying that the fields satisfy equations of motion.

Let us now consider the hypermultiplet in the presence of a local Yang-Mills symmetry. The covariant defining conditions now become

$$\sum_{(ij)} \mathcal{D}_{\alpha i} \phi_j = 0 = \sum_{(ij)} \mathcal{D}_{\dot{\beta} i} \phi_j \,, \tag{1.23}$$

where $\mathcal{D}_{\alpha i} = D_{\alpha i} - A_{\alpha i} \cdot Y$, etc.

It follows from these equations that

$$\sum_{(ijk)} \{ \mathcal{D}_{\alpha i}, \mathcal{D}_{\beta j} \} \phi_k = 0 = - \sum_{(ijk)} F_{\alpha i \beta j} \cdot Y \phi_k$$

$$\sum_{(ijk)} \{ \mathcal{D}_{\alpha i}, \mathcal{D}_{\dot{\beta} j} \} \phi_k = 0 = - \sum_{(ijk)} F_{\alpha i \dot{\beta} j} \cdot Y \phi_k$$

$$\sum_{(ijk)} \{ \mathcal{D}_{\dot{\alpha} i}, \mathcal{D}_{\dot{\beta} j} \} \phi_k = 0 = - \sum_{(ijk)} F_{\dot{\alpha} i \dot{\beta} j} \cdot Y \phi_k \,. \tag{1.24}$$

These equations in turn imply that

$$\sum_{(ij)} F_{\alpha i \beta j} = 0 = \sum_{(ij)} F_{\alpha i \dot{\beta} j} = \sum_{(ij)} F_{\dot{\alpha} i \dot{\beta} j} \,. \tag{1.25}$$

Just as before, we find that demanding the existence of irreducible representations in the presence of the Yang-Mills symmetry leads to covariant constraints on the field strengths. In fact, the constraints of equation (1.25) are sufficient to ensure the existence of all irreducible representations in the local case.

Examination of the field strengths shows that we can impose only one conventional constraint. The field strength is given by

$$F_{\alpha}{}^{i}{}_{\dot\beta j} \cdot Y = + D_{\alpha}{}^{i} A_{\dot\beta j} \cdot Y + D_{\dot\beta j} A_{\alpha}{}^{i} \cdot Y - \{ A_{\alpha}{}^{i} \cdot Y, A_{\dot\beta j} \cdot Y \} + 2i\, \delta_j^i (\sigma^c)_{\alpha\dot\beta} A_c \tag{1.26}$$

and consequently the constraint

$$F_{\alpha}{}^{i}{}_{\dot\beta i} = 0 \tag{1.27}$$

defines A_c in terms of $A_{\alpha}{}^{i}$ and $A_{\dot\beta j}$.

The set of constraints we have arrived at, given by equations (1.25) and (1.27), are just those which were found by R. Grimm, M. Sohnius and J. Wess [14] to describe the N = 2 Yang-Mills theory. These authors also demanded that the field strengths have zero central charge. They found the action to be

$$I = \int d^4x \, \mathrm{Tr} \{ \bar{\mathcal{D}}^i \gamma_5 \mathcal{D}_j \, \bar{\mathcal{D}}^j \gamma_5 \mathcal{D}_i \, (F_{\alpha i}{}^{\alpha i})^2 + h.c. \}. \tag{1.28}$$

The central charge condition is not demanded for the preservation of representations. In fact, as will be discussed later, it is possible to adopt an alternative condition and arrive at a different off-shell formulation.

Before leaving N = 2 supersymmetry it is instructive to examine another irreducible representation. Consider the irreducible representation described by the superfield ϕ subject to the equations

$$D_{\alpha i}\, \phi = 0 \tag{1.29}$$

$$\sum_{(ij)} \{ D_{\alpha i} D^{\alpha}{}_{j}\, \phi^* + D_{\dot\beta i} D^{\dot\beta}{}_{j}\, \phi \} = 0 . \tag{1.30}$$

It is an obvious consequence of equation (1.29) that $D^{\alpha i} D_{\alpha i} \phi = +4ia D_z \phi = 0$. That is, ϕ does not have a central charge.

Let us now consider this representation in the presence of the Yang-Mills symmetry. The defining condition of equation (1.29) generalizes to

$$\mathcal{D}_{\alpha i}\, \phi = 0 . \tag{1.31}$$

Were we to assume that the central charge were absent in the Yang-Mills case we would find that

$$\{ \mathcal{D}_{\alpha i}, \mathcal{D}_{\beta j} \} \phi = 0 = - F_{\alpha i \beta j} \cdot Y \phi .$$

Applying equation (1.25) yields the equation

$$F_{\alpha i}{}^{\alpha i} \cdot Y\, \phi = 0 . \tag{1.32}$$

For an arbitrary (up to its defining conditions) superfield ϕ this would imply that $F_{\alpha i}{}^{\alpha i} = 0$. This constraint, together with those of equation (1.25) are too strong; for one can show, using the Bianchi identities, that all the field strengths would vanish and consequently the theory would be pure gauge.

This example demonstrates the folly of omitting the central charge generator from the algebra. Clearly, ϕ in the local case possesses a central charge and we must subsequently covariantize the central charge, in accord with all the other derivatives. The consequence of equation (1.31) then reads

$$\{ \mathcal{D}_{\alpha i}, \mathcal{D}_{\beta j} \} \phi = -F_{\alpha i \beta j} \cdot Y \phi + 2ia \epsilon_{\alpha\beta} \epsilon_{ij} \mathcal{D}_z \phi = 0 \tag{1.33}$$

which is then just an equation that defines the action of the central charge on ϕ. This argument strongly suggests that the gauge fields themselves are not inert under the action of the central charge generator. To do otherwise would be to make the arbitrary and unnatural assumption that

$$D_z A_{\alpha i} = 0 .$$

Although we must, according to our programme, insist on the preservation of a general representation, ϕ, we can, as a special case, consider the preservation of a particular representation. We could then solve equation (1.32) by identifying ϕ with $F_{\alpha i}{}^{\alpha i}$. A necessary consequence of this is that $\mathcal{D}_z F_{\alpha i}{}^{\alpha i} = 0$. This possibility is acceptable since it leads to a set of constraints on F_{AB} consistent to those of equation (1.25). Although this leads to the Yang-Mills formulation given in reference [14] it is not, as will be explained later, the only formulation of this theory.

One can apply the above programme to the N = 4 Yang-Mills theory. To do so requires a knowledge of the representations of N = 4 supersymmetry. However, in this case the Yang-Mills multiplet itself is the simplest representation. Nevertheless, this does not prevent us from demanding that all representations of rigid N = 4 supersymmetry be preserved in the local case.

As an example, consider the representation defined by

$$D_\alpha{}^i \phi = 0, \tag{1.34}$$

supplemented with other conditions involving more derivatives. Covariantizing this condition yields the result

$$\mathcal{D}_\alpha{}^i \phi = 0 . \tag{1.35}$$

This implies that

$$\{ \mathcal{D}_\alpha{}^i , \mathcal{D}_\beta{}^j \} \phi = 0 = - F_\alpha{}^i{}_\beta{}^j \cdot Y\phi - 2(a^\ell)^{ij} \epsilon_{\alpha\beta} \mathcal{D}_{z_\ell} \phi , \tag{1.36}$$

which in turn implies that

$$\sum_{(i,j)} F_\alpha{}^i{}_\beta{}^j = 0 \ . \tag{1.37}$$

Similarly we can deduce that

$$\sum_{(i,j)} F_{\dot\alpha i \dot\beta j} = 0 \ .$$

One could find other rigid representations by writing down a set of conditions on a superfield which have non-trivial solutions and do not lead to field equations and so can be taken to be representation defining conditions. Preservation of all these representations would lead, with the addition of the one conventional constraint, $F_\alpha{}^i{}_{\dot\beta i} = 0$, to a set of N = 4 Yang-Mills constraints. Given this set of constraints we could discover whether the Yang-Mills fields have a central charge or not. For example, if these constraints, under the assumption of no central charge, imply that Yang-Mills fields are on-shell we would be forced to conclude that the Yang-Mills fields possess a central charge.

Before the discovery of the off-shell N = 4 supersymmetric theory and the subsequent recognition of the importance of central charges, one of the most puzzling features was that the natural generalization of the N = 2 Yang-Mills constraints to the N = 4 case (namely

$$\sum_{(ij)} F_\alpha{}^i{}_\beta{}^j = 0 = \sum_{(ij)} F_{\dot\alpha i \dot\beta j} = F_{\alpha}{}^{i}{}_{\dot\beta i}$$

) seemed to lead to field equations. The above argument makes it clear that _if_ the programme leads to such constraints then one would have to conclude that the N = 4 Yang-Mills fields possess a central charge. The fact that the preservation of representations to the Yang-Mills case automatically introduces central charges, which we then must gauge covariantize, as for the N = 2 case, also makes the possibility of the Yang-Mills fields carrying a central charge very natural.

It was these arguments that first lead us to strongly suspect that the N = 4 Yang-Mills field carried a central charge. In the next section we will present different arguments to support this suspicion and the actual off-shell result which confirms this conclusion.

To summarize, we have found that the formulation of supersymmetric theories in superspace involves superspace constraints.

The constraints for N = 1 and 2 Yang-Mills theory can be found from a systematic programme once the representations of rigid supersymmetry are known. This programme has also been shown to work for N = 1 and 2 supergravity [10,11]. Given the constraints, we can calculate the auxiliary fields of the corresponding Minkowski space formulation.

We saw how the action of the central charge is specified by the defining conditions of the representation. As such, we cannot assume a priori that fields are inert under the central charge; in fact the opposite is more often the case.

1.3 Ghost Excitations and the Off-Mass-Shell Formulation of Supersymmetric Gauge Theories

1.3.1 Auxiliary fields in Minkowski space

In the previous section we showed how knowing the irreducible representations we could deduce the superspace constraints, which were necessary for the formulation of supersymmetric theories in superspace. Theories formulated in superspace carry, of neccessity, a representation of the supersymmetry algebra. As such they must include the auxiliary fields. In this section we present an alternative method of finding auxiliary fields which works entirely in Minkowski space. This method also requires as input a knowledge of the irreducible representations of supersymmetry in Minkowski space. That the two methods have the same starting point is not surprising when one remembers that the formulation in Minkowski space with auxiliary fields and the superspace formulations are completely equivalent. We apply the method to find the auxiliary fields of the N = 2 and N = 4 supersymmetric Yang-Mills theories.

Unfortunately, very little is known about the representations of extended supersymmetry. However, it is a relatively simple matter to calculate the on-shell representations of supersymmetry, that is, to find the sets of states which form a representation of supersymmetry provided their classical equations hold. At first sight, this would seem of little help in the search for auxiliary fields, for they are just those fields which vanish on-shell and so are not included in the on-shell representations. However, we shall now discuss how a knowledge of the different on-shell representations of supersymmetry can indeed lead to information about the auxiliary fields.

Consider a supersymmetric theory for which we do not know the auxiliary fields. We take this theory, for sake of argument, to be an Abelian gauge theory,

$$\mathcal{L} = -\frac{1}{4} F_{\mu\nu}^{2} \quad + \text{ supersymmetric extension} \qquad (1.38)$$

Let us now consider placing the operator ∂^2 in between all the component fields to obtain the Lagrangian

$$\mathcal{L}^G = -\frac{1}{4} F_{\mu\nu}{}^2 - \frac{1}{4m^2} F_{\mu\nu}\, \partial^2 F^{\mu\nu} + \text{ supersymmetric extension.} \tag{1.39}$$

We will refer to this procedure [15] as a d'Alembertian insertion into the theory. Since ∂^2 is a supersymmetric operation it follows that this new theory is also supersymmetric. Consequently, the particles of this theory must belong to supersymmetric representations; in particular the on-shell states must belong to the on-shell supersymmetric representations that can easily be calculated.

The auxiliary fields appear in the original Lagrangian $\mathcal{L}$ in the generic form H^2 and thus do not correspond to on-shell states. However, in the Lagrangian $\mathcal{L}^G$ they appear in the generic form $H^2 + \frac{1}{m^2} H \partial^2 H$ and so do correspond to on-shell states. The operation ∂^2 introduces no new fields, so by knowing the content of the on-shell states of $\mathcal{L}^G$ and the contribution to these states from the physical fields of $\mathcal{L}$, we can deduce what must be the contribution from the auxiliary fields of $\mathcal{L}$. In this way we gain information about the auxiliary fields of $\mathcal{L}$. It should be emphasised that the Lagrangian $\mathcal{L}^G$ is an entirely theoretical model to enable us to discover auxiliary fields from a very limited knowledge of the representations. We are not in any way suggesting that it be taken seriously as a model for physics.

These considerations will be spelt out in particular models, but first we discuss the on-shell representations.

1.3.2 On-Shell Irreducible Representations of Supersymmetry

These representations may be calculated from the algebra using an extension of Wigner's method of induced representations. We refer the reader to reference [16] for the details of this simple construction and record here only the results. The nature of the on-shell representation, of course, depends on the on-shell characteristics being considered, namely, on whether the multiplet is massive or massless, has zero or non-zero central charge, and on the number N of spinorial symmetry generators.

If the multiplet is massless and has no central charge on-shell, then the irreducible representation contains states with helicities from a maximum helicity λ to a minimum helicity $\lambda - N/2$. The demand of a CPT invariant theory requires us to add to this the irreducible representation with helicities from $-\lambda$ to $-(\lambda - N/2)$ if these helicities are not already contained in the first representation. For a N = 4 theory with a maximum helicity 1 we find $\lambda - N/2 = -1$. Consequently, if N were to exceed 4 we would require particles of spin greater than one. This leads us to the well-known statement that the N = 4 supersymmetric Yang-Mills theory is the maximally extended supersymmetric Yang-Mills theory.

If the multiplet is massive but has no central charge on-shell, then the irreducible representation with the lowest maximal spin contains a maximal spin of $N/2$. For the case of N = 4 we get a maximal spin of two.

If the multiplet is massive but does possess an on-shell central charge, then the irreducible representation with the lowest maximal spin has a maximal spin of $\frac{N+1}{4}$, if N is odd and N/4 if N is even. For the case of N = 4 this maximal spin is one. The ability of the central charge to reduce the maximal spin of representations will prove essential in our later considerations.

It is also possible to calculate the number of on-shell states with each helicity. The table below gives the result for the Yang-Mills theories.

spin \ N	1	2	4
spin 1	1	1	1
spin ½	1	2	4
spin 0		2	6

(1.40)

1.3.3 Ghost Excitations

To prepare the way for a discussion of the consequences of the insertion of d'Alembertian in supersymmetric theories, we first calculate the spectrum from such insertions into the Lagrangian. Let us study the effects for one field at a time.

The Lagrangian for the scalar with the d'Alembertian insertion is

$$\mathcal{L}^G_\phi = +\frac{1}{2}(\partial_\mu\phi)^2 + \frac{1}{2m^2}\partial_\mu\phi\,\partial^2\partial^\mu\phi \,. \tag{1.41}$$

To extract the spectrum we introduce the field B to ensure that the Lagrangian contains only second derivatives

$$\mathcal{L}^G_\phi = +\frac{1}{2}(\partial_\mu\phi)^2 + \frac{1}{2}B^2 + \frac{1}{m}B\partial^2\phi \,. \tag{1.42}$$

Making the field redefinitions $C = B/m$, $A = \phi + C$ we find

$$\mathcal{L}^G_\phi = +\frac{1}{2}(\partial_\mu A)^2 - \frac{1}{2}(\partial_\mu C)^2 + \frac{1}{2}m^2C^2 \,. \tag{1.43}$$

This expression makes it clear that the spectrum is that of a massless scalar and that of one massive spin-zero ghost.

The Lagrangian for the fermion with the insertion of the d'Alembertian is

$$\mathcal{L}^G_\lambda = -\frac{i}{2}\bar{\lambda}\not{\partial}\lambda - \frac{i}{2m^2}\bar{\lambda}\not{\partial}\partial^2\lambda . \quad (1.44)$$

The system is treated in a similar way to the scalar case, except that we must now introduce two fields to reduce the Lagrangian to an expression containing only first derivatives. The spectrum is found to be a massless spin-$\frac{1}{2}$ and two massive spin-$\frac{1}{2}$ ghosts. The two ghosts correspond to the two additional fields introduced.

The Lagrangian for a vector with the insertion of the d'Alembertian is,

$$\mathcal{L}^G_{A_\mu} = -\frac{1}{4}F_{\mu\nu}^2 - \frac{1}{4m^2}F_{\mu\nu}\partial^2 F^{\mu\nu} . \quad (1.45)$$

The resulting spectrum is that of one massless spin-1 particle and one massive spin-1 ghost.

A spin-0 particle can also be described by an antisymmetric tensor field [17]. The Lagrangian is $-\frac{1}{2}(\partial_\nu {}^*A^{\mu\nu})^2$ and it possesses the gauge invariance

$$A_{\mu\nu} \rightarrow A_{\mu\nu} + \partial_\mu\Lambda_\nu - \partial_\nu\Lambda_\mu .$$

Making the d'Alembertian insertion we find the Lagrangian

$$\mathcal{L}^G_{A_{\mu\nu}} = -\frac{1}{2}(\partial_\nu {}^*A^{\nu\mu})^2 - \frac{1}{2m^2}\partial_\nu {}^*A^{\nu\mu}\partial^2\partial_\kappa {}^*A^\kappa{}_\mu . \quad (1.46)$$

It has a spectrum of a massless spin-0 and a massive spin-1 ghost! It is a remarkable fact, essential to future arguments, that different representations of the spin-0 lead to different ghost spectra once the d'Alembertian insertions are made.

Finally, we consider the auxiliary fields, inserting a d'Alembertian gives

$$\mathcal{L}^G_H = +\frac{1}{2}H^2 + \frac{1}{2m^2}H\partial^2 H . \quad (1.47)$$

The spectrum is that of one massive ghost.

We can now test out the insertion of a d'Alembertian into a supersymmetric theory. Let us consider the N = 2 supersymmetric theory [14] which has one spin-1, A_μ, two spinors, $\lambda_{\alpha i}$, two spin-0's in the usual representation, A, B, and three auxiliary fields $\vec{C}$. Making the d'Alembertian insertion in this theory leads to a massless set of on-shell states which are in the same representation as those on-shell states in the original theory, and also to a supersymmetric representation of massive ghost states. According to the previous considerations we get on-shell states of mass m as follows: one vector, four spinors and five scalars. The five scalars arise in two ways: two come from the two scalars and three come from the three auxiliary fields. Had we not known that there were three auxiliary fields in the original theory, we could have deduced this from the fact that the on-shell massive ghosts lie in the supersymmetric representation which has one vector, four spinors and five scalars.

The concrete construction underlying these considerations is the following Lagrangian with the d'Alembertian insertion:

$$\mathcal{L}^G = -\tfrac{1}{4} F_{\mu\nu}^2 + \tfrac{1}{2}(\partial_\mu A)^2 + \tfrac{1}{2}(\partial_\mu B)^2 - \tfrac{i}{2}\bar{\lambda}^i \not{\partial} \lambda_i + \tfrac{1}{2}\vec{C}^2 + \tfrac{1}{m^2}\Big(F_{\mu\nu}\partial^2 F^{\mu\nu} + \tfrac{1}{2}\partial_\mu A\, \partial^2 \partial^\mu A + \tfrac{1}{2}\partial_\mu B\, \partial^2 \partial^\mu B - \tfrac{i}{2}\bar{\lambda}^i \not{\partial} \partial^2 \lambda_i + \tfrac{1}{2}\vec{C}\, \partial^2 \vec{C}\Big).$$

1.3 Central Charges and N ≥ 4 Supersymmetric Theories

It was shown by Ferrara and de Wit [18] that the structures of the off-shell $N \geq 4$ representations are different in character from those for lower N. This is demonstrated in the context of the N = 4 supersymmetric theory [1]. This theory has one spin-1 and four spin-$\frac{1}{2}$'s and six spin-0's which we shall, for the moment, assume to be represented by six scalars. If we make the d'Alembertian insertion into this theory then the on-shell ghost states with mass m must belong to an N = 4 supersymmetric representation. If we now assume that this multiplet does not possess a central charge we must, according to the previous discussion, accept that it contains states of spin two. Such states can only come from the auxiliary fields and hence we are forced to conclude that some of the auxiliary fields of the N = 4 Yang-Mills have spin two. The idea of spin-2 auxiliary fields seems rather unpalatable, but there is to our knowledge no concrete objection to them.

Let us now examine the assumptions of this argument to see if we can avoid the introduction of spin-2 auxiliary fields into these theories [1]. The principal assumption is that the massive ghosts have a central charge. This assumption is entirely equivalent to the requirement that the original massless gauge multiplet possesses a central charge that vanishes on-shell. To accept this possibility one must free one self from certain prejudices about the nature of multiplets that have central charges.

It is not true that representations with a central charge must have a complex highest spin; we will present an example of such a representation shortly. If a representation possesses a central charge, then we may act on the fields of that representation with the central charge generator, i.e., $z\phi$, $z^2\phi$ etc. In order for each applixation of z not to produce an infinite tower of fields, we require a relation of the form $z^n\phi = r\phi$ where r is an object of dimension $(m)^r$. The only object out of which r can be constructed is p^2. In a massive free theory $p^2 = m^2$ so we can have an on-shell central charge. However it is perfectly possible to have in a massless theory an off-shell central charge which vanishes on-shell, that is when $p^2 = 0$. Finally, with regard to the particular theory under study, it is not consistent to demand both that the spin-zero's sit in a $\underline{6}$ of SU (4) or a $\underline{3} \oplus \underline{3}$ of O (4) and also that the representation has a central charge. This is because the groups SU(4) or O(4) will rotate the spinorial charges and must form an automorphism group of the supersymmetry algebra. Such an action is not consistent with the existence of central charges in the algebra.

In the rest of this section we will explore the possibility that the extended Yang-Mills multiplets possess an off-shell central charge.

1.3.4 N = 2 Yang-Mills Theory with Central Charge

In a previous section we saw how the on-shell states of the N = 2 Yang-Mills theory (one spin -1, 2 spinors - $\frac{1}{2}$, 2 spin-0) could be represented off-shell by one vector, 2 spinors, 2 scalars and 3 scalar auxiliary fields. We examined how an insertion of a d'Alembertian led to a supersymmetric representation of on-shell massive ghost states containing one vector, four spinors and five scalars. This supermultiplet is known not to have a central charge.

In the light of the analysis of the previous section we will now examine the possibility that there is an alternative off-shell representation of the N = 2 Yang-Mills theory into which an insertion of a d'Alembertian will lead to on-shell massive ghosts which do possess a central charge [1]. If we can find an N = 4 formulation of the Yang-Mills theory with central charge then we can by contraction find such a formulation in N = 2 theory. The only massive N = 2 central charge representation which includes one vector has as its on-shell states two spin-1's, four spin-$\frac{1}{2}$'s and two spin-0's. This immediately poses a problem: where do the two spin-1's come from ? One spin-1 arises from the vector A_μ in the original theory, the other can only come from the spin-0's or the auxiliary fields of the original theory. Let us suppose that it arises from the spin-0's. This is possible if one of the spin-0's was represented by an antisymmetric tensor field, $A_{\mu\nu}$ The other, in order not to create a third spin-1 will have to be represented by a scalar ϕ. The four spin-$\frac{1}{2}$ states of the massive on-shell ghost states are accounted for by the two spin-$\frac{1}{2}$ fields $\lambda_{\alpha i}$ of the original theory. We are finally left with the problem of accounting for the two massive, spin-0 on-shell ghost states. One comes from the scalar of the original theory, ϕ, and since we have now accounted for all the ghosts arising from the physical states of the theory before the d'Alembertian insertion, the other scalar must come from a single auxiliary field present in the original theory, H. Thus we are led to propose the existence of an alternative N = 2 Yang-Mills theory constructed out of the off-shell representation of one vector A_μ, two spinors $\lambda_{\alpha i}$, two spin zero's represented by one scalar ϕ and one antisymmetric tensor field $A_{\mu\nu}$, and one scalar auxiliary field H [1]

As a check on the validity of this representation we can test whether it has the same number of boson and fermion off-shell fields. Counting the boson states we get 4-1 = 3 for A_μ, 1 for ϕ, 6-3 = 3 for $A_{\mu\nu}$, and 1 for H; totalling 8. In this count it is necessary to subtract the gauge degrees of freedom, 1 for A_μ and 3 for $A_{\mu\nu}$. There are 8 fermion states from $\lambda_{\alpha i}$, and the numbers match. We refer the reader to reference [1,2] for the construction of the Lagrangian and field transformations for this representation. These results are contained as a subset of the analogous N = 4 case which is given in the next section.

1.3.5 The N = 4 Yang-Mills Theory

We now repeat the above analysis for the N = 4 supersymmetric Yang-Mills theory [1]. The on-shell states of this theory are one spin-1, four spin-$\frac{1}{2}$'s and six spin-0's. We shall assume that an insertion of a d'Alembertian in this theory leads to massive on-shell ghost states which possess a central charge. The only such representation has two spin-1's, eight spin-$\frac{1}{2}$'s, and ten spin-0's, and so avoids the spin-2's which occur in massive N = 4 multiplets without a central charge. One of the two massive spin-1 ghost states is provided by the spin-1 vector A_μ, the other, assuming it does not come from an auxiliary field, is provided by demanding that one of the spin-0's of the theory before the d'Alembertian insertion is described by an antisymmetric tensor $A_{\mu\nu}$. The five remaining spin-zero's must be in the usual scalar representation, ϕ. The eight massive on-shell ghost spinors are provided by the four spin-$\frac{1}{2}$ fields $\lambda_{\alpha i}$. Five of the ten massive on-shell ghost states of spin zero are provided by the five spin-0 scalars and the other five must be accounted for by five scalar auxiliary fields. Hence we conclude that the off-shell N = 4 Yang-Mills theory has a representation consisting of one spin-1 vector, of four spinors and six spin-zero's, which are represented by one antisymmetric tensor and five scalar fields, and of five auxiliary scalar fields. This multiplet has the same number of fermion as boson fields, provided we do not count the gauge degrees of freedom.

In order to calculate the transformation laws of these fields, it is necessary to know what the internal symmetry group is, i.e. the group which acts on the i -index of the spinorial charges $Q_\alpha{}^i$. The presence of the one central charge Z implies that the anticommutator of two spinorial charges is given by

$$\{ Q_\alpha{}^i , Q_\beta{}^j \} = 2 \epsilon_{\alpha\beta} \Omega^{ij} Z , \tag{1.48}$$

where Ω^{ij} is an antisymmetric matrix. For this relation to be preserved under the action of the internal symmetry group, this group must preserve the metric Ω^{ij} and so by definition must be Sp(4) or a subgroup of Sp(4). We note that the central charge vanishes on-shell and so the group enlarges to SU(4) on-shell. Let us consider for a moment the question of a representation for the five auxiliary fields. These fields occur in the variation of the spinors in the following way

$$\delta \lambda_i = \ldots + H_i{}^j \zeta_j . \tag{1.49}$$

The five auxiliary fields are contained in the four by four matrix $H_i{}^j$ whose indices must transform under Sp(4) or a subgroup of Sp(4). Using the metric $\Omega^{ij} = -\Omega^{ji}$ of Sp(4) we can lower the upper index on $H_i{}^j$ and then, demanding it to be antisymmetric, $H_{ij} = -H_{ji}$, and traceless $H_{ij}\Omega^{ij} = 0$, we are just left with five elements, which we take to be the five auxiliary fields. A similar argument can be applied to the five scalar fields ϕ_{ij}. These belong to a second rank antisymmetric traceless representation of Sp(4).
The four spinors belong to the fundamental representation of Sp(4). The antisymmetric tensor $A_{\mu\nu}$ and the vector A_μ are scalars under Sp(4). We now consider only the Abelian theory. The (interacting) Yang-Mills case will be treated in the subsequent chapters.

To calculate the transformations of these fields we just add all terms of the correct dimension with arbitrary coefficients to the variations of the fields. We must add only gauge-covariant terms to the variation of gauge-covariant fields. Some of the arbitrary coefficients can be chosen by field redefinitions, the rest are fixed by demanding that the algebra close with one central charge. The parity of the fields is also fixed by the demand that the algebra close. The field transformations are [1]

$$\begin{aligned}
\delta A_\mu &= -i \bar{\zeta}^i \gamma_\mu \lambda_i + \omega\, \partial^\nu {}^*A_{\nu\mu} + \partial_\mu \Lambda \\
\delta \lambda_i &= -\sigma^{\mu\nu} F_{\mu\nu} \zeta_i + \gamma_5 \gamma^\mu \partial_\nu {}^*A^\nu{}_\mu \zeta_i - 2 \not\partial \phi_i{}^j \zeta_j \\
&\quad + 2\gamma_5 H_i{}^j \zeta_j + \omega \gamma_5 \not\partial \lambda_i \\
\delta \phi_{ij} &= +i \bar{\zeta}_i \lambda_j - i \bar{\zeta}_j \lambda_i - \frac{i}{2} \Omega_{ij} \bar{\zeta}^k \lambda_k - \omega H_{ij}
\end{aligned} \tag{1.50}$$

$$\delta H_{ij} = -i\bar{\zeta}_i\gamma_5\not{\partial}\lambda_j + i\bar{\zeta}_j\gamma_5\not{\partial}\lambda_i - \frac{i}{2}\Omega_{ij}\bar{\zeta}^k\gamma_5\not{\partial}\lambda_k - \omega\partial^2\phi_{ij}$$

$$\delta A_{\mu\nu} = -2i\,\bar{\zeta}^k\sigma_{\mu\nu}\lambda_k - \omega F_{\mu\nu} + \partial_\mu\Lambda_\nu - \partial_\nu\Lambda_\mu, \qquad (1.50)\ \text{cont.}$$

where the parameters of supersymmetry, central charge, the vector gauge invariance and the antisymmetric tensor gauge invariance are ζ^i, ω, Λ and Λ_μ respectively.

This algebra closes. The commutator for two supersymmetry transformations [1] is

$$[\delta_S(\zeta_1), \delta_S(\zeta_2)] = +2i\,\bar{\zeta}_2^j\gamma^\mu\zeta_{1j}\partial_\mu + 2i\,\bar{\zeta}_2^k\gamma_5\zeta_{1k} + \text{gauge transformations.}$$

The commutator of a central charge and a super-symmetry transformation is

$$[\delta_S(\zeta), \delta_Z(\omega)] = \text{field dependent gauge transformations} \qquad (1.51)$$

The occurrence of the field-dependent terms in the commutator is a well known phenomenon connected with the Wess-Zumino gauge choice [19] in a superfield treatment.

The invariant Lagrangian which describes the N = 4 Abelian theory [1] is

$$\mathcal{L} = -\frac{1}{4}F_{\mu\nu}^2 + \frac{1}{2}(\partial_\mu\phi_{ij})^2 - \frac{1}{2}(\partial_\nu{}^*A^{\nu\mu})^2 - \frac{i}{2}\bar{\lambda}^i\not{\partial}\lambda_i + \frac{1}{2}(H_{ij})^2. \qquad (1.52)$$

This theory is clearly Sp(4) invariant. The non-Abelian theory will be given in the remaining chapters.

A Lagrangian with a set of supersymmetry transformations which describe an N = 4 Abelian theory has been found by a dimensional reduction procedure [20]. The transformations, although they leave the action invariant, do not form a representation of the supersymmetry algebra. The Lagrangian, apart from containing no auxiliary fields, differs from the Lagrangian given in Eq. (1.52) in that it is SU(4) invariant. The six spin zero's are represented by six scalar A_{ij} fields belonging to a <u>6</u> of SU(4), $A_{ij} = -A_{ji}$, $A_{ij} = \frac{1}{2}\epsilon_{ijkl}A^{kl}$. The existence of the SU(4) symmetry ensures that the fields of this theory do not possess a central charge.

Although the two formulations differ off-shell, they have the same equations of motion and their field transformations agree once these field equations are enforced. It is unknown whether the SU(4) invariant formulation can be extended to include auxiliary fields that would lead to a closing algebra. Should such an extension exist, it would have to contain high spin auxiliary fields. On the other hand, if an extension were not possible it is not clear that the symmetries of the classical equations of motion would be inherited by the quantum theory.

This raises the broader enquiry, are these two formulations equivalent at the quantum level, or more precisely, do they lead to the same on-shell S-matrix elements ? One of the major obstacles to answering this question is that the non-Abelian formulation of equations (1.50) and (1.52) involves a constraint. This non-Abelian formulation and possible methods for implementing the constraint will be discussed in the remaining chapters. Even so, at the quantum level, with the introduction of path integrals and ghosts, many subtle mechanisms could occur.

We could also ask what is the physics of off-shell central charges ? Naïvely it could appear that a theory with an off-shell central charge would have different internal symmetries at the quantum and classical levels. For the N = 4 Yang-Mills theory these would be Sp(4) and SU(4), respectively. There is then the possibility that such an effect could make itself apparent in the appearance of bound states which belong to representations of Sp(4) whereas the classical theory possesses an SU(4).

The presence of these two formulations and the fact that the one found in reference [20] came from a dimensional reduction of a ten-dimensional theory leads us to ask if there is not a ten dimensional Yang-Mills theory which could lead, via dimensional reduction, to the four-dimensional theory presented here.

To answer this question we need to observe that in ten dimensions the spinorial generators of supersymmetry, denoted Q_α, $\alpha = 1 \to 32$ have no internal index. Consequently it is not possible to have a central charge in the ten dimensional algebra. Dimensional reduction as so far practiced has, in the case of massless theories, meant setting the dependence of the field on the extra dimensions to zero. Such a procedure cannot give rise to a central charge in the resulting four dimensional algebra! Hence if the manifestly supersymmetric formulation given above is to be obtained from a ten dimensional theory , then the dimensional reduction procedure used will have to be different from that so far practised. In particular, it will have to introduce a central charge into the lower dimensional theory. Since the existence of a central charge generator is equivalent to the presence of an additional coordinate in the theory, such a dimensional reduction procedure would have to introduce a non-trivial dependence of the fields on the extra dimension. This new dimensional reduction procedure and the resolution of the problems of obtaining invariant actions in the presence of extra coordinates are given in the next chapter.

To summarize, we have seen how a knowledge of on-shell representations of supersymmetry can lead, via a consideration of ghost excitations, to off-shell representations of supersymmetry. The difference between these two representations being the presence of auxiliary fields. We saw how it was possible to formulate the N = 4 supersymmetric Yang-Mills theory in a manifestly supersymmetric fashion, without the occurrence of high spin auxiliary fields. This was achieved by introducing the concept of an off-shell central charge. That is, the gauge supermultiplet must have a central charge which vanishes on shell. Clearly, the arguments presented here could equally well be applied to the N = 8 supergravity theory, in order to discover its off-shell formulation.

2. OFF-SHELL STRUCTURE OF SUPERSYMMETRIC YANG-MILLS THEORIES*)

As has been emphasized in the preceding chapter, in order for one to have a true representation of the supersymmetry algebra on fields, auxiliary (i.e. non-propagating) fields must be included. While it is clear at an algebraic level why this need arises, the geometrical and dynamical origin of the auxiliary fields has remained obscure. We shall see in this chapter that once the role of the central charge is properly understood in relation to the theory's dynamics, a very natural interpretation for the auxiliary fields can be made.

Besides the origin of the auxiliary fields, some other long-standing problems in supersymmetric theories can now be resolved. The major one is how to extend the results presented in chapter one for Abelian supersymmetric gauge theories to the general non-Abelian case. In this discussion, we shall see the consequences of non-linear coupling of spinless fields when formulated in non-standard representations such as antisymmetric tensors or vectors. The method that we shall use to derive the non-Abelian theories is one familiar to every physicist from Hamiltonian mechanics: the Legendre transformation. This method of Legendre transformation provides a new way to perform dimensional reduction, and we shall examine the differences between it and the straightforward dimensional reduction of reference [20]. In this way, it is possible to understand both the origin of the rigid internal symmetries of the extended supersymmetric theories and the differences between the symmetries of the action and the symmetries of the field equations. This subject will be treated fully in chapter three.

*) This chapter corresponds to the lecture given by K.S.S.

2.1 Dimensional Reduction by Legendre Transformation

There are two options for the method of realizing the central charge in the supersymmetry algebra. This algebra includes the anti-commutation relation, which we may write as

$$\{Q_{\alpha i}, \bar{Q}^{\beta j}\} = 2\delta_i^{\ j} \left((\gamma^\mu)_\alpha^{\ \beta} P_\mu - (\gamma_5)_\alpha^{\ \beta} Z \right) . \tag{2.1}$$

The first option is to realize Z as an ordinary U(1) phase transformation on complex fields. Since Z has dimensions of mass, its non-trivial action on a field requires the existence of a parameter with dimensions of mass. This massive parameter is the eigenvalue of a given multiplet for Z. In many cases its value turns out to be that of the physical mass of the multiplet.

The second option is to treat Z on a similar footing to P_μ i.e. by allowing the fields of the theory to depend upon an extra space-time dimension x^5. The action of Z is then simply to perform a shift in the x^5 coordinate. Of course, the physical states of the theory must be independent of x^5, and we will find that the action of the central charge upon physical states must be the same in a given theory regardless of the option chosen for realizing its action upon off-shell fields. The virtue of the second method of realization is that it permits a non-trivial action upon fields even when there is no massive parameter present.

In considering the algebra (2.1), it should be borne in mind that the parity-odd character of the fifth dimension is not an essential feature of the central charge. One could equally well have chosen the central charge to be parity-even, but it will prove to be convenient for us to consider the example given, since it enables us to deal with ordinary Dirac gamma matrices γ_μ, γ_5. In this connection, we note that in the form (2.1), the i,j indices on the spinorial generators belong to the maximal rigid internal symmetry group of the supersymmetry algebra with a central charge. Since this will be discussed in detail in chapter three, we will not dwell upon it here. We merely note that for i = 1....2m, the internal symmetry group is Sp(2m), and the supercharges $Q_{\alpha i}$ satisfy the Sp(2m)-Majorana condition.

$$Q_{\alpha i} = i(\gamma_5 C)_{\alpha\beta}\,\Omega_{ij}\,\overline{Q}^{\beta j} \tag{2.2}$$

where C is the charge-conjugation matrix and Ω_{ij} is the Sp(2m) metric.

As a concrete example of a realization of the central charge through the use of a fifth dimension, we take the "scalar hypermultiplet" which is an N=2 supermultiplet [12] . The expression of the multiplet as a superfield [13] has already been presented in chapter one. The complex superfield

$$\phi^i(x^\mu, \theta_{\alpha j};\, x^5) \qquad\qquad i,j = 1,2$$

is an SU(2) = Sp(2) isospinor, and is subject to the constraint

$$D_{\alpha i}\phi^j = \frac{1}{2}\,\delta_i^{\;j}\, D_{\alpha k}\phi^k \;, \tag{2.3}$$

where all spinors are now SU(2)-Majorana. The superspace action for the multiplet is

$$I_{(4)} = \int d^4x\; \bar{D}\tau^r D\, (\bar{\phi}\,\tau^r \frac{\overleftrightarrow{\partial}}{\partial x^5}\,\phi) \tag{2.4}$$

When written out in terms of the components of the superfield, which are defined over x^μ and x^5, the action is

$$I_{(4)} = \int d^4x\; (\partial^\mu A^i \partial_\mu A_i^* - \frac{i}{4}\,\bar{\psi}\,\gamma^\mu \overleftrightarrow{\partial}_\mu \psi + G^i G_i^*) \tag{2.5}$$

A^i and G^i are complex scalar-isospinor fields and ψ is a Dirac spinor. In terms of real fields, and taking into account the behaviour under parity, A^i contains two scalars and two pseudo-scalars.

In order for the action (2.4),(2.5) to be invariant under central charge transformations, we must have

$$\frac{d}{dx^5} I_{(4)} = 0. \tag{2.6}$$

How is this to be achieved, when $I_{(4)}$ involves an integration only over d^4x? The central charge invariance of (2.4) follows from its invariance under supersymmetry transformations, as is required by the anticommutation relation (2.1). The supersymmetric invariance of (2.4) is guaranteed by the fact that the constraint (2.3) implies that $\bar{D}\tau^r D\,(\bar{\phi}\,\tau^r \frac{\partial}{\partial x^5}\,\phi)$ varies by a total derivative ∂_μ under supersymmetry transformations. Our task here is to unravel this rather complicated web of conditions and isolate the essential feature that makes $\frac{d}{dx^5} I_{(4)}$ vanish, and thereby establish rules for construction of central charge invariant actions in general.

Once the question of central charge invariance is asked in the form, "How does one assure invariance under translations in a coordinate that is not integrated over?" the answer can be lifted directly from any mechanics textbook. In a mechanical system, the Hamiltonian

$$H_{(3)} = \int d^3x\, \mathcal{H}(p_i, q_i) \tag{2.7}$$

is conserved in time, $\frac{dH}{dt} = 0$, provided that the equations of motion hold :

$$\frac{dq_i}{dt} = \frac{\delta H}{\delta p_i}\ , \quad \frac{dp_i}{dt} = -\frac{\delta H}{\delta q_i}\ . \tag{2.8}$$

In the present circumstances, we must construct the Hamiltonian for translations in a fifth dimensional "time". We shall see, however, that in order to maintain positivity of the physical potential of the theory, the fifth dimension must actually be spacelike. We start from a 4+1 dimensional theory and construct the Hamiltonian by Legendre transformation from the (4+1)-dimensional Lagrangian $\mathcal{L}_{(5)}$:

$$H_{(4)} = \int d^4x \, \left(\frac{\partial \mathcal{L}_{(5)}}{\partial(\partial_5\psi_i)} \partial_5\psi_i - \mathcal{L}_{(5)}(\partial_5\psi, \partial_\mu\psi, \psi) \right). \tag{2.9}$$

This object is guaranteed to satisfy $\frac{d}{dx^5} H = 0$ provided the five-dimensional equations of motion hold:

$$\frac{\partial}{\partial x^5} \frac{\partial \mathcal{L}(5)}{\partial \, (\partial_5\psi_i)} + \partial_\mu \frac{\partial \mathcal{L}(5)}{\partial(\partial_\mu\psi_i)} - \frac{\partial \mathcal{L}_{(5)}}{\partial \, \psi_i} = 0 \tag{2.10}$$

The essential point of the construction is that the five-dimensional equations of motion do not imply field equations in four dimensions as long as we continue to allow the fields to depend on x^5. It is at this point that the technique of dimensional reduction by Legendre transformation differs essentially from the usual dimensional reduction [20], where the dependence on the dimension being reduced is required to be trivial. For a free scalar field like A^i, we require that the five-dimensional d'Alembertian vanish:

$$\lozenge \, A^i \equiv \left(\frac{-\partial^2}{(\partial x^5)} + \Box \right) A^i = 0 \tag{2.11}$$

But this is exactly what is required by the constraint (2.3) on the hypermultiplet superfield, which iterates to give

$$\lozenge \, \phi^i = 0 \quad . \tag{2.12}$$

Our procedure is thus as follows: starting from a theory in five dimensions we construct the Hamiltonian and impose the five dimensional field equations as constraints. The construction of the Hamiltonian may be done by performing a Legendre transformation starting from the five-dimensional Lagrangian. Even if this were not known, however, the construction of the Hamiltonian could in principle be carried out by finding a first integral of the field equations. The action in four dimensions is then taken to be minus the Hamiltonian, in order to give the four-dimensional kinetic terms the correct sign:

$$I_{(4)} = - H_{(4)} \tag{2.13}$$

In considering the Hamiltonian as the action of a four-dimensional theory, it is necessary to decide what fields it is taken to be a functional of. Up to now we have only focused upon the question of invariance under shifts in x^5, for which it is sufficient to construct the "energy". Since in the end we must make contact with four-dimensional field theory, there must not be any explicit derivatives with respect to x^5 appearing in the Hamiltonian. This is of course the standard situation, for in mechanics one replaces time derivatives in the Hamiltonian by independent fields which are the canonical momenta. In the present situation, this means that we are introducing extra fields that are the canonical momenta for the five-dimensional theory.

The five-dimensional canonical momenta are nothing other than the <u>auxiliary fields</u> of the four dimensional theory.

To see how all this works in practice, we return to the example of the scalar hypermultiplet. We start from a formulation of the theory in five dimensions, without any auxiliary fields, since we shall be imposing the five-dimensional field equations as constraints. The physical fields are A^i and ψ. Their field equations may be derived from the manifestly SU(2) invariant Lagrangian

$$\mathcal{L}_{(5)} = \partial^a A^i \partial_a A^*_i - \frac{i}{4}\bar{\psi}\gamma^a \overleftrightarrow{\partial}_a \psi , \tag{2.14}$$

where $a = 0,1,2,3;5$, $i=1,2$, and the γ^a matrices satisfy the five-dimensional Dirac algebra

$$\{\gamma^a,\gamma^b\} = 2\eta^{ab} \quad , \quad \eta_{ab} = \mathrm{diag}\ (+1,-1,-1,-1;-1) .$$

Although A^i and ψ do not form a complete representation of the N=2 supersymmetry algebra, they do contain all the physical states, and their field equations possess an invariance under the supersymmetry transformations

$$\begin{aligned} \delta A^i &= i\bar{\zeta}^i \psi \\ \delta\psi &= 2\gamma^a \zeta_i \partial_a A^i \end{aligned} \tag{2.15}$$

where the anticommuting spinorial parameter ζ_i satisfies the SU(2)-Majorana condition

$$\zeta_i = i(\gamma_5 C)\, \varepsilon_{ij} (\bar{\zeta}^j)^T \quad . \tag{2.16}$$

The transformations (2.12) are also an infinitesimal invariance of the action given by (2.11), although this infinitesimal invariance cannot be integrated to a finite transformation without involving other spurious transformations containing the field equations for A^i and ψ. This restriction does not apply to the invariance of the field equations, however, for the term that would spoil the closure of the supersymmetry algebra is proportional to the field equation for ψ. Thus, subject to the field equations

$$0 = \gamma^a \partial_a \psi = \Box A^i \equiv \partial^a \partial_a A^i \quad , \tag{2.17}$$

the transformations (2.12) close.

The next step is to construct the Hamiltonian. The canonical momenta conjugate to the fields A^i and ψ are, from (2.11),

$$\Pi(A^i) = \frac{\partial \mathcal{L}_{(5)}}{\partial\, \partial_5 A^i} = -\partial_5 A_i^* \equiv G_i \tag{2.18a}$$

$$\Pi(\psi) = \frac{\partial \mathcal{L}_{(5)}}{\partial\, \partial_5 \psi} = \frac{i}{4} \bar{\psi}\gamma_5 \quad . \tag{2.18b}$$

The field G^i is the auxiliary field of the theory. Now we perform the Legendre transformation to obtain the Hamiltonian density $\mathcal{H}_{(4)}$ and identify $-\mathcal{H}_{(4)} = \mathcal{L}_{(4)}$ as the four-dimensional Lagrangian density :

$$\mathcal{L}_{(4)} = \partial^\mu A^i \partial_\mu A_i^* - \frac{i}{4} \bar{\psi}\, \gamma^\mu \partial_\mu \psi + G^i G_i^* \tag{2.19}$$

This is the correct off-shell Lagrangian density for the scalar hypermultiplet.

The invariance of the four-dimensional action under supersymmetry and central charge transformations holds subject to the five-dimensional Hamiltonian equations :

$$\partial_5 A^i = -G^i \tag{2.20a}$$

$$\partial_5 \psi = \gamma^\mu \gamma_5 \partial_\mu \psi \tag{2.20b}$$

$$\partial_5 G^i = -\partial^\mu \partial_\mu A^i \quad . \tag{2.20c}$$

These equations have the effect of restricting the independent data needed to determine a field defined over five dimensions to just that contained in an ordinary field over four dimensions. In other words, they restrict the expansion of ϕ^i in a Taylor series in x^5 to depend only on the components A^i, ψ and G^i evaluated at $x^5=0$, together with their four-space derivatives. Thus, the effect of a central charge transformation is now to rotate A^i into $-G^i$, G^i into $-\Box A^i$ and ψ into $-\gamma_5 \partial\!\!\!/ \psi$.

Note that no fermionic auxiliary fields have been generated by the above procedure. This is because the field ψ satisfies a first order field equation, and so already contains its own canonical momenta, as can be seen from (2.20b). Consequently, when the five-dimensional field equations are imposed, all terms in the expansion of ψ in powers of x^5 are given by four-space derivatives of $\psi(x^5=0)$.

The supersymmetry transformations that leave the action given by (2.19) invariant are read straight off from the five-dimensional on-shell transformations (2.12). It is only necessary to eliminate the $\partial/\partial x^5$ derivatives using (2.20).

The resulting four-dimensional off-shell transformations are :

$$\delta A^i = i\bar{\zeta}^i \psi \tag{2.21a}$$

$$\delta\psi = 2\gamma^\mu \zeta_i \partial_\mu A^i + 2\gamma_5 \zeta_i G^i \tag{2.21b}$$

$$\delta G^i = -i\, \bar{\zeta}^i \gamma^\mu \gamma_5 \partial_\mu \psi \tag{2.21c}$$

These transformations form a closed algebra without use of the four-dimensional equations of motion :

$$\left[\delta(\zeta),\delta(\eta)\right] = 2i\, \bar{\eta}^i \gamma^\mu \zeta_i \partial_\mu - 2i\bar{\eta}^i \gamma_5 \zeta_i \, Z \tag{2.22}$$

In the second term on the right hand side, we have replaced the partial derivative ∂_5 by the central charge generator Z, whose action is given by (2.20).

The central charge is active in the off-shell formulation of the scalar hypermultiplet even though no massive parameter has been introduced. Since there is no massive parameter, however, there can be no effect of the central charge upon physical states. This is seen to be the case from (2.20), because the central charge transformations of all the fields are just the four-dimensional field equations. In fact, the four-dimensional field equations may be very simply summarized by

$$\partial_5 \phi^i = 0 \quad . \tag{2.23}$$

In the reduction from five to four dimensions by Legendre transformation, the original five-dimensional Lorentz invariance is abandoned, just as the usual Hamiltonian of a four-dimensional theory possesses manifest three dimensional rotational invariance only. Abandoning the five-dimensional Lorentz invariance has the effect of permitting chiral spinor fields to be defined, since γ_5 is now a singlet. Thus, irreducible spinors in five dimensions become reducible under the four-dimensional Lorentz algebra. This is true in particular of the SU(2)-Majorana spinor parameter ζ_i of the supersymmetry transformations. In five dimensions, this spinor is irreducible under the Lorentz group alone, without regard to its SU(2) covariance properties. Thus it describes an N=1 supersymmetry. In four dimensions, it can be reduced into two two-component Weyl spinors with an SU(2) isospinor index. Consequently, in four dimensions we have an N=2 supersymmetry. A full discussion of the lineage of internal symmetries and of the number of independent supersymmetries will be given in chapter three.

2.2 Supersymmetric Yang-Mills Theories

The method of dimensional reduction by Legendre transformation permits us to obtain in a straightforward fashion the off-shell formulation of supersymmetric Yang-Mills theories. This is a problem that has heretofore been an important obstacle to the development of these theories, particularly regarding their quantum behaviour.

The present discussion is devoted to the maximally extended N=4 supersymmetric Yang-Mills theory [20] , which holds some promise of being completely scale-invariant to all loop orders. We obtain the off-shell formulation of this theory in four dimensions by a two stage dimensional reduction:

ordinary dimensional reduction $10 \longrightarrow 5$
followed by
Legendre transformation $5 \rightsquigarrow 4$.

As will be explained in chapter three, there are many other equivalent paths of dimensional reduction,all including one stage of Legendre transformation. We adopt the above because it is the simplest, and as we did for the hypermultiplet, we take the gamma matrix for the fifth dimension to be the Dirac γ_5.

The field equations for the N=1 supersymmetric Yang-Mills theory in 10 dimensions can be derived from the Lagrangian density ("on-shell formalism", without auxiliary fields):

$$\mathcal{L}_{(10)} = \mathrm{Tr}\,(-\frac{1}{4}F_{mn}F^{mn} - \frac{i}{4}\bar{\lambda}\,\Gamma^m \overleftrightarrow{\mathcal{D}}_m \lambda) \tag{2.24}$$

where m,n run through 0,1,2,3;5....10 and Tr denotes a trace over the Yang-Mills indices which are not explicitly written. The Lie-algebra-valued gauge field appears in the covariant derivative $\mathcal{D}_m \equiv \partial_m - A_m$. In order to have N=1 supersymmetry in 10 dimensions, the spinor λ must be in an irreducible representation of the Lorentz group. Thus λ is a simultaneously chiral and Majorana 32-spinor:

$$(1 - i\Gamma_{11})\lambda = 0 \tag{2.25a}$$

$$\lambda = C_{(10)}\,\bar{\lambda}^T \tag{2.25b}$$

where $C_{(10)}$ is the charge conjugation matrix for ten dimensions. Consequently, λ has 16 real components, which is correct for an object that must reduce to four irreducible four-dimensional spinors.

In order to perform the ordinary dimensional reduction to five dimensions, we adopt an appropriate representation of the Γ^m matrices, which will be given in chapter three, and then demand that none of the fields depend upon $x^6 \ldots x^{10}$. In this case, the Yang-Mills parameter cannot depend upon $x^6 \ldots x^{10}$ either, and consequently the gauge components $A_6 \ldots A_{10}$ become a set of five Lie-algebra-valued scalars. As will be shown in chapter three, the ordinary dimensional reduction breaks the Lorentz group in ten dimensions down into $SL(2,c) \times Sp(4)$. Thus, the above five scalars actually fall into the five-dimensional representation of Sp(4):

$$A_{5+s} = \phi_s \quad ; \quad s = 1....5. \tag{2.26}$$

The spinor λ also breaks up into a set of Sp(4)-Majorana spinors

$$\lambda_{\alpha i} = i(\gamma_5 C)_{\alpha\beta}\,\Omega_{ij}\,\bar{\lambda}^{\beta j} . \tag{2.27}$$

The resulting five-dimensional Lagrangian density is

$$\mathcal{L}_{(5)} = \text{Tr}\ (-\frac{1}{4}F_{ab}F^{ab} + \frac{1}{2}\mathcal{D}_a\phi_s\mathcal{D}^a\phi_s - \frac{i}{4}\bar{\lambda}^i\gamma_a\overleftrightarrow{\mathcal{D}}^a\lambda_i$$
$$+ \frac{1}{2}\bar{\lambda}^i(t_s)_i{}^j[\lambda_j,\phi_s] + \frac{1}{4}[\phi_r,\phi_s]\ [\phi_r,\phi_s]\ ) \tag{2.28}$$

where $a,b = 0....5$, $s = 1....5$ and $i,j = 1....4$. The matrices $(t_s)_i{}^j$ are the "gamma matrices" for the compact group Sp(4) and will be discussed in chapter three. The Lagrangian (2.28) gives rise to an action that is invariant under the following ("on-shell") supersymmetry transformations :

$$\delta A_a = i\bar{\zeta}^i\gamma_a\lambda_i \tag{2.29a}$$

$$\delta\phi_s = i\bar{\zeta}^i(t_s)_i{}^j\ \lambda_j \tag{2.29b}$$

$$\delta\lambda_i = -\sigma_{ab}\zeta_i F^{ab} + \gamma_a(t_s\zeta)_i\ \mathcal{D}^a\phi_s - \frac{i}{2}(t_r t_s\zeta)_i[\phi_r,\phi_s]\ . \tag{2.29c}$$

The five-dimensional field equations following from (2.28) are

$$\mathcal{D}^b F_{ab} + \frac{1}{2}\ \{\bar{\lambda},\gamma_a\lambda\} - i\left[\phi_s,\mathcal{D}_a\phi_s\right] = 0 \tag{2.30a}$$

$$\mathcal{D}^a\mathcal{D}_a\phi_s - \frac{1}{2}\ \{\bar{\lambda},t_s\lambda\} + \left[\phi_r,\ \left[\phi_r,\phi_s\right]\right] = 0 \tag{2.30b}$$

$$\gamma^a\mathcal{D}_a\lambda\ + i\left[t_s\lambda,\phi_s\right] = 0, \tag{2.30c}$$

where spinorial Sp(4) indices are suppressed:

The Legendre transformation from the Lagrangian (2.28) to the Hamiltonian is carried out in standard fashion. The canonical momenta are :

$$\Pi(A^\mu) = F_{5\mu} \equiv V_\mu \tag{2.31a}$$

$$\Pi(\phi^s) = -\mathcal{D}_5\phi_s \equiv H_s \tag{2.31b}$$

$$\Pi(\lambda) = \frac{i}{4}\bar{\lambda}\gamma^5 \tag{2.31c}$$

$$\Pi(A^5) = 0\quad . \tag{2.31d}$$

The Hamiltonian density can now be constructed, and its negative is the off-shell Lagrangian density for the four-dimensional theory:

$$\mathcal{L}^{N=4}_{(4)} = \mathrm{Tr}\,\Big(-\frac{1}{4}F_{\mu\nu}F^{\mu\nu} - \frac{1}{2}V_\mu V^\mu + \frac{1}{2}\mathcal{D}_\mu\phi_s\mathcal{D}^\mu\phi_s + \frac{1}{2}H_sH_s$$

$$- \frac{i}{4}\bar{\lambda}\gamma^\mu\overleftrightarrow{\mathcal{D}}_\mu\lambda + \frac{1}{2}\bar{\lambda}\,t_s[\lambda,\phi_s] + \frac{1}{4}[\phi_r,\phi_s][\phi_r,\phi_s]\Big)\ . \qquad (2.32)$$

Rewriting the five-dimensional supersymmetry transformations (2.30) to replace the terms in $\mathcal{D}_5$ by canonical momenta, we have

$$\delta A_\mu = i\bar{\zeta}\gamma_\mu\lambda \qquad (2.33a)$$

$$\delta\phi_s = i\bar{\zeta}t_s\lambda \qquad (2.33b)$$

$$\delta\lambda = -\sigma_{\mu\nu}\zeta\ F^{\mu\nu} + t_s\gamma^\mu\zeta\,\mathcal{D}_\mu\phi_s - \gamma^\mu\gamma_5\zeta V_\mu \qquad (2.33c)$$

$$+ t_s\gamma_5\zeta H_s - \frac{i}{2}t_rt_s\zeta\,[\phi_r,\phi_s]$$

$$\delta V_\mu = -2i\bar{\zeta}\,\sigma_{\mu\nu}\gamma_5\mathcal{D}^\nu\lambda - \bar{\zeta}\gamma_\mu\gamma_5t_s[\lambda,\phi_s] \qquad (2.33d)$$

$$\delta H_s = -i\bar{\zeta}t_s\gamma^\mu\gamma_5\mathcal{D}_\mu\lambda - \frac{1}{2}\bar{\zeta}\gamma_5\,[t_s,t_r]\,[\lambda,\phi_r] \qquad (2.33e)$$

where again, spinorial Sp(4) indices are suppressed.

In order for the action (2.32) to be invariant under the supersymmetry and central charge transformations, the equations of motion (2.30) must be imposed together with the relations between momenta and x^5-derivatives given in (2.31). Most of these equations just limit the dependence of the fields upon x^5, and so limit the action of the central charge to transformations among the physical fields and momenta defined over x^μ only, just as we found for the scalar hypermultiplet.

A distinctly new feature arises in the Yang-Mills theory, however, as a consequence of its gauge invariance. This is seen in two ways. The equation (2.31d) does not relate $\Pi(A^5)$ to an x^5-derivative, but rather says that A^5 does not have a conjugate momentum. In fact, A^5 is not a dynamical field at all; it is actually a Lagrange multiplier whose field equation is a constraint upon the purely x^μ - dependence of the other fields. Combining (2.31a) with the a=5 component of (2.30a), we have an equation that does not contain any x^5-derivatives:

$$\mathcal{D}^\mu V_\mu + \frac{1}{2} \{\bar{\lambda}, \gamma_5 \lambda\} - i[H_s, \phi_s] = 0. \tag{2.34}$$

In the form that we have written (2.32), this constraint has already been taken into account.

At this stage we recall the analogous feature in the Hamiltonian treatment of electrodynamics. The Maxwell action

$$I^{Max}_{(4)} = -\frac{1}{4} \int d^4x \, F_{\mu\nu} F^{\mu\nu} \tag{2.35}$$

yields the field equation, obtained by varying A^o,

$$\partial_i F^{io} = 0 \quad . \tag{2.36}$$

In the Hamiltonian formalism, F^{io} is treated as an independent field $\mathcal{E}^i$, just as we have treated V^μ as an independent field. $\mathcal{E}^i$ is of course just the electric field and (2.36) is Gauss' law in free space. Since (2.36) contains no time derivatives, it is an equation of constraint. It eliminates from the formalism the part of the canonical momentum $\mathcal{E}^i$ that would be conjugate to the longitudinal gauge part of A_i. The existence of such a constraint is a direct consequence of the gauge invariance of the theory.

In the present discussion, where the Hamiltonian of the five-dimensional theory is taken as the action of a four-dimensional theory, the presence of real $\partial/\partial x^4$ time derivatives in (2.34) has the effect of making the field V_μ propagate, even though it appears without derivatives in the action (2.32). The field V_μ actually describes a propagating spin zero particle.

In an Abelian theory, the constraint reduces simply to

$$\partial_\mu V^\mu = 0. \tag{2.37}$$

This can be solved very simply by setting

$$V^\mu = \partial_\nu {}^*A^{\mu\nu} \quad . \tag{2.38}$$

Then

$$V_\mu V^\mu = (\partial_\nu {}^*A^{\mu\nu}) \ (\partial_\rho {}^*A_\mu{}^\rho) \tag{2.39}$$

is the Lagrangian density for an antisymmetric tensor field describing a spin-zero particle. In this way, we make contact with the formulation of the Abelian N=4 supersymmetric Yang-Mills theory presented in chapter one.

The simple solution to the Abelian constraint (2.37) does not easily generalize to the non-Abelian case. A straightforward iterative solution of (2.34) starting from (2.38) gives an expression for V^μ in terms of $A^{\mu\nu}$ that is non-local. This situation is under further study, and here we will restrict ourselves to indicating an outline of the resolution of this problem by demonstrating in a non-supersymmetrically-covariant fashion the equivalence of the present formulation to the original on-shell formalism.

We may obtain the same dynamics as the system (2.32),(2.34) by relaxing the constraint on V^μ and re-introducing A_5 as a Lagrange multiplier to include the constraint in the action. The terms involving H_s and V^μ are then

$$\mathrm{Tr}\Big(-\frac{1}{2} V_\mu V^\mu + \frac{1}{2} H_s H_s + A_5 (\mathcal{D}^\mu V_\mu + \frac{1}{2}\{ \bar{\lambda},\gamma_5\lambda\} - i[H_s,\phi_s])\Big) \tag{2.40}$$

Note that this step is not supersymmetrically covariant because the constraint term has been introduced without any supersymmetric partners. Put otherwise, the field content of the theory has been restricted throughout our discussion to what in superspace is known as a "Wess-Zumino gauge", where the Yang-Mills gauge parameter does not have any superpartners. Since the constraints in a gauge theory are in one-to-one relationship to the gauge parameters, the constraint in the present formalism does not have any partners. In order to relax the constraint without harming the manifest supersymmetry, we would need to work from the start in a general superspace gauge.

It is clear from (2.40) that the fields V^μ and A_5 describe a spin zero particle, for in this expression we have a standard covariant first order formulation of a spin zero field. The fields V^μ and H_s satisfy algebraic equations of motion that can be re-substituted back into the action :

$$V_\mu = -\mathcal{D}_\mu A_5 \tag{2.41}$$

$$H = i[\phi_s, A_5] \ . \tag{2.42}$$

We may also make the SU(4) rigid internal symmetry of the theory manifest at this stage by combining $\phi_{ij} = \frac{1}{2}(t_s)_i{}^k \Omega_{jk}\phi_s$ and $-\frac{i}{2}\Omega_{ij} A_5$ into a six-dimensional representation of SU(4) :

$$S_{ij} \equiv \phi_{ij} - \frac{i}{2} \Omega_{ij} A_5 \tag{2.43}$$

where

$$S_{ij} = -\frac{1}{2} \varepsilon_{ijk\ell} \bar{S}^{k\ell} . \tag{2.44}$$

In order to make the SU(4) manifest, we must also introduce the chiral projection

$$\chi_i = \frac{1}{2} (1 + i\gamma_5) \lambda_i , \tag{2.45}$$

with the resulting on-shell form of the Lagrangian density

$$\begin{aligned} \mathcal{L}^{N=4}_{\text{on-shell}} = \mathrm{Tr} \Big(&- \frac{1}{4} F_{\mu\nu} F^{\mu\nu} + \frac{1}{2} \mathcal{D}^\mu S_{ij} \mathcal{D}_\mu \bar{S}^{ij} - \frac{i}{2} \bar{\chi}^i \gamma_\mu \overleftrightarrow{\mathcal{D}}^\mu \chi_i \\ &+ \bar{\chi}^i C \left[\bar{\chi}^{jT}, S_{ij} \right] + \chi_i{}^T C^{-1} \left[\chi_j, \bar{S}^{ij} \right] \\ &+ \frac{1}{4} \left[S_{ij}, S_{k\ell} \right] \left[\bar{S}^{ij} , \bar{S}^{k\ell} \right] \Big). \end{aligned} \tag{2.46}$$

This is the on-shell form in which the N=4 supersymmetric Yang-Mills theory was originally presented. In it the SU(4) symmetry is made manifest at the expense of manifest supersymmetry. The usefulness of one formalism or another will depend upon which symmetry is more powerful in controlling the structure of the theory at the quantum level. In this respect, the off-shell supersymmetric version would clearly seem to be favoured.

3. SUPERSYMMETRIC YANG-MILLS THEORIES AND THEIR INTERNAL SYMMETRIES*)

We have seen, in the previous chapter, how the technique of dimensional reduction can be modified to yield off-shell supersymmetric theories from higher-dimensional on-shell ones. The essential step was the replacement of the compactification of one of the extra dimensions by a Legendre transformation. This procedure abolishes the on-shell condition on the theory, namely that the algebra of the field transformations closes only when the fields are subjected to the equations of motion. The on-shell condition is replaced by constraints on the fields which are to be fulfilled off-shell and which resemble the original equations of motion of the higher-dimensional theory.

The invariance under translations in the one spatial dimension which has been reduced by Legendre transformation survives in the form of an invariance under central charge transformations. The constraints of the theory determine the behaviour of the fields under these transformations, just as equations of motion describe the time evolution of a system.

In this chapter, we shall see how all known supersymmetric Yang-Mills theories with particle spins ≤ 1 (i.e. with $N\leq 4$) can be understood as originating from the same theory in ten space-time dimensions. Various combinations of compactifications, Legendre transformations, eliminations of auxiliary fields ("going on-shell"), and truncations (i.e., the discarding of some of the supersymmetries) give us the known Yang-Mills and matter multiplets in four dimensions (for N=1,2,4), as well as the Lagrangians for all these theories.

Moreover, careful analysis of what happens to that part of the ten-dimensional Lorentz-group which acted on the six reduced dimensions gives us a good understanding of the global internal symmetries of these theories. With only a few diagrammatical rules, we can develop a "family tree" of supersymmetric Yang-Mills theories, and we can see, for instance, how the lineage of the N=4 theory provides the origins of the respective Sp(4) and SU(4) invariances of the off-shell and the on-shell versions of that theory; we can also see why the Grimm-Sohnius-Wess [14] off-shell formulation of the N=2 theory shares its internal symmetry group, SU(2)×U(1), with the on-shell version, while the more recent alternative formulation of ref. [1] has only an SU(2), besides a central charge.

Finally, in the last subsection of this chapter, we will indicate how the results, i.e., the various off-shell theories in

---

*) This chapter corresponds to the lecture given by M.F.S.

four dimensions, translate into superspace. Indeed, we will simply use the lessons that Legendre-transformations have taught us (not to be afraid of higher-dimensional equations of motion and of central charges) and reduce the ten-dimensional on-shell constraints to four dimensions.

However, before dealing with the main body of this chapter, which has been outlined so far, we will have another look at the analysis in Ref. 5 of supersymmetry algebras with central charges. We will show how compact symplectic groups naturally arise as internal symmetry groups in the presence of central charges.

3.1. Central charges and symplectic groups

Ref. [5] was heavily based upon earlier work by S. Coleman and J. Mandula [3] , who had studied the general structure of the largest possible Lie-group of symmetries of a non-trivial S-matrix in a relativistic theory with non-zero masses. Their result can be roughly summarized by a formula for that maximal symmetry group $\mathcal{G}$:

$$\mathcal{G} = \mathcal{P} \times (\mathcal{S} \times \mathcal{A}) \tag{3.1}$$

$\mathcal{P}$ is the Poincaré-group, $\mathcal{S}$ is a semisimple Lie-group, $\mathcal{A}$ is an Abelian Lie-group, and both $\mathcal{S}$ and $\mathcal{A}$ are compact. It is the direct-product structure which earned these results the name "no-go theorem" : an internal symmetry (from $\mathcal{S}$ or $\mathcal{A}$) can change neither spin nor mass.

Ref. [5] studied the restrictions on supersymmetries which are a consequence of the fact that the bosonic part of the algebra must resemble the structure (3.1). The results are reasonably well known, except, we feel, the exact properties which were derived for the so-called "central charges". We shall therefore review those properties and their derivation.

The relevant algebraic relations are

$$\begin{aligned}
\{Q_{\alpha i}, Q_{\beta j}\} &= 2\varepsilon_{\alpha\beta}\,(\Omega^e)_{ij}\, Z_e \\
\{Q_{\alpha i}, \bar{Q}_{\dot{\beta}}^{\;j}\} &= 2(\sigma^\mu)_{\alpha\dot{\beta}}\,\delta_i^j\, P_\mu \\
[Q_{\alpha i}, T_a] &= (f_a)_i^{\;j}\, Q_{\alpha j}
\end{aligned} \tag{3.2}$$

The Q's are, of course, the spinorial charges, the T_a are the internal symmetries, and the Z_e are a priori just a subset of these :

$$Z_e \varepsilon \{T_a\} \tag{3.3}$$

The $(f_a)_i{}^j$ and $(\Omega^e)_{ij}$ are numbers which turn out to be severely restricted by Jacobi-identities. In particular, the following relations must hold :

(i) The matrices f_a must generate a representation of the internal symmetry group.

(ii) δ_i^j must be a numerical invariant. This restricts the internal symmetry group to be unitary.

(iii) The commutator $[Z_e, T_a]$ must be a linear combination of the Z_e, i.e., the Z_e form an invariant subalgebra.

(iv) The commutator $[Z_e, Z_k]$ must vanish, i.e., the invariant subalgebra is moreover Abelian. This implies that none of the Z_e can belong to the semisimple part $\mathcal{S}$ of the internal symmetry group. They must be from the Abelian part $\mathcal{A}$, and therefore commute not only with each other, but also with all other T_a, due to the direct-product structure of eq. (3.1).

(v) The Z_e must commute with the Q. Now we are in the position to say : the Z_e commute with everything, they are <u>central</u> charges.

(vi) $$f_a \Omega^e + \Omega^e f_a{}^T = c_{ka}{}^e \Omega^k .$$

This matrix equation used to be quite puzzling. The right-hand side is actually zero, since $[Z_e, T_a] = 0$ and thus all structure constants $c_{ea}{}^b$ vanish. From the unitarity of the internal symmetry group follows the Hermiticity of f_a, and thus

$$f_a \Omega^e = -\Omega^e f_a^*$$

This result has been interpreted to mean that the representation f_a is equivalent to its complex conjugate, with at least one antisymmetric intertwining matrix Ω. Thus, e.g., the fundamental representations of SU(N) and O(N) were excluded (except for SU(2)).

However, it seems now more appropriate to read eq. (vi) in the form

$$(f_a)_i^{\ i'} \, (\Omega^e)_{i'j} + (f_a)_j^{\ j'} (\Omega^e)_{ij'} = 0 \tag{3.4}$$

This simply says that there are to be a number of antisymmetric numerical invariants of the representation f_a.

In the case of just one central charge and N supersymmetries, the maximal group to act on the Q's and to respect (3.4) is the symplectic group Sp(N). The particular representation f_a will generate a compact version of Sp(N) (since the f_a are Hermitian). It is relation (3.4) which intimately connects Sp-groups with central charges. In the absence of the latter, $\Omega^e = 0$, Eq. (3.4) is trivial, and U(N) is the maximal internal symmetry group consistent with the algebra.

3.2 The internal symmetries generated in dimensional reduction.

In this subsection we shall set up the rules by which internal symmetries get generated in particular dimensional reduction processes.

The effect of dimensional reduction by compactification [20] is illustrated by yet another look at the reduction from ten to five space-time dimension (10 ⟶ 5), which has been presented in chapter 2.

Compactification of five spatial dimensions obviously breaks down the Lorentz group SO(1,9) into SO(1,4) × O(5). Accordingly, the Yang-Mills potential A_m of the ten-dimensional theory ($m = 0,1,2,3;5,\ldots,10$) splits into the Yang-Mills potential A_a for the five-dimensional theory ($a = 0,1,2,3;5$) and five Lorentz-scalars (i.e. scalars under SO(1,4)):

$$\phi_s \equiv A_{5+s} \qquad (s = 1,\ldots,5) \quad . \tag{3.5}$$

Conversely, under O(5), the A_a are scalars, and the ϕ_s are a five-vector. The compactification procedure implies that no field depends on the coordinates $x^6 \ldots x^{10}$, thus the ϕ_s are gauge-covariant fields, since the inhomogeneous term in their transformation law vanishes :

$$\frac{\partial}{\partial x^{5+s}} \Lambda = 0 \qquad (\Lambda = \text{gauge parameter}) \; . \tag{3.6}$$

The Yang-Mills field-strengths F_{mn} break up into the parts

$$
\begin{aligned}
F_{ab} &= \partial_a A_b - \partial_b A_a + i\,[A_a, A_b] \\
F_{a,5+s} &= \partial_a \phi_s + i\,[A_a, \phi_s] = \mathcal{D}_a \phi_s \\
F_{5+s,5+r} &= i\,[\phi_s, \phi_r]
\end{aligned}
\tag{3.7}
$$

which are, of course, covariant under $SO(1,4) \times O(5)$.

The treatment of the spinor is somewhat more tricky. First, we have to go on a little excursion into higher-dimensional Dirac-matrices. We use the freedom to define them so that the dimensional reduction will be easy. In view of the reduction $10 \longrightarrow 5$, we define the ten Γ_m and Γ_{11} as :

$$
\begin{aligned}
\Gamma_a &= \begin{bmatrix} 0 & \gamma_a \otimes 1 \\ \gamma_a \otimes 1 & 0 \end{bmatrix} , \quad a = 0,1,2,3;5 \\
\Gamma_{5+s} &= \begin{bmatrix} 0 & -1 \otimes t_s \\ 1 \otimes t_s & 0 \end{bmatrix} ; \quad s = 1,\ldots,5 \\
\Gamma_{11} &\equiv -i\Gamma_o \ldots \Gamma_{10} = \begin{bmatrix} -i & 0 \\ 0 & i \end{bmatrix} .
\end{aligned}
\tag{3.8}
$$

Here γ_a and t_s are the respective Dirac-matrices for $O(1,4)$ and $O(5)$. They are 4×4 matrices, so that the Γ_m are indeed 32×32 matrices. The 32×32 charge conjugation matrix $C_{(10)}$ has the properties

$$
C^{-1}_{(10)} \Gamma_m C_{(10)} = - \Gamma_m^T \; ; \quad C^T_{(10)} = - C_{(10)} \tag{3.9}
$$

and can be built from the charge conjugation matrices $C_{(5)}$ for γ_a and Ω for t_s in the following way :

$$C_{(10)} = \begin{bmatrix} 0 & C_{(5)} \otimes \Omega \\ -C_{(5)} \otimes \Omega & 0 \end{bmatrix} \tag{3.10}$$

The defining properties of $C_{(5)}$ and Ω are

$$C^{-1}_{(5)} \gamma_a C_{(5)} = +\gamma_a^T \; ; \quad C^T_{(5)} = -C_{(5)}$$
$$\Omega^{-1} t_s \Omega = +t_s^T \; ; \quad \Omega^T = -\Omega \; . \tag{3.11}$$

We can give $C_{(5)}$ in terms of the usual C-matrix of four-space as $C_{(5)} = i\gamma_5 C$. Ω can be chosen to be real.

The spinorial representations of $O(1,4)$ and $O(5)$ are generated by $\frac{i}{4}[\gamma_a, \gamma_b]$ and $\frac{i}{4}[t_s, t_r]$, respectively, and from Eqs. (3.11) it can be derived that $C_{(5)}$ and Ω are invariants of these representations: the covering group in each case is thus $Sp(4)$, since real antisymmetric 4×4 matrices are left invariant.

Let us now return to the gluino field λ of the ten-dimensional supersymmetric Yang-Mills theory. It is constrained by both a Weyl- and a Majorana - condition :

$$(1 - i\Gamma_{11})\lambda = 0$$
$$\lambda = C_{(10)} \bar{\lambda}^T \tag{3.12}$$

which reduce a priori 32 complex components of a spinor in ten dimensions to 16 real ones, corresponding to eight on-shell states. This fits in nicely with the 10-2=8 on-shell bosonic states corresponding to the gauge-field A_m.

Since Γ_{11} is diagonal in our representation, the chirality condition just means that we can write λ in the form $\lambda_{\alpha i}$ $(\alpha, i=1,\ldots,4)$, where the γ_a act on the α-index and the t_s on the i-index. The Majorana condition then reads

$$\lambda_{\alpha i} = C_{(5)\alpha\beta}\, \Omega_{ij}\, \bar{\lambda}^{\beta j} \; . \tag{3.13}$$

"Dimensional reduction of the spinor" now means nothing more than that Eq. (3.13) is not to be seen anymore as a particular non-covariant way of writing the ten-dimensional conditions, Eqs. (3.12), but as a genuine condition for an $SO(1,4) \times Sp(4)$ spinor $\lambda_{\alpha i}$. We call it an Sp(4)-covariant reality (or Majorana-)condition. The dimensional reduction of an expression like $\Gamma^m \mathcal{D}_m \lambda$ is of course

$$\gamma^a \mathcal{D}_a \lambda_i - i(t_s)_i{}^j [\phi_s, \lambda_j] ,$$

since again $\lambda_{\alpha i}$ is supposed not to depend on the coordinates $x^6 \ldots x^{10}$. The only surviving part of $\mathcal{D}_{5+s} \lambda$ is then $i[\phi_s, \lambda]$.

Combining the results so far presented in this chapter, we get from the ten-dimensional Lagrangian

$$\mathcal{L}_{(10)} = \mathrm{Tr}\,(-\frac{1}{4} F_{mn} F^{mn} - \frac{i}{4} \bar{\lambda} \Gamma^m \overleftrightarrow{\mathcal{D}}_m \lambda) \tag{3.14}$$

to the five-dimensional one

$$\mathcal{L}_{(5)} = \mathrm{Tr}\,(-\frac{1}{4} F_{ab} F^{ab} + \frac{1}{2} \mathcal{D}_a \phi_s \mathcal{D}^a \phi_s - \frac{i}{4} \bar{\lambda}^i \gamma^a \overleftrightarrow{\mathcal{D}}_a \lambda_i + \frac{1}{2} \bar{\lambda}^i (t_s)_i{}^j [\lambda_j, \phi_s] + \frac{1}{4} [\phi_r, \phi_s][\phi_r, \phi_s]). \tag{3.15}$$

So far we have only reproduced earlier work by Gliozzi, Olive, Scherk, Brink, and Schwarz [20] , but with particular emphasis on the internal symmetry group, here Sp(4). $\mathcal{L}_{(5)}$ is indeed covariant under Sp(4), and so are the supersymmetry transformations under which the theory is invariant (see chapter 2).

Let us now abstract the first diagrammatical rule from the given example :

RULE A: Compactification of n space-like dimensions produces as internal symmetry group the universal covering group of O(n).

The covering groups for O(n) with n = 2,3,4,5,6 are respectively: U(1), SU(2), SU(2)×SU(2), Sp(4),SU(4). Indeed, we know that the N=4 on-shell theory has SU(4) as internal symmetry. This is in accordance with Rule A, since it can be seen as the result of the compactification 10 → 4.

The next question we ask is: which internal symmetries are generated by a dimensional reduction by Legendre transformation? The answer is simply :

<u>RULE B:</u> Dimensional reduction by Legendre transformation retains a translational invariance in the reduced dimension (central charge transformation), but does not generate any additional internal symmetry.

This rule, tested on the various examples, appears quite justified: generally a Legendre transformation mediates between different formulations of the same theory (Lagrangian or Hamiltonian formalism), so we do not expect anything dramatic to happen.

A dimensional reduction by Legendre transformation takes the supersymmetric theory off-shell. We have seen in the previous chapter that we can go back on-shell, i.e., eliminate the auxiliary fields, by means of their equations of motion, and the other canonical momenta by a suitable Lagrange multiplier technique. The result will be the same as if we had never performed the Legendre transformation in the first place and compactified right down to four dimensions. Therefore we can formulate the effect of "going on-shell" indirectly through the following

<u>RULE C:</u> The operations compactification ($\longrightarrow$), Legendre-transformation ($\leadsto$), and going on-shell ($\cdots\cdots\blacktriangleright$) commute.

So, for instance, the Sp(4) - covariant N=4 off-shell theory in four dimensions will become SU(4)-covariant if taken on-shell: the chiral projections of λ form a $\underline{4}$ of SU(4), and the $\underline{5}$ of Sp(4), namely ϕ_{sj} combines with the single spin-0 described by $\overline{V}_\mu$ to form a $\underline{6}$ of SU(4).

Actually, Rule C states more than what we just have made plausible. It states that, e.g., the chain of operations $10 \longrightarrow 5 \leadsto 4$ leads to the same N=4 off-shell theory as $10 \longrightarrow 7 \leadsto 6 \longrightarrow 4$ or $10 \longrightarrow 6 \leadsto 5 \longrightarrow 4$. Using Rules A and B, we would get the respective internal symmetry groups SP(4), SU(2)×U(1), SU(2) × SU(2) for the three presumably equivalent paths, in defiance of Rule C. We therefore have to augment our rules with

<u>RULE D:</u> If several possible paths between the same theories would naively generate different internal symmetry groups (according to Rules A,B), then it is the largest one of these groups which is actually realized.

Explicit calculation justifies this rule.

The following diagram shows two reduction paths, 10 ⟶ 5 ⇝ 4 and 10 ⟶ 7 ⇝ 6 ⟶ 4, which lead to the same N=4 off-shell theory in four dimensions (described in more detail in chapter 2). It also contains 10 4, which gives the on-shell N=4 theory directly. The latter, of course, can also be reached by going on-shell from the N=4 off-shell version.

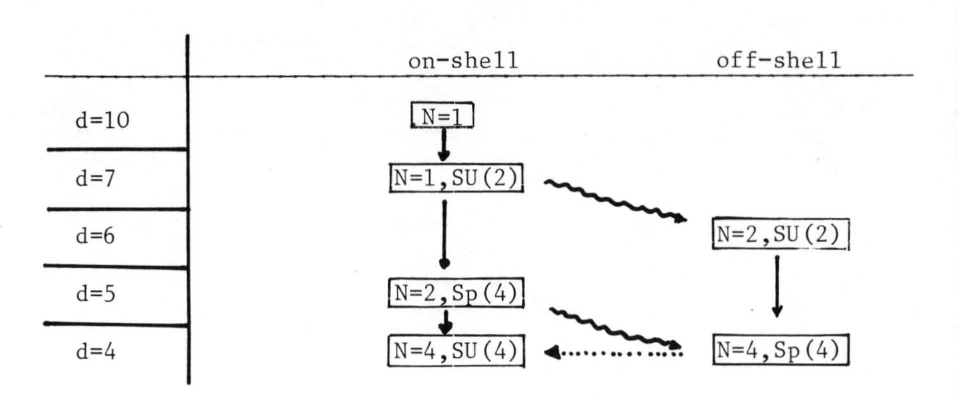

DIAGRAM 1. Reduction scheme for the maximally extended supersymmetric Yang-Mills theory.

We have indicated the internal symmetry group for each theory.

The label N denotes the total number of real spinorial charge components in the theory, divided by the smallest number n of real components which a Lorentz spinor must have in the respective dimension:

TABLE 1. The smallest possible number n of real components of a Lorentz covariant spinor in d-dimensional Minkowski space.

d	4	5	6	7	8	9	10	11
n	4	8	8	16	16	16	16	32

Thus the N=1 theory in ten dimensions, when reduced through seven, six, and five to four dimensions, stays N=1 for d=7, becomes N=2 for d=6 and d=5, and becomes N=4 for d=4.

This definition of N is consistent with the spirit of ref. [5] , where N is the "number of supersymmetries", although it is not always intuitively obvious. For example, the Sp(4) covariant Majorana condition

$$Q_{\alpha i} = i(\gamma_5 C)_{\alpha\beta}\, \Omega_{ij}\, \bar{Q}^{\beta j}$$

allows Q to have 16 real components, and therefore defines N=4 in four dimensions, but only N=2 in five dimensions, in spite of the fact that i takes values i=1,...,4. The reason behind this slight paradox is the impossibility of a Majorana condition for a single spinor field in five dimensions.

3.3 Truncation

From the very definition of N at the end of the previous subsection, we can see that for N>1 it is possible to impose further conditions on the spinors of such a theory without violating Lorentz covariance.

In particular, we can discard half the number of supersymmetry transformations by imposing some condition on the parameters. This we call a "truncation", e.g. from N=2 to N=1:

$$\boxed{N=2} \; - - - - - \rightarrow \; \boxed{N=1}$$

Let us consider an example. For an Abelian gauge group*), the chain 10→7⇝6 leads to the following off-shell N=2 theory in six dimensions (m,n=0,1,2,3;5,6).

$$\mathcal{L}_{(6)} = -\frac{1}{4} F_{mn} F^{mn} - \frac{1}{4} V_m V^m + \frac{1}{2}\partial_m \vec{\phi}\cdot\partial^m \vec{\phi} + \frac{1}{2}\vec{H}\cdot\vec{H}$$

$$- \frac{i}{4}\, \bar{\lambda}^i\, \gamma^m \overleftrightarrow{\partial}_m \lambda_i \qquad (3.16)$$

which is invariant under SU(2) and under the following set of supertransformations

$$\delta A_m = i\,\bar{\zeta}\gamma_m \lambda$$

$$\delta\vec{\phi} = i\,\bar{\zeta}\,\vec{\tau}\,\lambda$$

$$\delta\lambda = -\sigma_{mn}\zeta\, F^{mn} + \vec{\tau}\gamma^m \zeta\partial_m \vec{\phi} - \gamma_m\gamma_7 \zeta V^m + \vec{\tau}\gamma_7 \zeta \vec{H}$$

$$\delta V_m = -2i\bar{\zeta}\sigma_{mn}\gamma_7 \partial^n \lambda \qquad (3.17)$$

$$\delta\vec{H} = -i\bar{\zeta}\vec{\tau}\gamma_m\gamma_7\partial^m \lambda$$

if the constraint $\partial^m V_m = 0$ is imposed. λ_i and ζ_i are 16-component

*) see ref. [2] for the non-Abelian case

spinors with the SU(2)-Majorana condition [21] :

$$\zeta_{\alpha i} = C^{(6)}_{\alpha\beta}\,\varepsilon_{ij}\,\bar{\zeta}^{\beta j} \qquad \alpha,\beta = 1,\dots,8 \qquad i,j = 1,2 \quad . \tag{3.18}$$

Note that the reduction scheme gives this version of an N=2 theory in six dimensions, which is SU(2)-covariant, and not, e.g., one which has an unconstrained Dirac spinor.

We can now truncate by imposing a chirality condition

$$(1 - i\gamma_7)\zeta_i = 0 \tag{3.19}$$

which is compatible with (3.18), because $\gamma_7 C_{(6)} = -C_{(6)}\gamma_7^T$. As a result of (3.19), the multiplet splits up into a Yang-Mills multiplet, composed of A_μ, $(1+i\gamma_7)\lambda_i$, and $\vec{H}$, and a matter multiplet (in the adjoint representation of the gauge group), composed of $\vec{\phi}$, $(1-i\gamma_7)\lambda_i$, and V_μ. This matter multiplet describes four spin-0 and four spin-½ states and is the six-dimensional equivalent of the linear multiplet [22].

In this particular case, the SU(2) internal symmetry has survived the truncation, but the central charge invariance has not. This is, however, not generally true for all truncations, and we have to formulate

RULE E: The fate of internal symmetries and central charges under truncation is determined by the details of the particular case.

Or, in short, there is no rule for what truncations do.

The truncated N=1, SU(2) theory in six dimensions is, of course, the starting point for further dimensional reductions. Since it is already off-shell, we can compactify to four dimensions straight away. This splits off two scalar fields M and N from A_m, does not introduce a new central charge (the formerly present central charge related to x^7 has disappeared in the truncation), and leaves us with the Grimm-Sohnius-Wess [14] version of the N=2 off-shell supersymmetric gauge-theory,

$$\mathcal{L}_{(4)} = -\frac{1}{4}F_{\mu\nu}F^{\mu\nu} + \frac{1}{2}\partial_\mu M\partial^\mu M + \frac{1}{2}\partial_\mu N\partial^\mu N - \frac{i}{4}\bar{\lambda}^i\gamma^\mu\overleftrightarrow{\partial}_\mu\lambda_i + \frac{1}{2}\vec{H}\cdot\vec{H} \tag{3.20}$$

which has SU(2)×U(1) as internal symmetry (Rule A) and differs from its on-shell version only by the $\vec{H}^2$ term.

Another alternative is to take the theory back on-shell in six dimensions and then reduce to four via 6 ⟶ 5 ⤳ 4. The result is the version of the N=2 off-shell gauge theory which has a constrained vector V_μ(i.e. an antisymmetric tensor for an Abelian gauge group) and which has been described in Chapter 1. Its internal symmetry is only SU(2), since 6 ⟶ 5 ⤳ 4 cannot pick up additional internal symmetry, but it has a central charge, related to x^5. If we take this theory back on-shell, we end up with the same "physics" as for the off-shell version of ref. [14] without a central charge. In particular, we recover the larger symmetry SU(2)×U(1) according to Rule D.

Both alternatives of the N=2 off-shell theory could have been derived by truncation of the N=4 off-shell theory in four dimensions. The two truncations would break Sp(4) in different ways, one into SU(2)×U(1), one into SU(2). The central charge would survive in the second case, but not in the first.

By further truncation, each of the N=2 off-shell versions in four dimensions gives the N=1, U(1) theory of Ferrara-Zumino-Salam-Strathdee [7] .

The manipulations described in this subsection can be combined into Diagram 2. The left-most column of this diagram connects it with Diagram 1.

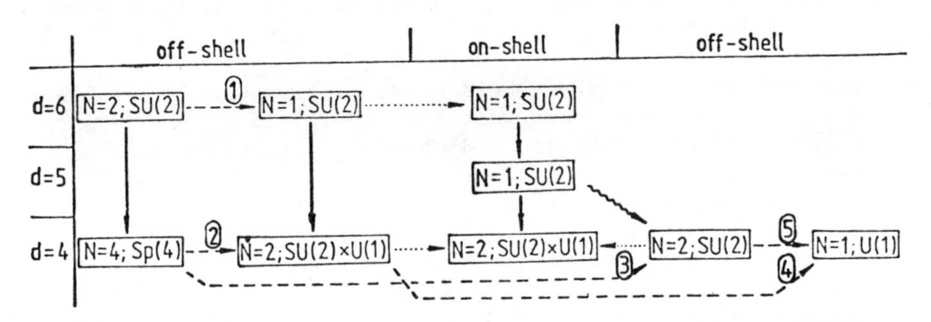

DIAGRAM 2. Reduction scheme for N=2 type Yang-Mills theories

The five different truncations in Diagram 2 produce as "side products" the various matter multiplets of supersymmetric Yang-Mills theories :

1. the six-dimensional linear multiplet.
2. the N=2 linear multiplet in four dimensions.
3. the scalar hypermultiplet.
4. the Wess-Zumino chiral multiplet.
5. the linear multiplet.

The first three of these describe each four spin-0 and four spin-$\frac{1}{2}$ states, the last two of these describe each two spin-0 and two spin-$\frac{1}{2}$ states.

3.4. Superspace

The off-shell version for the N=4 theory contains a vector field V_μ which is constrained by a differential condition (see chapter 2). If the gauge group is Abelian, the constraint can be explicitly solved in terms of an antisymmetric tensor field. But the Abelian theory is non-interacting. So far no solution for the general constraint is known.

Since the original reason behind the search for off-shell supersymmetric theories was to facilitate quantum calculations with supermultiplets, a non-algebraic differential constraint of this type is not acceptable. The constraint has to be solved either explicitly or implicitly, by use of a Lagrange multiplier. The Lagrange multiplier procedure that we have indicated does indeed give us the proper on-shell theory upon elimination of V_μ, but it is not supersymmetric, and therefore again does not meet the aim to make our off-shell theory workable. Eventually one will have to introduce a whole multiplet of Lagrange-multipliers, which will correspond to a multiplet of constraints, only one of which will be of a non-algebraic nature. To find the full set of constraints one will have to use a full superspace treatment, where the gauge parameter is a superfield itself.

This program has not yet been carried through. We do, however, know and want to present here what the superfield constraints on the Yang-Mills field strengths are and how the Lagrangians look in superspace:

In ten dimensions, we are dealing with a theory that contains only A_m and λ_α. The constraint must therefore be

$$F^{(10)}_{\alpha\beta} = 0 \quad ; \quad \alpha,\beta = 1.....32 \quad , \tag{3.21}$$

since otherwise $F^{(10)}_{\alpha\beta}$ would correspond to a covariant bosonic field of dimension 1. This constraint implies field equations by means of the Bianchi-identities. This is as expected, since we know that we are dealing with an on-shell theory.

$F^{(10)}_{\alpha\beta}$ is chiral in both indices and has the reality properties of a product of two Majorana spinors. We can use the notation of subsection 3.2 and write the constraint in the form

$$F^{(10)}_{\alpha i\beta j} = 0 \quad ; \quad \alpha,\beta,i,j = 1,.....,4. \tag{3.22}$$

The curvature in ten dimensions contains, of course, the usual $(\Gamma^m C^{(10)})_{\alpha\beta} A_m$ counterterm due to the torsion. In the four-dimensional notation, we then have

$$F^{(10)}_{\alpha i \beta j} = D_{\alpha i} \mathcal{A}_{\beta j} + D_{\beta j} \mathcal{A}_{\alpha i} + i\{\mathcal{A}_{\alpha i}, \mathcal{A}_{\beta j}\} \tag{3.23}$$

$$- 2(\gamma^\mu \gamma_5 C)_{\alpha\beta}\, \Omega_{ij} A_\mu - 2\, C_{\alpha\beta} \Omega_{ij} A_5 + 2(\gamma_5 C)_{\alpha\beta} (t_s \Omega)_{ij}\, \phi_s \, .$$

The algebra of our four-dimensional N=4 theory, however, contains only the four-momentum and one central charge on the right-hand side. Therefore the curvature in four dimensions is

$$F^{(4)}_{\alpha i \beta j} \equiv D_{\alpha i} \mathcal{A}_{\beta j} + D_{\beta j} \mathcal{A}_{\alpha i} + i\{\mathcal{A}_{\alpha i}, \mathcal{A}_{\beta j}\} \, . \tag{3.24}$$

$$- 2\, (\gamma^\mu\, \gamma_5 C)_{\alpha\beta}\, \Omega_{ij}\, A_u - 2\, C_{\alpha\beta}\, \Omega_{ij} A_5$$

and the constraint which corresponds to $F^{(10)}_{\alpha\beta} = 0$ is

$$F^{(4)}_{\alpha i \beta j} = - 2(\gamma_5 C)_{\alpha\beta}\, (t_s \Omega)_{ij} \phi_s \, . \tag{3.25}$$

We know that this constraint implies equations of motion [23] if there is no central charge. But now we also know that it does not imply equations of motion if we keep the central charge. The action is

$$I_{(4)} \sim \int d^4x\ \bar{D}^i \gamma_5\, D_j\ \mathrm{Tr}\ \bar{\lambda}^j \gamma_5 \lambda_i \tag{3.26}$$

where λ is defined through

$$(t_s)_i{}^j\ \lambda_j = D_i \phi_s \, . \tag{3.27}$$

In the case of six dimensions,

$$F^{(6)}_{\alpha\beta} = 0 \tag{3.28}$$

does not imply field equations. In the absence of a central charge $(6 \longrightarrow 4)$, this corresponds to

$$F^{(4)}_{\alpha i\beta j} = \varepsilon_{ij} \; ((A_5+\gamma_5\phi)C)_{\alpha\beta} \tag{3.2}$$

and describes the theory of ref.[14] with the action

$$I_{(4)} \sim \int d^4x \;\; \bar{D}^i\gamma_5 D_j \bar{D}^j \gamma_5 D_i \; \mathrm{Tr}\; (\phi^2 - A_5^2). \tag{3.3}$$

If we go on-shell in six dimensions, and then perform $6 \longrightarrow 5 \rightsquigarrow 4$ we keep a central charge,and get

$$F^{(4)}_{\alpha i\beta j} = \varepsilon_{ij}(\gamma_5 C)_{\alpha\beta}\phi \; , \tag{3.3}$$

and, as a constraint, the remnant of the six-dimensional field equation

$$(\bar{D}_i D_j \; + \bar{D}_j D_i)\;\phi = \; 0. \tag{3.3}$$

The Lagrangian is formally identical to (3.26), only that now $i,j = 1,2$ only, and $\lambda_i \equiv D_i\phi$.

4. CONCLUSION

The techniques presented in this article for finding the off-shell formulations of supersymmetric Yang-Mills theories can equally well be applied to supergravity theories. The off-shell formalism which avoids high-spin auxiliary fields through the action of a central charge has a direct analogue for extended supergravity theories. In analogy with the Yang-Mills case, the gauge invariances in supergravity will give rise to constraints. These constraints will make the canonical momenta propagate, just as we have found for the momentum V_μ, conjugate to the Yang-Mills field. A method of implementing these constraints is that of the Lagrange multiplier. However, in order to maintain an off-shell supersymmetric formalism, we must implement a whole supermultiplet of constraints, using a supermultiplet of Lagrange multipliers. The problem remains to find a systematic method of finding such multiplets.

It is clear that such Lagrange multiplier supermultiplets will contain spinor fields which will appear as auxiliary fields in the final formulation. Such spinor auxiliary fields are of course alrea known in the off-shell formulation of N=2 supergravity [24] and we expect them to be a general feature of all extended supergravity theories.

REFERENCES

1. M.F. Sohnius, K.S. Stelle, P.C. West, "Off-Mass-Shell Formulation of Extended Supersymmetric Gauge Theories", to be published in Physics Letters B, (1980).
2. M.F. Sohnius, K.S. Stelle, P.C. West, "Dimensional Reduction by Legendre Transformation Generates Off-Shell Supersymmetric Yang-Mills Theories", to be published in Nucl. Phys. B, (1980).
3. S. Coleman and J. Mandula, Phys. Rev. 159 (1967) 1251.
4. Y.A. Gol'fand and E.P. Likhtman, J.E.T.P. Lett. 13 (1971) 323; D.V. Volkov and V.P. Akulov, Phys. Letters 46B (1973) 109; J. Wess and B. Zumino, Nucl. Phys. B70 (1974) 39.
5. R. Haag, J.T. Lopuszanski and M.F. Sohnius, Nucl. Phys. B88 (1975) 257.
6. A. Salam and J. Strathdee, Nucl. Phys. B76 (1974) 477, Phys. ReV. D11 (1975) 1521.
7. S. Ferrara and B. Zumino, Nucl. Phys. B79 (1974) 413; A. Salam and J. Strathdee, Phys. Letters 51B (1974) 353.
8. Article by S. Ferrara and P. van Nieuwenhuizen in Supergravity edited by P. van Nieuwenhuizen and D.Z. Freedman (North Holland, Amsterdam, 1979), p. 221; P. van Nieuwenhuizen and P.C. West, "From Conformal Supergravity in Ordinary Space to Its Superspace Constraints", to be published in Nuclear Physics B, (1980). Further references are contained in these articles.
9. J. Wess, "Supersymmetry-Supergravity", in Lecture Notes in Physics Vol. 77 (Springer, Berlin, 1978), p. 81.
10. Articles by S.J. Gates, Jr., and by K.S. Stelle and P.C. West in Supergravity edited by P. van Nieuwenhuizen and D.Z. Freedman (North Holland, Amsterdam, 1979) p. 215 and 133 respectively; S.J. Gates, Jr., K.S. Stelle and P.C. West, "Algebraic Origins of Superspace Constraints in Supergravity", to be published in Nuclear Physics B, (1980).
11. K.S. Stelle and P.C. West, Phys. Letters 90B (1980) 393.
12. P. Fayet, Nucl. Phys. B113 (1976) 135.
13. M.F. Sohnius, Nucl. Phys. B138 (1978) 109.
14. R. Grimm, M. Sohnius and J. Wess, Nucl. Phys. B133 (1978) 275.
15. K.S. Stelle, Gen. Rel. and Grav. 9 (1978) 353; S. Ferrara, M.T. Grisaru and P. van Nieuwenhuizen, Nucl. Phys. B138 (1978) 430; B. de Wit and J.W. van Holten, Nucl. Phys. B155 (1979) 530.
16. A. Salam and J. Strathdee, Nucl. Phys. B80 (1974) 499; W. Nahm, Nucl. Phys. B135 (1978) 149.
17. H. Hayashi, Phys. Letters 44B (1973) 497; M. Kalb, P. Ramond, Phys. Rev. D9 (1974) 2279; E. Cremmer, J. Scherk, Nucl. Phys. B72 (1974) 117;

Y. Nambu, Phys. Rep. 23C (1976) 250;
P.K. Townsend, Phys. Letters 88B (1979) 97.
18. B. de Wit and S. Ferrara, Phys. Letters 81B (1979) 317.
19. J. Wess and B. Zumino, Nucl. Phys. B78 (1974) 1.
20. F. Gliozzi, J. Scherk and D. Olive, Nucl. Phys. B133 (1978) 253;
L. Brink, J. Schwarz and J. Scherk, Nucl. Phys. B121 (1977) 77.
21. P. Breitenlohner and A. Kabelschacht, Nucl. Phys. B148 (1979) 96.
22. W. Siegel, Phys. Letters 85B (1979) 333.
23. M. Sohnius, Nucl. Phys. B136 (1978) 461.
24. B. de Wit and J.W. van Holten, Nucl. Phys. B155 (1979) 530;
A.S. Fradkin and M.A. Vasiliev, Phys. Letters 85B (1979) 47;
P. Breitenlohner and M. Sohnius, Nucl. Phys. B165 (1980) 483;
B. de Wit, J.W. Holten and A. Van Proeyen, Nucl. Phys. B167 (1980) 186.

WHAT IS SUPERGRAVITY AND WHICH OF ITS GOALS HAVE BEEN REACHED?

P. van Nieuwenhuizen

Institute of Theoretical Physics
State University of New York
at Stony Brook, N.Y. 11794

DEDICATION

This contribution is dedicated to my friend Joel Scherk, whose last scientific activity was this conference and who died two months later.

Joel was, in my opinion, the most creative supergravity practitioner. His work in extended supergravity and dimensional reduction was unusually inventive and elegant. His death is a personal blow for many of us. It is a setback for the Ecole Normale in Paris and for CERN, where he would have started a six-year appointment in September. For supergravity the loss is irreplaceable.

1. INTRODUCTION

It is a relief to show you the beautiful vistas of supergravity now that we have been engulfed here on top of this mountain in fog and rain for more than a week. Local Fermi-Bose symmetry (= local supersymmetry) can only be implemented in curved spacetime. Thus supergravity is born by joining particle physics and relativity. To promote contact and further offspring this school was organized.

This contribution consists of three disjoint parts. In section 2 we summarize in how far the goals of supergravity as formulated in 1976 have been reached. In section 3 we show how a particle physicist would discover supergravity if he just went ahead and repeated what one does in electromagnetism where one starts with, say, charged pions, and promotes the global phase invariance to a local invariance by introducing the photon gauge field. In section 4 I will put a series of new results into an old perspective.

2. AIMS VERSUS RESULTS

Supergravity is the gauge theory of supersymmetry. Its two main goals were and are

(i) the unification of gravity with the strong, electromagnetic and weak interactions (super unified as opposed to grand unified theories)

(ii) to be a completely finite quantum theory of gravity.

The progress made since the discovery of supergravity (ref. 1,2) in the beginning of 1976 is impressive but we are nowhere near a complete solution. Let us summarize.

The most interesting model of supergravity is the N=8 model (ref. 3) with eight gravitini. The gravitino is the fermionic counterpart of the graviton and has spin 3/2. Initially massless and real, it can acquire a mass through spontaneous symmetry breaking (see below). In this way all gravitini and spin 1/2 particles become massive. The results are shown in Figure 1 (ref. 4). In addition to massless gluons and the photon, there is a tenth massless vector field which is either the Z° (which then should acquire a mass through further spontaneous symmetry breaking) or the antigraviton proposed by J. Scherk. There are up, down and charmed quarks with the correct charges (charges are predicted, not put in by hand!), but absent as elementary particles are: $W^{+}, W^{-}, \mu^{-}, \tau^{-}, \nu_e, \nu_\mu, \nu_\tau$ and the strange quarks which have been eaten by the gravitini (which then become massive as a result of this super-Higgs effect). The central problem for the future is to show that these missing particles appear as bound states.

As far as quantum supergravity is concerned, note that in theories with a dimensional coupling constant the counterterms ΔL have different functional form from the action L. Thus one cannot renormalize and absorb the infinities by rescaling of the physical

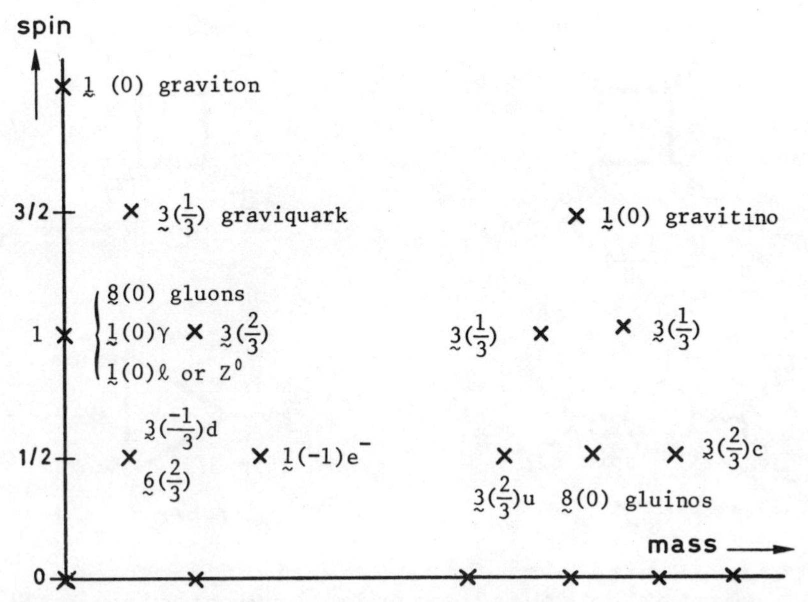

Figure 1. Spectrum of the spontaneously broken N=8 model with SU(3) x U(1) x U(1) invariance. The SU(3) representations are underlined (ex: 3); the bracket indicates the electric charge.

parameters in the original action, and one either has incurable divergences or otherwise finiteness due to "miraculous" cancellations. In supergravity the Fermi-Bose symmetry leads to such miraculous cancellations in all one- and two-loop S-matrix elements. The first time this was found in the case of photon-photon scattering in the N=2 model (ref. 5); see Figure 2. However, when one couples to matter (N-extended matter with N global supersymmetry invariances coupled to the pure N-extended gauge action) finiteness is lost. In N=1 (and N=2 also, see deWit's talk) one can prove with the tensor calculus that at 3, 4, 5, etc. loop level there exist dangerous off-shell invariants which are candidates for ΔL and which would destroy finiteness if their coefficients were nonzero. For N>2 this is not so clear; in particular constructions up to order κ^2 are not sufficient to prove this. The recent cancellations in N>4 theories (see Duff's talk) are encouraging: they herald the discovery of new symmetries which might serve to exclude the 3 and higher loop counterterms.

These results are entirely equivalent to what one

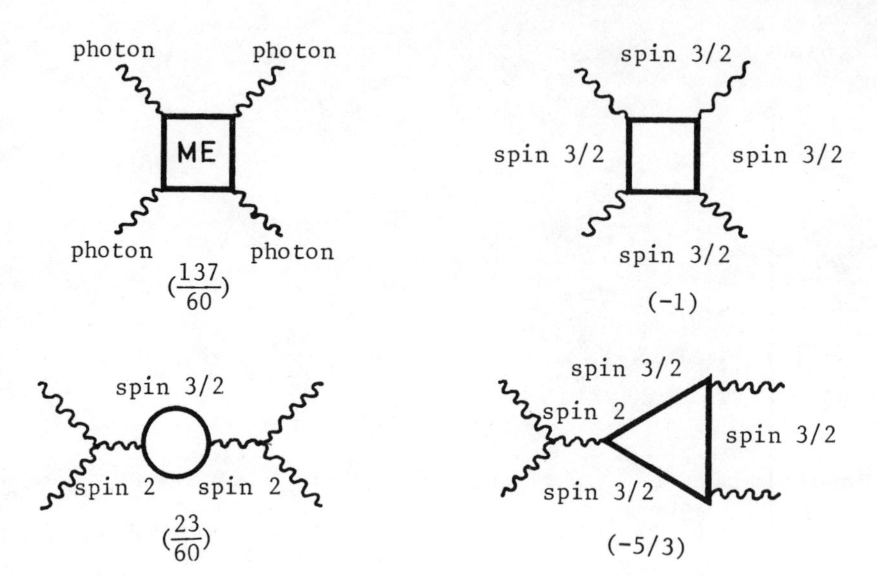

Figure 2. Finiteness of the S-matrix for photon-photon scattering in N=2 extended supergravity. All infinities are proportional to the energy-momentum tensor squared. The first diagram consolidates all diagrams with only photons and gravitons. Adding the diagrams with gravitini, the sum of the coefficients of the infinities cancels.

has in ordinary Einstein gravity, where processes with only gravitons are one-loop finite but where finiteness is lost whenever one couples to matter. The only difference is that, due to the extra Bose-Fermi symmetry, one can prove that supergravity conserves helicity and this enables one to prove two-loop finiteness (ref. 6). It is at present not known whether also ordinary quantum Einstein gravity preserves helicity, but if it does it will also be two-loop (and 3, 4, 5 loop) finite as was shown six years ago (ref. 7).

We mentioned the new kind of spontaneous symmetry breaking (ref. 8). Let us explain how it works by considering a scalar field $\varphi(x^1, \cdots x^5)$ in five dimensions. By choosing a special class of fields $\varphi(x^1,\cdots x^5) = \hat{\varphi}(x^1,\cdots x^4) \exp iMx^5$ it is clear that a scalar which is massless in five dimensions, can acquire mass in four dimensions (using Stelle's five dimensional d'Alembertian)

$$\pentagon \varphi = (\Box - M^2)\,\varphi = 0$$

The basic idea is that one can introduce in a natural way a massive parameter because coordinates have the dimension of $(\text{mass})^{-1}$.

In the N=8 model one introduces masses in the following way (ref. 9). First one writes down this model as N=1 theory in d=11 dimensions (it contains then an 11 x 11 vielbein $e^m{}_\mu$, an 11 x 32 gravitino ψ^a_μ and an antisymmetric tensor $A_{\mu\nu\rho}$). Then one reduces to N=8 theory (i.e. with 8 gravitini) in d=5 dimensions. This theory is massless and its obvious symmetry group is O(6) local (from O(11) → O(3,1) x O(6)) and SL(6,R) global (due to general coordinate transformations with parameters $\xi^\alpha = M^\alpha{}_\beta y^\beta$ with constant traceless M where α and β refer to the (compactified) six dimensional space). However, this symmetry group can be extended to Sp(8) local ⊗ E(6) global. One now introduces masses as in (1) but M is now an Sp(8) matrix, and since Sp(8) has rank 4, one introduces in this way four arbitrary masses.

In this way one arrives at N=8 d=4 massive supergravity by reduction of the N=8 d=5 massless theory down to four dimensions. The crucial question now is whether this theory is still finite, in other words whether the symmetry breaking is so soft that infinities still cancel in the S-matrix. This question is under study by E. Szesgin and myself at Stony Brook. Note that if we would start from an N<8 theory in five dimensions, one would end up in four dimensions with a matter coupled system and such systems are already infinite when there are no masses present. This again stresses the exceptional role the N=8 model plays.

All particles in pure N-extended supergravity form an irreducible representation of a new kind of Lie algebra, namely super-Lie algebras. The easiest way to understand what a super-Lie algebra is, is to consider canonical transformations which mix the creation and absorption operators of K bosons and L fermions (ref. 10). These transformations define of course a group, and the generators define the ortho-symplectic superalgebra OSp(K,K|2L). Contact with field theory is made by gauging such algebras. Just as general relativity is obtained by gauging the de Sitter algebra O(3,2) ≃ Sp(4), N=1 extended supergravity is obtained (ref. 11) by gauging OSp(1/4) and the N=2 case results are recovered (ref. 12) when one gauges OSp(2/4). (For an approach to gauging using the group manifold, see Fré's talk.)

Following a discussion at this conference regarding the N=8 theory in d=11 dimensions (where its only fields are e^m_μ, ψ^α_μ and $A_{\mu\nu\rho}$ as we mentioned above), Nicolai, Townsend and the author tried to reobtain this action (ref. 13) by gauging OSp(1/32). (Spinors in d=11 have 32 components.) The generators of Sp(32) are Γ^m, Γ^{mn} and Γ^{mnrst} (where the eleven Γ^m satisfy $\{\Gamma^m, \Gamma^n\} = 2\delta^{mn}$ and are 32 x 32 Dirac matrices). Clearly the gauge fields of Γ^m are $e^m{}_\mu$ and of Γ^{mn} are the spin connection $\omega_\mu{}^{mn}$, but if one associates with Γ^{mnrst} a six index photon $A_\mu{}^{mnrst}$ then it has only the same number of states as $A_{\mu\nu\rho}$ if it is completely antisymmetric in all its indices. In that case counting shows that one might only need one auxiliary field, namely an antisymmetric two-index field $t^{\mu\nu}$!!(With $A_{\mu\nu\rho}$ one needs much more auxiliary fields.) However, although group theory straightfowardly leads to a globally supersymmetric action, one cannot extend it to a locally supersymmetric action ! (This is thus a counter-example to the often stated claim that this can always be done.) In fact, one only runs into problems at the point where the $A_{\mu\nu\rho}$ formulation needs the gauge invariant coupling in the action

$$\epsilon^{\mu_1\cdots\mu_{11}} F_{\mu_1\cdots\mu_4} F_{\mu_5\cdots\mu_8} A_{\mu_9\cdots\mu_{11}}$$

This failure is exciting since apparently much new geometry must be discovered before group theory can reproduce all actions in supergravity.

3. FROM GLOBAL TO LOCAL SUPERSYMMETRY (ref. 14)

Consider as a globally supersymmetric model two free massless bosons and fermions, namely a scalar A, a pseudoscalar B and a spinor λ^α (α = 1,4). The spinor describes only two states since it is a Majorana spinor,* which means that its Dirac conjugate

*A Majorana conjugate spinor $\bar\lambda^M$ is required to transform in the same way as the Dirac conjugate spinor $\bar\lambda^D$. This leads to the condition that $C\sigma^{mn}$ be symmetric. For a Majorana spinor one requires in addition that $\bar\lambda^M$ satisfies the same Dirac equation as $\bar\lambda^D$. This leads to the stronger condition that

$\bar{\lambda}_\beta = \lambda^{\dagger}_{\alpha} (\gamma_0)^{\alpha}{}_{\beta}$. equals its Majorana conjugate $\bar{\lambda}_\beta = \lambda^{\alpha} C_{\alpha\beta}$. The action is the sum of Klein-Gordon and Dirac actions and is invariant under

$$\delta A = \bar{\epsilon}\lambda \ , \ \delta B = -\frac{i}{2}\bar{\epsilon}\gamma_5\lambda \ , \ \delta\lambda = \frac{1}{2}(\partial\!\!\!/(A - i\gamma_5 B))\epsilon$$

$$\mathcal{L}^0 = -\frac{1}{2}(\partial_\mu A)^2 - \frac{1}{2}(\partial_\mu B)^2 - \frac{1}{2}\bar{\lambda}\partial\!\!\!/\lambda \qquad (1)$$

Covariance of the generic law δ(boson) = fermion times ε, of which (1) gives two examples, requires that ε be (i) a Majorana spinor, (ii) anticommuting, (iii) have spin 1/2 and (iv) have dimension (-1/2). Indeed, since the dimension of A equals [A] = 1 and $[\lambda] = 3/2$, one has $[\varepsilon] = -1/2$, see again (1). Thus in the reverse generic law δ(fermion) = boson times ε there is a gap of one unit of dimension. In flat spacetime without masses there is only one object which can fill this gap: a derivative. Indeed, in $\delta\lambda$ in (1) a derivative is present. Making two successive supersymmetry transformations on A, one easily derives

$$[\delta(\epsilon_1) \ , \delta(\epsilon_2)] A = \frac{1}{2}(\bar{\epsilon}_2\gamma^\mu\epsilon_1)\partial_\mu A \qquad (2)$$

Thus two internal Fermi-Bose rotations lead to a translation, and one has here a unification of internal and spacetime symmetries.

For B the same result (2) is found. For λ one finds extra terms proportional to $\partial\!\!\!/\lambda$. Thus only on-shell (2) holds for λ, but one can extend the validity of (2) for λ off-shell by introducing two extra fields F and G, whose variations are proportional to $\partial\!\!\!/\lambda$. Adding to L° a term $\frac{1}{2}(F^2+G^2)$ and to $\delta\lambda$ a term $\frac{1}{2}(F + i\gamma_5 G)\epsilon$, one sees that F and G are auxiliary fields, and that one now has as many bosonic as fermionic fields, namely four λ^α and four A,B,F,G. For all eight of them (2) holds.

$C\gamma^m$ be symmetric. The matrix C is the charge conjugation matrix. It is antisymmetric and unique up to a constant.

An interesting alternative (ref. 15) is to use as field representation for the scalar field not A but the antisymmetric tensor $t^{\mu\nu}$ with action $(\partial_\mu t^{\mu\nu})^2$. This field is a gauge field (the action is invariant under $\delta t^{\mu\nu} = \epsilon^{\mu\nu\rho\sigma}\partial_\rho \Lambda_\sigma$), and has thus 3 field components, namely 6 minus the 3 parameters Λ_σ. (The fourth component $\Lambda_\sigma = \partial_\sigma \Lambda$ cancels in δt.) Together with B and λ one has equal numbers of bosonic and fermionic fields as well as states, and one can easily find transformation laws which leave the action invariant. These transformations satisfy (2) again uniformly. We conjecture that one can always eliminate auxiliary fields by choosing an appropriate field representation.

Let us now go to local supersymmetry. One expects in (2) for local $\epsilon^\alpha(x)$ translations ∂_μ over distances $d^\mu = \frac{1}{2}\bar{\epsilon}_2(x)\gamma^\mu\epsilon_1(x)$ which differ from point to point, hence general coordinate transformations and hence gravity. Let us suppress this thought for a moment.

If one varies the action I° using local parameters in (1), one finds on general grounds

$$\delta I^0 = \int d^4x \left(\partial_\mu \bar{\epsilon}_\alpha\right)\left(j^{\mu,\alpha}_N\right) \tag{3}$$

where $j^{\mu,\alpha}_N$ is the Noether current which is a spinorial vector and which is as always conserved on-shell

$$j^\mu_N = \frac{1}{2}\left(\not{\partial}\left(A + i\gamma_5 B\right)\right)\gamma^\mu\lambda \tag{4}$$

We cancel, as in Yang-Mills theory, δI^0 by introducing a gauge field which varies into the derivative of the parameter plus more, and coupling this gauge field to the Noether current

$$\mathcal{L}^N = \kappa\bar{\psi}_\mu j^\mu_N \, , \delta\psi^\alpha_\mu = \frac{1}{\kappa}D_\mu\epsilon^\alpha + \text{more} \tag{5}$$

<u>Thus the gauge field of supersymmetry is a vectorial spinor (or a spinorial vector) and contains spin 3/2.</u> Since $[\psi] = -3/2$, the constant κ has dimension, $[\kappa] = -1$. Let us resist the idea that this will be

the gravitational constant.

Varying $I^{\circ} + I^{N}$ under (1) and (5), all order κ° terms cancel, and to order κ one finds for the terms bilinear in A and B the following result

$$\delta\left(I^{\circ}+I^{N}\right) = \int d^4x \left(\frac{\kappa}{2}\bar{\Psi}_{\mu}\gamma_{\nu}\epsilon\right)\left(T^{\mu\nu}(A)+T^{\mu\nu}(B)\right) + \kappa\left(\bar{\Psi}_{\mu}\gamma_5\gamma_{\nu}\epsilon^{\mu\nu\rho\sigma}\right)\left(i\gamma_5\epsilon\right)\left(\partial_{\rho}A\,\partial_{\sigma}B\right) \tag{6}$$

The symbol $T^{\mu\nu}(A)$ is the energy-momentum tensor of A, defined by $\partial^{\mu}A\partial^{\nu}A - \frac{1}{2}\eta^{\mu\nu}(\partial_{\lambda}A)^2$. These terms can be cancelled as follows

(i) Since $T^{\mu\nu}(A)$ is the Noether current of translations, we repeat our strategy and couple L° in (1) to the gauge field of translations. If one chooses its transformation law appropriately, the $T^{\mu\nu}(A)$ and $T^{\mu\nu}(B)$ terms in (6) are cancelled simultaneously. Since 1915 we know how to do this to all orders at once: we put L° in curved space. There is no other way to cancel $T^{\mu\nu}(A)$. Thus gravity unavoidably enters the stage. All previous results go over into their curved space analogues. For example, the Noether current becomes (det e) times $\gamma^{\lambda}(\partial_{\lambda}(A+i\gamma_5 B))\gamma^{\mu}\lambda$ where $\gamma^{\mu} = e_m{}^{\mu}\gamma^m$ with constant γ^m. The symbol $e_m{}^{\mu}$ is the vierbein field. It is the square root of the metric $g_{\mu\nu}$, and is needed since fermions are present. Thus det e $= (g)^{1/2}$ with g = det $g_{\mu\nu}$. If one now requires that

$$\delta e^m{}_{\mu} = -\frac{\kappa}{2}\bar{\Psi}_{\mu}\gamma^m\epsilon, \quad e^m{}_{\mu}e^{\mu}{}_n = \delta^m{}_n \tag{7}$$

then the AA and BB terms in (6) cancel. Hence we deduced the vierbein variation law.

(ii) Partially integrating $\partial_{\rho}A$ and $\partial_{\sigma}B$ in (6) (one half of each; any other combination would work at the order κ level as well, but would lead to variations of order κ^2 which one cannot cancel*) one finds two kinds of terms, namely terms with $\partial_\rho \epsilon$ and terms with $\partial_\rho \bar{\psi}_\mu$. The former are easily cancelled by adding a seagull-type of term to the action, of the form $\mathcal{L}^4 \sim \kappa^2 \bar{\psi}\psi A \partial B$. Indeed, using $\delta\psi_\mu = \kappa^{-1}\partial_\mu \epsilon$ in $\mathcal{L}^4$ one can cancel the $\partial_\rho \epsilon$ terms. More interesting are the $\partial_\rho \bar{\psi}_\mu$ terms. They are cancelled by requiring that

$$R^\sigma = D_\rho \bar{\psi}_\mu \gamma_5 \gamma_\nu \epsilon^{\mu\nu\rho\sigma} \tag{8}$$

be the classical gravitino field equation and by adding a new term to the gravitino transformation law

$$\delta\psi_\mu = \kappa^{-1} D_\mu \epsilon + \frac{\kappa}{2}(i\gamma_5 \epsilon)(A \overleftrightarrow{\partial}_\mu B) \tag{9}$$

From (8) <u>we deduce the gravitino action</u>

$$\mathcal{L}^{3/2} = -\frac{1}{2}\epsilon^{\mu\nu\rho\sigma} \bar{\psi}_\mu \gamma_5 \gamma_\nu D_\rho \psi_\sigma \tag{10}$$

and indeed, if we substitute the last term in (9) into (10) we cancel the second term in (6). This possibility of deriving the gauge action is a feature not present in Yang-Mills theory. (Indeed, a perverse gauge action $(G_{\mu\nu}^{\ a})^2$ times $(G_{\rho\sigma}^{\ b})^2$ cannot be ruled out if one only requires invariance of the action.)

Since we are in curved space, we have replaced $\partial_\mu \epsilon$ in (9) by $D_\mu \epsilon = \partial_\mu \epsilon + \frac{1}{2}\omega_\mu^{\ mn}\sigma_{mn}\epsilon$. The spin connection $\omega_\mu^{\ mn}$ will always be expressed in terms of other fields by solving its field equation $\delta I(\text{total})/\delta\omega_\mu^{\ mn} = 0$. This means three things: (i) there is torsion induced by gravitinos, (ii) the transformation law for $\omega_\mu^{\ mn}$ follows from the chain

*This should be a warning for those people who analyze counter terms only to a given order in κ and then claim that they exist. In fact, we believe quite the contrary: extra symmetries will rule out such dangerous counter terms.

rule and (iii) in the action one never needs to vary the spin connection, since its variation is always multiplied by $\delta I/\delta\omega = 0$. This last observation is known as 1.5 order formalism and is so obvious that some people never understand it. From now on we will use 1.5 order formalism and forget about the spin connection.

To order κ there remain variations in $\delta(I^{\circ} + I^{N})$ of the generic form $\kappa \bar{\psi} \epsilon \bar{\lambda} \lambda$. They come from (7) into δI° and (1) into δI^{N}. All these terms can be cancelled by adding again extra terms to the action of order κ^2 and with two gravitini, and extra terms to $\delta\psi_{\mu}$ of order κ of the form $\kappa \bar{\lambda}\lambda \epsilon$, and (again a new feature not present in Yang-Mills theory) extra terms in $\delta\lambda$ of the form $\kappa \bar{\psi} \lambda \epsilon$.

At this point we must pursue two tracks. First of all we must take care of the $\delta\psi_{\mu} = \kappa^{-1} D_{\mu}\epsilon$ variation in $I^{3/2}$. This will complete the gauge action of supergravity. Then we must proceed to higher orders in κ iteratively to obtain our invariant action to all orders in κ.

Varying $\delta\psi_{\mu} = \kappa^{-1} D_{\mu}\epsilon$ in $I^{3/2}$ in (10) one finds a commutator of covariant derivatives (due to the ε symbol) and hence a (Riemann) curvature. The generic form of this variation is $\kappa^{-1} \bar{\psi}_{\tau}\epsilon R_{\mu\nu\rho\sigma}$ but since there are not enough indices available, only the contracted curvatures appear. Thus $\delta I^{3/2}$ is proportional to the Einstein tensor $G^{\mu\nu}$. The only way to cancel this term is to choose the Hilbert action $I^2 = \kappa^{-2}\int d^4x \, \det e \, R(e,\omega)$ as bosonic counterpart of $I^{3/2}$, since δI^2 is proportional to $G^{\mu\nu}$. Since we need not vary $\omega_{\mu}{}^{mn}$ (1.5 order formalism), a consistency test develops: does $\delta(I^2 + I^{3/2})$ cancel to order κ^{-1} if one uses (7)? It does. In fact, $I^2 + I^{3/2}$ is already invariant to all orders in κ under (7) and $\delta\psi_{\mu} = \kappa^{-1}D_{\mu}\epsilon$. Hence we derived that the bosonic part of the gauge action is the Hilbert action. This had better be true, because if we had found instead, say, $I^2 = \det e \, R^2$, then ω would have been propagating and 1.5 order formalism would not have been possible with all attendant algebraic miseries.

One can now go on, and make also the system (gauge action + matter action) invariant under local supersymmetry variations to all orders in κ, by adding terms to the action and the transformation laws (of the fermions). This has been done (ref. 16).

One thus finds matter fields in the transformation law of the gauge fields (namely of ψ_μ, see (9)). This precludes the possibility to add two invariant matter actions such that the sum is again invariant. Instead, each time one adds a matter system to the gauge action, one has to carry the Noether yoke and must painfully adjust the transformation rules such that the final system is invariant. (One also needs terms in the action bilinear in the fields of both matter systems.) The way out of this hornets' nest is to introduce Lagrange multipliers which are called in supergravity auxiliary fields.

We illustrate the procedure by means of the example in (9). If one replaces (9) by

$$\delta\psi_\mu = \kappa^{-1} D_\mu \epsilon + \tfrac{1}{2} A_\mu (i\gamma_5 \epsilon) \tag{11}$$

and adds to the action two more terms

$$\mathcal{L}(A_\mu) = -\tfrac{1}{3} \det e\, A_\mu^2 + \tfrac{2\kappa}{3} A^\mu (A \overleftrightarrow{\partial}_\mu B) \tag{12}$$

(all normalizations are arbitrary) then one would recover (9) if one were to solve the A_μ field equation and substitute the result in (11). Instead we consider A_μ as an independent field. Now there is no $\delta\psi \sim \kappa(A\partial B)\gamma \epsilon$ term to cancel the last term in (6) but if we require that δA_μ is proportional to the gravitino field equation in (8), then the same results are obtained as far as cancellation of the terms in (6) are concerned. Up to this point we merely introduced a new field A_μ and defined new couplings and δA_μ. Now, however, a consistency test develops. Do the variations of the form $A_\mu \partial\psi \epsilon$ due to (11) in (10) and to δA_μ in the first term of (12), cancel? Indeed they do, but this was not guaranteed from the beginning. In fact, it is very hard to find in general the auxiliary fields.

If one goes on to higher orders in κ, one finds two more auxiliary fields, namely a scalar S and a pseudoscalar P. All fields transform only into themselves, but no longer into matter fields. Also the commutator algebra of general coordinate, local Lorentz and local supersymmetry transformations (the local extension of (2)) closes off-shell uniformly on all fields (ref. 17). With auxiliary fields there are again equal numbers of fields

$16e_\mu^m$ - 6 Lorentz invar. - 4 gen. coord. = 6 bosonic

$16\psi_\mu^a$ - 4 local supersymmetries = 12 fermionic

S, P and A_μ (which is not a gauge field) = 6 bosonic

The gauge action reads

$$I^{gauge} = I^2 + I^{3/2} + \frac{1}{3} \det e \left(S^2 + P^2 - A_\mu^2\right) \tag{13}$$

In terms of these gauge fields and any set of matter fields, a tensor calculus has been developed (ref. 18). It is entirely equivalent to superspace approaches, but somewhat easier to understand and to work with. Using this tensor calculus, all previous matter coupling models are reproduced, but now without any work (in particular without the Noether method). Also it has been shown that at 3, 4, 5, etc. loop order dangerous counter terms really exist (ref. 19) (to all order in κ and ψ , not merely to second order). Also it was found that extending the gravitational topological invariants to superinvariants, following the rules of this tensor calculus, the same invariants (and no ψ corrections) were recovered (ref. 20).

For the extended supergravities, the approach we sketched above might also be used. Another approach to the auxiliary fields (and from there to a tensor calculus) for N>1 global supersymmetry is presented at this conference by Sohnius, Stelle and West. For N=2 supergravity, Fradkin and Vassiliev, and also de Wit and van Holten, and also Breitenlohner and Sohnius found the auxiliary fields (see their talks). It seems a long way to N = 8, but who knows?

4. NEW RESULTS FOR OLD PROBLEMS

Quantizing N=1 supergravity by path-integral techniques, one fixes the local supersymmetry and local spacetime symmetries (namely local Lorentz rotations and general coordinate transformations--otherwise stated: rotations and shifts of freely falling lifts). In this way one finds

(i) a complex anticommuting vector ghost C^{ν} from the general coordinate transformations;

(ii) a complex anticommuting antisymmetric tensor ghost C^{mn} from the local Lorentz rotations;

(iii) a complex commuting spin 1/2 ghost C^{α} from the local supersymmetry.

The kinetic terms are

$$\mathcal{L} = - C^{*\nu} \square C^{\nu} - \bar{C} \not{\partial} C - C^{*mn} \left(C^{mn} + \partial^{m} C^{\nu} \delta^{n}_{\nu} \right)$$

so that these ghosts are not gauge fields. Note that although the antisymmetric vierbein is nonpropagating since its kinetic matrix is given by the Lorentz gauge fixing term $(e_{m\mu} - e_{\mu m})^2$, it is incorrect to state that one may drop the Lorentz ghosts. It is C^{*mn} and $C^{mn} + \frac{1}{2}(\partial^m C^n - \partial^n C^m) \equiv \tilde{C}^{mn}$ which are nonpropagating. Replacing C^{mn} by $\tilde{C}^{mn} - \frac{1}{2}(\partial^m C^n - \partial^n C^m)$ one may everywhere drop C^{*mn} and $\tilde{C}^{mn}$, but extra interactions are introduced by this substitution which one would have missed if one simply had dropped C^{*mn} and C^{mn}. We found this result when we tried to show that, just as for the graviton, also the gravitino selfenergy is transversal (ref. 21).

In this Ward identity one needs that C^{ν} anticommutes with itself, and C^{α} commutes with itself. One does not need any rule for $C^{\nu} C^{\alpha} = \pm C^{\alpha} C^{\nu}$. Recently Ne'eman and Thierry-Mieg proposed a $Z_2 \times Z_2$ grading and claimed that only with the - sign does unitarity hold (ref. 22). Usually fields (anti) commute as given by a Z_2 grading

$$\phi(a_1,a_2)\,\phi(b_1,b_2) = (-)^{(a_1+b_1)(a_2+b_2)}\,\phi(b_1,b_2)\,\phi(a_1,a_2)$$

where $a_1 = 0\,(1)$ if ϕ has integer (half-integer) spin and $a_2 = 0\,(1)$ if ϕ is a physical (ghost) field. They proposed however as sign $a_1b_1 + a_2b_2$. Actually, as we have shown, these signs are unobservable, and unitarity holds in both cases (ref. 23). The Z_2 grading is simpler but the $Z_2 \times Z_2$ grading is not less correct.

Suppose one puts a gravitino in a background gravitational field. One adds to the path integral "unity" twice. Namely, once one fixes the gauge and finds in this way a Faddeev-Popov superdeterminant which, when exponentiated, yields C^α. Then one smears $\delta(\gamma.\psi - a)$ by an 't Hooft averaging functional

$$1 = N\int[da^\alpha]\,e^{\frac{1}{4}\bar{a}_\alpha \not{D}^\alpha{}_\beta a^\beta}\ , \quad N = \frac{1}{(\det \not{D})^{1/2}}$$

The normalization factor N is background field dependent, and exponentiating it, one gets a new, Nielsen-Kallosh ghost (ref. 24)

$$\mathcal{L}^{NK} = -\frac{1}{2}\,\bar{N}\not{D}N$$

where N^α is real (the bar is a Majorana bar) since a^α is real, and commuting in order that one gets $(\det \not{D})^{1/2}$ in the denominator. A problem for some time has been that for real commuting spinors $\bar{N}\not{\partial}N$ is a total derivative (as already observed by Majorana himself). In fact the solution of this puzzle is very simple. Replace

$$(\det \not{D})^{-1/2} = \frac{(\det \not{D})^{1/2}}{\det \not{D}}$$

in which case one gets: a real anticommuting and a complex commuting NK ghost. No problems with statistics now arise since for complex λ the action $\bar{\lambda} \not{D} \lambda$ is not a total derivative. This new ghost is needed to get, for example, the correct axial anomaly. (Duff and Christensen used an approach which is equivalent to working in the unweighted gauge $\gamma.\psi = 0$. Thus they did not need extra NK ghosts.)

Let us now turn to antisymmetric tensor fields coupled to gravity. (Examples: the antisymmetric $e_{m\mu} - e_{\mu m}$ vierbein, $A_{\mu\nu\rho}$ in d=11 N=1 supergravity, the Lorentz ghosts, etc.) We consider first the classical action

$$\mathcal{L} = F_{\mu\nu\rho} F_{\sigma\tau\upsilon}\, g^{\mu\sigma} g^{\nu\tau} g^{\rho\upsilon} \sqrt{g}, \quad F_{\mu\nu\rho} = \partial_\mu A_{\nu\rho} + \text{cyclic}$$

The field equation reads $D^\mu F_{\mu\nu\rho} = 0$ and since $F_{\mu\nu\rho} = \epsilon_{\mu\nu\rho\sigma} \varphi^\sigma$ in four dimensions, $\varphi_\sigma = \partial_\sigma \varphi$. Substituting this result back, one sees that $A_{\mu\nu}$ describes a scalar field φ. If one does not like substituting field equations with derivatives back into the action, one can start from a first order approach with $A_{\mu\nu}$ and $F_{\mu\nu\rho}$ as independent fields and arrive at the same conclusions. However, φ need not be regular (for example, the Dirac monopole) and this leads at the quantum level to an interesting phenomenon. Calculating the one loop divergences of $A_{\mu\nu}$ and φ coupled to gravity, and substituting the $A_{\mu\nu}$ and φ field equations in the result, one finds (ref. 25, 26).

$$\Delta\mathcal{L}(A_{\mu\nu} + \text{gravity}) - \Delta\mathcal{L}(\varphi + \text{gravity}) = \chi$$

where χ is the Euler number (properly normalized

as $(32\pi^2)^{-1}\ {}^*R^*_{\mu\nu\rho\sigma} R^{\mu\nu\rho\sigma}$). In fact, one finds that the trace anomaly due to an $A_{\mu\nu}$-loop minus that due to a φ-loop is nonzero, although $A_{\mu\nu}$ and φ are equal at the classical level ! The extra contribution comes from the singularity in φ which we mentioned above.

Even more interesting is what happens when one considers an antisymmetric tensor $A_{\mu\nu\rho}$. Now $F_{\mu\nu\rho\sigma} = \epsilon_{\mu\nu\rho\sigma}\varphi$ and $\partial_\sigma \varphi = 0$. Thus φ = constant. In other words, substituting this result back: <u>one obtains the cosmological constant from a gauge principle</u>. Calculating again the one loop divergences, one finds

$$\Delta L(A_{\mu\nu\rho} + \text{gravity}) - \Delta L(\text{gravity} + \text{cosm. constant}) = -2\chi$$

It is tempting to speculate that also for fermions dramatic results are at hand: different field representations for spin 1/2 fermions might lead to different axial anomalies, and require a different set of quarks in order that anomalies cancel (ref. 26).

A few words about the covariant quantization of antisymmetric tensorial gauge fields coupled to gravity. If one fixes the gauge by $\delta(D^\mu A_{\mu\nu} - b_\nu)$, then b_ν satisfies $D^\nu b_\nu = 0$ since $D^\mu D^\nu A_{\mu\nu} = R^{\mu\nu} A_{\mu\nu}$ = 0 (since $A_{\mu\nu}$ is antisymmetric but $R^{\mu\nu}$ is symmetric). Thus, as shown by Siegel (ref. 27), one must treat the 't Hooft averaging functional (which depends on b_ν) as the Maxwell action, and add extra gauge fixing delta functions and a ghost action for this smearing function as well! Smearing this extra delta function, the secondary smearing function has a normalization factor which is background field dependent and leads to a Nielsen-Kallosh ghost. In addition there are the standard Faddeev-Popov ghosts, but, as shown by Namazie and Storey, and Townsend (ref. 28), these ghosts are gauge fields and hence one needs ghosts for ghosts (for ghosts for . . .).

When one calculates quantum corrections, one must use a regularization scheme which preserves supersymmetry. There is now such a scheme, introduced and subsequently rejected by W. Siegel (ref. 29). (I believe that the "inconsistencies" he found are just the usual ambiguities due to the unavoidable axial anomaly). The scheme is based on the observation that if one continues spacetime down to n<4, then all supersymmetry manipulations go through unchanged if one keeps all field components as in four dimensions. Thus not only λ^{α} remains four-dimensional but also A_{μ}! One might think that this violates gauge Ward identities, but not so. An explicit two-loop computation of the Yang-Mills selfenergy showed that it remains transversal. Also the β function comes out unchanged (as expected: a different regularization technique should not lead to different physical results). And, most importantly, supersymmetry Ward identities remain valid at the one-loop level, both in global supersymmetry and in supergravity. These results prove that the scheme works (Ref. 30).

Having such a scheme proves that supergravity is indeed two-loop finite. Previous proofs assumed that ΔL had the same supersymmetry as L, but one did not know whether anomalies might contradict this assumption.

Let us give an example how this modified scheme of regularization by dimensional reduction works. In order to prove the invariance of the supersymmetric Yang-Mills action, one needs the identity

$$(\bar{\lambda}^{a}\gamma_{\mu}\lambda^{b})(\bar{\lambda}^{c}\gamma^{\mu}\epsilon)f_{abc} = 0$$

Clearly, this identity remains valid if the coordinates are in n-dimensions since it is only a rearrangement of a (fixed) number of components. Since the classical action remains invariant, one can now use path-integral techniques to derive Ward identities, and these have been verified explicitly.

Another example is QED. One can decompose $A_{\mu} = \{A_i, A_{\sigma}\}$ with $1 \leq i \leq n$ and $n \leq \sigma \leq 4$. Thus the action becomes

$$\mathcal{L} = -\frac{1}{4}F_{ij}^{2} - \bar{\lambda}\gamma^{j}(\partial_{j} - ieA_{j})\lambda$$
$$-\frac{1}{2}(\partial_{i}A_{\sigma})^{2} + ie(\bar{\lambda}\gamma^{\sigma}\lambda)A_{\sigma}$$

In non-supersymmetric theories the equivalence of two coupling constants (such as e in this example) is probably not maintained at the quantum level, so that this new scheme might be applicable only in supersymmetric theories.

In products such as $\gamma^\mu \partial_\mu$ the range of μ in γ^μ is $1 \leq \mu \leq 4$ but in ∂_μ the index μ runs from $1 \leq \mu \leq n$. In true democratic fashion the smallest wins and $\gamma^\mu \partial_\mu = \gamma^i \partial_i$. One can view the scalars A_σ as extra compensating fields which restore the supersymmetry. These fields have new couplings, namely Yukawa couplings. In ordinary dimensional regularization, no such extra fields are produced, and this is the reason that in ordinary dimensional regularization supersymmetry is not preserved.

REFERENCES

1. D. Z. Freedman, P. van Nieuwenhuizen and S. Ferrara, Phys. Rev. D 13, 3214 (1976).

2. S. Deser and B. Zumino, Phys. Lett. 62 B, 335 (1976).

3. E. Cremmer and B. Julia, Phys. Lett. 80 B, 48 (1978) and Nucl. Phys. B 159, 141 (1979).
   For earlier work see:
   E. Cremmer, B. Julia and J. Scherk, Phys. Lett. 76 B, 409 (1978).
   B. de Wit and D. Z. Freedman, Nucl. Phys. B 130, 105 (1977).
   B. de Wit, Nucl. Phys. B 158, 189 (1979).

4. Taken from J. Scherk in "Proceedings of the Supergravity Workshop at Stony Brook," P. van Nieuwenhuizen and D. Z. Freedman, eds, North-Holland (1979), p. 43.

5. Taken from D. Z. Freedman and P. van Nieuwenhuizen, Scientific American, February 1978.

6. M. T. Grisaru, Phys. Lett. 66 B, 75 (1977).

7. M. T. Grisaru, P. van Nieuwenhuizen and C. C. Wu, Phys. Rev. D 12, 1563 (1975).

8. J. Scherk and J. Schwarz, Phys. Lett. 82 B, 60 (1979) and Nucl. Phys. B 153, 61 (1979).

9. E. Cremmer, J. Scherk and J. Schwarz, Phys. Lett. 84 B, 83 (1979).

10. I thank Prof. Berezin for showing me this.

11. S. MacDowell and F. Mansouri, Phys. Rev. Lett. 38, 739 (1977).

12. P. K. Townsend and P. van Nieuwenhuizen, Phys. Lett. B 67, 439 (1977).

13. H. Nicolai, P. K. Townsend and P. van Nieuwenhuizen, Cern preprint TH.2844 (1980).

14. This approach was first developed by the author in "Proceedings of the Mathematical Physics Conference 1979," Lausanne (to be published).

15. B. de Wit and J. W. van Holten, Nucl. Phys. B 155, 530 (1979).

16. S. Ferrara, F. Gliozzi, J. Scherk and P. van Nieuwenhuizen, Nucl. Phys. B 117, 333 (1977). The same authors plus D. Z. Freedman and P. Breitenlohner, Phys. Rev. D 15, 1013 (1977).

17. S. Ferrara and P. van Nieuwenhuizen, Phys. Lett. 74 B, 333 (1978). K. S. Stelle and P. C. West, Phys. Lett. 74 B, 330 (1978).

18. S. Ferrara and P. van Nieuwenhuizen, Phys. Lett. 76 B, 404 (1978), idem with M. T. Grisaru, Nucl. Phys., B 138, 430 (1978), idem with B. de Wit, Nucl. Phys. B 139, 216 (1978). K.S. Stelle and P.C. West, Nucl. Phys. B140, 285 (197 and B145, 175 (1978); Phys. Lett. 77B, 376 (1978).

19. S. Ferrara and P. van Nieuwenhuizen, Phys. Lett. 78 B, 573 (1978).

20. P. K. Townsend and P. van Nieuwenhuizen, Phys. Rev. D 19, 3592 (1979).

21. M. K. Fung, D. R. T. Jones and P. van Nieuwenhuizen, Phys. Rev. D, to be published.

22. Y. Ne'eman and J. Thierry-Mieg, Proc. 2nd. Marcel Grosman meeting.

23. P. van Nieuwenhuizen, Journal of Math. Phys., to be published.

24. N. K. Nielsen, Nucl. Phys. B 140, 499 (1978).
R. Kallosh, Nucl. Phys. B 141, 141 (1978).

25. E. Sezgin and P. van Nieuwenhuizen, Phys. Rev. D, to be published.

26. M. Duff and P. van Nieuwenhuizen, Imperial College preprint.

27. W. Siegel, Phys. Lett. B, to be published.

28. M. A. Namazie and D. Storey, Nucl. Phys. B 157, 170 (1979).
P. K. Townsend, see ref (4).

29. W. Siegel, Phys. Lett. 84 B, 193 (1979) and to appear in Phys. Lett. B.

30. D.M. Capper, D.R.T.Jones and P. van Nieuwenhuizen, Nucl. Phys.B, to appear.

QUARK-LEPTON UNIFICATION AND PROTON DECAY

Jogesh C. Pati *

International Centre for Theoretical Physics
P.O.B. 586, Miramare
Trieste, Italy

and

Department of Physics
University of Maryland
College Park, Maryland 20742
U.S.A.

ABSTRACT

Complexions for proton decay arising within a maximal symmetry for quark-lepton unification, which leads to spontaneous rather than intrinsic violations of B, L and F, are considered. Four major modes satisfying $\Delta B = -1$ and $\Delta F = 0, -2, -4$ and -6 are noted. It is stressed that some of these modes can coexist in accord with allowed solutions for renormalization group equations for coupling constants of a class of unifying symmetries. None of these remarks is dependent on the nature of the quark charges. It is noted that if quarks and leptons are made of constituent preons, the preon binding is likely to be magnetic.

* Supported in part by the U.S. National Science Foundation and in part by the John Simon Guggenheim Memorial Foundation Fellowship.

1. INTRODUCTION

The hypothesis of grand unification[1-3] serving to unify all basic particles - quarks and leptons - and their forces - weak, electromagnetic as well as strong - stands at present primarily on its aesthetic merits. It gives the flavour of synthesis in that it provides a rationale for the existence of quarks and leptons by assigning the two sets of particles to one multiplet of a gauge symmetry G. It derives their forces through one principle-gauge unification.

With quarks and leptons in one multiplet of a local gauge symmetry G, baryon and lepton number conservations cannot be absolute. This line of reasoning had led Salam and myself to predict in 1973 that the lightest baryon - the proton - must ultimately decay into leptons[2]. Theoretical considerations suggest a lifetime for the proton in the range of 10^{28} to 10^{33} years[2-5]. Its decay modes and corresponding branching ratios depend in general upon the details of the structure of the symmetry group and its breaking pattern. What is worth noticing at this juncture is that proton decay modes and the value of the weak angle[6] $\sin^2 \theta_W$ may be the only effective tools we would have for some time to probe into underlying design of grand unification.

Experiments[7] are now underway to test proton stability to an accuracy one thousand times higher than before[8]. In view of this, I shall concentrate primarily on the question of expected proton decay modes within the general hypothesis of quark-lepton unification, and on the question of intermediate mass scales filling the grand plateau between 10^2 and 10^{15} GeV which influence proton decay. At the end I shall indicate some new features, which may arise if quarks and leptons are viewed - perhaps more legitimately - as composites of more elementary objects - the "preons".

Much of what I have to say arises in the context of maximal quark-lepton unifying symmetries of the type proposed before[9] *). I specify such symmetries in detail later. One characteristic feature worth noticing from the beginning is that within such symmetries a linear combination of baryon and lepton numbers as well as fermion number F are locally gauged and are therefore conserved in the basic Lagrangian. They are violated spontaneously, and unavoidably as the associated gauge particles acquire masses. The purpose of my talk would be many-fold:

---

*) The present talk is based largely on a forthcoming paper by J.C. Pati and Abdus Salam[10].

1) First to restress in the light of recent developments that within maximal symmetries the proton may in general decay via four major modes*), which are characterized by a change in the baryon number by -1 and a change in the fermion number defined below by 0, -2, -4 and -6 units **):

$$\left.\begin{array}{l} p \to 3\nu + \pi^+ \\ p \to 2\nu + e^- + \pi^+\pi^+ \text{ etc.} \end{array}\right\} \quad \left.\begin{array}{l} \Delta F = 0 \\ \Delta B_q = -3,\ \Delta L = +3 \end{array}\right\} \Delta(B-L) = -4 \tag{1}$$

$$\left.\begin{array}{l} p \to e^-\pi^+\pi^+,\ \nu\pi^+,\ e^-K^+\pi^+ \\ p \to e^+\nu_1\nu_2,\ e^+e^-e^-\pi^+\pi^+ \end{array}\right\} \quad \left.\begin{array}{l} \Delta F = -2 \\ \Delta B_q = -3,\ \Delta L = +1 \end{array}\right\} \Delta(B-L) = -2 \tag{2}$$

$$\left.\begin{array}{l} p \to e^+\pi^0,\ \bar{\nu}_e\pi^+,\ \mu^+K^0 \\ p \to \mu^+\pi^0,\ \bar{\nu}_\mu\pi^+,\ e^+e^+e^- \end{array}\right\} \quad \left.\begin{array}{l} \Delta F = -4 \\ \Delta B_q = -3,\ \Delta L = -1 \end{array}\right\} \Delta(B-L) = 0 \tag{3}$$

$$\left.\begin{array}{l} p \to 3\bar{\nu} + \pi^+,\ e^+\bar{\nu}_1\bar{\nu}_2 \\ p \to e^+e^+\bar{\nu} + \pi^- \end{array}\right\} \quad \left.\begin{array}{l} \Delta F = -6 \\ \Delta B_q = -3,\ \Delta L = -3 \end{array}\right\} \Delta(B-L) = +2 \tag{4}$$

---

*) That protons may decay via all these four modes (ΔF = 0, -2, -4 and -6) was to my knowledge first observed in Ref. 9. The modes ΔF = 0 and -4 occur within specific models of Refs 2 and 3, respectively, while the questioning of baryon number conservation (Ref. 2) is based on more general grounds and is tied simply to the idea of quark-lepton gauge unification.

**) I am not listing decay modes satisfying ΔB = -1 and ΔF = +2, 4, etc. corresponding to p → (5 or 7 leptons) + π's. These appear to be suppressed compared to those listed within the unification hypothesis.

(Here B_q denotes quark number which is +1 for all quarks, -1 for antiquarks and 0 for leptons. The familiar baryon number*) B, which is +1 for the proton, is one third of quark number ($B \equiv B_q/3$). L denotes lepton number which is +1 for $(\nu_e, e^-, \mu^-, \nu_\mu, \ldots)_{L,R}$, -1 for their antiparticles and 0 for quarks. Fermion number F is the sum of B_q and L: $F \equiv B_q + L$. Since for each case of proton decay, quark number must change by a fixed amount $B_q = -3$, there is a one-one relationship between the change of fermion number F and that of any other linear combination of B_q and L, for example of $[(B_q/3) - L] = B - L$ for the proton decay modes as exhibited in (1)-(4)

2) To point out that these four alternative decay modes cannot only exist but in general even coexist.

3) To stress that observation of any of the three decay modes satisfying $\Delta F = 0$, -2 and -6 would unquestionably signal the existence of one or several intermediate mass scales filling the plateau between 10^2 and 10^{15} GeV. [Existence of such intermediate mass scales is a feature which naturally rhymes with (a) maximal symmetries of B, L and F violations, and (b) partial quark-lepton unification at moderate energies 10^4 - 10^6 GeV.][2,11]

4) To point out that the complexions of proton-decay selection rules alter if one introduces intrinsic left-right symmetry in the basic Lagrangian thereby permitting the existence**) of ν_R's parallelling ν_L's and

5) to stress that none of the remarks (1)-(4) is tied to the nature of quark charges. These remarks hold for integer as well as fractional quark charges with SU(3) colour symmetry either being softly spontaneously broken or remaining exact.

To motivate these remarks let me first specify what I mean by "maximal" symmetry. Maximal symmetry[9] corresponds to gauging all fermionic degrees of freedom with fermions consisting of quarks and leptons. Thus with n two-component left-handed fermions F_L plus n two-component right-handed fermions F_R, the symmetry G is

*) The reader may note that in our previous papers we had by convention chosen to call quark number B_q the baryon number B. Therefore the 15th generator of $SU(4)_{col}$ (Ref. 2) which was written as (B - 3L) stood for (B_q - 3L). With the more conventional definition of baryon number, as adopted in the present note, $B_q - 3L = 3(B - L)$.

**) Note that in this case one must assume that ν_R and ν_L combine to form a light four-component Dirac particle and that $m_{W_R} \gg m_{W_L}$. (To conform with astrophysical limits one may[12] need $m_{W_R} \gtrsim 50\, m_{W_L}$). Alternatively the neutrinos may acquire Majorana masses.

$SU(n)_L \times SU(n)_R$. One may extend the symmetry G by putting fermions F_L and antifermions F_L^c (as a substitute for F_R) in the same multiplet. In this case the symmetry G is SU(2n), which is truly the maximal symmetry of 2n two-component fermions. As an example, for a single family of two flavours and four colours including leptonic colour, n = 8 and thus G = SU(16). One word of qualification: such symmetries generate triangle anomalies, which are avoided, however, by postulating that there exist a conjugate mirror set of fermions $F^m_{L,R}$ supplementing the basic fermions $F_{L,R}$ with the helicity flip coupling represented by the discrete symmetry ($F_{L,R} \leftrightarrow F^m_{R,L}$). Thus by "maximal" symmetries I shall mean symmetries which are maximal up to discrete symmetries as above*).

Though old, it is now useful to recall the argument leading to violations of B, L and F. If all quark-lepton degrees of freedom are gauged locally as in a maximal symmetry G specified above, then fermion number $F \equiv B_q + L = 3B + L$ as well as an independent**) linear combination of baryon and lepton numbers (B + xL) are among the generators of the local symmetry G. Now if all gauge particles***) with the exception of the photon and possibly the octet of gluons acquire masses spontaneously, then both fermion number F and (B + xL) must be violated spontaneously, as the associated gauge particles acquire their masses. The important remark here is that even though B, L and F are conserved in the basic Lagrangian, they are inevitably and unavoidably violated spontaneously****).

---

*) Later such discrete symmetries may include family discrete symmetry such as $e \leftrightarrow \mu \leftrightarrow \tau$. These would not materially alter our discussions of B, L and F violations except for the imposition of the constraints that $B = B_e + B_\mu + B_\tau$, $L = L_e + L_\mu + L_\tau$ and $F = F_e + F_\mu + F_\tau$.

**) For the case of lepton number being the fourth colour this generator is $\propto B_q - 3L = 3(B - L)$.

***) From the limits on Eötvös type experiments, one knows that no massless gauge particle coupled to B, L or F can lead to an effective four fermion coupling $\gtrsim G_{Newton} \times 10^{-8}$.

****) The generality of the argument demands violations of only the linear combinations F = (3B + L) and (B + xL), i.e., at least either B or L must be violated. Simple models of quark-lepton unification (see for example Refs 2,3, and 9) lead, however, to violations of B and L. Several authors have remarked that it is possible to introuuce global quantum numbers within the quark-lepton unification hypothesis, which would preserve proton stability (see Ref. 13)). But as stated above all simple models end up with an unstable proton.

Instead of the maximal symmetry G, one may of course choose to gauge a subgroup $\mathcal{G} \subset G$. But as long as the subgroup $\mathcal{G}$ assigns quarks and leptons into one irreducible multiplet, there are only two alternatives open. Either the subgroup $\mathcal{G}$ still possesses effectively a fermion number F and/or (B + xL) among its generators*), in which case these must be violated spontaneously for the reasons stated above; or the subgroup $\mathcal{G}$ arises through such "squeezing" of gauges of the maximal symmetry G that one and the same gauge particle couples for example to the diquark ($\bar{q}^c q$) as well as to the quark-lepton ($\bar{q}\ell^c$) currents**). In this case, baryon, lepton and fermion numbers are violated intrinsically in the basic Lagrangian. But, one way or another, be the violations spontaneous or intrinsic, some linear combinations of B and L must be violated; the single underlying reason in either case being quark-lepton unification.

Spontaneous and intrinsic violations of B, L and F can in general lead to similar predictions for proton decay. But the two cases would differ characteristically from each other at superhigh temperatures, where intrinsic violation would acquire its full strength with the superheavy gauge masses going to zero, but where spontaneous violation would in fact vanish***).

2. MODELS OF GRAND UNIFICATION

It is useful to see the inter-relationships between different types of unification models. The simplest realization of the idea of quark-lepton unification is provided by the hypothesis that lepton number is the fourth colour[2]. For a single family of (u,d) flavours, the corresponding multiplet is

$$(F_e)_{L,R} = \begin{bmatrix} u_r & u_y & u_b & u_\ell = \nu_e \\ d_r & d_y & d_b & d_\ell = e^- \end{bmatrix}_{L,R} \tag{5}$$

*) For example $[SU(4)]^4$ and $[SU(6)]^4$ contain (B-L) as a local symmetry, but no F (though F is a global symmetry in the basic Lagrangian of these models). SU(16) operating on 16-plets of e, μ and τ family fermions (see Table I) is a subgroup of $[SU(16)]^3$ and SU(48). It contains B-L and F with $B = B_e + B_\mu + B_\tau$ and likewise for L and F.

**) Examples of this type are SU(5) (Ref. 3) and SO(10) (Ref. 14). SU(5) does not contain B-L or F as local symmetries. SO(10) contains (B-L) but not F. Both SU(5) and SO(10) violate B, L and F intrinsically in the basic Lagrangian.

***) Alternatively, spontaneous violation may pass through a phase transition and increase[15] beyond the critical temperature T_c.

with r, y, b and ℓ denoting red, yellow, blue and lilac colours, respectively. The corresponding local symmetry is

$$\mathcal{G} = SU(2)_L \times SU(2)_R \times SU(4)'_{L+R} \tag{6}$$

where $SU(2)_{L,R}$ operates on the flavour indices $(u,d)_{L,R}$ and $SU(4)_{L+R}$ operates on the four colour indices (r,y,b,ℓ). It is the SU(4) colour symmetry which intimately links quarks and leptons. The symmetry $\mathcal{G}$ has three notable features. i) It is the simplest subunification model of all containing the low energy symmetry $SU(2)_L \times U(1) \times SU(3)'_{L+R}$ on the one hand and quark-lepton unification on the other. It gauges through the SU(4) colour symmetry which links quarks and leptons, with the combination $(B_q - 3L) = 3(B - L)$ as a local symmetry. ii) It is non-Abelian and thus provides a simple rationale for the quantization of charges. iii) Its gauge structure is left-right symmetric[16]. (Indeed the idea of lepton number being the fourth colour requires that neutrinos must be introduced with left and right helicities*) and thereby the basic matter multiplet must be left-right symmetric given that quarks enter into the basic Lagrangian with both helicities.)

Table I

The symmetries $[SU(n)]^4$ and SU(16) require the presence of mirror fermions for cancelling anomalies (see discussions in Section I). The quark fields $(u,d,u^c$ and $d^c)$ come in three different colours.

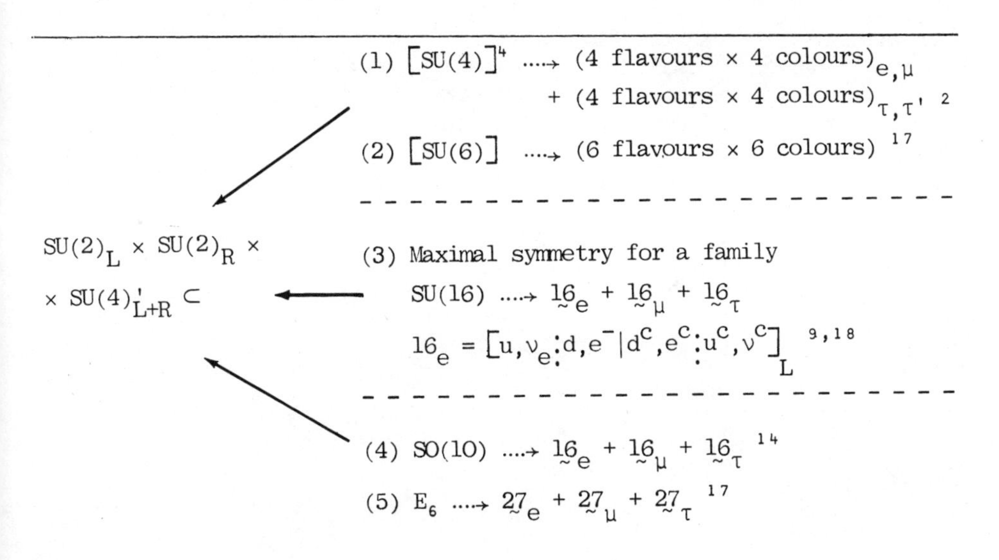

*) With ν_R being distinct from $\bar{\nu}_R$.

The symmetry $\mathcal{G}$, because of its simplicity, might be the right stepping stone towards grand unification, It should of course be viewed as a subunification symmetry as it contains at least two gauge coupling constants - one[16] for $SU(2)_{L,R}$ and the other for $SU(4)'_{L+R}$. In other words, it should be regarded as part of a bigger unifying symmetry G possessing a single gauge coupling constant. There are a number of candidates for G, which do contain the sub-unification symmetry $\mathcal{G}$. Table I provides a list of some of these symmetries.

All the unifying symmetries listed In Table I are left-right symmetric. By constrast one may consider the left-right asymmetric model[3] SU(5), which is the smallest grand unifying symmetry of all with the multiplet structure

$$(\bar{5}+10)_e + (\bar{5}+10)_\mu + (\bar{5}+10)_\tau$$

in which the right-handed neutrinos $(\nu_{e,\mu,\tau})_R$ are missing. Note that by contrast to the models listed in Table I the multiplet structure of SU(5) $(\bar{5} + 10)$ is reducible within one family.

Maximal symmetry and spontaneous violations of B, L and F

A maximal symmetry SU(2n), which puts n left-handed quarks and leptons as well as their charge conjugate fields within the same multiplet $F = [q, \ell | q^c, \ell^c]_L$ generates the sets of gauge fields shown in Fig. 1 and thus these quantum numbers are conserved in the basic Lagrangian. The violations of these quantum numbers arise, however, as the gauge particles, barring the photon and possibly the octet of gluons, acquire masses through spontaneous breakdown of the local symmetry G. The violations come about in two distinct ways.

(a) Through spontaneously induced mass mixing[9] of gauge particles carrying different sets of values of B, L and F. For instance $Y \leftrightarrow Y'$ mixing will lead to effective violations of the form $\Delta B_q = -3$, $\Delta L = -1$ and $\Delta F = -4$, which will lead to proton decays of the type

$$p \rightarrow \bar{\ell} + \text{pions} \tag{7}$$

while $Y \leftrightarrow \bar{X}$ mixing will induce effective violations of the form

$$\Delta B_q = -3, \ \Delta L = +1 \text{ and } \Delta F = \Delta(B_q + L) = -2$$

leading to proton decays of the type

$$p \rightarrow \ell + \text{pions} \tag{8}$$

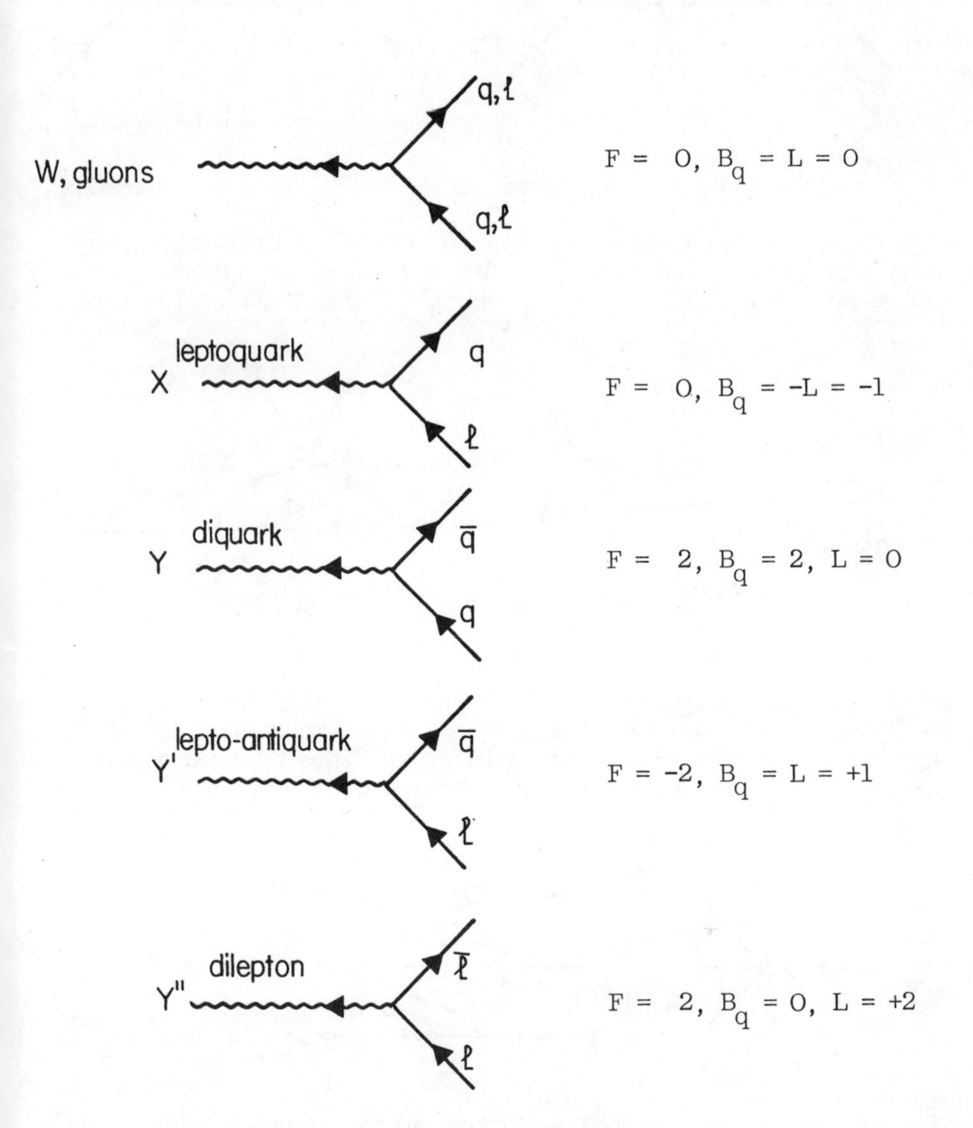

Fig.1 : Gauge particles within a maximal symmetry. Here B_q, L and $F \equiv B_q + L$ denote quark, lepton and fermion numbers, respectively, as defined in Section 1.

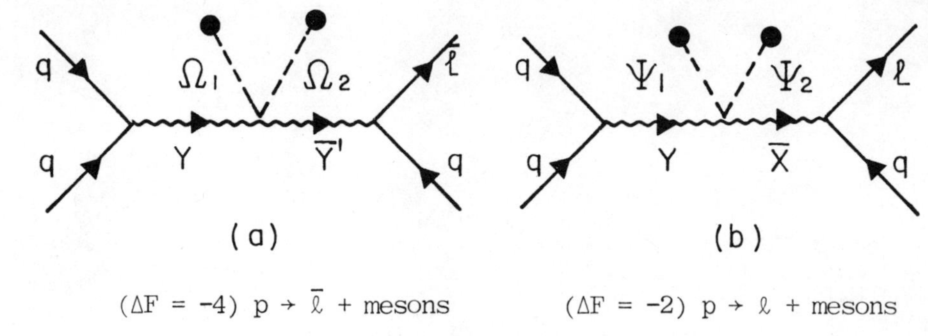

($\Delta F = -4$) $p \to \bar{\ell}$ + mesons ($\Delta F = -2$) $p \to \ell$ + mesons

Figs 2a,b: Spontaneous violations of B,L and F in a maximal symmetry leading to gauge mixings. These induce, for example, $\Delta F = -4$ and -2 proton decays. The $\Omega_{1,2}$ and $\Psi_{1,2}$ are Higgs fields (see text

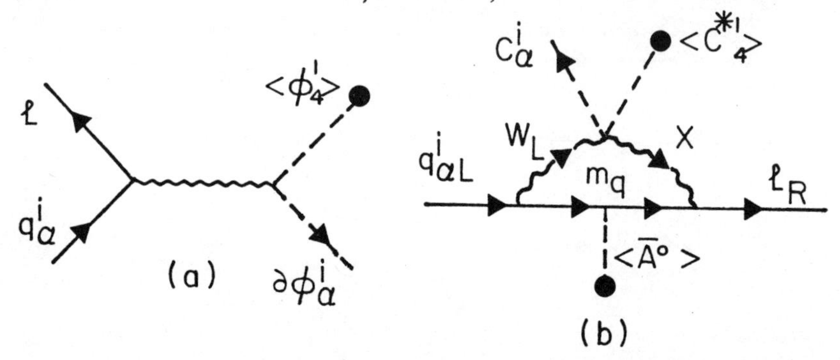

Figs 3a,b: Spontaneous violations of B,L (and in general F) leading to virtual Yukawa transitions: $q^i_\alpha \to \ell + \phi^i_\alpha$. These transitions induce $\Delta F = 0$ proton decays in third order: $p \to 3\ell$ + mesons (and analogously $\Delta F = -6$ decays: $p \to 3\bar{\ell}$ + mesons), (see text).

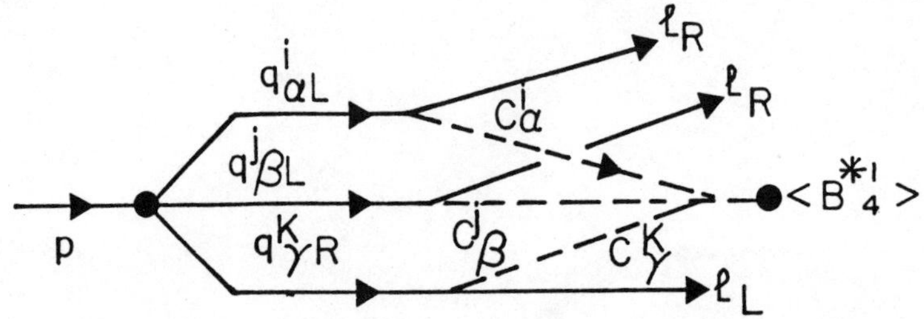

Fig. 4: $\Delta F = 0$ proton decays through spontaneous violations of B and L. These utilize the virtual Yukawa transitions of Fig. 3 thrice. Note that the mechanism of Figs 2, 3 and 4 apply to integer as well as fractionally charged quarks, hereafter denoted by ICQ and FCQ respectively.

These two kinds of mixings and the corresponding transitions are exhibited in Figs 2a and 2b, respectively. For the case of the "maximal" symmetry SU(16) defined earlier, $Y \leftrightarrow \bar{Y}'$ mixing of the sort mentioned above would arise from vacuum expectation values VEV of a Higgs field $\Omega^{\{AB\}}_{\{CD\}}$, where {AB} and {CD} are symmetric combinations of two 16-plets and two 16*-plets, respectively[18]. Such VEV's violate B, L and F, but respect (B-L). In fact they may be chosen to be invariant under the full subunification symmetry $SU(2)_L \times SU(2)_R \times SU(4)'_{L+R}$. Thus the $Y \leftrightarrow \bar{Y}'$ mixing mass squared may typically be superheavy, i.e., of the same order as $m_Y^2 \sim m_{Y'}^2$, where Y and Y' are regarded to be among the heaviest gauge particles.

On the other hand, the $Y \leftrightarrow \bar{X}$ mixing leading to $\Delta F = -2$ transitions arises from the VEV of a field Ψ transforming like $[16 \times 16]_{\text{Antisymmetric}} \times [16^* \times 16^*]_{\text{Antisymmetric}}$ tensor, which violates $SU(2)_L$ while still conserving $SU(3)'_{L+R}$ colour. Hence the $(\bar{Y} \leftrightarrow X)$ mixing mass squared is limited by $m_{W_L}\, m_X$.

(b) The violations of B, L and F may also arise through spontaneously induced three-point Yukawa transitions of the type $q \rightarrow \ell + \phi$ (see Figs 3a and 3b)

$$
\begin{aligned}
q^i_{\alpha L} &\rightarrow \nu_R + C^i_\alpha + \langle C^{*}{}^1_4 \rangle + \langle \bar{A}^c \rangle \\
q^i_{\alpha R} &\rightarrow \nu_L + B^i_\alpha + \langle B^{*}{}^1_4 \rangle + \langle A^c \rangle \\
d_{\alpha L} &\rightarrow e^-_R + C^1_\alpha + \langle C^{*}{}^1_4 \rangle + \langle A^c \rangle \\
d_{\alpha R} &\rightarrow e^-_L + B^1_\alpha + \langle B^{*}{}^1_4 \rangle + \langle \bar{A}^c \rangle
\end{aligned}
\tag{9}
$$

Here i and α denote flavour and SU(3) colour indices, respectively. The fields A,B and C, which are identical to those introduced in Ref. 2, transform as $(2,2,1)$, $(1,2,\bar{4})$ and $(2,1,\bar{4})$ respectively under $SU(2)_L \times SU(2)_R \times SU(4)'_{L+R}$. Under SU(16), A belongs to a $[16 \times 16]_{\text{symmetric}}$ representation, while C and $B^\dagger$ together make a a 16-plet. The fields C^1_4 and $\overset{*}{B}{}^1_4$ have the same quantum numbers within a 16-plet as ν_{eL} and ν^c_{eL} respectively, while A^0 possesses $I_{3L} = -I_{3R}$ $-1/2$. The VEV $\langle C^*_4\rangle$ and $\langle A^0\rangle$ are of order (m_{W_R}/g), while $\langle \overset{*}{B}{}^1_4\rangle$ is of order (m_{W_R}/g) or (m_X/g), whichever is lower. The effective Yukawa transitions (9), used thrice, induce F = 0 proton decays (see Fig. 4 of the type [18,20].

$$p \rightarrow \nu_L \nu_R \ell_{L,R} + \pi\text{'s} \tag{10}$$

where $\ell_{L,R}$ denotes either the charged or the neutral lepton. These transitions are made possible through quartic scalar interactions which permit for example $(C^1_1 + C^2_2 + B^2_3)$ to make a transition into $\overset{*}{B}{}^1_4$ and thereby disappear into the vacuum through $\langle B^{*1}_4\rangle \neq 0$ (see discussions later). Fermi statistics together with the colour singlet nature of the proton inhibits all three leptons in the final state from having the same helicity[18] (see (10)).

An analogous mechanism induces the Yukawa transitions $q^i_\alpha \rightarrow$ $\rightarrow \bar{\ell} + \chi^i_\alpha$ which in third order induce proton decays satisfying $\Delta F = -6$

$$p \rightarrow 3q \rightarrow (3\bar{\ell} + \pi\text{'s}) + 3\chi \rightarrow (3\bar{\ell} + \pi\text{'s}) + \langle \overset{*}{\chi}{}^1_4 \rangle \tag{11}$$

The above mechanisms show (in accord with the observation in Ref. 9) that all four modes for proton decay satisfying $\Delta F = 0, -2, -4$ and -6 can arise within a maximal symmetry G. Their relative rates would depend upon the associated gauge masses, which in turn depend upon the parent symmetry G as well as upon its breaking pattern. I will discuss later how at least some of them can have competing rates.

Two common features of these mechanisms are worth noting:

i) They utilize only <u>spontaneous</u> rather than intrinsic violations of B, L and F. None of them would be operative if the vacuum expectation values of all the afore-mentioned Higgs fields were set to zero.

ii) None of these mechanisms is tied to the nature of quark charges. They hold for quark charges being either integral or fractional, with SU(3) colour local symmetry either being broken spontaneously and softly or remaining exact.

[I would like to make a small digression here. The suggestion of Salam and myself on quark-lepton unification has been misrepresented in this regard - as though it is tied to integer charges for quarks (ICQ). Here a bit of history is relevant. During the years 1972-74 almost everyone accepted fractional charges (FCQ) and absolute confinement. We, on the other hand, noticed that both possibilities - fractional as well as integer charges for quarks - arise within the same unification hypothesis. For example, the hypothesis "lepton number is the fourth colour" permits both charge patterns depending only upon the nature of spontaneous symmetry breaking, which we noted in our paper (Ref. 2). We therefore took up the task of building up the theory of integer charges for quarks and possible "quark liberation" so that it can meaningfully be confronted with experiments. As far as I know there does not exist any theoretical or experimental argument as yet providing unambiguous evidence for one quark charge pattern in favour of the other*). We therefore still keep our options open regarding the nature of quark charges and wait for experiments to settle this question**), but bearing in mind all the time that the suggestions[1,2] of quark-lepton unification and consequent baryon number violation are not tied in any way to the nature of quark charges. They are more general. So much for the digression.]

*)Recent arguments of Okun, Voloshin and Zakharov[21] favouring fractional charges for quarks do not take into account the facts that (a) variations of electric charges for ICQ and FCQ as functions of momentum are governed by different renormalization group equations due to the presence of the colour component in the former which is absent in the latter, and (b) that for a partially confining theory there exist singularities in the variable mass parameters in time-like regions even near the origin without requiring the existence of physical particles at such points. This is to be elaborated in a forthcoming preprint by myself and Abdus Salam[22]. The second recent argument[23] based on $\eta' \to 2\gamma$ also favouring FCQ is subject to the uncertain PCAC extrapolation from 1 GeV^2 to zero for the case of ICQ. There is a third argument[24] based on deep inelastic Compton scattering $\gamma p \to \gamma + X$ which by contrast to the previous two favours ICQ over FCQ. This argument is uncertain as well to the extent that the P_T involved in present experiments is not high enough to permit a legitimate use of the parton model. We must wait to settle the question of quark charges. See footnote below.

**)The two photon experiments $e^-e^+ \to e^-e^+$ + hadrons now in progress are likely to provide a definitive answer in the near future.

Spontaneous versus intrinsic violations of B, L and F

It is now instructive to compare violations of B, L and F, which are spontaneous in origin (as outlined above) to those which are intrinsic. The latter arise in general *if* one chooses to gauge special subgroups of the maximal symmetry defined by the fermion content. As mentioned in the introduction, examples of such subgroups are SU(5) and SO(10) and (lepto-quark) current $(q\bar{\ell}^c)$ coupling to distinct gauge particles Y and $\bar{Y}'$, respectively, the two currents couple to one and the same gauge particle Y_s in the basic Lagrangian. This is equivalent to "squeezing" the two gauges associated with the two distinct currents mentioned above so that $Y_s \sim (Y + \bar{Y}')/\sqrt{2}$ coupling to the sum of the currents is present in the basic Lagrangian, but $Y_a \sim (Y - \bar{Y}')/\sqrt{2}$ coupling to the orthogonal combination is absent. (Equivalently, Y_a is assigned an *infinite* mass.) The exchange of the Y_s particle thus leads to a violation of B, L and F in the second order of the basic gauge interactions (see Fig. 5) and induces the $\Delta F = -4$ decays

$$p \longrightarrow \bar{\ell} + \text{mesons} \tag{12}$$

We see that intrinsic violations of B, L and F arising through squeezed gauging can in general lead to similar consequences for proton decay as the case of spontaneous violation arising for a maximal symmetry (compare Fig. 5 with Fig. 2a).

As a matter of *personal attitude*, however, it appears to me that squeezed gauging arises effectively only through spontaneous descent of a maximal symmetry, rather than as a symmetry of the basic Lagrangian, as otherwise the choice of the specific subgroup of the maximal symmetry yielding the squeezing appears to be rather arbitrary*). But admittedly this is only a personal attitude. The two cases - spontaneous versus intrinsic violations of B, L and F - appear to possess an absolute distinction from each other only at high temperatures. In this case they would be distinguishable in absolute terms only through cosmology.

*) For example, such a choice does not seem to follow under any simple prescription such as the imposition of discrete symmetries.

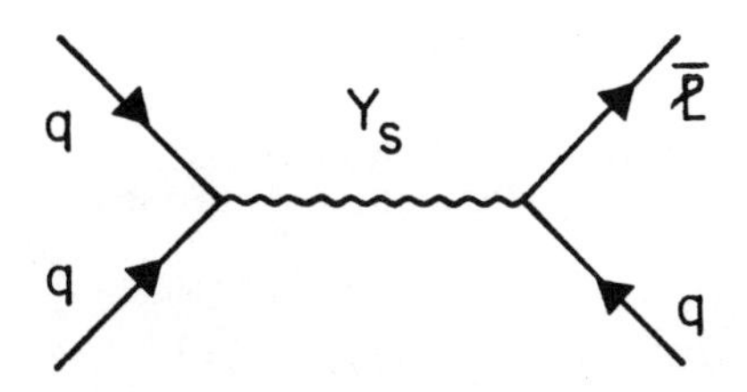

Fig. 5: Intrinsic violations of B, L and F through gauge squeezing. Here for example, $Y_S \equiv (Y + \bar{Y}')''/\sqrt{2}$ is a gauge particle of the basic Lagrangian but the orthogonal combination $(Y - \bar{Y}')/\sqrt{2}$ is absent or effectively has infinite mass. Such gauge squeezings occur in SU(5) and SO(10).

3. CONDITIONS FOR RELEVANCE OF ALTERNATIVE PROTON DECAY MODES

I now proceed to obtain the "necessary" and sufficient conditions for alternative proton decay modes satisfying $\Delta F = 0$, -2, -4 and -6 to be relevant for either the forthcoming or the second generation proton decay searches. For this purpose I shall consider only those mechanisms for proton decay which arise within a maximal symmetry, outlined in the previous section. Depending upon the decay modes*) the experimental searches are expected to be sensitive to proton lifetimes varying between 10^{30} to 10^{33} years.

Now let us first observe the restrictions which arise from the effective low energy symmetry being $SU(2)_L \times U(1) \times SU(3)_{colour}$. Weinberg and Wilczek and Zee[25] have shown that the effective proton decay interactions based on operators of lowest dimension, which is six**), automatically conserve B-L, if they are constrained to satisfy the low energy symmetry $SU(2)_L \times U(1) \times SU(3)_{colour}$. Based

---

*) For the two-body modes such as $p \to e^+\pi^0$ and $n \to e^+\pi^-$ satisfying $\Delta F = -4$ as well as $n \to e^-\pi^+$ satisfying $\Delta F = -2$, the forthcoming experiments may be sensitive to proton lifetimes $\lesssim 10^{33}$ years. For the multiparticle modes such as $p \to e^- + 2\nu + \pi^+\pi^+$ and $n \to e^- + 2\nu + \pi^+$ satisfying $\Delta F = 0$, the sensitivity might be two or three orders of magnitude lower, while for $\Delta F = -6$ modes such as $p \to e^+ + \bar{\nu}_1 + \bar{\nu}_2$, the sensitivity may lie in between (see Ref. 7 for details).

**) Thus these can include in general only operators of the form $qqq\ell$ and $qqq\bar{\ell}$ which, respectively, induce only $\Delta F = -4$ (e.g., $p \to e^+ + \pi^0$) and $\Delta F = -2$ (e.g., $p \to e^- + \pi^+\pi^+$) decays. The $\Delta F = 0$ and $\Delta F = -6$ decays would involve in any case higher dimensional operators with a minimum of six fermion fields (dimension $\gtrsim 9$).

on this observation they had concluded that proton decay should be dominated by the $\Delta F = -4$ modes (e.g., $p \to e^+\pi^0$, $\bar{\nu}\pi^+$ etc.) which conserve B-L. In drawing this conclusion they were motivated by the assumption that the theory possesses essentially only two mass scales $m_{W_L} \sim 100$ GeV and*) $M_Y \sim 10^{14} - 10^{15}$ GeV, in which case the alternative decay modes $F = 0, -2, -6$ - requiring higher dimensional operators and/or violation of $SU(2)_L \times U(1)$ - would be damped at least by a factor $\sim (M_{W_L}/M_Y)$ compared to the "allowed" (B-L) conserving decay modes in the amplitudes.

There are, however, good reasons why one may consider departures from the aforementioned assumption.

The most important one is that there do exist grand unification models such as $[SU(4)]^4$, $[SU(6)]^4$ and their extended maximal versions involving fermion number gauging (such as SU(32), SU(48) or the smaller tribal group $[SU(16)]^3$), which do permit intermediate mass scales filling the gap between 10^2 and 10^{15} GeV with the lightest leptoquark gauge particle X being as light as perhaps $10^4 - 10^5$ GeV [2,11]. It is precisely because of the existence of these intermediate mass scales that the models of Refs 2 and 9 have permitted all along alternative proton decay modes**). This I elaborate below.

The second important reason why such intermediate mass scales are worthy of consideration is purely experimental. They provide the scope for discovery of new physics through tangible evidence for quark-lepton unification in the conceivable future, especially if there exist leptoquark (X) gauge particles in the 10-100 TeV region, and thus should be of interest to our experimental colleagues.

---

*) Here Y is used in the generic sense to denote a superheavy gauge particle coupling to different sorts of $F = \pm 2$ currents.

**) Several authors (Ref. 26) have recently considered the possibility of intermediate mass scales as in Refs 2 and 9, permitting Higgs rather than gauge particles to acquire such masses and introducing Yukawa interactions to induce new complexions for proton decay. We are, on the other hand, pursuing the consequences of only the gauge interactions subject to spontaneous symmetry breaking. Recently, Weinberg has extended[27] his analysis, permitting intermediate mass scales. I understand that H.A. Weldon and A. Zee have made a similar analysis, though I have not seen their preprint.

The third important reason is that if these intermediate mass scales do exist they would permit $\Delta F = 0$, -2 and -6 modes, whose rates may in general even exceed the rate of the so-called allowed $\Delta F = -4$ mode. Experiments must therefore be designed to look for such modes. It is proton decay and not any particular decay mode, which is important for the idea of quark-lepton unification.

The other reason for the existence of intermediate mass scales (with successive steps perhaps differing by powers of α or α^2) is that it may make it easier to understand the problem of the gauge hierarchy*). And finally the existence of intermediate mass scales may also account for the departure**) of present experimental $\sin^2 \theta_W \approx 0.23 \pm 0.01$ from the "canonical" theoretical value of ≈ 0.20.

With these to serve as motivations for the existence of intermediate mass scales let me first present a scenario for the hierarchy of gauge masses. This is depicted in Fig. 6. I discuss later that such a scenario can be obtained within maximal symmetries in accord with renormalization group equations for the gauge coupling constants. The characteristic feature of this scenario is that the leptoquark gauge particles (X) are rather light, characterizing the fact that they belong to the lower subunification symmetry $SU(2)_L \times SU(2)_R \times SU(4)'_{L+R}$. The masses of the fermion number ± 2 gauge particles Y, Y' and Y" defined before may range between 10^{10} and 10^{15} GeV, with all possible mutual orderings including the possibility that they may all be nearly degenerate.

I now wish to argue that the $\Delta F = 0$ and -2 modes involving the decays $p \to 3\ell$ + pions and $p \to (e^-$ or $\nu)$ + pions would be relevant to forthcoming proton decay searches for the following set of values of the X and Y gauge particles:

$$\Delta F = 0 \cdots\cdots\to \text{Need} \quad m_X \approx 10^4 - 10^5 \text{ GeV}$$

$$\Delta F = -2 \cdots\to \text{Need} \begin{cases} m_Y \approx 10^{10} - 10^{12} \text{ GeV} \\ m_X \approx 10^4 - 10^5 \text{ GeV} \end{cases} \tag{13}$$

(Later I show that these requirements are met within a class of maximal symmetries.)

---

*) This is only a conjecture at present and needs to be investigated.

**) The weight of this remark is dependent upon further refinements in the measurements of $\sin^2 \theta_W$.

$F = \pm 2$	$\ell\ell$	Y''	$10^{10} - 10^{15}$ GeV
	$(q\ell)$	Y'	
	qq	Y	
$F = 0$	$V + A$	$W_R^\pm$	$10^4 - 10^6$ GeV
$F = 0$	$(\bar{\ell}q)$	X's	
$F = 0$	$V - A$	$W_L^\pm$	100 GeV
$F = 0$	photon	gluons	$m = 0$

Fig. 6: A scenario for gauge masses arising within a maximal symmetry. The masses of Y, Y' and Y'' range between $10^{10} - 10^{15}$ GeV with all possible mutual orderings including the possibility that they are degenerate. $W_R^\pm$ can be heavier or lighter than X's.

$\Delta F = 0$ modes ($p \to 3\ell + \pi$'s)

These decays occur for either integer or fractionally charged quarks as follows. Each quark makes a virtual transition to a lepton and a Higgs field ϕ_α^i; the three Higgs fields generated thereby combine to annihilate into the vacuum through a VEV $\langle \overset{*}{B}{}_4^1 \rangle \neq 0$ (see Figs 3 and 4 and discussions in the previous section). The preferred configuration corresponds to one quark being right-handed, which emits a B_α^1 field, and the other two being left-handed, emitting appropriate components of C_α^i fields*). Furthermore it turns

*) Recall that B and C transform as $(1,2,\bar{4})$ and $(2,1,\bar{4})$ respectively under $SU(2)_L \times SU(2)_R \times SU(4)$. Together C and $B^\dagger$ make a 16-plet of SU(16). The C and B fields have the same quantum numbers as the fermion fields F_L and F_R. Thus only C_4^1 and B_4^1 possess non-zero VEV which give masses to $W_L^\pm$ and $W_R^\pm$ respectively (see Section 2).

out, owing to selection rules, that one of the quarks must proceed via a tree (Figs 3a) and the other two via loops (Fig. 3b)*). The corresponding amplitudes (suppressing spinors) are given by

$$M^{L,R}_{tree}\ (\text{Fig.3a}) = (f_X^2/m_X^2)(m_q)(\langle \overset{*}{C}{}_4^1\rangle\ ,\ \langle \overset{*}{B}{}_4^1\rangle)$$

$$M^{L,R}_{loop}\ (\text{Fig.3b}) = (f_X^2/m_X^2)\left(\frac{g^2 m_q}{8\pi^2}\right)(\langle \overset{*}{C}{}_4^1\rangle , \langle \overset{*}{B}{}_4^1\rangle)$$
$$\times \ln(m_X^2/m_{W_{L,R}}^2) \qquad (14)$$

where the superscripts L and R and the sets of parameters ($\langle \overset{*}{C}{}_4^1\rangle$, m_{W_L}) and ($\langle \overset{*}{B}{}_4^1\rangle$, m_{W_R}) go with the transitions of left and right-handed quarks, respectively.

Thus the low energy $3q \to 3\ell$ amplitude (Fig. 4) with two quarks being left-handed and one quark being right-handed (which is the preferred configuration) is

$$M(q_L q_L q_L \to \ell_R \ell_R \ell_L) \approx \frac{(M^L_{tree}\, M^L_{loop}\, M^R_{loop})\, 4\langle \overset{*}{B}{}_4^1\rangle\, \lambda_D}{m^4(C^i_\alpha)\ m^2(B^i_\alpha)} \qquad (15)$$

Here λ_D denotes the scalar quartic coupling constant, entering into the triple Higgs annihilation vertex (see Fig. 4). We shall take Higgs masses to be given by $\mu_\phi^2 \sim \lambda \langle\phi\rangle^2$ where λ is the associated scalar quartic coupling constant**). Furthermore, we shall take

---

*) This is under the assumption that leptoquark gauge particles (X') coupling to cross-currents of the form ($\bar{e}u$) and ($\bar{\nu}d$) are much heavier than those coupling to ($\bar{\nu}u$) and ($\bar{e}d$), which are denoted by X. This situation emerges automatically if the unifying symmetry G descends to low energy symmetries via $SU(2)_L \times \times SU(2)_R \times SU(4)'_{L+R}$. See Ref. 20 for a more general discussion.

**) In general λ may be a combination of several λ's. We use it in a generic sense.

$\lambda \sim \lambda_D \sim O(\alpha^2)$ to $O(\alpha)$, which appears to be a reasonable range for quartic coupling constants. Note that the amplitude (15) is proportional to $\langle \overset{*}{B}{}^1_4 \rangle^2 \, [m^2_*(B^i_\alpha)]^{-1} \simeq 1/\lambda$, which is independent of the rather unknown quantity $\langle \overset{*}{B}{}^1_4 \rangle$ or correspondingly $m_{W_R^\pm}$. Now, for concreteness, take $\lambda \approx \lambda_D = (\alpha/\pi^2)$, $m_X \approx (3 \text{ to } 10) \times 10^4$ GeV, $m_q = (1/3)$ GeV, $f_X^2 \sim g^2 \approx 4e^2$ and $\langle C^1_4 \rangle \sim (m_{W_L^+}/g) \sim 200$ GeV. Folding the $3q \to 3\ell$ amplitude into the proton wave function and taking the characteristic mass scale for proton decay to be $\simeq 1$ GeV, the proton lifetime (assuming that $\Delta F = 0$ modes are dominant) is given by

$$\tau_p \approx 10^{28} - 10^{34} \text{ years} \tag{16}$$

Admittedly the calculation is sensitive to some of the parameters such as the λ's and also the nature of the proton wave function. But the main point worth noting is that for reasonable values of the choice of λ, the $\Delta F = 0$ modes become important if the leptoquark gauge boson X has a mass $\approx (10^4 - 10^5$ GeV).

$\Delta F = -2$ modes ($p \to e^- \pi^+ \pi^+$, etc.)

These decays arise through $\bar{X} \leftrightarrow Y$ mass mixing (see Fig. 2b), which in turn is induced through VEV of the Higgs field $\Psi^{[AB]}_{[CD]}$ (see Section 2). Since such a mixing violates $SU(2)_L \times U(1)$, the corresponding mixing mass squared denoted by Δ^2_{XY} must be proportional to a VEV $\leq m_{W_L}$. Taking $\Delta^2_{XY} = m_{W_L} m_X$ (or $m^2_{W_L}$), and $m^2_X \approx 10^5$ GeV (as before) we see that the $\Delta F = -2$ proton decay interaction viewed as an effective four fermion interaction would have a strength $(\Delta^2_{XY})/(m^2_X\, m^2_Y)$, which would exceed the canonical value $\approx 10^{-29}$ GeV^{-2} for $m_Y \approx 10^{13}$ GeV (or 3×10^{11} GeV). This would lead to a proton lifetime in the range of $10^{31} - 10^{32}$ years.

We thus see that for $m_X \approx 10^4 - 10^5$ GeV and $m_Y \approx 3 \times 10^{11} - 10^{13}$ GeV (depending upon Δ^2_{XY}) $\Delta F = 0$ as well as $\Delta F = -2$ can coexist with comparable rates and be relevant to forthcoming proton decay searches.

$\Delta F = -4$ modes ($p \to e^+ \pi^0$, etc.)

These arise through $Y \leftrightarrow \bar{Y}'$ mass mixing (see Fig. 2a), which is induced through the VEV of a Higgs field $\Omega_{\{ABCD\}}$ (see Section 2). Denoting the mixing mass squared by $\Delta^2_{YY'}$, the corresponding amplitude is $\Delta^2_{YY'}/(m^2_Y - m^2_{Y'})$. As explained in Section 2, $\Delta^2_{YY'}$ can be as large as $m_Y^2 \sim m_{Y'}^2$, but of course it can be smaller*) than m_Y^2 and $m_{Y'}^2$ as well. Thus if $\Delta_{YY'}^2$ has its maximum value $\approx m_Y^2 \approx m_{Y'}^2$, then the $\Delta F = -4$ amplitude would have a strength**) $\approx 1/m_Y^2$ and a proton lifetime $\approx 10^{30}$ years would require $m_Y \simeq 10^{14}$GeV. But with $\Delta_{YY'}^2$ a few orders of magnitude smaller than m_Y^2, there is the interesting possibility that a proton lifetime of 10^{30} years can be compatible with Y and Y' being much lighter. For example, if $\Delta_{YY'}^2 = (10^{10}\ \text{GeV})^2$ then a proton lifetime of 10^{30} years would be compatible with $m_Y \approx m_{Y'} \sim 10^{12}$ GeV. As noted earlier, for such values of the Y mass (and with $m_X \approx 10^4 - 10^5$ GeV), the $\Delta F = -2$ and the $\Delta F = 0$ modes becomes relevant as well. Thus we see that a gauge mass pattern

$$\begin{gathered} m_X \sim 10^4 - 10^5\ \text{GeV},\ m_Y \approx m_{Y'} \approx 10^{12}\ \text{GeV} \\ (\Delta_{YY'}^2)^{1/2} \approx 10^{10}\ \text{GeV and } (\Delta_{XY}^2)^{1/2} \approx m_{W_L} m_X \end{gathered} \tag{17}$$

would permit the possibility that $\Delta F = 0$, -2 and -4 modes can co-exist and be relevant to present searches. The possible coexistence of the $\Delta F = -6$ mode depends upon further considerations which I shall not enter into here.

The task now is to show that such mass patterns as outlined in (17) can be realized within unifying symmetries in accord with renormalization group equations as well as the observed values of $\sin^2 \theta_W$ and α_S.

4. SOLUTIONS TO HIERARCHY EQUATIONS FOR A CLASS OF UNIFYING SYMMETRIES

Perturbative renormalization group equations for the running coupling constants of a spontaneously broken unifying symmetry permit in general solutions for the gauge masses[6] which exhibit a hierarchy. I refer to these equations as hierarchy equations and

---

*) This would be the case if Y and Y' receive their principal masses from the VEV of a Higgs field such as the adjoint $\underline{255}$ of SU(16), which does not mix Y and $\bar{Y}'$. The adjoint $\underline{255}$ breaks SU(16) to $SU(8)_L \times SU(8)_R \times U(1)_F$.

**) Note that in this case $\Delta F = -4$ would supersede $\Delta F = -2$ by a factor $\gtrsim (m_X/m_{W_L})$ in the amplitude.

ask: do there exist solutions to these equations within some class of unifying symmetries which permit:

$$\begin{aligned} &(a)\quad M_X \sim 10^4 - 10^5 \text{ GeV} \\ &(b)\quad M_Y \sim 10^{11} - 10^{15} \text{ GeV} \\ &(c)\quad \sin^2\theta_W \approx 0.23 \text{ and} \\ &(d)\quad \alpha_s(m_W) \approx 0.14 \end{aligned} \tag{18}$$

We know that the answer is negative for SU(5) and SO(10).

Now, to see how a "light" leptoquark gauge particle X with a mass $\approx 10^4 - 10^5$ GeV can be realized in the first place, it is useful to recall the case of the $[SU(4)]^4$ model, which possesses a single gauge coupling constant because of discrete symmetry between the four SU(4) factors. This symmetry, depending upon the nature of spontaneous symmetry breaking can break via two alternative chains

$$[SU(4)]^4 \begin{cases} \xrightarrow{M_1} SU(2)_L^{I+II} \times U(1) \times SU(3)'_{L+R} \\ \xrightarrow[M_1]{} SU(2)_L^{I+II} \times U(1) \times SU(3)'_L \times SU(3)'_R \end{cases} \tag{19}$$

Here $SU(4)^{flavour}_{L,R}$ acts on $(u,d,c,s)_{L,R}$ flavours, $SU(2)^I$ and $SU(2)^{II}_L$ act on $(u,d)_L$ and $(c,s)_L$ doublets respectively, and $SU(2)^{I+II}_L$ is their diagonal sum. The gauge particles of $SU(2)^{I+II}_L$ are related to those of $SU(2)^{I,II}_L$ by $W^{I+II}_L = (W^I_L + W^{II}_L)/\sqrt{2}$. Thus, if g is the symmetric gauge coupling constant of each SU(4) factor, the coupling constant g_2 of the low energy symmetry $SU(2)^{I+II}_L$ would approach $g/\sqrt{2}$ in the symmetric limit. Likewise the coupling constant g_3 for vector colour $SU(3)_{L+R}$ would also approach $g/\sqrt{2}$, since it is obtained by diagonal summing of $SU(3)'_L$ and $SU(3)'_R$. By contrast the coupling constant g_3^c for chiral colour*) $SU(3)_L \times SU(3)_R$ (relevant to the lower chain) would approach g in the symmetric limit. Thus,

$$\begin{aligned} &\text{Vector Colour } g_2 = g_3^V = \frac{g}{\sqrt{2}} \text{ (Symmetric limit)} \\ &\text{Chiral Colour } g_2 = \frac{g}{\sqrt{2}} ; \; g_3^c = g \text{ (Symmetric limit)} \end{aligned} \tag{20}$$

This difference of a factor of $\sqrt{2}$ between flavour versus colour coupling constants, after it is squared and after some arithmetic translates it into a factor ≈ 2.5, multiplies the logarithm of M_1/μ and thus alters drastically the determination of the unification mass M_1, so that one obtains[11]

$$\begin{aligned} &\text{For vector colour} \begin{cases} M_1 \approx 10^{15} \text{ GeV} \\ \sin^2\theta_W \approx 0.20 \end{cases} \\ &\text{For chiral colour} \begin{cases} M_1 \approx 10^{6} \text{ GeV} \\ \sin^2\theta_W = \frac{2}{7} + \frac{10}{21\alpha_s} \approx 0.30 \end{cases} \end{aligned} \tag{21}$$

The mass of X is about a factor of 10 lower than M. Thus for $M_1 \simeq 10^6$ GeV, m_X is $\simeq 10^5$ GeV, as desired. However, the case of $[SU(4)]^4$ descending via chiral colour is now excluded experimentally, since it yields too high a value for $\sin^2\theta_W$ ($\simeq 0.30$) compared to the experimental value of $\simeq 0.23$.

Nevertheless the above example provides the clue for low mass unification. The idea is to create through spontaneous descent a dichotomy between low energy flavour versus colour coupling constants such that the former is lower than the latter in the

---

*) The chiral colour symmetry must break to vectorial colour eventually. But if this breaking takes place by a mass scale $< m_{W_L}$, one can ignore the effect of such a breaking for studies of renormalization group equations at momenta $> m_{W_L}$.

symmetry limit*). This is best illustrated by the symmetry[17] $[SU(6)]^4$, which operators on six flavours (u,d,c,s,t,b) and six colours. There are three leptonic colours rhyming with three quark colours (r,y and b). There are the six observed leptons plus twelve unobserved heavy leptons in the model**). Here the low energy flavour $SU(2)_L$ is obtained by diagonal summing of three SU(2)'s, which respectively act on the (u,d), (c,s) and (t,b) doublets. Thus $g_2 = g/\sqrt{3}$ in the symmetric limit. In this case, even if the low energy colour symmetry is vectorial $SU(3)'_{L+R}$ and therefore the corresponding coupling constant $g_3^V = g/\sqrt{2}$ in the symmetric limit, still $g_2 = g/\sqrt{3} < g_3^V$ in the symmetric limit. This, together with the fact that the bare value of the weak angle[28] $\sin^2\theta_0 = 9/28$ (rather than 3/8), leads again to a low unification mass***) for the descent $[SU(6)]^4 \overset{M_1}{\to} SU(2)_L^{I+II+III} \times U(1) \times SU(3)'_{L+R}$.

$$M_1 \approx 10^6 \text{ GeV}, \text{ i.e. } M_X \approx 10^5 \text{ GeV} \tag{22}$$

In this case one furthermore obtains a desirable value for the weak angle[28]

$$\sin^2\theta_W = \frac{5}{24} + \frac{19}{36}\frac{\alpha}{\alpha_s} \approx 0.235 \tag{23}$$

We thus see that quark-lepton unification can be visible through leptoquark gauge interactions at an energy scale 10^5 or even as low as 10^4 GeV.

---

*) This ingredient hastens the "meeting" of the colour and the SU(2) flavour coupling constants. There is a second ingredient which can speed up the "meeting" of SU(2) and U(1) coupling constants. This is realized through a lowering of the bare value of the weak angle $\sin^2\theta_0$ from the canonical value 3/8. In some models this is possible. See discussions later.

**) This is without counting the mirror fermions.

***) Note that had we chosen the descent via chiral colour, i.e., $[SU(6)]^4 \to SU(2)_L^{I+II+III} \times U(1) \times SU(3)'_L \times SU(3)'_R$, we would have obtained a still further reduction in M_1.

What about the masses of the fermion number F = ±2 gauge particles Y, Y' and Y" arising within a maximal symmetry? To obtain a scenario in which the masses of these gauge particles lie in the range of 10^{10} - 10^{15} GeV, while the X's are as light ≃ 10^4 - 10^5 GeV, let us proceed as follows*). Assume (following the illustrations for $[SU(4)]^4$ and $[SU(6)]^4$) that each individual family defines a distinct SU(2) within the parent symmetry G. These distinct SU(2)'s combine (or following a terminology used before, they are "squeezed") through spontaneous symmetry breaking by a relatively heavy mass scale**) M_L to yield a single SU(2), which is the SU(2) of low energy electroweak symmetry. Thus allowing for q left-handed families we envisage the descent

$$[SU(2)_L]^q \xrightarrow[SSB]{M_L} SU(2)_L \tag{24}$$

Recall that for $[SU(4)]^4$, q = 2, while for $[SU(6)]^4$, q = 3. If the theory is left-right symmetric, there would be the corresponding "squeezing" of $[SU(2)_R]^q$ into a single $SU(2)_R$ or even $U(1)_R$ through a heavy mass scale M_R

$$[SU(2)_R]^q \xrightarrow{M_R} SU(2)_R \text{ or } U(1)_R \tag{25}$$

In general the parent symmetry G may contain distinct SU(4) colour symmetries***) as well which are distinguished from each other either through the helicity of the fermions on which they operate, or through the family attribute, or both. For generality assume that there are p SU(4) colour symmetries within G. To be specific I shall furthermore assume that these are vectorial L+R symmetries. The generalization to chiral SU(4) colour is straightforward.

---

*) The discussions to follow are based on a forthcoming paper[20] by D. Deo, J.C. Pati, S. Rajpoot and Abdus Salam.

**) To realize the known universality of different families in electroweak interactions and to preserve the GIM mechanism up to its know accuracy, M_L should exceed about 10^5 GeV.

***) With the fourth colour being lepton number.

These p $SU(4)_{L+R}$ symmetries are "squeezed" through SSB to a single*) $SU(4)_{L+R}$ by a heavy mass scale M_4. The single $SU(4)_{L+R}$ subsequently descends also spontaneously to $SU(3)^{colour}_{L+R} \times U(1)_{L+R}$ via a heavy mass scale M_3. The leptoquark gauge particles X' receive their mass through M_3 with $M_X \approx M_3/10$. Thus the color sector may break as follows

$$[SU(4)_{L+R}]^p \xrightarrow{M_4} SU(4)_{L+R} \xrightarrow{M_3} SU(3)_{L+R} \times U(1)_{L+R} \tag{26}$$

In short, the scenario which we are led to consider for the sake of obtaining intermediate mass scales and thereby signals for grand unification at moderate energies is this: the families define distinct SU(2)'s and possibly even distinct SU(3) or SU(4) colour symmetries at the level of the parent symmetry. The distinction is lost and thereby the universality of families defined by discrete symmetries $e \leftrightarrow \mu \leftrightarrow \tau$ emerges at low energies due to spontaneous symmetry breaking. In other words, taking only two families e and μ, there are two distinct W's (W_e and W_μ) in the basic Lagrangian. Due to hierarchical SSB $(W_e - W_\mu)/\sqrt{2}$ acquires a heavy mass $\gtrsim 10^5$ GeV, but $(W_e + W_\mu)/\sqrt{2}$ acquires a mass only of order 100 GeV, hence the low energy $e \leftrightarrow \mu$ universality. Such a picture is logically feasible, since tests of $e \leftrightarrow \mu$ universality in weak interactions extend at best up to 10 to 30 GeV of centre-of-mass energies. Do there exist W's and Z's which couple to differences of e and μ currents rather than their sums? Tests of such family universality should provide an important motivation for building high energy accelerators in the 1-100 TeV region.

Such family distinctions are not realized within smaller symmetries such as SU(5), SO(10) and SU(16), if these are to be viewed as parent symmetries. But they do exist within symmetries such as $[SU(4)]^4$, $[SU(6)]^4$, $[SU(5)]^3 = SU(5)_e \times SU(5)_\mu \times SU(5)_\tau \subset SU(15)$ and likewise $[SO(10)]^3$ and $[SU(16)]^3$ or the still bigger symmetry SU(48).

We are aware that the symmetries of the latter kind are gigantic. But then Nature appears to be proliferated anyway beyond one's imagination at the quark-lepton level. The fermion contents of the symmetries mentioned above are essentially no more that the

---

*) Alternatively $[SU(4)]^p$ may descend first to $[SU(3)]^p \times [U(1)]^p$, which subsequently descends to $[SU(3)] \times U(1)$. This is considered in Ref. 20.

ones already observed. At the present stage of our ignorance it appears prudent therefore not to limit ourselves to smaller symmetries which permit no family distinctions, and to appeal to experiments to provide guidance in this matter. The answer to proliferation at the quark-lepton level both as regards the fermion content and perhaps the associated symmetry structure must be sought for elsewhere. This is what leads one to believe that quarks and leptons are composites of more elementary objects - the PREONS. I shall return to this notion at the end.

With these remarks to serve as motivations, we consider the possibility that the parent symmetry G breaks spontaneously to low energy components as follows[20]:

$$
\begin{array}{ccccccc}
G \xrightarrow{M} & [SU(2)_L]^q & \times & [SU(2)_R]^q & \times & [SU(4)_{L+R}]^p & \times\, U(1) \\
& \downarrow M_L & & \downarrow M_R & & \downarrow M_4 & \\
& SU(2)_L & & U(1)_R & & SU(4)_{L+R} & \\
& & & & & \downarrow M_3 & \\
& & & & & SU(3)_{L+R} \times U(1)_{L+R} & \\
& & & \downarrow m_{W_L} & & & \\
& & & SU(2)_L \times U(1) \times SU(3)_{L+R} & & &
\end{array} \tag{27}
$$

Such a hierarchy leads to the following two equations via the renormalization group equations for the coupling constants:

$$
\frac{\sin^2\theta_0 - \sin^2\theta_W}{\alpha \cos^2\theta_0} = -\frac{11 \ln_e A}{6\pi \cos^2\theta_0} \xrightarrow[M]{\text{Single}} \frac{11}{3\pi} \ln \frac{M}{\mu}
$$

$$
\frac{q}{\alpha_s} - \frac{p}{\alpha} \sin^2\theta_W = -\frac{11 \ln B}{6\pi} \xrightarrow[M]{\text{Single}} -\left(\frac{3q-2p}{2}\right) \frac{11}{3\pi} \ln \frac{M}{\mu} \tag{28}
$$

$$A \equiv D^{-1} \left[\left(\frac{M}{M_L}\right)^{2q} \left(\frac{M_L}{m_{W_L}}\right)^{2} \left(\frac{M}{M_R}\right)^{2q} \left(\frac{M}{M_4}\right)^{\frac{8p}{3}} \left(\frac{M_4}{M_3}\right)^{\frac{8}{3}} \right]^{\sin^2\theta_0}$$

$$B \equiv D^{-1} \left[\left(\frac{M}{M_4}\right)^{4p} \left(\frac{M_4}{M_3}\right)^{4} \left(\frac{M_3}{m_{W_L}}\right)^{3} \right] \qquad \text{(28) cont.}$$

$$D = \left(\frac{M}{M_L}\right)^{2q} \left(\frac{M_L}{m_{W_L}}\right)^{2}$$

We now ask: are there solutions to these equations for some p and q, which would satisfy the constraints on gauge masses as well as $\sin^2\theta_W$ and α_S as listed in Eq. (19)? (For these purposes chiral colour will correspond to p = 2.) We find[20] that there is no solution satisfying constraint (19) for p = q = 1 (corresponding for example to SU(5) and SO(10)), for p = 2, q = 1 (corresponding to SU(16)) etc. But there do exist desired solutions for p = q = 2; p = 2, q = 3 and p = q = 3, which could be obtained for example within $[SO(10)]^3$, $[SU(16)]^3$ and their extensions. These solutions and the corresponding coexistence of alternative proton decay modes are listed below:

$$p = q = 2 \quad \begin{cases} M \sim 10^{15}\ \text{GeV},\ M_4 \sim M_L \sim M_R \sim 10^{12}\ \text{GeV} \\ M_3 \sim 10^{5.5}\ \text{GeV} \Rightarrow M_X \sim 10^{4.5}\ \text{GeV} \\ \text{Here } \Delta F = 0 \text{ and } \Delta F = -4 \\ \text{Can Coexist, but } \Delta F = -2 \text{ is} \\ \text{suppressed} \end{cases} \qquad (29)$$

$$p = 2,\ q = 3 \quad \begin{cases} M \approx M_4 \approx 10^{12}\ \text{GeV},\ M_R \approx 10^{10}\ \text{GeV}, \\ M_L \approx 10^{7}\ \text{GeV},\ M_3 \approx 10^{5}\ \text{GeV} \Rightarrow \\ M_X \approx 10^{4}\ \text{GeV} \\ \text{Here } \Delta F = 0, -2, \text{and } -4 \text{ Can Coexist} \end{cases}$$

$$p = q = 3 \begin{cases} M \approx M_R \sim 10^{12}\ \text{GeV},\ M_4 \approx 10^9\ \text{GeV}, \\ M_L \approx 10^6\ \text{GeV},\ M_3 \approx 10^5\ \text{GeV} \Rightarrow \\ M_X \sim 10^4\ \text{GeV} \\ \text{Here } \Delta F = 0,\ -2 \text{ and } -4 \text{ can Coexist} \end{cases} \tag{29 cont.}$$

We thus see that within maximal symmetries permitting intrinsic family distinctions the proton can decay through alternative decay modes as claimed in the introduction.

5. A SUMMARY OF THE FIRST PART

I raised two questions:

i) Is it conceivable that the basic idea of quark-lepton unification may be tested tangibly through the manifestation of exotic quark-lepton interactions in the conceivable future? I have answered this in the affirmative. A number of unification models permit at least the leptoquark X gauge particles to possess a mass in the 10-100 TeV region. We need Isabelle, ISR and the immediate successors thereof, as well as improved cosmic ray studies, to see X effects. These may be seen for example through enhanced lepton pair production in pp and $\bar{p}p$ processes.

ii) Can the variety of complexions for proton decay outlined in Section 1 exist and coexist? This question has been answered in the affirmative. The answer here is in part correlated to that in i).

To summarize the first and the major part of my talk:

1) The idea of quark-lepton unification is basic. It is not tied to the nature of quark charges.

2) Proton decay is central to the hypothesis of quark lepton unification, but not any particular decay mode. It is a reasonable expectation within most models that the lifetime of proton should lie within the range of 10^{28} - 10^{33} years.

3) Proton decay modes can provide an important clue to the underlying design of grand unification. For example, observation of $\Delta F = 0$, -2 or -6 mode at any level within the conceivable future will signal the existence of intermediate mass scales, which in turn will reflect upon the nature of the parent symmetry G. In particular, the observation of the $\Delta F = 0$ mode will strongly suggest the existence of new interactions in the 10 to 100 TeV region. Thus, a search for such decay modes, if need to be through second and third generation experiments, would be extremely important in that such searches would have implications for the building of high energy accelerators.

4) The observation of proton decay will strongly support the idea that quark and leptonic matters are ultimately of one kind.

5) This is not to say, however, that they need to be the final constituents.

This leads to the second part of my talk where I indicate the directions in which some of the changes might occur for the unification hypothesis, if quarks and leptons are viewed as composites of more elementary objects - the preons - and also how the preons may bind.

6. PREONS*)

To resolve the dilemma of quark-lepton proliferation it was suggested in 1974 that quarks and leptons may define only a stage in one's quest for elementarity[30,31]. The fundamental entities may more appropriately correspond to the truly fundamental "attributes" (charges) exhibited (or yet to be exhibited) by Nature. The fields carrying these fundamental attributes are named "PREONS". Quarks and leptons**) may be viewed within this picture as composites of a set of preons consisting, for example, of m elementary "flavons" (f_i) plus n elementary "chromons" (C_α). The flavons carry only flavour but no colour, while the chromons carry only colour but no flavour. If both flavons and chromons carry spin-1/2, one needs to include a third kind of spin-1/2 attribute (or attributes) in

---

*) This section is based on a recent paper[29] by the present author

**) For simplicity let us proceed with the notion that lepton number is the fourth colour (Ref. 2). In this case the composite structure is as follows: $(q_{u,r,y,b}) = u + (r,y \text{ or } b) + \zeta$, while $\nu = u + \ell + \zeta$, etc. Within the preon idea leptons may, however, differ from quarks by more than one attribute. For example, we may have $\nu = u + \ell + \zeta'$ where $(\zeta' \neq \zeta)$. Such variants will be considered.

the preon-set, which for convenience we shall call "spinons" (ζ_ν); these serve to give spin-1/2 to quarks and leptons*) but may in general serve additional purposes. The quarks and leptons are in the simplest case composites of one flavon, one chromon and one spinon plus the "sea". If the μ and τ families are viewed to differ from the e-family only in respect of an "excitation quantum number" or degeneracy quantum number, which is lifted by some "fine" or "hyperfine" interaction, then only seven preons consisting of (u, d,r,y,b,ℓ and ζ) suffice to describe the 24 quarks and leptons of three families, and even more if they are to be discovered.

For this reason, the preon idea appears to be attractive. But can it be sustained dynamically? The single most important problem which confronts the preon hypothesis is this: what is the nature and what is the origin of the force which binds the preons to make quarks and leptons?

Our first observation is that ordinary "electric" type forces**) - Abelian or non-Abelian - arising within the grand unification hypothesis are inadequate to bind preons to make quarks and leptons unless we proliferate preons much beyond the level depicted above.

The argument goes as follows: since quarks and leptons are so point-like, their sizes are shorter than 10^{-16} cm as evidenced (especially for leptons) by the (g - 2) experiments - it follows that the preon binding force F_b must be strong or superstrong at short distances $r \leq 10^{-17} - 10^{-18}$ cm corresponding to running momenta $Q \gtrsim 1$ to 10 TeV. (Recall for comparison that the chromodynamic forces generated by the $SU(3)_{colour}$-symmetry are strong ($\alpha_c \gtrsim 1$) only at distances of order 1 Fermi, which correspond to the sizes of the known hadrons.) This says that the symmetry generating the preon binding force must lie outside the familiar $SU(2) \times U(1) \times \times SU(3)_{col}$ symmetry.

Now, consistent with our desire to adhere to the grand unification hypothesis, we shall assume that the preon binding force F_b derives its origin either intrinsically or through the spontaneous breakdown of a grand unifying symmetry G. Thus either the basic symmetry G is of the form $G_k \times G_b$ with G_k generating the known electroweak-strong forces and G_b generating the preon-binding forces, in which case G_k and G_b are related to each other by a discrete symmetry so as to permit a single gauge coupling constant, or the unifying symmetry G breaks spontaneously as follows:

*) With the spinon present the flavons and chromons can carry integer spin 0 or 1.

**) By "electric" type forces we mean forces whose effective coupling strength is of order $\alpha \simeq 1/137$ at the unification point M.

$$G \xrightarrow{SSB} G_K \times G_b \times [\text{possible } U(1) \text{ factors}] \tag{30}$$

In the second case G_k need not be related to G_b by discrete symmetry. But in either case G_k contains the familiar $SU(2)_L \times U(1)_{EW} \times SU(3)_{colour}$-symmetry and therefore the number of attributes (N_k) on which G_k operates needs to be at least 5. This corresponds to having two flavons (u,d) plus three chromons (r,y,b). To incorporate the leptonic chromon ℓ and possibly also the spinon ζ, N_k may need to be at least 7, but for the present we shall take conservatively $N_k \geq 5$.

Now consider*) the size of G_b. On the one hand the effective coupling constant $\bar{g}_b$ of the binding symmetry G_b is equal to the effective coupling constant $\bar{g}_c$ of the familiar SU(3)-colour symmetry (up to embedding factors[11] like $1/\sqrt{2}$ or $1/\sqrt{3}$ etc.) at the unification mass scale $M >> 10^4$ GeV. On the other hand, $\bar{\alpha}_b \equiv \bar{g}_b^2/4\pi$ needs to exceed unity at a momentum scale $\mu_b \geq 1$ to 10 TeV, where the chromodynamic coupling constant $\bar{\alpha}_c << 1$. It therefore follows (assuming that the embedding factor mentioned above is unity) that G_b is much larger than SU(3)**). Using renormalization group equations for variations of the coupling constants $\bar{\alpha}_b$ and $\bar{\alpha}_c$, one may verify that the minimal G_b is SU(5) and correspondingly the minimal dimension N_b of the space on which G_b operates is 5.

Now the preons $\{P_i\}$ which bind to make quarks and leptons must be non-trivial with respect to both G_k and G_b. Since each of G_k and G_b requires for their operations a space, which is minimally five dimensional, it follows that the number of preons N_p needed (under the hypothesis alluded to above) is minimally $N_k \times N_b \geq 25$.

$$N_p \geq N_K \times N_b \geq 5 \times 5 = 25 \tag{31}$$

*) In these considerations I assume all along that conventional perturbative renormalization group approach applies to the variations of running coupling constants down to such momenta where they are small ($g_i^2/4\pi < 0.3$ say) (see Ref. 6).

**) This incidentally excludes the possibility that G_b is Abelian.

We may consider relaxing the assumption that the embedding factor is unity. This would permit the ratio $[\bar{g}_b(\mu)/\bar{g}_c(\mu)]_{\mu=M}$ to be a number like $\sqrt{2}$ or $\sqrt{3}$ for example. In turn this can result in a reduction in the size of G_b. But simultaneously such a step necessitates an increase in the size of G_k or effectively of the number N_k with the result that the minimal number of preons needed $N_p \geq N_k \times N_b$ is not reduced below 21.

This number 25 (or 21) representing the minimal number of preons needed already exceeds or is close to the number of quarks and leptons which we need at present, which is 24. And if we include, more desirably, the leptonic chromon ℓ and the spinon ζ in the preonic degrees of freedom, the number of preons needed would increase to 35 (or 27).

Such a proliferation of preons defeats from the start the very purpose for which they were introduced - economy. In turn, this poses a serious dilemma. On the one hand giving up the preon idea altogether and living with the quark-lepton system as elementary runs counter to one's notion of elementarity and is thus unpalatable. On the other hand, giving up the grand unification hypothesis is not aesthetically appealing.

Noting this impasse, we are led to believe that there exists yet a new kind of force, which operates on preons and binds them. The only force which strikes us as missing so far, which can serve the purpose of preon binding without demanding a proliferation, and which is aesthetically appealing, is the magnetic counterpart of the electric forces. We are thus led to suggest that the preons carry not only electric but also magnetic charges and that their binding force is magnetic in nature. The two types of charges are related to each other by the familiar Dirac-like quantization conditions[32,33] for charge-monopole or dyon systems, which imply that the magnetic coupling strength $\alpha_m \equiv g_m^2/4\pi$ is $O(1/\alpha_e) \simeq O(137)$ and thus is superstrong. In other words, the magnetic force can arise through an Abelian U(1)-component within the unification hypothesis (as remarked further at the end) and yet it can be superstrong. This is what gives it the power to bind preons into systems of small size without requiring a proliferation. Quarks and leptons do not exhibit this superstrong force because they are magnetically neutral (see remarks below).

I first discuss the consistency of this idea with presently known phenomena from a qualitative point of view and later indicate the possible origin of this magnetic force.

1) Since the electric fine structure constant $\alpha_e = e^2/4\pi$ varying with running momentum remains small $\lesssim 10^{-2}$ almost everywhere (at least up to momenta $\sim 10^{14}$GeV and therefore down to distances $r \sim 10^{-28}$ cm), the magnetic "fine structure" constant

$\alpha_m \equiv g^2/4\pi$ related to α_e by the reciprocity relations in superstrong even at distances as short as 10^{-28} cm (if not at $r \to 0$). It is this strong short-distance component of the magnetic force which makes quarks and leptons so point-like with sizes $r_0 << 10^{-16}$ cm. Their precise size would depend upon the dynamics of the superstrong force which we are not yet equipped to handle. For our purposes we shall take r_0 to be as short as perhaps $M_{Planck}^{-1} \sim 10^{-33}$ cm, but perhaps as large as 10^{-18} cm (i.e., $r_0 < 10^{-18}$ cm).

2) Quarks and leptons do not exhibit even a trace of the superstrong interactions of their constituents because they are magnetically neutral composites of preons and their sizes are small compared to the distances $r \geq 10^{-16}$ cm which are probed by present high energy experiments.

3) We mention in passing that had we assumed, following Schwinger[33], that quarks (rather than preons) carry magnetic charges, we would not understand why they interact so weakly at short distances as revealed by deep inelastic ep-scattering.

4) Due to their extraordinarly small sizes, it can also be argued[29] that low energy parameters such as the $(g - 2)$ of leptons would not show any noticeable departures from the normal expectations. A similar remark applies to the P and T violations for quarks and leptons, which would be severely damped despite large P and T violations for preons carrying electric and magnetic charges.

I now end this talk by indicating the possible origin of the magnetic charges of preons. The origin could perhaps be topological[34,35]. Spontaneous breaking of the non-Abelian preonic local symmetry G_p to lower symmetries may generate monopoles or dyons. Such a picture would be attractive if in particular it could generate spin 1/2 monopoles (in addition to spin 0 and spin 1) and assign electric and magnetic charges to the originally introduced spin 1/2 fields and their topological counterparts.

There is a second alternative, which is the simplest of all in respect of its gauge structure. Assume that the basic Lagrangian of the preons is generated simply by the Abelian symmetry $U(1)_E \times U(1)_M$. The $U(1)_F$ generates "electric" and $U(1)_M$ the "magnetic" interactions of preons. Subject to subsidiary conditions, the theory generates only one photon coupled to electric as well as magnetic charges[36]. The charge are constrained by the Dirac quantization condition. In this model the basic fields are only the spin 1/2 preons and the spin-1 photon. The strong magnetic force binds preons to make spin 1/2 quarks and leptons as discussed earlier. Simultaneously it makes spin-1 and spin-0 composites of even numbers of preons (including antipreons), which also have very small sizes like the quarks and leptons as well as among themselves. The use of a recently prevalent theorem[37] would then suggest that their effective interactions must be generated from a local non-Abelian symmetry of the Yang-Mills type which is broken spontaneously, in order that they may be renormalizable. The spin-0 composites will now play the role of Higgs fields. It is amusing that if this picture can be sustained, the proliferated non-Abelian quark-lepton gauge structure $G_{(q,\ell)}$ with the associated spin 1/2, spin-1 as well as spin-0 quanta may have its origin in the simplest interaction of all: electromagnetism defined by the Abelian symmetry $G_P = U(1)_E \times U(1)_M$.

If the suggestion of Harari and Shupe[38] for generating the colour degree of freedom at the composite level can be sustained dynamically, it would be an attractive economical picture. To me it seems that (a) one may need to go to the preon or even to the pre-preon-basis for developing this economical basis, and (b) the binding of such preons or pre-preons should be magnetic in origin for reasons outlined here.

The idea of the magnetic binding of preons and its origin needs to be further developed. What I have argued here is that within the unification context a magnetic binding of preons appears to be called for if we are not to proliferate preons unduly.

REFERENCES

1. J.C. Pati and Abdus Salam, Phys. Rev. D8 (1973) 1240.
2. J.C. Pati and Abdus Salam, Phys. Rev. Letters 31 (1973) 661; Phys. Rev. D10 (1974) 275; Phys. Letters 58B (1975) 333.
3. H. Georgi and S.L. Glashow, Phys. Rev. Letters 32 (1974) 438.
4. J.C. Pati, Proc. of Symp. on Elementary Particle Physics, Seoul 1978, ed. J. Kim, P.Y. Pac and H.S. Song, (National University, Seoul, 1979).
5. T. Goldman and D. Ross, Caltech preprint. CALT-68-759 (1980).
6. H. Georgi, H. Quinn and S. Weinberg, Phys. Rev. Letters 33 (1974) 451.
7. See L. Sulak, these proceedings, p. 641.
8. F. Reines and M.F. Crouch, Phys. Rev. Letters 32 (1974) 493.
9. J.C. Pati, Abdus Salam and J. Strathdee, Nuovo Cimento 26A (1975) 77;
   J.C. Pati, Proceedings of the Second Orbis Scientiae, Coral Gables, Florida 1975, ed. B. Kursunoglu, A. Perlmutter and L.F. Scott (Plenum Press, New, 1975):
   J.C. Pati, S. Sakakibara and Abdus Salam, ICTP, Trieste preprint IC/75/93 (1975), (unpublished).
10. J.C. Pati and Abdus Salam, ICTP, Trieste preprint IC/80/72 (1980) .
11. V. Elias, J.C. Pati and Abdus Salam, Phys. Rev. Letters 40 (1978) 920.

12. G. Steigman, these proceedings, p. 495.
13. See M. Gell-Mann, P. Ramond and R. Slansky, Rev. Mod. Phys. 50 (1978) 721;
P. Langacker, G. Segrè and H.A. Weldon, Phys. Letters 73B (1978) 87.
14. H. Fritzsch and P. Minkowski, Ann. Phys. (N.Y.) 93 (1975) 193;
H. Georgi, "Particles and fields - 1974", (APS/DPF, Williamsburg), ed. C.E. Carlson (AIP, New York, 1975, p. 575 (1974).
15. See for example R.N. Mohapatra and G. Senjanovic, Phys. Rev. D20 (1979) 3390.
16. J.C. Pati and Abdus Salam, Phys. Rev. D10 (1974) 275;
R.N. Mohapatra and J.C. Pati, Phys. Rev. D11 (1975) 566;
G. Senjanovic and R.N. Mohapatra, Phys. Rev. D12 (1975) 1502.
17. J.C. Pati and Abdus Salam, unpublished work (1975);
J.C. Pati, Proceedings of the Scottish Univ. Summer School, 1976 ed. I.M. Barbour and A.T. Davies (SUSSP publications, Edinbur 1977).
18. J.C. Pati and Abdus Salam, forthcoming preprint, ICTP Trieste preprint IC/80/72 (1980).
19. F. Gürsey, P. Ramond and P. Sikivie, Phys. Letters 60B (1976) 177. For other references see B. Stech, R. Barbieri and D.V. Nanopoulos, these proceedings, pp. 23, 17, 435.
20. B. Deo, J.C. Pati, S. Rajpoot and Abdus Salam, preprint in preparation;
Abdus Salam, Proceedings of the EPS Conference, Geneva (CERN, 1979) p. 853.
21. L. Okun, M.B. Voloskin and V.I. Zakharov, ITEP, Moscow preprint 79-79 (1979).
22. J.C. Pati and Abdus Salam, ICTP, Trieste preprint in preparation (1980).
23. M.S. Chanowitz, Phys. Rev. Letters 44 (1980) 59.
24. H.K. Lee and J.K. Kim. Phys. Rev. Letters 40 (1978) 485;
J.K. Kim and H.K. Lee, Korea Advanced Institute of Science, Seoul, preprint (1979).
25. S. Weinberg, Phys. Rev. Letters 43 (1979) 1566;
F. Wilczek and A. Zee, Phys. Rev. Letters 43 (1979) 1571.
26. See F. Wilczek and A. Zee, preprint UPR-0135 T (1980);
R.N. Mohapatra and R. Marshak, preprints VPI-HEP-80/1,2 (1980);
S. Glashow, unpublished.
27. S. Weinberg, preprint HUTP-80/A023 (1980);
H.A. Weldon and A. Zee, U. Pennsylvania preprint (1980).
28. V. Elias and S. Rajpoot, ICTP, Trieste preprint IC/78/159 (1978).
29. J.C. Pati, University of Maryland preprint TR 80-095 (1980).

30. J.C. Pati and Abdus Salam, Phys. Rev. D10 (1974) 275; Proceedings of the EPS International Conference on High Energy Physics, Palermo, June (1975), p. 171, Ed. A. Zichichi; J.C. Pati, Abdus Salam and J. Strathdeed, Phys. Letters 59B (1975) 265.
31. See H. Terazawa, University of Tokyo preprint INS-Rep. 351 (1979) for a partial list of references on work of other authors.
32. P.A.M. Dirac, Proc. Roy. Soc. (London) A133 (1931) 60; Phys. Rev. 74 (1948) 817.
33. J. Schwinger, Phys. Rev. 144 (1966) 1087; D12 (1975) 3105; Science 165 (1969) 757; D. Zwanziger, Phys. Rev. Letters 40 (1978) 147.
34. G. 't Hooft, Nucl. Phys. B79 (1974) 276; A.M. Polyakov, JETP Letters 20 (1974) 194.
35. C. Montonen and D. Olive, Phys. Letters B72 (1977) 117; P. Goddard, J. Nuyts and D. Olive, Nucl. Phys. B125 (1977) 1.
36. The formalism may follow that of D. Zwanziger, Phys. Rev. D3 (1971) 880.
37. M. Veltman (unpublished); See J. Ellis, M.K. Gaillard, L. Maiani and B. Zumino, these proceedings, p. 69.
38. H. Harari, Phys. Letters 86B (1979) 83; M. Shupe, Phys. Letters 86B (1979) 87.

SUBCOMPONENT MODELS FOR QUARKS AND LEPTONS : DIFFICULTIES AND PERSPECTIVES *

R. Gatto

Département de Physique Théorique
Université de Genève
1211 Genève 4, Switzerland

When learning that I am working on subcomponents of quarks and leptons, my friends usually ask me for my possible motivations in spending my time in such intriguing business. Frankly I reply that the best answer I can give is that my motivations are mostly personal. When you are at a certain age, with all the various burdens etc, you can only devote the little time you are left with for research, either to a less competitive subject, such as for instance history of physics, or to something exotic such as subcomponents. But there are still other reasons, more closely related to the present developments in particle theory. For instance, there is the proliferation of quarks and leptons. Also, as it has become customary to discuss what may happen up to 10^{16} GeV or even to 10^{19}GeV, one cannot avoid the thought that, rigorously, leptons are known to be pointlike only up to $10^{-16}\div10^{-17}$cm, or in energy, up to say, 1TeV. There are still many, many orders of magnitudes up to the Planck mass, so one asks why one excludes a priori that there may be many more levels of substructure, or at least one more. Other reasons have to do with unification with gravity if it ever will be done. Maybe such unification will not be done in terms of our quarks and leptons but rather in terms of other more elementary objects. This has been already discussed, at this meeting, by M.K. Gaillard[1] with conclusions not entirely encouraging. In any case I shall have nothing to say on this subject and actually

---

* Supported in part by the Swiss National Science Foundation.

deliberately exclude that supergravity should limit in any way my freedom in inventing models. Another motivation for my work has been that after having seen some papers on the subject I started wondering why should one not try the most unimaginative and orthodox approach; just copying, almost, what was once done to go from hadrons to quarks. This time, to go from quarks and leptons to subcomponents.

The state of the art on this subject is summarized in the summary talk by Abdus Salam at the 1979 EPS Conference in Geneve[2]. For instance in the approach by Harari[3] and Shupe[4], color and flavor arise only at the composite level. Specifically in that approach, states such as TTV, TVT, and VTT, formed out of the two subcomponents T,V are supposed to be independent and degenerate. Clearly, unorthodox ideas are required to develop such interesting idea[5] We had also, Casalbuoni and I, worked out another unorthodox model based on the use of a Clifford algebra of order eight, which is related by a projective transformation to the Clifford algebra of order ten defining the spinor representation of O(10). In such a lowe Clifford algebra the 4 complex degrees of freedom are 3 fermi oscillators for colors and one related to electric charge, so that a physical quasi-particle subconstituent picture becomes available, easily extendable to include generations[6]. I consider this approach not as a subcomponent picture, but rather as the introduction of Grassmann coordinates for a superfield description of unified quarks and leptons. We are still developing such an approach in its dynamical consequences, but I shall not consider it here, as it is not really an orthodox subcomponent description. For such a reason indeed we started to work at the concrete subcomponent picture[7]. We do not see, at this time, how the two descriptions may converge.

Listing motivations for subcomponents, even if of limited applicability and not of general validity, a subconstituent scheme may be of pedagogical usefulness. For instance, reactions among quarks and leptons might be looked at as exchanges of subcomponents, as in reactions among nuclei with rearrangements of nucleons. At a more serious theoretical level, it must be stressed that a main motivation for composite (or semicomposite) models lies on the possibility they may offer to solve the problem of naturalness in gauge theories. Bound scalar states of elementary fermions would act as Higgs in providing for spontaneous symmetry breaking. This suggestion was most rigorously advanced and developed by Susskind and his collaborators. In the original technicolor schemes by Dimopoulos and Susskind the quarks and leptons were considered as elementary but additional "techniquarks" were added, strongly confined by a technicolor force. In all such cases we may expect, in some phase (if there is more than one) of the confining theory, fermion-antifermion condensates to occur, reproducing an effective mechanism of spontaneous breaking without elementary Higgs fields.

Let me first say that whatever I am going to say is only to be considered as a first preliminary look at the problem. In fact, most probably, the schemes we shall construct will have serious problems. I may point out immediately what seem to me to be possible serious difficulties. I shall present in detail the construction of the composite states. The set of composite states will contain, in addition to ordinary families of quarks and leptons, states which do not appear in the present phenomenology. They may be regarded as higher states, lying at higher masses. However a suitable mass-splitting mechanism must be devised for such a purpose. It may be difficult to obtain that such a mechanism be natural. The problem is connected to that of the mass scale for the subcomponent theory. It may be that such mass scale is much beyond the phenomenological limit for pointlike leptons, which is 1÷100 TeV . In that case our quarks and leptons are perhaps to be regarded as effectively massless. Also some of the additional states could be effectively massless, even if not yet found experimentally. The problem then arises of what could guarantee masslessness, and certainly our models do not seem to offer any easy solution. This subject has been recently the object of serious efforts by 'tHooft(8) and by Susskind and his group(9). In particular one demands that the massless sector be again described by a local field theory, which must be free of anomalies. 't Hooft also demands additional limitations related to Appelquist-Carazzone decoupling and concludes that it is difficult to construct an acceptable theory. Perhaps the exact requirement of massless quarks and leptons is too strict, and one would then have to face problems of large ratios of scales, of the type that are plaguing gauge theories since their start. One could envisage the possibility that the mass scale we are considering is <u>not</u> far beyond the present phenomenological limit for pointlike leptons. Then again our models below appear unsufficient, since, as we shall see, they all imply violation of baryon number, and we would like to have proton decay only from the extremely large unification mass scale.

It is needless to point out the tremendous dynamical problems which a model of composite quarks and leptons will have to face. The first problem that appears is one of energy scales. Take for instance the electron or the muon. As far as we know they are pointlike Dirac particles of small mass, minimally coupled to photons. For instance the anomalous magnetic moment of the muon has been measured at CERN to one part in 10^8. Its value completely agrees, to the same precision, with the 6-th order QED calculations including hadronic vacuum polarization. If we want the muon to be an almost pointlike object, containing confined subcomponents, we must think of some very small length scale Λ_{sub}^{-1}, much smaller, by many orders of magnitude, than the length scale Λ^{-1} of QCD. But in QCD the masses of the nucleons do not differ very much from the scale Λ and their magnetic moment anomalies also do not require completely different mass scales. Here instead we would be faced by a situation where the composites are almost massless, with respect to the scale Λ_{sub}, and their anomaly in magnetic moment is well understood entirely by the radiative corrections to a pointlike Dirac particle, rather then being controlled by the large Λ_{sub}. So, whatever the origin of the confinement is, it does not seem to bear much resemblance to the confining QCD we are familiar with (or, at least, to which we have got accustomed). If, however, one could think of chiral symmetry remaining exact in a confining subcolor theory, then one might hope to have vanishing masses and magnetic moment anomalies for some composite states. I have unsuccessfully tried to figure out how would such a theory look like, still possessing a large scale Λ_{sub}. Any intuition based on classical or semiclassical approximations (such as loop expansions, etc.) would most probably be unable to describe a system of very large Compton wavelength which is a compound of tightly bound objects, so that it appears as pointlike over a longe range of momentum transfers. In this sense thinking of the neutrino, for instance, as a bag of confined subcomponents appears not only as repugnant, but also perhaps as unacceptable within our semiclassical notion of a bag. The same difficulty one finds with the other light leptons and quarks. For the proton, the kaon, even for the pion, the situation is instead different, as the Compton wavelengths of such particles do not exceed (or are comparable to) their sizes and also their magnetic moment anomalies are big and compatible with such sizes.

In QCD, confinement and spontaneous breaking of chiral symmetry appear to be interrelated. The relations one usually employs are the low momentum limit of the Ward identity for the axial current vertex

$$\lim_{K\to 0} K^{\mu}\Gamma^{5}_{\mu\alpha}(p + K, p) = [S^{-1}(p),\gamma_5]_{+}\frac{\tau_\alpha}{2}$$

and the definition

$$<\pi_\alpha(K)|J^5_{\mu\beta}|0> = i\delta_{\alpha\beta}K_\mu F_\pi$$

In the Goldstone mode one finds all the current algebra relations and using (in a rather forced way) a quark propagator $S^{-1}(p) = \not{p} - \sum(p^2)$ one can relate F_π (we recall that F_π in the σ-model in tree approximation is the vacuum expectation value of the σ-field) to an integral over the self-energy $\sum(p^2)$

$$F_\pi^2 = \frac{-3i}{(2\pi)^2}\,\frac{1}{\pi^2}\int d^4p\,\frac{\sum(p^2)}{(p^2-\sum(p^2))^2}\left(1 - \frac{1}{2}\,\frac{d}{\log p^2}\right)\sum(p^2)$$

A rough, tentative, estimate by Pagels and Stokar[10] gives $F_\pi \sim \frac{1}{3}m_D$, where m_D gives the (expected) behaviour of $\sum(p^2)$ for large negative p^2 ($\sum \to 4m^3_D/p^2$). On the other hand it is also expected that F_π be, apart from factors, the Regge slope of the linear confining potential, so that the whole picture may be expected to have consistency. Unfortunately complete proofs are lacking. Nevertheless QCD seems to be essentially characterized by one scale, with pure number ratios widely different but not a priory ununderstandable. The consistency of the picture in the confining phase does not of course prove the dogma "confinement $\to$ chiral symmetry breaking".

In the theory of superconductivity (which has so much inspired elementary particle theory) it is known how the binding of Cooper pairs destabilizes the Wigner vacuum. For two electrons, on opposite points near the Fermi surface and with opposite spins, the Fröhlich term due to phonon exchange can dominate over the Coulomb term. The new many-electron superconducting state, as obtained non-perturbatively through a Bogoliubov transformation, has an energy (and thermodynamic potential) lower than for the normal state, by a non-analytic function of the density on the Fermi surface and of the binding potential. The instability of the Fermi surface, implying spontaneous symmetry breaking, follows from whatever small attraction of the electrons to give Cooper pairs. Recently, Casher[11] has studied the Dirac-vacuum instability in relation to binding, when the effective two-body interaction V between massless fermion and antifermion is non-chiral and (in a ladder approximation) has shown that the existence of an s-wave bound state of $|2p| + V$ is a necessary and sufficient condition for such instability. He thus argues that chiral symmetry is automatically broken by the confinement of massless quarks.

In a recent work, remarkable for its depth and generality, 't Hooft has investigated the possibility of gauge models of confined elementary fermions for which chiral symmetry remains, at least partially, exact. His conclusions are negative, provided a number of, perhaps inescapable, assumptions are made in defining the physical models. For his reasoning 't Hooft assumes a color gauge theory on a set of anomaly free representations. Such a set may also include the trivial representation (spectator fermions) with multiplicity such that a gauging of the implicit global flavor-symmetry would also result in an anomaly-free flavor-gauge theory. At low energies the massless chiral bound states would appear as minimally coupled to the flavor gauge fields. The anomalies of such couplings must be cancelled by the spectator anomalies, which however were just such as to give overall vanishing anomalies on the elementary fields. The anomalies in the bound-state representation must then be equal to those of the constituents. The negative result of 't Hooft comes from adding another requirement, very natural and probably unavoidable, the Appelquist-Carazzone requirement (and, in general, also another technical requirement, n-independence). One imagines that a heavy mass term can be added for an elementary colored constituent, thus leaving us a smaller flavor symmetry. The bound states containing such constituents must form vector-like representations of the smaller symmetry in order to become simultaneously very massive. This poses a strong set of conditions, which, together with the anomaly requirement, do not allow any solutions, at least in the simplest case discussed by 't Hooft. In the general case 't Hooft conjectures that the addition of n-independence (of the indices, which give the minimal number of massless chiral states, from the numbers n_i giving the multiplicities of the colored representations) would again not allow for any solution.

In spite of all the difficulties, the evident ones and those hidden, of which we are not even aware, I thought it may still be useful to present some constructions of bound states possibly representing quarks and leptons, at least as a first preliminary approach. What will be exemplified here is simply the elementary group-theoretic construction according to the rules of unitary groups and SL(2,c). Some physical consequences, such as the implied baryon non-conservation and the appearance of B-L as the only possibly additional conserved quantity, apart from Q, may be more general. One can also hope that only a limited number of solutions will be acceptable (here we find essentially SU(8) or SU(7) or, less preferred, SU(6)), just from the viewpoint of family phenomenology.

As the simplest confining gauge group, which would give quarks and leptons as three-particle bound states (plus sea, etc.), we take SU(3). To construct possible models we introduce a set of N left-handed subconstituents described by 3N left-handed Weyl spinors $\psi_{a,A,\alpha}$ (a = 1,2,3 is an index of $SU(3)_{sc}$, the subscript sc standing for subcolor; A = 1,2,...,N is an SU(N) index; α = 1,2 is a Weyl spinor index). The spinor $\psi_{a,A,\alpha}$ transforms like $(\underline{3},\underline{N})$ under $SU(3)_{sc} \times SU(N)$. The gauge group $SU(3)_{sc}$ of "subcolor" is assumed to remain exact, and, through a confinement mechanism, provide for subcolor singlet bound states. The group SU(N) is required to contain the SU(5) of Glashow and Georgi, as identified in the reduction $SU(N) \to SU(5) \times SU(N-5) \times U(1)$ (for $N > 7$, whereas $SU(6) \to SU(5) \times U(1)$). For the righthanded subconstituents we introduce a corresponding set of righthanded Weyl spinors $\chi^{\dot\alpha}{}_{a,A}$ again transforming as $(\underline{3},\underline{N})$ under $SU(3)_{sc} \times SU(N)$. Effectively the complete symmetry is thus the entire $SU(3)_{sc} \times SU(N) \times SU(N)$, and an additional U(1) would not affect our considerations below.

We want to think of quarks and leptons as $SU(3)_{sc}$-singlet bound states of three subcomponents (plus a possible sea, etc.). We require that the total antisymmetry be realized by complete symmetry in SU(N) and Weyl indices together with the antisymmetry in subcolor. The Weyl spinor notation is useful in the construction and is summarized for convenience in Appendix A. In Appendix B we give the $SU(N) \times SU(2)_L \times SU(2)_R$ decomposition of the totally symmetric Young tableau. We require that the families (or $\underline{5}$ + $\underline{10^*}$) belong to definite representations of the group SU(N-5). For the choice (i) it is then necessary (although in principle not sufficient) that either N = 6 (from (N-5)(N-4)/2 = N-5) or N = 8 (from (N-6)(N-5)/2 = N-5), that is, SU(6) or SU(8). Both cases work : in fact for N = 8 (and only then) the SU(N-5)-representation ⊟ (for the $\underline{5}$'s of SU(5)) coincides with the complex conjugate of the □ of SU(N-5) (for the 10's of SU(5)).

With SU(6) one has for the decomposition (i) of (B5) : $\underline{70}$ = $\underline{40} + \underline{10} + \underline{15} + \underline{5}$, and taking into account helicities, as from (B2), the full set of left-handed states is $2[(\underline{5})_L + (\underline{10^*})_L] + 2[(\underline{5^*})_L + (\underline{10})_L] + 2[(\underline{15})_L + (\underline{15^*})_L + (\underline{40})_L + (\underline{40^*})_L]$. It contains two standard families $(\underline{5})_L + (\underline{10^*})_L$. A suitable splitting mechanism must give large masses to the states $(\underline{5^*})_L + (\underline{10})_L$, which would have V + A couplings. To account for the b-quark, the τ, and its neutrino one must look into the $SU(3)_c \times SU(2)$ decompositions of the SU(5) representations $\underline{15}$ and $\underline{40}$ and their complex conjugates. For $\underline{15}$ one has : $\underline{15} = (\underline{6},\underline{1}) + (\underline{3},\underline{2}) + (\underline{1},\underline{3})$; while for $\underline{40}$ one has : $\underline{40} = (\underline{8},\underline{1}) + (\underline{3^*},\underline{2}) + (\underline{6},\underline{2}) + (\underline{3},\underline{1}) + (\underline{3},\underline{3}) + (\underline{1},\underline{2})$. One sees that there would in any case be no room for the standard assignment $[b_L \sim (\underline{3},\underline{2}),\ \bar{b}_L \sim (\underline{3^*},\underline{1}),\ (\tau,\nu_\tau)_L \sim (\underline{1},\underline{2}),\ (\tau)_L \sim (\underline{1},\underline{1})]$. Non-standard assignments could be devised. For instance a non-vector $\underline{15}_L + \underline{40}^*_L$ with standard b_L and $\bar{b}_L$, but with SU(2) triplet leptons and doublet antileptons, etc. Vector assignments with singlet, or doublet, b and doublet leptons are of course possible within $\underline{40}$ + $\underline{40^*}$ but are most probably unacceptable.

With the second possible choice one has SU(8), and the decomposition (i) of (B5) according to SU(5) x SU(3) is : $\underline{168} = (\underline{40},\underline{1}) + (\underline{10},\underline{3}) + (\underline{15},\underline{3}) + (\underline{5},\underline{3}^*) + (\underline{5},\underline{6}) + (\underline{1},\underline{8})$. The left-handed states, as from B2, are $2[(\underline{5} + \underline{10}^*,\underline{3}^*)_L + 2[(\underline{5}^* + \underline{10},\underline{3})_L] + 2[2(\underline{1},\underline{8})_L + (\underline{5},\underline{6})_L + (\underline{5}^*,\underline{6}^*)_L + (\underline{15},\underline{3})_L + (\underline{15}^*,\underline{3}^*)_L + (\underline{40},\underline{1})_L + (\underline{40}^*,\underline{1})_L]$. We see that one can account for two triplets of families of the standard type $\underline{5} + \underline{10}^*$. As we have already explained, the internal constituent structure is however different for the two triplets. One might envisage two possibilities. The first possibility is that the families come in pairs (for instance the e-family together with the μ-family) corresponding to the two interval structures. The second possibility is that one triplet comes first (e-,μ-,τ- families) and the second triplet after.

We now come to the totally antisymmetric representation in (B4) (see B6). We have two possibilities: either N = 7, (from $(N-6) \times (N-5)/2 = 1$) in which case one family $\underline{5} + \underline{10}^*$ comes from the same tensor; or N = 8 (from $(N-6)(N-5)/2 = N-5$) where the families come from the tensor and its conjugate. The case N = 7, SU(7), leads to a $\underline{35}$ with SU(5) x SU(2) decomposition $\underline{35} = (\underline{10}^*,\underline{1}) + (\underline{10},\underline{2}) + (\underline{5},\underline{1})$ and the set of left-handed states is $(\underline{5} + \underline{10}^*,\underline{1})_L + (\underline{5}^* + \underline{10},\underline{1})_L + (\underline{10} + \underline{10}^*,\underline{2})_L$. For N = 8, we have SU(8), and the antisymmetric $\underline{56}$ of SU(8) decomposes under SU(5) x SU(3) as $\underline{56} = (\underline{10}^*,\underline{1}) + (\underline{10},\underline{3}) + (\underline{5},\underline{3}^*) + (\underline{1},\underline{1})$. The set of left handed states (see B2) is

$$(\underline{5} + \underline{10}^*,\underline{3}^*)_L + (\underline{5}^* + \underline{10},\underline{3})_L + 2(\underline{1},\underline{1})_L + (\underline{10} + \underline{10}^*,\underline{1})_L$$

This theory would thus allow for three standard families transforming as $\underline{3}^*$ of SU(3). The model, like all the other models we have discussed, is of course anomaly free. It does not stay anomaly free however at the SU(5) x SU(3) level if the three compound families are treated as elementary. Any non-vector mechanism for anomaly cancellation, even allowing other states to come down in mass, appears impossible. We have therefore to face a situation where anomalies are cancelled only at the subcomponent level, but not by taking only a partial, low mass, set of composite states and treating then as elementary. Such situation is perhaps unsatisfactory[8].

A subcomponent model giving quarks and leptons belonging to a $\underline{56}$ of SU(8) has been proposed by T.L. Curtright and P.G.O. Freund[10] without introducing any subcolor. In such a case however one would need a totally symmetric state in the spinor indices, for total symmetry in configuration space. The $SU(2)_L \times SU(2)_R$ content of such a state would be $(\underline{4},\underline{1}) + (\underline{1},\underline{4}) + (\underline{3},\underline{2}) + (\underline{2},\underline{3})$. The only spin $-\frac{1}{2}$ states would belong to the same irreducible Lorentz representation $(\underline{3},\underline{2}) + (\underline{2},\underline{3})$ of a spin $\frac{3}{2}$. Such a situation would clearly be unacceptable.

A more general approach to the family problem would be to choose for the family group only a subgroup SU(n) of SU(N-5). The decomposition of the $\underline{10}^*$ and $\underline{5}$ SU(5) - states (i) in B5, according to SU(N) → SU(5) x SU(n) x SU(N-5-n) can be readily found :

$$(\underline{10},\underline{N-5})^* = (\underline{10},\underline{n},\underline{1})^* + (\underline{10},\underline{1},\underline{N-n-5})^*$$

$$(\underline{5},\underline{\tfrac{1}{2}(N-6)(N-5)}) = (\underline{5},\underline{\tfrac{1}{2}n(n-1)},\underline{1}) + (\underline{5},\underline{n},\underline{N-n-5}) + (\underline{5},\underline{1},\underline{\tfrac{1}{2}(N-n-5)(N-n-6)})$$

$$(\underline{5},\underline{\tfrac{1}{2}(N-5)(N-4)}) = (\underline{5},\underline{\tfrac{1}{2}n(n+1)},\underline{1}) + (\underline{5},\underline{n},\underline{N-n-5}) + (\underline{5},\underline{1},\underline{\tfrac{1}{2}(N-n-5)}\,\underline{(N-n-4)})$$

A non-trivial multiplet will contain $(\underline{10}^*,\underline{n}^*,\underline{1})$ and n ⩾ 4 is clearly excluded by the Young diagram which would require 3 or more antisymmetric indices, and no $\underline{5}$ of SU(5) is available of such a kind. For n = 3 we obtain $(\underline{10}^*,\underline{3}^*,\underline{1}) + (\underline{5},\underline{3}^*,\underline{1})$ as the only possibility, for whatever N, since the only $\underline{3}^*$ is the $\underline{\tfrac{1}{2}n(n-1)}$. So one is indeed forced to trivial tensor properties with respect to SU(N-5-n) = SU(N-8) and indeed one only has an uninteresting (from the viewpoint of families) extension of the SU(8) we have found (with the $\underline{168}$). For n = 2 one has $(\underline{10}^*,\underline{2},\underline{1})$, and the $\underline{5}$ can be taken from both $(\underline{5},\underline{2},\underline{N-7})$, appearing in the above decompositions of the $\underline{5}$. The tensor properties with respect to SU(N-5-n) = SU(N-7) are non-trivial and different from those of the $(\underline{10}^*,\underline{2},\underline{1})$ except for N = 8. So any SU(N), N > 8, is possible, but for N > 8 the net results are (N-8) extra $\underline{5}$'s, not needed for families of $\underline{10}^* + \underline{5}$ type. For N = 8 one has SU(8) → SU(5) x SU(2) and two families transforming as $(\underline{5} + \underline{10}^*,\underline{2})$ (actually 4 considering the two internal structures in Eq. B3). Exactly the same reasoning applies to the states (ii) in B6. Also one would have obtained the same conclusion even by directly looking for a subgroup $SU_c(3) \times SU(2) \times U(1) \times SU(n)$, without passing through SU(5). This is related to our previous discussion where it was found that no standard assignment would be possible for the b quark within $(\underline{15} + \underline{15}^* + \underline{40} + \underline{40}^*)$.

In a uniform description for all SU(N) (>6), we assign names to the subcomponents belonging to the fundamental representations of $SU(3)_{sc} \times SU(N)$ as follows

$$(3,N)_L \equiv \{\psi_{a,A,\alpha}\}; \quad (3,N)_R \equiv \{\chi^{\dot{\alpha}}{}_{a,A}\}$$

(a = index of $SU(3)_{sc}$, A of SU(N), $\alpha,\dot{\alpha}$ undotted and dotted spinor indices), where (i = 1,2,3 color index, n = 6,7...N, SU(N-5)-index)

$$\psi_{a,A,\alpha} \equiv [\bar{D}_{a,i,\alpha}; N_{a,\alpha}; E_{a,\alpha}; G_{a,n,\alpha}] \equiv$$

$$\equiv [(\bar{D}_i)_L; (N)_L; (E)_L; (G_n)_L]$$

$$(i\sigma_2)^{\dot{\alpha}\dot{\beta}}(\psi_{a,A,\beta})^* \equiv [(D_i)_R; (\bar{N})_R; (\bar{E})_R; (\bar{G}_n)_R]$$

$$\chi^{\dot{\alpha}}{}_{a,A} \equiv [\bar{D}^{\dot{\alpha}}_{a,i}; N^{\dot{\alpha}}_a; E^{\dot{\alpha}}_a; G^{\dot{\alpha}}_{an}] \equiv [(\bar{D}_i)_R; (N)_R; (E)_R; (G_n)_R]$$

$$(i\sigma_2)_{\alpha\beta}(\chi^{\dot{\beta}}{}_{a,A})^* \equiv [(D_i)_L; (\bar{N})_L; (\bar{E})_L; (\bar{G}_n)_L]$$

The subcomponents G_n we call "genons", as they may generate families. In the following we shall always omit the subcolor index a. The subcomponents transform as follows under $SU(3)_c \times SU(2) \times U(1)$ $(\bar{D}_i)_L \in (\underline{3}^*,\underline{1})$, $Q(\bar{D}_i) = \frac{1}{3}$; $(N,E)_L \in (\underline{1},\underline{2})$, $Q(N) = 0$, $Q(E) = -1$; $G_n \in (\underline{1},\underline{1})$, $Q(G_n) = 0$. We still stress, later on, that a consistent assignment of B-L is possible, but not of B and L separately.

The states $(\underline{5})_L + (\underline{10^*})_L$ for the various cases we have found are shown in Table C1, for models with singlet families, and C2, for models with triplets of families. In the tables the states are denoted by typical terms : the full expression is just obtained by adding permutations according to the corresponding Young tableau. If one abstracts from the helicity content one finds two possible composite structures. In all models where the $\underline{10^*}$ is taken from the complex conjugate representation one has for the $\underline{10^*}$ the structure $(d,u,\bar{u},e^+)_L = \bar{G}(\bar{N}D,\bar{E}D,DD,\bar{N}\bar{E})$. Only for $\underline{SU(7)}$, where the $\underline{10^*}$ is taken from the same representation as the $\underline{5}$, one has $(d,u,\bar{u},e^+)_L = (E\bar{D}\bar{D},N\bar{D}\bar{D},NE\bar{D},\bar{D}\bar{D}\bar{D})$. The $\underline{5}$ is always given by $(\bar{d},\nu,e^-)_L = GG(\bar{D},N,E)$. Baryon and lepton-number non-conservation have a very intuitive origin in our subcomponent models. They reside in the impossibility of simultaneously assigning these numbers to both constituents and subconstitutents. If we assign baryon number B to the constituents, as usual, we see that in any of the above models, we must have: $-\frac{1}{3} = B(\bar{d}) - B(\nu) = B(\bar{D}) - B(N)$, but also: $\frac{2}{3} = B(d) - B(\bar{u}) = B(\bar{D}) - B(N)$. This shows the impossibility of assigning B to the subcomponents. On the other hand it is easy to verify that the quantum number $Z = B - L$ can be consistently defined by giving $Z = -\frac{1}{3}$ to D,N,E,G. More completely, it is straightforward to demonstrate that Q and $Z = B - L$ are the only additive non-chiral color- (and subcolor-) singlet quantum numbers that can be consistently assigned. To see this let $H(\bar{X})$ be an additive quantum number assigned to the state X, such that $H(X) = -H(X)$ (for instance T_3 of weak SU(2) is not of this kind). This condition implies, in any of the models, for the subcomponents $H(G) = H(N)$, $H(D) = \frac{2}{3}H(N) + \frac{1}{3}H(E)$ and for the components

$$H(u) = \frac{1}{3}H(\nu) - \frac{2}{3}H(e^-)$$

$$H(d) = -\frac{2}{3}H(\nu) + \frac{1}{3}H(e^-)$$

where $H(\nu)$ and $H(e^-)$ satisfy

$$H(e^-) = 2H(N) + H(E)$$

$$H(\nu) = 3H(N)$$

In other words it is enough to assign H(N) and H(E), or equivalently $H(e^-)$ and $H(\nu)$. Any such quantum number H(X) can thus be written as

$$H(X) = [H(\nu) - H(e^-)]Q(X) - H(\nu)Z(X) =$$

$$= [H(N) - H(E)]Q(X) - 3H(N)Z(X)$$

that is as definite linear combination of the charge and of $Z = B - L$. All these equations hold independently of the fact that the $\underline{10}^*$ be taken in the direct or in the complex-conjugate representation.

Let us define as N_D, N_N, N_E, N_G the numbers of D minus $\bar{D}$, N minus $\bar{N}$, etc. We then have

$$Z \equiv B\text{-}L = -\frac{1}{3}(N_D + N_N + N_E + N_G)$$

(Note that Z is never among the generators except for N = 6) and

$$Q = -\frac{1}{3}N - N_E$$

whereas for the weak-chiral isospin T_3 we have

$$T_3 = \frac{1}{2}(N_N - N_E)$$

We can think of a physical quark, for instance the u-quark, as given by a "valence" term $\bar{G}D\bar{E}$ plus a "sea" of pairs, etc., for instance

$$u_{physical} = \bar{G}D\bar{E}[1 + \epsilon(D\bar{D} + N\bar{N} + E\bar{E} + G\bar{G}) + \ldots] =$$
$$= u + \epsilon'\bar{u}e^+\bar{d} + \ldots$$

Therefore, in addition to the component u with $B = \frac{1}{3}$, $L = 0$, one has an additional component with $B = -\frac{2}{3}$ and $L = -1$. If we think of the proton as composite of physical quarks it will then have to disintegrate.

In terms of our subcomponents the basic reactions for proton decay are

$$D + \bar{N} \rightarrow \bar{N} + D \qquad \text{(a)}$$

$$D + \bar{D} \rightarrow N + \bar{N} \qquad \text{(b)}$$

By adding to (a) two spectator $\bar{G},\bar{E}$ and two spectators $\bar{G}$,D, one obtains $u_L + d_L \rightarrow e_L^+ + \bar{u}_L$. Analogously by adding to (b) spectators $\bar{G},\bar{E}$ and G,$\bar{D}$, one obtains $u_L + u_R \rightarrow e_L^+ + \bar{d}_R$.

In all the models one has, in addition to the gauge bosons of SU(5) (gluons, γ,W,Z, leptoquarks, transforming under SU(5) x x SU(N-5) like $(\underline{24},\underline{1})$), gauge bosons transforming like $(\underline{5},(\underline{N-5})^*)$ (of the type $\bar{G}D,\bar{G}N,\bar{G}E$) and $(\underline{5}^*,\underline{N-5})$ (of the type $DG,\bar{N}G,\bar{E}G$), plus bosons transforming like $(\underline{1},\underline{(N-5)^2-1})$ (i.e. $\bar{G}_mG_n$) and like $(\underline{1},\underline{1})$ (i.e. $\sum\bar{G}_mG_m$). The $(\underline{5},(\underline{N-5})^*)$ and $(\underline{5}^*,\underline{N-5})$ produce transitions between different representations of SU(5) x SU(3). In the SU(8) model based on $\underline{168}$ the $(\underline{5},\underline{3}^*)$ generators lead from the $(\underline{5} + \underline{10}^*,\underline{3}^*)_L$ families to the right-handed families $(\underline{5}^* + \underline{10},\underline{3})_L$ and also to $(\underline{5}^*,\underline{6}^*)_L$ and $(\underline{15},\underline{3})_L$; the $(\underline{5}^*,\underline{3})$ generators lead from $(\underline{5} + \underline{10}^*,\underline{3}^*)_L$ to $(\underline{1},\underline{8})_L$ and $(\underline{40}^*,\underline{1})_L$. In the second SU(8) model based on $\underline{56}$, the $(\underline{5},\underline{3}^*)$ generators give $(\underline{5} + \underline{10}^*,\underline{3})_L \to (\underline{5}^* + \underline{10},\underline{3})_L$ and $(\underline{5}^*,\underline{3})_L$ give $(\underline{5} + \underline{10}^*,\underline{3}^*)_L \to (\underline{1},\underline{1})_L + (\underline{10},\underline{1})_L$. Note that the generators transforming like DG and $\bar{D}\bar{G}$ (chromogenons) also carry color and B-L, similarly to leptoquarks; anyway, as we have seen, these bosons do not contribute to transitions between ordinary quarks and leptons of the various families (apart from possible mixings). The generators $(\underline{1},\underline{1})$ and $(\underline{1},\underline{8})$, of the type $\bar{G}G$, produce instead transitions within $(\underline{5} + \underline{10}^*,\underline{3}^*)$, including family-exchange reactions such as $\mu^+ + u \to e^+ c$, through basic subcomponent processes $G_n + G_m \to G_m + G_n$. All these considerations can be repeated for the $\underline{35}$ of SU(7) with the interesting difference that one has found $(\underline{5} + \underline{10}^*,\underline{1})_L$, singlet under SU(2), so that the $(\underline{5},\underline{2})$ and $(\underline{5}^*,\underline{2})$ generators both lead to $(\underline{10},\underline{2})$- and never to the $(\underline{5}^* + \underline{10},\underline{1})_L$ - (V + A) states. Similarly, because of the singlet character, the $\bar{G}_m G_n$ generators of $(\underline{1},\underline{3})$ are unaffective on $(\underline{5} + \underline{10}^*,\underline{1})_L$. Finally in the $\underline{70}$ of SU(6) one has transitions from $(\underline{5} + \underline{10}^*)_L$ to $(\underline{5}^* + \underline{10})_L$, and to $\underline{15}_L$ via the $\underline{5}$ generators $\bar{G}_6\bar{D}$, $\bar{G}_6 N$, $\bar{G}_6 E$ and to $\underline{10}_L + \underline{40}^*_L$ via the $\underline{5}^*$, with $\bar{G}_6 G_6$ keeping it in the family.

Mass breaking within a family group requires a coupling $(\underline{5} + \underline{10}^*,\underline{f})_L \times (\underline{5} + \underline{10}^*,\underline{f})_L$ where $\underline{f}$ is a representation of the family group, so that no masses can be given without removing the SU(5) degeneracy (one cannot have a tower of families degenerate with SU(5) and of different mass). Even more, by looking at the $SU_c(3) \times SU(2) \times U(1)$ level the masses are within the self-coupling of $[(\underline{3},\underline{2})_L + (\underline{1},\underline{2})_L + 2(\underline{3}^*,\underline{1})_L + (\underline{1},\underline{1})_L,\underline{f}]$ so that it is inevitable to break SU(2) x U(1). When, for instance, SU(8) is directly broken into SU(5) x SU(2) one has in place of $(\underline{10},\underline{3}^*)$ in the states (i), (B5), $(\underline{10}^*,\underline{2}) + (\underline{10}^*,\underline{1})$ and, in place of $(\underline{5},\underline{3}^*)$, $(\underline{5},\underline{2}) + (\underline{5},\underline{1})$, thus allowing for a separate classification for the 3rd family. However the explicit mass breaking cannot be constructed without violating SU(2) x U(1).

The most general mixing allowed from charge conservation is among the subcomponents G and N, $\bar{G}$ and $\bar{N}$ of the same helicity. Assuming also (B-L) conservation one is only left with mixings among the G's and N, and among the $\bar{G}$'s and $\bar{N}$. Unitary transformations in SU(8) among the G's act simultaneously on $(u^n)_L$ and $(d^n)_L$ giving a unit Kobayashi-Maskawa matrix. Note that mixing between G and N breaks weak SU(2).

One may speculate that subcolor may give vacuum expectation values to bilinears of subcomponents $\underline{6} \times \underline{6}^*$, $\underline{7} \times \underline{7}^*$, $\underline{8} \times \underline{8}^*$, in SU(6),

SU(7), SU(8) analogously to what is done in technicolor models(13). However since subcolor is assumed to be exact there will not be vacuum expectation values for bilinears $\underline{6}\times\underline{6}$ or $\underline{6}^*\times\underline{6}^*$, etc, simply because they are not subcolor singlets ($\underline{3}\times\underline{3} = \underline{3}^* + \underline{6}$). So there is no B-L violation from bilinears of subcomponents (the bilinears which carry B-L also carry color). Indeed, because of exact subcolor, B-L violation will come from higher terms such as (for instance for SU(8)).

$$\epsilon_{abc}\epsilon_{def}\epsilon_{nml}\epsilon_{pqr} < G_{a,n}G_{b,m}G_{c,l}G_{d,p}G_{l,q}G_{f,r} >_o$$

giving Δ(B-L) = 2, but preserving both SU(5) and the family group SU(3).

It was a pleasure for me, during my stay at Erice, to be able to discuss on these subjects with Leonard Susskind, and to verify identities of viewpoints and the awareness of the same problems and difficulties.

Appendix A - SL(2,c) notations

Undotted spinors, $(\frac{1}{2},0)$, transform as

$$\psi'^{\alpha} = a^{\alpha}{}_{\beta}\psi^{\beta} \tag{A1}$$

where a is 2 x 2 and unimodular (a SL(2,c)). The invariant bilinear is

$$\phi^{\alpha}\epsilon_{\alpha\beta}\psi^{\beta} = \phi_{\alpha}\psi^{\alpha} = -\phi^{\alpha}\psi_{\alpha} \tag{A2}$$

where

$$\psi_{\alpha} = \psi^{\beta}\epsilon_{\beta\alpha}, \qquad \psi^{\alpha} = \epsilon^{\alpha\beta}\psi_{\beta} \tag{A3}$$

and ψ_{α} transforming as

$$\psi'_{\alpha} = (a^{T-1})_{\alpha}{}^{\beta}\psi_{\beta} \equiv a_{\alpha}{}^{\beta}\psi_{\beta} \tag{A4}$$

Dotted spinors, $(0,\frac{1}{2})$, transform as

$$\psi^{\dot{\alpha}\,'} = (a^{\alpha}{}_{\beta})^{*}\psi^{\dot{\beta}} \equiv a^{\dot{\alpha}}{}_{\dot{\beta}}\psi^{\dot{\beta}} \tag{A5}$$

and $\psi_{\dot{\alpha}} = \psi^{\dot{\beta}}\epsilon_{\dot{\beta}\dot{\alpha}}$ (A6)

with $\epsilon_{\dot{\alpha}\dot{\beta}} = \epsilon_{\alpha\beta} = \epsilon^{\dot{\alpha}\dot{\beta}}$ (A7)

A general spinor

$$\psi^{\alpha_1 \ldots \alpha_p ; \dot{\beta}_1 \ldots \dot{\beta}_q} \tag{A8}$$

transforms as p-times $(\frac{1}{2},0)$ and q-times $(0,\frac{1}{2})$ and can be reduced by the standard procedure (Young tableaux).

For instance $\psi^{\alpha\beta}$ reduces into the scalar (0,0)

$$\psi^{[\alpha,\beta]} = \frac{1}{2}(\psi^{\alpha\beta} - \psi^{\beta\alpha}) = \epsilon^{\alpha\beta}\mathbf{A} \quad (\mathbf{A} = \frac{1}{2}\psi^{\alpha\beta}\epsilon_{\alpha\beta}) \tag{A9}$$

and into (1,0) given by $\psi^{\{\alpha,\beta\}}$. The important rule is that antisymmetrizing two indices acts as contraction. The $(0,\frac{1}{2}) \times (\frac{1}{2},0)$ spinor $\psi^{\alpha\beta}$ corresponds to a four-vector

$$v_{\mu} = \psi^{\alpha\dot{\beta}}(\sigma_{\mu})_{\dot{\beta}\alpha} \tag{A10}$$

$$\psi^{\alpha\dot{\beta}} = \frac{1}{2}(\tilde{\sigma}_{\mu})^{\alpha\dot{\beta}}v^{\mu} \tag{A11}$$

where

$$(\sigma_{\mu})_{\dot{\alpha}\beta} \equiv (\sigma_{o}, -\vec{\sigma}) \tag{A12}$$

$$(\tilde{\sigma}_{\mu})^{\alpha\dot{\beta}} \equiv (\sigma_{o}, \vec{\sigma}) \tag{A13}$$

and it follows that v_μ transforms as

$$v'_\mu = \Lambda_{\mu\nu}(a) v^\nu \tag{A14}$$

where

$$\Lambda_{\mu\nu}(a)\sigma^\nu = a^\dagger \sigma_\mu a \tag{A15}$$

A Dirac spinor can be thought of as a four component ψ

$$\begin{bmatrix} \psi_{\dot{\alpha}} \\ \phi^\alpha \end{bmatrix} \tag{A16}$$

and the bilinear invariant is $\bar{\psi}\psi = \psi^\dagger \gamma^o \psi$, or

$$\psi^{\dot{\alpha}} \phi_{\dot{\alpha}} + \psi_\alpha \phi^\alpha \tag{A17}$$

where γ^o is antidiagonal.

The matrix (charge conjugation)

$$\begin{bmatrix} \epsilon^{\dot{\alpha}\dot{\beta}} & o \\ o & \epsilon_{\beta\alpha} \end{bmatrix} \tag{A18}$$

transforms (A16) into the charge-conjugate

$$\begin{bmatrix} \psi^{\dot{\alpha}} \\ \phi_\alpha \end{bmatrix} \tag{A19}$$

The parity representation (where γ^o is diagonal) is instead

$$\frac{1}{\sqrt{2}}(\psi_{\dot{\alpha}} + \phi^\alpha), \ \frac{1}{\sqrt{2}}(-\psi_{\dot{\alpha}} + \phi^\alpha), \tag{A20}$$

It is not covariant under SL(2,c), but only under an SU(2) subgroup.

Appendix B

SU(N) and Weyl indices can be read off from the SU(N) x x $SU(2)_L \times SU(2)_R$ decomposition ($SU(2)_L \times SU(2)_R$ from the Lorentz SL(2,C)) of the totally symmetric Young tableau. The decomposition is as follows :

$$
\begin{aligned}
\square\square\square = & \Big[\, \square\square\square \,,\; (\square\square\square, \cdot) + (\cdot, \square\square\square) \\
& (\square\square, \square) + (\square, \square\square) \Big] + \qquad \text{(B1)} \\
& + \Big[\, \begin{smallmatrix}\square\square \\ \square\;\;\end{smallmatrix} \,,\; 2(\cdot, \square) + 2(\square, \cdot) + (\square\square, \square) + (\square, \square\square) \Big] + \\
& + \Big[\, \begin{smallmatrix}\square \\ \square \\ \square\end{smallmatrix} \,, (\cdot, \square) + (\square, \cdot) \Big]
\end{aligned}
$$

where the dot stands for the scalar representation. The quark and lepton states must transform like $(\cdot, \square)$ or $(\square, \cdot)$ under $SU(2)_L \times SU(2)_R$. We are thus left with the states

$$
\begin{aligned}
& 2\left(\begin{smallmatrix}\square\square \\ \square\;\;\end{smallmatrix}, \cdot, \square\right) + 2\left(\begin{smallmatrix}\square\square \\ \square\;\;\end{smallmatrix}, \square, \cdot\right) + \\
& + \left(\begin{smallmatrix}\square \\ \square \\ \square\end{smallmatrix}, \;, \square\right) + \left(\begin{smallmatrix}\square \\ \square \\ \square\end{smallmatrix}, \square, \cdot\right) \qquad \text{(B2)}
\end{aligned}
$$

The multiplicity of two occurring in Eq. (1) does not mean exact duplication but rather expresses, for instance for a left-handed composite, the two possibilities

$$
[1] = \begin{array}{|c|c|}\hline L & L \\ \hline L & \\ \cline{1-1}\end{array} \,, \qquad [2] = \begin{array}{|c|c|}\hline R & L \\ \hline R & \\ \cline{1-1}\end{array} \qquad \text{(B3)}
$$

whereas only one antisymmetric state exists

$$\begin{array}{|c|}\hline R \\ \hline R \\ \hline L \\ \hline\end{array} \qquad \text{(B4)}$$

The situation should be contrasted with what happens in the non-relativistic SU(6) where the states (B3) coincide and (B4) disappears.

The two SU(N) representations which occur in Eq. (B2) one of mixed symmetry, one totally antisymmetric can be reduced with respect to SU(5) xSU(N-5) xU(1), as shown below :

(i) Mixed symmetry

$$\frac{N(N^2-1)}{3} \rightarrow (40,1) + (10,N-5) + (15,N-5) + \qquad \text{(B5)}$$

$$+ \left(5,\frac{1}{2}(N-6)(N-5)\right) + \left(5,\frac{1}{2}(N-5)(N-4)\right) + \left(1,\frac{1}{3}(N-6)(N-5)(N-4)\right)$$

(ii) Antisymmetry

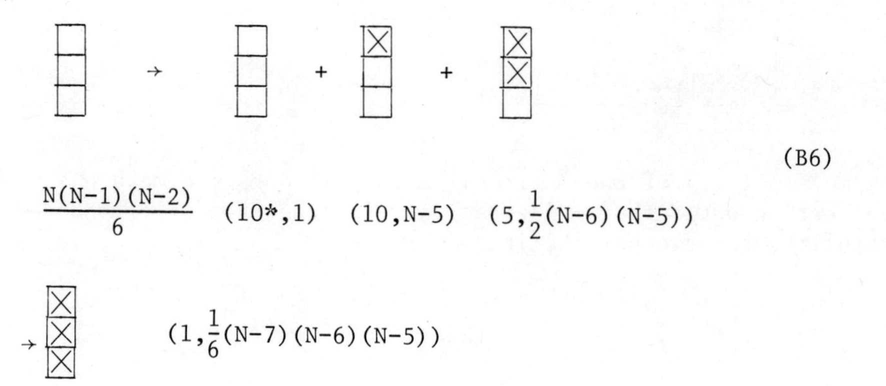

$$\frac{N(N-1)(N-2)}{6} \rightarrow (10^*,1) + (10,N-5) + \left(5,\frac{1}{2}(N-6)(N-5)\right) \qquad \text{(B6)}$$

$$+ \left(1,\frac{1}{6}(N-7)(N-6)(N-5)\right)$$

In (B5) and (B6) a cross stands for an index >5; below each tableau we have reported the dimensions of the corresponding SU(5) x SU(N-5) representations. According to Eq. (B2) the SU(N) x SU(N-5) states (i), Eq. (B5), are obtained both as left- and as right-handed and with multiplicity of two; whereas the states of (ii) Eq. (B6) are again obtained both as left- and right-handed but with multiplicity one. Left-handed $\underline{5}$ of SU(5) are contained in (i) and also in (ii). Left-handed $\underline{10}^*$ must be taken from the conjugates of the right-handed states (except for SU(7)).

Appendix C

The quark and lepton states in terms of the subcomponents are given in Tables C1 and C2, for the different cases, that is : SU(6) with two standard families in its $\underline{70}$, SU(7) with one standard family in its $\underline{35}$, SU(8) with two triplets of families in its $\underline{168}$, or with one triplet in its $\underline{56}$.

Table C1 : Composite quark and lepton states for the SU(6) and SU(7) models

SU(5)		SU(6):$\underline{70}_{[1]}$ (Young tableau: L L / L)	SU(6):$\underline{70}_{[2]}$ (Young tableau: R L / R)	SU(7):$\underline{35}$ (Young tableau: R / R / L)
$\underline{5}$	$(\bar{d}_i)_L$	$(G_6)_L(G_6)_L(\bar{D}_i)_L$	$(G_6)_R(G_6)_L(\bar{D}_i)_R$	$\epsilon_{mn}(G_m)_R(G_n)_R(\bar{D}_i)_L$
	$(\nu)_L$	$(G_6)_L(G_6)_L(N)_L$	$(G_6)_R(G_6)_L(N)_R$	$\epsilon_{mn}(G_m)_R(G_n)_R(N)_L$
	$(e^-)_L$	$(G_6)_L(G_6)_L(E)_L$	$(G_6)_R(G_6)_L(E)_R$	$\epsilon_{mn}(G_m)_R(G_n)_R(E)_L$
$\underline{10^*}$	$(d_i)_L$	$(\bar{G}_6)_L(\bar{N})_L(D_i)_L$	$(\bar{G}_6)_R(\bar{N})_L(D_i)_R$	$(E)_L(\bar{D}_j)_R(\bar{D}_k)_R\,\epsilon_{ijk}$
	$(u_i)_L$	$(\bar{G}_6)_L(\bar{E})_L(D_i)_L$	$(\bar{G}_6)_R(\bar{E})_L(D_i)_R$	$(N)_L(\bar{D}_j)_R(\bar{D}_k)_R\,\epsilon_{ijk}$
	$(\bar{u}_i)_L$	$(\bar{G}_6)_L(D_j)_L(D_k)_L\,\epsilon_{ijk}$	$(\bar{G}_6)_R(D_j)_L(D_k)_R\,\epsilon_{ijk}$	$(N)_L(E)_R(\bar{D}_i)_R$
	$(e^+)_L$	$(\bar{G}_6)_L(\bar{N})_L(\bar{E})_L$	$(\bar{G}_6)_R(\bar{E})_L(\bar{N})_R$	$(\bar{D}_i)_L(\bar{D}_j)_R(\bar{D}_k)_R\,\epsilon_{ijk}$

Table C2 : Composite quark and lepton states for the SU(8) models

SU(3)	SU(5)		SU(8): $\underline{168}_{[1]}$ ([L L / L])	SU(8): $\underline{168}_{[2]}$ ([R L / R]); $\underline{56}$ ([R / R / L])
$\underline{3^*}$ (n=1,2,3)	$\underline{5}$	$(\bar{d}_i)^n_L$	$\epsilon_{lmn}(G_l)_L(G_m)_L(\bar{D}_i)_L$	$\epsilon_{lmn}(G_l)_R(G_m)_R(\bar{D}_i)_L$
		$(\nu)^n_L$	$\epsilon_{lmn}(G_l)_L(G_m)_L(N)_L$	$\epsilon_{lmn}(G_l)_R(G_m)_R(N)_L$
		$(e^-)^n_L$	$\epsilon_{lmn}(G_l)_L(G_m)_L(E)_L$	$\epsilon_{lmn}(G_l)_R(G_m)_R(E)_L$
	$\underline{10^*}$	$(d_i)^n_L$	$(\bar{G}_n)_L(\bar{N})_L(D_i)_L$	$(\bar{G}_n)_R(\bar{N})_L(D_i)_R$
		$(u_i)^n_L$	$(\bar{G}_n)_L(\bar{E})_L(D_i)_L$	$(\bar{G}_n)_R(\bar{E})_L(D_i)_R$
		$(\bar{u}_i)^n_L$	$(\bar{G}_n)_L(D_j)_L(D_k)_L\,\epsilon_{ijk}$	$(\bar{G}_n)_R(D_j)_L(D_k)_R\,\epsilon_{ijk}$
		$(e^+)^n_L$	$(\bar{G}_n)_L(\bar{N})_L(\bar{E})_L$	$(\bar{G}_n)_R(\bar{N})_L(\bar{E})_R$

REFERENCES

1. J. Ellis, M.K. Gaillard, L. Maiani and B. Zumino, these proceedings, p. 69.
2. Abdus Salam, in Proc. of the International Conference on High Energy Physics, CERN, Geneva, Vol. 2, p. 868 (1980).
3. H. Harari, Phys. Letters 86B (1979) 83.
4. M.A. Shupe, Phys. Letters 86B (1979) 87.
5. There is a recent preprint by S.L. Adler (Princeton Institute for Advanced Study, 1979) in this direction.
6. R. Casalbuoni and R. Gatto, Phys. Letters 88B (1979) 306. ibid. 90B (1980) 81.
7. R. Casalbuoni and R. Gatto, UGVA-DPT 1980/02-235, preprint (1980) (to be published).
8. G. 't Hooft, Lectures given at the Cargèse Summer Institute, August, 1979..
9. S. Raby, S. Dimopoulos and L. Susskind (to be published). I would like to thank Prof. Susskind for informing me on this work and also of related work by S. Coleman (to be published).
10. H. Pagels and S. Stokar, Phys. Rev. D20 (1979) 2947.
11. A. Casher, Phys. Letters 83B (1979) 395.
12. T.L. Curtright and P.G.O. Freund, EFI 79/25, Enrico Fermi Institute, Chicago (1979).
13. S. Weinberg, Phys. Rev. D13 (1976) 974;
    S. Dimopoulos and L. Susskind, Nuclear Phys. B155 (1979) 237.

FORMULATIONS OF N = 2 SUPERGRAVITY THEORIES

B. de Wit

NIKHEF-H, Amsterdam

The Netherlands

We discuss various formulations of extended supergravity based on the N = 2 super-Weyl and super-Poincaré algebra. Representations of these algebras define a calculus for the construction of invariant actions. We also present a new version of Poincaré supergravity with local SU(2) symmetry. We exhibit the relation between various formulations, and give explicit examples of supersymmetry multiplets and invariant actions.

1. INTRODUCTION

The question whether supergravity plays a rôle in Nature is difficult to answer. Certainly no experimental indications are presently available to support a confirmative answer. On the other hand supergravity theories provide a unique principle for the unification of particles and their interactions, and have already been shown to exhibit a number of remarkable properties[1]. This is particularly the case for extended supergravity. In this talk we will discuss the simplest extended supergravity theory, namely supergravity with two independent supersymmetry transformations (N = 2). Most of the material presented here is based on work done in collaboration with M. de Roo, J.W. van Holten and A. Van Proeyen. Work along similar lines will be discussed at this conference by P. Breitenlohner.

The original approach to supergravity was based on attempts to construct invariant actions order-by-order in the gravitational coupling constant, starting from a Lagrangian that was invariant under rigid supersymmetry[2]. Apart from being cumbersome this rather heuristic procedure does not lead to much insight in the underlying structure of the theory. Instead it is preferable to start from the symmetry structure as defined by the supersymmetry algebra, and then find representations of this algebra on the basis of which one can construct invariant field theories. This procedure is rather straightforward for gauge theories of internal symmetries, where the algebra and its representations are simple. But local supersymmetry implies gravity, so that its gauge algebra will have at least the complexity of the combined algebra of general coordinate and internal gauge transformations. In addition the field representations of supersymmetry contain fields that are mutually different. Not only do both fermions and bosons occur in one representation, but the various components play rather different rôles when considered in the context of an invariant action. Some fields correspond to dynamical modes of the theory, whereas others are only auxiliary, i.e. they can be eliminated from the action by means of their field equations without changing the dynamical content of the theory. These auxiliary fields are only required to form complete representations of the supersymmetry algebra, or, stated differently, to close the algebra under (anti)commutation.

The problem is how to find a complete algebra of supersymmetry transformations, and the corresponding (gauge) field representation. This has been solved for $N=2$ supergravity, first in linearized form[3,4], and subsequently to all orders in the gravitational coupling constant by iteration[5,6]. In principle it is then straightforward to implement the full algebra on known representations of rigid supersymmetry. A component of such a representation that transforms into a total derivative, which defines an action invariant under rigid supersymmetry, can then be generalized into an invariant action under local supersymmetry. If this is possible one has found an invariant density, and obtained a calculus for the construction of supergravity actions. In principle one has the option of implementing different supersymmetry algebras. The superconformal algebra has the highest degree of invariance. We will argue that it is advantageous to classify field representations according to that algebra. After having found superconformal representations it is always possible to restrict the gauge algebra to a subalgebra, and to construct actions within that context.

In section 2 we discuss the N = 2 superconformal algebra, and its relation to the Poincaré theory. The implementation of this algebra on chiral superfields is given in section 3. From these results we obtain the corresponding calculus for the Poincaré subalgebra in section 4. In section 5 we present several examples. In particular we derive a minimal field representation of the Poincaré algebra by means of an appropriate gauge choice for the superconformal algebra. This representation has a local SU(2) symmetry, but does not lead to a meaningful Poincaré action. Another Poincaré formulation with local SU(2), which is gauge equivalent to the original representation presented in refs. 3,4, is introduced in section 6. We give our conclusions in section 7.

2. THE N = 2 POINCARE MULTIPLET AND THE SUPERCONFORMAL THEORY

As we have mentioned in the introduction the complete multiplet of N = 2 Poincaré supergravity is known[3,4]. It consists of the following 40 + 40 (fermionic + bosonic) degrees of freedom*):

$$\begin{array}{lll} \textit{physical fields} & : & e_\mu{}^a\,,\ \psi_\mu{}^i\,,\ B_\mu{}^{ij}\,, \\ \textit{auxiliary boson fields} & : & A_a,\ V_a{}^i{}_j\,,\ S^{ij}, \\ & & t_{ab}{}^{ij}\,,\ V_a,\ M^{ij}, \\ \textit{auxiliary fermion fields:} & & \chi^i,\ \lambda^i. \qquad (i,j=1,2) \end{array} \tag{2.1}$$

We use a chiral SU(2) notation throughout, with left- and righthanded chiral components $\psi_\mu{}^i$, χ^i, λ_i, and $\psi_{\mu i}$, χ_i, λ^i, respectively. The vierbein field $e_\mu{}^a$, the gravitino field $\psi_\mu{}^i$, and the SO(2) gauge field $B_\mu{}^{ij}$, generate the massless SO(2) supermultiplet (2, 3/2, 3/2, 1)[7]. The field $B_\mu{}^{ij}$, the antisymmetric Lorentz tensor $t_{ab}{}^{ij}$ and the complex spinless field M^{ij} are in the anti-symmetric SU(2) singlet representation. Also the vector and axial vector V_a and A_a are SU(2) singlets, whereas the vector and axial vectors contained in the anti-Hermitean traceless tensor $V_a{}^i{}_j$ transform in the adjoint representation. The complex symmetric Lorentz scalar S^{ij} is subject to a reality condition

---

*) In terms of the components defined in refs. 4,6 we have $S^{ij} = S\delta^{ij} - iP^{ij}$, $M^{ij} = M^{ij} - iN^{ij}$, $V_a{}^i{}_j = V_a{}^i{}_j - iA_a{}^i{}_j$

$$S_{ij} \equiv (S^{ij})^* = \varepsilon_{ik}\varepsilon_{jl} S^{kl} . \tag{2.2}$$

The full transformation rules of the Poincaré multiplet as well as the supersymmetry algebra are rather complicated. They have been given in the literature[5,6,8].

To clarify the transformation rules and the algebra we make use of the superconformal theory[6,9]. This procedure has also been used succesfully for N = 1 supergravity[10]. The use of superconformal concepts may have certain limitations, like in the presence of central charges. But once we return to the Poincaré theory we can always deal with central-charge multiplets in that context.

The superconformal theory is based on the submultiplet of the Poincaré fields that contains the highest-spin components. The fields of this so-called Weyl-multiplet, which has 24 + 24 components, are[6,9]:

$$\begin{aligned}
&\textit{gauge fields} &&: e_\mu{}^a ,\ \psi_\mu{}^i ,\ A_\mu \equiv e_\mu{}^a A_a ,\\
& && \ V_\mu{}^i{}_j \equiv e_\mu{}^a V_a{}^i{}_j + \kappa(2\bar\psi_\mu{}^i\lambda_j - \delta^i{}_j\bar\psi_\mu{}^k\lambda_k - \text{H.c.}) ,\\
&\textit{matter fields} &&: T_{ab}{}^{ij} \equiv t_{ab}{}^{ij} - \sqrt{2}\,\hat F_{ab}{}^{ij} ,\\
& && \ \chi_C{}^i \equiv (\chi - \tfrac{1}{3}\gamma\cdot R^P)^i ,\\
&\textit{auxiliary field} &&: D_C \equiv D^P\cdot V - \tfrac{1}{3}\kappa^{-1}R^P\\
& && \quad -\tfrac{1}{4}\kappa(V_a{}^i{}_j V_a{}^j{}_i + 2V_a{}^2 + |M^{ij}|^2)\\
& && \quad +\kappa(\bar\lambda_i(2\chi^i + 2\not{D}^P\lambda^i - \tfrac{1}{2}\gamma\cdot R^{iP} + i\kappa\not{A}\lambda^i\\
& && \qquad -\tfrac{1}{2}\kappa\sigma\cdot T^{-ij}\lambda_j) + \text{H.c.}) ,
\end{aligned} \tag{2.3}$$

where R denotes the Riemann curvature scalar, $R_\mu{}^i$ the Rarita-Schwinger field equation, and $T^\pm$ the selfdual combinations $T^\pm = T \pm \tilde T$. The index "P" indicates that the corresponding quantity has been covariantized with respect to super-Poincaré transformations. The quantity $\hat F_{ab}{}^{ij}$ is the covariantized field strength of the gauge field $B_\mu{}^{ij}$.

To eliminate lower-spin components the Weyl multiplet has a higher degree of gauge invariance. The gauge symmetries are those of the graded conformal algebra, SU(2,2|2), which include a local internal U(2) invariance, whose gauge fields are $V_\mu{}^i{}_j$ and A_μ. The only other gauge fields that are explicitly contained in the Weyl multiplet (2.3) are those of general coordinate transformations and Q supersymmetry, $e_\mu{}^a$ and $\psi_\mu{}^i$ respectively. The gauge fields of the remaining symmetries of the superconformal algebra do not occur as elementary fields in the Weyl multiplet, but are expressed in terms of the Weyl components*). They are denoted by $\omega_\mu{}^{ab}$, $f_\mu{}^a$, b_μ, and $\phi_\mu{}^i$, and correspond to, Lorentz transformations, conformal boosts, dilatations and a second kind of supersymmetry transformation called S supersymmetry, respectively. It has been shown that all gauge fields can be understood from the graded algebra subjected to certain constraints. Derivatives and curvatures that are covariantized with respect to the full conformal group, and thus depend on all these gauge fields, then appear naturally in the transformation rules of superconformal multiplets[6].

An important result is that Poincaré supersymmetry transformations can be decomposed in terms of the superconformal transformations. This decomposition, which holds uniformly, takes the form

$$[Q]^{\text{Poincaré}} = Q + S + SU(2) , \qquad (2.4a)$$

where the parameters of Q- and S-supersymmetry and of the local SU(2) transformation are expressed in terms of the Poincaré transformation parameter ε^i by

$$\begin{aligned} \varepsilon_Q{}^i &= \varepsilon^i , \\ \varepsilon_S{}^i &= \kappa(S^{ij} + \tfrac{1}{2}\sigma\cdot t^{ij})\varepsilon_j + (\not{b} + i\kappa\not{A})\varepsilon^i , \qquad (2.4b) \\ \Lambda_{SU(2)}{}^i{}_j &= -\kappa(2\bar{\varepsilon}^i\lambda_j - \delta^i{}_j\bar{\varepsilon}^k\lambda_k - \text{H.c.}) . \end{aligned}$$

It is this decomposition rule which allows us to return to Poincaré supergravity. This will be discussed in section 3.

Because of the constraints the superconformal algebra will no longer coincide with the SU(2,2|2) algebra. The commutator of two Q-supersymmetry transformations has changed, and is now given by (for fields inert under conformal boosts):

---

*) An exception is the dilatational gauge field b_μ, which remains independent, but cancels in the Poincaré transformations[6].

$$[\delta_Q(\varepsilon_1), \delta_Q(\varepsilon_2)] = 2(\bar{\varepsilon}_2{}^k \gamma^a \varepsilon_{1k} + \text{H.c.}) D_a^C + \delta_M(\varepsilon^{ab}) + \delta_S(\eta^i) \quad , \tag{2.5a}$$

where the superconformal derivative D_a^C denotes the effect of a supercovariant translation[11,12], and the parameters of the extra Lorentz and S-supersymmetry transformation are given by

$$\begin{aligned} \varepsilon^{ab} &= \kappa(\bar{\varepsilon}_1{}^i T^{+ab}{}_{ij}\, \varepsilon_2{}^j + \text{H.c.}) \quad , \\ \eta^i &= 3\kappa\, \bar{\varepsilon}_1{}^{[i}\, \varepsilon_2{}^{j]}\, \chi_{Cj} \quad . \end{aligned} \tag{2.5b}$$

The superconformal commutator algebra is much simpler than the super-Poincaré algebra, which makes its implementation on field representations much easier. We now discuss this for the case of $N = 2$ chiral superfields.

3. N = 2 CHIRAL SUPERFIELDS

On the basis of $N = 2$ chiral superfield a large variety of models can be constructed, such as Poincaré supergravity, Weyl supergravity and the supergravity version of Yang-Mills theory. A (lefthanded) chiral superfield[13,14] can be decomposed on the basis of (lefthanded) chiral anticommuting coordinates θ^i in terms of functions of the complex space-time variable

$$z^\mu = x^\mu + \bar{\theta}^i \gamma^\mu \theta_i \quad ,$$

with

$$\gamma_5 \theta^i = \theta^i \quad , \quad \gamma_5 \theta_i = -\theta_i \quad . \tag{3.1}$$

The decomposition reads

$$\begin{aligned} \Phi^{(+)}(z^\mu, \theta^i) = {} & A(z) + \bar{\theta}^i \Psi_i(z) + \tfrac{1}{2} \bar{\theta}^i \theta^j B_{ij}(z) \\ & + \frac{1}{2}\, \varepsilon_{ij} \bar{\theta}^i \sigma^{ab} \theta^j\, F^-_{ab}(z) \\ & + \frac{1}{3}\, \varepsilon_{ij} \bar{\theta}^i \sigma^{ab} \theta^j\, \bar{\theta}^k \sigma_{ab} \Lambda_k(z) \\ & + \frac{1}{12}\, (\varepsilon_{ij} \bar{\theta}^i \sigma_{ab} \theta^j)^2\, C(z) \quad . \end{aligned} \tag{3.2}$$

In the context of the superconformal algebra such fields are characterized by a Weyl and chiral weight factor w, which determines the behaviour under dilatations and chiral U(1) transformations according to

$$\Phi^{(+)}(z^\mu, \theta^i) \rightarrow e^{w\xi^*} \Phi^{(+)}(z^\mu, e^{-\frac{1}{2}\xi}\theta^i) \quad , \tag{3.3}$$

where the real and imaginary parts of ξ are the transformation parameters of the dilatation and chiral transformation, respectively. The product of two equal-chirality fields is again a chiral field, with a weight factor that equals the sum of the separate weights.

It is straightforward to give the rigid supersymmetry transformations for the various components of (3.2). To obtain the full transformation rules one implements the algebra (2.5) on these components thereby allowing modifications of the linearized results. The advantage of implementing the superconformal algebra rather than the super-Poincaré algebra then becomes clear. The higher degree of symmetry in the first case severely restricts these modifications. For example the low-θ components have no modifications possible in their transformation rules, other than a supercovariantization of the derivatives. We now give the full transformations under S- and Q-supersymmetry, and refer to ref. 15 for further details.

$$\begin{aligned}
\delta_S A &= 0 \, , \\
\delta_S \Psi_i &= 2wA\eta_i \, , \\
\delta_S B_{ij} &= (1-w)\bar{\eta}_{(i}\Psi_{j)} \, , \\
\delta_S F^-_{ab} &= -(1+w)\varepsilon^{ij}\bar{\eta}_i\sigma_{ab}\Psi_j \, , \\
\delta_S \Lambda_i &= -(1+w)B_{ij}\varepsilon^{jk}\eta_k + (1-w)\sigma\cdot F^-\eta_i \, , \\
\delta_S C &= 2w\varepsilon^{ij}\bar{\eta}_i\Lambda_j \, .
\end{aligned} \tag{3.4}$$

$$
\begin{aligned}
\delta_Q A &= \bar{\varepsilon}^i \Psi_i \,, \\
\delta_Q \Psi_i &= 2 \not{D}^C A \varepsilon_i + B_{ij} \varepsilon^j + \sigma \cdot F^- \varepsilon_{ij} \varepsilon^j \,, \\
\delta_Q B_{ij} &= \bar{\varepsilon}_{(i} \not{D}^C \Psi_{j)} - \bar{\varepsilon}^k \Lambda_{(i} \varepsilon_{j)k} \,, \\
\delta_Q F^-_{ab} &= \varepsilon^{ij} \bar{\varepsilon}_i \not{D}^C \sigma_{ab} \Psi_j + \bar{\varepsilon}^i \sigma_{ab} \Lambda_i \,, \\
\delta_Q \Lambda_i &= -\sigma \cdot F^- \overleftarrow{\not{D}}^C \varepsilon_i - \not{D}^C B_{ij} \varepsilon_k \varepsilon^{jk} + C \varepsilon^j \varepsilon_{ij} \\
&\quad + \tfrac{1}{2}\kappa \left((\not{D}^C A) T^+_{ij} \cdot \sigma + w A \not{D}^C T^+_{ij} \cdot \sigma \right) \varepsilon_k \varepsilon^{jk} \\
&\quad - \tfrac{3}{2}\kappa (\bar{\chi}_{C[i} \gamma_a \Psi_{j]}) \gamma_a \varepsilon_k \varepsilon^{jk} \,, \\
\delta_Q C &= -2\varepsilon^{ij} \bar{\varepsilon}_i \not{D}^C \Lambda_j - 6\kappa \bar{\varepsilon}_i \chi_{Cj} B_{kl} \varepsilon^{ik} \varepsilon^{jl} \\
&\quad - \tfrac{1}{2}\kappa \bar{\varepsilon}_i \left((w-1) \sigma \cdot T^+_{jk} \overleftarrow{\not{D}}^C \Psi_l + \sigma \cdot T^+_{jk} \not{D}^C \Psi_l \right) \varepsilon^{ij} \varepsilon^{kl}
\end{aligned}
\tag{3.5}
$$

where η^i and ε^i are the spinorial transformation parameters of S- and Q-supersymmetry, respectively.

Chiral superfields may be subject to an additional constraint for $N = 2$ [16]. In the superconformal context this is only possible for $w = 1$. The constraint has the following form in rigid supersymmetry.

$$
(\varepsilon_{ij} \bar{D}^i \sigma_{ab} D^j)^2 (\Phi^{(+)})^* = \mp 96 \,\Box\, \Phi^{(+)} \,, \tag{3.6}
$$

where both signs lead to equivalent results, and are related through the substitution $\Phi^{(+)} \to i\,\Phi^{(+)}$. For local supersymmetry (3.6) is modified, and implies for the components

$$
\begin{aligned}
\Lambda_i &= \mp \varepsilon_{ij} \not{D}^C \Psi^j \,, \\
B_{ij} &= \pm \varepsilon_{ik} \varepsilon_{jl} B^{*kl} \,, \\
C &= \mp (2\Box^C A^* + \tfrac{1}{4}\kappa T^+_{abij} F^+_{ab} \varepsilon^{ij} + 3\kappa \bar{\chi}_{Ci} \Psi^i \\
&\qquad - 2\kappa (1-a) D_C A^*) \,, \\
D_a^C F^-_{ab} &= \pm D_a^C F^+_{ab} + \tfrac{1}{4}\kappa D_a^C (\varepsilon^{ij} T^+_{abij} A \mp \varepsilon_{ij} T^{-ij}_{ab} A^*) \\
&\quad - \tfrac{3}{4}\kappa (\bar{\chi}_{Ci} \gamma_b \Psi_j \varepsilon^{ij} \mp \bar{\chi}^i_C \gamma_b \Psi^j \varepsilon_{ij}) \,.
\end{aligned}
\tag{3.7}
$$

Conformal invariance ensures that these equations have no dependence on the dilatational gauge field b_μ. The last equation of (3.7) is a Bianchi identity, which implies that F_{ab} (or $\tilde{F}_{ab}$, depending on the signs) can be expressed as a field strength in terms of a vector potential. The solution of this Bianchi identity (assuming the minus sign in (3.6)) is

$$\begin{aligned} F_{\mu\nu} &\equiv F^{+}_{\mu\nu} + F^{-}_{\mu\nu} \\ &= \partial_\mu V_\nu - \partial_\nu V_\mu - \tfrac{1}{2}\kappa(\bar{\psi}^{i}_{[\mu}\gamma_{\nu]}\Psi^{j}\varepsilon_{ij} + \text{H.c.}) \\ &\quad -\kappa^2(A\bar{\psi}_{\mu i}\psi_{\nu j}\varepsilon^{ij} + \text{H.c.}) - \tfrac{1}{4}\kappa(AT^{+}_{\mu\nu ij}\varepsilon^{ij} + \text{H.c.}), \end{aligned} \tag{3.8}$$

with V_μ a vector gauge field inert under S supersymmetry. Under Q supersymmetry V_μ transforms according to

$$\delta_Q V_\mu = \bar{\varepsilon}_i\gamma_\mu\Psi_j\varepsilon^{ij} + 2\kappa\bar{\varepsilon}_i\psi_{\mu j}\varepsilon^{ij}A + \text{H.c.}\,. \tag{3.9}$$

Of course, commutation closure is only guaranteed up to a gauge transformation on V_μ. Indeed we find

$$\begin{aligned} [\delta_Q(\varepsilon_1),\delta_Q(\varepsilon_2)]V_\mu &= 2(\bar{\varepsilon}_2^{\,i}\gamma_a\varepsilon_{1i} + \text{H.c.}) \\ &\quad \times(F^{+}_{\mu a} + \tfrac{1}{4}\kappa AT^{+}_{\mu a\, ij}\varepsilon^{ij} + \tfrac{1}{2}\kappa\bar{\psi}_{\mu i}\gamma_a\bar{\Psi}_j\varepsilon^{ij} + \text{H.c.}) \\ &\quad - 4\partial_\mu(A^*\bar{\varepsilon}_1^{\,i}\varepsilon_2^{\,j}\varepsilon_{ij} + \text{H.c.})\,, \end{aligned} \tag{3.10}$$

where the first term is precisely a covariant translation[11,12], and the second term is a gauge transformation which occurs as a field-dependent central charge. This discussion can also be extended to non-Abelian vector gauge fields. In that case all components of the chiral superfield are in the adjoint representation of the gauge group, and the central charge acts by means of a non-Abelian gauge transformation. Apart from the extra covariantization $-g\,\underset{\sim}{V}_\mu \times \underset{\sim}{X}$ in derivatives $D_\mu\underset{\sim}{X}$, the supersymmetry algebra requires the following modifications in the transformation rules.

$$\begin{aligned} \delta_g\underset{\sim}{\Psi}_i &= -2g(\underset{\sim}{A}\times\underset{\sim}{A}^*)\varepsilon_{ij}\varepsilon^{j}\,, \\ \delta_g\underset{\sim}{B}_{ij} &= 2g(\bar{\varepsilon}_{(i}\underset{\sim}{A}\times\underset{\sim}{\Psi}^{k}\varepsilon_{j)k} - \varepsilon_{(ik}\bar{\varepsilon}^{k}\underset{\sim}{A}^*\times\underset{\sim}{\Psi}_{j)})\,, \end{aligned} \tag{3.11}$$

with g the gauge coupling constant.

What remains is to find an invariant density for chiral multiplets. The obvious choice is to start from the C component of the chiral multiplet, which transforms under rigid supersymmetry into a total derivative. If $w = 2$, C has Weyl weight 4, and is a chiral singlet, which makes it a suitable candidate for a superconformally invariant density. Indeed a straightforward calculation leads to the following Lagrangian[15].

$$e^{-1}\mathcal{L} = C - \frac{1}{16}\kappa^2 A (T^+_{abij}\varepsilon^{ij})^2 - \kappa\varepsilon^{ij}\bar{\psi}_i\cdot\gamma\Lambda_j + \text{further terms with } \psi_{\mu i} . \tag{3.12}$$

With this density formula it is now straightforward to find the invariant action for the vector gauge multiple Namely, we take the square of a reduced chiral field, which is again a chiral field, but now with $w = 2$. Its components can then be inserted directly into (3.11), and we give the most important terms

$$\begin{aligned} e^{-1}\mathcal{L}_{\text{vector}} = {} & A\Box^C A^* - \tfrac{1}{4}(F^-_{ab})^2 - \tfrac{1}{2}\bar{\Psi}_i \not{D}^C \Psi^i + \tfrac{1}{8}|B_{ij}|^2 \\ & - (1-a)\kappa D_C |A|^2 + \tfrac{1}{8}\kappa A F^+_{ab} T^+_{abij}\varepsilon^{ij} \\ & + \tfrac{3}{2}\kappa\bar{\chi}_{Ci}\Psi^i + \tfrac{1}{64}\kappa^2 A^2 (T^+_{abij}\varepsilon^{ij})^2 \\ & + \psi_\mu\text{- dependent terms} . \end{aligned} \tag{3.13}$$

4. FROM SUPER-WEYL TO SUPER-POINCARE

Having established a calculus for multiplets and invariants in superconformal gravity it is straightforward to obtain the analogous results in Poincaré supergravity[10,17]. One first limits the invariance transformations to those of the Poincaré theory, with Poincaré supersymmetry transformations defined by the decomposition rule (2.4). After that the transformation rules will still depend on the Weyl weight w, which should lose its relevance here. Indeed, it turns out that the w-dependent terms can be absorbed into the various fields by making suitable field redefinitions. This then defines a realization of the super-Poincaré algebra.

We prefer to make these field redefinitions such that Poincaré and Weyl components will coincide for $w=0$[17]. Because products of $w=0$ superfields have still zero Weyl weight, this choice has the advantage that the multiplication rules of Poincaré and Weyl multiplets will be the same, whereas the Weyl multiplications are already known to be those of rigid supersymmetry[15]. We now give the Poincaré components $[A,\Psi_i,B_{ij},F^-_{ab},\Lambda_i,C]^P$ expressed in terms of the conformal components $(A,\Psi_i,B_{ij},F^-_{ab},\Lambda_i,C)$.

$$\begin{aligned}
[A]^P &= A \quad , \\
[\Psi_i]^P &= \Psi_i \quad , \\
[B_{ij}]^P &= B_{ij} + 2\kappa w A S_{ij} \quad , \\
[F^-_{ab}]^P &= F^-_{ab} + \tfrac{1}{4}\kappa w A t^-_{abij}\,\varepsilon^{ij} \quad , \qquad\qquad (4.1)\\
[\Lambda_i]^P &= \Lambda_i - \kappa w(S_{ik}\varepsilon^{kl}\Psi_l + \tfrac{1}{8}\sigma\cdot t^-_{jk}\,\varepsilon^{jk}\Psi_i + 2A\varepsilon_{ij}\varphi^j)\,, \\
[C]^P &= C - \kappa w(B_{ij}S^{ij} - \tfrac{1}{4}F^-_{ab}\,t^-_{abij}\,\varepsilon^{ij} - 2\bar{\varphi}^i\Psi_i) \\
&\quad -\kappa w A(2G + \kappa(w+1)|S_{ij}|^2 - \tfrac{1}{32}\kappa(w+1)(t^-_{abij}\varepsilon^{ij})^2)\,.
\end{aligned}$$

The newly introduced quantities φ^i and G form together with S^{ij} and a Lorentz vector $E_a{}^{ij}$ the tensor gauge multiplet that was discussed in refs. 4 and 6 as a sub-multiplet of the Poincaré multiplet*); $E_a{}^{ij}$ is simply the supercovariant field strength of the tensor gauge field. The chiral spinor φ^i and the complex Lorentz scalar G are defined by

$$\begin{aligned}
\varphi^i &= \chi^i - \tfrac{1}{2}\gamma\cdot R^{iP} \\
G &= D^P\cdot(V + iA) - \tfrac{1}{2}\kappa^{-1}R^P \\
&\quad -\tfrac{1}{4}\kappa(V_a{}^i{}_j V_a{}^j{}_i + 2V_a{}^2 - 4A_a{}^2 + |M^{ij}|^2 - \tfrac{1}{4}t^+_{ab}{}^{ij}T^+_{ijab}) \\
&\quad +\kappa(\bar{\lambda}_i(2\chi^i + 2\not{D}^P\lambda^i - \tfrac{1}{2}\gamma\cdot R^{iP} + i\kappa\not{A}\lambda^i - \tfrac{1}{2}\kappa\sigma\cdot T^{-ij}\lambda_j) \\
&\qquad + \text{H.c.}) \qquad\qquad (4.2)
\end{aligned}$$

*) This representation is called the linear multiplet in ref. 8.

To calculate the Poincaré supersymmetry transformations implied by (2.4) on the Poincaré components (4.1) is in principle straightforward. We refer to ref. 17 for further details, but we mention here that the Poincaré supersymmetry transformations are considerably more complicated than the superconformal ones. Furthermore, we note that neither M^{ij} nor V_a occurs in the transformation rules, whereas the field λ^i occurs only through the field-dependent SU(2) transformations, specified by (2.4).

The invariant density for chiral superfields in Poincaré supergravity can be directly obtained from the superconformal result (3.12). Since the latter is an invariant density under the full superconformal symmetry for $w = 2$, we can simply translate (3.12) into Poincaré components for this value of w. Since the Poincaré density is independent of w this leads to the required result.

$$\begin{aligned} e^{-1} L = {} & C + 2\kappa B_{ij} S^{ij} - \tfrac{1}{2}\kappa F^-_{ab}\, t^-_{abij}\, \varepsilon^{ij} - 4\kappa \bar{\varphi}^i \Psi_i \\ & + A\left(4\kappa G - 2\kappa^2 |S|^2 + \frac{1}{16}\kappa^2 (t^-_{abij}\varepsilon^{ij})^2 - \frac{1}{16}\kappa^2 (T^+_{abij}\varepsilon^{ij})^2\right) \\ & + \text{ terms with } \psi_{\mu i} \quad . \end{aligned} \qquad (4.3)$$

5. APPLICATIONS; THE MINIMAL FIELD REPRESENTATION

The results of the previous sections can be applied to various supergravity models. We discuss four examples.

a. Poincaré supergravity

As mentioned in ref. 18 this theory can be obtained from the unit chiral superfield. In the superconformal theory this is the multiplet with only the A-component non-vanishing and equal to 1. Obviously a constant multiplet must have $w = 0$, so that the Poincaré unit multiplet has the same form. Using (4.3) one finds that the corresponding action leads to the $N = 2$ Poincaré theory.

b. Yang-Mills-Einstein supergravity

We can add the action for one or several vector gauge multiplets (3.11) to the Poincaré supergravity action. In this way we obtain the model constructed in ref. 19 in terms of dynamic fields. The non-polynomial modifications occur as a result of substituting the Poincaré expressions for $T_{ab}{}^{ij}$, $\chi_C{}^i$, and D_C (see 2.3), and subsequently eliminating the auxiliary fields. We refer to ref. 15 for further details. This model can also be extended with an invariant selfinteraction by using the Poincaré calculus discussed in the previous section.

c. The minimal field representation

Precisely as for N = 1 we can obtain the Poincaré theory from a superconformal system by choosing appropriate gauge conditions to eliminate those conformal gauge invariances that are not contained in the Poincaré algebra[20]. We apply this idea to the vector multiplet, and choose as gauge conditions

$$A = \kappa^{-1} \quad , \quad \Psi_i = 0 \quad . \tag{5.1}$$

The first condition fixes the gauge for dilatations and chiral U(1) transformations, the second one for S supersymmetry. Poincaré supersymmetry transformations are defined by the particular field-dependent linear combination of Q and S supersymmetry which leaves Ψ_i invariant. Hence, we find the decomposition rule

$$[Q]^{\text{Poincaré}} = Q + S \ , \tag{5.2a}$$

with the parameters of Q and S supersymmetry expressed in the Poincaré supersymmetry parameter ε^i according to

$$\varepsilon_Q^{\ i} = \varepsilon^i ,$$

$$\varepsilon_{Si} = -\tfrac{1}{2}\kappa(B_{ij} + \sigma\cdot F^- \varepsilon_{ij})\varepsilon^j + (\not{b} - i\kappa\not{A})\varepsilon_i \ . \tag{5.2b}$$

Apart from the SU(2) transformation this result coincides precisely with (2.4). The field B_{ij} satisfies the same constraint (2.2) as S_{ij}, whereas F^-_{ab} satisfies a Bianchi identity which implies here (see (3.8))

$$\begin{aligned} F_{\mu\nu} &\equiv F^+_{\mu\nu} + F^-_{\mu\nu} \\ &= \partial_\mu V_\nu - \partial_\nu V_\mu - \kappa(\bar{\psi}_{\mu i}\psi_{\nu j}\varepsilon^{ij} + \text{H.c.}) \\ &\quad - \tfrac{1}{4}(T^+ + T^-)_{\mu\nu ij}\varepsilon^{ij} \ . \end{aligned} \tag{5.3}$$

Hence, we may identify $-\frac{1}{2}B_{ij}$ as the auxiliary field S_{ij}, and $-\frac{1}{2}\sqrt{2}V_\mu\varepsilon^{ij}$ as the gauge field $B_\mu^{\ ij}$. Note that both the transformation rule (3.9) and the central-charge term in the commutator (3.10) are consistent with this identification. The field strength (5.3) is then defined as a linear combination of the covariant SO(2) field strength $\hat{F}_{ab}^{\ ij}$ and the tensor field $T_{ab}^{\ ij}$, which corresponds to the auxiliary field $t_{ab}{}^{ij}$.

Therefore the gauge conditions (5.1) leave us with a locally SU(2) invariant set of fields:

$$\begin{aligned} &\textit{gauge fields} \quad : e_\mu{}^a,\ \psi_\mu{}^i,\ V_\mu{}^i{}_j,\ B_\mu{}^{ij}, \\ &\textit{auxiliary fields:}\ A_a,\ \chi_C{}^i,\ D_C,\ t_{ab}{}^{ij},\ S_{ij}. \end{aligned} \tag{5.4}$$

The multiplet (5.3) has only 32 + 32 components; it has still a closed supersymmetry algebra since a gauge choice generally leaves the closure property unaffected. Therefore this is a smaller field representation for Poincaré supersymmetry, which can be compared to the multiplet of currents constructed in ref. 21.

However, this minimal field representation is too small to allow a meaningful invariant Poincaré Lagrangian. This can be seen most directly by considering the term $-\kappa e D_C |A|^2$ of the Lagrangian (3.13) (The a-dependent term should cancel[15]). If we first solve the field equation for D_C we find $|A| = 0$, which will no longer allow the gauge choice (5.1). Alternatively, if we first fix the gauge according to (5.1) then the Lagrangian has a term $-\kappa^{-1} e D_C$, which leads to an inconsistent field equation for D_C (assuming a nonsingular vierbein). Hence the minimal field representation does not allow a consistent super-Poincaré Lagrangian. This negative conclusion agrees with that of Breitenlohner and Sohnius[8,18]. It brings us back to the fields (2.1), originally proposed in refs. 3 and 4. This set is known to be the smallest representation that allows a consistent Lagrangian (see the arguments presented in ref. 4).

d. Weyl supergravity

The curvatures of the superconformal algebra are contained in a chiral multiplet with external Lorentz indices. But unlike the case of N = 1 this multiplet has additional matter fields. The superfield turns out to be a selfdual antisymmetric tensor, with $T_{ab}^{-\,ij}$ as its first component. Most of the other components have been identified to all orders in terms of fully covariantized curvatures with certain modifications[6]. This chiral multiplet is also subject to constraints of the type (3.7), and has therefore 24 + 24 components. The fully contracted square of the Weyl multiplet is a scalar chiral superfield with w = 2, so that application of the action formula (3.11) will lead to the Weyl supergravity action. As for N = 1 this scalar multiplet can also be used to construct invariant counterterms for Poincaré supergravity[18,22].

6. POINCARE SUPERGRAVITY WITH LOCAL SU(2)

In the preceding sections we have argued that having a theory with a larger degree of gauge invariance is helpful in finding explicit representations of the supersymmetry algebra. For that reason we have used the superconformal theory as an intermediate step, from which the (more complicated) Poincaré results could be obtained rather directly. The superconformal theory does not allow the presence of field-independent central charges. These charges can at most coexist with a local SU(2) group acting on the supersymmetry generators, which is also the symmetry exhibited by the minimal field formulation (5.4). This suggests the possibility of constructing a locally SU(2) symmetric version of the Poincaré theory. Such a formulation would be closer to the superconformal theory and to (5.4), but could have a more general applicability.

Local SU(2) would simplify the transformation rules, since the SU(2) component of Poincaré supersymmetry, as given by the decomposition rule (2.4), would then decouple and become an independent invariance. This would thus eliminate most of the terms depending on the spinor field λ^i which occur mainly through field-dependent SU(2) transformations. For the Poincaré supersymmetry transformations of chiral fields this would imply that the transformation rules would only involve 32 + 32 supergravity components. One might also hope to simplify the supersymmetry algebra, which is rather complicated because of the presence of field-dependent supersymmetry transformations. A further motivation for attempting to construct a locally SU(2) invariant formulation is that N = 8 supergravity has a formulation with local SU(8) invariance[23] (but not with a closed algebra). This SU(8) invariance has aroused some interest, because of the conjecture that the gauge fields of this symmetry may become dynamical at the quantum level. Such a phenomenon is known to happen in two-dimensional models[24], and could have implications for phenomenological applications of N = 8 supergravity[25]. Therefore it is of interest to bring the formulation of N = 2 supergravity more in line with that of ref. 23.

The construction that we follow employs the fact that the Poincaré theory has already a rigid SU(2) chiral invariance. We start by introducing a new field $\Phi^I{}_i(x)$, which is an element of the group SU(2). The rigid SU(2) group acts on the right of Φ, and thus on the indices $i, j, k, \ldots$, etc. But in addition Φ transforms under a local SU(2) group on the left; hence local SU(2) acts on the indices $I, J, K, \ldots$, etc. We can now use Φ to redefine all fields with rigid SU(2) indices to fields with local indices. Note that this redefinition amounts to a local SU(2) transformation itself. In this way Φ is somewhat analogous to the vierbein in general relativity, which is an element of GL(4) that connects local Lorentz to world indices.

By introducing Φ we have not changed the number of degrees of freedom of the theory, since we have assumed a new local SU(2) invariance at the same time. In other words, we could always choose the gauge where Φ is the identity element. This would simply bring us back to the previous formulation. With this gauge condition imposed we then expect to recover the field-dependent SU(2) transformation as a component of Poincaré supersymmetry (see eq. (2.4)). This suggests that a supersymmetry transformation of Φ is precisely related to that field-dependent SU(2) transformation, which leads us to assume for the supersymmetry variation

$$\delta\Phi^{I}{}_{i} = \kappa(2\bar{\varepsilon}^{I}\lambda_{J} - \delta^{I}{}_{J}\,\bar{\varepsilon}^{K}\lambda_{K} - \text{H.c.})\Phi^{J}{}_{i} \quad . \tag{6.1}$$

With this choice the fields with local rather than rigid SU(2) indices will no longer have a field-dependent SU(2) component in their supersymmetry transformation, as one can verify by explicit calculation.

The gauge fields of the newly introduced local SU(2) are related to the SU(2) gauge fields $V_{\mu}{}^{i}{}_{j}$ of the superconformal theory. But again we prefer to write these fields with local SU(2) indices, which implies that we apply a local SU(2) transformation Φ:

$$V_{\mu}{}^{i}{}_{j} \rightarrow V_{\mu}{}^{I}{}_{J} = \Phi^{I}{}_{i}\, V_{\mu}{}^{i}{}_{j}\, \Phi^{j}{}_{J} + \frac{2}{\kappa}\, \Phi^{I}{}_{i}\, \partial_{\mu}\Phi^{i}{}_{J} \quad , \tag{6.2}$$

where $\Phi^{i}{}_{I}$ is the inverse of $\Phi^{I}{}_{i}$:

$$\Phi^{i}{}_{I}\,\Phi^{I}{}_{j} = \delta^{i}{}_{j} \quad , \quad \Phi^{I}{}_{i}\,\Phi^{i}{}_{J} = \delta^{I}{}_{J} \quad , \quad \Phi^{i}{}_{I} = \varepsilon^{ij}\varepsilon_{IJ}\Phi^{J}{}_{j} \, . \tag{6.3}$$

We can now use (6.2) to express the Poincaré auxiliary fields $V_{a}{}^{i}{}_{j}$ in the form

$$V_{a}{}^{i}{}_{j} = \frac{2}{\kappa}\, \Phi^{i}{}_{I}\, \hat{D}_{a}\Phi^{I}{}_{j} \quad , \tag{6.4}$$

where from now on D_a denotes a derivative covariantized with respect to Lorentz and local SU(2) transformations, the latter with chiral gauge fields $V_{\mu}{}^{I}{}_{J}$. The symbol "^" denotes a covariantization with respect to supersymmetry.

Hence, we have obtained a formulation of $N = 2$ supergravity with $40 + 40$ components, and an $SU(2)_{local} \times SU(2)_{rigid}$ invariance group. This leads to two simplifications in the transformation rules. First we can now use chirally covariant derivatives, and secondly certain λ_i dependent terms will be absent. We now give the full transformation rules.

$$\delta e_\mu{}^a = \kappa(\bar\varepsilon^I \gamma_a \psi_{\mu I} + \text{H.c.}) \quad ,$$

$$\delta B_\mu{}^{ij} = -\sqrt{2}\, \varepsilon^{ij} (\bar\varepsilon^I \psi_\mu{}^J \varepsilon_{IJ} + \text{H.c.}) \quad ,$$

$$\delta \psi_\mu{}^I = 2\kappa^{-1} D_\mu \varepsilon^I - \tfrac{1}{4}\sigma\cdot T^{-IJ} \gamma_\mu \varepsilon_J - \gamma_\mu (S^{IJ}\varepsilon_J + i\not{A}\varepsilon^I + \tfrac{1}{4}\sigma\cdot t^{+IJ}\varepsilon_J) \quad ,$$

$$\delta V_\mu{}^I{}_J = 2\bar\varepsilon^I \gamma_\mu \chi_J - 2\bar\varepsilon^I \hat R_{\mu J} + 2\kappa \bar\varepsilon^{(I} \psi_\mu{}^{K)} S_{KJ} - (\text{H.c.; traceless}) ,$$

$$\delta A_a = i\bar\varepsilon^I \gamma_a \chi_I + i\bar\varepsilon^I \sigma_{ab} \hat R_{bI} + \text{H.c.} \quad ,$$

$$\delta \Phi^I{}_i = \kappa(2\bar\varepsilon^I \lambda_J - \delta^I{}_J \bar\varepsilon^K \lambda_K - \text{H.c.})\Phi^J{}_i \quad ,$$

$$\begin{aligned}\delta\lambda^I = {} & S^{IJ}\varepsilon_J + (iA_a + \kappa\bar\lambda^J \gamma_a \lambda_J)\gamma^a \varepsilon^I \\ & + (\tfrac{1}{4}t^{+\,IJ}_{ab} - 2\kappa\bar\lambda^I \sigma_{ab}\lambda^J)\sigma^{ab}\varepsilon_J + \kappa^{-1}\Phi^I{}_i \hat{\not{D}}\Phi^i{}_J \varepsilon^J \\ & - \tfrac{1}{2}\not{V}\varepsilon^I - \tfrac{1}{2}M^{IJ}\varepsilon_J - 2\kappa\lambda^I(\bar\lambda^J \varepsilon_J + \text{H.c.}) \quad ,\end{aligned}$$

$$\begin{aligned}\delta\chi^I = {} & -2\hat{\not{D}} S^{IJ}\varepsilon_J + 2i\hat{\not{D}}\not{A}\varepsilon^I + \tfrac{1}{2}\hat{\not{D}}\sigma\cdot t^{+IJ}\varepsilon_J + \hat D\cdot V\varepsilon^I \\ & + \sigma\cdot\hat F^I{}_J(V)\varepsilon^J - \tfrac{1}{2}i\kappa\sigma\cdot T^{-IJ}\not{A}\varepsilon_J \\ & - 4\kappa(S^{IJ} - \tfrac{1}{8}\sigma\cdot T^{-IJ})(S_{JK}\varepsilon^K - i\not{A}\varepsilon_J + \tfrac{1}{4}\sigma\cdot t^-{}_{JK}\varepsilon^K) \\ & + \kappa\varepsilon^I(\kappa^{-2}|\hat D_a \Phi^J{}_j|^2 - 2A_a^2 - \tfrac{1}{2}V_a^2 - \tfrac{1}{4}|M^{JK}|^2 - \tfrac{1}{8}t^{+JK}_{ab} T^{+\,ab}_{JK}) \\ & + \kappa\varepsilon^I(\bar\lambda_J(2\chi^J + 2\hat{\not{D}}\lambda^J - \tfrac{1}{2}\gamma\cdot\hat R^J + i\kappa\not{A}\lambda^J - \tfrac{1}{2}\kappa\sigma\cdot T^{-JK}\lambda_K \\ & \qquad + 2\Phi^J{}_i \hat{\not{D}}\Phi^i{}_K \lambda^K) + \text{H.c.}) \quad ,\end{aligned}$$

$$\delta S^{IJ} = \bar\varepsilon^{(I}(\chi - \tfrac{1}{2}\gamma\cdot\hat R)^{J)} + \varepsilon^{IK}\varepsilon^{JL}\bar\varepsilon_{(K}(\chi - \tfrac{1}{2}\gamma\cdot\hat R)_{L)} \quad ,$$

$$\delta t^{+\,IJ}_{ab} = \varepsilon^{IJ}(-4\varepsilon_{KL}\bar\varepsilon^K \sigma_{ab}\chi^L + 2\varepsilon_{KL}\bar\varepsilon^K \gamma^d \sigma_{ab}\hat R_d{}^L + 2\varepsilon^{KL}\bar\varepsilon_K \sigma_{ab}\gamma\cdot\hat R_L)$$

$$\delta M^{IJ} = 2\bar{\varepsilon}^I(\chi^J + 2\hat{\not{D}}\lambda^J - \tfrac{1}{2}\kappa\sigma\cdot T^{-JK}\lambda_K + \kappa M^{JK}\lambda_K - \kappa(\not{V} - i\not{A})\lambda^J$$
$$+ 2\Phi^J{}_i\,\hat{\not{D}}\Phi^i{}_K\,\lambda^K) - (I \leftrightarrow J) \quad ,$$

$$\delta V_a = \bar{\varepsilon}_I\gamma_a(\chi^I + 2\hat{\not{D}}\lambda^I - \tfrac{1}{2}\kappa\sigma\cdot T^{-IJ}\lambda_J + \kappa M^{IJ}\lambda_J - \kappa(\not{V} - iA)\lambda^I$$
$$+ 2\Phi^I{}_i\,\hat{\not{D}}\Phi^i{}_J\,\lambda^J) - 2\bar{\varepsilon}^I(\hat{D}_a - \tfrac{1}{2}i\kappa A_a)\lambda_I$$
$$+\kappa\bar{\lambda}_I(-\tfrac{1}{4}\sigma\cdot T^{-IJ}\gamma_a\varepsilon_J - \gamma_a(S^{IJ}\varepsilon_J + i\not{A}\varepsilon^I + \tfrac{1}{4}\sigma\cdot t^{+IJ}\varepsilon_J))$$
$$+ \text{H.c.}, \tag{6.5}$$

where $\hat{F}_{ab}{}^I{}_J(V)$ denotes the supercovariantized chiral SU(2) field-strength tensor.

It is important to consider the commutator of two supersymmetry transformations on $\Phi^I{}_i$. It turns out that the result is a covariant translation (now also covariantized with respect to chiral SU(2)!) and a field-dependen SU(2) transformation. The SU(2) modifications will therefore occur uniformly in the commutators of other fields that carry local SU(2) indices. But a more important modification of the algebra is caused by the fact that we will assume ε^I to be the basic field-independent transformation parameter, rather than ε^i. This leads to additional supersymmetry transformations in the commutato algebra with parameters depending on the field λ^I. However, these terms cancel precisely similar terms in the original supersymmetry algebra[5,6], which then simplifies. The result is

$$[\delta(\varepsilon_1),\delta(\varepsilon_2)] = 2(\bar{\varepsilon}_2{}^I\gamma^a\varepsilon_{1I} + \text{H.c.})\hat{D}_a$$
$$+ \delta_M(\varepsilon^{ab}) + \delta_{SU(2)}(\Lambda^I{}_J) \tag{6.6a}$$

with

$$\varepsilon^{ab} = \kappa\bar{\varepsilon}_1{}^I\,\varepsilon_2{}^J\,(T^{+\,ab}_{IJ} - t^{-\,ab}_{IJ}) + 4\kappa\bar{\varepsilon}_1{}^I\,\sigma_{ab}\bar{\varepsilon}_2{}^J\,S_{IJ}$$
$$+2i\kappa\bar{\varepsilon}_1{}^I\,\gamma_d\varepsilon_{2I}A_c\varepsilon^{abcd} + \text{H.c.}$$

$$\Lambda^I{}_J = -2\kappa(\bar{\varepsilon}_1{}^M\,\varepsilon_2{}^N\,\varepsilon_{MN} + \text{H.c.})\varepsilon^{IK}S_{KJ} \tag{6.6b}$$
$$+4i\kappa(\bar{\varepsilon}_1{}^I\,\not{A}\varepsilon_{2J} - \tfrac{1}{2}\delta^I{}_J\,\bar{\varepsilon}_1{}^K\,\not{A}\varepsilon_{2K} - \text{H.c.})$$
$$-\tfrac{1}{2}\kappa(\bar{\varepsilon}_1{}^{(I}\,\sigma\cdot t^-_{JK}\varepsilon_2{}^{K)} - \text{H.c.}) \quad .$$

It is rather straightforward to reformulate previous results in terms of the local SU(2) theory. As an example we give the Poincaré supergravity action

$$
\begin{aligned}
e^{-1}\mathcal{L}^{\text{Poincaré}} = & -\tfrac{1}{2}\kappa^{-2} R(e,\omega) - \tfrac{1}{2}(\bar{\psi}_\mu{}^I R^\mu{}_I + \text{H.c.}) - \frac{1}{8}(F_{ab}{}^{ij}(B))^2 \\
& - \frac{1}{16}\sqrt{2}\,\kappa(\bar{\psi}_a{}^I \psi_b{}^J \varepsilon_{IJ}(F(B) + \tilde{F}(B) + \hat{F}(B) + \tilde{\hat{F}}(B)_{ab}{}^{ij}\varepsilon_{ij} + \text{H.c.}) \\
& -\tfrac{1}{2}|S|^2 + \frac{1}{8}(t_{ab}{}^{IJ})^2 + A_a{}^2 + \kappa^{-2}|\hat{D}_a \Phi^I{}_i|^2 - \tfrac{1}{2}V_a{}^2 - \tfrac{1}{4}|M^{IJ}|^2 \\
& +4(\tfrac{1}{4}\varepsilon^{abcd}\bar{\psi}_{bI}\gamma_c\psi_d{}^J + \bar{\lambda}_I\sigma_{ab}\psi_b{}^J - \bar{\lambda}^J\sigma_{ab}\psi_{bI} + \bar{\lambda}_I\gamma_a\lambda^J)\Phi^I{}_i\hat{D}_a\Phi^i{}_J \\
& +(\bar{\lambda}^I(2\chi_I + 2\not{D}\lambda_I - 2\kappa S_{IJ}\gamma\cdot\psi^J + 2i\kappa\gamma^\mu\not{A}\psi_{\mu I} - \tfrac{1}{2}\kappa\gamma^\mu\sigma\cdot t^-_{IJ}\psi_\mu{}^J \\
& \qquad \kappa V_a\psi_{aI} - i\kappa\not{A}\lambda_I - \tfrac{1}{2}\kappa\sigma\cdot T^+_{IJ}\lambda^J) + \text{H.c.}) \\
& -2\kappa^2(\bar{\psi}_\mu{}^I\lambda_I + \text{H.c.})(\bar{\lambda}^J\sigma^{\mu\nu}\psi_{\nu J} + \text{H.c.}) \\
& + \kappa^2 e^{-1}\varepsilon^{\mu\nu\rho\sigma}\bar{\psi}_\mu{}^I\gamma_\rho\psi_{\nu I}\bar{\lambda}^J\gamma_\sigma\lambda_J\,. \qquad (6.7)
\end{aligned}
$$

7. CONCLUSIONS

We have discussed N = 2 supergravity theories within the context of different supersymmetry algebras. In each of these formulations the Poincaré theory can be reobtained, either by imposing the decomposition rule (2.4), or upon an appropriate gauge choice. In general it is advantageous to classify field representations according to the algebra that has the highest degree of symmetry. The higher symmetry will restrict the modifications of the linearized transformation rules, and the structure of these modifications will be more transparent. For the construction of invariant actions only a subalgebra may be relevant, but as we have shown this poses no important problems. On the other hand the applicability of larger symmetry algebras may be less general.

The general strategy is to impose a local supersymmetry algebra on a representation of rigid supersymmetry. However, it should be emphasized that the success of this procedure is a priori not guaranteed. For instance the algebra for the non-Abelian gauge multiplet was introduced in section 3 on the basis of the reduced chiral superfield in the adjoint representation of the gauge group that contains the field strength (3.8). But it turns out that the same algebra cannot be realizedd on any other chiral multiplet that is not a singlet under the gauge group (and therefore not inert under the central charge).

We have already indicated that field-independent central charges can only be discussed in a Poincaré context. As is well known the Poincaré algebra has itself a central charge, which acts as a gauge transformation for the vector field $B_\mu{}^{ij}$. This central charge can be understood as arising from a de Sitter supersymmetry algebra after a Wigner-Inönü group contraction[26]. Before this contraction there is no central charge, and the SO(2) gauge transformations will act on the supersymmetry generators. The SO(2) gauge field will therefore couple minimally to all SO(2) covariant quantities. This gauging has been achieved for all SO(N)-extended supergravity theories with $N \leq 4$ in formulations without auxiliary fields[27]. In principle it is not clear whether the Poincaré auxiliary field representation will suffice to realize a closed de Sitter algebra as well. In fact we have not succeeded to extend the Poincaré algebra to the de Sitter case on the basis of the auxiliary fields of SO(2) Poincaré supergravity. These aspects of extended de Sitter supergravity clearly deserve further study.

We hope to have shown that the tensor calculus for $N = 2$ supergravity theories is well understood. It opens the way for a more systematic study of extended supergravity models. This is clearly needed to gain further insight in the implications of local supersymmetry for the unification of the fundamental forces in Nature.

REFERENCES

1. For references, see "Recent developments in gravitation", Summer Institute Cargèse, 1978 eds. M. Lévy and S. Deser (Plenum Press, New York 1978); "Supergravity", proc. of the Stony Brook Supergravity Workshop, 1979, eds. P. van Nieuwenhuizen and D.Z. Freedman. (North-Holland, Amsterdam, 1979); talks presented at this conference.
2. S. Ferrara, D.Z. Freedman, and P. van Nieuwenhuizen, Phys. Rev. D13:3214 (1976); B. Zumino and S. Deser, Phys. Lett. 62B:335 (1976).
3. E.S. Fradkin and M.A. Vasiliev, Lett. Nuovo Cim. 25:79 (1979).
4. B. de Wit and J.W. van Holten, Nucl. Phys. B155:530 (1979).
5. E.S. Fradkin and M.A. Vasiliev, Phys. Lett. 85B:47 (1979).
6. B. de Wit, J.W. van Holten, and A. Van Proeyen, Nucl. Phys. B167:186 (1980).

7. S. Ferrara and P. van Nieuwenhuizen, Phys. Rev. Lett. 37:1669 (1976).
8. P. Breitenlohner and M. Sohnius, Nucl. Phys. B165: 483 (1980).
9. B. de Wit, in "Supergravity", proc. of the Stony Brook Supergravity Workshop, 1979, eds. P. van Nieuwenhuizen and D.Z. Freedman (North-Holland, Amsterdam, 1979): A. Van Proeyen, in proc. of the 17th Winter School of Theoretical Physics, Karpacz, (to be published).
10. S. Ferrara and P. van Nieuwenhuizen, Phys. Lett. 76B:404 (1978).
11. B. de Wit and D.Z. Freedman, Phys. Rev. D12:2286 (1975).
12. P. Breitenlohner, Phys. Lett. 67B:49 (1977).
13. A. Salam and J. Strathdee, Nucl. Phys. B76:477 (1979), Phys. Rev. D11:1521 (1975): S. Ferrara, J. Wess and B. Zumino, Phys. Lett. 51B:239 (1974).
14. P.H. Dondi and M. Sohnius, Nucl. Phys. B81:317 (1974).
15. M. de Roo, J.W. van Holten, B. de Wit, and A. Van Proeyen, Nucl. Phys. B, to be published.
16. R.J. Firth and J.D. Jenkins, Nucl. Phys. B85:525 (1975).
17. B. de Wit, J.W. van Holten, and A. Van Proeyen, Leiden preprint, in preparation.
18. P. Breitenlohner, in "Supergravity", proc. of the Stony Brook Supergravity Workshop, 1979, eds. P. van Nieuwenhuizen and D.Z. Freedman (North-Holland, Amsterdam, 1979), and this volume p. 349.
19. J.F. Luciani, Nucl. Phys. B132:325 (1978).
20. A. Das, M. Kaku, and P.K. Townsend, Phys. Rev. Lett. 40:1215 (1978).
21. M. Sohnius, Phys. Lett. 81B:8 (1979).
22. S. Ferrara and P. van Nieuwenhuizen, Phys. Lett. 78B:573 (1978).
23. E. Cremmer and B. Julia, Phys. Lett. 80B:48 (1978); Nucl. Phys. B159:141 (1979).
24. A. D'Adda, M. Lüscher and P. Di Vecchia, Nucl. Phys. B146:73 (1978); B152:125 (1979); E. Witten, Nucl. Phys. B149:285 (1979).
25. T.L. Curtright and P.G.O. Freund, in "Supergravity", proc. of the Stony Brook Supergravity Workshop, 1979, eds. P. van Nieuwenhuizen and D.Z. Freedman (North-Holland, Amsterdam, 1979); M.K. Gaillard, this volume p. 69.
26. P.K. Townsend and P. van Nieuwenhuizen, Phys. Lett. 67B:439 (1977).
27. D.Z. Freedman and A. Das, Nucl. Phys. B120:221 (1977); E.S. Fradkin and M.A. Vasiliev, Lebedev Institute preprint (1976); D.Z. Freedman and J.H. Schwarz, Nucl. Phys. B137:333 (1978).

TENSOR CALCULUS FOR N = 2 EXTENDED SUPERGRAVITY

Peter Breitenlohner*

Max-Planck-Institut für Physik und Astrophysik

D-8000 München 40, Germany

ABSTRACT

We present most elements of tensor calculus for N = 2 extended (Poincaré) supergravity. Starting from the off-shell algebra with the minimal set of 32 + 32 field components, we study various types of abstract multiplets (linear, chiral, Yang-Mills, matter) and give an action formula for the linear as well as the chiral one. As an application, we construct invariant actions for supergravity, an arbitrary Yang-Mills multiplet, a gauged non-trivial SU(2) and the matter (hyper) multiplet. We give the candidate for an invariant three-loop counterterm.

The work presented here was mostly done together with M. Sohnius and A. Kabelschacht; part of it is published[1], other parts are prepared for publication[2]. Our work is closely related to that of Fradkin and Vasiliev[3] and of De Roo, De Wit, van Holten and van Proeyen[4].

1. OFF-SHELL VERSUS ON-SHELL ALGEBRA

For the benefit of the GUTs people in the audience, let me start with an explanation of these terms, since they have not been explained during this conference so far.

---

* Present address: Institut für Theoretische Physik, FB 20/WE 4 Freie Universität Berlin, D-1000 Berlin 33, Germany.

An off-shell realization of supersymmetry or supergravity consists of a set of fields together with their transformation laws. These transformation laws are independent of the parameters of models such as masses, coupling constants and wave function normalization constants. The commutator of two infinitesimal supersymmetry transformations acting on any field has a universal geometrical interpretation; in general it will include field dependent gauge transformations. The action for a specific model is given as a sum of terms which are separately invariant with the parameters of that model as coefficients. Tensor calculus allows one to construct such invariant terms.

A peculiarity of supersymmetry is the presence of auxiliary fields, i.e., fields which satisfy algebraic equations of motion. Note, however, that only the action for a model determines which fields are auxiliary ones.

We can use these equations of motion, which will, in general, depend on the parameters of the model, to eliminate all the auxiliary fields from the action and from the transformation laws. In this way we obtain the on-shell version for a specific model. Here the transformation laws do depend on the parameters of the model and only the total action is invariant. The commutator of two infinitesimal supersymmetry transformations has the same geometrical meaning as before plus additional terms which vanish only if all equations of motion are satisfied (on-shell algebra). The advantage is that all unphysical non-propagating fields have been eliminated.

Clearly an off-shell formulation is necessary for tensor calculus, where we study abstract multiplets and their properties. An off-shell formulation is very desirable, though not necessary, for the Faddeev-Popov quantization procedure or for the study of renormalization properties.

Unfortunately it is by no means simple to reintroduce the auxiliary fields, or even to guess their structure, once they have been eliminated. The complete solution is known for non-extended ($N = 1$) and $N = 2$ extended supersymmetry and supergravity only.

2. THE GEOMETRY OF SUPERSPACE

Starting from the affine geometry of a curved superspace[5] we will impose constraints on torsion and curvature components and thus obtain a minimal set of auxiliary fields.

We use an SU(2) notation and almost all our spinors will be SU(2) doublets, satisfying an SU(2) invariant Majorana condition[6]. Consequently, given two such spinors α and β, the quantity $i\bar{\alpha}\beta =$ $= i\bar{\beta}\alpha$ is real and the term $\overline{\alpha}\vec{t}\beta\vec{t}$ corresponds to $\bar{\alpha}^i\beta^i S + \bar{\alpha}^i\gamma_5\beta^j P^{ij}$ in SU(2) notation[3,4].

Our superspace has co-ordinates $z^M = (x^m, \Theta^\mu)$ where Θ^μ has 2 × 4 real components for N = 2. The affine structure is based on the vielbein superfield $E_M{}^A(z)$ and its inverse $E_A{}^M(z)$ which relate curved space tensors with indices M,N,P, ... (m,n,p ..., respectively ν,μ,π, ... for the bosonic respectively fermionic part) to tangent (or flat) space tensors with indices A,B,C, ... (a,b,c, ... respectively α,β,γ, ...). The structure group acting on tangent space does not mix the bosonic and fermionic part of vectors, such that the decomposition $U_A = (U_a, U_\alpha)$ is invariant; the corresponding decomposition $U_M = (U_m, U_\mu)$ is not invariant, U_m and U_μ transform into each other under general co-ordinate transformations of superspace. We consider the (graded) Lie algebra with generators

$$P_A = (P_a, Q_\alpha),\ L_{ab},\ \vec{I},\ Z,\ I_i \tag{1}$$

and corresponding parameters

$$\xi^A = (\xi^a, \xi^\alpha),\ \lambda^{ab},\ \vec{\lambda},\ \alpha,\ \lambda^i \tag{2}$$

P_A generates superspace translations (i.e., P_a the ordinary translations and Q_α supersymmetry transformations), the others generate the structure group. Lorentz transformations (L_{ab}) act on P_a and Q_α: the non-trivial SU(2) ($\vec{I}$) acts on Q_α; the central charge (Z) and the arbitrary internal symmetry (I_i) (Yang-Mills group) do not act on P_a but may both act on matter superfields.

It is convenient to use the Lie algebra valued parameter superfields

$$\lambda(z) = \tfrac{1}{2}\lambda^{ab}(z)L_{ab} + \vec{\lambda}(z)\vec{I} + \alpha(z)Z + \lambda^i(z)I_i \tag{3}$$

$$\xi(z) = \xi^A(z)P_A = \xi^a(z)P_a + i\,\bar{\xi}(z)Q \tag{4}$$

(the extra i in $i\bar{\xi}Q$ is due to the SU(2) notation) and gauge fields

$$V_M(z) = v_M(z) + E_M(z) \tag{5}$$

$$v_M(z) = \frac{1}{2}\omega_M{}^{ab}(z)L_{ab} + \vec{B}_M(z)\vec{I} + A_M(z)Z + B_M^i(z)I_i \tag{6}$$

$$E_M(z) = E_M{}^A(z)P_A = E_M{}^a(z)P_a + i\bar{\Psi}_M Q \tag{7}$$

where we have used the vielbein $E_M{}^A$ as the gauge field corresponding to translations.

The curvatures and torsions

$$R_{MN} = r_{MN} + T_{MN} \tag{8}$$

$$r_{MN} = \frac{1}{2}R_{MN}{}^{ab}L_{ab} + \vec{G}_{MN}\vec{I} + F_{MN}Z + G_{MN}^i I_i \tag{9}$$

$$T_{MN} = T_{MN}{}^A P_A \tag{10}$$

are given as

$$R_{MN} = \partial_M V_N - \partial_N V_M - [V_M, V_N] + [E_M, E_N] \tag{11}$$

or as tangent space tensors

$$R_{AB} = E_B{}^N E_A{}^M R_{MN} = (R_{ab}, R_{\alpha b}, R_{\alpha\beta}) \, . \tag{12}$$

where extra signs due to the grading are never written explicitly but are understood to be present.

The action of an infinitesimal structure group transformation is the obvious one,

$$\delta(\lambda)V_M(z) = \partial_M \lambda(z) + [\lambda(z), V_M(z)] = D_M \lambda(z) \qquad (13)$$

for the gauge fields and

$$\delta(\lambda)\phi(z) = \rho(\lambda(z))\phi(z) \qquad (14)$$

for all covariant fields, where $\rho(\lambda)$ is a (matrix) representation of the structure group.

The action of the translations on covariant fields is <u>defined</u> to give the covariant derivative

$$\delta(\xi)\phi(z) = \xi^A(z) D_A \phi(z) \qquad (15)$$

where

$$D_M \phi = E_M{}^A D_A \phi = \partial_M \phi - \rho(\omega_M)\phi \qquad (16)$$

Everything else follows from this contribution, in particular the Ricci-identities

$$[\delta(\xi_1), \delta(\xi_2)] = \delta(\xi_1{}^B \xi_2{}^A R_{AB}) \qquad (17)$$

and the action of the covariant translation on the gauge fields

$$\delta(\xi)V_M(z) = D_M \xi(z) + E_M{}^B(z)\xi^A(z) R_{AB}(z) \qquad (18)$$

where [with the same abuse of notation as in the analogous Eq. (13)]

$$D_M \xi(z) = \partial_M \xi(z) + [\xi(z), \omega_M(z)] \qquad (19)$$

Using the definitions (11) and (12) we find that $\delta(\xi)$ can be reinterpreted as a covariant Lie derivative with respect to $\xi^M = \xi^A E_A{}^M$, i.e., covariantized with respect to the structure group, a universal geometrical interpretation which covers Eqs (15) and (18).

We can specialize Eq. (18) to

$$\delta(i\bar{\xi}Q)e_m{}^a = i e_m{}^c \bar{\xi}^\beta T_{\beta c}{}^a + \bar{\xi}^\beta T_{\beta\gamma}{}^a \psi_m{}^\gamma \tag{20}$$

$$\delta(i\bar{\xi}Q)\bar{\psi}_m{}^\alpha = D_m\bar{\xi}^\alpha + i e_m{}^c \bar{\xi}^\beta T_{\beta c}{}^\alpha + \bar{\xi}^\beta T_{\beta\gamma}{}^\alpha \psi_m{}^\gamma \tag{21}$$

and similar equations for $\delta(i\,\bar{\xi}\,Q)v_m$. Note that until now everything is a superfield and that we can either compute $\delta(i\,\bar{\xi}\,Q)V_m$ once we know $R_{\alpha\beta}$ and $R_{\alpha b}$ or read off these quantities from $\delta(i\,\bar{\zeta}\,Q)V_m(z)$. All this applies to $\omega_m{}^{ab}$, regardless whether $\omega_m{}^{ab}$ is an independent field or can be expressed as a function of $e_m{}^a$, and $\bar{\psi}_m{}^\alpha$ (second order formalism).

Our strategy will be to use only $V_m(z)\,\partial_m$ but not $V_\mu(z)$ and ∂_μ. Clearly this means that we cannot use $E_A{}^M$ anymore; likewise we have no way of directly computing what $\delta(i\,\bar{\xi}\,Q)$ or D_α are.

The Bianchi identities are the integrability conditions which allow us to re-introduce the gauge dependent quantities $V_\mu(z)$ and ∂_μ if we wish to do so. We will use them, together with the Ricci identities (17), to learn as much as possible about $R_{\alpha\beta}$, $R_{\alpha b}$ and the action of D_α. When the spinorial derivative $D_\alpha\phi$ of a covariant superfield ϕ is left undetermined by Bianchi and Ricci identities, $D_\alpha\phi$ defines a new superfield; eventually, having introduced as many superfields as there are component fields, this procedure will terminate.

The covariant derivative

$$D_m\phi = \partial_m\phi - \rho(\omega_m)\phi = E_m{}^A D_A\phi = e_m{}^a D_a\phi + i\bar{\psi}_m{}^\alpha D_\alpha\phi \tag{22}$$

of a covariant field $\phi(z)$ and the curvature components

$$\begin{aligned} R_{mn} &= \partial_m V_n - \partial_n V_m - [V_m, V_n] + [E_m, E_n] = \\ &= E_n{}^B E_m{}^A R_{AB} = e_n{}^b e_m{}^a R_{ab} + 2i\, e_{[n}{}^b \bar{\psi}_{m]}{}^\alpha R_{\alpha b} + \bar{\psi}_m{}^\alpha R_{\alpha\beta}\psi_n{}^\beta \end{aligned} \tag{23}$$

can be computed in terms of $V_m(z)$ and ordinary derivatives $\partial_m = \partial/\partial x^m$. Once we know $D_\alpha\phi$, respectively $R_\alpha{}^\beta$, and $R_{\alpha b}$ we can use the inverse $e_a{}^m(z)$ of $e_m{}^a(z)$ (not a component of $E_A{}^M(z)$!) and solve for the supercovariant derivative $D_a\phi$ respectively supercovariant curvature R_{ab}.

Having already identified the covariant derivative $i\bar{\xi}^\alpha D_\alpha$ with the supersymmetry variation $(i\ \bar{\xi}^\alpha\ Q_\alpha)$ we can now replace all superfields $\phi(z)$ by the corresponding ordinary field

$$\phi(x) = \phi(z)\big|_{\theta=0} \tag{24}$$

and similarly replace $\lambda(z)$, $\xi(z)$ by $\lambda(x)$, $\xi(x)$. Since, on the other hand, every superfield $\phi(z)$ is uniquely determined by the corresponding ordinary field $\phi(x)$, there is no real need to distinguish them.

The process of reading off $R_{\alpha\beta}(x)$ and $R_{\alpha b}(x)$ from $(i\ \bar{\xi}\ Q)\ V_m(x)$, constructing the supercovariant curvature $R_{ab}(x)$ and declaring that all these fields become superfields is sometimes called gauge completion in the literature.

3. THE MINIMAL SET OF 32 + 32 FIELDS

The off-shell formulations of N = 2 supergravity given previously[1,3,4], which are all equivalent, suffer from the presence of a dimension 1/2 field, say χ, in $T_{\alpha\beta}{}^\gamma$. This is no problem in principle but leads to the majority of the non-linear terms in the transformation laws.

We present here another equivalent formulation in which χ is absent from all curvatures and torsions. This is achieved through suitable field redefinitions (including redefinition of gauge fields) and was possible only with the inclusion of SU(2) into the structure group. We will see that this does not necessarily imply propagating SU(2) gauge fields.

One of our constraints is $T_{ab}{}^c = 0$ which means that we use second order formalism. Choosing suitable constraints we find the following set of fields

Table 1.

dimension	fields		number of components		
			bosonic	fermionic	gauge (generator)
0	vierbein	$e_m^{\ a}$	6		$4 + 6(P_a + L_{ab})$
	Z-gauge field	A_m	3		1 (Z)
1/2	Q-gauge field	$\bar{\psi}_m$		24	$8\ (Q_\alpha)$
1	axial vector	j_a	4		$3(\vec{I})$
	antisymm. tensor	v_{ab}	6		
	isovector	$\vec{t}$	3		
	SU(2) gauge field	$\vec{B}_m$	9		
3/2	spinor	ρ		8	
2	scalar	I	1		
	together		32 +	32	

Table 2.

dimension	fields		number of components		
			bosonic	fermionic	gauge (generator)
1	gauge field	B_m^i	3		1 (I_i)
	scalar	M^i	1		
	pseudoscalar	N^i	1		
3/2	spinor	λ^i		8	
2	isovector	$\vec{D}^i$	3		
	together		8 +	8	per generator

Some of the torsions and curvatures are

$$T_{\alpha\beta}{}^{c} = -i(\gamma^{c})_{\alpha\beta}\,, \quad T_{\alpha b}{}^{c} = T_{ab}{}^{c} = T_{\alpha\beta}{}^{\gamma} = 0$$

$$F_{\alpha\beta} = i(1)_{\alpha\beta} \;\{\text{actually } (C\gamma_5\tau_2)_{\alpha\beta}\}\,, \quad F_{\alpha b} = 0$$

$$T_{\alpha b}{}^{\gamma} = [\tfrac{1}{2}\gamma_b \vec{\tau}\vec{t} - \tfrac{i}{4}\gamma_b\gamma^{cd}X_{cd} - \tfrac{i}{4}\gamma^{cd}\gamma_b v_{cd} - \tfrac{1}{2}\gamma_{bc}\gamma_5 j^{c}]_{\alpha}{}^{\gamma}$$

$$R_{\alpha\beta}{}^{ab} = [\tfrac{i}{2}\gamma^{c}\gamma^{ab}\gamma^{d}X_{cd} + \tfrac{i}{4}\{\gamma^{ab},\gamma^{cd}\}v_{cd} - \gamma^{ab}\vec{\tau}\vec{t} + \tfrac{1}{2}\{\gamma^{c},\gamma^{ab}\}\gamma_5 j_c]_{\alpha\beta}$$

$$\vec{G}_{\alpha\beta} = [2i\vec{t} - 2i\gamma^{a}\gamma_5\vec{\tau} j_a + \gamma^{ab}\vec{\tau} v_{ab}]_{\alpha\beta}$$

$$G_{\alpha\beta}^{i} = [iM^{i} + \gamma_5 N^{i}]_{\alpha\beta}$$

where $X_{ab} = T_{ab} + v_{ab}$ is a convenient combination.

Given the off-shell algebra we can ask for various types of multiplets, i.e., for linear realization of the algebra ($\{D_\alpha, D_\alpha\}$, $[D\ , D_b]$, $[D_a, D_b]$). We have already encountered the Yang-Mills multiplet

$$\mathcal{Y}^{i} = (B_m^{i}, M^{i}, N^{i}; \lambda^{i}; \vec{B}^{i})$$

with 8 + 8 components per generator; the realization of the algebra on this multiplet is linear up to terms with structure constants.

4. THE LINEAR MULTIPLET AND THE ACTION FORMULA

The linear multiplet (or superfield) with 8 + 8 components

$$\mathbb{K} = (\vec{K}\,;\, \varphi\,;\, S, P, V_a) \tag{25}$$

consists of an isovector $\vec{K}$, a spinor φ, a scalar s, a pseudoscalar P and a vector V_a and can be constructed from the defining relation

$$D_\alpha \vec{K} = -\tfrac{i}{2}(\vec{\tau}\varphi)_\alpha \tag{26}$$

It follows from the algebra that V_a is partially conserved

$$D_a V^a = \text{non-linear terms} \tag{27}$$

A linear multiplet may transform under an arbitrary matrix representation ρ_i of the Yang-Mills group

$$\delta(I_i)\mathbb{K} = \rho_i \mathbb{K} \tag{28}$$

(i.e., each of the 8 + 8 components transform under the same representation).

A linear multiplet may transform under the central charge; in this case we have a whole sequence of linear multiplets $\mathbb{K} = \mathbb{K}^{(0)}, \mathbb{K}^{(1)}, \mathbb{K}^{(2)}, \ldots$

$$\delta(Z)\mathbb{K}^{(n)} = \mathbb{K}^{(n+1)} \tag{29}$$

For rigid supersymmetry van Proeyen[7] has given such a sequence where only the first 16 + 16 components are independent. The importance of the linear multiplet comes from the fact, that in N = 2 rigid supersymmetry <u>every</u> action can be written as

$$I = \int d^4x\, S(x) = \tfrac{i}{12}\int d^4x\, (\bar{D}\vec{\tau}D\vec{K})\big|_{\theta=0} \tag{30}$$

with a suitable multiplet $\mathbb{K}$ with $\rho_i = 0$, but not necessarily $\mathbb{K}^{(1)} = 0$, the action

$$I(\mathbb{K}) = \int d^4x\, \det(e_m{}^a(x))\, \mathcal{L}(x) \tag{31}$$

$$\begin{aligned}\mathcal{L} = {} & S + \tfrac{i}{2}\bar{\psi}\cdot\gamma\varphi + \tfrac{1}{2}\bar{\psi}_a\gamma^{ab}\vec{\tau}\psi_b\vec{K} + \vec{t}\,\vec{K} \\ & + (V^a + \tfrac{i}{2}\bar{\psi}_b\gamma^{ab}\varphi - \tfrac{i}{2}\varepsilon^{abcd}\bar{\psi}_b\gamma_c\gamma_5\vec{\tau}\psi_d\vec{K})\, e_a{}^m A_m\end{aligned} \tag{32}$$

is invariant under all gauge transformations, in spite of the explicit appearance of the gauge field A_m. This somewhat unconventional form of the action is the price for using such a small multiplet as the linear one in the action formula. In most cases V_a is the divergence of an antisymmetric tensor we can integrate by parts.

We find that the gravitational fields

$$\vec{K} = \vec{t}, \quad \varphi = \rho, \quad S = I, \quad P = D_a j^a - \tfrac{1}{2}\varepsilon^{abcd} X_{ab} v_{cd}, \quad V_a = D^b (X_{ab} + v_{ab}) \tag{33}$$

form such a linear multiplet. If we apply the action formula and integrate by parts we obtain the supergravity Lagrangian

$$\mathcal{L}_{SG} = \frac{1}{\kappa^2}\Big[I + \vec{t}\vec{t} + \tfrac{i}{2}\bar{\psi}\cdot\gamma\rho + \tfrac{1}{2}\bar{\psi}_a\gamma^{ab}\vec{\tau}\psi_b\vec{t} + \tfrac{1}{2}(F_{ab} + i\bar{\psi}_a\psi_b)(F^{ab} + 2v^{ab} - \tfrac{1}{2}\varepsilon^{abcd}\bar{\psi}_c\gamma_5\psi_d)\Big] \tag{34}$$

As long as I is an independent field this action gives rise to the meaningless equation of motion det $e_m{}^a = 0$. It is this fact which forces us to reintroduce the field χ in order to express I (and ρ) in terms of other fields. The difference from previous formulations is the absence of χ from torsions and curvatures, i.e., from the right-hand side of the Ricci-identities (17) which govern the transformation laws of multiplets.

5. THE PRECURVATURE

We introduce the spinor χ of dimension 1/2 as precurvature. This means that supersymmetry variations of χ are curvature (and torsion) components but χ itself is not. In fact we just take the field χ of Ref. 1) and apply all field redefinitions to obtain

$$\begin{aligned} D\bar{\chi} = {} & \chi\bar{\chi} - \vec{\tau}\chi\bar{\chi}\vec{\tau} - \tfrac{1}{2}\vec{\tau}\vec{t} + \tfrac{1}{2}\gamma^a\gamma_5 j_a + \tfrac{i}{4}\gamma^{ab}v_{ab} \\ & + \tfrac{i}{2}(M - i\gamma_5 N) + \tfrac{i}{2}\gamma^a C_a \\ & + \tfrac{1}{4}(M - i\gamma_5 N)\vec{\tau} + \tfrac{1}{4}\gamma^a\vec{\tau}\vec{t}_a \end{aligned} \tag{35}$$

The supersymmetry variations of the new fields $M, N, C_a, \vec{M}, \vec{N}, \vec{t}$ involve further spinors $\lambda, \vec{\lambda}$ an isovector $\vec{D}$ and an isotensor $\tilde{\vec{D}}$ and we obtain the relations

$$\rho = \gamma \cdot R + 4\gamma^a D_a \chi + 2\lambda + i\vec{\tau}\vec{\lambda} + \ldots \tag{36}$$

$$\begin{aligned} I = & -\tfrac{1}{2}R + 2D_a C^a + \tilde{\vec{D}} - i\bar{\chi}(\rho + 2\lambda - i\vec{\tau}\vec{\lambda}) \\ & + 2(M^2 + N^2) - 2C_a C^a - \tfrac{1}{2}(\vec{M}^2 + \vec{N}^2) + \tfrac{1}{2}\vec{t}_a\vec{t}^a \\ & + j_a j^a - X_{ab} v^{ab} \end{aligned} \tag{37}$$

$$D_a \vec{t}^a = 3\vec{D} + \ldots \tag{38}$$

We see that χ replaces ρ and that the longitudinal parts of C_a and $\vec{t}_a$ are given in terms of other fields.

The new field components can be grouped into two Yang-Mills type multiplets $(C_a, M, N; \lambda; \vec{D})$ with 8 + 8 components and $(\vec{t}_a, \vec{M}, \vec{N}; \vec{\lambda}; \vec{D})$ with 3 ×(8 + 8) components.

If we substitute I and ρ in the supergravity Lagrangian, we find the Fayet-Iliopoulos term

$$\frac{1}{\kappa^2}\left[\tilde{\vec{D}} - \tfrac{1}{2}\bar{\psi}\cdot\gamma\vec{\tau}\vec{\lambda}\right] \tag{39}$$

as part of the supergravity Lagrangian, a situation known from non-extended supergravity[8].

We can consistently set

$$\vec{M} = \vec{N} = \vec{\lambda} = \tilde{\vec{D}} = 0 \tag{40}$$

and find as a consequence that $\vec{B}_m$ is SU(2) gauge equivalent to

$$-e_m{}^a \vec{t}_a - 2\bar{\psi}_m\vec{\tau}\chi \tag{41}$$

We will do this for the moment and are left with 8 + 8 components in addition to the minimal 32 + 32 ones.

6. PURE GRAVITY

With these 40 + 40 fields we get the usual N = 2 supergravity action. All field equations are contained in the superfield equation

$$\chi = 0 \tag{42}$$

The higher components are

$$M = N = C_a = \vec{t}_a = j_a = \nu_{ab} = \vec{t} = 0 \tag{43}$$

$$R^a \equiv \frac{i}{2} \varepsilon^{abcd} \gamma_b \gamma_5 R_{cd} = 0 \qquad (R_{cd}{}^{\alpha} = T_{cd}{}^{\alpha}) \tag{44}$$

$$G_{ab} + T_{ab}(\psi, A) = 0 \tag{45}$$

$$D^b F_{ab} = 0. \tag{46}$$

The lowest dimensional covariant field which does not vanish due to field equations is X_{ab}. On-shell $X_{ab} = F_{ab}$ but X_{ab} turns out to be more convenient off-shell. The superfield X_{ab} contains the Weyl spinor $W_{\alpha\beta\gamma}$, $\bar{W}_{\dot{\alpha}\dot{\beta}\dot{\gamma}}$ and the Weyl tensor $C_{\alpha\beta\gamma\delta}$, $C_{\dot{\gamma}\dot{\beta}\dot{\alpha}\dot{\delta}}$.

7. THE MATTER MULTIPLET

The 8 + 8 components of the matter multiplet[9]

$$\mathbb{A} = (A; \psi; F) \tag{47}$$

are two SU(2) doublets of scalars A_i and F_i and an SU(2) singlet Dirac spinor ψ. The relation

$$D_\alpha^i A_j = \delta_j^i \psi \tag{48}$$

defines the whole multiplet (superfield). The matter multiplet transforms under central charge

$$
\begin{aligned}
&\delta(z) A = F \\
&\delta(z)\psi = -\gamma^a D_a \psi \qquad + \text{non-linear terms} \qquad (49) \\
&\delta(z) F = D_a D^a A \qquad + \text{non-linear terms}
\end{aligned}
$$

and since central charge and supersymmetry transformations commute we find

$$D_\alpha^i F_j = \delta_i^j \, \delta(z)\psi \tag{50}$$

We can therefore construct a new multiplet with lowest component F

$$T(A) = \delta(z)\mathbb{A} = (F; \ldots) \tag{51}$$

As in rigid supersymmetry we can construct a linear multiplet. Due to Eq. (48)

$$D(A^+\vec{\tau}A) = \vec{\tau}(\cdots) \tag{52}$$

and there are two linear multiplets

$$\mathbb{K}_m = (A^+\vec{\tau}A; \ldots) \tag{53}$$

$$\mathbb{K}_K = (i(F^+\vec{\tau}A - A^+\vec{\tau}F); \cdots) \tag{54}$$

which give rise to the mass term and kinetic term of the action

$$I(\mathbb{K}_K + m\,\mathbb{K}_m) \tag{55}$$

The field equations are

$$\delta(z)\mathbb{A} + i m \mathbb{A} = 0 \tag{56}$$

8. THE CHIRAL MULTIPLET AND THE ACTION FORMULA

A chiral multiplet (or superfield)[10]

$$\mathbb{C} = (M, N; \lambda; T_{ab}, \vec{F}, \vec{G}; \mu; X, Y) \tag{57}$$

with 16 + 16 components is determined (up to field redefinitions) through the relation

$$DM = i\gamma_5 DN \qquad (= -\tfrac{1}{2}\lambda) \tag{58}$$

Clearly, given two chiral multiplets $\mathbb{C}_1$, $\mathbb{C}_2$ we can form a product multiplet

$$\mathbb{C}_1 \mathbb{C}_2 = (M_1 M_2 - N_1 N_2, M_1 N_2 + N_1 M_2; \cdots) \tag{59}$$

and there are scalar and pseudoscalar unit multiplets

$$\begin{aligned} \mathbb{1}_S &= (1, 0; 0; 0, 0, 0; 0; 0, 0) \\ \mathbb{1}_P &= (0, 1; 0; 0, 0, 0; 0; 0, 0) \end{aligned} \tag{60}$$

with properties

$$\mathbb{1}_S \mathbb{C} = \mathbb{C} \quad ; \quad \mathbb{1}_P \mathbb{1}_P = -\mathbb{1}_S . \tag{61}$$

Moreover we can form a kinetic multiplet

$$\begin{aligned} \mathbb{T}(\mathbb{C}) = (&X + \tfrac{8}{3}\vec{t}\vec{F} + \tfrac{4}{3}\omega^{ab}T_{ab} , \\ &-Y - \tfrac{8}{3}\vec{t}\vec{G} - \tfrac{2}{3}\varepsilon^{abcd}\omega_{ab}T_{cd} ; \cdots) \end{aligned} \tag{62}$$

A chiral multiplet $\mathbb{C}$ is reducible and contains two linear ones

$$\begin{aligned} \mathbb{K}_S(\mathbb{C}) &= (\vec{F} + \vec{t}M ; \ldots) \\ \mathbb{K}_P(\mathbb{C}) &= \mathbb{K}_S(-\mathbb{1}_P \cdot \mathbb{C}) = (\vec{G} + \vec{t}N ; \cdots) \end{aligned} \tag{63}$$

We can therefore apply the action formula for the linear multiplet and obtain one for the chiral one

$$I(\mathbb{C}) = I(\mathbb{K}_s(\mathbb{C})) \tag{64}$$

After integration by parts

$$\begin{aligned} I(\mathbb{C}) = \int d^4x\, e \Big\{ & \tfrac{1}{2}X - \tfrac{2}{3}v^{ab}T_{ab} + \tfrac{8}{3}\vec{t}\vec{F} + \tfrac{1}{2}\bar{\rho}\lambda \\ & -[I + \vec{t}\vec{t} + \tfrac{1}{2}X^{ab}X_{ab} - \tfrac{1}{2}v^{ab}v_{ab}] M \\ & -[D_a j^a + \tfrac{1}{4}\varepsilon^{abcd}(X_{ab} - v_{ab})(X_{cd} - v_{cd})] N \\ & + \text{ the terms with } \bar{\Psi}_m \Big\} \end{aligned} \tag{65}$$

We see that the supergravity action is just

$$I_{SG} = \frac{1}{\kappa^2} I(\mathbb{1}_s). \tag{66}$$

9. THE YANG-MILLS ACTION

A chiral multiplet may transform under the Yang-Mills group and in fact we can form a chiral multiplet transforming under the adjoint representation

$$\mathbb{C}^i = (M^i, N^i; \lambda^i; \ldots) \tag{67}$$

from the components of the Yang-Mills multiplet Y^i. We find that

$$\mathbb{K}_P(\mathbb{C}^i) = 0 \tag{68}$$

which expresses $\vec{G}, \mu$, X and Y in terms of the other components and gives the Bianchi identity for G^i_{ab}.

Using the action formula (64,65) we can construct the Yang-Mills action

$$I_{YM} = \frac{1}{e^2} I(-\tfrac{1}{2}\mathbb{C}^i\mathbb{C}^i) \tag{69}$$

The Yang-Mills field equations are

$$\mathbb{K}_s(\mathbb{C}^i) = 0 \tag{70}$$

where the linear multiplet $\mathbb{K}_s(\mathbb{C}^i)$ has the components

$$\begin{aligned} &\vec{K} = \vec{D}^i, \quad \varphi = -\gamma^a D_a \lambda^i + \ldots \\ &S = D_a D^a M^i + \ldots, \quad P = D_a D^a N^i + \ldots \\ &V^a = D_b G^{abi} + \ldots \end{aligned} \tag{71}$$

10. PROPAGATING SU(2) GAUGE FIELDS

In sections 6-9 we had put the fields $\vec{M}, \vec{N}, \vec{\lambda}, \vec{\tilde{D}}$ equal to zero. We can, instead construct an action for these fields. At the same time $\vec{B}_m$ will become independent of $\vec{t}_a$ and start propagating.

$$\vec{\mathbb{C}} = (\vec{M}, \vec{N}; \vec{\lambda}; \ldots) \tag{72}$$

transform almost as a chiral multiplet such that

$$\vec{\mathbb{C}}\,\vec{\mathbb{C}} = (\vec{M}\vec{M} - \vec{N}\vec{N}, 2\vec{M}\vec{N}; \ldots) \tag{73}$$

is in fact a chiral multiplet (or superfield). The SU(2) action is

$$I_{SU(2)} = \frac{1}{g^2} I(-\tfrac{1}{2}\vec{\mathbb{C}}\,\vec{\mathbb{C}}) \tag{74}$$

and contains the terms

$$\frac{1}{g^2}\left(\tfrac{1}{2}\vec{\tilde{D}}\,\vec{\tilde{D}} + \tfrac{1}{2} D_a \vec{M} D^a \vec{M} + \tfrac{1}{2} D_a \vec{N} \vec{D}^a N\right) \tag{75}$$

Together with the Fayet-Iliopoulos term (30) this gives rise to a vacuum expectation value for $\vec{\tilde{D}}$ and a cosmological constant of order $g^2/\varkappa^4$. Contrary to the corresponding situation for N = 1 [8] $\vec{M}$ (or $\vec{N}$) picks up a vacuum expectation value as well. Therefore supersymmetry and SU(2) are broken and both gravitinos obtain (different) masses via the super-Higgs effect.

11. CONSTRUCTION OF HIGHER INVARIANTS

We can now combine various elements of tensor calculus and construct higher invariants in close analogy of the N = 1 case[11]. We can construct a chiral multiplet with external Lorentz indices from X_{ab}

$$\mathbb{C}_{ab} = (X_{ab}, \tfrac{1}{2}\varepsilon_{ab}{}^{cd}X_{cd}; \ldots) \tag{76}$$

and observe that its square

$$\mathbb{C}_W = \mathbb{C}_{ab}\,\mathbb{C}^{ab} \tag{77}$$

contains the square of the Weyl tensor in its highest components. Consequently the invariant

$$I(\mathbb{C}_W(T(\mathbb{C}_W))^n) \tag{78}$$

contains the 2(n + 1)'st power of the Weyl tensor and is a candidate for a 2n + 1 loop counterterm. It is known that the one loop counterterm vanishes on-shell due to a Gauss-Bonnet type identity. Instead we can construct two topological invariants

$$I(\mathbb{C}_W) \qquad \text{and} \qquad I(\mathbb{1}_P\,\mathbb{C}_W) \tag{79}$$

corresponding to the Euler and Pontrjagin number.

REFERENCES

1. P. Breitenlohner and M.F. Sohnius, Nucl. Phys. B165 (1980) 483;
P. Breitenlohner, in "Supergravity", Proc. of the Supergravity workshop at Stony Brook, Sept. 1979, ed. by P. van Nieuwenhuizen and D.Z. Freedman (North Holland, Amsterdam, 1979) which also contains a fairly complete compilation of supergravity references;
P. Breitenlohner, "Superfield Formulation, Auxiliary Fields and Tensor Calculus for supergravity", presented at the 17th Winter School of Theoretical Physics, Feb. 22-March 6, 1980, Karpacz, Poland.
2. P. Breitenlohner and M.F. Sohnius, MP5 preprint, to appear (1980).
3. E.S. Fradkin and M.A. Vasiliev, Lett. Nuovo Cimento 25 (1979) 79; Phys. Letters 85B (1979) 47.
4. B. De Wit and J.W. van Holten, Nucl. Phys. B155 (1979) 530;
B. De Wit, in "Supergravity", Ref. 1);
B. De Wit, J.W. van Holten and A. van Proeyen, Leuven preprint KUL-TF-79/034 (1979);
M. De Roo, J.W. van Holten, B. De Wit and A. van Proeyen, "Chiral Superfields in N = 2 Supergravity", Leiden preprint (1980).
5. J. Wess and B. Zumino, Phys. Letters 66B (1977) 361;
J. Wess, "Supersymmetry-Supergravity", Lecture Notes in Physics 77 (1978) 81.
6. A. Salam and J. Strathdee, Nucl. Phys. B80 (1974) 499.
7. A. van Proeyen, "N = 2 Supergravity Multiplets", presented at the 17th Winter School of Theoretical Physics, Feb. 22-March 6 1980, Karpacz, Poland.
8. D.Z. Freedman, Phys. Rev. D15 (1977) 1173;
P. van Nieuwenhuizen and B. De Wit, Nucl. Phys. B139 (1978) 216.
9. P. Fayet, Nucl. Phys. B113 (1976) 135;
C. Zachos, Phys. Letters 76B (1978) 329.
10. P. Dondi and M.F. Sohnius, Nucl. Phys. B81 (1974) 317.
11. P. van Nieuwenhuizen and P.K. Townsend, Phys. Rev. D19 (1979) 3592.

QUANTIZATION OF SUPERGRAVITY WITH A COSMOLOGICAL CONSTANT

M.J. Duff

Physics Department
Imperial College
London. U.K.

In O(N) extended supergravity theories, the usual arguments for one-loop finiteness (modulo topological terms) cease to apply when the internal symmetry is gauged because of the appearance of a cosmological constant, Λ, and spin-3/2 mass parameter, m, related to the gauge coupling e. For $N \leq 4$, we find that infinite renormalizations are required. Remarkably, the particle content of theories with $N > 4$ results in a cancellation of these infinities implying, in particular, a vanishing one-loop $\beta(e)$ function. The topological significance of these results is also briefly discussed.

INTRODUCTION

I would like to describe some recent work on the quantization of gravity and supergravity in the presence of a cosmological constant. The pure gravity results were obtained in collaboration with Christensen[1,2]; those on supergravity with Christensen, Gibbons and Rocek[3].

Now when Einstein first introduced a cosmological constant into gravitational field equations, he was guided by various prejudices about the static nature of the universe which he later abandoned in the face of growing evidence for an expanding universe. Since Einstein's time the cosmological constant has had its ups and downs (both empirically and metaphorically), and current estimates put its physical value as very small and possibly zero ($< 10^{-57}$ cm^{-2}).

As far as quantum gravity is concerned, almost all discussions to date have focused their attention on the case of vanishing cosmological constant $\Lambda = 0$. This is not without reason. Whatever one's attitude to the value of Λ at the classical level, attempts to build a consistent quantum theory for non-vanishing Λ present new difficulties over and above the already formidable problems present when $\Lambda = 0$. Unfortunately both grand unified and super unified theories of Nature each predict, for one reason or another, an enormous cosmological constant, and in the opinion of the author, there is to date no satisfactory means by which the cosmological constant may be "argued away". The one arising from spontaneous symmetry breakdown may be cancelled by an ad hoc addition to the Lagrangian, but this will only result in a non-vanishing Λ in the early stages of the universe when the symmetry is restored through high-temperature effects. In general, moreover, a vanishing Λ at the tree level will not prevent its reappearance through closed loop effects. A review of the status of the cosmological constant and the problems involved may be found in Ref.(1). In the present context therefore, the words of Zeldovitch and Novikov seem particularly appropriate: "After a genie is let out of a bottle (i.e. now that the possibility is admitted that $\Lambda \neq 0$) legend has it that the genie can be chased back in only with the greatest difficulty". For the moment, then, let us not attempt to chase the genie back into the bottle: let us admit a non-vanishing Λ and see whether we can cope with it.

There are many questions one might ask of a quantum field theory with a cosmological constant, and I shall not attempt to answer all of them, but a very important one concerns ultraviolet divergences and renormalizability. Here the analysis turns out to be reasonably straightforward and in Ref (1) explicit results for one-loop counterterms and anomalous scaling behaviour were given both for pure gravity and for gravity plus matter fields of spin 0, $\frac{1}{2}$ and 1 in the presence of a cosmological constant. Related work may be found in Ref. (2). Summarized below are the calculations of a forthcoming publication[(3)] in which these techniques are generalized to spin-3/2 fields and hence to supergravity,[(4)] with dramatic results for the O(N) extended models.[(5)]

These developments may appear as something of a luxury: if ordinary quantum gravity is non-renormalizable without a cosmological constant, it is not likely to become so with the additional complication of a non-vanishing Λ. However, such arguments require drastic revision in extended supergravity where the gauging of the O(N) symmetry[(6)] requires a (huge) cosmological constant $\Lambda = -6e^2/\kappa^2$ and gravitino mass parameter m, with $\Lambda = -3\,m^2$. (e is the gauge coupling constant and $\kappa^2 = 8\pi \times$ Newton's constant). Thus it is plausible that the ultra-violet behaviour of these models at higher loops, by virtue of their extra local symmetry, may even be an improvement over theories without a cosmological constant. Moreover, the strong empirical evidence in favour of a vanishing cosmological constant may be only an apparent discrepancy between theory and experiment if, as suggested in Hawking's "space-time foam,"[7] one reinterprets Λ as a measure of the average small scale curvature of space-time.

ONE-LOOP COUNTERTERMS

Let us first recall the pure gravity results of Refs. (1) and (2). If, at the classical level, we take the Einstein action

$$S = -\frac{1}{2\kappa^2}\int d^4x\sqrt{g}\ (R - 2\Lambda) \tag{1}$$

then, using the background field method,[8] the one-loop counter-terms will be a linear combination of $R_{\mu\nu\rho\sigma}R^{\mu\nu\rho\sigma}, R_{\mu\nu}R^{\mu\nu}, R^2, \Lambda R$ and Λ^2 but with gauge dependent coefficients. Gauge invariance is achieved by use of the field equations $R_{\mu\nu} = \Lambda g_{\mu\nu}$. Alternatively, terms which vanish with the field equations may be removed by gauge-dependent field redefinition.[9] Either way, the resulting counter-term ΔS may then be written

$$\Delta S = -\frac{1}{\varepsilon}\gamma \tag{2}$$

where $\varepsilon = n - 4$ is the dimensional regularization parameter. γ is given by

$$\gamma = A\chi + B\delta \tag{3}$$

where A and B are numerical coefficients and where

$$\chi \equiv \frac{1}{32\pi^2}\int d^4x\sqrt{g}\ {}^*R_{\mu\nu\rho\sigma}{}^*R^{\mu\nu\rho\sigma} \tag{4}$$

$$\delta \equiv \frac{1}{12\pi^2}\int d^4x\sqrt{g}\ \Lambda^2 = -\frac{\kappa^2\Lambda}{12\pi^2}\ S \tag{5}$$

The star denotes the duality operation, and $\sqrt{g}\ {}^*R_{\mu\nu\rho\sigma}{}^*R^{\mu\nu\rho\sigma} = \sqrt{g}\ (R_{\mu\nu\rho\sigma}R^{\mu\nu\rho\sigma} - 4R_{\mu\nu}R^{\mu\nu} + R^2)$ is a total divergence which is sometimes discarded. Its integral over-all space, however, yields the Euler number χ, a topological invariant which takes on integer values in spaces with non-trivial topology.[10] The explicit calculations of Ref. (1) yield A = 106/45 and B = -87/10. Thus, in contrast to the case $\Lambda = 0$,[11] pure gravity with a Λ term is no longer one-loop "finite" (in the non-topological sense) because $B \neq 0$.

One may now repeat the exercise for simple supergravity with a gravitino mass term:[12]

$$S = \int d^4x (\det e^a_\mu) \left[-\frac{1}{2\kappa^2} R + \frac{1}{2} \varepsilon^{\mu\nu\rho\sigma} \psi_\mu \gamma_5 \gamma_\nu D_\rho \psi_\sigma + m\psi_\mu \sigma^{\mu\nu} \psi_\nu + \frac{1}{3}(s^2 + p^2 - A^\mu A_\mu) + \frac{2m}{\kappa} s \right] . \quad (6)$$

Elimination of the auxiliary field s yields a cosmological constant $\Lambda = -3\,m^2$. The one-loop counterterms will now be given by the appropriate supersymmetric completion[13] of those encountered in pure gravity, i.e., ΔS is again given by $-\varepsilon^{-1}(A\chi + B\delta)$ with $\delta = -\kappa^2\Lambda(12\pi^2)^{-1} S$ (on-shell) but where S is now given by Eq. (6). The topological invariant χ, on the other hand, acquires no extra terms.[14] The coefficients A and B will now receive contributions both from the graviton and the gravitino (with its appropriate mass parameter). Explicit calculations in Ref. (3) yield $A = 41/24$ and $B = -77/12$ and, once again in contrast to the case $\Lambda = 0 = m$,[15] simple supergravity is no longer one-loop finite.

EXTENDED SUPERGRAVITY

Having obtained the contribution to A and B for spins 2 and 3/2, it is now tempting to combine these results with those of Ref. (1) for spins 1, ½, and 0, and apply them to the extended O(N) theories with gauged internal symmetry, especially because the renormalization of Λ takes on a new significance: by supersymmetry the coefficient B also determines the renormalization of the gauge coupling constant e.

The supersymmetric completion of δ now contains the spin-1 gauge field contribution $e^2 \, \mathrm{Tr}\, F_{\mu\nu}F^{\mu\nu}$. Note that this arises from two different sources : in addition to the usual charge renormalization effects, there will also be one-loop counterterms of the form $\kappa^2 R \, \mathrm{Tr}\, F_{\mu\nu}F^{\mu\nu}$. On using the field equations $R = 4\Lambda + \ldots$ with $\kappa^2\Lambda = -6e^2$ this is converted into an extra $e^2 \, \mathrm{Tr}\, F_{\mu\nu}F^{\mu\nu}$ term.[16]

Before displaying our results, some qualifications are required. Although the construction of consistent O(N) supergravity Lagrangians has been successfully achieved for all N up to N = 8,[5] the corresponding Lagrangians with gauged internal symmetry have, to date been written down explicitly only for N = 2,3[6] and N = 4[17,18]. It is thus an assumption on our part that such

Lagrangians exist for N = 5, 6, 7 and 8. As far as we are aware, there are no theoretical reasons preventing such a construction since the appropriate supersymmetry algebras for $N > 4$ are perfectly respectable.[19] (We refrain from going beyond N = 8 for the usual reason of requiring no spin higher than 2). The crucial observation, however, is that by restricting our attention to the gravitational part of the on-shell counterterms at the one-loop level, the details of the interaction terms in such Lagrangians are not relevant: all that is required to determine the coefficients A and B is the pure spin-2 Lagrangian itself together with that part of the remaining Lagrangian quadratic in the lower-spin fields. Having calculated the gravitational contribution to ΔS on shell, the remainder is determined by the sypersymmetry which guarantees that (with $\kappa^2\Lambda = -6e^2$)

$$\Delta S = \frac{-1}{\varepsilon}\left[A\chi + B\frac{e^2}{2\pi^2}S\right] \tag{7}$$

where S is the classical action. The signal for asymptotic freedom is $B > 0$.[9]

Only the kinetic terms are needed to fix the contributions to A from fields of different spin. These have been calculated before.[20] To calculate B we also require knowledge of the mass terms. All particles must be massless for all N if, as we are assuming, supersymmetry is not spontaneously broken. For $N > 4$ there is an "apparent mass" parameter m for the gravitinos given by $\Lambda = -3\ m^2$ which we assume to remain the same for $N > 4$. Similarly we assign no such parameters to the spin-1 and spin-$\frac{1}{2}$ fields for $N > 4$ since they are absent for $N \leq 4$. The scalar fields, which first make their appearance at N = 4 require greater care. The spin-2, spin-0 coupling in the N = 4 model is known to be of the form [17,18].

$$L = -\frac{1}{2\kappa^2}\sqrt{g}\,(R - 2\Lambda) + \frac{1}{2}\sqrt{g}\,\phi^i\left[-\Box + \frac{2\Lambda}{3}\right]\phi^i + O(\phi^3) \tag{8}$$

i.e., minimal coupling with a mass term. However, one could equally well use

$$L = -\frac{1}{2\kappa^2}\sqrt{g}\,(R - 2\Lambda) + \frac{1}{2}\sqrt{g}\,\phi^i\left[-\Box + \frac{R}{6}\right]\phi^i + O(\phi^3) \tag{9}$$

i.e., conformal coupling with no mass term.[21] The equivalence is seen by making a Weyl rescaling in the Lagrangian (9) of the form

$$g_{\mu\nu} \to \Omega^2 g_{\mu\nu}, \quad \phi^i \to \Omega^{-1}\phi^i ; \quad \Omega^2 = 1 + \frac{\kappa^2}{6}\phi^i\phi^i \tag{10}$$

which yields the Lagrangian (8). Both versions yield the same B coefficient on mass shell since the field equations imply $R = 4\Lambda + \ldots$. We therefore adopt a conformal coupling with no mass term for all $N \geqslant 4$. With the above assumptions, the calculations of Ref. (3) give the contributions to A and B shown in Table I.

Table I

S	360A	60B
0	4	-1
1/2	7	-3
1	-52	-12
3/2	-233	137
2	848	-522

The combined results for O(N) supergravity then follow from the well-known particle content shown in Table II.

Table II

N \ S	2	3/2	1	1/2	0	A	B
1	1	1				41/24	-77/12
2	1	2	1			11/12	-13/3
3	1	3	3	1		0	-5/2
4	1	4	6	4	2	-1	-1
5	1	5	10	11	10	-2	0
6	1	6	16	26	30	-3	0
7	1	8	28	56	70	-5	0
8	1	8	28	56	70	-5	0

The most remarkable feature is clearly the vanishing of the B coefficient for all $N > 4$, though the integral value of A for all

$N > 2$ is not without interest. We now discuss the implications of these results.

RENORMALIZABILITY

Apart from the topological χ counterterm, the $N > 4$ theories are seen to remain one-loop finite on shell even when the internal symmetry is gauged and $\Lambda \neq 0$.[22] In particular, the one-loop contribution to the renormalization group $\beta(e)$ function vanishes! This is reminiscent of the $N = 4$ Yang-Mills multiplet in flat space, whose β function is known to vanish to two-loop order.[23] The vanishing β function in $N > 4$ gauged supergravity is no less mysterious than in $N = 4$ Yang-Mills and, at the time of writing, is understood only as a "miraculous" cancellation of numerical coefficients. There have been earlier speculations[24] that $N > 4$ theories might show improved ultra-violet behaviour, but we do not know their connection, if any, with the concrete calculations presented here. These cancellations can hardly be accidental however, and provide something of an a posteriori justification for our previous assumptions on $N > 4$ theories.

For $N \leq 4$, we do not have one-loop finiteness but rather one-loop renormalizability. Moreover, the negative value of B indicates that these theories are not asymptotically free (inasmuch as asymptotic freedom is meaningful for theories which may not be renormalizable at higher loops). One will find two-loop renormalizability for all N when $\Lambda \neq 0$ for the same reason one finds two-loop finiteness when $\Lambda = 0$,[25] and it would be interesting to know the β-function. Three loops and beyond is still a mystery.[26]

TOPOLOGY

Another remarkable feature peculiar to $N > 4$ models is that $\gamma = A\chi + B\delta$ = integer (since A is an integer, B is zero and χ takes on integer values). If previous experience is any guide, this may be indicative of a new "Super Index Theorem"[27] for $N > 4$. Let us recall the significance of γ[28,27]. At the one-loop level γ counts the total number of eigenmodes (boson minus fermion) of the differential operators whose determinants govern the one-loop functional integral. (It is closely related to the anomalous trace of the energy-momentum tensor.) The number of zero-eigenvalue modes will be finite and given by an integer; the number of non-zero modes is formally infinite. After regularization (e.g. by the zeta-function method), this number is rendered finite but not necessarily an integer. In certain circumstances, however, there may be a mutual cancellation of the non-zero modes between the bosons and fermions, in which case γ = integer. Such a cancellation does indeed take place in $\Lambda = 0$ supergravity[28] when the space is self-dual, i.e., $R_{\mu\nu\rho\sigma} = \pm {}^*R_{\mu\nu\rho\sigma}$

(which implies $R_{\mu\nu} = 0$). If, in addition, the space is compact with spin structure (i.e., fermions can be globally defined) then χ = integer × 24. Consistent with this is the result in Table II that A = integer/24 for all N. We do not know whether any similar mechanism can take place when $\Lambda \neq 0$ and χ is not so restricted, but our results indicate that $N > 4$ models are the most likely candidates. One might also ask whether there is anything special, from a topological point of view, about gauged supergravity when $N > 4$. There is one remark, brought to our attention by C.J. Isham: On a four-dimensional manifold with non-trivial topology, SO(N) gauge theories are "topologically stable"[(29)] only for $N > 4$, in the sense that there is a natural one-to-one correspondence between the SO(N) bundles (or, equivalently, between the topological sectors of the gauge theory). Again, we do not yet know whether this is mere coincidence.

Finally, we note that the signs of A and B in simple and extended supergravity reinforce the conclusions concerning "space-time foam" reached in the context of pure gravity.[(1)] If these one-loop results are taken seriously, the sign of γ would seem to imply that spacetime becomes "foamier and foamier" the shorter the length scale, in contrast to the picture of "one unit of topology per Planck volume" expected if γ were positive definite.[(7)]

NOTE ADDED

It has recently been discovered[30] that different field representations for particles of given spin and number of degrees of freedom, although naively equivalent[31], lead to different quantum effects in spaces with non-trivial topology. Thus the gauge theory of an antisymmetric rank-2 tensor $\phi_{\mu\nu}$ (with 1 degree of freedom) differs from that of a scalar ϕ; and that of an antisymmetric rank-3 tensor $\phi_{\mu\nu\rho}$ (with 0 degrees of freedom) differs from nothing. Since the effects are purely topological they show up only in the A coefficient of the Euler number which appears in the counterterms and trace anomalies, but not in the B coefficient. In fact (with minimally coupled ϕ)

$$A\left[\phi_{\mu\nu}\right] = A\left[\phi\right] + 1$$

$$A\left[\phi_{\mu\nu\rho}\right] = -2$$

The possible significance of these results for supergravity are presently being investigated, especially since (a) such representations seem to appear dynamically in one version[32] of the auxiliary fields (b) they also occur naturally in "Super Index Theorems"[27] and (c) the rank-3 tensors provide a gauge principle interpretation of the cosmological constant[30].

REFERENCES.

1. S.M. Christensen and M.J. Duff, "Quantizing Gravity with a Cosmological Constant," University of California at Santa Barbara Report ITP-79-01 (October 1979).
2. G.W. Gibbons and M.J. Perry, Nucl. Phys. B146, 90 (1978). This paper contains numerical errors corrected in Ref. (1).
3. S.M. Christensen, M.J. Duff, G.W. Gibbons and M. Rocek "One-loop effects in supergravity with a cosmological constant" (in preparation).
4. D.Z. Freedman, P. van Nieuwenhuizen, and S. Ferrara, Phys. Rev. D13, 3214 (1976); S. Deser and B. Zumino, Phys. Lett. 62B, 335 (1976).
5. See, for example, J. Scherk in Recent Developments in Gravitation: 1978 Cargese Summer School Lectures (Plenum Press, New York, 1979).
6. D.Z. Freedman and A. Das, Nucl. Phys. B120, 221 (1977); E.S. Fradkin and M. Vasiliev, Lebedev Institute Preprint N 197 (1976).
7. S.W. Hawking, Nucl. Phys. B144, 349 (1978). See also P.K. Townsend, Phys. Rev. D15, 2795 (1977).
8. B.S. DeWitt in Dynamical Theory of Groups and Fields (Gordon and Breach, New York, 1965).
9. G. 't Hooft, Nucl. Phys. B62, 444 (1973).
10. This will prove particularly important when one comes to the euclidean spacetime foam interpretation. See Ref. (7). The quantity $\delta = \Lambda^2 V/12\pi^2$ (in compact spaces with volume V) equals $c_1^2/6$ in Einstein-Kahler manifolds where c_1 is the first Chern number but is not in general a topological invariant.
11. G. 't Hooft and M. Veltman, Ann. Inst. Henri Poincare 20, 69 (1974).
12. P. van Nieuwenhuizen in Recent Developments in Gravitation: 1978 Cargese Summer School Lectures (Plenum Press, New York, 1979). Previous discussions of supergravity with a cosmological constant may be found in S.W. MacDowell and F. Mansouri, Phys. Rev. Lett. 38, 739 (1977); S. Deser and B. Zumino, Phys. Rev. Lett. 38, 1433(1977); P.K. Townsend, Phys. Rev. D15 2802 (1977).
13. We do not anticipate any problems in this respect of the kind discussed by B.de Wit and M.T. Grisaru, Phys. Rev. D20, 2082 (1979), since our calculations are insensitive to the elimination of the auxiliary fields.

14. P.K. Townsend and P. van Nieuwenhuizen, Phys. Rev. D19, 3592 (1979).
15. M. Grisaru, P. van Nieuwenhuizen and J.A.M. Vermaseren, Phys. Rev. Lett. 37, 1662 (1976).
16. Note that even in non-supersymmetric theories the β function will change when one allows for gravity and a non-vanishing Λ. Thus (modulo renormalizability problems) a theory which is asymptotically free in flat space might be converted into one which is not and vice-versa.
17. A. Das, M. Fischler and M. Rocek, Phys. Rev. D16, 3427 (1977).
18. D.Z. Freedman and J.H. Schwarz, Nucl. Phys. B137, 333, (1978).
19. W. Nahm, Nucl. Phys. B135, 149 (1978).
20. See S.M. Christensen and M.J. Duff, Phys. Lett. 76B, 571 (1978) and references therein.
21. In neither version does the scalar potential contain terms linear in ϕ. In this respect, and also in that it truncates consistently to the O(3) and O(2) models, the O(4) model differs from the alternative chiral SU(2) x SU(2) gauged supergravity of Ref, (18), which will not be discussed here. Both theories apparently suffer from a scalar potential $V(\phi)$ which is not bounded below. Since $V(\phi)$ is intimately connected with Λ, however, it may be that the criterion for stability is different when $\Lambda \neq 0$.
22. Note that, a priori, our results have nothing to do with those of E. Cremmer, J. Scherk and J. Schwarz, Phys. Lett. 84B, 83 (1979) who found that the Λ induced by one-loop quantum corrections was finite in the spontaneously broken version of ungauged extended supergravity which has zero Λ at the classical level.
23. D.R.T. Jones, Phys. Lett. 72B, 199 (1977); H. Pendleton and E. Poggio, Phys. Lett. 72B, 200 (1977). See also M. Grisaru, W. Siegel and M. Rocek, Nucl. Phys. B159 (1979) 429.
24. B. deWit and S. Ferrara, Phys. Lett. 81B, 317 (1979).
25. M. Grisaru, Phys. Lett. 66B, 75 (1977).
26. S. Deser, J. Kay and K. Stelle, Phys. Rev. Lett. 38, 527 (1977)
27. S.M. Christensen and M.J. Duff, Nucl. Phys. B154, 301 (1979).
28. S.W. Hawking and C.N. Pope, Nucl. Phys. B146, 381 (1978).
29. This means that for $N \geq 5$ the gauge symmetry may be spontaneously broken from SO(N+1) down to SO(N) in a unique way without meeting any topological obstructions. For $N \leq 4$, it may, depending on the space-time topology, break into inequivalent topological sectors, and/or the spontaneous symmetry breakdown may be completely inhibited. (C.J. Isham, private communication).
30. M.J. Duff and P.van Nieuwenhuizen (to appear).
31. P. van Nieuwenhuizen, these proceedings, p.245.
32. K.S. Stelle and P.C. West, Phys. Lett. B74, (1978) 330.

A SHORT APPRECIATION OF JOEL SCHERK AND HIS WORK

Joel Scherk died on 16 May, 1980 at the age of 33. His friends and collaborators miss him deeply. In spite of this relatively short life his impact on theoretical physics was great: he was a very fertile and original thinker. His name is connected with the beginnings of a number of ideas, and we would like to mention here a few of those which seem to have led to the most important further developments.

After some work on current algebra, Joel devoted his interest to dual theories and their connection with field theories. He collaborated in the first work in which the singularities appearing in the unitarization of the dual model were isolated and renormalized, an essential step in promoting the model to a theory. He was also at the origin of the discovery of the connection between dual theories and conventional field theories by the zero slope limit. In particular, the connection between dual models and Einstein's gravity led to the suggestion that the dual model could provide a renormalizable theory of gravity. Joel was also involved in the modern revival of the idea that internal symmetries are related to compactified additional space-time dimensions. He and his collaborators developed this into the powerful technique of dimensional reduction which has proved especially useful in the construction of field theories with extended supersymmetry.

Joel's recent scientific activity was mainly devoted to supersymmetry and supergravity. He has collaborated in a number of papers on matter couplings to supergravity, on the relation between supergravity and dual models and on the construction of extended supersymmetric theories. Using the dimensional reduction technique the $N = 4$ supersymmetric Yang-Mills theory was constructed, a theory remarkable for its symmetry and convergence. He also contributed to the discovery of a simple Lagrangian for supergravity in eleven space-time dimensions. A refined dimensional reduction of this theory gave N = 8 supergravity in four dimensions with spontaneous breaking of supersymmetry, a theory which exhibits a form of antigravity.

Joel Scherk's very recent interest, gravity at short range or "antigravity", formed the main subject of his lecture at this workshop. At the time of his death he had written only the first part of it, which follows. For the rest we include a reproduction of the handwritten transparencies which he used during the lecture. They are impressive as a model of clarity, and we hope that they will serve to inform the reader about the development of Joel's ideas at the end of his tragically short life.

GRAVITATION AT SHORT RANGE AND SUPERGRAVITY

J. Scherk

Ecole Normale Supérieure, Paris

ABSTRACT

Extended supergravity models provide examples of theories of gravity which are generally covariant but where gravity has not only a tensorial, but also vectorial and scalar components. Phenomenological models having these characteristics can be constructed and have testable predictions at laboratory distances. It is suggested that improvements of existing laboratory experiments on gravitation would provide more stringent limits on the free parameters occurring in these models.

1. INTRODUCTION

At the present time, the relation between extended supergravity models and physics is far from obvious. Supersymmetric theories have been constructed with N = 1, ... 8 spinorial generators. While N = 1 supersymmetric models compatible with the usual theory of weak, electromagnetic and strong interactions have been constructed, and have testable predictions in high-energy physics, it has not yet been possible to devise models of extended (N > 1) supersymmetry which are flexible enough to accommodate the real world.

Nevertheless, extended supergravity theories are a fascinating laboratory of theoretical physics because they unify Einstein's theory of general relativity with lower spin fields: as soon as N reaches 2, the gravitational sector includes a spin-1 particle; when it reaches 3, a spin-½ particle; for N = 4, it includes also scalars; finally, when N is bigger than 4, only the gravitational sector exists and one has fully-fledged unified field theories incorporating all spins. Further, these models, of which the biggest is the N = 8 model, have natural geometrical interpretations in the superspace framework.

If these models are to be applicable to the physical world it is likely that this will occur only when the detailed mechanism of symmetry breaking is understood. In the meantime, nevertheless, one can study whether these models have some invariant features which may survive the breaking of supersymmetry, and would be detectable in gravitational experiments, aside from high-energy physics.

The most obvious features of extended supergravity models is that gravity in general will proceed not only through a tensorial exchange, but also through vectorial and scalar exchanges. In two quite different supergravity models where masses are introduced, this is indeed the case. The first one is the N = 2 supergravity model coupled to a massive scalar multiplet, but where supersymmetry is unbroken. The second one is the N = 8 supergravity model where the masses are introduced as a soft breaking of supersymmetry. In both cases one observes that an Abelian vector couples to the massive particles of the model with a coupling constant g of order κm, where $\kappa^2 = 4\pi G_N$ and m is the mass of the particle to which it is coupled. Such a vectorial particle can be called a graviphoton. In these two models one finds that in the static limit the graviphoton exchange creates a repulsive force between two particles which cancels in one case the attractive force of gravity, and in the other case the combined attraction due to gravity and to the exchange of a scalar particle. On the other hand, between a particle and an antiparticle pair, the net force is attractive. The idea that gravity could distinguish between particles and antiparticles has been called "antigravity" but so far it has had little theoretical background. Also, the earlier idea of "antigravity" was that like particles attracted each other, while unlike repulsed, while in this framework, this is just the opposite.

One can construct phenomenological models of gravitation incorporating these features of supergravity theories, namely the existence of graviphotons and graviscalars. In order for these models to be compatible with experiment, one needs to assume that the vectorial and scalar particles acquire a mass, so that the associated force is short-range. An order of magnitude estimate gives a range of 1 km or less. However, such an estimate is at best tentative.

Models have been built in the past where gravity includes a

scalar component, and are within experimental limits if the couplings are weak enough, or the range sufficiently short. The rather novel aspect of this model is to include also a vector component, which yields a repulsive force at short range between particle pairs. Also, while the weak equivalence principle is not necessarily broken in a tensor-scalar theory, it is certainly broken in a tensor-vector theory and this gives further restrictions on the free parameters occurring in the model.

The main remaining theoretical problem is to understand the origin of the mass term for graviscalars and graviphotons, as this has not been done so far. It is not impossible that the recent advances in tensor calculus in extended supergravity theories may provide an answer to this question.

2. THEORETICAL FRAMEWORK

The first model where a vector occurs naturally in association with the metric tensor is the N = 2 model of supergravity[1]. It contains a graviton (J = 2), two Majorana gravitinos (J = 3/2) and a vector (J = 1), all massless. The algebra of supersymmetry contains a piece which is a gauge transformation on the vector field A_μ.

In the N = 2 model, the spectrum consists of a vierbein $V_\mu{}^r$ two gravitinos $\psi_\mu{}^i$ (i = 1,2) and the vector field A_μ, whose transformations are given by:

$$\delta V_\mu{}^r = -i\kappa\, \bar{\varepsilon}^i \gamma^r \psi_\mu{}^i \tag{1}$$

$$\delta \psi_\rho{}^i = \frac{1}{\kappa} \mathcal{D}_\rho(\hat{\omega})\, \varepsilon^i - \frac{i}{2} \varepsilon^{ij} \sigma^{\mu\nu} \hat{F}_{\mu\nu} \gamma_\rho \varepsilon^j \tag{2}$$

$$\delta A_\mu = -\varepsilon^{ij} \bar{\varepsilon}^i \psi_\mu{}^j \tag{3}$$

The commutator of two supersymmetry transformations contains field-independent transformations which do not vanish when the fields are set to zero, and field-dependent transformations which do vanish when the fields are set to zero. If we keep only the field-independent transformations, one finds a co-ordinate transformation with parameter

$$\xi^\mu(x) = i\, \bar{\varepsilon}_2{}^i \gamma^\mu \varepsilon_1{}^i \tag{4}$$

and also a gauge transformation of $A_\mu(x)$ with parameter

$$\frac{1}{\kappa} \varepsilon^{ij} \bar{\varepsilon}_2^{\,i} \varepsilon_1^{\,j}$$

This means[2] that $A_\mu(x)$ gauges a central charge, which can occur in the algebra

$$\{ Q_\alpha^i , \bar{Q}_\beta^j \} = i (\not{P})_{\alpha\beta} \delta^{ij} + \delta_{\alpha\beta} \varepsilon^{ij} (Z + i\gamma_5 Z') \tag{5}$$

In the case of N = 2 supergravity, there is no axial vector to gauge the axial charge Z' and also since the central charges have the dimension of a mass, they vanish for pure supergravity. This situation changes, however, if one couples the supergravity multiplet to a matter multiplet having an explicit mass term. Such a multiplet contains two Majorana spin-½ fields χ^i, two scalars A^i, and two pseudoscalars B^i, all having a common mass m.

If one denotes collectively these fields by ϕ^a, the commutator of two supersymmetry transformations in flat space-time contains, in addition to the usual translation piece, a global rotation of the two multiplets into one another

$$[\delta_2, \delta_1] \phi^a = i \bar{\varepsilon}_1^{\,i} \gamma^\mu \varepsilon_2^{\,i} \partial_\mu \phi^a + m \varepsilon^{ij} \bar{\varepsilon}_1^{\,i} \varepsilon_2^{\,j} \varepsilon^{ab} \phi^b \tag{6}$$

When one couples this matter multiplet to supergravity[3] the infinitesimal rotation parameter

$$\Lambda = m \varepsilon^{ij} \bar{\varepsilon}_1^{\,i} \varepsilon_2^{\,j}$$

becomes local and the vector field A_μ gauges the central charge of the scalar multiplet. Indeed, one finds that A_μ couples minimally through the covariant derivative

$$\mathcal{D}_\mu \phi^a = (\partial_\mu \delta^{ab} - \kappa m \varepsilon^{ab} A_\mu) \phi^b \tag{7}$$

If one introduces the complex fields $A^1 + iA^2$, $B^1 + iB^2$, $\chi^1 + i\chi^2$ the theory can be re-expressed in terms of two complex scalar fields and a Dirac spinor field coupled to N = 2 supergravity. Denoting collectively these fields by ψ^i, one notes that A_μ couples through

$$D_\mu \psi^i = \partial_\mu \psi^i - i \kappa m A_\mu \psi^i \tag{8}$$

The dimensionless coupling constant of the vector field with the matter multiplet is thus given by $g = \kappa m$. If one identifies A_μ with the electromagnetic field, this model unifies gravity and electromagnetism (Einstein's dream). However, if m is a typical hadronic mass, the coupling constant g is extremely tiny, of the order of 10^{-19}, and A_μ should rather be identified with an extra vectorial component of the gravitational field coupled to the number of particles, and proportionately to their mass, hence the name "graviphoton". The electromagnetic and other gauge fields should be added as members of N = 2 vector multiplets in which case mixing between the graviphoton and the true photon could occur. These will be left for a further study.

A remarkable phenomenon occurring in this model was discovered by Zachos[3]. In the weak field limit the potential between a pair of massive particles can be obtained from the Born diagrams where both a graviton and a graviphoton can be exchanged. If several scalar multiplets are introduced, with masses m_i and coupling constants g_i to the vectorial particle, one finds in the static limit the amplitude:

$$\mathcal{A} = 4 \kappa^2 \frac{m_i m'_i}{q^2} \left(m_i m'_i - \frac{g_i g'_i}{\kappa^2} \right) \tag{9}$$

The resulting potential in the static limit is given by:

$$V(r) = - \frac{\kappa^2}{4 \pi r} \left(m_i m'_i - \frac{g_i g'_i}{\kappa^2} \right) \tag{10}$$

In N = 2 supergravity the relation $g_i = \varepsilon \kappa m_i$ holds where $\varepsilon = +1$ for a particle and $\varepsilon = -1$ for an antiparticle. So one finds that

$$V(r) = - \frac{\kappa^2}{4 \pi r} m_i m'_i \left(1 - \varepsilon \varepsilon' \right) \tag{11}$$

which vanishes between a pair of particles or antiparticles while the gravitational attraction is doubled between a particle-antiparticle pair. This phenomenon can be called "Antigravity"[4], although the term was used previously to suggest that particles attracted each other while particle-antiparticle pairs repelled. Here it is the opposite, and this is due to the fact that the extra vectorial force can only give a repulsive force between like-particles and an attractive force between unlike-particles.

One notes that in such a model, although general covariance is maintained, the equivalence principle is maximally broken, since in the field of a particle, the acceleration of another particle will be zero while an antiparticle will fall according to Newton's law.

This model exhibits a cancellation between the tensorial and vectorial exchanges which is quite striking. It is, however, not the only case where cancellations occur between fields of different spin. The first example was found in the SO(3) gauge theory coupled to a triplet of Higgs fields[5]. After spontaneous symmetry breaking, the spectrum consists of a massless vector A_μ, two massive charged vector bosons $W_\mu^\pm$ of mass $m_W = g\langle\phi_3\rangle$, where $\langle\phi_3\rangle$ is the vacuum expectation value of the Higgs field in the third direction, and a neutral scalar field σ of mass $m_\sigma = \sqrt{2\lambda}\ \langle\phi_3\rangle$. There is a trilinear coupling of σ to the massive vector bosons given by $2g\ m_W\ W_\mu{}^\dagger W_\mu$. So this coupling is proportional to the mass, as is the case for the graviphoton. In the Prasad-Sommerfield limit $(\lambda \to 0)$, m_σ vanishes and two long-range forces are competing: the one due to the photon exchange which is either attractive or repulsive and the one due to the scalar exchange which is always attractive. In the static limit one finds that the potential between two W^+ or W^- particles vanishes as the scalar attraction balances exactly the vectorial repulsion, while between a W^+W^- pair the potential is given by

$$V(r) = -\frac{2g^2}{4\pi r}$$

This suggests an analogy between the N = 2 supergravity model and the SO(3) gauge theory, namely:

J = 2 graviton $\leftrightarrow$ J = 1 photon

J = 1 graviphoton $\leftrightarrow$ J = 0 Higgs scalar.

At the same time, it suggests that to rescue antigravity experimentally, one should give a mass to the graviphoton, just as this is the case in the SO(3) model for the Higgs scalar, so that the cancellation would hold only at short distances.

Before leaving the N = 2 model, one can also investigate whether the vector-tensor cancellation holds if the full non-linearities of Einstein's theory are taken into account. One can keep the static limit, but relax the weak field approximation. At the same time one can treat the massive particles of the N = 2 scalar multiplet as point-like sources, which is legitimate if $r \gg \hbar/mc$, and look for solutions of the coupled Maxwell-Einstein equations with the information that $Q^2 = \kappa^2 m^2$. A general isotropic, static solution with Q, m unrelated is given by the Reissner-Nordström solution

$$A^0 = \frac{Q}{4\pi r} \qquad A^i = 0 \tag{12}$$

$$ds^2 = B(r)\,dt^2 - \frac{dr^2}{B(r)} - r^2\left(d\theta^2 + \sin^2\theta\, d\phi^2\right) \tag{13}$$

where

$$B(r) = 1 - \frac{2\kappa^2 m}{4\pi r} + \frac{\kappa^2 Q^2}{(4\pi)^2 r^2} \tag{14}$$

B(r) has two zeros if $\kappa^2 m^2 > Q^2$ is regular everywhere if $\kappa^2 m^2 < Q^2$ and is a perfect square if $Q^2 = \kappa^2 m^2$ which is the case selected by N = 2 supergravity. This limiting case was studied in detail by B. Carter who showed that a complete analytic continuation of the metric could be found. Also one notes that in accordance with general theorems, a singularity exists at r = 0 and a horizon exists at $r_0 = \kappa^2 m/4\pi$. Thus the validity of the vector-tensor cancellation in the case of like particles holds only for distances much greater than r_0.

Let us now turn towards the N = 8 supergravity model with masses, which provides another example of such a cancellation. To start with, let us note that the N = 2 and N = 8 models are quite distinct because the N = 2 model is not a unified model. We must add to it massive scalar multiplets whose masses are put in "by hand", while supersymmetry is kept exact. On the other hand, the N = 8 model is a unified field theory in which there is no matter supermultiplet which can be introduced. Four mass parameters m_i ($i = 1, \ldots 4$) are introduced via dimensional reduction from five to four dimensions and break supersymmetry spontaneously in the sense that mass relations ensure that the one-loop corrections to the cosmological constant (which vanishes in the classical approximation) are finite[6,7].

The introduction of masses in the D = 4, N = 8 theory is similar to the Kaluza-Klein theory, so we shall illustrate it on the simple example of a scalar field coupled to gravity in D = 5. Starting with the Einstein-Klein-Gordon Lagrangian in five dimensions:

$$\mathcal{L}_5 = -\frac{1}{4K^2}\hat{V}\hat{R} + \hat{V} g^{\hat{\mu}\hat{\nu}} \partial_{\hat{\mu}} \phi^* \partial_{\hat{\nu}} \phi \tag{15}$$

one reduces it to four dimensions by splitting the co-ordinate $x^{\hat{\mu}} = (x^\mu, y)$ where $\mu = 0,1,2,3$ and y is the fifth co-ordinate. An ansatz which ensures that $\mathcal{L}_5$ is independent of y is

$$\hat{V}_{\hat{\mu}}{}^{\hat{r}}(x,y) = \hat{V}_{\hat{\mu}}{}^{\hat{r}}(x) \tag{16}$$

$$\phi(x,y) = \exp(imy)\,\phi(x) \tag{17}$$

where m has the dimension of a mass. The field content of the Einstein action can be read from the parametrization of the fünfbein:

$$\hat{V}_{\hat{\mu}}{}^{\hat{r}}(x) = \begin{bmatrix} \exp\left(-\frac{K\sigma(x)}{\sqrt{3}}\right) V_\mu^{\ r}(x) & 2K A_\mu(x)\exp\left(\frac{2K\sigma(x)}{\sqrt{3}}\right) \\ 0 & \exp\left(\frac{2K\sigma(x)}{\sqrt{3}}\right) \end{bmatrix} \tag{18}$$

This parametrization is chosen such that in D = 4 one gets the canonical Einstein action, a canonical kinetic term for the σ field, and is such that under co-ordinate transformations in the fifth direction A_μ transforms as an Abelian gauge field

$$\delta A_\mu = \frac{1}{2K}\,\partial_\mu \xi^5(x) \quad ; \quad \delta\sigma = \delta V_\mu^r = 0 \tag{19}$$

The resulting Lagrangian in four dimensions coming from the Einstein action is given by:

$$\mathcal{L}_4^E = -\frac{1}{4K^2} V R - \frac{1}{4} V \exp\left(2\sqrt{3}\,K\sigma\right) F_{\mu\nu}F^{\mu\nu} + \frac{1}{2} V g^{\mu\nu}\partial_\mu\sigma\,\partial_\nu\sigma \tag{20}$$

So it contains a graviton, a graviphoton and a graviscalar. Under co-ordinate transformations in the fifth direction, the scalar field $\phi(x)$ transforms as

$$\delta\phi(x) = i m\, \xi^5(x)\, \phi(x) \tag{21}$$

Thus, in the coupling of the graviphoton to the scalar field ϕ the covariant derivative $\partial_\mu - 2i\kappa m A_\mu$ will arise and one finds indeed the reduced Lagrangian:

$$\mathcal{L}_4^s = V\left\{ g^{\mu\nu}\left(\partial_\mu + 2iKmA_\mu\right)\phi^*\left(\partial_\nu - 2iKmA_\nu\right)\phi - m^2\phi^*\phi\, \exp\left(-2K\sqrt{3}\,\sigma\right)\right\} \tag{22}$$

Now the cancellation occurs between the gravitational field, the graviphoton field A_μ, and the graviscalar field σ. The static potential between two particles is given by:

$$V(r) = -\frac{K^2}{4\pi}\,\frac{mm'}{r}\left(1 - 4\varepsilon\varepsilon' + 3\right) \tag{23}$$

where the factors in the brackets arise from the spin-2, spin-1 and spin-0 exchanges, respectively. Again the cancellation of static forces occurs between particle or antiparticle pairs.

In the N = 8 theory with breaking of supersymmetry, masses are introduced in a similar way[6,7]. One starts with the D = 5, N = 8 theory and one reduces it to four dimensions. The spectrum of the model in five dimensions has a global Sp(8) invariance and consists of one graviton, 8 spin-3/2, 27 spin-1, 48 spin-½ and 42 spin-0 fields. Instead of the U(1) factor exploited in the previous case, one can exploit the Sp(8) group in the dimensional reduction. Sp(8) being of rank 4, four mass parameters can be introduced in the model, by setting

$$V_{\hat{\mu}}{}^{\hat{r}}(x,y) = V_{\hat{\mu}}{}^{\hat{r}}(x) \quad ; \quad \psi_{\hat{\mu}}{}^{\hat{a}}(x,y) = (\exp \mathcal{M} y)^{\hat{a}}{}_{\hat{b}}\, \psi_{\hat{\mu}}{}^{\hat{b}}(x)$$

and so on. $\mathcal{M}$ is a matrix of Sp(8) which can be brought to the form

$$\mathcal{M} = \begin{pmatrix} 0 & 1 \\ -1 & 0 \end{pmatrix} \otimes \mathrm{diag}(m_1, m_2, m_3, m_4)$$

and the four mass parameters m_i determine the masses of all the states arising in the model. In particular, the 8 Majorana spin-3/2 become four Dirac massive spin-3/2 of masses m_i by absorbing 8 spin-½ fields through the super-Higgs mechanism. Similarly, 24 of the spin-1 fields become complex massive fields through the Higgs mechanism and have masses $|m_i \pm m_j|$ $(i < j)$.

In such a dimensional reduction the vector and the scalar coming from the reduction of the fünfbein play a special role as in the previous example, in that they couple proportionately to the mass. In particular, the graviphoton field A_μ couples to the mass of each of the states of the model with coupling constant $g = \pm 2\kappa |m|$.

IV N = 8 SUPERGRAVITY AND ANTIGRAVITY

N = 2	N = 8
Not unified	Unified
(SG + Matter)	(1 multiplet only)
only one m	Several mass scales m_i (i = 1, ... 4)
Supersymmetry exact	Supersymmetry broken

Still, the cancellation persists, now between tensor + vector + + scalar.

Construct N = 8 theory m = 0:

$$J=2 \quad (1) \quad , J=\tfrac{3}{2} \quad (8) \quad , \quad 1 \quad (28) \quad , \tfrac{1}{2} \quad (56) \quad , 0 \quad (70)$$

Start with N = 1 SG in D = 11, reduce it to D = 4 (Cremmer-Julia)

Construct the N = 8 theory with masses and spontaneous breaking of supersymmetry:

Start with N = 8 SG in D = 5, m = 0, reduce it to 4. (Cremmer, S., Schwartz)

$$D=5 \qquad V_\mu^r \qquad \psi_\mu^a \qquad A_\mu^{ab} \qquad \chi^{abc} \qquad \phi^{abcd}$$

$$1 \qquad 8 \qquad 27 \qquad 48 \qquad 42$$

indices a, b, ... a = 1, ... 8 Sp(8)

Use generalized dimensional reduction:

$$\psi_\mu^a(x,y) = (\exp \mathfrak{M}\, y)_a^b\, \psi_\mu^b(x)$$

$\mathfrak{M}$: Sp(8) matrix

$$\mathfrak{M} = \begin{pmatrix} 0 & 1 \\ -1 & 0 \end{pmatrix} \otimes \begin{pmatrix} m_1 & & & 0 \\ & m_2 & & \\ & & m_3 & \\ 0 & & & m_4 \end{pmatrix}$$

4 mass parameters are introduced.

analogy: $\phi(x,y) = e^{imy}\phi(x) \qquad \Box_5\phi = 0 \Rightarrow (\Box_4 + m^2)\phi = 0$

Mass spectrum N = 8, D = 4 supergravity

Spin	Mass	Degeneracy
2	0	1
3/2	$\|m_i\|$	2
1	0 $\|m_i \pm m_j\|$ $i < j$	4 2
1/2	$\|m_i\|$ $\|m_i \pm m_j \pm m_k\|$ $i < j < k$	4 2
0	0 $\|m_i \pm m_j\|$ $i < j$ $\|m_1 \pm m_2 \pm m_3 \pm m_4\|$	6 2 2

Remarkable properties

$$G_r \operatorname{Tr}(\mathcal{M})^{2q} = \sum_{J=0}^{2} (-1)^{2J} (2J+1) \operatorname{Tr}(\mathcal{M}_J)^{2q} = 0$$

for q = 0, 1, 2, 3 (Ferrara-Zumino)

one loop correction to cosmological constant finite even in the presence of breaking.

Interesting cases:

$m_1 = 0$	N = 8 unbroken
$m_1 = m_2 = 0$	N = 4 "
$m_1 = m_2 = \ldots = m_4 = M$	SU(4) × U(1) unbroken; SS broken
$m_1 = m_2 = m_3 = m$	SU(3) × U(1) × U(1) unbroken
$m_4 = M$	SS broken

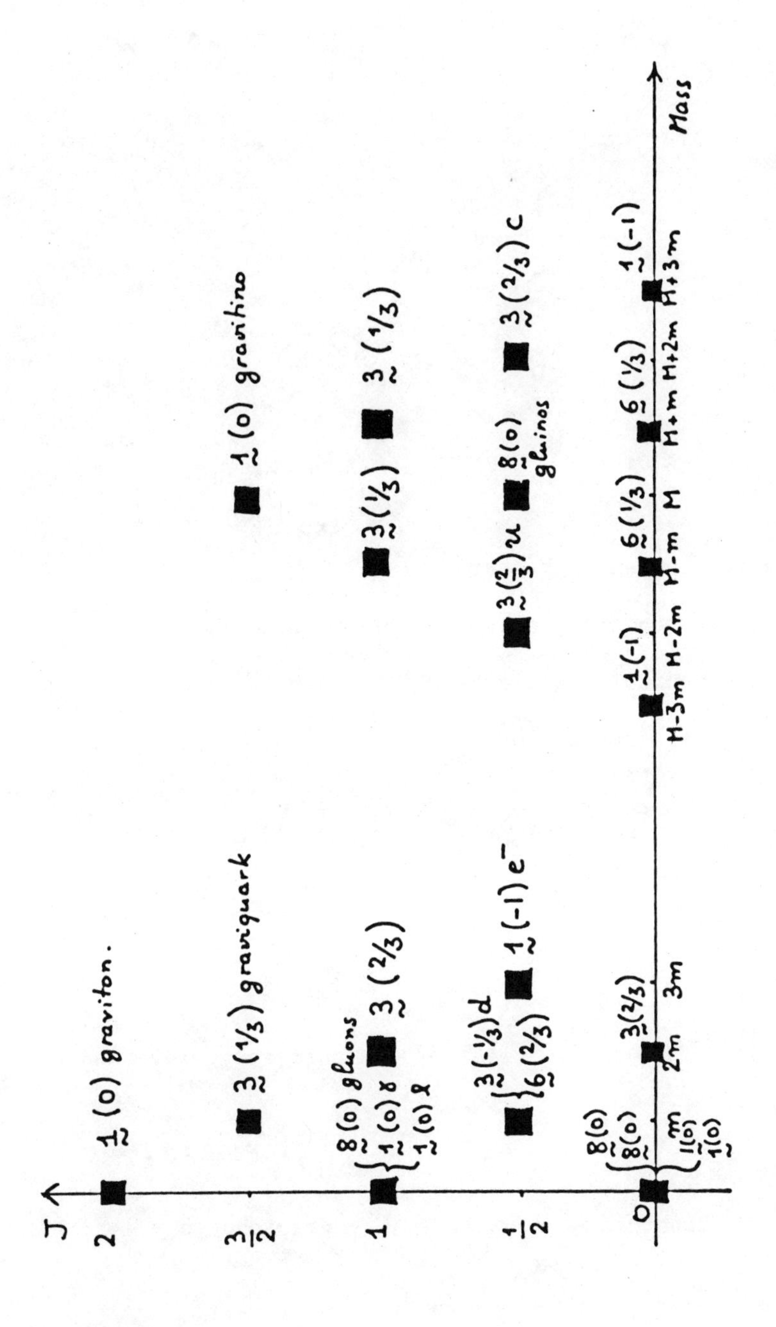

Fig.1 : Spectrum of the spontaneously broken, N = 8 model with SU(3) ⊗ U(1) ⊗ U(1) invariance. The SU(3) representations are underlined (ex : $\underset{\sim}{3}$) ; the bracket indicates the electric charge according to the Gell-Mann scheme.

Simple example for the introduction of a mass term (Kaluza-Klein)

$$\mathcal{L}_5 = -\frac{1}{4\kappa^2}\hat{V}\hat{R} \qquad + \qquad \hat{V} g^{\hat{\mu}\hat{\nu}} \partial_{\hat{\mu}}\phi^* \partial_{\hat{\nu}}\phi$$

gravity + massless scalar in D = 5

$$x^{\hat{\mu}} = (x^\mu, y)$$

Ansatz:

$$\hat{V}^r_\mu(x,y) = \hat{V}^r_\mu(x)$$

$$\phi(x,y) = e^{imy}\phi(x)$$

write the fünfbein as:

$$\hat{V}^r_\mu(x) = \begin{pmatrix} \exp -\frac{\kappa\sigma}{\sqrt{3}} V^r_\mu(x) & 2\kappa A_\mu \exp\frac{2\kappa\sigma}{\sqrt{3}} \\ 0 & \exp\frac{2\kappa\sigma}{\sqrt{3}} \end{pmatrix}$$

Parametrization chosen such that:

- get the canonical Einstein's action in D = 4
- get a canonical kinetic term for the σ
- under $\xi^5(x)$ translations:

$$\delta V^{\hat{r}}_{\hat{\mu}} = \xi^{\hat{\rho}}\partial_{\hat{\rho}} V^{\hat{r}}_{\hat{\mu}} + \partial_{\hat{\mu}}\xi^{\hat{\rho}} V^{\hat{r}}_{\hat{\rho}}$$

one has for $\xi^5(x)$:

$$\delta V^r_\mu = 0 \quad ; \quad \delta\sigma = 0$$

$$\delta A_\mu = \frac{1}{2\kappa}\partial_\mu \xi^5(x) \qquad \text{local gauge invariance}$$

Calculation gives:

$$\mathcal{L}_5 \to \mathcal{L}_4 = -\frac{1}{4\kappa^2} VR \qquad - \qquad \frac{V}{4}\exp 2\sqrt{3}\,\kappa\sigma\, F_{\mu\nu}F^{\mu\nu}$$

- Einstein -   - graviphoton -

$$+ \frac{1}{2} V g^{\mu\nu}\partial_\mu\sigma\partial_\nu\sigma$$

- graviscalar -

Why is A_μ a graviphoton, σ a graviscalar ?

Set $\phi(x,y) = \exp imy\ \phi(x)$

under ξ^5 translations $\delta\phi(x,y) = \xi^5 \partial_5 \phi(x,y) = im\xi^5(x)\phi(x)$

and thus:

$$\begin{cases} \delta\phi(x) = im\,\xi^5(x)\,\phi(x) \\ \delta A_\mu(x) = \frac{1}{2\kappa}\,\partial_\mu \xi^5(x) \end{cases}$$

So the combination:

$$(\partial_\mu - 2i\kappa m A_\mu)\phi$$

will appear naturally.

Indeed:

$$\mathcal{L}_5 \to \mathcal{L}_4 = V\{g^{\mu\nu}(\partial_\mu + 2i\kappa m A_\mu)\phi^*(\partial_\nu - 2i\kappa m A_\nu)\phi$$
$$- m^2\phi^*\phi \exp -2\kappa\sqrt{3}\sigma\}$$

In the weak field static limit:

$$V_{J=2}(r) = -\frac{\kappa^2}{4\pi}\frac{mm'}{r}$$

$$V_{J=1}(r) = +\frac{4\kappa^2}{4\pi}\frac{mm'}{r}\epsilon\epsilon' \qquad \epsilon = \pm 1$$

$$V_{J=0}(r) = -\frac{3\kappa^2}{4\pi}\frac{mm'}{r}$$

Cancellation occurs for particle pairs
true for any theory derived from a massless D = 5 theory
where $m = p_5$.

$$\hat{k} = (k_\mu, \epsilon m) \qquad \epsilon = \pm 1$$
$$\hat{k}^2 = k_\mu^2 - m^2 = 0$$
$$\hat{p} = (p_\mu, \epsilon' m')$$
$$\hat{p}^2 = p_\mu^2 - m'^2 = 0$$

Graviton exchange D = 5:

$$\mathcal{A}^{(5)} \sim \frac{G}{q^2 + i\epsilon} (\hat{k}\cdot\hat{p})(\hat{k}'\cdot\hat{p}')$$

Static limit
$$\hat{k} = \hat{k}' = m(1, \vec{0}, \epsilon)$$
$$\hat{p} = \hat{p}' = m'(1, \vec{0}, \epsilon')$$

$$\boxed{\mathcal{A}^{(5)} \sim \frac{G}{q^2 + i\epsilon} m^2 m'^2 (1 - \epsilon\epsilon')^2}$$

Conclusion:

N = 2, 8 exhibit cancellation of long range gravitational force through massless graviphotons ($g \sim km$) and graviscalar ($\lambda \phi^* \phi \sigma$ coupl. $\lambda \sim km^2$)

Can anything like that survive in the real world?

V MOTIVATIONS FOR A PHENOMENOLOGICAL MODEL

Take as a hint from supergravity models that in addition to $g_{\mu\nu}$, one has also a vector field A_μ with $g \sim$ km m: typical hadronic mass, $g \sim 10^{-18}$, and also a scalar field σ.

σ : already introduced for $m = 0$ by Jordan, Thirry, Brans, Dicke. Known to be weakly coupled to gravity.

A_μ: occurs in a model of Lee, Yang coupled to j_B^μ. Known also to be weakly coupled/gravity if $m = 0$.

If one wants to keep the idea of short range cancellation of the gravitational force, need to assume that A_μ, σ become massive.

Estimation of the mass (qualitative) assume Higgs scalar ϕ to give a mass A_μ, and A_μ coupled gravitationally to ϕ:

$$m_V = g \langle \phi \rangle = \kappa \mu \langle \phi \rangle$$

μ, $\langle \phi \rangle$ hadronic or 10^2 GeV ...

$$m_V \sim 10^{-19} \text{GeV}^{-1} \mu \langle \phi \rangle$$

$$\lambda_V \lesssim 10^3 \text{ m}$$

If μ, $\langle \phi \rangle$ much bigger, then all effects disappear.

VI PHENOMENOLOGICAL MODEL: IMPLICATIONS

Phenomenological model of gravity containing a vector field:

$$\mathcal{L} = -\frac{1}{4\kappa^2} V V^{\mu a} V^{\nu b} R_{\mu\nu ab} - \frac{1}{4} V g^{\mu\rho} g^{\nu\sigma} F^{\ell}_{\mu\nu} F^{\rho\sigma\ell}$$

$$+ m^2_{\ell} A^{\ell}_{\mu} A^{\ell}_{\nu} g^{\mu\nu} V + \mathcal{L}_M$$

$\mathcal{L}_M$ will contain A^{ℓ}_{μ} through:

$$\mathcal{D}_\mu \chi = \partial_\mu \chi_i - i\alpha\kappa m_i A^{\ell}_{\mu} \chi_i$$

α expected to be of order 1.

$\alpha = 1 \quad N = 2$

$\alpha = 2 \quad N = 8$

Uncertainties in the model:

- range m_ℓ, m_σ ?
- should one apply it blindly to nucleons, or to elementary constituents ? N = 8 suggests to apply it to leptons and quarks.
- how to reconcile it with the Higgs mechanism ?

Clearer if assumes a coupling of A_μ to conserved or semi-conserved currents

$$A_\mu \left[g_B J^\mu_B + g_c J^\mu_{Lc} + g_s J^\mu_s + \dots \right]$$

where the dimensionless constants g_B, g_c, g_S are roughly given by

$$g_B \sim \kappa M_p \quad \text{or} \quad \kappa m_L$$

$$g_c \sim \kappa m_c \;, \;\ldots$$

Features of the model.

a) deviations from Newton's law at short distances: true for A_μ, σ. Test:
Cavendish experiment.
Difference A_μ, σ:
A_μ: repulsive, σ attraction.

b) Generally covariant, but breaks the equivalence principle: few "neutral" particles exist for gravity.
Only those which are self-conjugate: π^0, ψ, γ fall with same acceleration.
p, n, e fall with slightly different accelerations.
Test:
Eötvös experiment.
Typical of A_μ, since σ could couple exactly to $T_{\mu\mu}$.

c) Apparent CPT violation in the field of the Earth:
p, $\bar{p}$, K^0, $\bar{K}^0$ do not have quite same weight:
test $K^0 - \bar{K}^0$ weight differences.

d) Long-range effects negligible (planets, ...)
possible effect in neutron stars or cosmology.

a) Deviations from Newton's law

There is room for deviations from Newton's law. (Mikkelson, Newman, Phys. Rev. D16, 4 (1977) 919).

$$G(r) = F_{grav} \frac{r^2}{Mm}$$

Tests of constancy of G(r) $\begin{cases} \text{celestial mechanics.} & G_c \\ \text{laboratory exp.} & G_0 \end{cases}$

Celestial mechanics

$$V(r) = -\frac{G_c Mm}{r}\left(1+\alpha e^{-\mu r}\right)$$

$$G(r) = G_c\left[1+\alpha(1+\mu r)e^{-\mu r}\right] \quad \text{(Fujii, 1977)}$$

$$G_0 = G_c[1+\alpha]$$

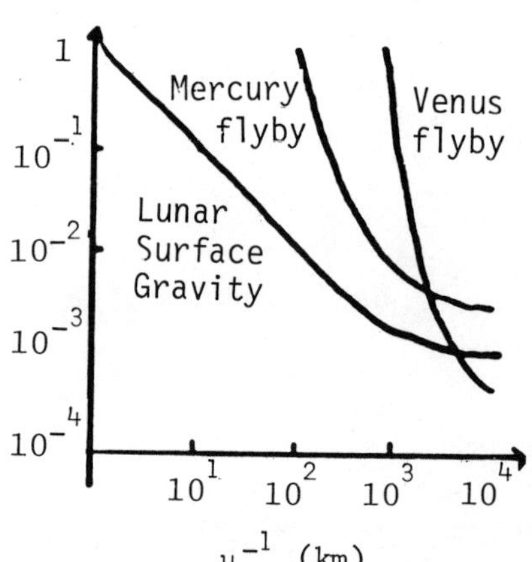

precession of planets:

$$G - G_c < 10^{-8} \qquad 0.3 . 10^8 < r < 9 . 10^8 \text{ km}$$

binary pulsars:

$$< 10^{-5} \qquad 7 . 10^5 < r < 3 . 10^6 \text{ km}$$

planetary mass determination;

$$< 3 . 10^{-2} \qquad 10^4 < r < 3 . 10^8 \text{ km}$$

neutron stars? probably exclude $\alpha \sim 1$ $\mu^{-1} \sim$ km.
Big Bang ?

Laboratory experiments

(Cavendish experiment) Long Phys. Rev. D9 (1974) 850
Nature 260 (1976, Apr. 1) 417.

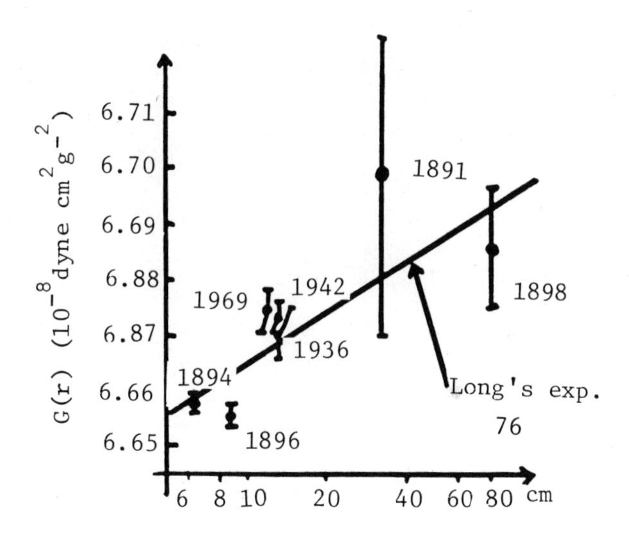

Parametrizes result as $G(r) = G_0\ (1 + 0.002\ \text{Ln}r)$
analysed in term of $(1 + \alpha e^{-\mu r})$

$$\boxed{\alpha < 0} \quad (!)$$

repulsive force at short distance.

$|\alpha| \geq 0.005$
(Long's experiment)

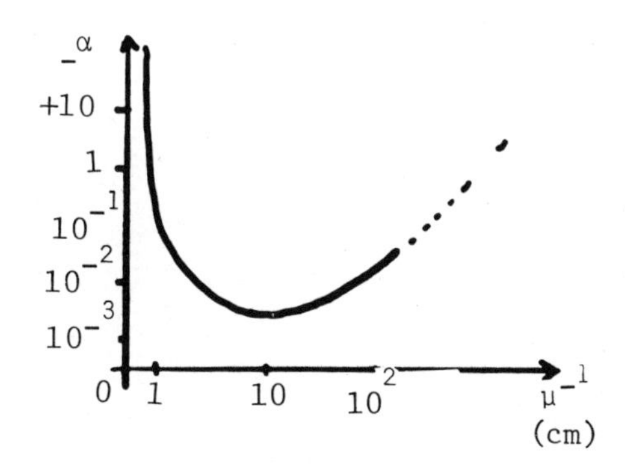

Needs confirmation. $\left.\begin{array}{l}\alpha \sim -5.\ 10^{-3} \\ \mu^{-1} \sim 10\ \text{cm}\end{array}\right\}$ possibly.

Datas on G(R)	$(10^{-18}$ dyne $cm^2/g^2)$	R(cm)
Boys : 1894	6.6576 ± 0.002	6.3
Braun : 1896	6.655 ± 0.002	8.6
Poynting : 1891	6.6984 ± 0.0029	32
Richarz et al : 1898	6.685 ± 0.011	80
Heyl : 1930	6.670 ± 0.005	13
Heyl Chranowsky : 1942	6.673 ± 0.003	13
Rose et al : 1969	6.674 ± 0.004	12

Gap between laboratory and celestral experiments

poor constraints on G(r)

$$10\ \text{m} < r < 1\ \text{km}$$

Models of earth, sun imply

$$0.50 < \frac{G_c}{G_0} < 1.32 \qquad \text{not more.}$$

Conclusion: room for deviations from Newtons law in the range

$$r \sim 1\text{m} - 1\ \text{km}.$$

$$\alpha \sim 10^{-2} - 1$$

D.R. Long's Experiment (Eastern Washington State College, Ph. Dept., Nature, 1976, April issue.)

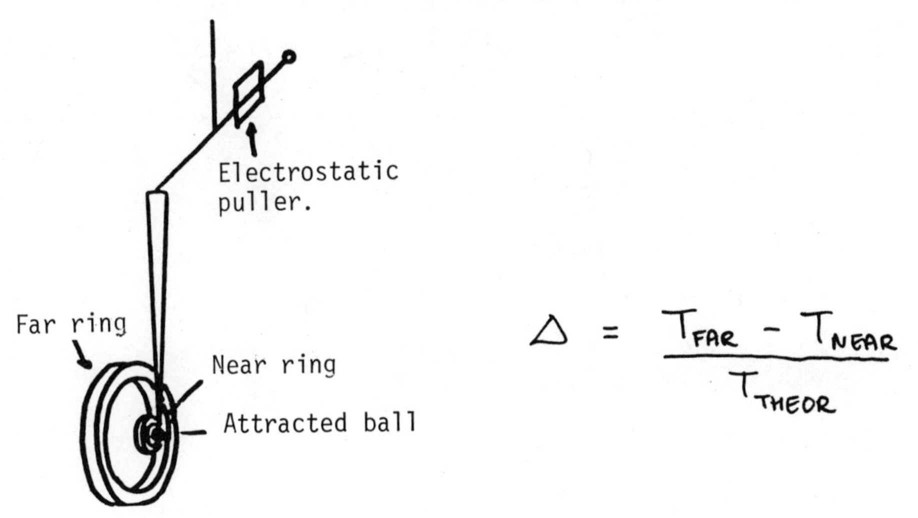

$$\Delta = \frac{T_{FAR} - T_{NEAR}}{T_{THEOR}}$$

$$\Delta_{TH} = 0.03807 \pm 0.0005$$

(position, masses, Newton's $1/r^2$ law, numerical integration)

$$\Delta_{EXP} = 0.04174 \pm 0.0004$$

$$\delta = \Delta_{EXP} - \Delta_{TH} = 0.0037 \pm 0.0007$$

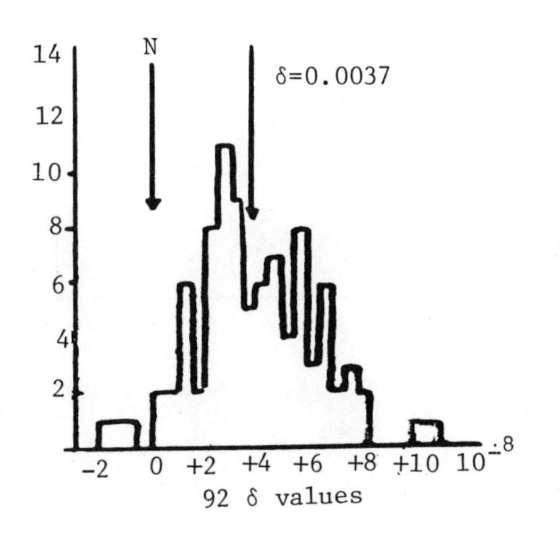

$$G(r) = G_0\left(1 + 2.10^{-3} \ln R\right)$$

a) Apparent CPT violation

$$\vec{F} = M_1 \vec{\gamma}_{\mp} = -\frac{G_N M_1 M_2}{r^2}\left[1 \mp \alpha^2\left(1+\frac{r}{\lambda_\ell}\right)e^{-r/\lambda_\ell}\right]$$

In the field of the Earth

u, d, s, e^-, ... fall with universal γ_-

$\bar{u}$, $\bar{d}$, $\bar{s}$, e^+, ... fall with universal γ_+

$\Rightarrow$ p, $\bar{p}$; e^-, e^+, K^0 - $\bar{K}^0$ have apparent different masses in the earth's field.

Comparison/experiment

- Earth not point-like
- ℓ couples to quarks rather than directly to baryons (debatable)

Non-pointlikeness of Earth

After integration

$$\vec{F} = -G_N \frac{M_1 M_\oplus}{r^2}\left(\frac{\vec{r}}{r}\right)\left[1 \mp \alpha^2\left(1+\frac{r}{\lambda_\ell}\right)\exp -\frac{h}{\lambda} f(x)\right]$$

$$h = r - R_\oplus \quad \text{(altitude)}$$

$$x = R_\oplus / \lambda_\ell$$

$$f(x) = \frac{3}{2x^3}\left[x\left(1+e^{-2x}\right) - \left(1-e^{-2x}\right)\right]$$

f(x): form factor $f(0) = 1$

in practice:

$$x = \frac{R_\oplus}{\lambda_\ell} \gg 1 \qquad f(x) \sim \frac{3}{2}\left(\frac{\lambda_\ell}{R_\oplus}\right)^2$$

- coupling to elementary constituents

J = 2 graviton couples to $T_{\mu\nu}$; sees quarks, but also gluons, e.m. fields etc... couples to $M = M_0 - \Delta M$ (binding)

J = 0 graviscalar also: couples to $T_{\mu\mu}$.

J = 1 graviphoton couples to $\sum_i m_i \chi_i \gamma^\mu \chi_i$
not necessarily to gluons, e.m. fields etc ...
tight algebraic constraints needed to make it couple to M rather than M_0.

$$\vec{F} = - \frac{G_N M_1 M_\oplus}{r^2}\left(\frac{\vec{r}}{r}\right)\left[1 \mp \alpha^2 \frac{M_1^0}{M_1}\frac{M_\oplus^0}{M_\oplus}\left(1+\frac{r}{\lambda}\right)\exp -\frac{h}{\lambda} f(x)\right]$$

$$M_p^{(0)} = 2m_u + m_d$$

$$M_n^{(0)} = 2m_d + m_u$$

$$M_e^{(0)} = m_e$$

$$M_{\pi^+}^{(0)} = m_u - m_d$$

$$M_\psi^{(0)} = m_c - m_{\bar{c}} = 0$$

$$M_\oplus^{(0)} = M_\oplus \frac{3m_u}{M_p}$$

uncertainties parametrized by m_u, m_d, ...

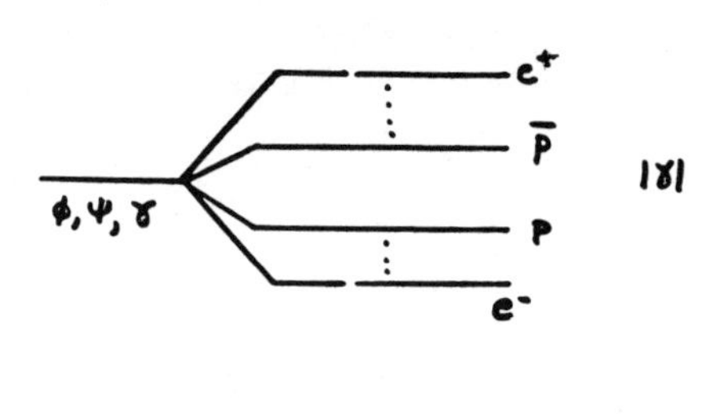

first order effect : distinguish particles-antiparticles

second order effect: distinguish particle types.

first order more reliable.

- CPT violation effects

$M_x = M(1 + V)$ weights of $K^0 - \bar{K}^0$ (M.L. Good), 1961.

$M_{\bar{x}} = M(1 - V)$

$$\boxed{\frac{M_x - M_{\bar{x}}}{M} = 3\, G_N \frac{M_\oplus}{R_\oplus} \alpha_x^2 \left(\frac{\lambda}{R_\oplus}\right)^2 \exp - \frac{h}{\lambda}}$$

$$\boxed{\alpha_x^2 = \alpha^2 \frac{M_x^0}{M_x} \cdot \frac{3m_u}{M_p}}$$

$$\boxed{\frac{M_x - M_{\bar{x}}}{M} \sim 2.10^{-9} \alpha_x^2 \left(\frac{\lambda}{R_\bullet}\right)^2} \quad \text{at } h=0$$

Experimentally:	μ	$-(2 \pm 5)10^{-6}$
	π	$-(2.4 \pm 4.7)10^{-5}$
poor tests	K	$-(1.3 \pm 1.1)10^{-4}$
	$\bar{p}$	$(7,6 \pm 3.8)10^{-5}$ ←

H. Poth, Phys. Letters 77B (1978) 321.

$K^0 - \bar{K}^0$ mass difference better known.

Steinberger analysis:

$$\left|\frac{M - \bar{M}}{M}\right| = \left|\frac{(\epsilon - \epsilon')}{1 - \epsilon\epsilon'}\right| \frac{M_S - M_L}{M_{K^0}} = \frac{|\epsilon - \epsilon'|}{|1 - \epsilon\epsilon'|} 7.08 \times 10^{-15}$$

$$\leq 2.8.10^{-17} \quad \leq 10^{-16}$$

$$\boxed{\alpha_{K^0} \lambda / R_\odot \leq 1.2 \times 10^{-4}}$$

$$\boxed{\lambda \leq \alpha_{K^0}^{-1}\, 7.6\; 10^2 \,m}$$

naive model:

$$\alpha_{K^0}^2 = 2 \left(\frac{m_S - m_d}{M_{K^0}}\right)^{\frac{1}{2}} \left(\frac{3m_u}{M_p}\right)^{\frac{1}{2}} \sim 2 \qquad \boxed{\lambda \leq 3.8\; 10^2 m}$$

The equivalence principle

$$\gamma = -\frac{G_N M_\oplus}{r^2}\left[1-\alpha^2\frac{M_0}{M}\frac{M^0_\oplus}{M_\oplus}\left(1+\frac{r}{\lambda}\right)\exp-\frac{h}{\lambda}f(x)\right]$$

$$f(x) \simeq \frac{3}{2}\left(\frac{\lambda}{R_\oplus}\right)^2$$

γ not universal since $M^0/M \neq 1$ for a vector

Limits: Dicke $\qquad \frac{\delta\gamma}{\gamma} < 10^{-11}$ $\qquad$ but use the Sun.

Vectorial piece negligible.

Eötvös $\qquad \frac{\delta\gamma}{\gamma} < 10^{-9}$ $\qquad$ Earth $\qquad h \sim 0$

$$\frac{\delta\gamma}{\gamma} = \left(\frac{M^0}{M}-\frac{M'^0}{M'}\right)\alpha^2\frac{M^0_\oplus}{M_\oplus}\frac{3}{2}\frac{\lambda}{R_\oplus} < 10^{-9}$$

gives

$$\boxed{\alpha^2\frac{M^0_\oplus}{M_\oplus}\left(\frac{M^0}{M}-\frac{M'^0}{M'}\right)\lambda < 4.10^{-1}\,\text{cm.}}$$

1) example

Lee-Yang model, Phys. Rev. 98 (1955) 1501.

$$J^\mu = \alpha\,\kappa\,M_p\,J^\mu_B$$

$$\frac{M^0_\oplus}{M_\oplus} \sim 1 \qquad \text{(neglect electrons)}$$

and:
$$\frac{M^0}{M}-\frac{M'^0}{M'} = \left(\frac{Z}{A}-\frac{Z'}{A'}\right)\frac{(M_p-M_n+M_e)}{M_p}$$
$$= \delta\left(\frac{Z}{A}\right) 8.4\,10^{-4} \sim \delta\left(\frac{Z}{A}\right)10^{-3}$$

Al: $\frac{Z}{A} = 0.4818$ $\qquad$ Cu $\frac{Z}{A} = 0.4564$ $\qquad \delta\left(\frac{Z}{A}\right) = 5.10^{-2}$

Au $\frac{Z}{A} = 0.4010$ $\qquad$ Pt $\frac{Z}{A} = 0.3995$

$$\boxed{\alpha^2\lambda < 10^2\,\text{m}}$$

2) example

$$J^\mu = \alpha\, \mathcal{K}\left(M_p \bar{\chi}_p \gamma^\mu \chi_p + M_n \bar{\chi}_n \gamma^\mu \chi_n + m_e \bar{\chi}_e \gamma^\mu \chi_e\right)$$

$$\left(\frac{M^0}{M}\right)_\oplus \simeq 1$$

$$\frac{M^0}{M} - \frac{M^{0\prime}}{M'} = \left(\frac{Z}{A} - \frac{Z'}{A'}\right) \times \frac{(M_p - m_n + m_e)\,\Delta E}{M_p^2}$$

$$= \left(\frac{Z}{A} - \frac{Z'}{A'}\right) \times 7.10^{-6} \simeq 3.5.10^{-7}$$

Equivalence principle less violated:

$$\boxed{\alpha^2 \lambda < 10^4 \text{ m}}$$

3) example

$$J^\mu = \alpha\, \mathcal{K}\left(m_u \bar{\chi}_u \gamma^\mu \chi_u + m_d \bar{\chi}_d \gamma^\mu \chi_d + m_e \bar{\chi}_e \gamma^\mu \chi_e\right)$$

$$m_u = 4.8 \text{ MeV} \qquad m_d = 7.5 \text{ MeV}$$

$$\frac{M^0_\oplus}{M_\oplus} = \frac{2m_u + m_d}{M_p} = 0.0182$$

$$\delta \frac{M^0}{M} = \delta\left(\frac{Z}{A}\right) \frac{\left[(M_n - \Delta E)(m_u - m_d + m_e) - (m_u + 2m_d)(M_p - M_n + m_e)\right]}{M_p^2}$$

$$= 0.0023\ \delta\left(\frac{Z}{A}\right) \sim 10^{-4}$$

$$\boxed{\alpha^2 \lambda < 2.10^3 \text{ m}}$$

Conclusion

Eötvös + CPT + absence of test 10m < r < 10^2 km leave a room for an effect with a range of 10^2 - 10^3m.

REFERENCES

1. S. Ferrara and P. van Nieuwenhuizen, Phys. Rev. Letters 37 (1976) 1669.
2. S. Ferrara, J. Scherk and B. Zumino, Nuclear Phys. B121 (1977) 393.
3. C. Zachos, Phys. Letters 76B (1978) 329; Ph.D. Thesis, Caltech (1979).
4. J. Scherk, Phys. Letters 88B (1979) 265;
   Lecture given at the International Conference on Mathematical Physics, Lausanne, 20-25 August 1979, preprint LPTENS 79/19.
   Lecture given at the Stony Brook Supergravity Workshop, in "Supergravity", eds P. van Nieuwenhuizen and D.Z. Freedman, North-Holland (1979), p.43.
   Lecture given at the 11th Summer School of Particle Physics, Gif-sur-Yvette, 3-7 September 1979, preprint LPTENS 79/23.
5. C. Montonen and D. Olive, Phys. Letters 72B (1977) 117.
6. J. Scherk and J.H. Schwarz, Phys. Letters 82B (1979) 60; Nuclear Phys. B153 (1979) 61.
7. E. Cremmer, J. Scherk and J.H. Schwarz, Phys. Letters 84B (1979) 83.

Additional references for the transparencies

- S. Weinberg, Gravitation and Cosmology, John Wiley & Sons (1972).
- M. Gell-Mann Lecture at the Washington Meeting of the American Physical Society, April 1977 (unpublished).
- M.L. Good, Phys. Rev. 121 (1961) 311.
- D.R. Mikkelsen and M.J. Newman, Phys. Rev. D16 (1977) 919.
- D.R. Long, Phys. Rev. D9 (1974) 850; Nature 260 (1976) 417.
- H. Poth, Phys. Letters 77B (1978) 321.
- T.D. Lee and C.N. Yang, Phys. Rev. 98 (1955) 1501.

SUPERGRAVITY AND GAUGE SUPERSYMMETRY

Pran Nath*

CERN, Geneva, Switzerland, and

R. Arnowitt

Department of Physics, Northeastern University
Boston, Massachusetts 02115, USA

1. INTRODUCTION

I would like to talk about the work which Richard Arnowitt and I have done recently regarding the structure of gauge supersymmetry and the relationship of the Riemannian space geometry of gauge supersymmetry to the non-Riemannian superspace geometry of supergravity[1]. Rather than start directly with a discussion of the Riemannian geometry of gauge supersymmetry it is perhaps worth while motivating the discussion by recalling the great amount of progress made in the superspace formulations of supergravity[2] and the nature of the questions that arise in the framework of such formulations. What the superspace formulations of supergravity show is that supergravity obeys a non-Riemannian geometry in superspace with torsion. The problem which we should like to address ourselves to, however, concerns the amount of unification that occurs in supergravity.

While supergravity theories show a remarkable unification of particles of different spins and statistics within their framework, have well defined particle content and are manifestly free of ghosts, a unification of interactions is thus far lacking. The lack of unification of interactions in supergravity arises essentially from the fact that the tangent space group of supergravity is a product group $H = O(3,1) \times O(N)$. An example of this lack of unification is that in the superspace formulations of extended supergravity the Yang-Mills connections and the Lorentzian connections arise in the formalism in an unrelated fashion.

---

*) On Sabbatical leave from Northeastern University, Boston, Massachusetts 02115, USA, presenter of this report.

There is a second novel aspect that appears in the superspace formulations of supergravity. This concerns the fact that the vacuum state of supergravity is postulated to obey the condition that the tangent space covariant derivative for the vacuum state obeys the graded algebra of OSp(N/4), i.e.

$$\left[D_A^{(0)}, D_B^{(0)}\right\} = D_C^{(0)} C^C_{AB} + \frac{1}{2} X^{mn} C_{mnAB} , \qquad (1.1)$$

where C^C_{AB} and C_{mnAB} are the structure constants of OSp(N/4). Equation (1.1) implies then that the vacuum expectation values of some of the components of torsion and curvature are non-vanishing. The assumption of Eq. (1.1) represents a novel aspect, because in the conventional gauge theory formulations one does not normally postulate non-vanishing vacuum expectation values. Rather, in conventional gauge theories one expects that the non-vanishing expectation values of dynamical objects would arise as a consequence of spontaneous or dynamical symmetry breaking. One may ask if such an approach is also possible in supergravity.

Since the tangent space group of extended supergravity is a product structure, it is natural to investigate if such a product structure is embedded in the unified tangent group of a larger theory. One might contemplate a situation anologous to the product group structure $SU(2) \times U(1) \times SU(3)_C$ of weak, electromagnetic, and strong interactions embedded in SU(5), SO(10) or some other grand unified theory. In analogy to the case of grand unified theories one may also expect that the spontaneous breakdown of the larger theory with a unified tangent space group $\mathcal{H}$ would result in the product group structure of supergravity $H = O(3,1) \times O(N)$. One may also hope that in the process of spontaneous breakdown, the vacuum expectation values of the various quantities such as supervielbeins, torsions and curvatures, postulated in the superspace formulations of supergravity would arise from a spontaneous breaking of such a theory.

It is from the viewpoint outlined above that we shall examine gauge supersymmetry, i.e. as a grand unified supergravity theory. In fact, we shall look in detail at the superspace geometry of supergravity as the contracted limit of the larger Riemannian theory whose tangent space group is the unified group $\mathcal{H} = OSp(3,1/4N)$ and contains as a subgroup the tangent space group $H = O(3,1) \times O(N)$ of supergravity. From the analysis of gauge supersymmetry and its spontaneous breakdown one arrives at a number of interesting results regarding supergravity. We list some of the more important results below.

First one finds that the vacuum state of supergravity is precisely deduced as a consequence of the spontaneous breakdown of gauge supersymmetry. One also finds that the Lorentzian and the Yang-Mills connections arise as two parts of a single unified connection of the OSp(3,1/4N) tangent space group. Since for the Riemannian theory a relation exists between the connection and the vielbein in a closed form, one also obtains closed form expressions for the connections of supergravity in terms of the vielbeins as the Riemannian geometry is contracted to the supergravity geometry[3,4]. We exhibit first the Lorentzian part of the supergravity connections. The Lorentzian tangent space connection $h_{mn\Lambda}(z)$ is given by*)

$$h_{mn\Lambda}(z) = h_{mnC}V^{C}{}_{\Lambda}(-1)^{C+\Lambda C} , \tag{1.2}$$

where the components h_{mns} and h_{mnai} of h_{mnC} are given by

$$h_{mns}(z) = \frac{1}{2}\left[\Omega_{mns}(z) + \Omega_{msn}(z) - \Omega_{nsm}(z)\right] \tag{1.3}$$

$$h_{mnai}(z) = \frac{1}{2}\left[\Omega_{main}(z) - \Omega_{naim}(z)\right] . \tag{1.4}$$

The components of $\Omega_{mBn}(z)$ appearing in Eqs. (1.3) and (1.4) are determined by the following relation in terms of the vielbein $V_{\Lambda}{}^{A}$:

$$\Omega_{mBn}(z) = V_{m}{}^{\Lambda}\left[V_{\Lambda,\Sigma}{}^{r} - (-1)^{\Lambda+\Sigma+\Lambda\Sigma}V_{\Sigma,\Lambda}{}^{r}\right](-1)^{B\Sigma}V_{B}{}^{\Sigma}\eta_{rn} . \tag{1.5}$$

---

*) Our notation is as follows. Superspace coordinates are denoted by $z^{A} = (x^{\mu},\theta^{\alpha i})$, where $[x^{\mu},x^{\nu}] = 0 = \{\theta^{\alpha i},\theta^{\beta i}\}$. Here α, β = 1, 2, 3, 4 are the Majorana indices and i, j = 1 ... N are the internal symmetry indices. In general we shall use Greek letters to denote global coordinates and Latin letters to denote the local or tangent space coordinates. The capital letters are assumed to run over the full array of Bose and Fermi indices of superspace. Further, early letters, such as α, β, ... or a, b, ... are used to label Fermi coordinates and late letters such as μ, ν, ... or m, n, ... are used to label the Bose coordinates. Factors such as $(-1)^{\Lambda}$ have values with Λ = (0,1) corresponding to whether the index is Bose or Fermi. η_{mn} denotes the Lorentz metric and η_{ab} (the Fermi metric) is related to the change conjugation matrix C by $\eta_{ab} = (-C^{-1})ab$.

For the O(N) Yang-Mills connections $h_{ij\Lambda}$ of supergravity one obtains

$$h_{ij\Lambda}(z) = \frac{1}{4}\,\bar{\omega}_{aibj\Lambda}\eta^{ab} \tag{1.6}$$

$$\bar{\omega}_{aibjm} = \frac{1}{2}\left[\bar{\Omega}_{aimbj} + \bar{\Omega}_{bjmai}\right] \tag{1.7}$$

$$\bar{\omega}_{aibjck} = \frac{1}{2}\left[\bar{\Omega}_{aibjck} - \bar{\Omega}_{aickbj} - \bar{\Omega}_{bjckai}\right], \tag{1.8}$$

where $\bar{\Omega}_{aicbj}$ is given in terms of the vielbein by the following relation

$$\bar{\Omega}_{aiCbj} = (-1)^{C+1} V_{ai}{}^{\Lambda}\left[V_{\Lambda}{}^{dj}{}_{,\Sigma} - (-1)^{\Lambda\Sigma} V_{\Sigma}{}^{dj}{}_{,\Lambda}\right](-1)^{C\Sigma} V_C{}^{\Sigma}\eta_{db}\,. \tag{1.9}$$

The closed form solutions of the connections in terms of the vielbeins given by Eqs. (1.2)-(1.9) hold for arbitrary values of N. It is understood that in these closed form solutions of the connections in terms of the vielbeins one must only use a gauge complete vielbein with respect to the supergravity transformation laws of an extended supergravity multiplet*). It is in the formulation of a gauge complete vielbein that the restriction $N \leq 8$ is expected to arise. We also note that closed form solutions of the connections in terms of the vielbeins have been obtained through explicit calculations for the case N = 1 by Gates and Siegel[5], though up to now these do not exist for general N. In the solutions that one obtains from the contraction of the Riemannian theory, one finds that the Lorentzian and the O(N) Yang-Mills connections arise from the different components of a single quantity Ω_{ABC}. We see here then a unification of these connections in the Riemannian geometry framework.

The reduction of the Riemannian geometry to the supergravity geometry leads to an apparent paradox. The apparent paradox arises from the fact that the Riemannian superspace geometry has a vanishing superspace torsion. Thus if the non-Riemannian supergravity geometry which has a vanishing torsion arises from the Riemannian geometry, one is led to ask where the non-vanishing

---

*) The technique of gauge completion was first introduced by P. Nath and R. Arnowitt in Ref. 2 in the superspace formulation of supergravity in the metric formalism. Extensive use of this technique has subsequently been made by many authors; see, for example, L. Brink et al., S. Ferrara and P. van Nieuwenhuizen, E. Cremmer and S. Ferrara in Ref. 2. See also S. Ferrara in these proceedings.

torsion of supergravity came from. The resolution of this apparent paradox is rather simple. While the torsion corresponding to the tangent space group OSp(3,1/4N) does indeed vanish in the Riemannian formulation, the torsion corresponding to its subgroup O(3,1) × O(N) does not. It is the non-vanishing O(3,1) × O(N) subpiece of the OSp(3,1/4N) torsion $\mathcal{T}_{AB}^{C}$ that goes over to the supergravity torsion in the limit*) $k \to 0$. Thus one may decompose the vanishing OSp(3,1/4N) torsion $\mathcal{T}_{AB}^{C}$ into two parts

$$\mathcal{T}_{AB}^{C} = \tilde{T}_{AB}^{C} + S_{AB}^{C} , \qquad (1.10)$$

where $\tilde{T}_{AB}^{C}$ represents the O(3,1) × O(N) subpiece of the OSp(3,1/4N) torsion and S_{AB}^{C} is the remainder. Equation (1.10) tells us that the vanishing torsion of the Riemannian theory polarizes into two equal and opposite parts. The supergravity torsion T_{AB}^{C} arises from $\tilde{T}_{AB}^{C}$ on contraction,

$$T_{AB}^{C} = \tilde{T}_{AB}^{C}\Big]_{k=0} . \qquad (1.11)$$

The topics which we should like to discuss in greater detail are the following. First we shall discuss the derivation of the Riemannian geometry in superspace in the supervielbein language. It is the vielbein rather than the metric formulation which is more appropriate for making contact with the supergravity geometry. We shall also discuss briefly how the vacuum state of supergravity arises as a consequence of the spontaneous breaking of gauge supersymmetry. The second topic will concern the reduction of the Riemannian space geometry to the supergravity geometry. In particular, we shall use the reduction of the Riemannian theory to derive the closed form solutions for the supergravity connections exhibited in Eqs. (1.2)-(1.9). Finally we shall discuss properties of gauge supersymmetry as such. These will include the ultraviolet finiteness of the S-matrix of gauge supersymmetry to all quantum loop orders and a discussion of the particle content of some prototype models of gauge supersymmetry.

2. SUPERSPACE GEOMETRY WITH TANGENT GROUP OSp(3,1/4N)

Superspace offers a natural framework for a discussion of the geometrical aspects of local supersymmetry. [For a discussion of superspace, see Zumino[7].] Here we want to discuss the superspace

*) This point was first made in Arnowitt and Nath[6]. (The parameter k arises in the spontaneous breaking solutions of the vacuum of gauge supersymmetry.)

geometry with the tangent space group OSp(3,1/4N), since the full discussion of this geometry represents the general case. We begin by introducing the supervielbein $V_\Lambda{}^A(z)$ which we assume possesses an inverse so that $V_B{}^\Lambda V_\Lambda{}^A = \delta_B{}^A$. Under the global coordinate space transformations $z^\Lambda \to z^{\Lambda'} + \xi^\Lambda(z)$ and the local tangent space transformations parametrized by $\varepsilon_B{}^A$, $V_\Lambda{}^A$ transforms so that*)

$$\delta V_\Lambda{}^A = V_\Lambda{}^A{}_{,\Sigma}\xi^\Sigma + (-1)^{(\Lambda+\Sigma)(A+1)} V_\Sigma{}^A \xi^\Sigma{}_{,\Lambda} + V_\Lambda{}^B \varepsilon_B{}^A \ . \quad (2.1)$$

One may also introduce the covariant derivative of $V_\Lambda{}^A$ by

$$V_\Lambda{}^A{}_{;\Sigma} = V_\Lambda{}^A{}_{,\Sigma} - (-1)^{\Sigma(A+\Delta)} \Gamma_\Lambda{}^\Delta{}_\Sigma V_\Delta{}^A + V_\Lambda{}^B \omega_B{}^A{}_\Sigma \ . \quad (2.2)$$

In Eq. (2.2) $\omega_B{}^A{}_\Sigma$ denotes the supervielbein connection and $\Gamma_\Lambda{}^\Delta{}_\Sigma$ stands for the global space affinity. From Eqs. (2.1) and (2.2) one finds that $\omega_A{}^B{}_\Sigma$ obeys the transformation law

$$\delta\omega_A{}^B{}_\Sigma = \varepsilon_A{}^B{}_{,\Sigma} - \varepsilon_A{}^C \omega_C{}^B{}_\Sigma + (-1)^{\Sigma(B+C)} \omega_A{}^C{}_\Sigma \varepsilon_C{}^B \ . \quad (2.3)$$

Further, one may always define a metric tensor $g_{\Lambda\Sigma}(z)$ of the global space given a vielbein $V_\Lambda{}^A$ by the introduction of a tangent space metric**) η_{AB}. Thus one defines

$$g_{\Lambda\Sigma}(z) = V_\Lambda{}^A \eta_{AB} (-1)^{\Sigma(1+B)} V_\Sigma{}^B \ . \quad (2.4)$$

It follows from Eq. (2.1) and the assumption of a constant tangent space metric η_{AB} that in order that $g_{\Lambda\Sigma}(z)$ obeys the correct transformation law under global coordinate transformations, $\varepsilon_A{}^B$ must obey the condition

$$\varepsilon_A{}^C \eta_{CB} + (-1)^{A+B+AB} \varepsilon_B{}^C \eta_{CA} = 0 \ . \quad (2.5)$$

Further, the general form of a constant tangent space metric which maintains Lorentz covariance and is consistent with its symmetry properties is

---

*) We use right derivatives as in Refs. 3 and 4.

**) The tangent space metric η_{AB} possesses the symmetry property $\eta_{AB} = (-1)^{A+B+AB}\eta_{BA}$.

$$\eta_{AB} = \begin{pmatrix} \eta_{mn} & 0 \\ 0 & k\eta_{ab} \end{pmatrix} . \tag{2.6}$$

We note that the transformation parameters $\varepsilon_B{}^A$ which describe the local tangent space transformations in Eq. (2.1) are not all arbitrary but must satisfy the conditions of Eq. (2.5). Actually the $\varepsilon_B{}^A$ as constrained by Eq. (2.5) generate the tangent space group transformations corresponding to the group OSp(3,1/4N). To see this more clearly, we consider the transformation of a contravariant vector $\phi^A(z)$ so that*)

$$\delta\phi^A = \phi^B \varepsilon_B{}^A . \tag{2.7}$$

We may rewrite Eq. (2.8) so that

$$\delta\phi^A = \phi^B \frac{1}{2} (Z^{MN})_B{}^A \varepsilon_{MN} , \tag{2.8}$$

where $(Z^{MN})_B{}^A$ is the contravariant vector representation of OSp(3,1/4N), i.e.

$$(Z^{MN})_B{}^A = (-1)^N \delta_B{}^M \eta^{NA} - (-1)^{M+MN} \delta_B{}^N \eta^{MA} \tag{2.9}$$

and $\varepsilon_{MN} \equiv \varepsilon_M{}^A \eta_{NA}$. Further, the Z^{MN} obey the graded algebra[8]),

$$\left[Z^{MN}, Z^{PQ} \right\} = f_{RS}^{MNPQ} Z^{RS} , \tag{2.10}$$

where f_{RS}^{MNPQ} represent the structure constants of OSp(3,1/4N):

$$f_{RS}^{MNPQ} = \left[\eta^{PQ} \frac{1}{2} (\delta_R^M \delta_S^Q - (-1)^{R+S+RS} \delta_S^M \delta_R^Q)(-1)^{N+Q+S} \right.$$
$$\left. - (-1)^{MN} (M \leftrightarrow N) \right] - (-1)^{PQ} [P \leftrightarrow Q] . \tag{2.11}$$

---

*) We note here that while there is only one contravariant vector $\phi^A(z)$ there are two types of covariant vectors $\chi_A = \eta_{AB}\chi^B$ and $\hat{\chi}_A = \chi^B \eta_{BA}$ depending on how the indices are lowered. The two types of covariant vectors are characterized by their different transformation properties, i.e. $\delta\chi_A = -\varepsilon_A{}^B \chi_B$ and $\delta\hat{\chi}_A = -(-1)^{A+AB} \hat{\chi}_B \varepsilon_{\hat{A}}{}^B$.

Here η^{PQ} appearing in Eq. (2.11) is the inverse of η_{AB} defined in Eq. (2.6) so that

$$\eta^{PQ} = \begin{pmatrix} \eta^{mn} & 0 \\ 0 & \frac{1}{k}\eta^{ab} \end{pmatrix} . \tag{2.12}$$

We have here exhibited η^{PQ} explicitly to emphasize the appearance of $O(1/k)$ factors in the structure constants of the group OSp(3,1/4N) of Eq. (2.11). This aspect of the structure constants will lead us to choose the correct basis in the Lie algebra of the tangent space group, so that we will take a smooth $k \to 0$ limit in order to make contact with the non-Riemannian geometry of supergravity.

In order to develop the geometry of superspace we need to introduce the definition of the tangent space covariant derivative. First let us introduce the covariant derivative $\mathfrak{D}_\Lambda$ of a tangent space vector ϕ^A. One has

$$\phi^A \overleftarrow{\mathfrak{D}}_\Lambda = \phi^A{}_{,\Lambda} - \phi^B \omega_B{}^A{}_\Lambda . \tag{2.13}$$

A more useful derivative for the definition of the geometry is obtained by projecting $\overleftarrow{\mathfrak{D}}_\Lambda$ into the tangent space as follows

$$\overleftarrow{\mathfrak{D}}_A = \overleftarrow{\mathfrak{D}}_\Lambda V^\Lambda{}_A . \tag{2.14}$$

The quantity $V^\Lambda{}_A$ appearing in Eq. (2.14) may be constructed explicitly by use of the inverse metric*)

$$V^\Lambda{}_A = g^{\Lambda\Sigma} V_\Sigma{}^B \eta_{BA} . \tag{2.15}$$

The reason that the tangent space projection $\mathfrak{D}_A$ of Eq. (2.14) is the useful quantity is that it is in terms of this projection that one defines the curvature and torsion of the space. Thus the curvatures and torsions are introduced through the action of a graded commutator on an arbitrary tensor ϕ as follows:

---

*) The definition of the inverse metric tensor easily follows from the existence of the inverse tangent space metric so that $g^{\Lambda\Pi} = V_A{}^\Lambda \eta^{AB} (-1)^{B(1+\Pi)} V_B{}^\Pi$, where $V_A{}^\Lambda$ is the inverse vierbein satisfying $V_A{}^\Lambda V_\Lambda{}^B = \delta_A{}^B$.

$$\phi\left[\overleftarrow{\mathfrak{D}}_A, \overleftarrow{\mathfrak{D}}_B\right\} = \phi\overleftarrow{\mathfrak{D}}_C \mathcal{T}^C_{AB} + \frac{1}{2}\phi Z^{MN}\mathcal{R}_{MNAB} , \tag{2.16}$$

where Z^{MN} is the representation of the OSp(3,1/4N) generators for the tensor ϕ. Further, computation of the graded commutator on the left-hand side of Eq. (2.16) allows one to determine the curvatures and torsions explicitly and one has

$$\mathcal{T}^C_{AB} = [V^C{}_\Pi \overleftarrow{\mathfrak{D}}_\Lambda - (-1)^{\Pi\Lambda} V^C{}_\Lambda \overleftarrow{\mathfrak{D}}_\Pi](-1)^{\Pi A} V^\Lambda{}_A V^\Pi{}_B \tag{2.17}$$

$$R_{MNAB} = \Big[\omega_{MN\Pi,\Lambda} - (-1)^{\Pi\Lambda}\omega_{MN\Lambda,\Pi} + \frac{1}{2}(-1)^{\Lambda(\Pi+R+S)} f^{PQRS}_{MN}\,\omega_{PQ\Lambda}\omega_{RS\Pi}\Big] V^\Lambda{}_A (-1)^{\Pi A} V^\Pi{}_B , \tag{2.18}$$

where we have defined

$$\omega_{MN\Lambda} \equiv \omega_M{}^P{}_\Lambda \eta_{NP} . \tag{2.19}$$

In the analysis carried out thus far we have treated the vielbein and the connections as independent quantities. However, in the Riemannian space the vierbein covariant derivative $V_\Lambda{}^A{}_{;\Sigma}$ vanishes and a relation between these quantities is thus obtained. Using Eq. (2.2) and the symmetry property of the affinity*) $\Gamma_\Lambda{}^\Delta{}_\Sigma$ one may obtain the following equation for $\omega_B{}^A{}_C$:

$$\omega_B{}^A{}_C - (-1)^{AB+BC+AC}\omega_C{}^A{}_B = V_B{}^\Lambda\Big[V_\Lambda{}^A{}_{,\Sigma} - (-1)^{A(\Lambda+\Sigma)+\Lambda+\Sigma+\Lambda\Sigma} V_\Sigma{}^A{}_{,\Lambda}\Big](-1)^{C\Sigma} V_C{}^\Sigma , \tag{2.20}$$

where $\omega_B{}^A{}_C \equiv \omega_B{}^A{}_\Lambda V^\Lambda{}_C$. Now $\omega_B{}^A{}_C$, etc., possess no symmetry properties in their indices and consequently one cannot directly solve

---

*) The global space affinity $\Gamma_\Lambda{}^\Delta{}_\Sigma$ obeys the following symmetry property

$$\Gamma_\Lambda{}^\Delta{}_\Sigma = (-1)^{\Lambda+\Sigma+\Lambda\Sigma+\Delta(\Lambda+\Sigma)}\Gamma_\Sigma{}^\Delta{}_\Lambda .$$

On using this symmetry to eliminate the affinity from Eq. (2.2), one arrives at Eq. (2.20).

for $\omega_B{}^A{}_C$ from Eq. (2.20). The essential step in the solution of ω's requires that one lower the upper middle index A by η_{DA} and use the symmetry property*)

$$\omega_{ABC} = -(-1)^{AB}\omega_{BAC} \tag{2.21}$$

to solve for ω_{ABC} from Eq. (2.20). One arrives then at the following expression for ω_{ABC} in terms of the vielbein $V_\Lambda{}^A$:

$$\omega_{ABC} = \frac{1}{2}\left[\Omega_{ABC} + (-1)^{BC}\Omega_{ACB} - (-1)^{A(B+C)}\Omega_{BCA}\right], \tag{2.22}$$

where

$$\Omega_{ABC} = V_A{}^\Lambda(-1)^{BC+C}\left[V_{\Lambda C,\Sigma} - (-1)^{C(\Lambda+\Sigma)+\Lambda+\Sigma+\Lambda\Sigma}V_{\Sigma C,\Lambda}\right](-1)^{B\Sigma}V_B{}^\Sigma . \tag{2.23}$$

One may note that the expression for Ω_{ABC} involves $V_{\Lambda C}$, which has its local index lowered by η_{DC}, i.e.

$$V_{\Lambda C} = V_\Lambda{}^D\eta_{DC} . \tag{2.24}$$

Because of the η_{DC} factor, $V_{\Lambda C}$ has now an explicit k dependence and thus Eq. (2.24) will play an important role when we later discuss the reduction of the Riemannian geometry to the supergravity geometry in the limit $k \to 0$.

Equations (2.22) and (2.23) represent the connections of a superspace Riemannian geometry. As in the Bose space Riemannian geometry, the connections involve three permutations of a single Ω_{ABC}. In order to make contact with the non-Riemannian geometry of supergravity, one must consider a contraction of the Riemannian space geometry. Before one can carry out the contraction, however, one must affect a change of basis in the Lie algebra of the Riemannian tangent space group $\mathcal{H} = OSp(3,1/4N)$. The reason for this is that it is only on the new basis that one achieves a smooth $k \to 0$ limit in the subsectors that connect to the supergravity geometry. The new basis X^{MN} is defined by

$$X^{MN} = C^{MN}_{RS}Z^{RS} . \tag{2.25}$$

---

*) Equation (2.21) follows from the requirement that Eq. (2.4) reproduces the correct formula for the covariant derivative for $g_{\Lambda\Sigma}$ when Eq. (2.2) is employed. The symmetries of Eq. (2.21) are consistent with those implied by Eqs. (2.3) and (2.5).

The O(3,1) × O(N) generators in the new basis are related to the components of Z^{RS} as follows*)

$$X^{mn} = Z^{mn} - \frac{i}{4}(k\eta\sigma^{mn})_{cd}Z^{ckdk} \tag{2.26}$$

$$Y^{ij} = Z^{aibj}k\eta_{ba}\ , \tag{2.27}$$

where X^{mn} present the generators of the O(3,1) Lorentzian group and Y^{ij} represent generators of the O(N) internal symmetry group. The connections in the new basis may also be defined so that

$$X^{MN}\tilde{\omega}_{MN\Lambda} = Z^{MN}\omega_{MN\Lambda}\ . \tag{2.28}$$

Hence the O(3,1) × O(N) connections in the new basis are**)

$$\tilde{\omega}_{mn\Lambda} = \omega_{mn\Lambda}\ ,\qquad \tilde{\omega}_{ij\Lambda} = \frac{1}{4}\omega_{aibj\Lambda}\frac{1}{k}\eta^{ab}\ . \tag{2.29}$$

Thus one may express the covariant derivative $\mathcal{D}_\Lambda$ on the new basis to exhibit explicity the O(3,1) × O(N) part of the OSp(3,1/4N) connections**)

$$\mathcal{D}_\Lambda = \partial_\Lambda - \frac{1}{2}\left[X^{mn}\tilde{\omega}_{mn\Lambda} + Y^{ij}\tilde{\omega}_{ij\Lambda}\right] + \left[X^{mai}\tilde{\omega}_{mai\Lambda} + \frac{1}{2}\bar{X}^{aibj}\tilde{\bar{\omega}}_{aibj\Lambda}\right]\ . \tag{2.30}$$

In a similar manner the curvature components on the new basis are defined by

$$X^{MN}\tilde{R}_{MNAB} = Z^{MN}R_{MNAB}\ . \tag{2.31}$$

---

*) The remaining set of generators of OSp(3,1/4N) on the new basis are given by

$$X^{mai} = Z^{mai}$$
$$X^{aibj} = \bar{X}^{aibj} + \frac{1}{4}\frac{1}{k}\eta^{ab}Y^{ij}$$
$$\bar{X}^{aibj} = Z^{aibj} - \frac{1}{4}Z^{cidj}\eta_{dc}\eta^{ab}\ .$$

**) The remaining set of connections of OSp(3,1/4N) on the new basis are

$$\tilde{\omega}_{mai\Lambda} = \omega_{mai\Lambda}$$
$$\tilde{\bar{\omega}}_{aibj\Lambda} = \omega_{aibj\Lambda} - \frac{1}{4}\eta_{ab}\omega_{cidj\Lambda}\eta^{dc} + \frac{i}{4}(k\eta\sigma^{mn})_{ab}\delta_{ij}\omega_{mn\Lambda}\ .$$

The curvature components corresponding to O(3,1) × O(N) parts are then

$$\tilde{R}_{mnAB} = \mathcal{R}_{mnAB} \tag{2.32}$$

$$\tilde{R}_{ijAB} = \frac{1}{4}\mathcal{R}_{aibjAB}\frac{1}{k}\eta^{AB} . \tag{2.33}$$

Here $\tilde{R}_{ijAB}$ are the Yang-Mills superfield strengths. From Eqs. (2.29), (2.32), and (2.33) we see that the Lorentzian and the Yang-Mills parts of connections as well as of curvatures are unified in the larger Riemannian geometry framework.

3. VACUUM SYMMETRIES

We now discuss how gauge supersymmetry deduces the vacuum state of supergravity as a consequence of spontaneous breakdown, in the limit $k \to 0$. Thus in this limit gauge supersymmetry and supergravity possess exactly the same vacuum state. The vacuum metric that results from the spontaneous breakdown of gauge supersymmetry has the form[4]*)

$$g^{(0)}_{\mu\nu}(z) = \eta_{\mu\nu} , \quad g^{(0)}_{\mu\alpha} = -i(\bar{\theta}\Gamma_\mu)_\alpha$$

$$g^{(0)}_{\alpha\beta} = k\eta_{\alpha\beta} + (\bar{\theta}\Gamma^\mu)_\alpha(\bar{\theta}\Gamma_\mu)_\beta , \tag{3.1}$$

where Γ_μ obeys the condition $(\eta\Gamma_\mu)^T = (\eta\Gamma_\mu)$. The full set of transformations that leave invariant the vacuum metric may now be obtained from the killing vectors $\xi^{\Lambda(0)}$ of the space defined by $g^{(0)}_{\Lambda\Pi}$, i.e.

$$0 = \delta g_{\Lambda\Pi}{}^{(0)} = \xi^{(0)}_{\Lambda|\Pi} + \xi^{(0)}_{\Pi|\Lambda} , \tag{3.2}$$

where "|" means the covariant derivative with respect to the vacuum metric $g^{(0)}_{\Lambda\Pi}$. The full set of $\xi^{\Lambda(0)}$ which obey Eq. (3.2) are[4]

---

*) Γ_μ are matrices in the Dirac and internal symmetry space and obey the relations

$$k^{-1}\Gamma_\mu\Gamma^\mu = -\frac{1}{2}(\lambda - K') , \quad k^{-1}\mathrm{Tr}(\Gamma_\mu\Gamma^\mu) = -4(\lambda - \lambda') ,$$

where the parameter λ enters in the action of gauge supersymmetry and λ' and K' arise from quantum loop corrections to the effective potential. The above equations assume a preserved global supersymmetry at the quantum loop level. Explicit solutions to spontaneous breakdown of gauge supersymmetry exist in the tree approximation and one hopes that the solutions to spontaneous symmetry breaking with quantum corrections included also exist.

$$\xi^{\mu(0)} = \left[\varepsilon^{\mu} + \varepsilon^{\mu\nu}x_{\nu}\right] + \{i\bar{\lambda}\Gamma^{\mu}\theta\} \ .$$

$$\xi^{\alpha(0)} = \left[\frac{i}{4}\varepsilon^{\mu\nu}(\sigma_{\mu\nu}\theta)^{\alpha i}\right] + \{\lambda^{\alpha i}\} - (M\theta)^{\alpha i} \tag{3.3}$$

$$\left[\Gamma_{\mu}, M\right] = 0 \ ; \quad M^{T} = -\eta M \eta^{-1} \ . \tag{3.4}$$

From Eqs. (3.3) we see that the preserved symmetries consist of the Poincaré transformations parametrized by ε^{μ} and $\varepsilon^{\mu\nu} = -\varepsilon^{\nu\mu}$, global supersymmetry transformations parametrized by $\lambda^{\alpha i}$, and a set of internal symmetry transformations generated by M obeying the conditions of Eq. (3.4). Further, in the vielbein language the spontaneous symmetry-breaking solutions may be expressed in the form of a vacuum vielbein*)

$$V_{\mu}{}^{m(0)} = \delta_{\mu}{}^{m} \ , \quad V_{\alpha}{}^{m(0)} = i(\bar{\theta}\Gamma^{m'})_{\alpha}$$

$$V_{\mu}{}^{a(0)} = 0 \ , \quad V_{\alpha}{}^{a(0)} = \delta_{\alpha}{}^{a} \ . \tag{3.5}$$

One may now proceed to determine the set of vacuum and tangent space transformations that leave the vacuum vielbein invariant. For $\xi^{\Lambda(0)}$ one obtains the same result as in Eq. (3.3), while for $\varepsilon_{A}{}^{B(0)}$ one obtains

$$\varepsilon_{m}{}^{n(0)} = \varepsilon_{m}{}^{n} \ , \quad \varepsilon_{ai}{}^{m(0)} = 0 = \varepsilon_{m}{}^{ai(0)}$$

$$\varepsilon_{ai}{}^{bj(0)} = -\frac{i}{4}\varepsilon_{mn}(\sigma^{mn})_{a}{}^{b}\delta_{i}{}^{j} - M^{bj}{}_{ai} \ . \tag{3.6}$$

From Eq. (3.6) we see that after spontaneous breaking the tangent space group invariance OSp(3,1/4N) is broken and the residual symmetries consist only of the set corresponding to O(3,1) × $\mathcal{G}$(N), where the group $\mathcal{G}$(N) corresponds to the internal symmetries generated by M of Eq. (3.4). It can also be easily shown[4] that as one takes the limit $k \to 0$, the group $\mathcal{G}$(N) expands to the group O(N), so that the residual tangent space symmetries in the limit $k \to 0$ consist of O(3,1) × O(N). These are precisely the tangent space symmetries of extended supergravity.

---

*) $\Gamma^{m'}$ plays the same role for the vielbein case that Γ^{m} does for the metric formulation.

4. REDUCTION OF RIEMANNIAN GEOMETRY

We discuss next the actual reduction of the non-Riemannian supergravity geometry. We have seen already that the unified tangent space group OSp(3,1/4N) of the Riemannian theory contains as subgroups the tangent space group O(3,1) × O(N) of extended supersymmetry, which turns out to be indeed the only preserved group of the tangent space of the Riemannian theory after spontaneous breaking in the limit $k \to 0$. The reduction of the Riemannian geometry to the supergravity geometry thus involves both the extraction of the correct O(3,1) × O(N) subgroup of the Riemannian tangent space as well as the limit $k \to 0$. To exhibit this more clearly we first carry out the decomposition of $\mathcal{D}_\Lambda$ into its O(3,1) × O(N) subpart and a remainder

$$\mathcal{D}_\Lambda = \tilde{D}_\Lambda + K_\Lambda \tag{4.1}$$

$$\tilde{D}_\Lambda = \overleftarrow{\partial}_\Lambda - \frac{1}{2}\left[X^{mn}\tilde{\omega}_{mn\Lambda} + Y^{ij}\tilde{\omega}_{ij\Lambda}\right] \tag{4.2}$$

$$K_\Lambda = -\left[X^{mai}\tilde{\omega}_{mai\Lambda} + \frac{1}{2}\bar{X}^{aibj}\tilde{\bar{\omega}}_{aibj\Lambda}\right], \tag{4.3}$$

where both $\tilde{D}_\Lambda$ and K_Λ are O(3,1) × O(N) covariant quantities. Thus in the formulation of a geometry with an O(3,1) × O(N) tangent space one may use either or both in a linear combination. The formulation of supergravity geometry however, requires that one uses only the reduced D_Λ in formulating this theory, i.e. the quantity

$$D_\Lambda = \tilde{D}_\Lambda\Big|_{k\to 0} . \tag{4.4}$$

D_Λ represents the tangent space covariant derivative for supergravity involving the Lorentzian connection $h_{mn\Lambda}$ and the Yang-Mills connection $h_{ij\Lambda}$:

$$D_\Lambda = \partial_\Lambda - \frac{1}{2}\left[X^{mn}h_{mn\Lambda}(z) + Y^{ij}h_{ij\Lambda}(z)\right], \tag{4.5}$$

where $h_{mn\Lambda}$ and $h_{ij\Lambda}$ are to be calculated. Since the connections for Riemannian space are known in terms of the vielbeins in a closed form, one also expects closed form solutions for the supergravity connections through the reduction of Eq. (4.4). The results of this reduction were displayed in Section 1. However, before we discuss the actual deduction of the closed forms, we should like to mention briefly the direct technique that one may always use for the construction of these connections. This technique is the gauge completion method and we illustrate it first for the construction of the vielbein $V_\Lambda{}^A$ that enters in the superspace formulations of supergravity.

The gauge completion method for the construction of the vielbein $V_\Lambda{}^A$ of supergravity consists in computing $V_\Lambda{}^A$ order by order in θ^α, so that the transformation laws of superspace Eq. (2.1) are consistent with the supergravity transformation laws*)

$$\delta e_\mu{}^m = i\bar{\psi}_\mu \gamma^m \lambda \ , \quad \frac{1}{2}\,\delta\psi_\mu = D_\mu \lambda \ . \tag{4.6}$$

Thus one starts with the boundary conditions on $V_\Lambda{}^A$ which are just the zeroth order terms in θ^α, i.e.

$$V_\mu{}^m(\theta = 0)^{(0)} = e_\mu{}^m(x) \ , \quad V_\mu{}^a(\theta = 0)^{(0)} = \frac{1}{2}\,\psi_\mu{}^a$$

$$V_\alpha{}^m(\theta = 0)^{(0)} = 0 \ , \quad V_\alpha{}^a(\theta = 0)^{(0)} = \delta_\alpha{}^a \ . \tag{4.7}$$

Similarly one assumes an expansion for $\xi^\Lambda(z)$ starting with

$$\xi^\mu(z) = i\bar{\lambda}\gamma^\mu\theta \ , \quad \xi^\alpha(z) = \lambda^\alpha \tag{4.8}$$

and allowing λ^α to become a function of x so that $\lambda^\alpha = \lambda^\alpha(x)$. Consistency of Eqs. (4.7) and (4.8) then demands specific higher order θ^α field dependent terms**).

A similar procedure can be carried out for the construction of the superconnections of supergravity through the gauge completion method. Thus in the analysis of Brink et al.[2] one uses the transformation equations for $h_{mn\Lambda}(z)$, i.e.

$$\delta h_{mn\Lambda}(z) = \varepsilon_{mn}(z)_{,\Lambda} + (-1)^{\Lambda+\Sigma\Lambda}\xi^\Sigma{}_{,\Lambda}\, h_{mn\Sigma} + h_{mn,\Sigma}\xi^\Sigma \tag{4.9}$$

---

*) D_μ appearing in Eq. (4.6) is defined by

$$D_\mu = \partial_\mu + \frac{i}{4}\,\sigma^{mn}\omega_{mn\mu}(x) \ ,$$

where $\omega_{mn\mu}(x)$ is the sum of the vierbein affinity and a spin 3/2 torsion (see Ref. 1).

**) We note also that gauge completion in the higher order in θ sectors without use of field equations would also require introduction of auxiliary fields.

and the boundary conditions at $\theta^{\alpha} = 0$, i.e.

$$h_{mn\mu}(x;\ \theta^{\alpha} = 0) = \omega_{mn\mu}(x) \qquad (4.10)$$

to construct $h_{mn\Lambda}(z)$ in a manner similar to the one described above to construct the vielbein. A similar procedure can also be used to construct the Yang-Mills connection $h_{ij\Lambda}(z)$, where the boundary conditions chosen are

$$h_{ij\mu}(x;\ \theta^{\alpha i} = 0) = eA_{\mu}{}^{ij}(x) \ , \qquad (4.11)$$

where $A_{\mu}{}^{ij}(x)$ are the Yang-Mills potentials and e is the parameter that gauges the internal symmetry in SO(N) supergravity models[9]. Next we discuss how the contraction procedure generates the closed form solutions for the supergravity connections.

From Eqs. (4.2), (4.4), and (4.5) we have the following expressions for the supergravity connections $h_{mn\Lambda}$ and $h_{ij\Lambda}$:

$$h_{mn\Lambda}(z) = \left[\tilde{\omega}_{mn\Lambda}(z)\right]_{k\to 0} \qquad (4.12)$$

$$h_{ij\Lambda}(z) = \left[\frac{1}{4}\,\tilde{\omega}_{aibj\Lambda}\,\frac{1}{k}\,\eta^{ba}\right]_{k\to 0} \ . \qquad (4.13)$$

From Eqs. (2.23) and (2.24) we find that

$$\Omega_{ABci} \sim O(k) \ , \quad \Omega_{ABm} \sim O(1) \ . \qquad (4.14)$$

Equations (4.12) and (4.14) then immediately lead to Eqs. (1.3) and (1.4). To verify that the connections of Eq. (1.3) and (1.4) are indeed the correct Lorentzian parts of the supergravity, we note that the $h_{mn\Lambda}$ of Eq. (4.12) obey the transformation law of Eq. (4.9). Further by a direct substitution of the gauge complete vielbein one can also verify that the boundary conditions of Brink et al.[2] for the cases N = 1, 2 of Eq. (4.10) are also reproduced.

In a similar fashion Eqs. (4.13) and (4.14) lead to Eqs. (1.6)-(1.8), though the limit is somewhat more subtle owing to the 1/k factor appearing in Eq. (4.13). Since one has a Majorana trace appearing in Eq. (4.13), the singular 1/k term actually cancels out at $\theta^{\alpha} = 0$ and thus the limit is smooth as $k \to 0$. Again the $h_{ij\Lambda}$ of Eqs. (4.13) or (1.6) satisfies the correct transformation laws appropriate for a Yang-Mills connection and the set of Eqs. (1.6)-(1.8) yield the correct boundary conditions*). Finally, a similar reduction procedure can also be employed to obtain the curvatures of supergravity. Thus the supergravity curvature R_{mnAB} and the Yang-Mills superfield strengths of supergravity are obtained directly from Eqs. (2.32) and (2.33). One has

$$R_{mnAB} = \tilde{R}_{mnAB}\Big]_{k\to 0} \,, \quad F_{ijAB} = \tilde{R}_{ijAb}\Big]_{k\to 0} \,. \tag{4.15}$$

5. PROPERTIES OF GAUGE SUPERSYMMETRY

The preceding discussion shows that the geometrical aspects of the supergravity theory arise in a natural and unified way from the reduction of the Riemannian space geometry of gauge supersymmetry. The Riemannian dynamics of gauge supersymmetry is essentially unique except for the presence of the parameter λ, which plays the role of a supercosmological constant in the theory (the theory is actually free of a true cosmological constant). A remarkable feature of the Riemannian dynamics of gauge supersymmetry is the ultraviolet finiteness of the theory. What actually has been shown is that the off-sheel Green's functions of gauge supersymmetry for $N \geq 2$ are ultraviolet finite to all orders in the quantum loop in the linearized harmonic gauge[10]. This result is then sufficient to establish the ultraviolet finiteness of the S-matrix of gauge supersymmetry.

The proof of the ultraviolet finiteness property of the Green's functions consists first in establishing that the zeroth order tensor propagator $\Delta_{\Pi\Lambda\Sigma\Delta}(z_1,z_2) = \langle T[h_{\Pi\Lambda}(z_1)h_{\Sigma\Delta}(z_2)]\rangle$, defined in the background field $g^{(0)}_{\Pi\Lambda}$ with $g_{\Pi\Lambda} = g^{(0)}_{\Pi\Lambda} + h_{\Pi\Lambda}$, has the decomposition in momentum space of the form**)

$$\Delta_{\Pi\Lambda\Sigma\Delta}(p;\,\theta_1,\theta_2) = e^{\bar{\omega}p_\mu\Gamma^\mu\xi}\sum_{n=1}^{4} F^{(n)}{}_{\Pi\Lambda\Sigma\Delta}(p;\,\omega^\alpha)P_n(\xi^\alpha) \,, \tag{5.1}$$

when the quantization of the theory is carried out in a manifestly globally supersymmetric manner. Here ω^α and ξ^α are the relative and centre-of-mass Fermi coordinates, i.e.

$$\omega^\alpha = (\theta_1{}^\alpha - \theta_2{}^\alpha) \,, \quad \xi^\alpha = \frac{1}{2}(\theta_1{}^\alpha + \theta_2{}^\alpha) \,. \tag{5.2}$$

---

*) As one goes to higher orders in θ^α, possible non-zero $1/k$ terms can be omitted since they represent irrelevant constants of integration.

**) A similar analysis holds for the ghost propagators.

Equation (5.1) contains the full content of the constraints of global supersymmetry on the propagators. The dynamics of gauge supersymmetry further requires that the zeroth order propagators of Eq. (5.1) obey superspace wave equations. An iteration procedure in θ can be used on these wave equations to compute the large Euclidean behaviour of the form factors $F^{(n)}{}_{\Pi\Lambda\Sigma\Delta}$. In the linearized harmonic gauge one finds for large Euclidean momenta the form

$$F^{(m)}(p;\,\omega^{\alpha}) \simeq \Sigma A_n p^{-(2+4N-n)} \omega^{\alpha_1} \ldots \bar{\omega}^{\alpha_n} \,. \qquad (5.3)$$

From Eq. (5.3) one notes that the form factors exhibit a dimensional rule where for every one less factor of ω^{α} the form factor falls off faster by a power of momentum.

Using the large Euclidean momentum behaviour of Eq. (5.3) one may next establish that the degree of divergence of an n-sided one-loop supergraph obeys the condition

$$d_{n,m} \leq m(4 - 4N) - 2\,(m - 1)\,. \qquad (5.4)$$

Then for $N \geq 2$ one has that the over-all degree of divergence is negative and all the subintegrations are finite. Finiteness of the off-shell Green's function in the linearized harmonic gauge then result from an application of Weinberg's theorem[11].

A question which naturally arises as a result of the ultraviolet finiteness of gauge supersymmetry concerns whether this theory possesses ghosts. A definitive answer to this question as well as to the broader question of the full particle content of gauge supersymmetry is still lacking. However, some insights have been gained through study of some prototype models which possess many of the properties exhibited by gauge supersymmetry. Thus the models investigated as prototypes share with gauge supersymmetry the properties that the quadratic part of their action is initially non-diagonal. They are invariant under global supersymmetry transformations and possess a mass term or a mass parameter. Here we shall discuss two such models briefly.

The first model investigated is a scalar model involving only spin-zero and spin-half fields. It is described by the Lagrangian density

$$\begin{aligned}\mathcal{L} = &-\Big(\partial_\mu A\partial^\mu A_1 + \frac{1}{2}A_1{}^2 + \partial_\mu B\partial^\mu B_1 + \frac{1}{2}B_1{}^2 \\ &+ \frac{1}{2}\partial_\mu F\partial^\mu F + \frac{1}{2}\partial_\mu G\partial^\mu G\Big) \\ &+ \bar{\psi}\frac{1}{i}\gamma^\mu\partial_\mu\psi_2 - \frac{1}{2}\bar{\psi}_1\frac{1}{i}\gamma^\mu\partial_\mu\psi_1 + \bar{\psi}_2\psi_1 \\ &- m^3\Big(AF + BG + \frac{1}{2}\bar{\psi}\psi\Big)\,. \end{aligned} \qquad (5.5)$$

Actually the model of Eq. (5.5) for the case m = 0 was investigated by Ferrara and Zumino[12] as a prototype of conformal supergravity[13]. In their analysis, Ferrara and Zumino found a set of dipole and tripole ghost states appearing in the theory. To exhibit their result we consider the A, A_1 sector of Eq. (5.5) when $m = 0$. Here one can introduce two creation operators $a^*(p)$ and $a_1{}^*(p)$ and the set of one-particle states $|p\rangle = a^*(p)|0\rangle$ and $|p\rangle_1 = a_1{}^*(p)|0\rangle$ obeying the normalization

$$\langle p|p'\rangle = 0 = {}_1\langle p|p'\rangle_1 \ , \ \langle p|p'\rangle_1 = \delta^3(p - p') \ . \qquad (5.6)$$

However, one cannot diagonalize the Hamiltonian in this case, as can be seen by noting that only one of the states introduced above is an eigenstate of energy. Thus defining $\omega \equiv |\vec{p}|$ one has

$$P^0|p\rangle_1 = \omega|p\rangle_1 \ , \quad P^0|p\rangle = \frac{1}{2\omega}|p\rangle_1 + \omega|p\rangle \ . \qquad (5.7)$$

This is a feature considered characteristic of dipole ghosts. A similar phenomena appears in the B, B_1 sector, while in the ψ, ψ_1, and ψ_2 sector one has a system of tripole ghost states appearing.

On the other hand, in gauge supersymmetry one has a mass parameter appearing in the theory after spontaneous breakdown, so that the situation characteristic of gauge supersymmetry in Eq. (5.5) would correspond to the case when $m \neq 0$. When $m \neq 0$, one finds that the A, A_1 sector is now coupled to the field F and one must diagonalize the quadratic action involving the fields A, A_1 and F in Eq. (5.5). The eigenmass spectrum of the system now arises from the condition

$$\det(K_{ij}p^2 + M_{ij}) = 0 \ , \qquad (5.8)$$

where K_{ij} and M_{ij} represent elements of the kinetic energy and mass matrices of the A, A_1, and F system. Solutions to Eq. (5.8) show that the eigenmasses are given by $p^2 = \mu^2{}_\lambda$, where

$$\mu^2{}_{\lambda=1} = m^2 \ , \quad \mu^2{}_{\lambda=2,3} = \mu^2{}_\pm = -m^2 \, e^{\pm\pi i/3} \ . \qquad (5.9)$$

Equation (5.9) exhibits a set of complex conjugate poles rather than the dipole ghost structure that emerged when $m = 0$. Further, one may carry out a creation and annihilation operator expansion of the field $\phi_i = (A, A_1, F)$ allowing for the complex energy eigensolutions in the expansion†). One may then establish that a set of creation operators $a^{\lambda *}$ and annihilation operators a^λ which exist for the coupled system obey the commutation relations

$$\left[a^\lambda(p), a^{\lambda'}(p')\right] = 0 \ , \quad \left[a^\lambda(p), a^{\lambda'}(p')^*\right] = \delta_{pp'} K^{-1}{}_{\lambda\lambda'} \ . \quad (5.10)$$

It is possible now to diagonalize the Hamiltonian of the A, A_1, and F system unlike the case of Ferrara and Zumino[12] and one has

$$H = \sum_{p,\lambda} \omega_\lambda(p)\left[N_\lambda(p) + \frac{1}{2}\right] , \quad (5.11)$$

where N_λ are a set of commuting operators such that

$$N_1 = a_1{}^* a_1 \ , \quad N_2 = a^{(-)*} a^{(+)} \ , \quad N_3 = a^{(+)*} a^{(-)} \ . \quad (5.12)$$

In fact the entire analysis now is equivalent to that of Lee and Wick[14]. An exactly identical phenomenon occurs in the remaining sectors of Eq. (5.5).

Several other prototype models have also been investigated involving features not contained in the simple model of Eq. (5.5) such as inclusion of vector particles and existence of residual gauge symmetries in the theory in addition to global supersymmetry. We mention here briefly only one such model described by

$$\mathcal{L} = -\frac{1}{4} V_{\mu\nu} V^{\mu\nu} - \frac{i}{2} \bar{\lambda}\gamma^\mu \partial_\mu \lambda + \frac{1}{2} D^2$$
$$- \frac{1}{4a^2} \Box^2 V_{\mu\nu} \Box^2 V^{\mu\nu} - \frac{i}{2} \Box^2 \bar{\lambda}\gamma^\mu \partial_\mu \Box^2 \lambda + \frac{1}{2a^2} (\Box^2 D)^2 \ . \quad (5.13)$$

In the vector meson sector the eigenmass spectrum now consists of

$$\mu^2{}_{\lambda=1} = 0 \ , \quad \mu^2{}_{\lambda=2,3} = \mu^2{}_\pm = \pm i|a| \ . \quad (5.14)$$

---

†) One has

$$\phi_i(x) = \sum_{\vec{p},\lambda} \left(\frac{1}{2\omega_\lambda V}\right)^{1/2} \left[u_i{}^\lambda(p) a^\lambda(p)\, e^{ip^\lambda x}\right.$$
$$+ \left(\frac{1}{2\omega_\lambda{}^* V}\right)^{1/2} \left[u_i{}^{\lambda *}(p) a^{\lambda *}(p)\, e^{-ip\lambda^* x}\right]\} ,$$

where the column vectors u_i (p) satisfy the condition

$$\left[K_{ij}p^2 + M_{ij}\right] u_j^{\lambda}(p) = 0 .$$

Thus once again one has a set of complex conjugate poles, but now with a massless gauge particle (a photon) remaining. This constitutes an interesting example, since gauge supersymmetry possesses specific residual gauge invariances in addition to global supersymmetry after spontaneous breaking.

One may recall that in a theory which exhibits the Lee-Wick phenomenon, one may define a unitary S-matrix by allowing the in and out states to involve only the normal states[15]. However, there is a loss of strict causality for a theory of this type[16]. For the Lee-Wick electrodynamics, the mass parameter of the heavy boson appearing in the theory is of the order of ~ 10 GeV and acausal effects are expected, in intervals $< 10^{-23}$ s. For gauge supersymmetry the size of the mass parameter is much larger, i.e. $\sim 10^{18}$ GeV, so that acausal effects would manifest themselves only in exceedingly small intervals, i.e. $< 10^{-40}$ s!

Acknowledgements

This research was supported in part by the National Science Foundation under Grant Nos. PHY78-09613 and PHY77-22864.

REFERENCES

1) D.Z. Freedman, P. van Nieuwenhuizen and S. Ferrara, Phys. Rev. D13:3214 (1976); S. Deser and B. Zumino, Phys. Lett. 62B:335 (1976).

2) P. Nath and R. Arnowitt, Phys. Lett. 65B:73 (1976); J. Wess and B. Zumino, Phys. Lett. 66B:361 (1977); V. Ogievetsky and E. Sokatchev, Nucl. Phys. B124:39 (1977); S.J. Gates, Phys. Rev. D16:1727 (1977); R. Grimm, J. Wess and B. Zumino, Phys. Lett. 73B:15 (1978); Phys. Lett. 74B:51 (1978); L. Brink, M. Gell-Mann, P. Ramond and J.H. Schwarz, Phys. Lett. 74B:336 (1978); Phys. Lett. 76B:417 (1978); R. Arnowitt and P. Nath, Phys. Lett. 78B:581 (1978); W. Siegel, Nucl. Phys. B142:301 (1978); W. Siegel and S.J. Gates, Nucl. Phys. B147:77 (1979); J.G. Taylor, Phys. Lett. 78B:577 (1978); Phys. Lett. 79B:399 (1978); Y. Ne'eman and T. Regge, Riv. Nuovo Cimento Ser. III

1:(No. 5) (1978); F. Mansouri, Proc. Integrative Conf. on Group Theory and Mathematical Physics, Austin, Texas, 1978, Springer-Verlag, N.Y. (1979). S. MacDowell, Phys. Lett. 80B: (1979); L. Brink and P. Howe, Phys. Lett. 88B:268 (1979); E. Cremmer and S. Ferrara, Ecole Normale Superior Preprint, 1980.

3) R. Arnowitt and P. Nath, Phys. Rev. Lett. 44:223 (1980).

4) P. Nath and R. Arnowitt, Supergravity geometry in superspace, Nucl. Phys. B165:462 (1980).

5) W. Siegel, Nucl. Phys. B142:301 (1978); W. Siegel and S.J. Gates, Nucl. Phys. B147:77 (1979). B. Zumino (private communication).

6) R. Arnowitt and P. Nath, in On the path of Albert Einstein, (eds. B. Kursunuglu, A. Perlmutter and L.T. Scott), Plenum, N.Y. (1979), p. 103.

7) B. Zumino, Invited talk in these proceedings. The earliest discussion regarding the relevance of the parameter k for superspace geometries is given in B. Zumino, Proc. Conf. on Gauge Theories and Modern Field Theory, Boston, 1975, MIT Press, Cambridge, Mass. (1976) (eds. R. Arnowitt and P. Nath), p. 256.

8) R. Arnowitt and P. Nath, 1975, unpublished; P.G.O. Freund, J. Math. Phys. 17:424 (1976).

9) D.Z. Freedman and A. Das, Nucl. Phys. B120:221 (1977). S.W. MacDowell and F. Mansouri, Phys. Rev. Lett 38:739 (1977)

10) P. Nath and R. Arnowitt, Phys. Rev. Lett. 42:138 (1979). R. Arnowitt and P. Nath, Nucl. Phys. B161:321 (1979).

11) S. Weinberg, Phys. Rev. 118:838 (1960).

12) S. Ferrara and B. Zumino, Nucl. Phys. B134:301-326 (1978).

13) M. Kaku, P.K. Townsend and P. van Nieuwenhuizen, Phys. Lett. 69B:304 (1977); Phys. Rev. Lett. 39:1109 (1977). S. Ferrara, M. Kaku, P.K. Townsend and P. van Nieuwenhuizen, Nucl. Phys. B129:125 (1979).

14) T.D. Lee and G.C. Wick, Nucl. Phys. B9:209 (1969); 10:1 (1969).

T.D. Lee, Quanta, University of Chicago Press (1970), p. 260.

15) R.E. Cutkosky, P.V. Landshoff, D.I. Olive and J.C. Polkinghorne, Nucl. Phys. B12:281 (1969).

16) S. Coleman, "Subnuclear Phenomena", Academic Press, New York (1970), p. 260. N. Nakanishi, Phys. Rev. D3:811 (1970).

THE FERMION SPECTRUM ACCORDING TO GUTs

D.V. Nanopoulos

CERN

1211 Geneva 23, Switzerland

PROLOGUE

Gauge theories seem to be nature's favourite candidate for describing correctly and elegantly the "observed" (fundamental?) interactions. Among other things gauge theories are supposed to shed light on and improve our understanding of many of the age-old "mysteries" of the observed fermion spectrum. Clearly such hopes have been at least partially fulfilled, especially with the advance of grand unified theories, but still a lot of "mysteries" remain unsolved. My purpose here is to report on our present status of understanding of the fermion spectrum including some old and some very recent work[1] on the subject.

1. QUESTIONS

In order to appreciate the strength of grand unified theories[2] (i.e., theories that unify strong and electro-weak interactions) let us see what kind of world is opened to us by the "standard" $SU(3)_C \times SU(2) \times U(1)$ ($\equiv G_1$) model. The well-known fermionic structure is as follows:

$$\text{generation} \equiv \begin{cases} \begin{pmatrix} u \\ d \end{pmatrix}_L & ; \quad u_R,\ d_R \\ \\ \begin{pmatrix} \nu_e \\ e \end{pmatrix}_L & ; \quad e_R \end{cases} \qquad (1)$$

with the understanding that quarks come in three colours (leptons are colour singlets) and there are at least two more generations with identical structure ((c,s,μ,ν_μ) ; (t,b,τ,ν_τ)). Definitely such a fermionic structure is consistent with all available "low energy" phenomenology ($E < 100$ GeV (?)) but certainly it leaves unanswered too many questions. Let us enumerate some of them:

1) Why is $Q_{proton} + Q_{electron} = O(< 10^{-20})$?

2) Why do leptons and quarks have similar weak interactions, e.g., why is there

$$\begin{pmatrix} \nu_e \\ e \end{pmatrix}_L \quad \text{and not} \quad \begin{pmatrix} e^+ \\ \bar{\nu}_e \end{pmatrix}_L \quad \left(\text{or equivalently} \quad \begin{pmatrix} \nu_e \\ e \end{pmatrix}_R \right) ?$$

3) Why are quarks heavier than leptons belonging to the same generation ?

4) How many generations are there ?

5) Why are the charged and neutral weak currents mixtures of V and A ? Or equivalently why do L- and R-fermionic components transform differently under SU(2) ?

6) Why do all the observed generations seem to have identical structure under G_1 ? (i.e., the so-called generation reproduction problem).

7) Can we understand the mass gaps between different generations ?

8) What is the origin of the Cabibbo-like angles and why are they small ?

9) Why is $t > b$ and $c > s$ but $u < d$? *)

10) Are the neutrinos really massless ?

We will show next that in the framework of GUTs most of these questions are satisfactorily answered.

2. ANSWERS ("THE CLASSICS")

The basic hypothesis of the "grand synthesis" is that there exists a simple group $G \supset G_1$, which is characterized by a single coupling constant g, and that all interactions are generated by G. Quarks and leptons are in general members of the same multiplet(s) of the

*) The names of quarks and leptons represent also their masses.

$$\frac{m_b(Q=Y)}{m_\tau} \simeq \left[\frac{\alpha_s(Q=Y)}{\alpha_s(Q=M_1)}\right]^{\frac{12}{33 - 2f}} \qquad (3)$$

where α_s is the strong coupling constant evaluated at the mass of Y, $M_1 \sim 10^{15}$ GeV and f is the number of flavours. Using experimental data to determine α_s and taking f = 6, we get[4] from Eq. (3) $m_b/m_\tau \simeq 2.7$-3 which is the observed value. So we are definitely on the right track.

4) A very important point to make here is that Eq. (3) is sensitive[4] to the number of flavours (f). Two loop corrections[5] to (3) determine uniquely[5] that f > 6 would imply an unaccepted m_b/m_τ value. So GUTs not only explain why quarks are heavier than leptons but also determine that there are only three generations[4,5]. This piece of news is most welcome to cosmologists, who have determined[6] independently that the observed He abundance of the universe implies not more than three light neutrinos, i.e., no more than three generations. We move next to some new applications of GUTs.

3. ANSWERS ("MODERN TIMES")

Before getting to the real meat of the new applications of GUTs, let me formulate a few principles that we should find helpful in the following. Undoubtedly physics is living through its GAUGE AGE and most of us are believers in the

Central Dogma of the gauge age

" a) All fundamental interactions are gauge interactions.

b) Gauge symmetries may be broken only spontaneously.

c) Exact symmetries in Nature are always local. "

I want to call attention to c) which says that that Nature does not like to have around exact global symmetries not coming from local symmetry*), which is of great importance in our case in discussing neutrino masses. Another principle of great importance is related to the symmetry breaking scheme as in Eq. (2). It is natural to assume that at each step of symmetry breaking, fermions that may receive masses that are invariant under the unbroken group indeed do so, and thus we are led to the

*) For instance non-perturbative gravitational effects break any global (but not local) symmetry[7].

group G. Then at some superhigh energy scale ($M_1 \sim 10^{15}$ GeV) G suffers a breaking down to G_1 and then around $M_2 \sim 100$ GeV G_1 breaks down in the usual way to $G_2 \equiv SU(3)_c \times U(1)_{E.M.}$:

$$G \xrightarrow[M_1(\sim 10^{15}\text{ GeV})]{} G_1\left(\equiv SU(3)_c \times SU(2) \times U(1)\right) \xrightarrow[M_2(\sim 10^{2}\text{ GeV})]{} \tag{2}$$

$$\longrightarrow G_2\left(\equiv SU(3)_c \times U(1)_{E.M.}\right)$$

If such a scenario is realized in nature then a few things follow immediately.

1) Since G is a simple group, electric charge is quantized. Actually the charge operator Q is a generator of the group G and thus traceless and so when it is acting on any representation of G containing both quarks and leptons it is bound to give some relation between quark and lepton charges. For example, in SU(5)[3] one finds that $Q_d = 1/3\, Q_{e^-}$, where the factor of three appears because there are just three colours! So there is no mystery anymore why $Q_{proton} + Q_{elect} = 0$.

2) Since quarks and leptons are sitting in the same representation(s) of the group G, it is natural to find correlations between their helicities and similar responses to weak interactions which are contained in the G-interactions. For example, in SU(5) one finds that $\binom{\nu}{e}_L$ and $\bar{d}_L$ are contained in the same multiplet ($\bar{5}$) and so d_R is not available to have SU(2) weak interactions. Only d_L is still available, and behold another multiplet (namely the 10_L) contains the $\binom{u}{d}_L$ doublet and there is no more mystery why

$$\begin{pmatrix} u \\ d \end{pmatrix}_L \quad \text{goes with} \quad \begin{pmatrix} \nu \\ e \end{pmatrix}_L \quad \text{and not} \quad \begin{pmatrix} \nu \\ e \end{pmatrix}_R \quad .$$

3) The fact that quarks and leptons share the same representation(s) of G, has another important consequence. Namely, there are relations between quark and lepton masses. At superhigh energies where G is a good symmetry, quark and lepton masses are more or less equal (modulo some possible Clebsch-Gordan coefficients). At lower energies though, when G is badly broken, higher order corrections due to the "standard" SU(3) × SU(2) × U(1) interactions modify these "symmetry relations" in an appropriate way: making quarks heavier than leptons, mainly because quarks have strong interactions and leptons do not. Quantitatively, one finds[4] that if $m_b = m_\tau$ at superhigh energies as it happens in many models, then

Survival hypothesis[1,8]: "Low mass fermions are those that cannot receive G_1-invariant masses"*).

Armed with the above principles we now try to answer the rest of the questionnaire.

5) According to the survival hypothesis, fermions may not have their L- and R-components transforming identically under SU(3) × SU(2) × × U(1), because then they will all have masses $M_1 \sim 10^{15}$ GeV. Since nature chooses to have $SU(3)_c \times U(1)_{E.M.}$ "vector-like" our only hope is that SU(2) is chiral. The simplest possibility for non-trivial weak interactions is to have

$$\begin{pmatrix} f_1 \\ f_2 \end{pmatrix}_L \; ; \; f_{1R}, \; f_{2R}$$

(or vice versa!) and that seems to be realized by nature. Putting things in triplets, etc., looks messy, uneconomical and ugly. So, accepting the survival hypothesis, GUTs provide a reason for chiral SU(2) weak interactions (i.e., charged and neutral weak currents are a combination of V and A) namely the very existence of "light" fermions.

6) Furthermore if a second or third or ... "light" generation has to exist, then again we should make sure that it is impossible to create G_1 invariant fermion masses, not only inside the same generation but also through inter-generation couplings. For instance, a world made of

$$\begin{pmatrix} u \\ d \end{pmatrix}_L \; ; \; u_R, \; d_R \;\Big|\; \begin{pmatrix} c \\ s \end{pmatrix}_R \; ; \; c_L, \; s_L$$

is unacceptable since by coupling

$$\begin{pmatrix} u \\ d \end{pmatrix}_L \quad \text{to} \quad \begin{pmatrix} c \\ s \end{pmatrix}_R$$

we may create G_1-invariant fermion mass terms, which will mean that at least some of the u,d,c,s quarks will be superheavy. Easiest way out: make every "new" generation a copy of the first one.

So we may begin to understand why weak interactions have the observed structure and why if there are two or more generations, they all need to be photocopies of the first one. Last but not least, let me briefly comment on the constraints that are

---

*) Sometimes G_1 may be enlarged to contain some other group beyond SU(3) × SU(2) × U(1).

imposed on the fermion representation content of the theory by demanding the absence of the Adler-Bell-Jackiw anomalies. We all agree that

a) there is strong experimental evidence that at least "low energy" physics is well described by renormalizable field theory and

b) the existence of anomalies spoils renormalizability,

so then we have to accept that the "effective" SU(3) × SU(2) × U(1) theory had better be anomaly free. It is quite remarkable, as found by Bouchiat-Iliopoulos and Meyer, that each generation [Eq. (1)] is anomaly free, with a cancellation between the anomalies of the quark and lepton sector. Here lies our main point: the Survival Hypothesis forbids the existence of "low energy" vector-like weak interactions which would not only be automatically anomaly-free for each generation [Eq. (1)], but furthermore each generation quark and lepton sector is anomaly-free separately. So again the simplest and maybe the most economic choice of an anomaly free fermion representation is that given by Eq. (1). Perhaps we would invent leptons (quarks) if we had only known about anomalies, and th quarks (leptons) exist.

Clearly the inter-relation between the existence of "low-mass" fermions, fermion representation content, and absence of anomalies is one of the most fascinating aspects of our "low-energy" theory. Before getting any further I will need some more input that I will provide next.

4. A NEW MECHANISM FOR "LIGHT" FERMION MASS GENERATION

We saw in the last section that "light" fermions are light because their L- and R-components transform differently under weak SU(2). Presumably by the same token "superheavy" fermions are superheavy because their L- and R-components transform identically under weak SU(2) (i.e., their weak interactions are "vector-like"). And here lies the clue of the new mechanism[1]. Suppose that in some unified theory of strong-electro-weak and possibly generation changing interactions there are "light" and "superheavy" fermions lying in the same representation of the unified group G' ($\supseteq$ G). Then in general there will be gauge bosons which couple "light" to "superheavy" fermions and presumably diagrams like those in Fig. 1 are allowed:

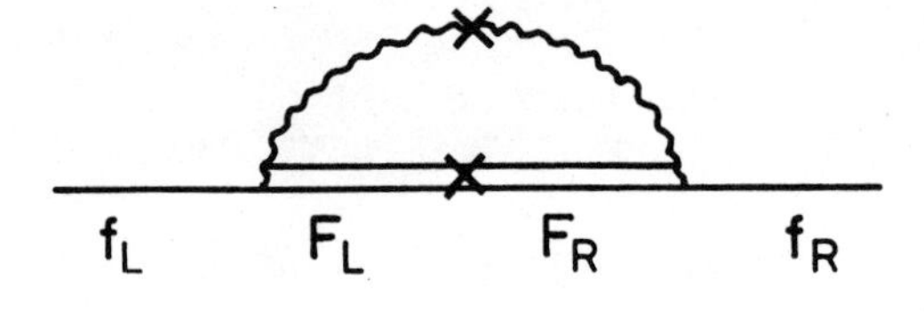

Figure 1.

Let us look closer at Fig. 1. Clearly the gauge boson that transforms $f_L \to F_L$ cannot be the same as the one which transforms $f_R \to F_R$. For reasons that we explained before, these gauge bosons should have different weak SU(2) transformation properties and so there should be a mixing of order M_W/M_1. But then we get a finite contribution δm of the form:

$$\delta m_f = \frac{\alpha}{\pi} \cdot M_W \cdot \frac{M_F}{M_1} \cdot O(1) \tag{4}$$

This is our main new result. Since we do not expect M_F/M_1 to be drastically different from O(1), we see that amazingly enough we have been able to create "light" fermion masses of the correct order of magnitude just by radiative corrections involving "superheavy" gauge boson exchanges and "superheavy" fermions. Needless to say that such a mechanism may be used iteratively if different symmetry breaking scales are present.

The natural appearance of such radiatively induced mass terms (diagonal and off-diagonal) of the "right" magnitude fill us with hope and joy that finally we can handle the main questions of Section 1 concerning the understanding of the mass generation gap and the smallness of the Cabibbo-like angles. Indeed one may envisage a situation where we start at tree level with only the third generation massive, and the first and second ones massless, with one-loop corrections then generating masses for the second generation $[\sim (\alpha/\pi)M_W]$ and two-loop corrections generating masses to the first generation $[\sim (\alpha/\pi)^2 M_W]$. Then we would get the correct mass generation gap plus small and calculable Cabibbo-like angles, through the off-diagonal mass terms. which are induced in the diagonal-Cabibbo matrix with which we started.

We will discuss next possible scenarios where such situations may be realized.

5. SCENARIOS OF "LIGHT" FERMION MASS GENERATION

5.1 Poor man's view

The simplest case of the proposed "radiative" mechanism is to assume that no "light" fermion gets a tree level mass (let us call it a "direct" mass to distinguish it from the radiatively induced mass: a "radiative" mass). All "light" masses are "radiative" and the observed fermion mass spectrum is nothing else but a "reflection" of the superheavy fermion mass spectrum; i.e., play with the M_F/M_1 ratio of Eq. (4). In other words Eq. (4) acts like a "transport" equation and the real game is played at superhigh energies. This approach has the virtue that we may start with only one mass parameter say M_{Planck} and get all other masses for free (radiatively).

Actually this has been already proposed[9] for the gauge bosons and Higgs particles ("light" and superheavy) in connection with the solution of the famous gauge-hierarchy problem. So there is a good chance that given M_{Planck} all other masses of all particles are created radiatively in a most agreable and natural way. It should be noticed that it is much easier to restrict the Yukawa couplings of "superheavy" fermions using discrete or continuous symmetries, involving superheavy multi-Higgs systems because these do not suffer from the well-known naturality constraints on "light" multi-Higgs systems[10].

5.2 Rich man's view

One may try to be a bit more imaginative. Thus suppose that we consider a unified theory of strong-electro-weak and generation changing interactions described by a gauge group $G' \supset G$($\equiv$ the strong and electro-weak interactions grand unified group). Let us assume now that fermions are thrown in different inequivalent representations ψ_i (i = 1,2 ...) of G', with only the constraint

$$\sum_{\substack{i \\ \text{anomalies}}} \psi_i \Big|_G = 0 \quad ^{*)} \tag{5}$$

and then presumably (if we are lucky) we will be left finally with only three-"light" generations of the "standard" model and everything else will be superheavy. Then it is not inconceivable to imagine that G'-invariance forbids at the tree level all but the third generation to get a "direct mass". This is possible because the different generations belonging in general to inequivalent reps of G'. Then one may imagine that a one-loop "radiative" mass is allowed for the second generation, but not for the first one, since again they belong to inequivalent reps of G' and the gauge transitions desired to complete the one-loop diagram for the first generation may not exist. Finally one may hope that in a natural way the first generation gets a two-loop-"radiative" mass. So we may get a satisfactory qualitative explanation of questions 7) and 8) but we need explicit and detailed calculations to answer 9). One may ask why nature has done such random thing ? i.e., why has she thrown different generations in inequivalent G'-reps ? Well, the obvious answer that comes to one's mind is that after all perhaps quarks and leptons (together with gauge bosons and Higgs particles) are not fundamental particles but composites of more basic object, preons (?). The energy scale where this compositeness is realized may be superhigh ($M_1 \sim 10^{15}$ GeV ; $M_{P\ell} \sim 10^{19}$ GeV ?) so we need not worry much about it yet. But if this is the case then we should also not care about nature's randomness because composite things may indeed fall into surprising representations. Actually, recent efforts[11] to unify grand unified theories with gravity through

*) In this case G' is very strongly broken and may only be useful as global classification symmetry.

supergravity, provide just such a picture. Then I strongly believe that the scenario proposed in this subsection when worked out in detail may solve the last "mysteries" of the fermion spectrum.

After all these speculations let me come to a more concrete proposal.

5.3 A Greek-Latin view[1]: AN EXCEPTIONAL GUT

A modest realization of the new ("radiative") mechanism[1] for "light" fermion mass generation is offered by the group E_6. Of course, this application is less ambitious than the scenario just described above, since E_6 claims to unify only strong and electro-weak interactions, and does not contain any generation changing interactions. Still, E_6 has enough structure to serve as a useful and simple example. E_6 has very nice and interesting properties and it is a serious candidate for being the correct group that unifies strong and electro-weak interactions. Here I am going to use those E_6 properties that serve my purposes, but the interessed reader should consult Ref. 1) for all the details of the specific E_6 model that I am using here. Fermions of each family lie in the fundamental $\underline{27}$ representation. The nice property of the $\underline{27}$ is that under $SU(\overline{3}) \times SU(2) \times U(1)$ it splits into the observed 15-plet (counting helicity states) of "light" fermions of Eq. (1) plus a self-conjugate 12-plet of fermions which will become superheavy in accord with our survival hypothesis[1,8]. The 12-plet of superheavy fermions transforms under the subgroup SU(5) as $\underline{5} + \underline{\bar{5}} + \underline{1} + \underline{\bar{1}}$; we call them (D_i, E^c, N^c), (D_i^c, E, N), ν^c and L. Of the last two neutral superheavy leptons, L is also an SO(10) singlet, whereas ν^c is the member of the 16-plet under SO(10) which is apparently missing from the "observed" spectrum. As to the Higgs representations we use the $\underline{351}_S$ and the $\underline{27}$, the first is capable of breaking $E_6 \to SU(3) \times SU(2) \times SU(1)$, and, giving $\underline{\text{superheavy}}$ masses to the self-conjugate 12-plet of fermions whereas the $\underline{27}$ breaks $SU(3) \times SU(2) \times U(1) \to$ $\to SU(3) \times U(1)_{E.M.}$, and it should at the same time give masses to the top generation. We assume at this level, that it will be possible for the charm and up-generation to remain massless in a natural way. What we have in mind is that eventually E_6 will be contained, together with some group of generation changing interactions in some bigger group G' so that the scenario presented in the preceding subsection will be fully realized. A relatively simple pattern of "radiative" masses emerges for the charm generation if one assumes a hierarchy in the "superheavy" fermion masses according to the suggestive cascade $E_6 \to SO(10) \to SU(5) \to SU(3) \times SU(2) \times U(1)$ (i.e., $m_L >> m_{\nu^c} >> m_{D,E}$). In this case a careful inspection shows that the dominant contribution to the μ- and the s-quark masses comes from L-exchange (Fig. 2).

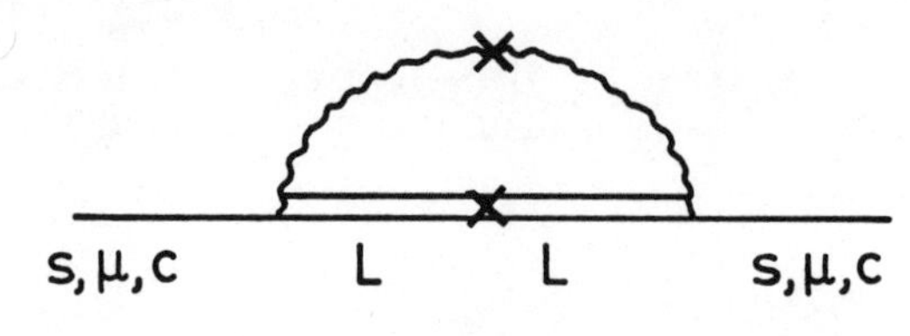

Figure 2.

whereas the charm quark receives contributions from L exchange and from ν^c exchange (Fig. 3)

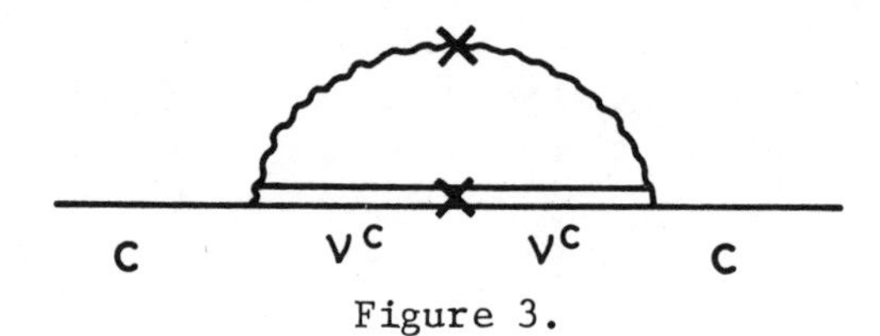

Figure 3.

The contributions are in general comparable, because the higher mass of L relative to ν^c is compensated by the higher masses of the gauge bosons exchanged in Fig. 2 relative to those of Fig. 3. This comes from the hierarchical pattern being postulated, which in turn suppresses all the diagrams with D_i or E exchanges. A similar suppression actually takes place for the contributions to the $Q = -1/3$ and $Q = -1$ masses coming from ν^c exchange, basically because they have superheavy partners unlike the $Q = 2/3$ quark. From the diagram of Fig. 2 $s \simeq \mu$ arises at superhigh energies whereas the additional contribution to the c quark mass from Fig. 3, althoug not possible to estimate numerically because of the extra parameters involved, gives a rather straightforward qualitative explanation*) why $c > s \simeq \mu$. All the charm generation masses produced in this way are $O((\alpha/\pi)M_W)$, which is what we find experimentally. Concerning the up-generation masses, one finds that if we stay within E_6 interactions (as we did for the charm generation masses), then we find zero masses to all orders in perturbation theory. Maybe this is an encouraging result, since we know that $u < d$ and $d \neq e$ in contrast with $c > s$, $s \simeq \mu$ and $t > b$, $b = \tau$. How then does the up-generation get mass ? Well, we would employ post E_6 interactions, which are there anyway (generation changing interactions, gravity, or ?[10]) and they may well generate masses of a few MeV (or$(\alpha/\pi)^2 M_W$ i.e., probably mass generation through two loops). So that is no big sweat. I hope to have shown with this simple example how we expect to get reasonable (and hopefully correct answers) to questions 7), 8) and 9) within the general framework described in this section.

*) I believe that this explanation survives in more complicated schemes than the one considered here.

Before embarking on the discussion of neutrino masses let me make a few more remarks on the structure of the charged fermion mass spectrum. The old dream that some masses in the fermion spectrum are "radiatively" generated seems to be realized. Namely, quark-lepton mass differences are generated[4,5] through radiative corrections involving the "standard" strong and electroweak interactions - essentially, quarks are heavier than leptons because quarks participate in strong interactions and leptons do not. On the other hand inter-generation mass differences - or quark doublet mass differences - may be explained[1] through radiative corrections involving "post" - SU(3) × SU(2)× U(1) - interactions, plus the fact that under the <u>unification group G'</u> of everything, different generations transform inequivalently and so gauge invariance of G' naturally prevents certain types of "direct" Yukawa couplings from appearing. A final comment concerning the mass of the as yet unseen top quark. In some grand unified theories like O(10), E_6, etc. suitably (and reasonably) constrained the following relation holds[1]:

$$\frac{t}{\tau} = \frac{c}{\mu} \quad . \tag{6}$$

At the toponium mass (T), i.e., taking into account finite renormalization effects when extrapolating from the charmonium region (ψ), this relation becomes

$$t(T) \approx \frac{\tau}{\mu} \cdot c(\psi) \cdot \left[\frac{\alpha_s(T)}{\alpha_s(\psi)}\right]^{\frac{4}{7}} \tag{7}$$

Using appropriate values for α_s and $c(\psi)$ we get[1]

$$t \simeq 18 - 22 \text{ GeV} \quad , \tag{8}$$

almost inside the energy range of PETRA or PEP. Thus, we should not be surprised by the discovery of the top quark very soon.

6. NEUTRINO MASSES (?)

Neutrinos, these mysterious and fascinating chargeless, massless (or nearly so), feebly interacting objects are now with us for exactly fifty years, after their "theoretical" discovery by Pauli, in 1930. Though until recently, we have not really had a satisfying answer to the question of their masses (question 10). In the "standard" SU(3) × SU(2) × U(1) model a direct Dirac mass term of the form $\bar{\nu}_R \nu_L$ is prohibited by making the neutrino a two component spinor. A Majorana mass term of the form $\nu_L \bar{\nu}_L$ is prohibited since it violates lepton number, which is supposed to be an exact global symmetry.

Concerning the absence of a Dirac mass term, the argument is clearly circular, and is certainly not a theoretical explanation. With the advance of GUTs the mechanism of L conservation that forbids a Majorana mass term cannot hold water. Lepton number is violated, as it should, because it is only a global symmetry and thus should be broken, according to the "central dogma of the gauge age". So we are left in the woods. This is not exactly the case, because in grand unified theories we have mechanisms[12] that may give an appropriate explanation of the neutrino masses. In general, there are particles around (N) that may serve as the right-handed missing partners of ν_L, but since they are $SU(3) \times SU(2) \times U(1)$ singlets, they acquire according to the survival hypothesis superheavy Majorana masses (M). So a general neutrino mass matrix will eventually look*) like:

$$\begin{array}{c c} & \begin{array}{cc} \nu & N \end{array} \\ \begin{array}{c} \nu \\ N \end{array} & \begin{pmatrix} 0 & m \\ m & M \end{pmatrix} \end{array} \tag{9}$$

where the off-diagonal terms (m) correspond to "normal" Dirac mass terms and thus they are naturally related to the energy scale that $SU(3) \times SU(2) \times U(1) \rightarrow SU(3) \times U(1)_{E.M.}$. Clearly (9) implies a neutrino mass:

$$m_{\nu_L} \simeq \frac{m^2}{M} \tag{10}$$

which may be anything between (10^{-5} eV → few eV), due to the uncertainties concerning the values of m and M used in (10), which in turn depend crucially on the specific model in use. This is the GUT answer for question 10). Certainly such values of neutrino masses are acceptable phenomenologically and they are very interesting from an experimental point of view, since the upper part of these predictions lies close to the presently available experimental upper bounds. Looking for effects of such masses, neutrino oscillations, distortion of the end-points of the energy spectrum in β decay, etc. may be a fruitful and worth while enterprise. I want to stress the point that Eq. (10) emerges naturally and so we understand[4] that corrections to the relation $M_Z \cos\theta_W = M_W$ are negligible, because the SU(2) symmetry breaking is predominantly through a Higgs doublet.

The point is that a Majorana mass term $\nu_L\nu_L$ may effectively break $SU(2) \times U(1)$ similarly to a Higgs triplet. Then the relation $M_Z \cos\theta_W = M_W$ gets spoiled if the breaking remains unchecked, but Eq. (10) assures us that GUTs take care of this problem in a very satisfying way[4]. Among grand unified theories, minimal SU(5)[3]

*) In fact the top left-hand entry should not be exactly zero: for reasons of naturalness[12] it will in general be $O(m^2/M)$.

uniquely predicts massless neutrinos in a straightforward way. The neutrinos do not have right-handed components because there is no space for them in the ($\bar{5}$ + 10). The neutrinos do not have Majorana mass terms because they violate B-L, which is a global (accidental) symmetry of minimal SU(5)[3]. Recently efforts have been made to undo both the above-mentioned arguments. It is possible that there are other representations of SU(5) around[11] (presumably self-conjugate under "SU(3) × SU(2) × U(1)" and thus possessing superheavy masses) which may contain particles which act as right-handed components of the "observed" neutrinos, or they may be "post" SU(5) interactions that break B-L and thus allow Majorana mass terms[13]. In both cases, though, it should be mentioned that the predictions for the neutrino masses should lie very close to the lower part of the range mentioned above (say, closer to 10^{-5} eV than to 1 eV). Other GUTs, like O(10), E_6, etc., since they are L-R symmetric contain naturally*) in the same representation both ν_L and ν_R and thus Eqs (9) and (10) apply directly to them, with the neutrino masses able to cover the whole range of 10^{-5} eV → few eV. So there seems to be a crucial distinction between the "extended" SU(5) model and other GUTs concerning the probable range of neutrino masses. It remains to be seen if nature uses her last chance to re-establish L-R symmetry at superhigh energies (which could perhaps be needed if we want to unify with gravity, which respects L-R symmetry) or if she prefers to remain always left-handed and so we have to swallow the fact that we are living in a lopsided world! A perhaps not unrelated matter concerns quark-lepton symmetry. In contrast with SU(5) where quarks and leptons are thrown randomly in ($\bar{5}$ + 10), in all other GUTs at superhigh energies $SU(4)_{colour}$[14] becomes a good symmetry, i.e., lepton-number serves as a fourth colour and so the observed quark-lepton symmetry looks less mysterious and finds a rather aesthetically and conceptually appealing explanation. On the other hand proponents of the compositeness of quarks and leptons may argue that the randomness of SU(5) should not worry us because since quarks and leptons are not fundamental they may find themselves sitting in rather random-looking representations. Clearly a lot of theoretical and experimental effort is needed before we decide which picture makes more sense, and clearly the neutrino mass spectrum may play a key role.

---

*) We recall that O(10) and E_6 are the only Lie-groups with the following desirable properties:

i) they are automatically anomaly free.

ii) they have useful complex representations and

iii) the fermions of each generation belong to an irreducible representation 16 [O(10)], 27 [E_6].

EPILOGUE

The most tantalizing predictions of GUTs concerning the fermion spectrum refers to the number of quarks (not more than six)[4,5] and the neutrino mass spectrum[12,13] (10^{-5} eV $\rightarrow$ few eV). In the next few years, experimentalists no doubt will put these predictions under scrutiny, but I hope that their results will not jeopardize my firm belief and convinction that grand unified theories are right.

ACKNOWLEDGEMENTS

I would like to thank R. Barbieri and J. Ellis for enjoyable discussions. Also I would like to thank the organizers, J. Ellis, S. Ferrara and P. Van Nieuwenhuizen for creating such a pleasant and successful meeting.

REFERENCES

1. R. Barbieri and D.V. Nanopoulos, Phys. Letters 91B (1980) 369 and CERN preprint TH-2870 (1980).
2. For a review of GUTs see: D.V. Nanopoulos, "Protons are not forever", in High Energy Physics in the Einstein Centenial Year, (Orbis Scientiae 1979) p. 91, Ed. by A. Perlmutter, F. Kransz and L. Scott, Plenum Press (1980);
   J. Ellis, Proceedings of the EPS Int. Conf. on High Energy Geneva (1979) (CERN 1979) p. 940.
3. H. Georgi and S.L. Glashow, Phys. Rev. Letters 32 (1974) 438.
4. A.J. Buras, J. Ellis, M.K. Gaillard and D.V. Nanopoulos, Nucl. Phys. B135 (1978) 66.
5. D.V. Nanopoulos and D.A. Ross, Nucl. Phys. B157 (1979) 273.
6. J. Yang, D.N. Schramm, G. Steigman and R.T. Rood, Astrop. J. 227 (1979) 697.
7. S. Hawking, Comm. Math. Phys. 43 (1975) 43;
   Ya.B. Zeldovich, Phys. Letters 59A (1976) 254;
   S.W. Hawking, D.N. Page and C.N. Pope, Phys. Letters 86 B (1979) 175.
8. R. Barbieri, D.V. Nanopoulos, G. Morchio and F. Strocchi, Phys. Letters 90B (1980) 91.
9. J. Ellis, M.K. Gaillard, D.V. Nanopoulos and C.T. Sachrajda, Phys. Letters 83B (1979) 339;
   J. Ellis, M.K. Gaillard, A. Peterman and C.T. Sachrajda, Nucl. Phys. B164 (1980) 253.
10. J. Ellis and M.K. Gaillard, Phys. Letters 88B (1979) 315.
11. J. Ellis, M.K. Gaillard, L. Maiani and B. Zumino, LAPP-TH-15/ CERN TH-2841 (1980);
    J. Ellis, M.K. Gaillard and B. Zumino CERN preprint TH-2842/ LAPP-TH-16 (1980).

12. H. Georgi and D.V. Nanopoulos, Nucl. Phys. B155 (1979) 52 and unpublished;
S. Weinberg, Phys. Rev. Letters 43 (1979) 1566;
M. Gell-Mann, P. Ramond and R. Slansky, unpublished.
R. Barbieri, D.V. Nanopoulos, G. Morchio and F. Strocchi, Ref. 8);
E. Witten, Phys. Letters 91B (1980) 81;
M. Magg and Ch. Wetterich, CERN preprint TH-2829 (1980).
13. R. Barbieri, J. Ellis and M.K. Gaillard, Phys. Letters 88B (1979) 315.
14. J.C. Pati and A. Salam, Phys. Rev. D10 (1974) 275.

MAGNETIC MONOPOLES AND GRAND UNIFIED THEORIES

D.I. Olive

Physics Department
Imperial College
London UK

The first part of my talk is devoted to explaining why magnetic monopoles, for so long a theoretical curiosity, arise naturally in the grand unified gauge theories, currently thought to be a step on the way to ultimate synthesis of physics. Then I discuss the theoretical interest of having a Higgs field in the adjoint representation, and further of it having vanishing self-interaction. I end by commenting on the masses of the monopoles and their possible production in the early Universe.

The fundamental forces are now thought to arise from gauge principles. The exact gauge groups appear to have the form "$U(1)_M \times K$" where $U(1)_M$ is the Maxwell gauge group, whose generator Q is the electric charge. Q is a colour singlet with respect to the colour group K, supposedly semisimple, and containing the familiar SU(3) colour but may be also "technicolour" or "colour prime". The K charges are for some reason invisible but the electric charges Q carried by the elementary particles display an interesting pattern: they are "quantised" in the sense of being integer multiples of an elementary unit of electric charge, that carried by the electron ($-q_o$), (except for the unseen quark which is thought to carry $^-q_o/_3$ or $^2q_o/_3$). Within conventional quantum electrodynamics there is no explanation of this.

Historically there have been three "explanations" of charge quantisation - the inverted commas indicate that the explanation is in terms of an alternate hypothesis rather than a fundamental principle. We shall see that these three "explanations", apparently so disparate, are, with modern insight, intimately related.

THE KALUZA-KLEIN EXPLANATION (1921)[1]

There exists a fifth, hitherto unsuspected, dimension of space with period L. The conjugate momentum P_5 is proportional to electric charge and has eigenvalues which are quantised as integer multiples of $2\pi/L$.

MAGNETIC CHARGE EXPLANATION (1931)[2]

A somewhat more serious explanation, due to Dirac[2], is that if $U(1)_{Maxwell}$ is the exact gauge symmetry and if there exist magnetic monopoles with charges g_i, then the consistency of quantum mechanics requires[2]

$$\exp\left(\frac{iq_j g_i}{\hbar}\right) = 1$$

for any electric charge q_j. It follows that if g_o is the smallest magnetic charge,

$$q_j = \frac{2\pi\hbar}{g_o} n_j \;, \quad n_j = 0, \pm 1, \pm 2, + 3 \ldots, \tag{1}$$

i.e. electric charge is quantized (and likewise magnetic charge).

If instead of $U(1)_M$ the exact gauge symmetry is $U(1)_M \times K$ one finds instead[3]

$$\exp\left(\frac{iq_j q_i}{\hbar}\right) = k \varepsilon K \tag{2}$$

Since Q is a colour (K) singlet so is the element k of K, the colour group occurring on the right-hand side. So it lies in what mathematicians call the centre of the colour group. Since K is semisimple this centre has a finite number of elements, and in the case of $K = SU(3)_{colour}$, precisely three, corresponding to the three cube roots of unity. Instead of (1) one finds

$$q_i = \frac{2\pi\hbar}{g_o} \left(n_j + \frac{t}{3}\right) \tag{3}$$

where t = 0, 1 or 2 is the SU(3) colour triality of the particle with charge q_i. Thus quarks have fractional charge. This explanation is linked essentially to the gauge nature of the colour group and has nothing to do with flavour.

THE UNIFICATION EXPLANATION (1954)

This is the most conventional explanation. When it was realized that SO(3) and higher non-Abelian groups could also be gauged[4] it was natural to choose Q as one of the generators. Any generator of SO(3) has quantized eigenvalues and hence so would Q. The remaining generators would correspond to some broken symmetry.

It is worth formulating this argument more precisely. It is not true that any generator of any semisimple group has quantised eigenvalues. For example, the ratios of the eigenvalues of $\frac{1}{2}(\lambda_3+\lambda_8)$ in SU(3) are irrational numbers because of the $\sqrt{3}$ in λ_8. In fact in SU(3), unlike SO(3), different generators generally have different eigenvalue. If one of these generators is to be Q, how is it selected, and what guarantees that it has quantised eigenvalues? We have learnt that a natural way of breaking a gauge symmetry without destroying the symmetry of the Lagrangian is to let the vacuum break the symmetry. The only explicit such mechanism is via a Higgs field[5], a scalar field with non-zero vacuum expectation value.

When this is done the exact gauge symmetry is the subgroup leaving the Higgs field invariant and this is necessarily compact. If it has the form $U(1) \times K$ with K semisimple it follows that the U(1) is compact and therefore generated by a charge Q with quantised eigenvalues.[6]

Now we notice a similarity between the monopole and unification arguments above. Both assume a $U(1) \times K$ exact gauge symmetry with K compact. Quantization of the Q generating the U(1) follows either if we assume magnetic monopoles or if we assume unification, i.e., that $U(1) \times K$ be the residue after vacuum breaking of a larger gauge symmetry G, assumed semisimple.

In fact it can be proved by topological arguments that in the second case there automatically exist field configurations satisfying the classical equation of motion and carrying a conserved U(1) magnetic charge. Thus magnetic monopoles automatically exist (in a classical sense) and Dirac's argument is applicable. This is the deep connection between magnetic monopoles and unification originally revealed by the work of 't Hooft and Polyakov[7].

We shall not delve into the topological argument but we can make the result plausible by making an extra assumption (which will link up with the Kaluza-Klein explanation).

The assumption is that the Higgs field ϕ lies in the adjoint representation of G. This seems to be the most natural way of selecting a direction Q, in the space of the generators of the unification group G. There are physical objections to this, which we discuss later.

Normalise the charge in the direction of the Higgs field as $Q = e\hbar\, \emptyset . T/_a$, where $a = \sqrt{\emptyset^2}$ in vacuo. The exact gauge symmetry is generated by the subalgebra of generators of G, commuting with Q. Obviously Q belongs to this set and commutes with all the others. Hence it generates an invariant U(1) subgroup of the exact gauge group and we have the desired situation providing K is semisimple. Comparing the conventional U(1) covariant derivators $\partial^\mu - iQA^\mu/\hbar$ and the covariant derivative $\partial^\mu - igT_iW_i{}^\mu$ we see that we have normalized Q appropriately for its eigenvalues to specify the U(1) charges of the elementary fields of the theory.

The G gauge particles not corresponding to exact gauge symmetries acquire mass from the vacuum by the Higgs-Kibble-Brout-Englert mechanism[5]. The result for the mass squared matrix is

$$M^2_{\alpha\beta} = g^2\hbar^2\, \emptyset\, T_\alpha T_\beta \emptyset$$

For $\emptyset$ in the adjoint representation the generators T are structure constants which can be chosen antisymmetric in all three indices. So $M^2 = a^2Q^2$ in matrix notation and the mass of any gauge particle is $a|q|$ where $|q|$ is the modulus of its U(1) charge. This is irrespective of the group G[8].

Now we derive a lower bound on the energy and hence the mass of a classical field configuration. Its energy is

$$H = \frac{1}{2}\int d^3a\,(E^2 + B^2 + (D^i\emptyset)^2 + (D^o\emptyset)^2 + V(\emptyset))$$

Each term is positive and $\frac{1}{2}(B^2+(D^i\emptyset)^2) = \frac{1}{2}(B_i \mp D_i\emptyset)^2 \pm B_i D^i\emptyset$ exceeds $|B_iD^i\emptyset| = |\nabla_i(B\,\phi)|$ using the Bianchi identity. So H exceeds $|\int dS_iB_i\emptyset| = a\,|g|$ where g is the U(1) magnetic flux out of a large surface containing the configuration, and hence its magnetic charge.

This bound implies that the mass $M \geqslant a\,|g|$.[9] Since g is quantised and so cannot vary there is a strong indication of an actual solution to the equations of motion satisfying this bound. This is the monopole solution. The bound can be satisfied exactly in the so-called "Prasad-Sommerfield" limit[11] in which $V(\emptyset)$ vanishes identically[9]. This apparently artificial limit is very interesting because it seems[12] that then all particle states, whether they be gauge particles, magnetic monopole solutions (or dyons carrying electric charge too) or Higgs particles satisfy the universal mass formula[10]

$$M = a\sqrt{q^2 + g^2} \tag{4}$$

This is invariant with respect to "duality" rotations between the electric and magnetic U(1) charges, q and g, and unifies all the particles irrespective of their field theoretic origin as classical solutions or quantum excitations. This Prasad-Sommerfield limit of vanishing Higgs self interaction appears to be an unusual sort of symmetry limit and as such is particularly interesting.

The mass formula (4) suggests that a q and a g form the fifth and sixth components of four-momentum and that this newly obtained six-momentum is massless. This is rather close to the original Kaluza-Klein explanation[1] of charge quantization mentioned above, and can be vindicated[13] if we recognise that the Higgs field $\emptyset$ be thought of as the fifth component of the gauge potential[14]. This is only possible because it lies in the adjoint representation, like the original gauge potentials. Further, the theory should be made supersymmetric in six dimensions[15]. The vanishing of the Higgs self-interaction would then be a consequence of this symmetry. The passage from six dimensions back to four dimensions is an example of the dimensional compactification[16] procedure mentioned by several speakers on supersymmetry at this meeting.

The scheme developed so far is quite pretty but unfortunately wrong. The success of the Salam-Weinberg theory[17] tells us that the electromagnetic U(1) is a subgroup of the electroweak $SU(2)\times U(1)_Y$. The electron charge is a linear combination of the weak hypercharge Y and an SU(2) generator. The symmetry breaking is via an SU(2) doublet. Because $SU(2)\times U(1)$ is not semisimple and therefore has two coupling constants true unification is lacking and there are no magnetic monopole solutions. Hence on either count there is no understanding of electric charge quantization.

In addition there is no unification with $SU(3)_{colour}$ so we have an intermediate symmetry $SU(3)\times SU(2)\times U(1)_Y$. But if we seek grand unification of this within a semisimple group G the appearance of an invariant U(1) subgroup, generated by weak hypercharge Y rather than electric charge Q, is again suggestive of symmetry breaking by a Higgs field in the adjoint representation which actually defines the hypercharge direction. Then G = SU(5)[18] but we can imagine slightly more complicated two-stage schemes, covering several of the models discussed at this meeting.

G : grand unified symmetry (of the Lagrangian)
broken $\downarrow$ via Higgs ϕ in the adjoint representation choosing the Y direction.

$U(1)_Y \times H$: the intermediate symmetries ($\supset$ Salam Weinberg x colour)
broken $\downarrow$ via Higgs ψ responsible for fermion masses

$U(1)_{EM} \times K_{colour}$: the exact gauge symmetry.

This seems to be the simplest possibility consistent with the known facts. It follows that

$$Q/q_o = Y + Y_\perp$$

where this defines the normalization of Y and $Y_\perp$ and $Y_\perp$ is a generator of H, assumed semisimple so that Y and $Y_\perp$ are mutually orthogonal. So if θ is the angle between Y and Q (the Weinberg angle), then Y is a projection of Q/q_o and $|Y| = |Q| \cos\theta/q_o$.

Again we can have classical monopole solutions but since the electromagnetic charge rather than the hypercharge is the charge finally conserved we shall seek to express the various masses in terms of that.

Just as before the monopole mass M exceeds $|\int dS_i \underline{\phi}.\underline{B}_i|$ since any contribution of the second Higgs field ψ is positive. Far away ϕ and ψ are covariantly constant which means that B_i viewed as a generator annihilates both ψ and ϕ and hence lies in the $U(1)_{EM} \times K_{colour}$ subalgebra:

$$B^i = \hat{Q}\, b^i + \beta^i_{colour}$$

with b_i the Maxwell magnetic field. Since $\phi.\hat{Q} = a\,\hat{Y}.\hat{Q} = a \cos\theta$ (hats denote unit vectors), we find the result of Scott[19] that the mass of a monopole M exceeds $a|g \cos\theta|$, where g is the Maxwell U(1) magnetic charge.

Let us likewise try to relate the masses of the G gauge particles to their Maxwell charges, this time electric. The adjoint representation of G carrying the gauge particles decomposes into irreducible multiplets of $U(1)_Y \times H$ of fixed Y and hence mass, but variable electric charge Q. As before the gauge particle masses m are given by

$$m = a\hbar g\, |\langle\hat{\phi}.T\rangle_H = a\hbar g|\langle\, \hat{Y}\, \rangle_H|$$

where $\langle\ \rangle_H$ means average over the irreducible H multiplet considered. Resolving the unit charge vector into its components

$$\hat{Q} = \hat{Y} \cos\theta + \hat{Y}_{\perp} \sin\theta$$

we see that $\langle\hat{Q}\rangle_H = \langle\hat{Y}\rangle_H \cos\theta$ since Y averages to zero as it is a generator of H which is assumed to be semisimple. But $Q = \hbar g \hat{Q}$, so finally the gauge particle masses m are given by

$$m = a \,|\, \langle Q\rangle_H \,| / \cos\theta$$

So again the Weinberg angle θ enters but in a different way.

The lightest magnetic monopole has presumably the smallest g_{EM} compatible with the generalised Dirac quantization condition (2), Acting on an electron state, which is a colour singlet, $\exp ig\, q_o/\hbar$ equals unity so the minimum possible g is simply $g = 2\pi\hbar/q_o$. Suppose that for a gauge particle multiplet $\langle Q\rangle_H = \beta q_o$, with β a number to be evaluated. Then the ratio of the lightest monopole mass to the gauge particle mass exceeds $|g \cos^2\theta / \langle Q\rangle_H|$ which equals $\cos^2\theta/2\alpha\beta$, where $\alpha = q_o^2/4\pi h$ is precisely the fine structure constant 1/137.

Let us evaluate β and hence this mass ratio in the SU(5) model [18]. All we need know is the charge assignment in the fermion $\underline{5}$ and its decomposition into H = SU(3) ⊗ SU(2) multiplets; $\underline{5} = (3,\bar{0}) \oplus (0,2)$ and $Q = q_o(-1/3, -1/3, -1/3, 0, 1)$. Hence $\overline{Y} = \langle Q\rangle_H/q_o = (-1/3, -1/3, -1/3, \tfrac{1}{2}, \tfrac{1}{2})$ and $\cos^2\theta = \langle Y^2\rangle_G / \langle (Q/q_o)^2\rangle_G = 5/8$. Since the adjoint 24 = 5 x 5 -1 we see $\beta = \tfrac{1}{2} + 1/3 = 5/6$ and so the mass ratio is $3/8\alpha$ which is about 50. Scott found the same number in different versions of the SO(10) model. In the simplest models this ratio is $1/\alpha$ [7], but what we have seen is that it gets reduced by an effect involving the Weinberg angle. So the monopoles are estimated to have a mass one order of magnitude heavier than the super-heavy gauge particles which themselves are thought to be enormously heavy (10^{15} GeV).

If these monopoles can be treated as ordinary particles in the quantum version of the grand unified theories then they can be pair created at high enough energy. In particular if the Universe was created in a big bang the thermal energies at very early times would be sufficient to pair produce them copiously. Estimates indicate that a large density would survive and affect both the observed helium abundance and rate of expansion of the Universe (because of the enormous individual masses), causing a significant divergence from the observed values. The result is a clash between different physical ideas but the weak point seems to be the estimate of the initial monopole density. The monopoles are not ordinary particles but rather composite objects which condense out at a phase transition as the Higgs potential changes shape with reducing temperature[20]. Roughly the monopoles have a radius r and a crude estimate for their initial density is that it is given by the density of closely packed spheres of this radius. The candidate for this monopole radius r is the larger of the Compton wavelengths r(h), r(w) of the Higgs or heavy gauge particles respectively. Usually they are assumed to be of the same order of magnitude and this leads to the paradox. But r(h) is essentially a free parameter inversely related to the Higgs self-coupling and therefore becomes very large in the Prasad-Sommerfield limit which as we have seen is a very interesting symmetry limit, manifesting dual and supersymmetry. In this limit (if the argument is still correct) the Universe has room for only one monopole and the paradox disappears. Of course the whole argument should be reexamined, and this has been recognized, but as far as I know the result is inconclusive.

Theoretically the point is very interesting. In two dimensions certain classical solutions have what is known as the soliton property; they cannot be pair created. Maybe in the Prasad-Sommerfield limit the monopole solutions are solitons in some modified sense, applicable in four dimensions, such that pair creation is inhibited.

I am grateful for discussions with T. Kibble and S. Rajpoot.

REFERENCES

1. T. Kaluza: Sitzungber. Preuss. Akad. Wiss. Berlin, Math. Phys. KA966 (1921).
O. Klein: Z. Physik. 37, (1920), 895.
2. P. Dirac: Proc. R. Soc. A133, (1931), 60.
3. E. Corrigan and D. Olive: Nucl. Phys. B110, (1976) 237
4. C.N. Yang and R.L. Mills: Phys. Rev. 96, (1954), 191.
R. Shaw: Ph.D Thesis, Cambridge University (1955).
5. P.W. Higgs: Phys. Rev. Lett. 12 (1964), 132
Phys. Rev. Lett. 13 (1964), 508
Phys. Rev. 145 (1966), 1156.
F. Englert and R. Brout: Phys. Rev. Lett. 13, (1964), 321
G.S. Guralnik, C.R. Hagen and T.W.B. Kibble: Phys.Rev. Lett. 13 (1964), 585
T.W.B. Kibble : Phys. Rev. 155,(1967), 1557.
6. C.N. Yang: Phys. Rev. D1 (1970) 2360.
7. G. 't Hooft: Nucl. Phys. B79 (1974) 276.
A.M. Polyakov. JETP Lett. 20, (1974),194.
8. D. Olive: Phys. Reports. 49 (1979) 165
F.A. Bais: Phys. Rev. D18 (1978) 1206

9. E.B. Bogomolny : Sov. J. Nucl. Physics 24,(1976) 449
S. Coleman, S. Parke, A. Neveu and C.M. Sommerfield
Phys. Rev. D15, (1977) 554.
10. B. Julia and A. Zee Phys. Rev. D11. (1975) 2227.
11. M.K. Prasad and C.M. Sommerfield: Phys. Rev. Lett. 35, (1975) 760.
12. C. Montonen and D. Olive: Phys. Lett. 72B, (1977) 117.
13. D. Olive: Nucl. Phys. B153. (1979) 1.
14. M. Lohe : Phys. Lett. 70B, (1977) 325.
15. A. D'Adda, R. Horsley and P. Di Vecchia Phys. Lett. 76B, (1978) 298.
E. Witten and D. Olive. Phys. Lett. 78B (1978), 97.
16. E. Cremmer and J. Scherk. Nucl. Phys.B103, (1976) 399; B108 (1976) 409, B118 (1977), 61.
17. A. Salam : Proc. 8th. Nobel Symp: Elementary Particle Theory ed. N. Svartholm (New York: Wiley) 337
S. Weinberg: Phys. Rev. Lett. 19, (1967) 1264.
18. H. Georgi and S.L. Glashow : Phys. Rev. Lett. 32, (1974) 438.
19. D.M. Scott: Imperial College and Cambridge University preprints.
20. T.W. Kibble : Imperial College preprint ICTP/79-80/23 (1980): This contains a full list of references. See also M. Einhorn, these proceedings, p. 569.

GRAND UNIFICATION AND COSMOLOGY *

John Ellis
CERN, Geneva, Switzerland

Mary K. Gaillard
LAPP, Annecy-le-Vieux, France

D.V. Nanopoulos
CERN, Geneva, Switzerland

1. A COSMIC CONNECTION?

Do cosmology and grand unified theories (GUTs) of elementary particle interactions have anything useful to say to each other? There is a great deal of theoretical work on GUTs unifying the strong, weak and electromagnetic interactions[1], now that many theorists perceive these individual interactions to be understood in principle. GUTs invoke energy scales of $O(10^{15})$ GeV which seem vertiginous to many physicists. It is not imaginable to reach these energies in laboratory experiments, and experimentalists are therefore forced to look for very indirect and feeble side-effects of grand unification such as proton decay. Even these valiant efforts may be brought to nought by a (logarithmically) modest increase in the grand unification mass-scale[2]. However, this mass-scale may be achieved directly in cosmological and astrophysical situations. For example, black hole explosions could in principle achieve temperatures up to the Planck temperature of 10^{32}°K corresponding to energies of 10^{19} GeV, while temperatures corresponding to particle energies of 10^{15} GeV or more are generally thought to have occurred very early in the Big Bang when the Universe was about 10^{-37} seconds old. It may indeed turn out that the Big Bang has been the only place to do direct experiments on GUTs (albeit rather uncontrolled, at least by us). In this paper we review progress on some interfaces between cosmology and grand unification, asking whether cosmology tells us important things about GUTs, and vice versa.

---

* Presented by John Ellis.

Table

Important Problems

Grand Unification	Cosmology
Fermion spectroscopy	Baryon to photon ratio, and lepton numbers
Grand unification mass m_X = ? and coupling constant α_{GUM} = ?	Homogeneity and isotropy
SU(5) or a more complicated GUT?	Galaxy formation
Mechanism of symmetry breaking	Simultaneity of expansion of causally separated parts of the Universe
Connection with gravitation?	How many black holes and/or monopoles in the Universe?

Shown in the Table are a few of the important problems in grand unification and cosmology: neither listing is exhaustive. Within the rubric of fermion spectroscopy we include the absolute number of fermion species and the masses of the apparently very light neutrinos as well as the possible existence of superheavy fermions. As far as the grand unification mass m_X is concerned, we have a phenomenological lower bound from the absence to date of observations of proton decay[3], and a suggestion[4] from the observed strength of the strong interactions that m_X may lie very close[5-7,2] to this lower bound: direct observation is, however, still lacking as yet. Simple-minded calculations[5-7] suggest that the strength $\alpha_{GUM} = g^2/4\pi$ of the gauge coupling at the grand unification mass of order 10^{15} GeV is around 1/40, but some authors have suggested that α_{GUM} may be O(1) at or near the Planck mass scale[8]. The biggest open problem in gauge theories is perhaps the mechanism of spontaneous symmetry breaking -- is it realized through explicit Higgs fields, or dynamical symmetry breaking, or a combination of the two? Eventually one must of course include gravitation in a truly unified theory of all the fundamental interactions[9], though present attempts at grand unification generally postpone this problem *sine die*.

One of the outstanding problems in cosmology has been to understand why there are just a few baryons in the Universe: $n_B/n_\gamma = 10^{-9\pm1}$, and why there seem to be no antibaryons[10], at least on the scale of our local galactic cluster. Related unknowns are the densities (chemical potentials) of other quantities not yet known to be violated, such as electric charge and the various lepton numbers (L_e, L_μ, L_τ). Another fundamental problem is to understand why the Universe is apparently so homogeneous and isotropic, for example in the distribution of luminous matter within it. The reader is actually living (?) evidence that some inhomogeneities exist, of which the most notable are galaxies, whose formation is not yet fully understood. But the Universe looks quite regular on a large scale, and a particularly striking and baffling aspect of this homogeneity and isotropy of the Universe is the apparent necessity within the standard Big Bang model for causally separated parts of the Universe to have chosen to behave in the same way at the same time. Presumably it was not coincidence, and the causality arguments must somehow be thwarted by some deviation from our "standard" ideas on particle interactions or on the evolutionary history of the Universe. Finally, on a more mundane level, our unified theories of the fundamental interactions predict the existence of massive classical solutions[11], grand unified monopoles, whose abundance is unknown[12] but must be strongly suppressed in order to avoid a conflict with the standard Big Bang model. This might prove to be a serious embarrassment for simple grand unified models, though we ourselves doubt it.

In this review we will not address, let alone answer, all the above problems, but will address ourselves to four topics where the cosmology-GUT interface seems particularly active. The first is the topic of fermion spectroscopy, in which we remind the reader that the successful calculation[5,13,14] of the bottom quark mass in GUTs imposes a phenomenological restriction to six of the number of quarks. This parallels nicely the restriction to three of the number of neutrinos due to nucleosynthesis[15] in the Big Bang. The second topic is baryon number generation in grand unified theories[16]. These provide a qualitative framework for generating a net baryon excess through C-, CP- and B-violating interactions dropping out of thermal equilibrium very early in the Big Bang. When one tries to put this scenario on a quantitative basis one discovers[17] that the grand unification mass should not be much lower than 10^{15} GeV, or α_{GUM} much larger than 1/40, and that a theory larger[18] than the minimal SU(5) GUT with just two Higgs representations ($\underline{5}$ and $\underline{24}$) is probably required[19]. The third topic is that of dissipative processes very early in the Universe, due to the relatively long mean free times of elementary particles possibly providing forms of viscosity and thermal conductivity. These suppress[20] some forms of inhomogeneity, but permit the growth of long wavelength compressional perturbations which have been suggested[21] as the origin of galaxy formation. The fourth topic is the cosmological density of grand unified monopoles[12]. It has been suggested[22] that this may be brought acceptably low by postulating a first order phase transition with large supercooling, or that one may have to postulate[23] a GUT larger than SU(5), or that symmetry breaking may have to be dynamical[24]. In fact, none of these may[25] be necessary.

It will be seen from the above synopsis that on many issues cosmology and grand unification do indeed have things to say to each other, though the utility of some of the connections is not yet clear.

2. GUTS AND FERMION SPECTROSCOPY

Before embarking on the amalgamation of grand unification and cosmology, perhaps we should first establish the credentials of GUTs for our cosmological friends -- we will, of course, take on trust their "standard" Big Bang model[26]. GUTs [1,27] have an immense aesthetic appeal in that they explain many previous baffling aspects of particle physics -- the quantization of electric charge $Q_e = -Q_p$ and the apparent "family resemblances" of quarks and leptons. While they have a unique gauge coupling g_{GUM} instead of the disunified $g_3 \neq g_2 \neq g_1$ of the low-energy SU(3) × SU(2) × U(1) gauge theory, they do however have many other free parameters, notably including those associated with the Higgs fields and their couplings. These include the Higgs-fermion-fermion Yukawa couplings which fix quark and lepton masses, mixing angles and CP-violating phases[28]. They also include the parameters of the Higgs potential which determine the largely unknown Higgs boson masses[5], and at least one of these parameters must be fixed[29,5] with a precision of $O(10^{-25})$ in order to yield the enormous hierarchical ratio

$$\frac{m_X}{m_W} \approx (10^{12} \text{ to } 10^{13}) \tag{1}$$

of vector boson masses expected in GUTs. Phenomenological estimates[2,5-7,30] of the mass m_X of the baryon-number violating gauge bosons X, on the basis of the observed strengths of the strong and electromagnetic interactions at low energies, suggest that $m_X \approx$ $\approx 6 \times 10^{14}$ GeV with substantial uncertainties[2,31] except in the minimal SU(5) GUT. With this value of m_X, protons and bound neutrons are expected to decay with lifetimes less than 10^{32} years which could be detected in forthcoming experiments[32]. However, proton decay has not yet been seen, and one would like to see at least some indirect evidence for GUTs in low-energy phenomena.

Two such pieces of evidence exist. The most celebrated is the SU(2) × U(1) mixing angle $\sin^2\theta_W$ which has been calculated[4,5,7] in simple GUTs such as SU(5) and SO(10). In these models one has

$$\sin^2\theta_W = \frac{\frac{3}{5}\, g_1^2}{g_2^2 + \frac{3}{5}\, g_1^2} \tag{2}$$

which is 3/8 in the symmetry limit at high energies where $g_2 = g_1$. As one comes down below m_X to low energies Q the couplings $g_1(Q)$, $g_2(Q)$ evolve differently so that the value of formula (1) also changes. When $Q \approx m_W$ one finds

$$\sin^2\theta_W\,(m_W) \approx 0.20 \text{ to } 0.21 \tag{3}$$

to be compared with the experimental value[33]

$$\sin^2\theta_W\,(Q \ll m_W) \approx 0.230 \pm 0.015\ . \tag{4}$$

To make a direct comparison between (3) and (4) requires a complete computation of radiative corrections at low energies which has not yet been done. However, the agreement between (3) and (4) for a number which *a priori* could have lain anywhere between 0 and 1 is surely impressive.

A second piece of evidence for GUTs comes from a calculation (Refs. 5, 13 and 14) of the bottom quark mass. In many GUTs there is a relation between this and the mass of the τ lepton:

$$m_b = m_\tau \tag{5}$$

valid in the high-energy symmetry limit. At low Q, where m_b and m_τ are actually measured, the prediction (5) gets renormalized[5,13,14] to

$$\frac{m_b}{m_\tau}\,(Q = 10\ \text{GeV}) \simeq \left[\frac{\alpha_s(Q)}{\alpha_X}\right]^{4/(11-\frac{2}{3}f)} \left[1 \pm O(10)\%\right]\ , \tag{6}$$

where α_X is the grand unified coupling at $Q \approx m_X$ and f is the number of quark flavours. The observed mass of the τ can be used with (6) to deduce that

$$m_b \approx (5 \text{ to } 5\tfrac{1}{2})\ \text{GeV} \tag{7}$$

if there are only six quark flavours. The number (7) is increased by 30% (or more) if there are eight (or more) quarks. If the basis of this successful calculation is accepted, it means that there can be at most six quarks. Since there are two quarks per neutrino in simple GUTs, this limit parallels very nicely the limit[15] of 3 on the number of (almost) massless neutrinos ($m_\nu \ll \ll 1$ MeV) coming from nucleosynthesis.

This parallelism is meaningful only if GUTs imply $m_\nu \ll 1$ MeV for all their neutrinos. Recall that the present laboratory limits are

$$m_{\nu_e} \lesssim 35 \text{ eV}, \; m_{\nu_\mu} \lesssim 600 \text{ keV}, \; m_{\nu_\tau} \lesssim 200 \text{ MeV} \quad . \tag{8}$$

Most fashionable GUTs do indeed predict $m_\nu \ll 1$ MeV, with a common tendency[34] to find

$$\frac{m_\nu}{m_W} \approx \frac{m_W}{m_X} \Rightarrow m_\nu \sim (10^{-5} \text{ to } 1) \text{ eV} \tag{9}$$

the second hierarchy emerging naturally given the (unnatural) first one (1). Neutrino masses in the lower end of the range (9) are best detected by astrophysical means such as oscillations in cosmic ray[32] or solar neutrinos[35], or time-delays in pulses from supernova collapses. Neutrino masses $\gtrsim 1$ eV fall within the sensitivity of present laboratory experiments[36], as well as having possible astrophysical implications through their aggregation in galactic haloes and/or clusters. The astrophysical and cosmological constraints on such neutrinos are reviewed by Steigman[7] at this meeting.

3. THE EVOLUTION OF THE VERY EARLY UNIVERSE

The Universe now contains many photons, principally in the 3°K microwave background radiation, and relatively few baryons by comparison:

$$n_B/n_\gamma = 10^{-9\pm 1} \quad . \tag{10}$$

Furthermore, there are no known concentrations of antimatter and the absence of high-energy photons which could be the products of B-$\bar{B}$ annihilation indicate that our local cluster of galaxies does not contain any significant fraction of antimatter[10]. The cosmic ray antiprotons recently observed[38] do not constitute evidence for cosmological antimatter, as their flux is consistent with that expected from secondary $\bar{p}$ production by primary nucleons. Before the advent of GUTs there was no satisfactory explanation of the ratio (10) or the apparent matter-antimatter asymmetry. Some held that the Universe is in fact symmetric and contains equal amounts of matter and antimatter, but failed to explain satisfactorily how their concentrations could have got separated in the present Universe[39]. Others hypothesized that the Universe had in fact started out with only baryons, and the large number of photons in (10) was generated subsequently by dissipative processes. Yet others just assumed that the Universe had always been slightly asymmetric. It now seems possible to believe that the matter-antimatter asymmetry and the small number (10) can be naturally explained[16] in GUTs. Three crucial ingredients are necessary in order to realize this trick. One is clearly that one has a theory with baryon number B-violating interactions. Another is that some component of these interactions violates both C and CP, as making either of these transformations reverses the sign of the net baryon number which must therefore be zero if the Universe evolves from an initially symmetric state. Finally, the B-, C- and CP-violating interactions must get out of thermal equilibrium, as in such a state the arrow of time is lost and CPT invariance guarantees that the net baryon number is zero[17,40]. We will now see how these three conditions are fulfilled with the expansion and cooling of the Universe introducing the necessary departure from thermal equilibrium while GUTs provide the necessary B-, C- and CP-violation.

We will be interested in epochs of the Universe's evolution when it was very hot with a temperature of $(10^{32}$ to $10^{23})^{o}$K equivalent to a particle energy $O(10^{19}$ to $10^{10})$ GeV. The upper limit is given by the Planck temperature T_P corresponding to a mass m_P related to Newton's gravitational constant by $m_P \equiv (G)^{-\frac{1}{2}} \approx 1.2 \times$ $\times\ 10^{19}$ GeV. The quantum gravitational interactions between particles at or above T_P or m_P are $O(1)$, conventional classical approximations break down, and we cannot even speculate sensibly any more. Below 10^{19} GeV quantum gravitational effects presumably decrease as a power of the temperature T, and we can use classical solutions of the Einstein equations, typically Robertson-Walker-Friedman (RWF) metrics, to describe the large scale structure of gravitational effects. The expansion of the Universe in the presence of curvature k is then described[26] by

$$\frac{\ddot{R}}{R} = -4\pi G\left(p + \frac{\rho}{3}\right) \tag{11a}$$

$$\left(\frac{\dot{R}}{R}\right)^2 = \frac{8\pi G\rho}{3} - \frac{k}{R^2} \tag{11b}$$

where R is the RWF scale parameter and ρ and p are the energy density and pressure of matter in the Universe. At very high temperatures when all particles are ultrarelativistic we have $p = {}^1\!/_3\rho$ and

$$\rho = g_T B\, T^4 , \tag{12}$$

where g_T is the effective number of particle helicity states (1 for a scalar, 2 for a massless vector, 7/4 for a two-component spinor, etc.) and B is a constant taking the value

$$B = \frac{\pi^2}{30} \tag{13}$$

if the particles are in kinetic equilibrium. Solving the evolution equations (11) with the particle energy density (12) yields an age-temperature relation for the Universe:

$$\frac{t(T)}{t_P} = \left(\frac{3}{32\pi g_T B}\right)^{\frac{1}{2}} \left(\frac{T_P}{T}\right)^2 , \tag{14}$$

where t_P is the Planck time $O(10^{-42})$ sec related to the Planck energy of 10^{19} GeV by the usual uncertainty relation.

The relativistic particles in the hot but cooling and expanding Universe have interactions of which the most important are $(2 \leftrightarrow 2)$ scattering processes (Fig. 1a) and $(1 \leftrightarrow 2)$ decays and inverse decays (Fig. 1b). Decays and inverse decays of a massive particle X will be characterized by a time-dilated version of the free decay time:

$$\tau \propto \frac{(T^2 + m_X^2)^{\frac{1}{2}}}{\alpha_X m_X^2} \tag{15}$$

which is very long in the limit that T >> the mass m_X. In fact, many or all of the particles in GUTs acquire their masses from spontaneous symmetry breaking, so that they are functions of temperature and are zero above some critical temperature T_c which is generally $\gtrsim m_X$. This entails a further lengthening of τ compared with (15). At low temperatures $T \ll m_X$ the density of massive particles is suppressed by a Boltzmann factor $e^{-m_X/T}$, and there is a similar reduction in the probability for two light particles to come together to make a massive particle. This means that a true measure of the mean free time $\tau_{1\leftrightarrow 2}$ between $1\leftrightarrow 2$ interactions is related to τ (15) by

$$\tau_{1\leftrightarrow 2} \approx \tau \times \exp - (m_X/T) . \tag{16}$$

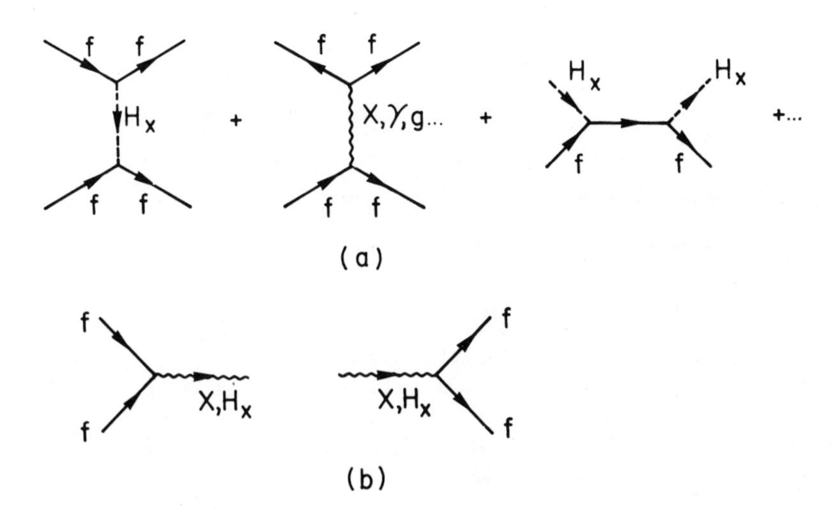

Fig. 1 Examples of important fundamental interactions: (a) ($2 \leftrightarrow 2$) scattering, (b) ($1 \leftrightarrow 2$) decays and inverse decays.

Dimensional arguments would suggest that in the ultrarelativistic limit $2 \leftrightarrow 2$ scattering in a GUT with a dimensionless coupling constant α_{GUM} would behave as

$$\sigma(2 \leftrightarrow 2) \approx \frac{\alpha^2_{GUM}}{T^2} \tag{17}$$

possibly modified by logarithmic factors, and for processes involving vector boson exchange the best guess[5,6] is that $\alpha_{GUM} \sim 1/40$. In fact, if the scattering particles were free, one would have from the t-channel exchange of a massive boson X

$$\sigma(2 \leftrightarrow 2) \approx \frac{\alpha^2_{GUM}}{m^2_X}$$

at high energies, but in the primordial plasma we are considering there is a reduction[41] of σ due to screening effects. These reduce the cross-section to at most $\alpha^2_{GUM}\ \lambda^2_D$, where the Debye length[41]

$$\lambda_D \approx \left(\frac{T}{32\pi^2 \alpha_{GUM} n} \right)^{\frac{1}{2}} \tag{18}$$

and n is some relevant number density:

$$n = g_T A\ T^3 \tag{19}$$

with $A \approx 0.12$ for a Boltzmann distribution. If (?) the relevant n corresponds to $g_T = 1$, then for $\alpha_{GUM} \approx 1/40$, $\lambda_D \approx 1/T$ and the dimensional cross-section formula (17) is applicable. If we consider ($2 \leftrightarrow 2$) scattering (Fig. 1a) via a boson whose mass m_X is comparable to the temperature T then the cross-section (15) is suppressed by propagator effects

$$\sigma(2 \leftrightarrow 2) \approx \frac{\alpha_{GUM}^2\ T^2}{(T^2 + m_X^2)^2} \tag{20}$$

which means that the scattering is greatly suppressed at temperatures $T << m_X$. From the cross-section (17) one can deduce a mean free time $\tau_{2\leftrightarrow 2}$ between collisions:

$$\tau_{2\leftrightarrow 2} = \frac{1}{n\sigma(2 \leftrightarrow 2)} \tag{21}$$

and we see from (17) and (19) that

$$\tau_{2\leftrightarrow 2} \propto \frac{1}{\alpha_{GUM}^2}\ \frac{1}{T} \tag{22a}$$

at very high temperatures. At very low temperatures $T << m_X$, Eqs (19) and (20) imply that for ($2 \leftrightarrow 2$) scattering through massive particle exchange Eq. (22a) is modified to become

$$\tau_{2\leftrightarrow 2} \propto \frac{1}{\alpha_{GUM}^2}\ \frac{m_X^4}{T^5}\ . \tag{22b}$$

In the neighbourhood of $T \sim m_X$ neither (22a) nor (22b) is appropriate. If the X exchange takes place in a crossed channel there will be a simple interpolating form for $\tau_{2\leftrightarrow 2}$. However, if the X exchange takes place in the <u>direct</u> channel, there is a possibility of enhancing $\sigma_{2\leftrightarrow 2}$, and reducing $\tau_{2\leftrightarrow 2}$ correspondingly, because of scattering through X bosons close to their mass shell[41]. It is convenient to explicitly remove these pole contributions from $\sigma_{2\leftrightarrow 2}$ and lump them together with the $(1\leftrightarrow 2)$ decays and inverse decays to which they are directly related by unitarity, leaving behind a reduced $(2\leftrightarrow 2)$ cross-section which leads to a smooth interpolation of $\tau_{2\leftrightarrow 2}$ between (22a) and (22b).

We can now determine approximately the epochs during which various interactions are in equilibrium by looking at the ratio between the expansion time t (14) and the various different mean free times τ. If $t/\tau < 1$, the corresponding interaction is out of equilibrium, whereas if $t/\tau > 1$ then the interaction is in equilibrium. We have several different cases to consider: the results are illustrated in Fig. 2. Hereafter unless explicitly stated to the contrary, we will use "natural" units in which Planck units (m_P, T_P, t_P, ...) are unity.

Quantum gravitational effects

If we assume that at $T < T_P = 1$, $\sigma_{grav} \leq G$ (and $\sigma_{grav} \approx G^2T^2$ seems likely to be a realistic estimate for $(2\leftrightarrow 2)$ scattering) then

$$\tau_{grav} \gtrsim \frac{1}{GT^3} \qquad \text{(for example } 1/G^2T^5\text{)}, \tag{23}$$

whereas $t \sim T^{-2}$ so that

$$\left.\frac{t}{\tau}\right|_{grav} \lesssim T \tag{24}$$

and quantum gravitational interactions are expected to be relatively unimportant and out of equilibrium for $T < 10^{19}$ GeV.

Strong, weak and electromagnetic interactions

Since the masses of particles mediating these interactions are negligible on the energy scales we are considering, the cross-section (17) should be applicable and we expect

$$(t/\tau)_{SWEM} \approx \frac{\alpha^2_{GUM}}{T} . \tag{25}$$

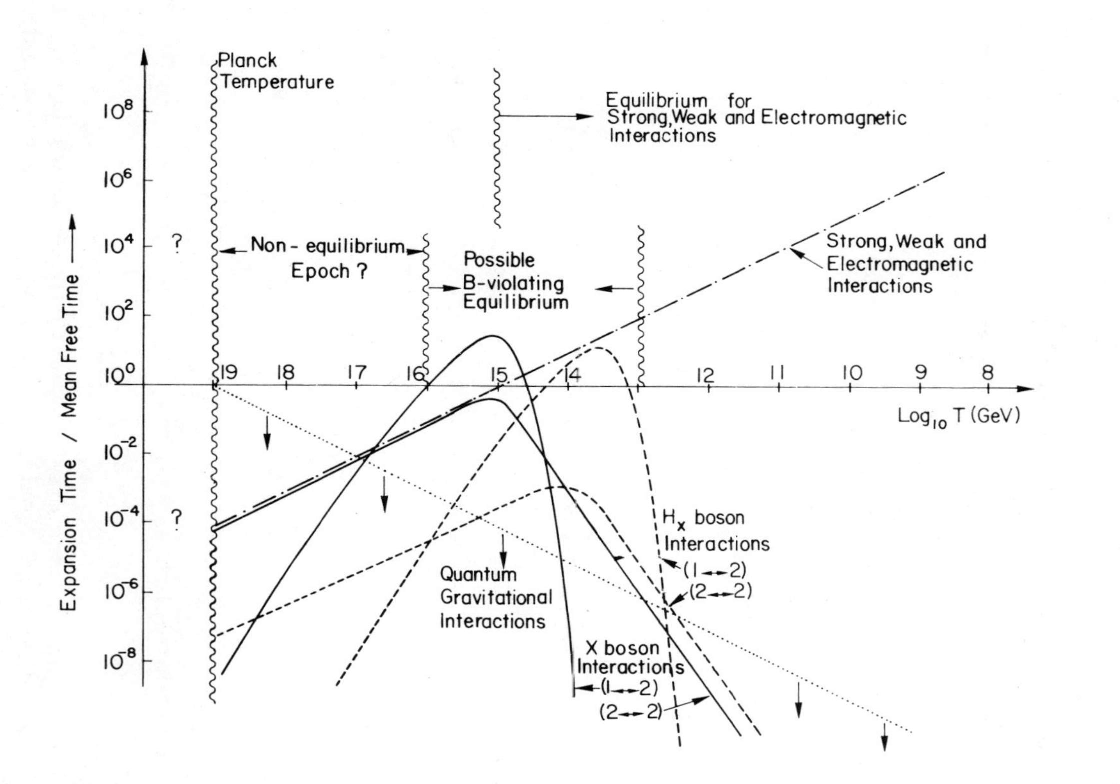

Fig. 2 The ratios of the universal expansion time (14) to the mean free times for different interactions[42] discussed in the text. Note the possibility that no known interactions were in equilibrium at 10^{19} GeV > T > 10^{15} GeV, the probable period of equilibrium for baryon number violating interactions at T $\sim 10^{15}$ GeV, and their subsequent non-equilibrium period.

This ratio is $O(10^{-4})$ at $T \sim 10^{19}$ GeV, meaning that these interactions should be out of thermal equilibrium at that temperature, a state of affairs which should persist until $T \sim \alpha^2 \sim 10^{15}$ GeV, after which strong, weak and electromagnetic interactions should be in equilibrium. This equilibrium state should be maintained until much lower temperatures, below those at which the weak interaction symmetry is spontaneously broken and the colour interactions are shielded by long-distance non-perturbative effects.

Grand unification interactions

By these we mean the interactions of (or involving exchanges of) the superheavy X bosons of a GUT. As far as the $(2\leftrightarrow2)$ interactions are concerned, we expect at $T >> m_X$ a behaviour like that of the strong, weak and electromagnetic interactions (25), while at temperatures $T << m_X$ we expect from (22b) that

$$\left(\frac{t}{\tau}\right)_{2\leftrightarrow2\ \mathrm{GUT}} \approx \frac{\alpha_X^2\ T^3}{m_X^4} \tag{26}$$

with a smooth interpolation between (25) and (26) at $T \sim m_X$. This function is therefore $O(1)$ at $T \sim m_X \sim \alpha_X^2$. For $(1\leftrightarrow2)$ interactions we expect from (15) and (16) that

$$\left(\frac{t}{\tau}\right)_{1\leftrightarrow2\ \mathrm{GUT}} \approx \frac{\alpha_X\ m_X^2}{(T^2 + m_X^2)^{\frac{1}{2}}\ T^2}\ e^{-m_X/T} . \tag{27}$$

This function is clearly $<< 1$ at $T >> m_X$ and at $T << m_X$, peaking for T near m_X at a value of order $(\alpha_X/m_X) \approx 10^2$. The ratios $(t/\tau)_{1\leftrightarrow2,\,2\leftrightarrow2\ \mathrm{GUT}}$ are plotted[42] in Fig. 2, and we see that they suggest that for an epoch where T was of order m_X the interactions involving X bosons may have been in equilibrium.

One can make similar estimates for the interactions of the superheavy Higgs bosons H_X expected in GUTs. Shown in Fig. 2 are the curves corresponding to $(2\leftrightarrow2)$ interactions via Higgs exchange and $(1\leftrightarrow2)$ Higgs decays and inverse decays, assuming a Higgs mass $\sim 10^{14}$ GeV and a Higgs coupling $\alpha_H \approx 10^{-3}$. Though these parameters are clearly subject to some uncertainty, we see a tendency to have a period in which H_X processes are in equilibrium.

Given the general qualitative features of the expansion time to mean free time ratios shown in Fig. 2, there are three qualitatively different scenarios[42] for generating the baryon asymmetry observed in the Universe, illustrated in Fig. 3.

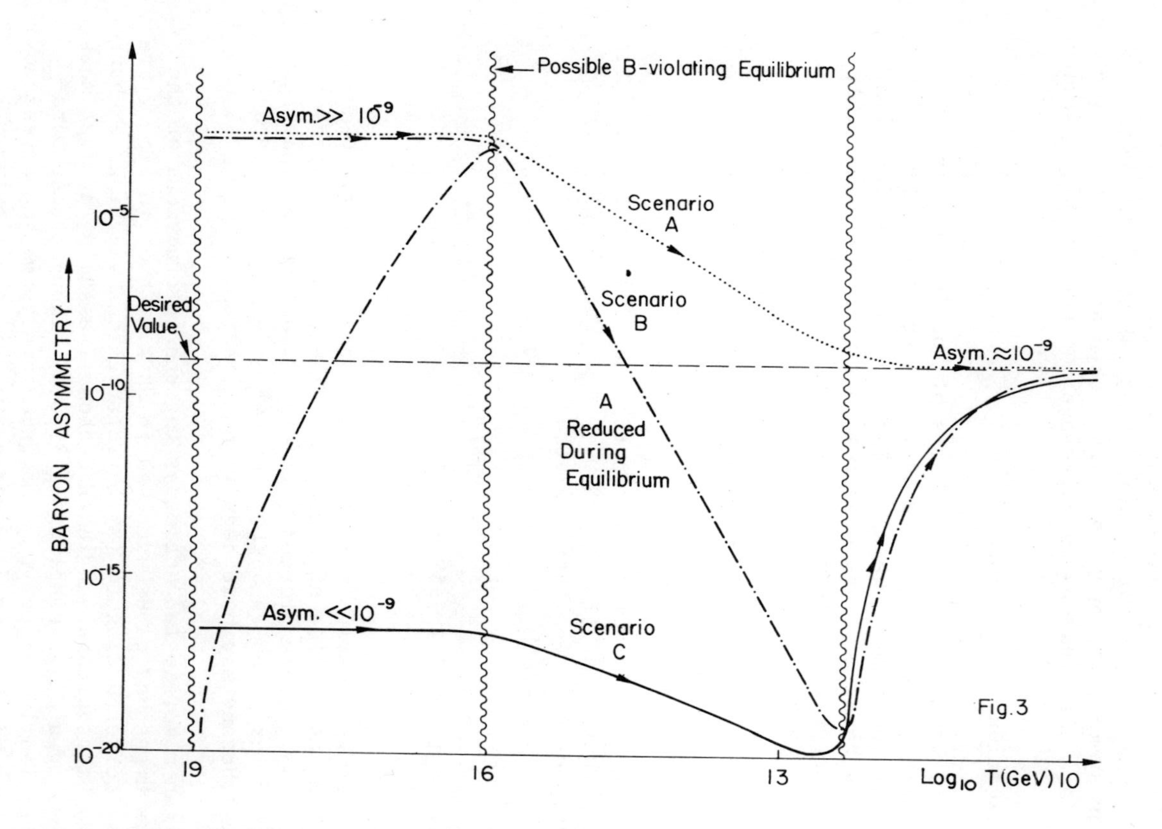

Fig. 3 The various scenarios[42] (A), (B) and (C) described in the text for baryon number generation in the context of GUTs

(A) The Universe might have started at $T \sim 10^{19}$ GeV with a baryon-antibaryon asymmetry $\gg 10^{-9}$, or such a large asymmetry might have been generated by interactions at $T > m_X \sim 10^{15}$ GeV. This baryon asymmetry would then have been reduced during the period of grand unified equilibrium around $T \sim 10^{15}$ GeV (in accord with general theorems[17,40]) to the value $10^{-9\pm1}$ observed today[43,17].

(B) Again one considers the possibility that the baryon asymmetry may have been large at temperatures $> m_X$, but that grand unified interactions in equilibrium may then have pushed the number much below $10^{-9\pm1}$, only to have X and H_X interactions in the post-equilibrium period regenerate the required number $10^{-9\pm1}$.

(C) Perhaps the Universe really did start off at $T > 10^{15}$ GeV with a baryon asymmetry which was either zero or very small $\ll 10^{-9\pm1}$. Then as X and H_X interactions drop out of thermal equilibrium after 10^{15} GeV the required baryon number of $10^{-9\pm1}$ per photon could have been generated.

Examining these different alternatives, we see that options (B) and (C) give us the virtue of naturalness: the final asymmetry is determined by GUT considerations independent of the initial conditions. In contrast, option (A) requires a very delicate fine-tuning[17] of the interaction rates if the total wash-out of the baryon number is to be avoided. It does not require a C- and CP-violating component in the B-violating GUT interactions, but we have strong reasons to believe that such a component exists. In the rest of this discussion we will discuss the baryon asymmetry[16] generated by C- and CP-violating interactions of GUTs in options (B) and (C).

4. MECHANISMS FOR BARYON NUMBER GENERATION

The generation of the net baryon number (10) may have been quite complicated, with several different baryon-number violating processes contributing and/or competing. However, it seems likely that the dominant driving mechanism may have been the decays and inverse decays of some superheavy boson[40,44] either a gauge vector X or a Higgs scalar H_X. In most simple[1,27] GUTs the dominant decays of these particles are expected (Fig. 4) to be into pairs of quarks or into antilepton and antiquark:

$$X \text{ or } H_X \to (q + q) \quad \text{or} \quad (\bar{q} + \bar{\ell}) \; . \tag{28}$$

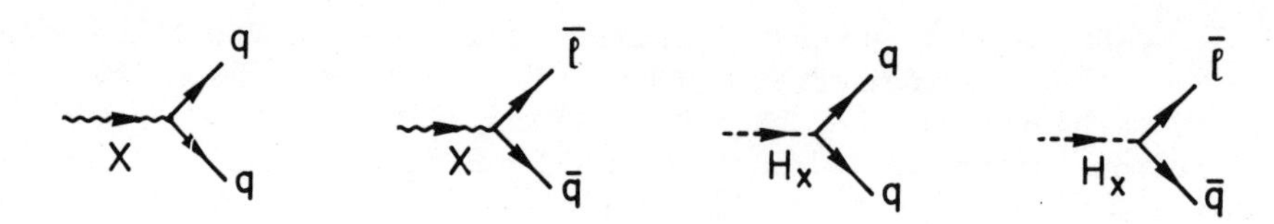

Fig. 4 Dominant decay modes of superheavy vector bosons X and Higgs bosons H_X .

The total decay rates of particles and antiparticles must be the same because of CPT:

$$\Gamma_{tot}(X) = \Gamma(X \to qq) + \Gamma(X \to \bar{q}\bar{\ell}) + \ldots$$
$$\Gamma_{tot}(\bar{X}) = \Gamma(\bar{X} \to \bar{q}\bar{q}) + \Gamma(\bar{X} \to q\ell) + \ldots \tag{29}$$

but the partial decay rates into conjugate channels may differ if C and CP are both violated:

$$B \equiv B(X \to \bar{q}\bar{\ell}) \equiv \frac{\Gamma(X \to \bar{q}\bar{\ell})}{\Gamma_{tot}(X)} \neq \bar{B} \equiv B(\bar{X} \to q\ell) \equiv \frac{\Gamma(\bar{X} \to q\ell)}{\Gamma_{tot}(\bar{X})} \ . \tag{30}$$

Then if one naively assumes that one starts with an identical population of X and $\bar{X}$ bosons (or of H_X and $H_{\bar{X}}$) their decays would generate a net baryon to photon number density ratio[44]

$$\frac{n_B}{n_\gamma} \backsimeq \left(\frac{g_X}{g_{tot}}\right)(\bar{B} - B) \approx (10^{-2})\ \Delta B \tag{31}$$

where g_X is the number of helicity states of X bosons and g_{tot} is the total number of helicity states. We will see later that this naïve decay mechanism is somewhat over-simplified[45], but it suffices for discussions of some qualitative features of the baryon generation problem.

It is clear that no difference ΔB in the branching ratios of particles and antiparticles can arise in lowest order. Such a non-zero ΔB can first arise in fourth order, through final state interactions (see Fig. 5) involving bosons which violate B conservation[44]. The resulting difference in branching ratios is of order

$$\Delta B \approx \text{Im Tr}(abcd)/\textstyle\sum_a \text{Tr}(aa^*) \ , \tag{32}$$

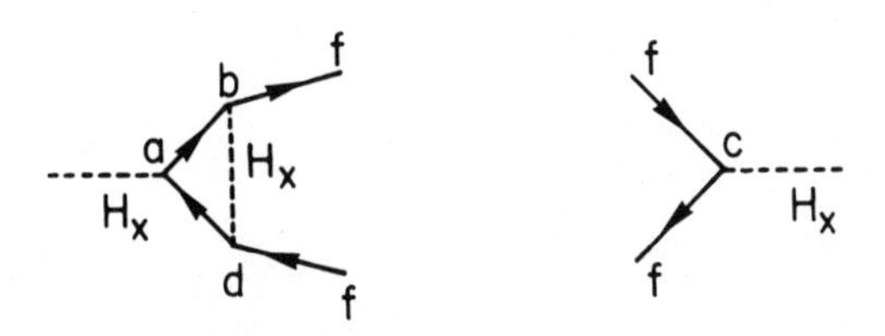

Fig. 5 A sample 4th order interference diagram which could (see Eq. (32)) contribute to ΔB and hence to the baryon asymmetry.

where a,b, c and d are the coupling matrices introduced in Fig. 5. In fact, in simple models the first C- and CP- violating ΔB often occurs only in higher orders. In a large class of GUTs all the CP violation originates from the fermion-fermion-Higgs Yukawa couplings. As in the simplest case (32), all leading CP-violating rate differences such as ΔB are related to products of Yukawa coupling matrices traced along fermion lines. These traces may arise from thermal averaging over initial states as well as from sums over possible final states, and occur for scattering processes as well as for $(1 \leftrightarrow 2)$ decays and inverse decays. It is then important in the context of any particular model to determine the lowest order trace of Yukawa couplings which violates C and CP. The lowest order contributions to the baryon asymmetry can then be got from this trace by connecting up the Yukawa vertices by Higgs lines in all possible ways, and then cutting the resulting diagrams in all possible ways corresponding to processes with different initial and final states and different possible arrangements of the final state interactions. As an example, we consider[42] the "minimal" SU(5) model with the usual $\underline{\bar{5}} + \underline{10}$ fermions and one $\underline{24}$ and one $(\underline{5} + \underline{\bar{5}})$ pair of Higgs representations. In this model the first CP-violating trace is 8th order in the Yukawa coupling -- see Fig. 6a. Many different ways of connecting up the Higgs vertices are possible, of which one example is shown in Fig. 6b. This diagram and its fellows may then be cut in many different ways corresponding to different physical processes, as indicated in Figs 6 c,d. Each one of these cuts corresponds to a direct CP-violating interaction rate ($(1 \leftrightarrow 2)$ or $(2 \leftrightarrow 2)$ etc.) which may contribute to the total baryon asymmetry generated. Suppose we consider the contribution of the cut in Fig. 6c which gives a CP-violating difference ΔB in decay branching ratios. Each Yukawa vertex is $O(m_f/m_W \times g)$ and the CP-violating effect is proportional to GIM-like products of differences in quark masses. If we assume there are just three generations of $\underline{\bar{5}} + \underline{10}$ fermions and that $m_t >> m_b >> m_c >>$ other fermion masses, then we have a CP-violating

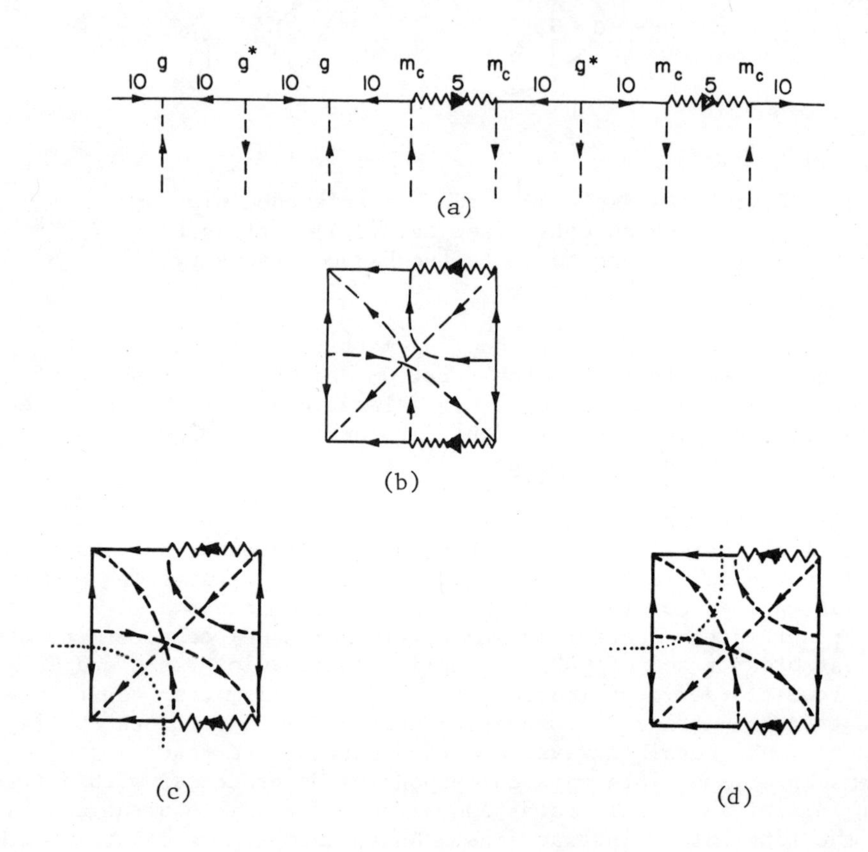

Fig. 6 (a) One of the lowest order C-,CP- and B-violating fermion traces in minimal SU(5). (b) The Higgs-fermion-fermion vertices of the trace connected up in one of many possible ways. (c) The previous diagram cut in such a way as to contribute a CP-violating ΔB in Higgs decays. (d) The same diagram cut so as to make a CP-violating difference between Higgs-fermion scattering cross-sections - cf. Fig. 1(a). The dashed lines are $\underline{5}$ Higgs bosons, the solid lines $\overline{\underline{10}}$ fermions, the zig zag lines $\underline{5}$ fermions, and the dotted lines are unitarity cuts.

$$\Delta B \lesssim \frac{1}{10}\,\alpha_X^4\,\frac{m_b^4\,m_t^3\,m_c}{m_W^8}\Bigg/(m_t^2/m_W^2)\,\alpha_X\,, \tag{33}$$

where the 1/10 is an upper limit on the product of mixing angles and phase factors, and we have neglected various numerical factors from phase space, etc., as well as combinatorial factors from the large number of similar diagrams. Putting in the known quark masses we find

$$\Delta B \lesssim 10^{-14}\,\frac{m_t}{m_W} \leq 10^{-14} \tag{34}$$

which Eq. (31) demonstrates to be too small to generate the observed baryon asymmetry of order $10^{-9\pm 1}$. A model more complicated than minimal SU(5) (more Higgs representations? superheavy fermions?) is necessary[18,44] to generate the observed number, and can easily be constructed. (Note that any process involving a gauge vector boson in this model, whether in an initial or final state or in a virtual interaction, has CP violation only in at least one order of α higher.)

Before discussing in more detail the mechanics of computing the baryon asymmetry in specific models it is perhaps worth making a few general remarks about the nature of the CP-violation necessary for generating non-zero baryon number. It is clear that one needs a "hard" form of CP-violation which persists up to high energies $\sim 10^{15}$ GeV, and therefore that models where CP-violation is generated by spontaneous breakdown on a weak interaction scale of order 100 GeV are not suitable. Furthermore, models where CP is violated spontaneously at $O(10^{15})$ GeV are also not acceptable. At these very early epochs causally connected domains in the Universe are very small, and in a scenario of spontaneous CP violation there would be no correlation between the signs and perhaps magnitudes of the baryon number generated in different domains. At subsequent epochs previously separated domains become connected, and the net baryon asymmetry we see today would tend to be a statistically averaged, and hence much diluted, version of the asymmetry possible in any individual domain. It therefore seems that a scenario of spontaneous CP-violation is difficult to maintain in the absence of acausal effects, and that some form of intrinsic CP-violation like that found[1,28] in the minimal SU(5) model is to be preferred. One might wonder whether the sign of the baryon asymmetry can be related to the sign of CP-violation in the K^0-$\bar{K}^0$ system. Quite apart from the computational problems of getting numbers for either of these physical quantities from parameters of a fundamental Lagrangian, it seems unlikely in general that the same CP-violating parameter controls both phenomena. For example, in the standard SU(5) with N_g (e.g. 3) generations the generalized Kobayashi-Maskawa matrix characterizing low-energy weak interactions has $(N_g-1)^2$ (e.g. 4) parameters of which $\frac{1}{2}(N_g - 1)(N_g - 2)$ (e.g. 1) are CP-violating phases "observable" via K^0-$\bar{K}^0$ phenomena, while at high energies there are[28,42] $(N_g - 1)$ (e.g. 2) additional CP-violating phases, one of which determines the dominating CP-violating rate (33). In fact, even in the minimal SU(5) model CP-violation and hence a net baryon asymmetry is possible with only four quarks, though the number generated increases if more quarks or other structure is introduced, and a realistic model probably needs[46] at least six quarks.

There are several obstacles remaining before a quantitative calculation of the baryon asymmetry can be made. The first is that because the minimal SU(5) model is too small, one must go[18] to a more complicated model with more free parameters, which hinders the attainment of numerical precision. Secondly, the naïve decay scenario outlined above is incomplete. This is because, as can be seen from Fig. 2, $(2\leftrightarrow2)$ interactions are still significant and near equilibrium around the time of the heavy particle decays. Since the net baryon asymmetry must be zero in equilibrium, this suggests that these $(2\leftrightarrow2)$ interactions have a tendency to reduce the net asymmetry generated by $(1\leftrightarrow2)$ interactions. This washing-out has been computed numerically by various groups[17,47,48] integrating up the Boltzmann equations using both $(1\leftrightarrow2)$ and $(2\leftrightarrow2)$ interactions. They find that the amount of washout is very sensitive both to the masses of the superheavy bosons and to the grand unified coupling α_X. We see from Fig. 7 that the efficiency of B generation via vector bosons X with $\alpha_X \sim 1/40$ is 100% efficient only for $m_X \gtrsim 10^{17}$ GeV, while for Higgs bosons H_X with $\alpha_{H_X} \sim 10^{-3}$ complete efficiency persists down to $m_{H_X} \approx 10^{15}$ GeV. If one requires an efficiency $\gtrsim 0.1\%$ in order to get a large enough n_b/n_γ, then one needs (see also the paper by Weinberg and the second paper by Yoshimura in Ref. 16):

$$m_X \gtrsim 10^{15}\ \text{GeV} \quad \text{or} \quad m_{H_X} \gtrsim 10^{13}\ \text{GeV}\ . \tag{35}$$

For comparison, the latest best estimates from low-energy phenomenology[2,6,7,30] are

$$m_X \approx 6 \times 10^{14}\ \text{GeV}\ , \quad m_{H_X}/m_X \approx 10^{0\pm1} \tag{36}$$

which are just consistent with the constraints (35). The greater amount of washing-out for vector boson processes is another reason (in addition to their intrinsically higher order of CP-violation) why one might expect the dominant contribution to the baryon asymmetry to come from processes involving superheavy Higgses. There

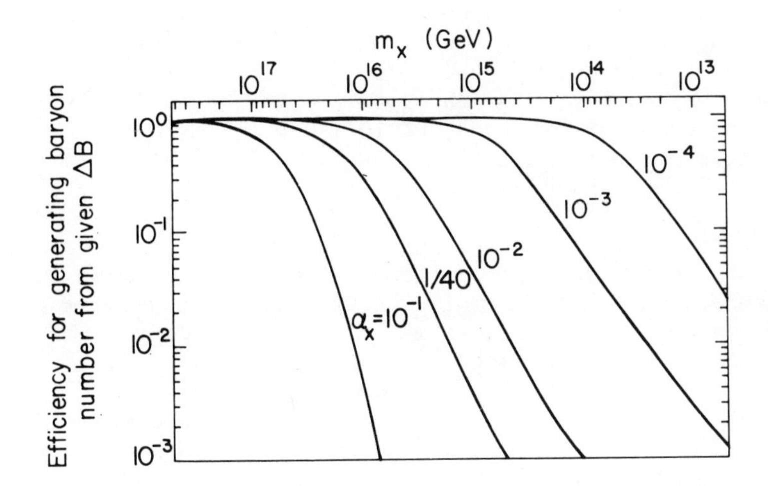

Fig. 7 A diagram from the work of Kolb and Wolfram[17] illustrating how the net baryon number obtained from a CP-violating difference in decay branching ratios may be less than expected due to the wash-out by (2↔2) interactions.

is one complication with these wash-out calculations that should be mentioned. It is that in models with reducible fermion representations for each generation of fermions, such as SU(5) with its $\underline{\bar{5}} + \underline{10}$ representations, the vector bosons which are the dominant (2↔2) interactions cannot[47] wash out all combinations of quark asymmetries because of helicity conservation. This means that in such "reducible" models the constraints (35) may be relaxed somewhat, though the degree of relaxation has not yet been computed. Another important problem is the interplay of washing out and CP-violation in a situation with several different varieties of superheavy boson (more than one stage of GUT symmetry breaking, $m_X \neq \neq m_{H_X}$, etc.). Under certain restrictions on α_X and m_X the heavier bosons may establish equilibrium for the lighter bosons[44], or the asymmetries from different sources may just add[47]. Other calculational complications arise from possible deviations from the simple isotropic and homogeneous, monotonically expanding and cooling Universe which we have been considering. For example, there may be significant inhomogeneities - primordial black holes (PBHs) - whose explosions at or after a time corresponding to a universal temperature of $\sim 10^{15}$ GeV might[48] contribute significantly to the net baryon asymmetry. (Conversely, a satisfactory computation of n_B/n_γ in a completely homogeneous and isotropic scenario might impose constraints on the spectrum of primordial black holes.) Related complications may arise from the observation[42] (see Fig. 2) that no known particle interactions may have been in equilibrium at temperatures between 10^{19} and 10^{15} GeV. This means that particles need not have been in a Boltzmann distribution in this epoch, and that dissipative phenomena may have been important, as discussed in the next section. Finally, as discussed in Section 6 of this paper, it has recently been suggested[22] in the context of grand unified monopole production early in the Universe that the Universe may have had a complicated thermal history during the grand unified phase transition. It may have supercooled in the symmetric phase down to a temperature $\ll 10^{15}$ GeV, and then been reheated by the latent heat released in a first order phase transition. This would complicate considerably the previous naïve discussion (see Fig. 2) of the expansion rate, interaction rates and thermal equilibrium. Because of all the complications mentioned here -- and doubtless others -- it seems reasonable to conclude that the GUT mechanism for generating the baryon asymmetry (10) is still in a qualitative stage, though the process of quantification is making significant advances.

5. NON EQUILIBRIUM IN THE VERY EARLY UNIVERSE?

It is interesting and perhaps significant that when the temperature of the Universe was between 10^{19} GeV and roughly 10^{15} GeV no "known" particle interactions may have been in equilibrium[42,49]: during that epoch one has approximately

$$\frac{\text{mean free time}}{\text{expansion time}} \approx \frac{T}{\alpha^2} > 1 \ . \tag{37}$$

This means that particles need not have been Boltzmann distributed, and that dissipative phenomena[20] may have been significant. What might the primordial distribution of particles have been? In a Robertson-Walker-Friedman Universe, when the temperature was of order 10^{19} GeV, the horizon diameter $D_H = 4t$ would have been of the same order as the Planck distance $\approx 1/m_P$. In the absence of acausal effects -- which should, however, perhaps be expected in the context of quantum gravity -- it is difficult to imagine how coherence could have been established over distances $> D_H$ with the same ease as over distances $< D_H$. (It is difficult to imagine particles with a wavelength size larger than the horizon size.) In this case, one might expect[49] particles with short wavelengths $\lambda < 1/m_P$ or high energies $E \gtrsim O(m_P)$ to have a spectrum distorted by comparison with a naïve Boltzmann distribution. As an extreme case, one could even imagine a cut-off of the low-energy tail of the spectrum at $E < O(m_P)$. If such an approximate cut-off occurred, one can estimate that $(1\leftrightarrow 2)$ production processes would have been unlikely to fill in the spectrum at $T >$ few $\times 10^{16}$ GeV, while $(2\leftrightarrow 3)$ interactions are unlikely[49] to have done the job before $T \approx 10^{16}$ GeV. Therefore, if the masses of the X bosons were significantly higher than the fashionable 10^{15} GeV, it is an open question whether their distributions were ever fully thermalized. In the absence of thermalization, various theorems[17,40] about the absence of a net baryon number become inapplicable, and the computation of n_B/n_γ becomes more complicated. This complication is unlikely to have arisen if m_X is as low as the expected 10^{15} GeV, but could be significant if m_X were significantly larger.

Equation (37) implies that during this very early epoch particles would have had mean free paths large compared with the horizon size, and therefore that dissipative phenomena[20] may have occurred as illustrated in Fig. 8. Generically, the shear viscosity η of an imperfect fluid is[50] of order

$$\eta \simeq T^4 \tau \tag{38a}$$

and taking the earlier estimates of the mean free time τ one gets[20]

$$\eta \approx (10^2 \text{ to } 10^3)\ T^3 \tag{38b}$$

when the temperature exceeds 10^{15} GeV. Similarly, the thermal conductivity χ of an imperfect fluid is of order

$$\chi \sim T^3 \tau \tag{39a}$$

which becomes[20]

$$\chi \simeq (10^3 \text{ to } 10^4)\ T^2 \tag{39b}$$

in this very early epoch. By contrast, the bulk viscosity is zero for a fluid of ultrarelativistic particles. The grand unified viscosity (38) and thermal conductivity (39) may have played a role in smoothing out primordial inhomogeneities or, perhaps, correcting initial anisotropies as illustrated in Fig. 8. Because of causality,

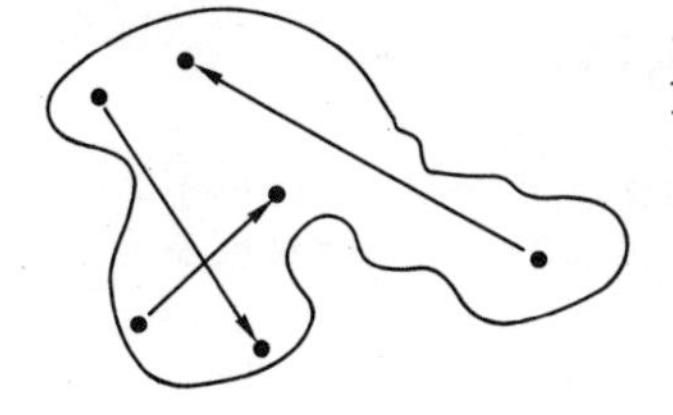

Fig. 8 Illustration of particles with long mean free paths transferring energy and momentum and hence possibly giving rise to dissipation through viscosity and/or thermal conductivity.

their effects should be limited to within the size of a horizon: this may be sufficient[20] to damp down high frequency modes of oscillation[26] (radiative, rotational or compressional) which might otherwise have contained an energy density comparable to that in elementary particles. These dissipative effects also tend to homogenize the Universe locally preparatory to the generation of baryon number: in particular they may[51] tend to hinder the production of primordial black holes $\leq$ the horizon size. However, viscous effects do not prevent the initiation and growth of very low frequency, large scale compressional inhomogeneities, which Press[21] has speculated may have originated at the time of the grand unified phase transition and subsequently grown in intensity and evolved into galaxies. In view of the logarithmically long growth time between 10^{15} GeV and the epoch at which galaxy formation is conventionally initiated, this may be a promising way[21] of avoiding the normal problems in explaining galaxy formation.

6. GRAND UNIFIED MONOPOLES

These appear[11] near the mass scale at which a primitive U(1) factor in the gauge group of the world gets absorbed into a simple group. For example, in the case of the minimal GUT SU(5): $SU(5) \rightarrow SU(3) \times SU(2) \times U(1)$ at around 10^{15} GeV, one expects to encounter topologically non-trivial and stable states (cf. Fig. 9a) called magnetic monopoles M with masses

$$m_M \approx \frac{m_X}{\alpha_X} \approx 10^{16} \text{ GeV} . \tag{40}$$

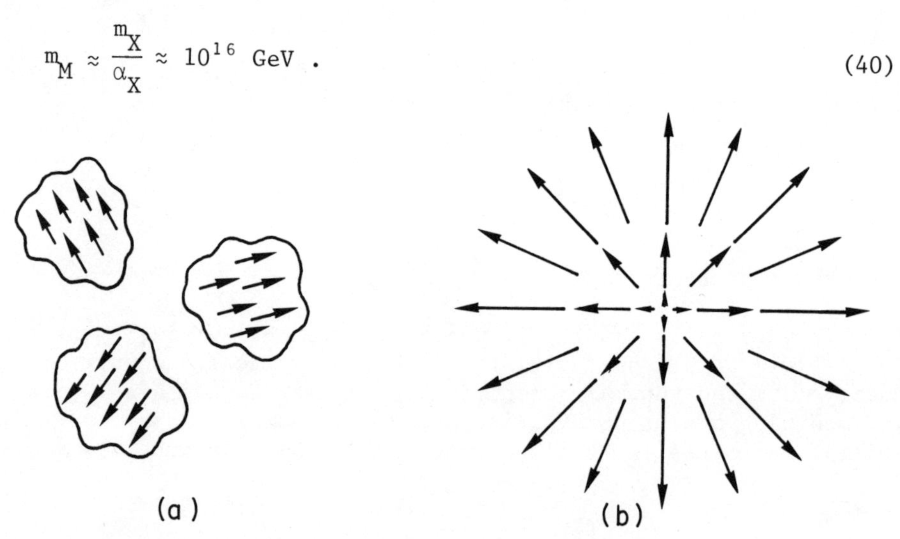

Fig. 9 Impressions of (a) Higgs fields in causally separated domains pointing in different (internal group) directions causing (b) a monopole to be formed where they meet: note that the value of the Higgs field goes to zero in the centre.

These monopoles may have been produced copiously in the very early Universe, and if so their topological stability may have caused them to survive to lower temperatures in unacceptably large numbers[12]. It is convenient to introduce a dimensionless density for these objects which is

$$r \equiv n/T^3 \tag{41}$$

and is basically the monopole-to-photon ratio. From the success of the conventional calculations[15] of nucleosynthesis at temperatures $T \sim 1$ MeV one deduces that monopoles did not dominate the energy density at that epoch, so that

$$r_{1\ \text{MeV}} < 10^{-19} . \tag{42a}$$

The absence of a collapse in the present Hubble expansion due to a high density of relic monopoles means that

$$r_{3^\circ K} < 10^{-25} . \tag{42b}$$

It has been estimated that conventional $M\bar{M}$ capture and annihilation in a homogeneous Universe could not have reduced r below 10^{-10}. The Universe now contains inhomogeneities (e.g. you, us and galaxies) but excess annihilation in these could not have explained the number (42a), and probably not the number (42b) either. We therefore conclude that <u>either</u> very few monopoles ($\lesssim$ Eq. (42)) were produced <u>and/or</u> the conventional capture and annihilation rates are not applicable.

The conventional picture[12,22] of monopole production in the early Universe goes somewhat as follows. At very high temperatures the Higgs fields have zero vacuum expectation value and tend to run around in random directions as in Fig. 10a. At low temperatures the Higgs fields will congeal with a fixed, non-zero v.e.v. but in general pointing in different internal group directions in different domains of space, as illustrated in Fig. 9b. Where these domains come together there may be a mismatch with the Higgses pointing in different directions so that the lowest energy configuration has a zero field value in the centre, as in Fig. 9a. Such a "hole" is a topologically stable monopole[11] for which one must pay the energetic price (40).

A popular line of argument[22] is to suggest that the number of "holes" is the same order[52] as the number of causally separated domains at the temperature where the monopoles freeze out. To get an acceptably low number (42) one of them tries to reduce as far as possible the number of different domains by setting this freeze-out at the lowest possible temperature, postulating a first-order phase transition and strong supercooling[22]. When the phase transition takes place the latent heat then reheats the Universe back to a temperature closer to 10^{15} GeV, after which it expands and cools in the normal way. Apart from the suppression of monopole production, an interesting feature[53] of this scenario is that the Universe expands exponentially during the supercooled epoch,

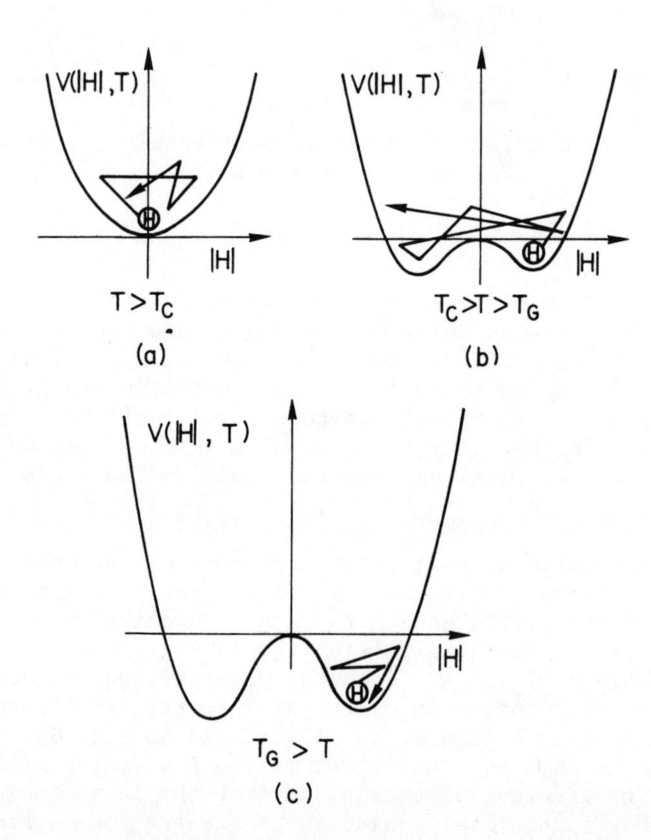

Fig. 10 Illustration of the Higgs field value as a function of temperature. (a) At high temperatures $T > T_C$, $|H|$ is expected to be zero. (b) For intermediate temperatures $T_C > T > T_G$ the minimum of the potential $V(|H|,T)$ is at $|H| \neq 0$, but the thermal fluctuations of $|H|$ are so large that it can pass over the central hump. (c) At low temperatures $|H| \neq 0$ and the Higgs field no longer fluctuates over the hump.

as the energy density and pressure are then dominated by the vacuum energy density. This has two[53] attractive implications, and one which is potentially catastrophic. One piece of good news is that the classic "simultaneity" problem is avoided because before this period of exponential expansion all the Universe we now see, and much more besides, was sufficiently close to be in causal contact and hence all expanding in the same way at the same time. Another piece of good news is that during this exponential expansion any initial curvature ($k \neq 0$ in the RWF solution (11)) would have got flattened out, rather like a balloon being blown up. This might explain why the present density of the Universe is so close to the critical density for eventual gravitational collapse. The bad news is that it is very difficult[53] to see how the transition to a broken phase of the GUT could ever have occurred. The basic problem is that while bubbles of asymmetric Universe expand at the speed of light, the Universe as a whole is expanding too fast for them to catch up. Unless this problem is solved, the scenario is doomed.

The necessity for a strongly first-order phase transition has in any case been criticized[25]. The concept of a Higgs field pointing in any given direction is gauge-dependent, and the concept of domains "pointing in a fixed direction" is not on as solid a basis as in a ferromagnet. A monopole is of course a gauge-independent object, and the proposal[25] is to calculate their density assuming a second-order or weakly first-order transition and the validity of thermal equilibrium arguments. In a second-order phase transition the lowest energy state starts having a non-zero value of the Higgs field at the critical temperature T_c (see Fig. 10a). However, for some range of temperatures below that, the thermal fluctuations are sufficient to knock the Higgs field out of its minimum and over the central "hump" in the Higgs potential, as indicated in Fig. 10b. Therefore, in this range of temperatures it does not even really make sense to talk of a conserved topological quantum number, or of monopoles. At a temperature $T_G < T_c$ the thermal fluctuations are no longer sufficient to jiggle the Higgs, and it congeals into the non-zero valley of the Higgs potential, as in Fig. 10c. This Ginzburg[54,12] temperature T_G is basically determined by the energy density in a correlation volume of radius $O(1/m_{Higgs})$ being less than the free energy density of order $(\langle 0|H|0\rangle)^4$:

$$T_G \approx (\text{geometrical factor}) \times \frac{1}{m_H^3} \times (\langle 0|H|0\rangle)^4 \ . \tag{43}$$

Down to this temperature the Higgs field is still moving around sufficiently fast

$$\tau \sim \frac{1}{m_H} < t \sim T_G^{-2} \tag{44}$$

that thermal equilibrium might be relevant and one would expect

$$r \approx O\left(\exp - \frac{m_M}{T_G}\right) = O\left(\exp - C\frac{m_H}{m_X}\right), \tag{45}$$

where[25] in the minimal SU(5) model

$$C \approx O(30) \ . \tag{46}$$

Bearing in mind the uncertainties in estimating C and the possibility that $m_H/m_X > 1$ (dynamical symmetry breaking?) this argument could well give r low enough to be compatible with the limits (42). Some people dislike[24] this argument on the grounds that since it gives much less than one monopole per causally connected volume, it seems to require the existence of some acausal correlation in the Higgs field. However, as remarked above, gauge invariance may require some of the arguments about causally unrelated domains to be relaxed.

A final remark about the present density of monopoles concerns the classic calculations of capture and annihilation rates which assumed no correlations in the spatial distributions of monopoles and antimonopoles. It has[23] been pointed out that in some GUTs some of the monopoles may be bound together by a string of Z^0 magnetic flux which enhances the annihilation rate. An example[55] of a model where this may play a role in suppressing monopoles is in SO(10) broken down through the chain SO(10) → SO(6) × SO(4) → SU(3) × × SU(2) → U(1) → SU(3) × U(1). For yet another mechanism for suppressing monopoles, see Ref. 56.

Our present feeling is that while none of the above arguments provides a totally convincing mechanism for disposing of the grand unified monopole problem, none of us proposes to lose any sleep worrying about it. But it would be worth improving the present experimental upper limits[57] on the relic monopole density by a few orders of magnitude: after all, perhaps ...?

7. CONCLUSIONS

There is indeed a flourishing cosmic connection between grand unification and cosmology, which has several aspects that are more or less strong. The principal points concern:

Neutrino masses and numbers

There are phenomenological reasons in GUTs to expect only three or four "light" neutrinos, and theoretical reasons to expect that they might have masses up to a few electron volts. The success of nucleosynthesis calculations seems to require three or four "light" neutrinos, and masses up to O(100) eV are allowed by cosmology. Such massive "light" neutrinos may even help alleviate the missing light problem of clusters of galaxies.

Baryon number generation

GUTs provide a framework and a qualitative mechanism for generating the baryon asymmetry observed in the Universe. For it to work, the masses of the superheavy bosons should not be too small, nor their coupling constants too large. Present GUTs are more or less consistent with these constraints, but it seems that a larger GUT than the minimal SU(5) may be necessary if one is to generate a large enough number.

Grand unified monopoles

First estimates of their production rates early in the Universe gave values much above the allowed limits. Ideas for getting rid of them have been proposed, one of which has other interesting cosmological implications (simultaneity, asymptotic flatness) but also has serious problems. It may well be that the grand unified monopole problem has a relatively banal solution. In the meantime: why not look for them?

Acknowledgements

We would like to thank F.A. Bais, J.D. Barrow, B.J. Carr, G. Cocconi, M.B. Einhorn, A. Guth, S. Rudaz, G. Steigman, M.S. Turner, S.-H.H. Tye, S. Weinberg and S. Wolfram for interesting conversations about the topics discussed here.

REFERENCES

1. J.C. Pati and A. Salam, Phys. Rev. Letters 31 (1973) 661 and Phys. Rev. D8 (1973) 1240 were the first proposers of grand unified gauge theories, but the type discussed in this paper was first proposed by
   H. Georgi and S.L. Glashow, Phys. Rev. Letters 32 (1974) 438.
2. J. Ellis, M.K. Gaillard, D.V. Nanopoulos and S. Rudaz - LAPP preprint TH-14/CERN preprint TH.2833 (1980) and references therein.

3. J. Learned, F. Reines and A. Soni, Phys. Rev. Letters 43 (1979) 907 and H.R. Steinberg, private communication (1980), quote limits on the nucleon lifetime of 2 and 3 $\times 10^{30}$ years, respectively, if the decay modes resemble those expected in the SU(5) model of Ref. 1.
4. H. Georgi, H.R. Quinn and S. Weinberg, Phys. Rev. Letters 33 (1974) 451.
5. A.J. Buras, J. Ellis, M.K. Gaillard and D.V. Nanopoulos, Nuclear Phys. B135 (1978) 66.
6. T. Goldman and D.A. Ross, Phys. Letters 84B (1979) 208 and Caltech preprint CALT-68-759 (1980).
7. W. Marciano, Phys. Rev. D20 (1979) 274 and Rockefeller Univ. preprint COO-2232B-195 (1980).
8. L. Maiani, G. Parisi and R. Petronzio, Nuclear Phys. 136B (1978) 115 and references therein.
9. J. Ellis, M.K. Gaillard, L. Maiani and B. Zumino, LAPP preprint TH-15/CERN preprint TH-2841 (1980), contribution to these proceedings; p.69.
   J. Ellis, M.K. Gaillard and B. Zumino, CERN preprint TH.2842/LAPP preprint TH-16 (1980) and references therein.
10. G. Steigman, Ann. Rev. Astron. and Astrophys. 14 (1976) 339.
11. G. 't Hooft, Nuclear Phys. B79 (1974) 276;
   A.M. Polyakov, JETP Letters 20 (1974) 194.
12. J.P. Preskill, Phys. Rev. Letters 43 (1979) 1365; see also Ya.B. Zeldovich and M.Y. Khlopov, Phys. Letters 79B (1979) 239.
13. M.S. Chanowitz, J. Ellis and M.K. Gaillard, Nuclear Phys. B128 (1977) 506.
14. D.V. Nanopoulos and D.A. Ross, Nuclear Phys. B157 (1979) 273.
15. J. Yang, D.N. Schramm, G. Steigman and R.T. Rood, Ap.J. 227 (1979) 697; and
   G. Steigman, contribution to these proceedings, p. 495.
16. A.D. Sakharov, Pis'ma Zh.Eksp.Teor.Fiz. 5 (1967) 32;
   A. Yu. Ignatiev, N.V. Krosnikov, V.A. Kuzmin and A.N. Tavkhelidze, Phys. Letters 76B (1978) 436;
   M. Yoshimura, Phys. Rev. Letters 41 (1978) 381; 42 (1979) 746(E).
   S. Dimopoulos and L. Susskind, Phys. Rev. D18 (1978) 4500;
   D. Toussaint, S.B. Treiman, F. Wilczek and A. Zee, Phys. Rev. D19 (1979) 1036;
   J. Ellis, M.K. Gaillard and D.V. Nanopoulos, Phys. Letters 80B (1979) 360, 82B (1979) 464(E),
   S. Weinberg, Phys. Rev. Letters 42 (1979) 850.
   A.D. Sakharov, Zh.Eksp.Teor.Fiz. 76 (1979) 1172;
   S. Dimopoulos and L. Susskind, Phys. Letters 81B (1979) 416;
   M. Yoshimura, Phys. Letters 88B (1979) 294.
17. E.W. Kolb and S. Wolfram, Phys. Letters 91B (1980) 217, and Caltech preprint OAP-579/CALT-68-754 (1979);
   J.N. Fry, K.A. Olive and M.S. Turner, Enrico Fermi Institute preprint 80-07 (198).

18. D.V. Nanopoulos and S. Weinberg, Phys. Rev. D20 (1979) 2484; S. Barr, G. Segrè and A. Weldon, Phys. Rev. D20 (1979) 2494; see also
J. Ellis, M.K. Gaillard and D.V. Nanopoulos, Ref. 16.
19. Another cosmological topic worth mentioning is the lepton number of the Universe. If lepton number L is violated as well as B, then one might expect a net n_L comparable with n_B. In many GUTs however, B-L is conserved or very nearly so, and in these models neither n_B nor n_L has good reason to be small unless the net n_B and n_L were both initially zero (a symmetric Universe). In more general theories, where (B-L) is violated, this problem does not arise. For discussions, see
S. Dimopoulos and G. Feinberg, Phys. Rev. D20 (1979) 1283;
D.V. Nanopoulos, D. Sutherland and A. Yildiz, Harvard Univ. preprint HUTP 79/A038 (1979), (to be published in Lettere al Nuovo Cimento).
D.N. Schramm and G. Steigman, Phys. Letters 87B (1979) 141.
20. J. Ellis, M.K. Gaillard and D.V. Nanopoulos, Phys. Letters 90B (1980) 253.
21. W.H. Press, Physica Scripta 21 (1980) 702.
22. A. Guth and S.-H.H. Tye, Phys. Rev. Letters 44 (1980) 631;
M.B. Einhorn, D.L. Stein and D. Toussaint, Univ. of Michigan preprint UM HE 80-1 (1980).
23. G. Lazarides and Q. Shafi, CERN preprint TH-2821 (1980).
24. For reviews of the cosmological monopole problem, see
M.B. Einhorn, contribution to these proceedings, p. 569, and
S.-H.H. Tye, Cornell University preprint "Monopoles, Phase Transitions and the Early Universe" (1980).
25. F.A. Bais, S. Rudaz, CERN preprint TH.2885 (1980).
26. S. Weinberg, "Gravitation and Cosmology" (Wiley, New York, 1972).
27. For GUTs based on bigger groups than SU(5), see
R. Barbieri, D.V. Nanopoulos, J.C. Pati, B. Stech, contributions to these proceedings.
28. J. Ellis, M.K. Gaillard and D.V. Nanopoulos, Phys. Letters 80B (1979) 320.
29. E. Gildener, Phys. Rev. D14 (1976) 1667; see also
J. Ellis, M.K. Gaillard, A. Peterman and C.T. Sachrajda, Nuclear Phys. B164 (1980) 253, and references therein.
30. P. Binétruy and T. Schücker, CERN preprint TH-2857 (1980).
31. K.T. Mahanthappa and M.A. Sher, Colorado preprint COLO-HEP 13 (1979).
32 L. Sulak and E. Bellotti, contributions to these proceedings, pp. 641, 675.
33. U. Amaldi, Proc. Neutrino 1979, Bergen, eds A. Haatuft and C. Jarlskog (Univ. of Bergen, 1979), p. 376;
F. Dydak, Proc. EPS Int. Conf. on High Energy Phys., Geneva (1979) (CERN, 1979) p. 25;
K. Winter, Proc. 1979 Int. Symp. on Lepton and Photon Interactions at High Energies, eds T.B.W. Kirk and H.D.I. Abarbanel (FNAL, Batavia, 1980), p. 258.

34. R. Barbieri, CERN preprint TH.2850 (1980), contribution to these proceedings, p. 17.
35. For a recent review, see
J.N. Bahcall, contribution to Int. Conf. on Astrophysics and Elementary Particles, Common Problems, Rome (1980).
36. Since this talk was given more information has become available about possible neutrino masses and oscillations:
F. Reines, H.W. Sobel and E. Pasierb, U.C. Irvine preprint "Evidence for Neutrino Instability" (1980).
E.F. Tretyakov et al., reportedly see a ν_e mass between 14 and 46 eV, while the ν_e/ν_μ induced event ratio in the 1979 CERN beam dump experiments was not unity; see also
H. Wachsmuth, Proc. 1979 Int. Symp. on Lepton and Photon Interactions at High Energies, eds T.B.W. Kirk and H.D.I. Abarbanel (FNAL, Batavia, 1980), p.541.
37. G. Steigman, contribution to these proceedings, p. 495.
38. R.L. Golden et al., Phys. Rev. Letters 43 (1979) 1196.
39. R.L. Omnès, Phys. Rev. Letters 23 (1969) 38 and Astron. and Astrophys. 10 (1971) 228.
40. See in particular D. Toussaint et al., and S. Weinberg, Ref. 16.
41. See E.W. Kolb and S. Wolfram, Ref. 17.
42. See J. Ellis, M.K. Gaillard and D.V. Nanopoulos, Ref. 16.
43. See the second paper of S. Dimopoulos and L. Susskind, Ref. 16.
44. D.V. Nanopoulos and S. Weinberg, Ref. 18.
45. For an estimate of the dilution of the baryon number generated due to subsequent entropy generation in a homogeneous and isotropic Universe, see
G. Steigman, K.A. Olive and D.N. Schramm, Phys. Rev. Letters 43 (1979) 239.
46. D.V. Nanopoulos, Phys. Letters 91B (1980) 67, discusses anthropocentric constraints on the baryon number asymmetry.
47. S.B. Treiman and F. Wilczek, Princeton preprint "Thermalization of Baryon Asymmetry" (1980).
48. M.S. Turner and D.N. Schramm, Nature 279 (1979) 303;
M.S. Turner, Nature 281 (1979) 549, and Phys. Letters 89B (1979) 155;
B.J. Carr and M.S. Turner, Enrico Fermi Institute preprint 80-09 (1980); see also D. Toussaint et al., Ref. 16.
49. J. Ellis and G. Steigman, Phys. Letters 89B (1980) 186;
J.J. Aly, Mon.Not.Roy.Astr.Soc. 189 (1979) 479 points out that the Universe could have been very chaotic initially, as long as it smoothed out before the baryon asymmetry[16] was generated.
50. J.M. Stewart, "Non-equilibrium Relativistic Kinetic Theory", Lecture Notes in Physics, Vol. 10 (Springer, Berlin, 1971);
C. Marle, Ann.Inst.H. Poincaré 10 (1969) 67, 127.
It should be noted that the numerical factors of order 1 in the formulae for η and χ used in Ref. 20 in fact depend on the precise angular form of the cross-section σ being used, and that some formulae given in the literature are not exactly correct. See for some clarification

N. Straumann, Helvetica Phys. Acta 49 (1976) 269.
Also the exponent in formula (24) of Ref. 20 should be a factor of 2 larger. We thank N. Straumann for correspondence and discussions on these points.
It has recently been emphasized by H. Sato, Kyoto University preprint RIFP-390 (1980) that the conventional viscosity approximation for discussing dissipative effects is only valid when $\tau < t$. In our case (38) this is only strictly valid when the temperature is in the lower part of the range 10^{19} GeV > T > 10^{15} GeV. This means that the results of Ref. 20 can only be regarded as qualitative. However, we believe that the basic physical picture of Fig. 8 remains valid even if $\tau > t$: the point is that particles are then able to "leak" out of an inhomogeneity.
51. A.F. Grillo, Frascati preprint LNF-80/21(P) (1980).
52. T.W.B. Kibble, J. Phys. A9 (1976) 1387 and
M.B. Einhorn, D.L. Stein and D. Toussaint, Ref. 22.
53. A. Guth, private communication (1980).
For a recent review of cosmological phase transitions, their problems and implications, see
T.W.B. Kibble, Imperial College London preprint ICTP/79-80/23 (1980).
For discussions of the nature of the grand unified phase transition, see
P. Ginsparg, CEN Saclay preprint DPh-T/80/27 (1980);
M. Daniel and C.E. Vayonakis, CERN preprint TH-2860 (1980).
54. V.L. Ginzburg, Fiz.Teor.Tela 2 (1960) 2031.
55. G. Lazarides, M. Magg and Q. Shafi, CERN preprint 2856 (1980).
56. P. Langacker and S.-Y. Pi, Princeton Institute for Advanced Study preprint "Magnetic Monopoles in Grand Unified Theories" (1980) propose to reduce the cosmological density of monopoles by having electromagnetic gauge invariance spontaneously broken at temperatures $T \geq 1$ TeV. Open problems in their scenario include its successful embedding in a grand unified theory, and the computation of the net charge density of the Universe due to C-, CP- and Q-violating effects out of thermal equilibrium. However, all other GUTs conserve Q and hence do not explain why the present charge density of the Universe is so small.
57. G. Barbiellini et al., DESY preprint 80/42 "Quarks and Monopoles at LEP" (1980) contains a review of monopole searches.

COSMOLOGY CONFRONTS GRAND UNIFICATION

Gary Steigman

Bartol Research Foundation of The Franklin Institute
University of Delaware, Newark, DE 19711, U.S.A.

INTRODUCTION

The euphoria accompanying the successes of Grand Unified Theories as models for particle physics at the very highest energies, is tempered by the emerging awareness of the limitations of conventional accelerators in testing and probing these theories. Increasing consideration is being given to alternate techniques for studying particle physics at its frontiers. One of the most promising of these approaches to elementary particle physics is via cosmology (Steigman, 1979 and references therein).

The early universe provided the ideal environment of a cosmic accelerator. The high temperatures and densities achieved during early epochs ensured frequent collisions at enormous energies--every experimentalist's dream. Unfortunately, no log books were kept. We must become detectives to learn from this cosmic accelerator, searching for and interpreting clues from the early evolution of the universe. Fortunately, our laboratory, the universe, is big enough to offer us some hope of success.

During the hot, dense epochs, all sorts of elementary particles, those known as well as those not yet dreamed of, were produced and came to equilibrium in the cosmic soup. As the universe cooled and expanded, the environmental impact was severe. Fortunately, a few relics survive to influence the subsequent evolution of the universe, providing evidence which may be unraveled to obtain information about the early evolution of the universe and about particle physics at the highest energies. Long-lived relics may be observed directly in terrestrial experiments or their presence inferred indirectly from astronomical observations. Indeed, since the present universe is the debris of the initial big bang, the past, present and future evolution of the universe is determined by this cosmic connection between particle physics and cosmology.

The efficacy of this approach to particle physics via cosmology will be illustrated by two examples of current interest. Neutrinos, light or heavy, would have been produced copiously in the cosmic accelerator and relics would have emerged from the cosmic soup in sufficient abundance to strongly affect the early evolution of the universe and, perhaps, to dominate the present universe and determine its ultimate fate. The constraints on particle physics and cosmology which may be derived from an analysis of neutrinos in big bang cosmology will be reviewed. By analogy, this same analysis has a more general application. New models of GUTs are likely to require the existence of new particles whose properties may be constrained by cosmological data.

RELIC NEUTRINOS

Attention here will be focussed on light neutrinos. For discussions of the cosmological role played by, and the constraints on, heavy neutrinos, see Lee and Weinberg (1977), Dicus et al. (1977), Gunn et al. (1978), Steigman et al. (1978). Light neutrinos were produced and maintained in equilibrium by neutral current weak interactions of the type,

$$e^{+} + e^{-} \leftrightarrow \nu_i + \bar{\nu}_i \; ; \quad i = e, \mu, \tau, --- . \tag{1}$$

At high temperatures, $T > m_\nu$, neutrinos (relativistic fermions) are comparable in importance to photons (relativistic bosons). Accounting for the spin states ($g_\gamma = 2$; $g_\nu = 2$ for each Majorana neutrino; $g_\nu = 4$ for each Dirac neutrino), the ratios of number and mass densities are,

$$\frac{n_\nu}{n_\gamma} = \frac{3}{4}\left(\frac{g_\nu}{g_\gamma}\right)\left(\frac{T_\nu}{T_\gamma}\right)^3 \; ; \quad \frac{\rho_\nu}{\rho_\gamma} = \frac{7}{8}\left(\frac{g_\nu}{g_\gamma}\right)\left(\frac{T_\nu}{T_\gamma}\right)^4 . \tag{2}$$

When the neutrinos are in equilibrium, the neutrino temperature and the photon temperature are equal. As the universe expands and cools, a critical temperature, T_* ($\approx$ 1 MeV), is reached, below which essentially no new neutrino pairs are created and annihilation virtually ceases. Below $T \approx m_e$ ($<T_*$), the electron-positron pairs annihilate heating the photons but not the decoupled neutrinos. Thus, for $T \equiv T_\gamma \ll m_e$,

$$\frac{T_\nu}{T_\gamma} = \left(\frac{4}{11}\right)^{1/3} \quad ; \quad \frac{n_\nu}{n_\gamma} = \frac{3}{22}\, g_\nu . \tag{3}$$

The above discussion clarifies the meaning of "light" as applied to neutrinos: relic neutrinos are comparable in abundance to relic photons if $m_\nu << T_* \approx 1$ MeV.

$$T > m_\nu \; : \; \frac{n_\nu}{n_\gamma} = \frac{3}{22}\, g_\nu \; , \quad \frac{\rho_\nu}{\rho_\gamma} = \frac{7}{16}\left(\frac{4}{11}\right)^{4/3} g_\nu \; ; \tag{4a}$$

$$T < m_\nu \; : \; \frac{n_\nu}{n_\gamma} = \frac{3}{22}\, g_\nu \; , \quad \frac{\rho_\nu}{\rho_\gamma} = \left(\frac{m_\nu}{2.7kT}\right)\left(\frac{n_\nu}{n_\gamma}\right) . \tag{4b}$$

During early "radiation dominated" (RD) epochs, relic neutrinos make an important contribution to the total density ($\rho_\nu \approx \rho_\gamma$) and, thereby, effect the early expansion rate ($t^{-1} \propto \rho^{1/2}$). Non-relativistic particles dominate the total density of the universe during more recent "matter dominated" epochs. Relic neutrinos will be seen to be much more abundant than nucleons ($n_\nu \gtrsim 10^9 n_N$) so that even very light ($m_\nu \gtrsim 10^{-9} m_N$) relic neutrinos could dominate the present universe. The consequences of these possibilities lead to constraints on neutrino physics and on cosmology which will now be reviewed.

RELIC NEUTRINOS AND PRIMORDIAL NUCLEOSYNTHESIS

During RD epochs, the age of the universe (or, its inverse, the expansion rate), the total density and the temperature are related by,

$$\frac{32\pi}{3} G\rho t^2 = 1 \quad ; \quad t_{sec} \approx \frac{2.4}{[g(T)]^{1/2}} \times \frac{1}{T_{MeV}^2} . \tag{5}$$

In (5), G is Newton's gravitational constant and g(T) is the effective number of degrees of freedom (spin states), defined by,

$$\rho = \frac{g}{2}\rho_\gamma \quad ; \quad g = g_B + \frac{7}{8} g_F . \tag{6}$$

In (6), B(F) stands for Bosons (Fermions).

It is clear from (5) and (6) that the more types of relativistic particles present, the higher the density (at a fixed temperature) and the faster the universe expands. In particular, the early expansion rate depends on the number of spin states contributed by light neutrinos; the more neutrino types, the faster the universe expands. Physical processes which probe the expansion rate can lead to constraints on the number of neutrino types. Primordial nucleosynthesis offers just such a probe (Steigman et al., 1977).

Primordial Nucleosynthesis

Starting our discussion at $T \sim 10$ MeV, we note that the usual charged current weak interactions ("beta decay"),

$$p + e^- \leftrightarrow n + \nu_e, \; n + e^+ \leftrightarrow p + \bar{\nu}_e, \; n \rightarrow p + e^- + \bar{\nu}_e, \tag{7}$$

are rapidly transforming neutrons into protons and vice-versa. The frequency of these reactions guarantees that the neutron to proton ratio is maintained at its equilibrium value,

$$\frac{n}{p} = \exp\left(\frac{-\Delta m}{T}\right). \tag{8}$$

As the universe cools, fewer neutrons are available.

At the same time, neutrons and protons occasionally collide producing deuterons.

$$n + p \leftrightarrow d + \gamma. \tag{9}$$

Photodissociation of deuterium is, however, so rapid (for $T \gtrsim 0.1$ MeV) that the deuterium abundance is negligibly small and there is no base on which the heavier elements may be built. All the while, fewer and fewer neutrons are surviving (c.f. eq. (8)).

When the temperature drops below a critical value, T_f, the weak interaction rate can no longer keep up with the expansion rate and the neutron to proton ratio "freezes-out". The surviving abundance of neutrons, therefore, depends on the competition between the weak interaction rate and the expansion rate, thereby providing a probe for the early expansion rate and, hence, the number of types of neutrinos. The more types of light neutrinos, the faster the universe expands leving behind more neutrons to be incorporated in nuclei when nucleosynthesis begins in earnest.

When the temperature drops below $\sim$0.1 MeV, the deuteron abundance increases leading to the buildup of the heavier elements. Very rapidly, in a sequence of two body nuclear reactions, ^{3}H, ^{3}He and ^{4}He are synthesized (for a review and references see Schramm and Wagoner, 1977). The lack of a stable nucleus at mass-5 results in a gap preventing the synthesis of significant amounts of heavier nuclei. The most tightly bound of the light nuclei, ^{4}He, incorporates virtually all the available neutrons. As a sensitive probe of the neutron abundance at nucleosynthesis, the primordial abundance of ^{4}He may be used to constrain the number of types of light neutrinos (Steigman et al., 1977; Steigman et al., 1979; Yang et al., 1979).

The nucleon abundance at nucleosynthesis (i.e.: the nucleon to photon ratio) determines the time at which the buildup of the heavier elements begins. The more nucleons, the earlier the photodissociation of deuterium is overcome as a barrier to nucleosynthesis. Earlier times correspond to more neutrons leading to the synthesis of more ^{4}He. This interplay between expansion rate, neutrino types and nucleon abundance permits a variety of constraints to be derived from the primordial abundance of ^{4}He.

For an abundance by mass of primordial ^{4}He of $Y_P \lesssim 0.25$ (Yang et al., 1979) and for at least three types of two component neutrinos (ν_e, ν_μ, ν_τ), the present ratio of nucleons to photons is restricted to (Yang et al., 1979),

$$\left(\frac{n_N}{n_\gamma}\right)_o \lesssim 4.2 \times 10^{-10} \tag{10}$$

If a lower limit to the nucleon to photon ratio is known, the same analysis leads to an upper limit to the number of neutrino types N_L ($2N_L \equiv \Sigma g_\nu$). It has already been noted that if light neutrinos have a small mass ($m_\nu \gtrsim 10^{-9} m_N$), relic neutrinos may dominate the mass in the present universe. Given this possibility (Schramm and Steigman, 1980a,b), it is very difficult to obtain a reliable lower limit to the nucleon abundance. Estimates from luminous (optical, x-ray) matter suggest (Schramm and Steigman, 1980a,b).

$$\left(\frac{n_N}{n_\gamma}\right)_o \gtrsim 10^{-10} \tag{11}$$

For a nuclear abundance at least this large and for $Y_P \lesssim 0.25$, at most four types of two component neutrinos are permitted.

$$N_L \lesssim 4. \tag{12}$$

This is an extremely strong constraint, eliminating, for example, right handed counterparts to the familiar left handed neutrinos. Perhaps all the leptons to be discovered, have been discovered!

A Neutrino Dominated Universe

The Hubble parameter describes the expansion of the universe.

$$H(t) = \frac{1}{a}\left(\frac{da}{dt}\right) \quad , \quad H_o = H(t_o). \tag{13}$$

In (13), a = a(t) is the scale factor; all objects (i.e.: galaxies, clusters, etc.) which expand along with the general cosmological expansion (i.e.: comoving) have their relative separations increase in proportion to the scale factor: (R(t) ∝ a(t). H_o is the present value of the Hubble parameter, often misleadingly called the Hubble "constant". There are large uncertainties in the numerical estimates of H_o (Sandage and Tammann, 1976; deVaucouleurs and Bollinger, 1979; Branch, 1979; Aaronson et al., 1979); $40 \lesssim H_o \lesssim 100$ kms^{-1} Mpc^{-1}. To deal with this large range of possible values, it is convenient to introduce h_o ($0.4 \lesssim h_o \lesssim 1$) where,

$$H_o = 100 h_o \text{ kms}^{-1} \text{ Mpc}^{-1}; \quad H_o^{-1} = 10 h_o^{-1} \times 10^9 \text{ yr}. \tag{14}$$

Since gravity slows the expansion of the universe, the present age of the universe (in the "standard" Robertson-Walker-Friedman models) is always less than H_o^{-1}. Data from globular clusters suggests the oldest stars may be at least 13 billion years old (Iben, 1974); this would argue for a relatively low value for H_o($h_o \lesssim 3/4$).

In terms of H_o there is a critical density,

$$\rho_c = \frac{3H_o^2}{8\pi G} \approx 2 \times 10^{-29} h_o^2 \text{ gcm}^{-3}, \tag{15}$$

which separates those models which expand forever ($\rho_o \leq \rho_c$) from those which eventually collapse ($\rho_o > \rho_c$). It is convenient to write the various contributions to the total density as a fraction of the critical density; introduce Ω_i where,

$$\rho_i = \Omega_i \rho_c \quad ; \quad \Omega_o = \sum_i \Omega_i. \tag{16}$$

For example, the density in nucleons may be written as,

$$\rho_N = m_N n_N = 2 \times 10^{-29} \Omega_N h_o^2 \text{ gcm}^{-3}. \tag{17}$$

Previously, we found, from primordial nucleosynthesis, an upper limit to the nucleon to photon ratio (eq. (10)). Since the present density of relic photons is,

$$n_{\gamma_o} \approx 400 \left(\frac{T_o}{2.7}\right)^3 ,$$

this leads to a constraint on Ω_N.

$$\Omega_N < \frac{0.014}{h_o^2}\left(\frac{T_o}{2.7}\right)^3 . \tag{18}$$

The present temperature of the microwave radiation is in the range $2.7 \lesssim T_o \lesssim 3.0$°K (Thaddeus, 1972; Hegyi et al., 1974; Woody et al., 1975; Danese and DeZotti, 1978); for $T_o < 3.0$°K and $h_o \gtrsim 0.4$,

$$\Omega_{N,MAX} < 0.12 . \tag{19}$$

There is evidence from the dynamics of clusters of galaxies (Faber and Gallagher, 1979 and references therein) and from the "cosmic virial theorem" (Peebles, 1979), that the total density of the universe exceeds this limit for the nucleon density ($\Omega_o \gtrsim 0.2$). The uncertainties, however, are large and $\Omega_o \lesssim 0.1$ may still be consistent with the observational data (Gott et al., 1974). If the current evidence suggesting $\Omega_o > \Omega_{N,MAX}$ should survive further critical investigations, the universe would be dominated by something other than ordinary matter (nucleons). Could it be relatively light, massive relic neutrinos?

Throughout the evolution of the universe and, in particular, at present, relic neutrinos and relic photons are of comparable abundance (c.f. eq. (4)). The present mass density in relic neutrinos whose mass is m_ν is

$$\Omega_\nu = \left(\frac{1}{100h_o^2}\right)\left(\frac{T_o}{2.7}\right)^3\left(\frac{g_\nu}{2}\right) m_\nu . \tag{20}$$

In (20) and subsequently, m_ν is in eV. The total contribution from all neutrinos is obtained by summing over all neutrino types. For convenience, we illustrate the potential importance of massive relic neutrinos by considering one, most massive, Majorana ($g_\nu = 2$) neutrino. By comparing (18) and (20), a limit to the relative contributions to the total mass in the universe from relic neutrinos and nucleons is obtained.

$$\frac{M_\nu}{M_N} = \frac{\Omega_\nu}{\Omega_N} \gtrsim \frac{m_\nu}{1.4\text{eV}} . \tag{21}$$

The present universe is neutrino dominated for $m_\nu \gtrsim 1.4$ eV. Notice that this result is unaffected by present uncertainties in H_o and T_o. It does, however, depend strongly on the adopted upper limit to the primordial abundance of ^{4}He($Y_p \lesssim 0.25$) and on the lower limit to the number of types of light (<< 1 MeV), two component neutrinos ($N_L \geq 3$, $\Sigma g_\nu \geq 6$).

It is difficult to use equation (20) to obtain a reliable estimate of m_ν since the result would depend on a combination of the uncertain parameters H_o, T_o, $\Omega_\nu(\lesssim\Omega_o)$. Given the large uncertainties, neutrino masses as large as ~100 eV may be permitted; a more likely upper limit is probably $m_\nu \lesssim 50$ eV (Schramm and Steigman, 1980b).

Neutrinos which cluster gravitationally with ordinary matter will, it must be noted, contribute to the dark (nonluminous) matter. Such neutrinos, therefore, provide an ideal solution to the "missing light" problem (c.f. Faber and Gallagher, 1979 and references therein). As suggested by Cowsik and McClelland (1972), relic neutrinos may dominate the dark mass in clusters of galaxies. In this case, (21) may be used to obtain an upper limit to m_ν which, while free of the uncertainties in Ω_o, H_o, T_o, is dependent on the estimate of the luminous mass and the total mass of clusters. Since $M_\nu < M_{TOT}$ and $M_N \gtrsim M_{Lum}$,

$$m_\nu \lesssim 1.4 \left(\frac{M_{TOT}}{M_{Lum}}\right)_{Clusters} . \tag{22}$$

Many clusters are sources of x-rays whose origin is the thermal bremsstrahlung radiation from a hot intracluster gas. Estimates of the amount of gas (Lea et al., 1973; Cavaliere and Fusco-Femiano, 1976; Malina et al., 1976) suggest

$$M_{TOT} \lesssim 10(2h_o)^{3/2} M_{Lum}, \tag{23}$$

so that

$$m_\nu \lesssim 40\ h_o^{3/2}\ \text{eV}. \tag{24}$$

If $t_o \gtrsim 13$ billion years and $\Omega_o \gtrsim 0.2$, H_o is restricted to be $\lesssim$70 kms^{-1} Mpc^{-1} (Schramm and Steigman, 1980b) so that $m_\nu \lesssim 23$ eV. Even this upper limit to m_ν may, however, be too high.

Tremaine and Gunn (1979) have shown that relic neutrinos with masses in excess of ~20 eV could cluster in galactic halos. But, such heavy neutrinos would, if clustered on such small scales, contribute too much dark matter. Similarly, if $m_\nu \gtrsim 12$ eV, relic neutrinos might add too much dark mass to binary galaxies and to small groups of galaxies. Notice, though, that to "solve" the missing light problem in clusters requires $m_\nu \gtrsim 4$ eV (Tremaine and Gunn,

1979). It is suggested, then, that the present universe may be dominated by relic neutrinos with masses in the range

$$4 \lesssim m_\nu \lesssim 10 \text{ eV}. \tag{25}$$

Although the evolutionary history of neutrino versus nucleon clustering may permit the upper limit of this estimate to be increased, it is unlikely that $m_\nu \gtrsim 40$ eV could be made consistent with the cosmological data.

Lighter neutrinos ($m_\nu \lesssim 4$ eV) will find it difficult to cluster and should be spread, rather homogeneously, throughout the universe. If their masses are not too small ($m_\nu \gtrsim 1$ eV), they still may make an important contriubtion to the total mass in the universe. Although there is no compelling evidence that $\Omega_o \gtrsim 1/2$, note that for $h_o \simeq 1/2$, $T_o \simeq 3.0°K$ and $\Sigma(g_\nu/2)m_\nu \approx 15$ eV, $\Omega_{\nu,TOT} \simeq 0.82$. Adding in the nucleon contribution could raise the total density perilously close to the critical density. Resolution of the crucial question of the fate of the universe must await better cosmological data (H_o, T_o) and laboratory measurements of the neutrino mass. Perhaps no where else is the cosmic connection more important.

CONSTRAINTS ON NEW PARTICLES

Many aspects of the previous analysis may be generalized to yield constraints on new particles other than neutrinos (Dover et al., 1979; Frampton and Glashow, 1980; Ellis et al., 1980). Particles yet to be dreamed of may be required by future GUTs. If, indeed they do exist, the cosmic accelerator would have produced them and they would have come to equilibrium in the cosmic soup of the early universe. Some of these exotic relics would have emerged from the big bang to influence the subsequent evolution of the universe. If sufficiently stable, some exotic relics would be present today.

Depending on their interactions, the abundance of exotic relics is likely to be significant (Dover et al., 1979; Ellis et al., 1980)

$$f = \frac{n}{n_N} \gtrsim 10^{-10}. \tag{26}$$

Since accelerator searches have failed to reveal new particles with $M \lesssim 10\ M_N$ (Cutts et al., 1978; Vidal et al., 1978), if they exist, these relics may make a non-negligible contribution to the total mass (of the universe or of terrestrial material).

$$\frac{\rho}{\rho_N} = f\left(\frac{m}{m_N}\right) \gtrsim 10^{-9}. \tag{27}$$

If long-lived ($\tau \gtrsim t_o$), the properties of such abundant relics may be constrained by a variety of terrestrial and astrophysical observations.

(i) Too much mass in the universe. Although quite weak, this very general constraint that, say, $\rho_{TOT} \lesssim 10\rho_N$, requires $f(m/m_N) \lesssim 10$.

(ii) Chemical versus physical masses. If associated with terrestrial matter, exotic relics would produce anomalous isotopes unless, say, $f(m/m_N) \lesssim 10^{-4}$. In directed searches for anomalous isotopes this limit may be improved enormously. For example, for $m \lesssim 100\ m_N$, Middleton et al. (1979) have failed to find evidence for new particles at a level $f \lesssim 10^{-16}$.

(iii) Proton decay type searches. The absence of anomalous isotopes may not require the absence of new particles. Heavy particles ($m \gtrsim 100\ m_N$) of low abundance ($f(m/m_N) \lesssim 10^{-4}$) may have escaped detection. If such long lived relics are ultimately unstable, they may reveal their presence in the underground searches for proton decay (see Physics Today, January 1980, p. 17). If B_μ is the number of muons per decay, the data of Reines et al. (1971, 1974) sets the constraint (Ellis et al., 1980)

$$\tau \gtrsim 10^{31}\ f\ B_\mu \left(\frac{m}{m_N}\right)\ \text{yr}. \tag{28}$$

(iv) Annihilation relics. During recent epochs when density inhomogeneities were amplified gravitationally, annihilation would have begun anew (it stopped in the early universe when the density became so low that the annihilation rate could not keep up with the expansion rate) for symmetric relics. Those relics which were not shielded from annihilation by having been incorporated in ordinary nuclei, would not have survived. Their absence terrestrially, therefore, would not be proof of their nonexistence. Of course, those which were shielded should be found terrestrially. Sufficiently abundant symmetric relics which annihilated during the recent past may have produced too many annihilation gamma rays. The observed gamma ray background above ~100 MeV (Fichtel et al., 1978) leads to the constraint (Ellis et al., 1980).

$$fB_{\gamma a} \lesssim 10^{-7}\ , \tag{29}$$

where $B_{\gamma a}$ is the gamma ray multiplicity in annihilation.

The fleeting existence of short-lived ($\tau < t_o$) relics is more easily hidden from view. Still, some constraints do exist. Particles which have decayed in the recent past, for redshifts $\lesssim 10^2$-10^3, may contribute their decay gammas to the gamma ray background. If $B_{\gamma d}$ is the gamma ray multiplicity in decay, intermediate lifetime particles must have been of low abundance (Ellis et al., 1980).

$$10^7 \text{ yr} \lesssim \tau \lesssim t_o \quad , \quad fB_{\gamma d} \lesssim 10^{-7} \quad . \tag{30}$$

The universe is optically thick to gamma rays from the decay of shorter-lived relics. Still, since the mass density in those relics may be comparable to the mass density in the relic photons, decay early on may distort the microwave radiation spectrum and/or produce too much entropy (Ellis et al., 1980). The time, t_*, when $\rho \approx \rho_\gamma$, may be related to the time, $t_{eq} \approx 10^6$ yr, when $\rho_N \approx \rho_\gamma$ (Ellis et al., 1980)

$$t_* \approx \left[f\left(\frac{m}{m_N}\right)\right]^{-2} t_{eq} \approx \left[f\left(\frac{m}{m_N}\right)\right]^{-2} \times 10^6 \text{ yr.} \tag{31}$$

For $t_* \lesssim t_{nuc} \approx 10^{-8}$ yr, there are no significant constraints on very short-lived relic particles. In between, the lifetime is limited to,

$$\tau < \text{Min } [10^7 \text{ yr}, t_*] \ , \tag{32a}$$

$$\tau < \text{Max } [t_{nuc}, t_*] \ , \tag{32b}$$

These general constraints, suitably adopted to specific cases, may prove of value in constraining new models of GUTs.

SUMMARY

There is a cosmic connection. Advances in our understanding of elementary particle physics influence the models we build to explore the early evolution of the universe. In pursuing the cosmological consequences of models of particle physics, we are led to astrophysical and terrestrial tests of those models. The feedback from cosmology to particle physics and, vice-versa, provides constraints on, and adds to our comprehension of, the macroscopic world and the microscopic world.

The weakness of the weak interaction ensures that neutrinos are exceedingly difficult to detect in the laboratory but guarantees that relic neutrinos survive in such abundance that they play a significant, even dominant, role in the past, present and future evolution of the universe. Neutrinos and neutrino physics provide an ideal illustration of the form and importance of the cosmic connection.

How many lepton flavors are there? Except for a few suggestive hints, GUTs have very little to say. Extra neutrino types corresponding to new lepton flavors would influence the early expansion of the universe. The abundance of ^{4}He produced in primordial nucleosynthesis constrains the early expansion rate, leading to limits which suggest $N_L \lesssim 4$. Perhaps all the leptons to be discovered, have been discovered.

Do neutrinos have a mass? The GUTs say, maybe, why not? For the cosmology, the implications of a neutrino mass are enormous. Relic neutrinos with greater than a billionth of the nucleon mass could dominate the universal density at present and during recent epochs and may determine the ultimate fate of the universe. Data on the distribution of dark matter in the universe suggests that the missing light problem could be solved by relic neutrinos with $4 \lesssim m_\nu \lesssim 10$ eV. More conservatively, the data could be consistent with the neutrinos having masses in the range $1 \lesssim m_\nu \lesssim 100$ eV or $3 \lesssim m_\nu \lesssim 13$ GeV (Steigman, 1980). The same data leads to the very strong constraint that intermediate mass neutrinos (100 eV $\lesssim m_\nu \lesssim$ 3 GeV) would have to be unstable (Gunn et al., 1978).

It must be noted that the cosmological evidence for neutrino masses is indirect. The dark matter in the universe need not be neutrinos and neutrinos need not be massive. If, however, laboratory experiments do reveal the neutrino to be massive, the astrophysical implications are clear and clearly important if $m_\nu \gtrsim 1$ eV.

The abundance, compared to photons, of more strongly interactio relics is expected to be small (c.f. Steigman, 1979). Compared to nucleons, however, such abundances may not be negligible ($f \gtrsim 10^{-10}$) Long-lived relics could be found in terrestrial experiments (anomalous isotopes), may contribute significantly to the present mass in the universe or, could reveal themselves through their decay or annihilation products. Short-lived relics could have had a significant influence on the earlier evolution of the universe and may, in the process of dying, have left clues to their existence.

The absence of evidence for exotic relics suggest that significant constraints exist on the properties of new particles. Relatively abundant relics ($f \gtrsim 10^{-8}$) must be short-lived ($\tau < t_o$); in general, their lifetimes are restricted to be $\tau < t_*$ where, 10^{-8} yr $\lesssim t_* \lesssim 10^7$ yr $<< t_o$. Very low abundance relics ($f \lesssim 10^{-8}$) could have escaped detection (if $m \gtrsim 100\ m_N$), but might reveal their presence through their decays.

In developing new models for GUTs, the cosmic connection must occupy a central role. The lessons from neutrino physics as well as the more general constraints outlined here must be kept center stage if the mysteries of the universe are to be revealed.

ACKNOWLEDGMENTS

Over the years, in the course of the work outlined here, I have benefitted from valuable discussions with many of my particle physics and astrophysics colleagues. At this conference I have especially profitted from conversations with R. Barbieri, M. Einhorn, J. Ellis, M. K. Gaillard, D. Nanopoulos, J. C. Pati, D. Ross, A. Salam, M. Sher, L. Sulak and L. Susskind. Thanks are due John Ellis, Sergio Ferrara and Peter Van Nieuwenhuizen for having organized a stimulating conference. This work is supported at Bartol by DOE Grant ER-78-8-02-5007.

REFERENCES

Aaronson, M., Mould, J., Huchra, J. Sullivan, W. T., Schommer, R. A., and Bothun, G. D., 1979, Ap. J. (in press).
Branch, D., 1979, MNRAS, 186:609.
Cavaliere, A. and Fusco-Femiano, R., 1976, Astron. & Astrophys., 49:137.
Cowsik, R. and McClelland, J., 1972, Phys. Rev. Lett., 29:669.
Cutts, D., et al., 1978, Phys. Rev. Lett., 41:363.
Danese, L. and DeZotti, G., 1978, Astron. & Astrophys., 68:157.
Dicus, D., Kolb, E. N., and Teplitz, V., 1977, Phys. Rev. Lett., 39:168.
Dover, C. B., Gaisser, T. K., and Steigman, G., 1979, Phys. Rev. Lett., 42:1117.
Ellis, J., Gaisser, T. K., and Steigman, G., 1980, CERN TH.2893.
Faber, S. M. and Gallagher, J. S., 1979, Ann. Rev. Astron. Astrophys., 17:135.
Fichtel, C. E., Simpson, G. A., and Thompson, D. J., 1978, Ap. J., 222:833.
Frampton, P. H. and Glashow, S. L., 1980, Preprint, HUTP-80/A007.
Gott, J. R., Gunn, J. E., Schramm, D. N., and Tinsley, B. M., 1974, Ap. J., 194:543.
Gunn, J. E., Lee, B. N., Lerche, I., Schramm, D. N., and Steigman, G., 1978, Ap. J., 223:1015.
Hegyi, D. J., Traub, W. A., and Carleton, N. P., 1974, Ap. J., 190:543.
Iben, I., 1974, Ann. Rev. Astron. & Astrophys., 12:215.
Lea, S. M., Silk, J., Kellogg, E., and Murray, S., 1973, Ap. J. (Lett.), 184:L105.
Lee, B. W. and Weinberg, S., 1977, Phys. Rev. Lett., 39:165.
Malina, R., Lampton, M., and Bowyer, S., 1976, Ap. J., 209:678.
Middleton, R., Zurmühle, R. W., Klein, J., and Kollarits, R. V., 1979, Phys. Rev. Lett., 43:429.
Peebles, P. J. E., 1979, Astron. J., 84:730.
Reines, F., et al., 1971, Phys. Rev., D4:80; ibid 99.
Reines, F. and Crouch, M. F., 1974, Phys. Rev. Lett., 32:493.
Sandage, A. and Tammann, G. A., 1976, Ap. J., 210:7.
Schramm, D. N. and Wagoner, R. V., 1977, Ann. Rev. Nucl. Part. Sci.

27:37.
Schramm, D. N. and Steigman, 1980a, First Prize Essay, Gravity Research Foundation (BA-80-17).
Schramm, D. N. and Steigman, 1980b, submitted to Ap. J. (BA-80-19).
Steigman, G., Schramm, D. N., and Gunn, J. E., 1977, Phys. Lett. B, 66:202.
Steigman, G., Sarazin, C. L., Quintana, H., and Faulkner, J., 1978, Astron. J., 83:1050.
Steigman, G., 1979, Ann. Rev. Nucl. Part. Sci., 29:313.
Steigman, G., Olive, K. A., and Schramm, D. N., 1979, Phys. Rev. Lett., 43:239.
Steigman, G., 1980, To appear in the Proc. of the Int. Meeting on Astrophys. and Elem. Part. (Rome, 21-23 Feb., 1980) (BA-80-18).
Thaddeus, P., 1972, Ann. Rev. Astron. & Astrophys., 10:305.
Tremaine, S. and Gunn, J. E., 1979, Phys. Rev. Lett., 42:407.
deVaucouleurs, G. and Bollinger, G., 1979, Ap. J., 233:433.
Vidal, R., et al., 1978, Phys. Lett. B, 77:344.
Woody, D. P., Mather, J. C., Nishioka, N., and Richards, P. L., 1975, Phys. Rev. Lett., 34:1036.
Yang, J., Schramm, D. N., Steigman, G., and Rood, R. T., 1979, Ap. J., 227:697.

SOLUTION OF THE BIANCHI IDENTITIES IN SU(2) EXTENDED SUPERSPACE WITH CONSTRAINTS

Richard Grimm

Institut für Theoretische Physik

Physikhochhaus, D-7500 Karlsruhe 1

Starting from the description of differential geometry in superspace as formulated by Wess and Zumino[1,2] , a solution of the Bianchi identities in SU(2) extended superspace in terms of some superfields and their covariant derivatives will be given.

The specific form of the solution is a consequence of the structure group which we take to be the direct product of Lorentz group and SU(2) and of constraints on the torsion which will be specified in detail below.

We proceed in close analogy to ordinary superspace[3] with notations and conventions carried over to extended superspace.

The Bianchi identities are:

$$\oint_{\mathcal{E}\mathcal{D}\mathcal{C}} \left(R_{\mathcal{E}\mathcal{D}\mathcal{C}}{}^{\mathcal{A}} - \mathcal{D}_{\mathcal{E}} T_{\mathcal{D}\mathcal{C}}{}^{\mathcal{A}} - T_{\mathcal{E}\mathcal{D}}{}^{\mathcal{B}} T_{\mathcal{B}\mathcal{C}}{}^{\mathcal{A}} \right) = 0 ,$$

and

$$\oint_{\mathcal{E}\mathcal{D}\mathcal{C}} \left(\mathcal{D}_{\mathcal{E}} R_{\mathcal{D}\mathcal{C}\mathcal{B}}{}^{\mathcal{A}} + T_{\mathcal{E}\mathcal{D}}{}^{\mathcal{F}} R_{\mathcal{F}\mathcal{C}\mathcal{B}}{}^{\mathcal{A}} \right) = 0 .$$

In the graded cyclic sum one has to take into account the bosonic or fermionic nature of the indices using the permutation rule

$$\mathcal{E}\mathcal{D} = -(-1)^{d(\mathcal{E})\,d(\mathcal{D})}\,\mathcal{D}\mathcal{E} \quad .$$

where $d(\xi)$ equals one if ξ is a fermionic and zero if it is a bosonic index.

Torsion and curvature have the properties

$$T_{\mathcal{C}\mathcal{B}}{}^{\mathcal{A}} = -(-1)^{d(\mathcal{C})d(\mathcal{B})}\,T_{\mathcal{B}\mathcal{C}}{}^{\mathcal{A}} \quad ,$$

$$R_{\mathcal{D}\mathcal{C}\mathcal{B}}{}^{\mathcal{A}} = -(-1)^{d(\mathcal{D})d(\mathcal{C})}\,R_{\mathcal{C}\mathcal{D}\mathcal{B}}{}^{\mathcal{A}} \quad .$$

In addition, the curvature takes values in the Lie algebra of the structure group with respect to the indices $\mathcal{B}$ and $\mathcal{A}$. Torsion and curvature determine the commutation relations of the covariant derivatives:

$$(\mathcal{D}_{\mathcal{C}}, \mathcal{D}_{\mathcal{B}}) = -R_{\mathcal{C}\mathcal{B}\cdot}{}^{\cdot} - T_{\mathcal{C}\mathcal{B}}{}^{\mathcal{F}}\,\mathcal{D}_{\mathcal{F}} \quad ,$$

$$(\mathcal{D}_{\mathcal{C}}, \mathcal{D}_{\mathcal{B}}) = \mathcal{D}_{\mathcal{C}}\mathcal{D}_{\mathcal{B}} - (-1)^{d(\mathcal{C})d(\mathcal{B})}\,\mathcal{D}_{\mathcal{B}}\mathcal{D}_{\mathcal{C}} \quad .$$

In extended superspace the covariant derivative, like any other supervector, has the components

$$\mathcal{D}_{\mathcal{A}} \sim \left(\mathcal{D}_a, \mathcal{D}^A_\alpha, \mathcal{D}^{\dot\alpha}_A\right) \quad ,$$

where the spinorial quantities now carry an additional index A for some internal group. We have chosen it to be SU(2). a, α and $\dot\alpha$ are the usual vector and spinor indices, $a = 0,1,2,3$ and $\alpha,\dot\alpha = 1,2$. Let me now specify the structure group and the constraints on torsion.

The structure group will be the product of the Lorentz group and SU(2). This means that in the Lie algebra valued quantity $R_{\mathcal{B}}{}^{\mathcal{A}}$ only the components

$$R_{\mathcal{B}}{}^{\mathcal{A}} \sim \left(R_b{}^a , R_{\beta A}^{B\alpha} , R_{B\dot{\alpha}}^{\dot{\beta} A} \right)$$

are different from zero and have the properties:

$$R_{ba} = -R_{ab} ,$$

$$R_{\beta A}^{B\alpha} = \delta_A^B R_\beta{}^\alpha + \delta_\beta^\alpha R^B{}_A ,$$

$$R_{B\dot{\alpha}}^{\dot{\beta} A} = \delta_B^A R^{\dot{\beta}}{}_{\dot{\alpha}} + \delta_{\dot{\alpha}}^{\dot{\beta}} R_B{}^A ,$$

$$R^B{}_A = - R_A{}^B ,$$

$$R_\alpha{}^\alpha = R^{\dot{\alpha}}{}_{\dot{\alpha}} = R_A{}^A = 0 .$$

Vector and spinor representations of the Lorentz group are related through

$$R_{\beta\dot{\beta}\,\alpha\dot{\alpha}} = -2\,\epsilon_{\beta\alpha} R_{\dot{\beta}\dot{\alpha}} + 2\,\epsilon_{\dot{\beta}\dot{\alpha}} R_{\beta\alpha} ,$$

$$R_{\beta\dot{\beta}\,\alpha\dot{\alpha}} \equiv \sigma^b{}_{\beta\dot{\beta}}\, \sigma^a{}_{\alpha\dot{\alpha}}\, R_{ba} .$$

The non-vanishing components of the torsion are

$$T_{\gamma}{}^{C}{}_{B}{}^{\dot{\beta}}{}^{a} = -2i\, \delta_B^C (\sigma^a \epsilon)_\gamma{}^{\dot{\beta}} ,$$

$$T_\gamma{}^C{}_b{}^\alpha{}_A = \delta_A^C T_{\gamma b}{}^\alpha ,$$

$$T^{\dot{\gamma}}{}_{c\,b}{}^A{}_{\dot{\alpha}} = \delta_C^A T^{\dot{\gamma}}{}_{b\dot{\alpha}} ,$$

$$T_\gamma{}^C{}_{b\dot{\alpha}}{}^A ,\quad T^{\dot{\gamma}}{}_{c\,b}{}^\alpha{}_A ,\quad T_{cb}{}^\alpha{}_A ,\quad T_{cb\dot{\alpha}}{}^A .$$

This set of constraints differs from others proposed earlier[4,5]. It is, however, very similar to the set of constraints used in the superspace formulation of ordinary supergravity[6].

Using this choice of structure group and constraints, the investigation of the first set of Bianchi identities shows that all components of torsion and curvature can be expressed in terms of the superfields

$$X^{\underset{\smile}{BA}} , \; X_{\underset{\smile}{\dot\beta\dot\alpha}} , \; Y_{\underset{\smile}{\beta\alpha}} \quad \text{and} \quad \bar X_{\underset{\smile}{BA}} , \; \bar X_{\underset{\smile}{\beta\alpha}} , \; \bar Y_{\underset{\smile}{\dot\beta\dot\alpha}} ,$$

which are symmetric in their indices, and $U_{\alpha\dot\alpha}$ and their covariant derivatives. Note that, up to sign factors $\bar X_{\underset{\smile}{BA}}$, $\bar X_{\beta\alpha}$, $\bar Y_{\dot\beta\dot\alpha}$ are the complex conjugates to $X^{\underset{\smile}{BA}}$, $X_{\dot\beta\dot\alpha}$, $Y_{\beta\alpha}$, respectively, and $(U_{\alpha\dot\alpha})^* \sim U_{\alpha\dot\alpha}$. Moreover, as a consequence of the Bianchi identities, these superfields are subject to covariant conditions.

$$\mathcal{D}^C_\gamma X_{\underset{\smile}{\dot\beta\dot\alpha}} = 0 ,$$

$$\oint_{ABC} \mathcal{D}^C_\gamma X^{\underset{\smile}{BA}} = 0 , \qquad \oint_{\alpha\beta\gamma} \mathcal{D}^C_\gamma Y_{\underset{\smile}{\beta\alpha}} = 0 ,$$

$$\mathcal{D}_{\alpha B} X^{\underset{\smile}{BA}} + \mathcal{D}^{\beta A} Y_{\underset{\smile}{\beta\alpha}} = 0 .$$

$$\mathcal{D}^{\dot\gamma}_C \bar X_{\underset{\smile}{\beta\alpha}} = 0 ,$$

$$\oint_{ABC} \mathcal{D}^{\dot\gamma}_C \bar X_{\underset{\smile}{BA}} = 0 , \qquad \oint_{\dot\alpha\dot\beta\dot\gamma} \mathcal{D}^{\dot\gamma}_C \bar Y^{\underset{\smile}{\dot\beta\dot\alpha}} = 0 ,$$

$$\mathcal{D}^{\dot\alpha B} \bar X_{\underset{\smile}{BA}} = \mathcal{D}_{\dot\beta A} \bar Y^{\underset{\smile}{\dot\beta\dot\alpha}}$$

$$\oint_{ABC} \mathcal{D}^{A}_{\dot\alpha} X^{\underset{\smile}{BC}} = 0 \,, \qquad \oint_{ABC} \mathcal{D}^{\dot\alpha}_{A} \bar{X}_{\underset{\smile}{BC}} = 0 \,,$$

$$\mathcal{D}^{D}_{\delta} U_{\gamma\dot\gamma} = -\frac{1}{4}\mathcal{D}^{D}_{\dot\gamma} Y_{\underset{\smile}{\delta\gamma}} + \frac{1}{4}\epsilon_{\delta\gamma}\left(\mathcal{D}^{\dot\delta D} X_{\underset{\smile}{\dot\delta\dot\gamma}} - \frac{1}{3}\mathcal{D}_{\dot\gamma C} X^{\underset{\smile}{CD}}\right),$$

$$\mathcal{D}_{\dot\varepsilon E} U_{\gamma\dot\gamma} = -\frac{1}{4}\mathcal{D}_{\gamma E}\bar{Y}_{\underset{\smile}{\dot\varepsilon\dot\gamma}} + \frac{1}{4}\epsilon_{\dot\varepsilon\dot\gamma}\left(\mathcal{D}^{\varepsilon}_{E}\bar{X}_{\underset{\smile}{\varepsilon\gamma}} - \frac{1}{3}\mathcal{D}^{C}_{\gamma}\bar{X}_{\underset{\smile}{CE}}\right),$$

and

$$\mathcal{D}^{\alpha A}\mathcal{D}^{\beta}_{A}\bar{X}_{\underset{\smile}{\alpha\beta}} + \mathcal{D}^{\dot\alpha}_{A}\mathcal{D}^{\dot\beta A} X_{\underset{\smile}{\dot\alpha\dot\beta}} = 4i\left(Y_{\underset{\smile}{\beta\alpha}}\bar{X}^{\underset{\smile}{\beta\alpha}} - X_{\underset{\smile}{\dot\beta\dot\alpha}}\bar{Y}^{\underset{\smile}{\dot\beta\dot\alpha}}\right).$$

As functions of this basic set of superfields the torsion components are expressed as follows:

$$T_{\gamma}{}^{C}{}_{\beta\dot\beta}{}^{A}{}_{\dot\alpha} = g^{CA}\left(\epsilon_{\gamma\beta} X_{\underset{\smile}{\dot\beta\dot\alpha}} + \epsilon_{\dot\beta\dot\alpha} Y_{\underset{\smile}{\gamma\beta}}\right) + \epsilon_{\gamma\beta}\epsilon_{\dot\beta\dot\alpha} X^{\underset{\smile}{CA}} \,,$$

$$T_{\dot\gamma C\,\beta\dot\beta\,\alpha A} = g_{CA}\left(\epsilon_{\dot\gamma\dot\beta}\bar{X}_{\underset{\smile}{\beta\alpha}} + \epsilon_{\beta\alpha}\bar{Y}_{\underset{\smile}{\dot\gamma\dot\beta}}\right) + \epsilon_{\dot\gamma\dot\beta}\epsilon_{\beta\alpha}\bar{X}_{\underset{\smile}{CA}} \,,$$

$$T_{\gamma}{}^{C}{}_{\beta\dot\beta}{}^{A}{}_{\dot\alpha} \equiv \sigma^{b}_{\beta\dot\beta}\, T_{\gamma}{}^{C}{}_{b}{}^{A}{}_{\dot\alpha} \,, \qquad T^{\dot\gamma}{}_{C\,\beta\dot\beta}{}^{\alpha}{}_{A} \equiv \sigma^{b}_{\beta\dot\beta}\, T^{\dot\gamma}{}_{C\,b}{}^{\alpha}{}_{A} \,.$$

$$T_{\gamma\,\beta\dot\beta\,\alpha} = \epsilon_{\beta\alpha} U_{\gamma\dot\beta} + \epsilon_{\beta\gamma} U_{\alpha\dot\beta} \,,$$

$$T_{\dot\gamma\,\beta\dot\beta\,\dot\alpha} = \epsilon_{\dot\beta\dot\alpha} U_{\beta\dot\gamma} + \epsilon_{\dot\beta\dot\gamma} U_{\beta\dot\alpha} \,,$$

$$T_{\gamma}{}^{C}{}_{\beta\dot\beta}{}^{\alpha}{}_{A} \equiv \sigma^{b}_{\beta\dot\beta}\, T_{\gamma}{}^{C}{}_{b}{}^{\alpha}{}_{A} \,, \qquad T^{\dot\gamma}{}_{C\,\beta\dot\beta}{}^{A}{}_{\dot\alpha} \equiv \sigma^{b}_{\beta\dot\beta}\, T^{\dot\gamma}{}_{C\,b}{}^{A}{}_{\dot\alpha} \,,$$

and

$$T_{\underbrace{\gamma\beta}\,\alpha A} = -\frac{i}{4}\mathcal{D}_{\alpha A}\bar{X}_{\underbrace{\gamma\beta}} - \frac{i}{12}\left(\epsilon_{\alpha\beta}\mathcal{D}_{\gamma}^{B} + \epsilon_{\alpha\gamma}\mathcal{D}_{\beta}^{B}\right)\bar{X}_{\underbrace{BA}},$$

$$T_{\underbrace{\dot\gamma\dot\beta}\,\alpha A} = -\frac{i}{4}\mathcal{D}_{\alpha A}\bar{Y}_{\underbrace{\dot\gamma\dot\beta}},$$

$$T_{\underbrace{\gamma\beta}\,\dot\alpha}{}^{A} = \frac{i}{4}\mathcal{D}^{A}_{\dot\alpha}Y_{\underbrace{\gamma\beta}},$$

$$T_{\underbrace{\dot\gamma\dot\beta}\,\dot\alpha}{}^{A} = \frac{i}{4}\mathcal{D}^{A}_{\dot\alpha}X_{\underbrace{\dot\gamma\dot\beta}} + \frac{i}{12}\left(\epsilon_{\dot\alpha\dot\beta}\mathcal{D}_{\dot\gamma B} + \epsilon_{\dot\alpha\dot\gamma}\mathcal{D}_{\dot\beta B}\right)X^{\underbrace{BA}}.$$

$$T_{\gamma\dot\gamma\,\beta\dot\beta}{}^{\alpha}{}_{A} \equiv \sigma^{c}{}_{\gamma\dot\gamma}\,\sigma^{b}{}_{\beta\dot\beta}\,T_{cb}{}^{\alpha}{}_{A},$$

$$T_{\gamma\dot\gamma\,\beta\dot\beta}{}^{\alpha}{}_{A} \equiv -2\epsilon_{\gamma\beta}T_{\underbrace{\dot\gamma\dot\beta}}{}^{\alpha}{}_{A} + 2\epsilon_{\dot\gamma\dot\beta}T_{\underbrace{\gamma\beta}}{}^{\alpha}{}_{A},$$

$$T_{\gamma\dot\gamma\,\beta\dot\beta}{}^{A}{}_{\dot\alpha} \equiv \sigma^{c}{}_{\gamma\dot\gamma}\,\sigma^{b}{}_{\beta\dot\beta}\,T_{cb}{}^{A}{}_{\dot\alpha},$$

$$T_{\gamma\dot\gamma\,\beta\dot\beta}{}^{A}{}_{\dot\alpha} \equiv -2\epsilon_{\gamma\beta}T_{\underbrace{\dot\gamma\dot\beta}}{}^{A}{}_{\dot\alpha} + 2\epsilon_{\dot\gamma\dot\beta}T_{\underbrace{\gamma\beta}}{}^{A}{}_{\dot\alpha}.$$

Here, spinor notation has been used in an obvious way.

Now from rather general arguments[7] we know that all components of the curvature are functions of torsion components and that the second set of Bianchi identities does not contain new information. Therefore, it is sufficient to analyze the first set of Bianchi identities which consists of thirty equations, labelled by different sets of indices ($\mathfrak{S}\mathfrak{D}\mathfrak{C}^{A}$). Due to the choice of structure groups and constraints, the equations corresponding to the combinations

$$\begin{pmatrix} E & D & C & A \\ \varepsilon & \delta & \gamma & \dot\alpha \end{pmatrix}, \quad \begin{pmatrix} E & D & C & a \\ \varepsilon & \delta & \gamma & \end{pmatrix}, \quad \begin{pmatrix} E & D & \dot\gamma & a \\ \varepsilon & \delta & C & \end{pmatrix},$$

$$\begin{pmatrix} \dot\varepsilon & \dot\delta & \dot\gamma & \alpha \\ E & D & C & A \end{pmatrix}, \quad \begin{pmatrix} \dot\varepsilon & \dot\delta & \dot\gamma & a \\ E & D & C & \end{pmatrix}, \quad \begin{pmatrix} \dot\varepsilon & \dot\delta & C & a \\ E & D & \gamma & \end{pmatrix},$$

of superindices $(\varepsilon\delta e^{\mathcal{A}})$ vanish identically.

Next, we look at equations which do not contain derivatives and are linear. We start with

$$R_{\delta\gamma E\dot\alpha}^{DC\dot\varepsilon A} - 2i\,\delta_E^D(\sigma^b\epsilon)_\delta{}^{\dot\varepsilon}\, T_{\gamma b\dot\alpha}^{C\ \ A} - 2i\,\delta_E^C(\sigma^b\epsilon)_\gamma{}^{\dot\varepsilon}\, T_{\delta b\dot\alpha}^{D\ \ A} = 0, \tag{1}$$

$$R_{\delta\gamma e}^{DC\ \ a} - 2i\,(\sigma^a\epsilon)_\gamma{}^{\dot\gamma}\, T_{\delta e\dot\gamma}^{D\ \ C} - 2i\,(\sigma^a\epsilon)_\delta{}^{\dot\delta}\, T_{\gamma e\dot\delta}^{C\ \ D} = 0, \tag{2}$$

$$R_{\varepsilon\delta\gamma A}^{EDC\alpha} + R_{\delta\gamma\varepsilon A}^{DCE\alpha} + R_{\gamma\varepsilon\delta A}^{CED\alpha} = 0 . \tag{3}$$

From (1) and (2) we obtain

$$T_{\gamma\beta\dot\beta\dot\alpha}^{C\ \ \ A} = \epsilon_{\gamma\beta}\epsilon_{\dot\beta\dot\alpha}\, X^{\underset{\smile}{CA}} + \epsilon_{\gamma\beta}\, g^{CA} X_{\underset{\smile}{\dot\beta\dot\alpha}} + \epsilon_{\dot\beta\dot\alpha}\, g^{CA} Y_{\underset{\smile}{\gamma\beta}} ,$$

$$R_{\delta\gamma E}^{DC\ \ A} = 2i\,\epsilon_{\delta\gamma}\left(\delta_E^D X^{\underset{\smile}{CA}} - \delta_E^C X^{\underset{\smile}{DA}}\right) - 2i\, Y_{\underset{\smile}{\delta\gamma}}\left(\delta_E^D g^{CA} + \delta_E^C g^{DA}\right),$$

$$R_{\delta\gamma\,\underset{\smile}{\dot\varepsilon\dot\alpha}}^{DC} = -2i\,\epsilon_{\delta\gamma}\, g^{DC} X_{\underset{\smile}{\dot\varepsilon\dot\alpha}} ,$$

$$R_{\delta\gamma\,\underset{\smile}{\varepsilon\alpha}}^{DC} = -2i\,\epsilon_{\delta\gamma}\, g^{DC} Y_{\underset{\smile}{\varepsilon\alpha}} + 2i\left(\epsilon_{\delta\varepsilon}\epsilon_{\gamma\alpha} + \epsilon_{\delta\alpha}\epsilon_{\gamma\varepsilon}\right) X^{\underset{\smile}{CD}} .$$

(3) is then identically satisfied and

$$\mathcal{D}_\delta^D\, T_{\gamma e\dot\alpha}^{C\ \ A} + \mathcal{D}_\gamma^C\, T_{\delta e\dot\alpha}^{D\ \ A} = 0 \tag{4}$$

implies

$$\mathcal{D}_\gamma^C X_{\dot\beta\dot\alpha} = 0 , \quad \mathcal{D}_{\alpha B} X^{BA} + \mathcal{D}^{\beta A} Y_{\beta\alpha} = 0 ,$$

$$\oint_{ABC} \mathcal{D}_\gamma^C X^{BA} = 0 , \quad \oint_{\alpha\beta\gamma} \mathcal{D}_\gamma^C Y_{\beta\alpha} = 0 .$$

In the same way, from the complex conjugate equations

$$R_{DC\varepsilon A}{}^{\dot\delta\dot\gamma E\alpha} - 2i\,\delta_D^E (\bar\sigma^b\epsilon)^{\dot\delta}{}_{\varepsilon} T^{\dot\gamma}_{C b A}{}^{\alpha} - 2i\,\delta_C^E (\bar\sigma^b\epsilon)^{\dot\gamma}{}_{\varepsilon} T^{\dot\delta}_{D b A}{}^{\alpha} = 0, \tag{5}$$

$$R_{DCe}{}^{\dot\delta\dot\gamma\, a} - 2i\,(\sigma^a\epsilon)_\gamma{}^{\dot\gamma} T^{\dot\delta}_{D e C}{}^{\gamma} - 2i\,(\sigma^a\epsilon)_\delta{}^{\dot\delta} T^{\dot\gamma}_{C e D}{}^{\delta} = 0, \tag{6}$$

$$R_{EDC\dot\alpha}{}^{\dot\varepsilon\dot\delta\dot\gamma A} + R_{DCE\dot\alpha}{}^{\dot\delta\dot\gamma\dot\varepsilon A} + R_{CED\dot\alpha}{}^{\dot\gamma\dot\varepsilon\dot\delta A} = 0 , \tag{7}$$

we obtain

$$T_{\dot\gamma C\,\beta\dot\beta\,\alpha A} = \epsilon_{\beta\alpha}\epsilon_{\dot\gamma\dot\beta}\bar X_{CA} + g_{CA}\epsilon_{\beta\alpha}\bar Y_{\dot\gamma\dot\beta} + g_{CA}\epsilon_{\dot\gamma\dot\beta} X_{\beta\alpha} ,$$

$$R_{DC}{}^{\dot\delta\dot\gamma E}{}_{A} = 2i\,\epsilon^{\dot\delta\dot\gamma}\left(\delta_D^E \bar X_{CA} - \delta_C^E \bar X_{DA}\right) + 2i\,\bar Y^{\dot\delta\dot\gamma}\left(\delta_D^E g_{CA} + \delta_C^E g_{DA}\right),$$

$$R_{DC}{}^{\dot\delta\dot\gamma}{}_{\varepsilon\alpha} = 2i\,\epsilon^{\dot\delta\dot\gamma} g_{DC} \bar X_{\varepsilon\alpha} ,$$

$$R_{DC}{}^{\dot\delta\dot\gamma}{}_{\dot\varepsilon\dot\alpha} = 2i\,\epsilon^{\dot\delta\dot\gamma} g_{DC} \bar Y_{\dot\varepsilon\dot\alpha} + 2i\left(\delta_{\dot\varepsilon}^{\dot\delta}\delta_{\dot\alpha}^{\dot\gamma} + \delta_{\dot\alpha}^{\dot\delta}\delta_{\dot\varepsilon}^{\dot\gamma}\right)\bar X_{DC} .$$

Clearly, (7) is again an identity with the results from (5) and (6), and

$$\mathcal{D}_D^{\dot\delta} T^{\dot\gamma}_{C e A}{}^{\alpha} + \mathcal{D}_C^{\dot\gamma} T^{\dot\delta}_{D e A}{}^{\alpha} = 0 \tag{8}$$

becomes

$$\mathcal{D}_C^{\dot\gamma} \bar X_{\beta\alpha} = 0 , \quad \mathcal{D}^{\dot\alpha B} \bar X_{BA} = \mathcal{D}_{\dot\beta A} \bar Y^{\dot\beta\dot\alpha} ,$$

$$\oint_{ABC} \mathcal{D}_C^{\dot\gamma} \bar X_{BA} = 0 , \quad \oint_{\dot\alpha\dot\beta\dot\gamma} \mathcal{D}_C^{\dot\gamma} \bar Y^{\dot\beta\dot\alpha} = 0 .$$

The solution of

$$R_{\delta c e}{}^{D\dot{\gamma}\,a} - 2i\,(\bar{\sigma}^a\epsilon)^{\dot{\gamma}}{}_{\gamma}\,T_{\delta e c}{}^{D\gamma} - 2i\,(\sigma^a\epsilon)_{\delta}{}^{\dot{\delta}}\,T_{c e \dot{\delta}}{}^{\dot{\gamma}\dot{D}} = 0, \tag{9}$$

$$\begin{aligned} &R_{C\varepsilon\delta A}{}^{\dot{\gamma}ED\alpha} + R_{C\delta\varepsilon A}{}^{\dot{\gamma}DE\alpha} - 2i\,\delta_C^D\,(\bar{\sigma}^b\epsilon)^{\dot{\gamma}}{}_{\delta}\,T_{\varepsilon b A}{}^{E\alpha} \\ &- 2i\,\delta_C^E\,(\bar{\sigma}^b\epsilon)^{\dot{\gamma}}{}_{\varepsilon}\,T_{\delta b A}{}^{D\alpha} = 0 , \end{aligned} \tag{10}$$

$$\begin{aligned} &R_{\gamma E D\dot{\alpha}}{}^{C\dot{\varepsilon}\dot{\delta}A} + R_{\gamma D E\dot{\alpha}}{}^{C\dot{\delta}\dot{\varepsilon}A} - 2i\,\delta_E^C\,(\sigma^b\epsilon)_{\gamma}{}^{\dot{\varepsilon}}\,T_{D b\dot{\alpha}}{}^{\dot{\delta}A} \\ &- 2i\,\delta_D^C\,(\sigma^b\epsilon)_{\gamma}{}^{\dot{\delta}}\,T_{E b\dot{\alpha}}{}^{\dot{\varepsilon}A} = 0 , \end{aligned} \tag{11}$$

is given by

$$T_{\delta\varepsilon\dot{\varepsilon}\alpha} = \epsilon_{\varepsilon\alpha}\,U_{\delta\dot{\varepsilon}} + \epsilon_{\varepsilon\delta}\,U_{\alpha\dot{\varepsilon}} ,$$

$$T_{\dot{\delta}\varepsilon\dot{\varepsilon}\dot{\alpha}} = \epsilon_{\dot{\varepsilon}\dot{\alpha}}\,U_{\varepsilon\dot{\delta}} + \epsilon_{\dot{\varepsilon}\dot{\delta}}\,U_{\varepsilon\dot{\alpha}} ,$$

$$R_{\delta\dot{\gamma}C\,\dot{\varepsilon}\dot{\alpha}}{}^{D} = 2i\,\delta_C^D\,(\epsilon_{\dot{\gamma}\dot{\alpha}}\,U_{\delta\dot{\varepsilon}} + \epsilon_{\dot{\gamma}\dot{\varepsilon}}\,U_{\delta\dot{\alpha}}) ,$$

$$R_{\delta\dot{\gamma}C\,\varepsilon\alpha}{}^{D} = 2i\,\delta_C^D\,(\epsilon_{\delta\alpha}\,U_{\varepsilon\dot{\gamma}} + \epsilon_{\varepsilon\delta}\,U_{\alpha\dot{\gamma}}) ,$$

$$R_{\delta\dot{\gamma}C}{}^{D}{}_{A}{}^{E} = 8i\left(\delta_A^D\,\delta_C^E - \tfrac{1}{2}\,\delta_A^E\,\delta_C^D\right) U_{\delta\dot{\gamma}} .$$

In the next step we investigate the equations linear in $T_{cb}{}^{A}{}_{\dot{\alpha}}$ and their complex conjugates. Let us begin with

$$R_{\delta e c \dot{\alpha}}^{D \quad \dot{\gamma} A} + \mathcal{D}_{\delta}^{D} T_{c e \dot{\alpha}}^{\dot{\gamma} \quad A} + \mathcal{D}_{c}^{\dot{\gamma}} T_{\delta e \dot{\alpha}}^{D \quad A} - \tag{12}$$

$$- 2i \, \delta_{c}^{D} (\sigma^{b} \epsilon)_{\delta}{}^{\dot{\gamma}} \, T_{be \dot{\alpha}}^{\quad A} = 0 \quad ,$$

$$- R_{\delta e c}^{D \quad a} + R_{\delta c e}^{D \quad a} - 2i (\sigma^{a} \epsilon)_{\delta}{}^{\dot{\delta}} \, T_{ec \dot{\delta}}^{\quad D} = 0 \; , \tag{13}$$

$$R_{\delta e \gamma A}^{D \quad C \alpha} + R_{\gamma e \delta A}^{C \quad D \alpha} + \mathcal{D}_{\gamma}^{C} T_{\delta e A}^{D \quad \alpha} + \mathcal{D}_{\delta}^{D} T_{\gamma e A}^{C \quad \alpha} = 0 . \tag{14}$$

Using the results from (12) and (13),

$$T_{\underset{\smile}{\gamma\beta} \dot{\alpha}}^{\quad A} = \frac{i}{4} \mathcal{D}_{\dot{\alpha}}^{A} Y_{\underset{\smile}{\gamma\beta}} \; ,$$

$$T_{\underset{\smile}{\dot{\gamma}\dot{\beta}} \dot{\alpha}}^{\quad A} = \frac{i}{4} \mathcal{D}_{\dot{\alpha}}^{A} X_{\underset{\smile}{\dot{\gamma}\dot{\beta}}} + \frac{i}{12} \left(\epsilon_{\dot{\alpha}\dot{\gamma}} \mathcal{D}_{\dot{\beta} D} + \epsilon_{\dot{\alpha}\dot{\beta}} \mathcal{D}_{\dot{\gamma} D} \right) X^{\underset{\smile}{DA}} \; ,$$

$$\mathcal{D}_{\delta}^{D} U_{\gamma\dot{\gamma}} = - \frac{1}{4} \mathcal{D}_{\dot{\gamma}}^{D} Y_{\underset{\smile}{\delta\gamma}} + \frac{1}{4} \epsilon_{\delta\gamma} \left(\mathcal{D}^{\dot{\delta} D} X_{\underset{\smile}{\dot{\delta}\dot{\gamma}}} - \frac{1}{3} \mathcal{D}_{\dot{\gamma} C} X^{\underset{\smile}{CD}} \right) ,$$

$$\oint_{ABC} \mathcal{D}_{\dot{\alpha}}^{A} X^{\underset{\smile}{BC}} = 0 \quad ,$$

$$R_{\delta e c a}^{D} = i (\sigma_{e} \epsilon)_{\delta}{}^{\dot{\delta}} T_{ca \dot{\delta}}^{\quad D} - i (\sigma_{c} \epsilon)_{\delta}{}^{\dot{\delta}} T_{ae \dot{\delta}}^{\quad D} - i (\sigma_{a} \epsilon)_{\delta}{}^{\dot{\delta}} T_{ec \dot{\delta}}^{\quad D} ,$$

$$R_{\delta \varepsilon \dot{\varepsilon} \quad A}^{D \quad C} = - \left(\delta_{A}^{D} \delta_{E}^{C} - \frac{1}{2} \delta_{A}^{C} \delta_{E}^{D} \right) \left(\mathcal{D}_{\dot{\varepsilon}}^{E} Y_{\underset{\smile}{\delta\varepsilon}} + \epsilon_{\delta\varepsilon} \left(\mathcal{D}^{\dot{\delta} E} X_{\underset{\smile}{\dot{\delta}\dot{\varepsilon}}} + \frac{1}{3} \mathcal{D}_{\dot{\varepsilon} D} X^{\underset{\smile}{DE}} \right) \right) ,$$

we see that (14) becomes an identity. In the same way we treat the complex conjugate equations

$$- R_{E d \gamma A}^{\dot{\varepsilon} \quad C \alpha} - \mathcal{D}_{E}^{\dot{\varepsilon}} T_{\gamma d A}^{C \quad \alpha} - \mathcal{D}_{\gamma}^{C} T_{E d A}^{\dot{\varepsilon} \quad \alpha} \tag{15}$$

$$+ 2i (\sigma^{b} \epsilon)_{\gamma}{}^{\dot{\varepsilon}} \, \delta_{E}^{C} \, T_{bd A}^{\quad \alpha} = 0 \quad ,$$

$$-R^{\dot\varepsilon}{}_{E\,dc}{}^{a} + R^{\dot\varepsilon}{}_{E\,cd}{}^{a} - 2i\,(\bar\sigma^{a}\epsilon)^{\dot\varepsilon}{}_{\varepsilon}\,T_{dc}{}^{\varepsilon}{}_{E} = 0, \qquad (16)$$

$$-R^{\dot\varepsilon}{}_{E\,dc}{}^{\dot\gamma A}{}_{\dot\alpha} - R^{\dot\gamma}{}_{c\,d}{}^{\dot\varepsilon A}{}_{E\dot\alpha} - \mathcal{D}^{\dot\varepsilon}_{E}\,T^{\dot\gamma}{}_{c\,d}{}^{A}{}_{\dot\alpha} - \mathcal{D}^{\dot\gamma}_{C}\,T^{\dot\varepsilon}{}_{E\,d}{}^{A}{}_{\dot\alpha} = 0, \qquad (17)$$

where now (17) becomes an identity with the results from (15) and (16):

$$T_{\underset{\smile}{\dot\gamma\dot\beta}}{}^{\alpha}{}_{A} = -\frac{i}{4}\,\mathcal{D}^{\alpha}_{A}\,\bar Y_{\underset{\smile}{\dot\gamma\dot\beta}},$$

$$T_{\underset{\smile}{\gamma\beta}}{}^{\alpha}{}_{A} = -\frac{i}{4}\,\mathcal{D}^{\alpha}_{A}\,\bar X_{\underset{\smile}{\gamma\beta}} - \frac{i}{12}\left(\delta^{\alpha}_{\gamma}\,\mathcal{D}^{D}_{\beta} + \delta^{\alpha}_{\beta}\,\mathcal{D}^{D}_{\gamma}\right)\bar X_{\underset{\smile}{DA}},$$

$$\mathcal{D}_{\dot\varepsilon E}\,U_{\gamma\dot\gamma} = -\frac{1}{4}\,\mathcal{D}_{\gamma E}\,\bar Y_{\underset{\smile}{\dot\varepsilon\dot\gamma}} + \frac{1}{4}\,\epsilon_{\dot\varepsilon\dot\gamma}\left(\mathcal{D}^{\varepsilon}_{E}\,\bar X_{\underset{\smile}{\varepsilon\gamma}} - \frac{1}{3}\,\mathcal{D}^{C}_{\gamma}\,\bar X_{\underset{\smile}{CE}}\right),$$

$$\oint_{ABC} \mathcal{D}^{\alpha}_{A}\,\bar X_{\underset{\smile}{BC}} = 0,$$

$$R^{\dot\varepsilon}{}_{E\,dca} = i\,(\bar\sigma_{d}\epsilon)^{\dot\varepsilon}{}_{\varepsilon}\,T_{ca}{}^{\varepsilon}{}_{E} - i\,(\bar\sigma_{c}\epsilon)^{\dot\varepsilon}{}_{\varepsilon}\,T_{ad}{}^{\varepsilon}{}_{E} - i\,(\bar\sigma_{a}\epsilon)^{\dot\varepsilon}{}_{\varepsilon}\,T_{dc}{}^{\varepsilon}{}_{E},$$

$$R_{\dot\varepsilon E\,\delta\dot\delta}{}^{C}{}_{A} = -\left(\delta^{C}_{E}\,\delta^{D}_{A} - \frac{1}{2}\,\delta^{C}_{A}\,\delta^{D}_{E}\right)\left(\mathcal{D}_{\delta D}\,\bar Y_{\underset{\smile}{\dot\varepsilon\dot\delta}} + \epsilon_{\dot\varepsilon\dot\delta}\left(\mathcal{D}^{\varepsilon}_{D}\,\bar X_{\underset{\smile}{\varepsilon\delta}} + \frac{1}{3}\,\mathcal{D}^{E}_{\delta}\,\bar X_{\underset{\smile}{ED}}\right)\right).$$

At this stage of our analysis we note that the non-linear equations, corresponding to the combinations.

$$\begin{pmatrix}\dot\varepsilon & & \alpha\\ E & dc & A\end{pmatrix},\quad \begin{pmatrix}E & & A\\ \varepsilon & dc & \dot\alpha\end{pmatrix},\quad \begin{pmatrix} & & \alpha\\ e & dc & A\end{pmatrix},\quad \begin{pmatrix} & & A\\ e & dc & \dot\alpha\end{pmatrix}$$

of super indices ($\varepsilon\,dc^{A}$) are identically satisfied, if one uses the results obtained so far.

We are therefore left with the last three equations:

$$R_{edca} + R_{dcea} + R_{ceda} = 0 \,, \tag{18}$$

$$-R_{dc\varepsilon A}{}^{E\alpha} + \mathcal{D}_{\varepsilon}^{E} T_{dcA}{}^{\alpha} - \mathcal{D}_{d} T_{\varepsilon c A}^{E}{}^{\alpha} + \mathcal{D}_{c} T_{\varepsilon d A}^{E}{}^{\alpha}$$
$$+ \left(T_{\varepsilon d}^{E}{}_{B}^{\beta} T_{\beta c A}^{B}{}^{\alpha} + T_{\varepsilon d \dot\beta}^{E}{}^{B} T_{B C A}^{\dot\beta}{}^{\alpha} - d \leftrightarrow c \right) = 0 \,, \tag{19}$$

$$-R_{dc E \dot\alpha}{}^{\dot\varepsilon A} + \mathcal{D}_{E}^{\dot\varepsilon} T_{dc\dot\alpha}{}^{A} - \mathcal{D}_{d} T_{E c \dot\alpha}^{\dot\varepsilon}{}^{A} + \mathcal{D}_{c} T_{E d \dot\alpha}^{\dot\varepsilon}{}^{A}$$
$$+ \left(T_{E d}^{\dot\varepsilon}{}_{B}^{\beta} T_{\beta c \dot\alpha}^{B}{}^{A} + T_{E d \dot\beta}^{\dot\varepsilon}{}^{B} T_{B C \dot\alpha}^{\dot\beta}{}^{A} - d \leftrightarrow c \right) = 0 . \tag{20}$$

The well-known consequence of (18) is

$$\chi_{\underset{\smile}{\delta\gamma}\,\underset{\smile}{\varepsilon\alpha}} = \chi_{\underset{\smile}{\varepsilon\alpha}\,\underset{\smile}{\delta\gamma}} \,, \quad \bar\chi_{\underset{\smile}{\dot\delta\dot\gamma}\,\underset{\smile}{\dot\varepsilon\dot\alpha}} = \bar\chi_{\underset{\smile}{\dot\varepsilon\dot\alpha}\,\underset{\smile}{\dot\delta\dot\gamma}} \,,$$

$$\chi_{\underset{\smile}{\delta\gamma}}{}^{\underset{\smile}{\delta\gamma}} = 6\lambda = \bar\chi_{\underset{\smile}{\dot\delta\dot\gamma}}{}^{\underset{\smile}{\dot\delta\dot\gamma}} \,, \quad \varphi_{\underset{\smile}{\delta\gamma}\,\underset{\smile}{\dot\varepsilon\dot\alpha}} = \varphi_{\underset{\smile}{\dot\varepsilon\dot\alpha}\,\underset{\smile}{\delta\gamma}} \,,$$

where we have already made use of the spinor decomposition of the curvature components:

$$R_{\delta\dot\delta\,\gamma\dot\gamma\,\varepsilon\dot\varepsilon\,\alpha\dot\alpha} \equiv \sigma^{d}_{\delta\dot\delta}\,\sigma^{c}_{\gamma\dot\gamma}\,\sigma^{e}_{\varepsilon\dot\varepsilon}\,\sigma^{a}_{\alpha\dot\alpha}\,R_{dcea} =$$

$$= 4\,\epsilon_{\dot\delta\dot\gamma}\,\epsilon_{\dot\varepsilon\dot\alpha}\,\chi_{\underset{\smile}{\delta\gamma}\,\underset{\smile}{\varepsilon\alpha}} - 4\,\epsilon_{\dot\delta\dot\gamma}\,\epsilon_{\varepsilon\alpha}\,\varphi_{\underset{\smile}{\delta\gamma}\,\underset{\smile}{\dot\varepsilon\dot\alpha}}$$

$$- 4\,\epsilon_{\delta\gamma}\,\epsilon_{\dot\varepsilon\dot\alpha}\,\varphi_{\underset{\smile}{\dot\delta\dot\gamma}\,\underset{\smile}{\varepsilon\alpha}} + 4\,\epsilon_{\delta\gamma}\,\epsilon_{\varepsilon\alpha}\,\bar\chi_{\underset{\smile}{\dot\delta\dot\gamma}\,\underset{\smile}{\dot\varepsilon\dot\alpha}}$$

$$R_{\delta\dot\delta\,\gamma\dot\gamma}{}^{E}{}_{A} \equiv \sigma^{d}_{\delta\dot\delta}\,\sigma^{c}_{\gamma\dot\gamma}\,R_{dc}{}^{E}{}_{A} =$$

$$= -2\,\epsilon_{\delta\gamma}\,\Phi_{\underset{\smile}{\dot\delta\dot\gamma}}{}^{E}{}_{A} + 2\,\epsilon_{\dot\delta\dot\gamma}\,\Phi_{\underset{\smile}{\delta\gamma}}{}^{E}{}_{A} \,.$$

Keeping in mind the relations obtained so far, the consequences of (18), (19) and (20) may be stated as follows:

$$\chi_{\underset{\smile}{\delta\gamma}\,\underset{\smile}{\varepsilon\alpha}} = \frac{1}{4!}\,\underset{\delta\gamma\varepsilon\alpha}{P}\left(-\frac{i}{8}\right)\left(\mathcal{D}^{E}_{\varepsilon}\mathcal{D}_{\alpha E} + 8i\,Y_{\underset{\smile}{\varepsilon\alpha}}\right)\bar{X}_{\underset{\smile}{\delta\gamma}} + \left(\epsilon_{\delta\varepsilon}\epsilon_{\gamma\alpha} + \epsilon_{\gamma\varepsilon}\epsilon_{\delta\alpha}\right)\lambda ,$$

$$\bar{\chi}_{\underset{\smile}{\dot\delta\dot\gamma}\,\underset{\smile}{\dot\varepsilon\dot\alpha}} = \frac{1}{4!}\,\underset{\dot\delta\dot\gamma\dot\varepsilon\dot\alpha}{P}\left(-\frac{i}{8}\right)\left(\mathcal{D}^{\dot\varepsilon}_{E}\mathcal{D}_{\dot\alpha}{}^{E} + 8i\,\bar{Y}_{\underset{\smile}{\dot\varepsilon\dot\alpha}}\right)X_{\underset{\smile}{\dot\delta\dot\gamma}} + \left(\epsilon_{\dot\delta\dot\varepsilon}\epsilon_{\dot\gamma\dot\alpha} + \epsilon_{\dot\gamma\dot\varepsilon}\epsilon_{\dot\delta\dot\alpha}\right)\lambda ,$$

$$6\lambda = 3\,U^{\alpha\dot\alpha}U_{\alpha\dot\alpha} - \frac{3}{2}X^{\underset{\smile}{BA}}\bar{X}_{\underset{\smile}{BA}} + \frac{1}{2}X_{\underset{\smile}{\dot\beta\dot\alpha}}\bar{Y}^{\underset{\smile}{\dot\beta\dot\alpha}} + \frac{1}{2}Y_{\underset{\smile}{\beta\alpha}}\bar{X}^{\underset{\smile}{\beta\alpha}}$$
$$+\frac{i}{16}\left(\mathcal{D}^{\dot\alpha}_{A}\mathcal{D}^{\beta A}X_{\underset{\smile}{\dot\alpha\beta}} + \mathcal{D}^{\dot\alpha}_{A}\mathcal{D}_{\dot\alpha B}X^{\underset{\smile}{AB}} - \mathcal{D}^{\alpha A}\mathcal{D}^{B}_{\alpha}\bar{X}_{\underset{\smile}{AB}} - \mathcal{D}^{\alpha A}\mathcal{D}^{\beta}_{A}\bar{X}_{\underset{\smile}{\alpha\beta}}\right),$$

$$\varphi_{\underset{\smile}{\gamma\alpha}\,\underset{\smile}{\dot\gamma\dot\alpha}} = \frac{1}{2}\bar{X}_{\underset{\smile}{\gamma\alpha}}X_{\underset{\smile}{\dot\gamma\dot\alpha}} - U_{\gamma\dot\gamma}U_{\alpha\dot\alpha} - U_{\gamma\dot\alpha}U_{\alpha\dot\gamma}$$
$$-\frac{i}{32}\left(\sum_{\dot\gamma\dot\alpha}\left(\mathcal{D}^{C}_{\dot\gamma}\mathcal{D}_{\dot\alpha C} + 4i\,\bar{Y}_{\underset{\smile}{\dot\gamma\dot\alpha}}\right)Y_{\underset{\smile}{\gamma\alpha}} + \sum_{\gamma\alpha}\left(\mathcal{D}^{C}_{\gamma}\mathcal{D}_{\alpha C} + 4i\,Y_{\underset{\smile}{\gamma\alpha}}\right)\bar{Y}_{\underset{\smile}{\dot\gamma\dot\alpha}}\right),$$

$$\Phi_{\underset{\smile}{\delta\gamma}}{}^{A}{}_{E} = -\frac{i}{8}\left(\delta^{A}_{B}\delta^{C}_{E} - \frac{1}{2}\delta^{A}_{E}\delta^{C}_{B}\right)\mathcal{D}^{\dot\alpha}_{C}\mathcal{D}^{B}_{\dot\alpha}\,Y_{\underset{\smile}{\delta\gamma}}$$
$$+\frac{1}{2}\left(Y_{\underset{\smile}{\delta\gamma}}\bar{X}^{A}{}_{E} - \bar{X}_{\underset{\smile}{\delta\gamma}}X^{A}{}_{E}\right),$$

$$\Phi_{\underset{\smile}{\dot\delta\dot\gamma}}{}^{A}{}_{E} = -\frac{i}{8}\left(\delta^{A}_{B}\delta^{C}_{E} - \frac{1}{2}\delta^{A}_{E}\delta^{C}_{B}\right)\mathcal{D}^{B}_{\alpha}\mathcal{D}^{\alpha}_{C}\,\bar{Y}_{\underset{\smile}{\dot\delta\dot\gamma}}$$
$$+\frac{1}{2}\left(\bar{Y}_{\underset{\smile}{\dot\delta\dot\gamma}}X^{A}{}_{E} - X_{\underset{\smile}{\dot\delta\dot\gamma}}\bar{X}^{A}{}_{E}\right),$$

$$\mathcal{D}^{\alpha A}\mathcal{D}^{\beta}_{A}\bar{X}_{\underset{\smile}{\alpha\beta}} + \mathcal{D}^{\dot\alpha}_{A}\mathcal{D}^{\beta A}X_{\underset{\smile}{\dot\alpha\beta}} = 4i\left(Y_{\underset{\smile}{\beta\alpha}}\bar{X}^{\underset{\smile}{\beta\alpha}} - X_{\underset{\smile}{\dot\beta\dot\alpha}}\bar{Y}^{\underset{\smile}{\dot\beta\dot\alpha}}\right).$$

The last relation follows from the reality property of the curvature scalar λ. The remaining properties of $R_{dcb}{}^{a}$ following from (18), however, are also consequences of (19) and (20) and relations obtained earlier.

We have constructed a complete solution of the Bianchi identities in terms of a few basic superfields and their covariant derivatives, as indicated at the beginning of the paper. In order to define component fields we introduce the notations:

$$\Sigma_\alpha{}^A = \mathcal{D}_{\alpha B} X^{\underset{\smile}{BA}} \quad , \quad \bar{\Lambda}^{\dot\alpha A} = \mathcal{D}^{\dot\alpha}_B X^{\underset{\smile}{BA}} \quad ,$$

$$\bar{\Sigma}^{\dot\alpha}_A = \mathcal{D}^{\dot\alpha B} \bar{X}_{\underset{\smile}{BA}} \quad , \quad \Lambda_{\alpha A} = \mathcal{D}^B_\alpha \bar{X}_{\underset{\smile}{BA}} \quad ,$$

$$\mathcal{M} = \mathcal{D}^\alpha_A \mathcal{D}_{\alpha B} X^{\underset{\smile}{BA}} \quad , \quad \bar{\mathcal{M}} = \mathcal{D}^A_{\dot\alpha} \mathcal{D}^{\dot\alpha B} \bar{X}_{\underset{\smile}{BA}}$$

$$\mathcal{D} = \mathcal{D}^{\dot\alpha}_A \mathcal{D}_{\dot\alpha B} X^{\underset{\smile}{BA}} + \mathcal{D}^A_\alpha \mathcal{D}^{\alpha B} \bar{X}_{\underset{\smile}{BA}} \; .$$

The covariant derivatives of $\mathcal{M}$, $\bar{\mathcal{M}}$ and $\mathcal{D}$ do not contain new independent quantities because of the properties of the basic superfields which are consequences of the Bianchi identities. Then using techniques developed in ordinary superspace by Wess and Zumino[8], we identify as component fields the values at $\theta = \bar\theta = 0$ of vielbein and connection superfields in a suitable gauge,

$$E_m{}^A(x,0,0) \sim \begin{pmatrix} e_m{}^a & \psi_m{}^\alpha_A & \bar{\psi}_m{}^A_{\dot\alpha} \\ 0 & \delta^\alpha_\mu \delta^M_A & 0 \\ 0 & 0 & \delta^{\dot\mu}_{\dot\alpha} \delta^A_M \end{pmatrix},$$

$$\phi_B{}^A(x,0,0) = dx^m \varphi_{mB}{}^A \quad ,$$

and of the covariant superfields

$$X_{\underset{\smile}{\beta\dot{\alpha}}},\ \bar{X}_{\underset{\smile}{\beta\alpha}},\quad X^{\underset{\smile}{BA}},\ \bar{X}_{\underset{\smile}{BA}},\ Y_{\underset{\smile}{\beta\alpha}},\ \bar{Y}_{\underset{\smile}{\dot{\beta}\dot{\alpha}}},$$

$$\Sigma^{A}_{\alpha},\ \Lambda_{\alpha A},\ \bar{\Sigma}^{\dot{\alpha}}_{A},\ \bar{\Lambda}^{\dot{\alpha}A},\ \mathcal{M},\ \bar{\mathcal{M}},\ \mathcal{D}.$$

By the same methods as in Ref. 8. differential geometry in superspace provides us with transformation laws and commutators of transformations

The lowest components of $X_{\underset{\smile}{\dot{\beta}\dot{\alpha}}}$ and $\bar{X}_{\underset{\smile}{\beta\alpha}}$ are related to the field strength of a spin one field, which is the supersymmetric partner of graviton and gravitinos. Taking the superfields $X^{\underset{\smile}{BA}}$, $\bar{X}_{\underset{\smile}{BA}}$, $Y_{\underset{\smile}{\beta\alpha}}$, $Y_{\underset{\smile}{\dot{\beta}\dot{\alpha}}}$ and $U_{\alpha\dot{\alpha}}$ to be zero implies equations of motion for the spin-2, spin-1 and spin-3/2 fields.

The chiral multiplet as well as the chiral density multiplet[8] have been constructed in this kind of extended superspace[9]. Other superspace densities which might be candidates for superspace actions are under investigation.

ACKNOWLEDGEMENT

Part of this and related work, in particular the investigation of the linearized version of N = 2 supergravity in superspace, has been done together with N. Dragon and J. Wess.

REFERENCES

1. J. Wess, Proceedings of the VIII G.I.F.T. International Seminar on Theoretical Physics, Salamance (June 1977), Lecture Notes in Physics (Springer-Verlag, Berlin, Heidelberg, New York 1978), vol. 77, p. 81.
2. B. Zumino, Proceedings of the NATO Advanced Study Institute on Recent Developments in Gravitation, Cargèse, Corsica (August 1978), NATO Advanced Study Institutes Series B (Plenum Press, New York, London) p. 405.
3. R. Grimm, J. Wess and B. Zumino, Nucl. Phys. B152 (1979) 255.
4. J. Wess, Proceedings of the EPS International Conference on High Energy Physics, Geneva (June 1979), CERN publication (1979), vol. 1, p. 462.
5. P. Breitenlohner and M.F. Sohnius, Nucl. Phys. B165 (1980) 483.
6. J. Wess and B. Zumino, Phys. Lett. 66B (1977) 361.
7. N. Dragon, Z. Phys. C2 (1979) 29.
8. J. Wess and B. Zumino, Phys. Lett. 79B (1978) 394.
9. J. Wess and R. Grimm, in preparation.

THE U(1) PROBLEM AND CURRENT ALGEBRA*

R. Arnowitt

Department of Physics, Northeastern University
Boston, Massachusetts 02115

Pran Nath†

CERN, Geneva, Switzerland

ABSTRACT

The procedures for applying current algebra methods to construct an effective Lagrangian for the physical particle states of strongly interacting systems is illustrated for the U(1) problem of QCD. The effective Lagrangian incorporates the Kogut-Susskind pole and the U(1) axial anomaly. Previous results on the U(1) problem obtained from the fundamental QCD Lagrangian are rederived simply from the effective Lagrangian along with a number of new results. The solutions of the different U(1) problems are seen to have a unified origin in the anomalous PCAC condition. All chiral and SU(3) breaking effects are included, and shown in certain circumstances to be considerable.

---

*Talk presented by R. Arnowitt

†On sabbatical leave from Northeastern University, Boston, Massachusetts 02115, U.S.A.

I. INTRODUCTION

I would like to talk about some work that Pran Nath and I have done on applying current algebra techniques to the U(1) problem. There are a number of different ways of implementing the current algebra conditions. The method that I will use here makes use of effective Lagrangians. While I will restrict the discussion to problems involving QCD only, we feel that the most interesting uses of effective Lagrangians may reside in dealing with "hypercolor" or "technicolor" interactions(1) as well. Thus this talk might be subtitled "The Uses of Effective Lagrangians in Strong Interactions" and the QCD analysis given here represents an interesting special case which illustrates some of the ideas and procedures which may have a wider scope of applicability.

Let me start by summarizing some of the advantages (and disadvantages) of the effective Lagrangian approach:

(1) First, the effective Lagrangian is a useful tool as it allows one to impose all the general principles one knows about a problem in a Lagrangian formalism. Particularly important for theories such as QCD or technicolor (where the particle spectrum is distantly related to the fundamental quark and gluon fields) these principles are imposed on fields that represent the <u>physical spectrum</u> of the states, and so the Lagrangian can be applied directly to physically observable phenomena.

(2) Symmetry breaking (e.g., SU(3), chiral, etc.) can be taken into account in a general way without resorting to perturbative methods. We will see that this can be important, as symmetry breaking may be large for some situations, and hence cannot always be treated as a small correction to the symmetric limit.

(3) Things that are difficult to calculate using the fundamental Lagrangian can often be characterized by a few phenomenological parameters in the effective Lagrangian. Thus by resorting to experiment to determine these parameters, one may extend the predictive range of the fundamental theory.

(4) One disadvantage of the approach is that the technique is generally restricted to "low energy" phenomena.(2) Furthermore, unitarity is generally difficult to impose.(3)(4)

What we will discuss in this talk, then, is the effective Lagrangian approach to the U(1) problem of QCD. In the process we will see how one extracts from the fundamental QCD Lagrangian the necessary information to build a useful effective Lagrangian. This analysis both helps to illuminate the conceptual understanding of the U(1) problem, as well as make feasible more realistic calculations of such processes as the $\eta \to 3\pi$ decay and the neutron electric dipole moment. One may expect similar advantages when the technique is extended to the higher energy domain of technicolor.

II. WHAT IS THE U(1) PROBLEM?

The "U(1) Problem" is actually a complex of different questions of which we list three:

(a) Why is the η' so heavy?

Experimentally $m_{\eta'} = 958$ MeV, nearly twice as heavy as the η. The origin of this additional mass creates a puzzle, as was first pointed out by Glashow[5] in 1968.

(b) Why doesn't the amplitude for $\eta \to \pi^o + \pi^+ + \pi^-$ vanish in the soft π^o limit?

Experimentally, in the region of the Dalitz plot close to the soft π^+ or π^-, the $\eta \to 3\pi$ amplitude is indeed small, in accord with current algebra. Yet the experimental amplitude is quite large near the soft π^o region.

(c) What is the relation of the above questions to the QCD θ-vacuum?

It is clear that such apparently disparate questions can be resolved only within an agreed upon theoretical framework. (Indeed, item (c) is more like an answer in search of a question!) Much of the debate about the U(1) problem has centered upon agreeing on this framework. So let me state next the "ground rules" for "solving" the U(1) problem. We assume therefore the following:

(1) Strong plus electroweak interactions at currently accessible energies are described by the standard $SU(3)_c \times SU(2)_L \times U(1)_Y$ gauge theory.

The QCD part of the Lagrangian is thus

$$\mathcal{L}_{QCD} = -\tfrac{1}{4}F_{\mu\nu}{}^A(x)F^{\mu\nu A}(x) - \bar{q}(x)\gamma^\mu(\tfrac{1}{i}\partial_\mu - gA_\mu{}^A(x)\lambda^A)q(x) - \theta\left[\frac{g^2}{32\pi^2} F^{\mu\nu A}(x)\,\tilde{F}_{\mu\nu}{}^A(x)\right] , \qquad (2.1)$$

where $A_\mu{}^A(x)$, $A = 1...8$ is the SU(3) color gluon field, $F_{\mu\nu}{}^A$ the corresponding field strength ($\tilde{F}_{\mu\nu}{}^A = \tfrac{1}{2}\epsilon_{\mu\nu\alpha\beta} F^{\alpha\beta A}$) and $q(x)$ a set of quark fields. The last term of Eq. (2.1) is the strong CP violating term due to the topologically non-trivial aspects of the θ vacuum. No bare mass is assumed to exist, and consequently the strong part of the Lagrangian, Eq. (2.1) possesses an "accidental" $U(N_f) \times U(N_f)$ global chiral invariance, where N_f is the number of quark flavors.

(2) Origin of Quark Masses

The origin of quark masses is assumed to arise from two sources:

(i) Spontaneous symmetry breaking of the electroweak interactions by a fundamental Higgs field (or by a dynamical symmetry breaking in the "family" or "sidewise" interactions of extended technicolor[1]).

Such breaking gives rise to the "current algebra" quark mass m_q.

(ii) Dynamical symmetry breaking of the chiral invariance in the strong QCD sector.

This breaking occurs when the interaction becomes strong enough for the condensate Λ to form:

$$\Lambda^3 = \langle 0|\bar{q}q|0\rangle \quad . \tag{2.2}$$

Experimentally, $\Lambda \approx 300$ MeV and gives rise to the "constituent" quark mass at energies $\leqslant \Lambda$.

(3) Goldstone Bosons

We assume that there are three light quarks with $m_q \ll \Lambda$: the up (u), down (d) and strange (s) quarks. Then the above dynamical breaking implies that the breaking of the chiral $U(N_\ell) \times U(N_\ell)$ ($N_\ell = 3$ is the number of light quarks) is realized in a Goldstone fashion, i.e., there are nine Goldstone bosons coupled to the nine weak axial currents $A_a{}^\mu$ such that

$$\langle 0|A_a{}^\mu|q,b\rangle = F_{ab} iq^\mu \quad ; \quad A_a{}^\mu \equiv \bar{q}\gamma^\mu\gamma^5 \frac{\lambda_a}{2}q, \quad a,b = 1...9 \tag{2.3}$$

where $|q,a\rangle$ are 0^- meson states. In Eq. (2.3), the "decay constants" F_{ab} are non-zero in the chiral limit when the current algebra masses $m_q \to 0$, and the Goldstone bosons become massless. For $m_q \neq 0$ (but $m_q \ll \Lambda$) one expects then nine light pseudo-Goldstone bosons. However, the ninth member of the nonet, the η', is heavy, which is the first U(1) problem.

(4) PCAC and the U(1) Anomaly

There exists an anomaly in the PCAC equation for the ninth axial current. Thus one may write

$$\partial_\mu A_a^{\ \mu} = \bar{q} i\gamma^5 \{\frac{\lambda_a}{2}, M\} q + \delta_{a9} 2N_\ell \partial_\mu K^\mu \quad , \tag{2.4}$$

where M is the light quark (current algebra) mass matrix: diag $M = (m_u, m_d, m_s)$. The anomaly in Eq. (2.4) is connected to the topological charge density Q(x)

$$Q(x) \equiv \partial_\mu K^\mu = \frac{g^2}{32\pi^2} F^{\mu\nu A} \tilde{F}_{\mu\nu}^{\ A} \quad , \tag{2.5}$$

where the "topological current density" K^μ is given by (C^{ABC} = SU(3) structure constants):

$$K^\mu(x) \equiv \frac{g^2}{32\pi^2} \epsilon^{\mu\nu\alpha\beta} A_\alpha^{\ A} (F_{\beta\gamma}^{\ A} - \frac{1}{3} C^{ABC} A_\beta^{\ B} A_\gamma^{\ C}) \quad . \tag{2.6}$$

As we will see, the solution of the U(1) problems centers around the existence of this anomaly, since Eq. (2.4) implies there are actually two currents in the ninth channel, the original current A_9^μ and the current

$$\tilde{A}_9^{\ \mu} \equiv A_9^{\ \mu} - 2N_\ell K^\mu \equiv \text{"symmetry current"} \quad . \tag{2.7}$$

These currents have complementary properties, i.e., A_9^μ is gauge invariant, is not conserved in the chiral limit $M \to 0$ (due to the presence of the anomaly) and does not generate the U(1) symmetry transformations. On the other hand, one has that $\tilde{A}_9^\mu$ is not gauge invariant (since K^μ is not gauge invariant), is conserved in the chiral limit (i.e., from Eqs. (2.4) and (2.7), $\partial_\mu \tilde{A}_9^\mu \to 0$ as $M \to 0$), and does in fact generate the U(1) symmetry transformation on operators and states, i.e.,

$$U(1) = e^{i\alpha \tilde{Q}_9} \quad ; \qquad \tilde{Q}_9 \equiv \int d^3x\, \tilde{A}_9^{\ o} \quad . \tag{2.8}$$

III. RESOLUTION OF THE U(1) PROBLEMS

We will conclude our introductory survey by briefly summarizing what are the current resolutions of the U(1) problems within the QCD framework.[6] We start first by examining the $\eta \to 3\pi$ puzzle. This decay which is electromagnetic in origin (see Fig. 1)

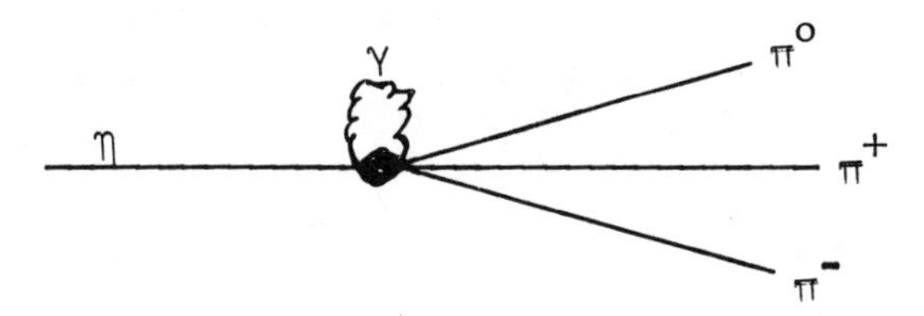

Fig. 1. Diagram for the $\eta \to \pi^+ + \pi^- + \pi^o$ decay.

is governed in the soft pion limit by the third component of the isotopic spin density, i.e.,

$$M_{\eta\to3\pi} \sim \langle\pi^o\pi^+\pi^-| \int dx\ \bar{q}(x)\lambda_3 q(x)|\eta\rangle \quad . \tag{3.1}$$

Contracting the π^o and going to the soft pion limit, one finds, by the usual PCAC analysis,

$$M_{\eta\to3\pi} \sim \frac{1}{F_\pi} \langle\pi^+\pi^-| \int dx\ \bar{q}\gamma^5 q|\eta\rangle$$

$$\sim \frac{1}{F_\pi m_q} \langle\pi^+\pi^-| \int dx\ \partial_\mu \tilde{A}_9{}^\mu|\eta\rangle \quad , \tag{3.2}$$

where m_q is an appropriate quark mass. Note that it is the <u>symmetry</u> current $\tilde{A}_9{}^\mu$ of Eq. (2.7) that enters in Eq. (3.2). We have then

$$M_{\eta\to3\pi} \sim [q_\mu \langle\pi^+\pi^-|\tilde{A}_9{}^\mu|\eta\rangle]_{q^\mu\to0} \tag{3.3}$$

where q^μ is the momentum transfer. Under normal circumstances, one would expect then that $M_{\eta\to3\pi}$ vanishes as $q^\mu \to 0$ (which is the Sutherland theorem[7]). However, Kogut and Susskind[8] suggested that actually $\tilde{A}_9{}^\mu$ couples to a massless ghost pole so that

$$\langle a|\tilde{A}_9{}^\mu|b\rangle = \frac{q^\mu}{q^2} \langle a|G|b\rangle + \text{smooth part} \tag{3.4}$$

where $\langle a|G|b\rangle$ is non-zero as $q^\mu \to 0$. Thus in the soft π^o limit one obtains now a <u>non-zero</u> amplitude

$$M_{\eta\to3\pi} \sim [\langle a|G|b\rangle]_{q^\mu=0} \quad . \tag{3.5}$$

The coupling of the non-physical ghost pole to $\tilde{A}_9{}^\mu$ is feasible since $\tilde{A}_9{}^\mu$ is not a gauge invariant quantity. If one postulates that $2N_\ell K^\mu$ has a similar pole with residue of <u>opposite</u> sign

$$2N_\ell\langle a|K^\mu|b\rangle = -\frac{q^\mu}{q^2} \langle a|G|b\rangle \tag{3.6}$$

then the <u>gauge invariant</u> (and hence observable) current $A_9{}^\mu$ would not couple to the ghost pole and hence no basic principle would be violated.[9] Kogut and Susskind showed that indeed such ghost pole couplings existed in the 2 dimensional Schwinger model and argued that a similar phenomena should arise ("seizing of the vacuum") in 4 dimensional QCD. We will assume in our analysis here that indeed the gluon dynamics of QCD does arrange itself to produce this "Kogut-Susskind ghost pole," and hence it is the non-zero residue at this pole which gives rise to the non-zero $\eta \to 3\pi$ amplitude and hence resolves this puzzle.

The remaining two U(1) problems of Sec. II were more recently examined within the framework of QCD by Witten[10] using the 1/N expansion.[11] What Witten pointed out was that while the U(1) axial anomaly is a 1/N correction to the leading terms, it actually can become important in the chiral limit, where leading terms may vanish. Thus the anomaly produces an additional growth to the η' mass, and hence the total mass has the form

$$m_{\eta'}^2 = \mu^2 + \frac{\text{const}}{N} \quad . \tag{3.7}$$

Here μ is the chiral contribution ($\mu^2 \sim m_q$) and const/N is the addition produced by the anomaly. Thus in the chiral limit $m_{\eta'}^2 \to \text{const}/N \neq 0$, and so the η' is not a Goldstone boson (except in the $N \to \infty$ limit). The additional anomaly contribution to the η' mass presumably accounts for its heaviness. Witten was also able to relate this additional piece to $E(\theta)$, the dependence of the vacuum energy on θ. Thus in the <u>chiral and SU(3) symmetric limit</u> Witten showed from QCD that

$$m_{\eta'}^2 \to \frac{4N_\ell}{F_\pi^2}\left(\frac{d^2E(\theta)}{d\theta^2}\right)^{\text{no quarks}}_{\theta=0} \quad , \tag{3.8}$$

where the "no quark" label means the gluon contribution to $E(\theta)$ (in the absence of quarks).

IV. EFFECTIVE LAGRANGIAN AND THE 1/N EXPANSION

The previous discussion has shown how the U(1) problems may be resolved within the framework of QCD. However, the analyses are qualitative in nature, not quantitative. Thus to actually calculate the $\eta \to 3\pi$ decay rate, one would need to know the matrix elements of the operator G in Eqs. (3.4), (3.5), i.e., the residue at the Kogut-Susskind pole. Similarly, the anomaly mass growth of the η' is discussed only in the chiral and SU(3) symmetric limit. Equally important is the conceptual gap that while all the arguments resolving the U(1) problems involve the U(1) axial anomaly, the interrelation among the various problems remains unclear. We turn now to the construction of effective Lagrangians for this situation which helps answer these questions.

Effective Lagrangians in terms of phenomenological fields describing the physically observed mesons and obeying all current algebra and PCAC conditions, but <u>not</u> including the U(1) anomaly have, of course, been known for over a decade[12],[13]. It is interesting to note that it is possible to relate these older ideas to more fundamental analyses of QCD involving the 1/N expansion.

Recently, Witten[14] has pointed out that if the 1/N expansion does indeed produce confinement, then QCD possesses many qualitatively correct results in the large N limit. In particular, the color singlet meson states have "narrow" widths with $\Gamma \sim 1/N$ and decay and scattering amplitudes proceed through tree and seagull diagrams involving, to leading order in 1/N, <u>only the physical color singlet meson</u> states. Matrix elements of any quark bilinear $J(x) \equiv \bar{q}(x)\Gamma q(x)$ (Γ = arbitrary matrix) between vacuum and 1 meson states $|m\rangle$ obey

$$\langle 0|J(x)|m\rangle \sim \sqrt{N} \tag{4.1}$$

while the different n-point mesic vertex coupling constants $g^{(n)}$ behave as indicated in Fig. 2. For an arbitrary low energy scattering problem, one may separate out from the total amplitude, T, the low mass s,t,u channel pole terms

$$T = \text{low mass poles} + R(s,t,u) \tag{4.2}$$

and approximate the remainder (coming from the higher mass meson exchanges) in the low energy domain by a polynomial: $R \cong a + bs + ct + du + \ldots$. In this approximation,[15] the 1/N expansion then implies that the scattering may be described by an effective Lagrangian (with vertices of Fig. 2) to be used to tree and seagull order.

This is as far as the 1/N expansion can go, since to explicitly determine the coupling constants $g^{(n)}$ would involve actually summing the leading 1/N diagrams. However, one may learn more by including the constraints of current algebra.[16] Since, in fact, the fundamental QCD Lagrangian obeys current algebra, any mesonic matrix elements must obey the current algebra conditions. But this is precisely how one constructs the effective Lagrangian in current algebra.[12,13] Thus the (a priori) unknown constants $g^{(n)}$ of the

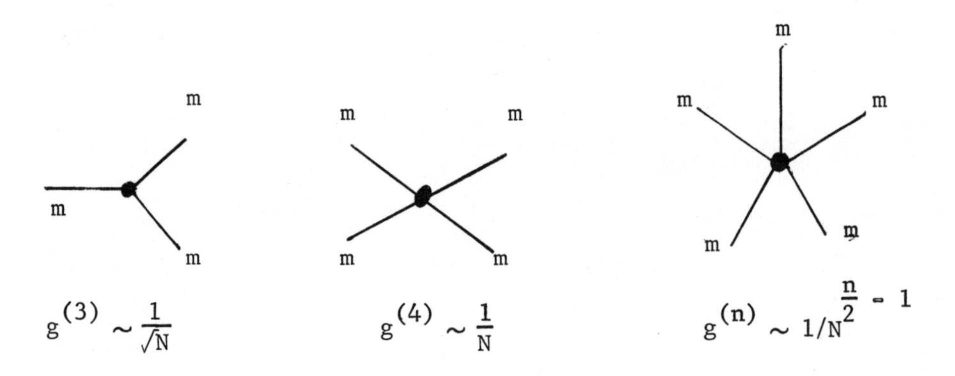

Fig. 2. Behavior of n-point vertices in the large N limit.

1/N expansion must be the ones already determined by current algebra![17] The only reason this might not be true is if the current algebra constraints were inconsistent with the 1/N approximation scheme (e.g., cuts across different orders in 1/N). That this is not the case can be seen from Refs. (12,13). For example, consider the PCAC condition on the axial current $A_a^\mu = \bar{q}\gamma^\mu \frac{1}{2}\lambda_a q$. One may use A_a^μ as an interpolating field for the set of 0^- pseudoscalar mesons $\varphi_a(x)$ and 1^+ axial mesons $a_a^\mu(x)$:

$$A_a^\mu = g_{ab} a_b^\mu(x) + F_{ab}\partial^\mu \varphi_a(x) \quad . \tag{4.3}$$

From Eq. (4.1) one has

$$\begin{aligned} \langle 0|\bar{q}\gamma^\mu \tfrac{1}{2}\lambda_a q|1^+ b\rangle &= g_{ab} \sim \sqrt{N} \\ \langle 0|\bar{q}\gamma^\mu \tfrac{1}{2}\lambda_a q|0^- b\rangle &= F_{ab} i q^\mu \sim \sqrt{N} \quad . \end{aligned} \tag{4.4}$$

For an arbitrary n-point vertex of the effective Lagrangian, $\mathcal{L}^{(n)} = g^{(n)}\varphi_1 a_2^\mu a_3^\nu \ldots \varphi_n$, PCAC requires[18]

$$F_{ab}\frac{\partial \mathcal{L}^{(n)}}{\partial \varphi_b} \sim \frac{\partial \mathcal{L}^{(n-1)}}{\partial \varphi}\varphi + \ldots \tag{4.5}$$

which may be viewed as an equation to determine $\mathcal{L}^{(n)}$ by iteration, i.e., $\mathcal{L}^{(n)} \sim F^{-1}\mathcal{L}^{(n-1)}\varphi \sim \mathcal{L}^{(n-1)}\varphi/\sqrt{N}$. Starting with the mass term, $\mathcal{L}^{(2)} = \frac{1}{2}\varphi_a\mu_{ab}\varphi_b$, which is O(1) in N, one has $\mathcal{L}^{(3)} \sim 1/N^{\frac{1}{2}}$, $\mathcal{L}^{(4)} \sim 1/N, \ldots, \mathcal{L}^{(n)} \sim 1/N^{(n/2-1)}$. This is precisely what was required by the 1/N expansion (Fig. 2). A similar result holds for all the current algebra constraints, showing that current algebra and the 1/N expansion are in agreement with each other at low energies.

V. EFFECTIVE LAGRANGIAN WITH ANOMALY

The previous analysis has discussed the conventional current algebra effective Lagrangian from the viewpoint of the 1/N expansion. It does not include the effects of the U(1) axial anomaly. In order to do this, it is necessary to introduce a phenomenological field $K^\mu(x)$ to represent the topological current density of Eq. (2.6). The general form of the effective Lagrangian which is in accord with the QCD principles of Sec. III is then the following[19]

$$\begin{aligned} \mathcal{L}_{eff} = {} & [-\varphi_a^\mu \partial_\mu \varphi_a + \tfrac{1}{2}\varphi_a^\mu \varphi_{\mu a} - \tfrac{1}{2}\varphi_a \mu_{ab}\varphi_b + \mathcal{L}_{C.A.}(\chi_A)] \\ & + \left[\frac{1}{2C}(\partial_\mu K_\nu)^2 + G(\chi_A)\partial_\mu K^\mu - \theta\partial_\mu K^\mu\right] \quad , \end{aligned} \tag{5.1}$$

where φ_a, $\varphi_a{}^\mu$, a = 1...9 represent the nonet of 0^- meson fields (we are using first order formalism), $\{\chi_A\}$ are the set of all meson fields ($0^\pm$, $1^\pm$ mesons), μ_{ab} are the chiral contributions to the meson masses ($\mu_{ab} \sim m_q$) and C is a constant to be determined. The first bracket represents the total current algebra Lagrangian of the preceding section, and is explicitly calculated in Refs. (12,13). The second bracket represents the new effects due to the topological charge and anomaly. Thus the term $-\theta\partial_\mu K^\mu$ is the CP violating θ-dependent term [i.e., from Eq. (2.5) it represents the last term of Eq. (2.1)]. The term $G(\chi_A)\partial_\mu K^\mu$ is a general expression representing the coupling of the mesic fields to the anomaly (and clearly is needed to account for, among other things, the anomaly mass growth of the η′ discussed by Witten[10] in his QCD analysis). Finally, the $(\partial_\mu K_\nu)^2$ term is precisely what is needed to account for the existence of the Kogut-Susskind[8] ghost pole.[20] To see this, one may vary Eq. (5.1) with respect to K^μ:

$$-\Box^2 K_\mu = C\partial_\mu G(\chi_A) \quad . \tag{5.2}$$

Hence for any two states

$$\langle a|K_\mu|b\rangle = i\,\frac{q_\mu}{q^2}\,C\langle a|G(\chi_A)|b\rangle \tag{5.3}$$

which exhibits the Kogut-Susskind behavior of Eq. (3.6).

Equation (5.3) shows that the function $G(\chi_A)$ of $\mathcal{L}_{eff}$ is the quantity that governs the residue at the Kogut-Susskind pole (and hence by Eq. (3.5) will govern the amplitude for the $\eta \to 3\pi$ decay). We now show that it is the anomaly in PCAC that determines this function. To see this we first recall that in QCD, the quark axial current obeys the PCAC relation Eq. (2.4). The chiral mass term involves the operators $\bar{q}i\gamma^5\lambda_a q$ which are interpolating fields for the pseudoscalar fields $\varphi_a(x)$. Thus in the effective Lagrangian, the PCAC relation must read

$$\partial_\mu A_a{}^\mu = F_{ac}\mu_{cb}\varphi_b + \delta_{a9}2N_\ell\partial_\mu K^\mu \quad , \tag{5.4}$$

where $\mu_{ab} \sim M$ is the 0^- chiral mass matrix (and hence $\mu_{ab} \to 0$ in the chiral limit). Inserting Eq. (4.3) for $A_a{}^\mu$ on the l.h.s., and using the field equations from the Lagrangian of Eq. (5.1) one obtains two equations

$$F_{ab}\,\frac{\partial\mathcal{L}_{C.A.}}{\partial\varphi_b} = -\left[Z_{1abc}\,\frac{\partial\mathcal{L}_{C.A.}}{\partial\chi_b}\,\chi_c + \ldots\right] - F_{ac}\mu_{cb}\varphi_b \tag{5.5a}$$

$$F_{ab}\frac{\partial G(\chi)}{\partial\varphi_b} = -\left[Z_{1abc}\frac{\partial \mathcal{L}_{C.A.}}{\partial\chi_b}\chi_c + \ldots\right] - 2N_\ell\delta_{a9} \quad , \qquad (5.5b)$$

where the + ... represents additional terms involving gradients of the meson fields and the $1^\pm$ vector meson contributions.[21] Equation (5.5a) (which is Eq. (4.5) written out in more detail) comes from equating the parts of Eq. (5.4) independent of the anomaly, while Eq. (5.5b) comes from the terms linear in $\partial_\mu K^\mu$. Equation (5.5a) is not new. It clearly determines the PCAC part of the conventional current algebra effective Lagrangian [and was first derived in Refs. (12,13)]. The new result is Eq. (5.5b) which determines the Kogut-Susskind residue function $G(\chi_A)$. One may iterate it to obtain a series solution

$$G(\chi) = -2N_\ell F^{-1}_{a9}\varphi_a + 2N_\ell F^{-1}_{ar}Z_{1rbc}F^{-1}_{b9}\varphi_a\sigma_c + \ldots \quad , \qquad (5.6)$$

where σ_a are a nonet of 0^+ mesons and + ... represents higher non-linear terms as well as additional structures involving gradient and vector meson couplings.[22]

We will see below that it is the linear term in Eq. (5.6) that gives rise to the anomaly mass growth of the η' discussed by Witten in his 1/N analysis.[10] The quadratic (and higher terms) contribute to the $\eta \to 3\pi$ matrix element of Eq. (3.5). Thus the single Kogut-Susskind residue function accounts for all the different U(1) problems, and the unifying principle that determines $G(\chi)$ is just the anomalous PCAC condition Eq. (5.5b). The form of Eq. (5.5b) shows that if there is an η' anomaly mass growth [i.e., a linear term in Eq. (5.6)] then there must also be a non-zero contribution to the $\eta \to 3\pi$ amplitude [i.e., non-linear terms in Eq. (5.6)]. The two effects are really part of a single item.

Thus, not only does the effective Lagrangian allow one to calculate $G(\chi)$ explicitly so that quantitative calculations can now be made, but it also produces a conceptual clarification in that the solutions of all the U(1) problems are explicitly shown to be unified into a single theoretical framework.

VI. QUADRATIC PARTS OF $\mathcal{L}_{eff}$

From Eqs. (5.1) and (5.6) the quadratic part of $\mathcal{L}_{eff}$ involving the pseudoscalar fields is

$$\mathcal{L}^{(2)}_{eff} = -\varphi_a{}^\mu\partial_\mu\varphi_a + \tfrac{1}{2}\varphi_a{}^\mu\varphi_{\mu a} - \tfrac{1}{2}\varphi_a\mu_{ab}\varphi_b + \frac{1}{2C}(\partial_\mu K_\nu)^2$$
$$- 2N_\ell F^{-1}_{a9}\varphi_a\partial_\mu K^\mu - \theta\partial_\mu K^\mu \quad . \qquad (6.1)$$

$\mathcal{L}^{(2)}_{eff}$ is of interest as it allows one to calculate the two point functions,

$$\Delta_{ab} = i\langle T(\varphi_a(x)\varphi_b(0))\rangle \ , \qquad \Delta_{\mu\nu} = i\langle T(K_\mu(x)K_\nu(0))\rangle$$

$$\tau \equiv i\langle T(\partial_\mu K^\mu(x)\partial_\nu K^\nu(0))\rangle \tag{6.2}$$

and so obtain some physical insights as to the effects of the anomaly. In particular, one may deduce all the results previously obtained from the fundamental QCD analysis[10] in a much simpler fashion, as well as a number of new results.

Equation (6.1) yields $\Delta_{ab} = \delta_{ab}(q^2 + m_a^2)^{-1}$ where

$$m_a^2\delta_{ab} = \mu_{ab} + 4N_\ell^2 C F_{a9}^{-1} F_{b9}^{-1} \quad . \tag{6.3}$$

Here $m_8 \equiv m_\eta$ and $m_9 \equiv m_{\eta'}$ are the physical meson masses. In the chiral and SU(3) limits, $\mu_{ab} \to 0$ and $F_{89}^{-1} \to 0$. Hence $m_\eta \to 0$ but

$$m_{\eta'}^2 \to 4N_\ell^2 C/(F_{99})^2 \neq 0 \quad . \tag{6.4}$$

Thus the η is a Goldstone particle but due to the presence of the anomaly, the η' is not. Note that since $F_{99} \sim \sqrt{N}$, the r.h.s. of Eq. (6.4) has the form const/N, in accord with the QCD result of Eq. (3.7). Equation (6.4) also implies that

$$C > 0 \quad . \tag{6.5}$$

The other two propagators of Eq. (6.2) are evaluated to be

$$\Delta_{\mu\nu}(q) = -C\frac{\eta_{\mu\nu}}{q^2} + \frac{q_\mu q_\nu}{q^4}4N_\ell^2 C^2 \sum_a \frac{(F_{a9}^{-1})^2}{q^2 + m_a^2} \ , \tag{6.6}$$

$$\tau(q) = 4N_\ell^2 C^2 \sum_a \frac{(F_{a9}^{-1})^2}{q^2 + m_a^2} \quad . \tag{6.7}$$

Equation (6.6) exhibits again the Kogut-Susskind pole and on account of Eq. (6.5) one sees that it is indeed a ghost pole. However, the <u>gauge invariant</u> propagator $\tau(q)$ possesses no such poles. In general, the Kogut-Susskind pole cancels from <u>all</u> gauge invariant quantities constructed from the $\mathcal{L}_{eff}$ of Eq. (5.1).

We consider next the relation between the anomaly and the θ dependence of the vacuum energy $E(\theta)$. First, within the framework of QCD, one may show that

$$\left(\frac{d^2E(\theta)}{d\theta^2}\right)_{\theta=0} = \left(\frac{g^2}{8\pi^2}\right)^2 \langle B^i{}_a(0)B^i{}_a(0)\rangle \quad ; \text{ QCD} \tag{6.8}$$

where $B^i{}_a = \frac{1}{2}\epsilon^{ijk}F_a{}^{jk}$ is the color magnetic field. Alternatively, one may show in QCD that Eq. (6.8) is equivalent to

$$\left(\frac{d^2E(\theta)}{d\theta^2}\right)_{\theta=0} = -[q^\mu q^\nu \Delta_{\mu\nu}(q)]_{q=0} \quad , \tag{6.9}$$

where $\Delta_{\mu\nu}$ is the propagator defined in Eq. (6.2).(23) One may also calculate $E(\theta)$ using the effective Lagrangian Eq. (5.1). One finds that Eq. (6.9) is again valid. Now, however, one may insert the explicit form for $\Delta_{\mu\nu}$ of Eq. (6.6) and so find(24)

$$\left(\frac{d^2E}{d\theta^2}\right)_{\theta=0} = C - 4N_\ell{}^2C^2 \sum_a \frac{(F_{a9}{}^{-1})^2}{m_a{}^2} \quad . \tag{6.10}$$

One is now in a position to understand the physical significance of the constant C. In the absence of quarks, i.e., $N_\ell = 0$, one has

$$C = \left(\frac{d^2E}{d\theta^2}\right)_{\theta=0}^{\text{no quarks}} \quad . \tag{6.11}$$

In passing we note that Eqs (6.4) and (6.11) automatically imply Eq. (3.8), which was the basic result obtained from QCD in Ref. (10). Alternatively, one may use Eq. (6.11) to relate C to the QCD gluon effects:

$$C = \left(\frac{g^2}{8\pi^2}\right)^2 \langle B^i{}_a B^i{}_a\rangle^{\text{no quarks}} \quad . \tag{6.12}$$

Finally the quantity

$$\tau(q=0) \equiv \int d^4x \; i\langle T(Q(x)Q(0))\rangle \quad ; \qquad Q \equiv \partial_\mu K^\mu \tag{6.13}$$

represents a measure of the "quantum fluctuations" of the topological charge.(6) From Eqs. (6.7) and (6.4) one finds that in the chiral limit

$$\tau(q=0) \rightarrow C \text{ (chiral limit)} \quad . \tag{6.14}$$

While in many of the discussions in this section we have made use of the chiral and SU(3) symmetric limits (to make contact with previous QCD results), it should be stressed that all the effective Lagrangian formulae hold without making such approximations. We will see in the next section that such approximations may sometimes be quite inaccurate.

VII. EVALUATION OF C AND F_{ab}

The effective Lagrangian discussed above depends upon the following quantities:

$$F_{88} , \quad F_{89} , \quad F_{98} , \quad F_{99} ; \quad C \tag{7.1}$$

It is important to know the numerical size of these constants. Current algebra furnishes three sum rules for the five constants[25]:

$$(F_{88}m_{\eta})^2 + (F_{89}m_{\eta'})^2 = \frac{4}{3}(F_K{}^2 m_K{}^2 + C_{\varkappa}) - \frac{1}{3}F_{\pi}{}^2 m_{\pi}{}^2 \quad , \tag{7.2}$$

$$F_{88}(F_{88} + \sqrt{2}\, F_{98})m_{\eta}{}^2 + F_{89}(F_{89} + \sqrt{2}\, F_{99})m_{\eta'}{}^2 = F_{\pi}{}^2 m_{\pi}{}^2 \quad , \tag{7.3}$$

$$(F_{88} + \sqrt{2}\, F_{98})^2 m_{\eta}{}^2 + (F_{89} + \sqrt{2}\, F_{99})^2 m_{\eta'}{}^2 = 3F_{\pi}{}^2 m_{\pi}{}^2 + \frac{16}{3}N_{\ell}{}^2 C \ , \tag{7.4}$$

where m_K and m_{π} are the K and π masses and $C_{\varkappa}$ governs the amount of SU(3) breaking. To proceed further one must appeal to experiment (see item (3) of Sec. I!) and we use here the η', η and π^o decays into two photons. Experimentally, the amplitudes for these decays are in the ratio

$$A_{\eta' \to 2\gamma} : A_{\eta \to 2\gamma} : A_{\pi^o \to 2\gamma} = 1.9 : 1.0 : 1.3 \tag{7.5}$$

Since the photon decay amplitudes can be expressed in terms of the F_{ab}, Eqs. (7.2)-(7.5) represent five equations for the five parameters of Eq. (7,1). We find[26]

$$F_{88} \cong 1.2F_{\pi} \ , \quad F_{89} \cong -0.25F_{\pi} \ , \quad F_{98} \cong -0.2F_{\pi} \ , \quad F_{99} \cong 1.3F_{\pi}$$

$$C \cong 1.35F_{\pi}m_{\pi}{}^2 \quad . \tag{7.6}$$

While it might appear from Eq. (7.6) that F_{89}, F_{98} and C are relatively small, what appears in PCAC are $F_{89}m_{\eta'}{}^2/F_{88}m_{\eta}{}^2 \simeq 2/3$ and $4N_{\ell}{}^2C \simeq F_{\pi}{}^2m_{\eta'}{}^2$. Thus the SU(3) symmetry breaking and the anomaly produce effects comparable to the normal terms.

As an application of the above results (which also exhibits the largeness of the SU(3) breaking effects) we consider the gluon content of the η and η' [27] which can be calculated from $\Delta_a \equiv i\langle T(\partial_{\mu}K^{\mu}(x)\varphi_a(0))\rangle$. Equation (6.1) yields the result $\Delta_a(q) = 2N_{\ell}C\ F_{a9}^{-1}(q^2 + m_a{}^2)^{-1}$. Since $\partial_{\mu}K^{\mu}$ represents the topological charge density of Eq. (2.5) one has that

$$\langle 0| \frac{g^2}{32\pi^2} F_{\mu\nu}{}^A \tilde{F}^{\mu\nu F} |\eta\rangle = 2N_{\ell}C\ F_{89}{}^{-1} \quad , \tag{7.7}$$

$$\langle 0|\frac{g^2}{32\pi^2} F_{\mu\nu}{}^A \tilde{F}^{\mu\nu A}|\eta'\rangle = 2N_\ell C\, F_{99}{}^{-1} \quad . \tag{7.8}$$

Only the η' has non-zero gluon content in the SU(3) limit where $F_{89}^{-1} \to 0$ (but F_{99}^{-1} remains non-zero). However, this SU(3) symmetric result is misleading. Thus a measure of the gluon content of the η and η' would be the ratio $R_a \equiv \langle 0|2N_\ell \partial_\mu K^\mu|a\rangle / F_{aa} m_a{}^2$ ($a \equiv \eta, \eta'$) and from Eq. (7.6) and Eqs. (7.7), (7.8) we find $R_\eta \simeq 1/3$ and $R_{\eta'} \simeq 1/2$. Thus with the physical symmetry breaking parameters, <u>both</u> the η and η' have a sizeable gluon content.

VIII. STRONG CP VIOLATION

We conclude our discussion with an examination of the strong CP violating effects produced by the θ dependent part of the Lagrangian. In QCD, the Lagrangian, including a quark current algebra mass matrix M, reads

$$\mathcal{L} = -\frac{1}{4} F_{\mu\nu}{}^A F^{\mu\nu A} - \bar{q}\left[\gamma^\mu\left(\frac{1}{i}\partial_\mu - gA_\mu{}^A\lambda^A\right) + M\right] q$$
$$- \theta\left(\frac{g^2}{32\pi^2} F_{\mu\nu}{}^A \tilde{F}^{\mu\nu A}\right) \quad . \tag{8.1}$$

As is well known,[28] in the chiral limit $M \to 0$, one can eliminate the θ-dependence in Eq. (8.1) by making a U(1) <u>symmetry</u> transformation

$$U(\theta) = \exp[i\theta(2N_\ell)^{-1}\tilde{Q}_9] \quad , \tag{8.2}$$

where $\tilde{Q}_9$ is given by Eqs. (2.8) and (2.7). For $M \neq 0$, the transformation Eq. (8.2) still eliminates the θ dependence from the gluon sector of $\mathcal{L}$, and transforms it into the quark sector. Using this technique, Baluni[29] showed that the CP violating interaction to linear order in θ had the form

$$\delta\mathcal{L}_{CP} = \theta\beta\bar{q}i\gamma^5 q \; ; \quad \beta \equiv m_u m_d m_s (m_u m_d + m_u m_s + m_d m_s)^{-1} \quad . \tag{8.3}$$

Note that $\delta\mathcal{L}_{CP}$ is correctly an SU(3) singlet as required by the theorems of Dashen and Nuyts.[30] Calculations of the neutron electric dipole moment D_n based on Eq. (8.3) have been done using the bag model[29] and the soft pion approximation.[31] The experimental upper limit[32] of $|D_n| \leqslant 10^{-24}$ cm then leads to an upper limit on θ:

$$\theta \leqslant 10^{-9} \quad . \tag{8.4}$$

Let us see now how one may carry out a similar analysis in the effective Lagrangian formalism, where we can make direct contact with the physical meson states. One may again make the transformation of Eq. (8.2) on the effective Lagrangian of Eq. (5.1). For small θ, the canonical commutation relations yield

$$U(\theta)\mathcal{L}_{eff}(\varphi_a, K_\mu)U^{-1}(\theta) = \mathcal{L}_{eff}(\varphi_a' - \theta(2N_\ell)^{-1}F_{9a}, K_\mu) \ , \qquad (8.5)$$

where[21]

$$\varphi_a' = \varphi_a - \theta(2N_\ell)^{-1} Z_{19ab}\sigma_b \ . \qquad (8.6)$$

The constant translation of the pseudoscalar fields precisely eliminates the $-\theta\partial_\mu K^\mu$ term in Eq. (5.1) (just as in QCD!) and reinserts the θ dependence into the chiral violating parts of the meson terms (the analogue of the quark mass terms). To obtain the CP violating interaction one must also make the SU(3) × SU(3) transformation that minimizes the quantity[30]

$$F(\beta_i) \equiv \langle 0|V^{-1}(\beta_i)H_{eff}V(\beta_i)|0\rangle \ , \qquad (8.7)$$

where $V(\beta_i) = \exp i\beta_i Q_i^5$, $i = 1...8$. One finds then to linear order in θ

$$\delta\mathcal{L}_{CP} = \theta\tilde{\beta}v_o \ ; \qquad v_o = (\sqrt{Z}\ \varphi)_9 \ , \qquad (8.8)$$

where

$$\tilde{\beta} = [2N_\ell(\sqrt{Z}\ \mu\ \sqrt{Z})^{-1}{}_{99}]^{-1}(F\ \tilde{\sqrt{Z}})_{99} \ . \qquad (8.9)$$

Here $(\sqrt{Z})_{ab}$ is the wave function renormalization matrix defined by Glashow and Weinberg,[33] F_{ab} are the η-η' interpolating constants [evaluated in Eq. (7.6)] and μ_{ab} is the chiral mass matrix [see Eq. (6.2)].

Equation (8.8) is a rigorous equation to linear order in θ for arbitrary current algebra interactions of the 0^+, 0^-, 1^+, 1^- mesons (including derivative couplings required by current algebra). Note that β is proportional to the chiral mass matrix μ_{ab} and so $\delta\mathcal{L}_{CP}$ correctly vanishes in the chiral limit, just as in the QCD formula Eq. (8.3). Also, as in QCD, $\delta\mathcal{L}_{CP}$ is an SU(3) singlet (as required by Dashen and Nuyts[30]). Thus, because of SU(3) breaking $\varphi_9(x) \equiv \eta'(x)$ is not an SU(3) singlet. However, the quantity $v_o = (\sqrt{Z})_{9a}\varphi_a(x)$ appearing in Eq. (8.8) does transform as a singlet under SU(3). Finally, equating the matrix element $\langle 0|[\tilde{Q}_9, \delta\mathcal{L}_{CP}]|0\rangle$ evaluated from QCD with that obtained from the effective Lagrangian yields (for $N_\ell = 3$)

$$[(\sqrt{Z}\ \mu\ \sqrt{Z})^{-1}{}_{99}]^{-1}[(F\ \tilde{\sqrt{Z}})_{99}]^2 = 12m_u m_d m_s(m_u m_d + m_u m_s + m_d m_s)^{-1} \times \langle 0|\bar{q}q|0\rangle \ . \qquad (8.10)$$

Equation (8.10) is an interesting sum rule relating the quark masses and the condensate $\langle\bar{q}q\rangle$ with the phenomenological η-η' parameters.

Perhaps the most important application of Eq. (8.8) would be using it to calculate the neutron electric dipole moment D_n. The above formalism allows for the inclusion of arbitrary chiral and SU(3) symmetry breakdown and for hard pion corrections. An accurate calculation of the neutron electric dipole moment may be of special importance as a reasonably large class of hypercolor interactions predict a spontaneous generation of CP breaking[34] producing contributions to D_n of order 10^{-24} cm, i.e., at the present experimental upper limit.

ACKNOWLEDGMENT: This research was supported in part by the National Science Foundation under Grant Nos. PHY78-09613 and PHY77-22864.

REFERENCES

1. S. Weinberg, Phys. Rev. D13, 974 (1976); Phys. Rev. D19, 1277 (1978); L. Susskind, Phys. Rev. D20, 2619 (1979); E. Eichten and K. Lane, Phys. Lett. 90B, 125 (1980).
2. For QCD, when some form of unitarity is imposed, low energy usually means $\lesssim$ 500 MeV-1 GeV (see e.g., Ref. 4). For technicolor, "low energy" means $\ll$ 1 TeV!
3. Approximate unitarity can be imposed using Breit-Wigner methods and has had reasonable success (see Ref. 4). Perturbative unitarity has been considered by S. Weinberg, Physica 96A, 327 (1979).
4. R. Arnowitt, Proc. of Conference on $\pi\pi$ and $K\pi$ Interactions, Argonne National Laboratory, 1969. P. Nath and S. S. Kere, Proc. of LAMPF Summer School on the Theory of π-Nucleus Scattering, Los Alamos Scientific Laboratory, 1973.
5. S. L. Glashow, "Hadrons and Their Interactions" (Academic Press, N.Y. 1968).
6. For a general review of the U(1) problems see R. J. Crewther, Revista Nuovo Cimento 2, No. 8, 63 (1979).
7. D. G. Sutherland, Phys. Lett. 23, 384 (1966).
8. J. Kogut and L. Susskind, Phys. Rev. D11, 3594 (1975).
9. Alternatively, one might reverse the logic and argue that if topologically non-trivial states exist, then $\langle a|\int dx\, Q(x)|b\rangle$ must be non-zero, and since $Q \equiv \partial_\mu K^\mu$, Eq. (3.6) must hold. Gauge invariance of $A_9{}^\mu$ then requires that a compensating pole exists for $\tilde{A}_9{}^\mu$, i.e., Eq. (3.4). Note, however, as discussed in Ref. (6), instantons produce contributions that are too small to resolve the U(1) problems.
10. E. Witten, Nucl. Phys. B156, 269 (1979).

11. G. 't Hooft, Nucl. Phys. B72, 461 (1974).
12. J. Wess and B. Zumino, Phys. Rev. 163, 1727 (1967); R. Arnowitt, M. H. Friedman and P. Nath, Phys. Rev. Lett. 19, 1085 (1967); C. G. Callan, S. Coleman and B. Zumino, Phys. Rev. 177, 2239, 2247 (1968).
13. R. Arnowitt, M. Friedman, P. Nath and R. Suitor, Phys. Rev. D3, 594 (1971).
14. E. Witten, Nucl. Phys. B160, 57 (1979).
15. Since the 1/N expansion resembles a dual amplitude in that it implies the existence of an infinite number of resonances (Ref. 14), it may not be possible to represent the rigorous amplitude by an effective Lagrangian. Note that the approximate treatment of the higher mass states by a polynomial in s, t and u is precisely what is done in current algebra effective Lagrangians (see e.g., Refs. 12 and 13).
16. Pran Nath and R. Arnowitt, "The U(1) Problem: Current Algebra and the θ-Vacuum," Northeastern University preprint NUB #2417 (1979).
17. Of course, current algebra does not determine all the n-point couplings, but only gives information on those vertices involving mesons that couple to the currents.
18. See, e.g., Eq. (6.2) of Ref. (13).
19. We follow here the general analysis of Ref. (16). There are several other analyses which also discuss effective Lagrangians with anomalies: C. Rosenzweig, J. Schechter and G. Trahern, Syracuse University preprint SU-4217-148 (1979); E. Witten, Harvard University preprint HUTP-80/A005 (1980); P. Di Vecchia and G. Veneziano, CERN preprint TH.2814 (1980). These papers deal only with the spin 0 mesons and with an SU(3) symmetric format (i.e., σ-model) and so cannot discuss a number of the phenomena described below.
20. One may add to Eq. (5.1) an additional term $-(\partial_\mu K^\mu)^2/2D$. All physical predictions remain unchanged provided $D/C \neq 1$ (i.e., the kinetic energy of the K^μ field is non-singular).
21. The constant Z_{1abc} is evaluated in Ref. (13), Eqs. (A.39), (A.40), (C.26). In the SU(3) and chiral limits it is just the SU(3) f_{abc} and d_{abc} constants.
22. If one neglects the gradient and vector meson couplings and the SU(3) breaking effects of F_{ab} and Z_{1abc}, Eq. (5.6) reduces to the σ-model results of the papers listed in Ref. (19). Note also that since $F_{ab}^{-1} \sim 1/\sqrt{N}$, the terms of Eq. (5.6) are in accord with what is expected for $G(\chi_A)$ in the 1/N expansion.
23. Equations (6.8) and (6.9) were calculated in the temporal gauge. The technique for arriving at these results is outlined in the Appendix of Ref. (10). Note that Eq. (6) of Ref. (10) [which reads $(d^2E/d\theta^2)_{\theta=0} = \tau(q = 0)$] should be replaced by Eq. (6.9). As will be seen below, $-q^\mu q^\nu \Delta_{\mu\nu}$ and τ have considerably different properties.

24. In the chiral limit $F_{89}^{-1}/m_\eta \to 0$, and so $(d^2E/d\theta^2)_{\theta=0}$ vanishes in that limit on account of Eq. (6.4). This is in accord with the requirement that the θ dependence of the vacuum should disappear in the chiral limit. (Note, however, from Eq. (6.7) that $\tau(q = 0)$ is non-zero in the chiral limit.)
25. Equations (7.2)-(7.4) without anomaly were first derived in Ref. (13) (see Eqs. C(3), C(16-18) there) and are related to equations subsequently used by S. Weinberg, Phys. Rev. D11, 3583 (1975) in his analysis of the η mass problem. Formulae equivalent to Eqs. (7.2)-(7.4) with the anomaly addition in Eq. (7.4) were first discussed by H. Goldberg, Phys. Rev. Lett. 44, 363 (1980).
26. In order to obtain a solution for F_{ab} and C one must insert in a value for F_K/F_π. In principle this is to be determined from the $K_{\ell 3}$ data [see e.g., R. Arnowitt, M.H. Friedman and Pran Nath, Nucl. Phys. B10, 578 (1969)] but the $K_{\ell 3}$ ξ parameter is still not accurately determined. We have chosen the value $F_K = 1.23F_\pi = 116$ MeV (corresponding to $C_\varkappa = 0$) which minimizes the SU(3) breaking effects (i.e., all other choices correspond to larger values of F_{89} and F_{98}). This predicts a value for ξ of -0.28 in agreement with the $K^\pm$, $K_{\mu 3}^+$ and $K_{\mu 3}^o$ data (which average to -0.27).
27. This question has also been considered by H. Goldberg, Ref. (25).
28. R. Jackiw and C. Rebbi, Phys. Rev. Lett. 37, 172 (1976); C. G. Callan, Jr., R. F. Dashen and D. J. Gross, Phys. Lett. 63B, 334 (1976).
29. V. Baluni, Phys. Rev. D19, 2227 (1979).
30. R. Dashen, Phys. Rev. D3, 1879 (1971); J. Nuyts, Phys. Rev. Lett. 26, 1604 (1971).
31. R. J. Crewther, P. Di Vecchia, G. Veneziano and E. Witten, Phys. Lett. 88B, 123 (1979).
32. W. B. Dress et al., Phys. Rev. D15, 9 (1977).
33. S. L. Glashow and S. Weinberg, Phys. Rev. Lett. 20, 224 (1968). These parameters were evaluated for the η-η' sectors in Ref. (13) [see Eqs. (A39, A42, C26)] in the absence of anomalies. The relevant equations may be easily extended to include the non-zero effects of C.
34. E. Eichten, K. Lane and J. Preskill, Harvard University preprint HUTP-80/A016 (1980).

SUPERFIELD PERTURBATION THEORY

Marcus T. Grisaru

Department of Physics
Brandeis University
Waltham, Massachusetts 02254

ABSTRACT

This is a review of recent work which uses superfield Feynman diagrams to carry out loop calculations in supersymmetric models. I summarize the status of a calculation to compute the 3-loop β-function in O(4) Yang-Mills theory, describe a calculation in the much simpler Wess-Zumino model, and discuss the calculation of anomalies.

INTRODUCTION

I will describe some recent work in superfield Feynman diagram calculations, carried out with M. Rocek and W. Siegel,[1] and L.F. Abbott.[2] Superfield Feynman rules and calculations are not new and in fact date back to the early days of supersymmetry [see ref. 1 and earlier work by Salam and Strathdee, Ferrara and Piguet, Capper, Delbourgo, Fujikawa, Lang, West, Honerkamp and others]. However, motivated by the possibility that the β-function in O(4) supersymmetric Yang-Mills theory[3] vanishes at the three-loop level, as it does at the one- and two-loop level[4] and the fact that a component field calculation seems prohibitively difficult compared to the corresponding superfield one, we have taken a new look at the methods and have come up with some improvements and streamlining of the techiques. We believe, and have partial results to support this belief, that these improved methods will also streamline and simplify calculations in supergravity.[1,5] Here in particular, a background field method which has been developed for the superfield Yang-Mills case[1] and can be extended to supergravity[5] is expected to allow higher-loop locally supersymmetric calculations. It has also allowed us to

*Work supported in part by NSF Grant No. PHY 79-20801.

compute the one-loop, four-particle S-matrix in supergravity.

The 0(4) three-loop β-function calculation is in principle simple: one is dealing with three chiral superfields interacting with a gauge (real) superfield and also self-interacting (with the same, gauge, coupling constant). The Lagrangian is

$$\mathcal{L} = tr\left[\int d^4\theta \, e^{-gV} \bar{\Phi}_i e^{gV} \Phi_i + \frac{1}{64g^2}\int d^2\theta \, W^\alpha W_\alpha \right.$$

$$\left. + \frac{ig}{3!}\int d^2\theta \, \varepsilon_{ijk} \Phi_i [\Phi_j, \Phi_k] + c.c. \right] \quad (1)$$

with

$$V = V^a G_a \,, \quad \Phi_i = \Phi_i^a G_a \qquad i = 1, 2, 3$$

$$W^\alpha = \bar{D}^2 (e^{-gV} D^\alpha e^{gV}) \quad (2)$$

and G_a are the Yang-Mills generators. The action is supersymmetric and also invariant under 0(4) transformations which mix the chiral fields and the gauge field. [In principle a formulation directly in terms of a single 0(4) superfield would be simpler. However such a formulation is not known at the present time.] Since by superfield power counting[1] (or Ward identities[6]) the chiral three-point function is finite, it suffices to show that the two-point function is finite in order to obtain the desired result. Using superfield methods the one- and two-loop vanishing of the β-function[4] can be demonstrated with deceptive ease.[1] However the three-loop calculation involves a large number of diagrams (albeit a fraction of the number required in a component calculation) and has not yet been completed. I shall instead describe, in order to illustrate the methods, the calculation of the three-loop β-function for the pure self-interacting massless chiral multiplet (massless Wess-Zumino model).[2] The whole calculation is simple, and can be carried out analytically.

FEYNMAN RULES

The Wess-Zumino Lagrangian is

$$\mathcal{L} = \int d^4\theta\, \bar{\Phi}\Phi - \frac{\lambda}{3!}\int d^2\theta\, \Phi^3 - \frac{\lambda}{3!}\int d^2\bar{\theta}\, \bar{\Phi}^3 \tag{3}$$

$$\Phi = e^{i\theta\sigma^\mu\bar{\theta}\partial_\mu}\left[\frac{A+iB}{\sqrt{2}} + \theta^\alpha\psi_\alpha + \frac{F-iG}{\sqrt{2}}\theta^2\right] \tag{4}$$

$$\bar{\Phi} = (\Phi)^* \tag{5}$$

$$\bar{D}_{\dot{\alpha}}\Phi = \left(\frac{\partial}{\partial\bar{\theta}^{\dot{\alpha}}} - i(\theta\sigma^\mu)_{\dot{\alpha}}\frac{\partial}{\partial x^\mu}\right)\Phi = 0 \tag{6}$$

[We use two-component spinor notation. See Ref. 1, Appendix A.] In terms of component fields

$$\begin{aligned}\mathcal{L} &= -\frac{1}{2}(\partial_\mu A)^2 - \frac{1}{2}(\partial_\mu B)^2 - \frac{1}{2}\bar{\Psi}\gamma\cdot\partial\Psi + \frac{1}{2}F^2 + \frac{1}{2}G^2 \\ &\quad + \frac{\lambda}{2\sqrt{2}}\left[-F(A^2-B^2) + 2GAB + \bar{\Psi}(A+i\gamma_5 B)\Psi\right]\end{aligned} \tag{7}$$

The Feynman rules are obtained from the generating functional

$$\begin{aligned}Z(J,\bar{J}) &= \int \mathcal{D}\Phi\,\mathcal{D}\bar{\Phi}\; e^{\int \bar{\Phi}\Phi - \frac{\lambda}{3!}\Phi^3 - \frac{\lambda}{3!}\bar{\Phi}^3 + J\Phi + \bar{J}\bar{\Phi}} \\ &= e^{-\frac{\lambda}{3!}\left[\int d^2\theta \frac{\delta^3}{\delta J^3} + \int d^2\bar{\theta}\frac{\delta^3}{\delta\bar{J}^3}\right]}\int \mathcal{D}\Phi\,\mathcal{D}\bar{\Phi}\; e^{\int \bar{\Phi}\Phi + J\Phi + \bar{J}\bar{\Phi}}\end{aligned} \tag{8}$$

Here $J(\bar{J})$ are chiral (antichiral) sources, so that

$$\frac{\delta}{\delta J(x,\theta)} J(x',\theta') = -\frac{1}{4}\bar{D}^2\delta^4(\theta-\theta')\,\delta^4(x-x') \tag{9}$$

with

$$\delta^4(\theta-\theta') = (\theta-\theta')^2(\bar\theta-\bar\theta')^2 \tag{10}$$

The relations

$$\int d^4x\, d^2\theta = -\frac{1}{4}\int d^4x \frac{\partial^2}{\partial\theta^2}$$

$$= -\frac{1}{4}\int d^4x\left[\frac{\partial}{\partial\theta_\alpha} - i(\sigma^\mu\bar\theta)^\alpha\frac{\partial}{\partial x^\mu}\right]\left[\frac{\partial}{\partial\theta^\alpha} - i(\sigma^\mu\bar\theta)_\alpha\frac{\partial}{\partial x^\mu}\right] = -\frac{1}{4}\int d^4x\, D^2 \tag{11}$$

[since the terms we have added are total derivatives] and

$$\bar D^2 D^2 \phi = 16\,\Box\,\phi \tag{12}$$

obtained by using the commutation relations $\{D_\alpha, \bar D_{\dot\alpha}\} = 2i\,\sigma^\mu_{\alpha\dot\alpha}\partial_\mu$ and $\bar D^{\dot\alpha}\phi = 0$, allow us to "complete squares" in eq. (8):

$$\int d^4\theta\,\bar\Phi\phi + \int d^2\theta\, J\phi + \int d^2\bar\theta\,\bar J\bar\Phi$$

$$= \int d^4\theta\left[\left(\bar\Phi - \frac{D^2}{4\Box}J\right)\left(\phi - \frac{\bar D^2}{4\Box}\bar J\right) - D^2 J\frac{1}{16(\Box)^2}\bar D^2\bar J\right] \tag{13}$$

By integration by parts and the antichirality of $\bar J$ the last term is simply $J\frac{1}{4\Box}\bar J$ so that, doing the functional integral in eq. (8) we obtain

$$Z(J,\bar J) = e^{-\frac{\lambda}{3!}\left[\int d^4x\, d^2\theta\frac{\delta^3}{\delta J^3} + \int d^4x\, d^2\bar\theta\frac{\delta^3}{\delta\bar J^3}\right]}\cdot e^{-\int d^4x\, d^4\theta\, J\frac{1}{\Box}\bar J} \tag{14}$$

The following Feynman rules are then obtained: (see Fig. 1)

Propagators:

$$\langle T\,\bar\Phi\phi\rangle = \frac{1}{p^2}\delta^4(\theta-\theta')$$

$$\langle T\,\phi\phi\rangle = 0$$

Figure 1

Vertex: Three point vertex with a $d^4\theta$ integral and factors of $-\bar{D}^2/4$ on any two of the chiral lines entering it [$-D^2/4$ on two of the antichiral lines entering it] except that one such factor is omitted for each external line. [This follows from eq. (9) and the fact that in eq. (14) one has only $\int d^2\theta\ \delta^3/\delta J^3$. One factor $-\bar{D}^2/4$ is used up in converting $\int d^2\theta$ to $\int d^4\theta$.]

In Fig. 2 we compare the old rules and new rules for chiral fields. The advantage is that there are no exponentials to manipulate but only δ-functions and derivatives that can be integrated by parts and "slid" along a propagator from one end to the other by using $D^\alpha_{\theta_1}\delta(\theta_1-\theta_2) = -\delta(\theta_1-\theta_2)\overleftarrow{D}^\alpha_{\theta_2}$. [The rules for the gauge multiplet are unchanged.]

The following useful facts should be kept in mind. Let $\delta_{12} = \delta^4(\theta_1-\theta_2)$. Then

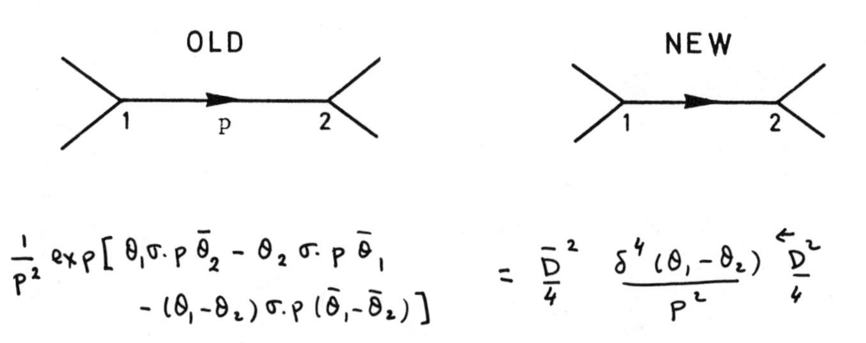

Figure 2

$$\delta_{12}\,\delta_{12} = \delta_{12}\,D^{\alpha}\,\delta_{12} = \delta_{12}\,D^{2}\,\delta_{12} = \delta_{12}\,D^{2}\bar{D}_{\dot{\alpha}}\,\delta_{12} = 0 \quad (15)$$

$$\delta_{12}\,D^{2}\bar{D}^{2}\,\delta_{12} = 16\,\delta_{12} \quad (16)$$

[For more than four D's, we use the commutation relations $\{D^{\alpha}, \bar{D}^{\dot{\alpha}}\} = 2\sigma^{\mu\alpha\dot{\alpha}}\,p_{\mu}$ to reduce their number to four or less.]

From this we have the following lemma: <u>Any loop can be shrunk to a point in θ-space</u>.

Proof: We integrate by parts the D's acting on some line and repeat the procedure until all the lines in the loop have no D's acting on them except one. For example from Fig. 3 we might obtain momentum factors and

$$\delta_{12}\,\delta_{23}\cdots\delta_{n-1,n}\,D^{2}\bar{D}^{2}\,\delta_{n,1} = \\ = 16\,\delta_{12}\,\delta_{23}\cdots\delta_{n-1,n} \quad (17)$$

Furthermore, <u>if any loop</u>, either from the beginning or as a result of integration by parts <u>has fewer than two D's and two $\bar{D}$'s, then it is zero</u>.

The following theorem is then an easy consequence: <u>Any graph is local in θ</u>. For example the effective action at any loop order depends on only one θ (and $\bar{\theta}$) and can be written as

$$\Gamma(\phi) = \int d^4x_1 \ldots d^4x_n\, G(x_1 - x_2, x_1 - x_3 \cdots x_1 - x_n) \\ d^4\theta\, \phi(x_1,\theta,\bar{\theta}) \ldots \phi(x_n,\theta,\bar{\theta}) \quad (18)$$

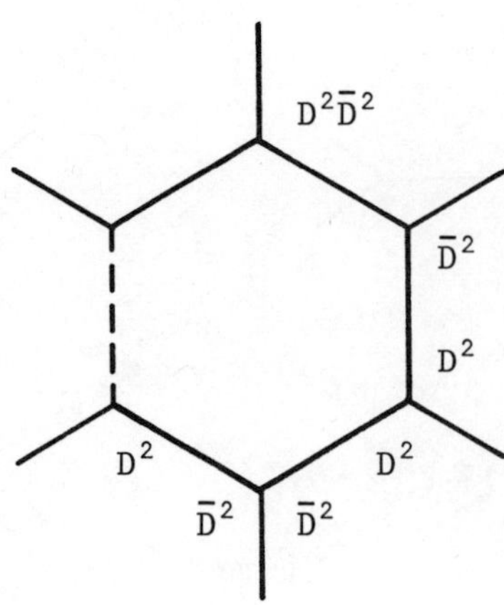

possibly with D_α, or $\bar{D}_{\dot\alpha}$ acting on some of the Φ's. This is because each loop can be shrunk successively to a point until only one point remains in θ-space.

A consequence of this theorem is that supersymmetry is manifest: The quantity above is trivially invariant under the translations

$$\theta \to \theta + \varepsilon \qquad \bar\theta \to \bar\theta + \bar\varepsilon$$
$$x_a \to x_a - i\,\theta\sigma_a\bar\varepsilon - i\,\varepsilon\sigma_a\bar\theta \tag{19}$$

[This is also true for composite operators: For example adding the source $\int \mathcal{Y}(y,\theta,\bar\theta)\,\phi^n(y,\theta)\,\bar\Phi^m(y,\bar\theta)\,d^4\theta\,d^4y$ gives the effective action

$$\Gamma(\phi,\ldots\mathcal{Y}) = \int d^4\theta\, \mathcal{Y}(y,\theta)\,\phi(x,\theta)\ldots \tag{20}$$

Therefore, as we shall discuss later, there is no problem with maintaining global supersymmetry if one regularizes momentum integrals in any manner which maintains the (x-space) translational invariance.]

THE THREE-LOOP β-FUNCTION

We describe now the calculation of the three-loop β-function for the massless Wess-Zumino model.[2] While it has no fundamental interest, it is a good pedagogical example both for superfield calculations and for general renormalization group properties. The calculation is done by doing dimensional regularization (after the θ-algebra has been performed) and using minimal subtraction on the resulting supersymmetric quantity. We have discussed other subtraction schemes in Ref. 2.

We start with

$$\mathcal{L} = \int d^4\theta\, \bar\Phi_0\,\phi_0 - \frac{\lambda_0}{3!}\int d^2\theta\,\phi_0^3 + h.c. \tag{21}$$

where the subscript denotes bare quantities. By power counting[1] (or Ward identities[6]) one knows that the three-point function $\Gamma(\phi,\phi,\phi)$ is finite. Hence, if we define

$$\lambda_0 = \mu^\varepsilon Z_\lambda \lambda \quad , \quad \phi_0 = Z_\phi^{1/2}\,\phi \tag{22}$$

then $\lambda_0\,\phi_0^3$ being finite implies

$$Z_\lambda = Z_\phi^{-3/2} \tag{23}$$

and with

$$\gamma = \frac{1}{2}\mu\frac{\partial}{\partial\mu}\ln Z_\phi$$
$$\beta = -\lambda\mu\frac{\partial}{\partial\mu}\ln Z_\lambda \tag{24}$$

we conclude[7]

$$\beta = 3\lambda\gamma \tag{25}$$

Hence it suffices to calculate Z_ϕ .

We write the two field part of the effective action, in momentum space with $\bar\Phi_0\phi_0 \to Z_\phi\bar\Phi\phi$, as

$$\Gamma(\phi,\bar\Phi) = \int d^4\theta\,\frac{d^4p}{(2\pi)^4}\,\bar\Phi(-p)\,\phi(p)\ .$$

$$\cdot\Big[\, Z_\phi + \frac{1}{2}\left(\frac{\lambda}{4\pi}\right)^2 \qquad + \left(\frac{\lambda}{4\pi}\right)^4 \qquad + \left(\frac{\lambda}{4\pi}\right)^6 \tag{26}$$

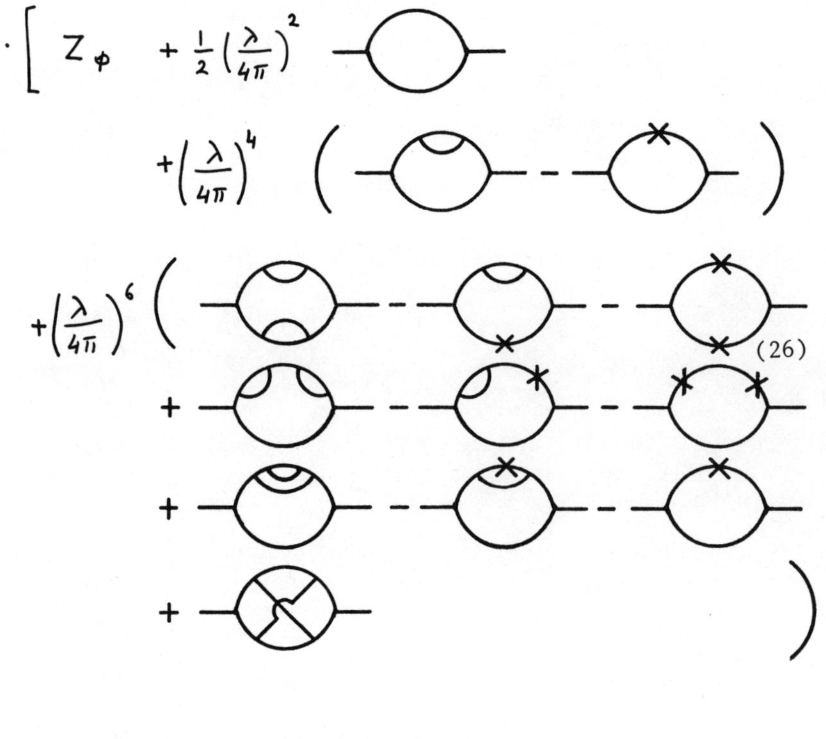

which includes all the non vanishing diagrams in this model. Z_ϕ is chosen so as to cancel all the divergences and the crosses indicate loop by loop subtractions of lower order divergences defined by the expansion

$$Z_\phi = 1 + \left(\frac{\lambda}{4\pi}\right)^2 Z_1 + \left(\frac{\lambda}{4\pi}\right)^4 Z_2 + \cdots \tag{27}$$

The one-loop calculation is trivial. The Feynman rules give, using eq. (16)

$$\Gamma(\text{ONE LOOP}) = \int \bar{\Phi}(-p,\theta)\left[\frac{D^2}{4}\,\frac{\delta^4(\theta-\theta')}{q^2}\,\frac{\overleftarrow{\bar{D}}^2}{4}\right]\frac{\delta^4(\theta'-\theta)}{(p-q)^2}\,\Phi(p,\theta')$$

$$= \int \bar{\Phi}(-p,\theta)\,\Phi(p,\theta)\int\frac{d^4q}{q^2(p-q)^2} \tag{28}$$

The integral, continued to n-dimensions, gives

$$\left(\frac{p^2}{4\pi\mu^2}\right)^{-\varepsilon}\frac{\Gamma(\varepsilon)[\Gamma(1-\varepsilon)]^2}{\Gamma(2-2\varepsilon)}$$

and

$$Z_1 = -\frac{1}{2\varepsilon} \tag{29}$$

For the two loop integral we have

$$\Gamma = \int \bar{\Phi}(-p,\theta)\,\frac{D^2}{4}\,\frac{\delta_{\theta 1}}{q^2}\,\frac{\overleftarrow{\bar{D}}^2}{4}\,\frac{\bar{D}^2}{4}\,\frac{\delta_{12}}{k^2}\,\frac{\overleftarrow{D}^2}{4}\,\frac{\delta_{12}}{(q-k)^2}\,\frac{D^2}{4}\,\frac{\delta_{23}}{q^2}\,\frac{\overleftarrow{\bar{D}}^2}{4}\,\frac{\delta_{\theta 3}}{(p-q)^2}\,\Phi(p,\theta_3) \tag{30}$$

Using integration by parts, "sliding" of D's along propagators and

$$D^2\bar{D}^2D^2\bar{D}^2 = -16\,q^2 D^2\bar{D}^2 \tag{31}$$

the two loop diagram reduces to doing an equivalent scalar loop diagram with one line shrunk to a point (because of the q^2 factor in eq. (31)) as depicted in Fig. 4.

Figure 4

We obtain then

$$Z_2 = -\frac{1}{4\varepsilon^2} + \frac{1}{4\varepsilon} \tag{32}$$

Similar manipulations for the higher loop diagrams lead to an effective action which is $\bar{\Phi}\Phi$ times equivalent scalar loop integrals where some lines have been shrunk to a point, as given by Fig. 5

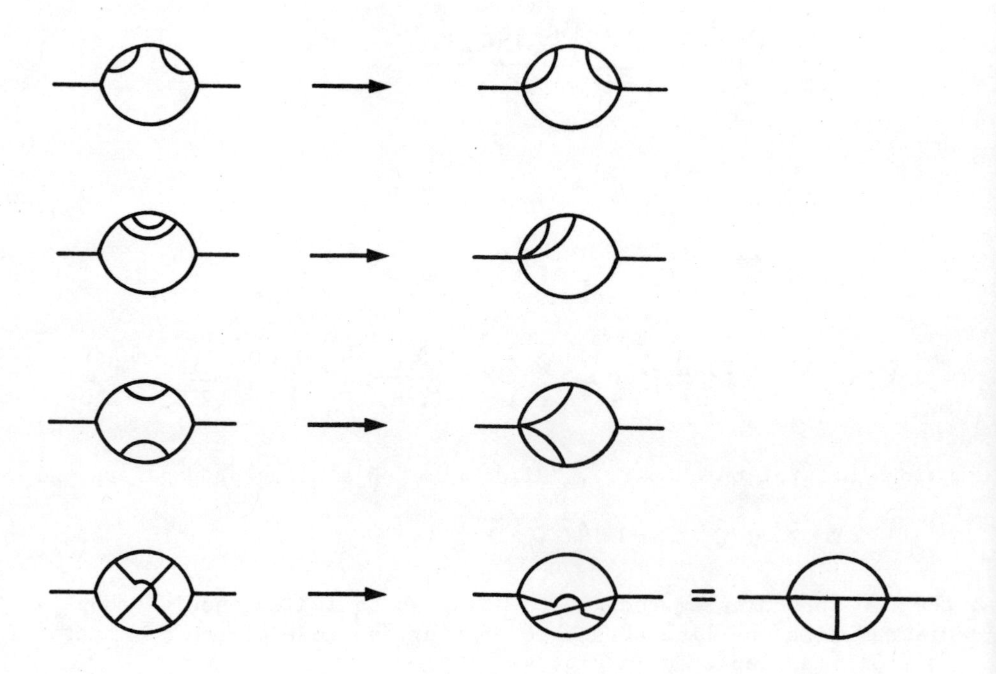

Figure 5

Details may be found in Ref. 2. We find

$$Z_3 = -\frac{1}{24\epsilon^3}\left[5 - 9\epsilon + (5 + 3\zeta(3))\epsilon^2\right] \tag{33}$$

$$\beta(\text{MIN. SUBTR}) = \lambda\left[\frac{3}{2}\left(\frac{\lambda}{4\pi}\right)^2 - \frac{3}{2}\left(\frac{\lambda}{4\pi}\right)^4 + \left(\frac{15}{8} + \frac{9}{8}\zeta(3)\right)\left(\frac{\lambda}{4\pi}\right)^6 + \mathcal{O}\left(\left(\frac{\lambda}{4\pi}\right)^8\right)\right] \tag{34}$$

The Riemann ζ-function comes from the last diagram in Fig. 5. Since everything can be done analytically it is also possible to investigate the dependence of the β-function on the subtraction scheme.

SUPERSYMMETRIC REGULARIZATION

In the above example we have imagined doing all the algebraic manipulations involving the spinor coordinates in four dimensions. Subsequently we have continued the momentum integrals to n-dimensions. If $n < 4$ this is equivalent to regularization by dimensional reduction.[8]

We should emphasize however that any other scheme which maintains (x-space) translational invariance will preserve supersymmetry once we have reduced the θ-integrals to one over a single θ. We can use BPHZ, analytic regularization, Pauli-Villars, or higher derivative. In fact it is very hard not to maintain global supersymmetry, if one does the calculations using superfields. Concerning supergravity, one would expect that any calculation which maintains general coordinate invariance (gravitational gauge invariance) and global supersymmetry, will also maintain local supersymmetry. This will also be the case in calculations done using the background field method[5]: Manifest local supersymmetry is then maintained for the background field functional as long as the regularization method maintains the invariance of the action under background transformations of the quantum fields. Since these latter transformations have essentially a global character [the quantum fields transform covariantly] one expects again that a regularization which respects global supersymmetry will be sufficient.

ANOMALIES

The status of anomalies in supersymmetric theories is still unclear.[9] It is known that the chiral current, supersymmetry current and energy momentum tensor fit into a current superfield so that one expects that in a calculation done supersymmetrically (e.g. with superfields and suitable regularization) the anomalies will also fit in a superfield. (If the system under consideration includes scalar multiplets this statement requires some modification since it is possible then to define more than one chiral current.) One also expects that the anomalies are associated with breaking of superconformal invariance so that any regulator which breaks this invariance will give anomalies.

In superfield language the relevant object is the current $J_{\alpha\dot{\alpha}}(x, \theta, \bar{\theta})$ (or J_{μ}) whose first component is the ordinary chiral current. Violations at the quantum level of the "classical" conservation equation $D^{\alpha} J_{\alpha\dot{\alpha}} = 0$ [or, more generally, if the classical current is not conserved, modifications of the classical equations] correspond to a multiplet of anomalies defined by the right hand side of $D^{\alpha}\langle J_{\alpha\dot{\alpha}} \rangle_{\text{quantum}} = \langle S_{\dot{\alpha}} \rangle_{\text{quantum}}$. In principle one need only calculate matrix elements of $J_{\alpha\dot{\alpha}}$, suitably regularized, and examine the result.

At the one-loop level several methods, and several regularization schemes, are available.

a) One can use Zimmerman normal product methods together with BPHZ techniques to define at the quantum level $J_{\alpha\dot{\alpha}}$, in a manifestly supersymmetric fashion. This has been done by Clark, et al[10] in a series of papers for the self interacting scalar multiplet, and for the scalar multiplet interacting with an Abelian vector multiplet (supersymmetric QED). They demonstrate the presence of multiplets of anomalies, as well as the existence of more than one chiral current for the scalar multiplet. In particular one which fits into a multiplet with the supercurrent and energy momentum tensor and whose anomaly is present at all orders, and another one, which has only a one-loop anomaly (Adler-Bardeen theorem). However the interesting case of (self-interacting) Yang-Mills multiplet has not yet been investigated.

b) One can do a conventional Feynman diagram calculation, with superfields, and regulators which maintain supersymmetry. With some care, supersymmetric dimensional regularization can be used and one should find a supersymmetric result. [Presumably the anomalies, and possibilities of shifting them from one current to another are connected with the "inconsistencies" in dimensional reduction regularization pointed out by Siegel.[11] However the situation could use further clarification.]

c) At the one-loop level at least, there exists, for global supersymmetry models an unambiguous way of regularizing, namely Pauli-Villars. That this can be done for loops of scalar multiplets is clear, since one can write down models for massive scalars. But it can also be done, at least at the one-loop level, for the vector multiplet (Yang-Mills) if one uses background field methods.[1] It is known that in such methods the quantum field (inside the loop) transforms covariantly (without inhomogeneous terms) under background (gauge) transformations so that one can add a mass term to the quantum field (hence a Pauli-Villars regulator field) without destroying the gauge invariance. Superconformal invariance is broken by the mass term and the expected multiplet of anomalies emerges. Whether one can carry this on to higher loops in unclear because of renormalization subtleties, but it would be one way to settle the clash between the Adler-Bardeen theorem (no renormalization of the axial anomaly) and the existence of the higher loop β-function (renormalization of the trace anomaly).

d) Other manifestly supersymmetric methods are in principle available: Adler-Rosenberg, if one can impose on the external fields some gauge invariance constraints and a method, proposed by Siegel, which consists in coupling the quantum system to a manifestly superconformally invariant form of supergravity and searching for quantum violations of the superconformal invariance. In any event, one would always expect to obtain a multiplet of anomalies (unless scalar multiplets are present which can modify the picture somewhat).

REFERENCES

1. M.T. Grisaru, M. Roček and W. Siegel, Nuclear Phys. B159 (1969) 429; see also M. Grisaru, "Supergravity" (eds P. van Nieuwenhuizen and D.Z. Freedman, North-Holland Publ. Co. 1979), p. 1.
2. L.F. Abbott and M.T. Grisaru, "The Three-Loop β-Function for the Wess-Zumino Model", Brandeis preprint, March 1978 (to be published in Nuclear Physics).
3. L. Brink, J.H. Schwarz and J. Scherk, Nuclear Phys. B121 (1977) 77.
4. E.C. Poggio and H.N. Pendleton, Phys. Letters 72B (1977) 200; D.R.T. Jones, Phys. Letters 72B (1977) 1991.
5. M.T. Grisaru and W. Siegel "Supergraphity" (in preparation).
6. J. Iliopoulos and B. Zumino, Nuclear Phys. B76 (1974) 310.
7. S. Ferrara, J. Iliopoulos and B. Zumino, Nuclear Phys. B77 (1974) 413.
8. W. Siegel, Phys. Letters 84B (1979) 193.
9. For a review, see, for example, M. Grisaru, "Anomalies in Supersymmetric Theories" in "Recent Developments in Gravitation" (Cargèse 1978), eds M. Lévy and S. Deser (Plenum Press).
10. T.E. Clark, O. Piguet and K. Sibold, "The Renormalized Supercurrents in Supersymmetric QED", Los Alamos preprint LA-UR-79-3439 (1979).
11. W. Siegel, "Inconsistency of Supersymmetric Dimensional Regularization", Princeton Institute of Advanced Study preprint (1980).

SUPERSYMMETRIC REGULARIZATION AND SUPERCONFORMAL ANOMALIES*

H. Nicolai and P.K. Townsend

CERN, Geneva, Switzerland

ABSTRACT

We review the status of dimensional regularization by dimensional reduction as a regularization scheme for supersymmetric field theories with particular emphasis on the calculation of the superconformal anomalies.

1. INTRODUCTION

When discussing quantum corrections to supersymmetric field theories the question arises: is supersymmetry preserved by these corrections? One way to answer this question is to find a regularization scheme that preserves supersymmetry and then prove its consistency. Alternatively one can attempt to establish by regulator independent means whether an anomaly is unavoidable, and if not to find a regularization prescription with the desired properties. The latter approach is rather less ambitious as we do not regard the regularization prescription itself as a proof that supersymmetry is preserved. In fact, Abbott, Grisaru, and Schnitzer[1] have provided general regularization independent arguments that we can have conservation of the supersymmetry current S_μ if we abandon its spin $3/2$ character, there being a clash between the two properties. That is $\partial_\mu S_\mu = 0$ implies $\gamma \cdot S \neq 0$ and vice versa. From this result we can expect that it will be possible to find a supersymmetric regularization scheme. For such a scheme to be consistent it follows that it must be able to reproduce the *known* anomalies in

* Talk given by P.K. Townsend

the superconformal currents. These currents are the energy momentum tensor, $\theta_{\mu\nu}$, the supersymmetry current S_μ, and the axial current, j^5_μ, which were shown to form part of a supermultiplet by Ferrara and Zumino[2]. Related to this is the "multiplet of anomalies" containing $\gamma \cdot S$, θ^α_α, and $\partial \cdot j^5$ which all vanish classically for a theory without dimensional parameters. As the name suggests however, they are non-zero in the quantum theory because of perturbation theory anomalies. For supersymmetry to be preserved the coefficients of these anomalies should also be related in a particular way as discussed by Curtright[3] and Lukierski[4].

Dimensional regularization was initially avoided because of a general belief in its inapplicability to supersymmetric theories. This was because verification of supersymmetry requires algebraic manipulations specific to four dimensions, such as Fierz rearrangements. Early evidence that dimensional regularization does *not* spoil supersymmetry was a calculation by Curtright and Ghandour[5] of the two loop renormalization constants of a model with one scalar, one pseudoscalar, and one Majorana spinor field. In the supersymmetric limit (the Wess-Zumino model) only one renormalization constant remained. More recently the supersymmetry Ward identities of this model were checked at two loops[6]. The results satisfied the Ward identities not just for the infinite and finite parts but also, surprisingly, for the parts that vanish in the $n \to 4$ limit! These results relied on some calculational rules which are

$$\mathrm{Tr}\,\mathit{1} = 4 \qquad \mathit{1} = \text{unit Dirac matrix}$$

$$\{\gamma_5, \gamma_\mu\} = 0 \qquad \forall\, \gamma_\mu \, .$$

The first of these rules is needed to ensure various boson/fermion cancellations. It expresses the equality of boson to fermion degrees of freedom. The second rule is needed to ensure various scalar/pseudoscalar cancellations typical of supersymmetric theories.

For supersymmetric gauge theories it is not so obvious how we are to maintain the equality of bosons to fermions because in ordinary dimensional regularization a field W_μ has n rather than four components. Siegel[7] proposed that an effective Lagrangian be obtained by dimensional reduction of the $n = 4$ Lagrangian to $n < 4$. By this procedure a vector W_μ becomes an n-component vector $W_{\hat\mu}$ and $(4-n)$ "ε-scalars" $W_{(\mu)}$, but the total number of components of $W_\mu = [W_{\hat\mu}, W_{(\mu)}]$ remains the same. For example, the effective super-Yang-Mills Lagrangian in Feynman-Wess-Zumino gauge becomes

$$\mathfrak{L} = \left\{ -\frac{1}{4}\left(G^a_{\hat\mu\hat\nu}\right)^2 - \frac{1}{2}\left(\partial \cdot W^a\right)^2 + C^{*a}\partial_{\hat\mu}\left(D_{\hat\mu}C\right)^a - \frac{1}{2}\bar\lambda^a\gamma^{\hat\mu}\left(D_{\hat\mu}\lambda\right)^a \right\}$$
$$+ \left\{ -\frac{1}{2}\left[D_{\hat\mu}W^a_{(\nu)}\right]^2 - \frac{1}{4}W^b_{(\mu)}W^c_{(\nu)}f^{abc} - \frac{1}{2}f^{abc}\bar\lambda^a\gamma_{(\mu)}\lambda^c W^b_{(\mu)} \right\} . \tag{1.1}$$

The important point is that this is equivalent to the original four-dimensional Lagrangian except that all derivatives are n-dimensional. This modified form of dimensional regularization is therefore equivalent to the prescription in which all γ-matrix algebra is done in four-dimensions while only the momenta are continued to n-dimensions. Capper, Jones, and van Nieuwenhuizen[8], and also Majumdar, Poggio, and Schnitzer[9] have recently used this prescription to calculate the one loop contributions to a supersymmetry Ward identity of the theory defined by the above Lagrangian. The finite parts satisfy the identity, but only in the dimensional reduction scheme, not if ordinary dimensional regularization is used.

Having established that dimensional reduction, now to be called supersymmetric dimensional regularization, preserves supersymmetry Ward identities, the next question is: can we use it to calculate the superconformal anomalies? Two problems immediately arise.

i) 't Hooft and Veltman[10] used a different γ_5 prescription from that of Eq. (1.1) for their calculation of the anomaly. They had $n > 4$ and a γ_5 which anticommutes with the first $\bar{4}$ γ_μ but *commutes* with the last $(n-4)$ γ_μ. This γ_5 prescription would generate spurious supersymmetry anomalies.

ii) The algebraic identity $\gamma_\mu \sigma_{\alpha\beta} \gamma_\mu = 0$, valid in four dimensions ensures that $\gamma \cdot S = 0$ classically. How then can we get $\gamma \cdot S \neq 0$ while using four-dimensional Dirac algebra? We shall argue later that the index μ in the above identity should be n-dimensional as in ordinary dimensional regularization. If $\sigma_{\alpha\beta}$ is also n-dimensional we have the n-dimensional identity $\gamma_\mu \sigma_{\alpha\beta} \gamma_\mu = (n-4)\sigma_{\alpha\beta}$ leading to the usual result for $\gamma \cdot S$. However, we have a four-dimensional $\sigma_{\alpha\beta}$ and therefore ε-scalar contributions to $\gamma \cdot S$ proportional to $\gamma_\mu \sigma_{\alpha(\beta)} \gamma_\mu = 2\sigma_{\alpha(\beta)} + O(n-4)$. This coefficient does not vanish as $n \to 4$, but the ε-scalar index (β) will produce a single factor of $(n-4)$. Therefore ε-scalars can contribute at one loop to $\gamma \cdot S$ and $\partial \cdot S$.

We have investigated both of these problems in a recent article[11]. The first one is not new. It has been discussed by Chanowitz, Furman, and Hinchliffe[12], whose ideas we have used in our analysis, although we differ with them in detail. The second problem is new and particular to supersymmetric dimensional regularization. We will now consider these two problems in turn.

2. THE AXIAL ANOMALY IN TWO DIMENSIONS

For simplicity in this talk we will consider only the two-dimensional anomaly $\partial_\mu j^5_\mu \propto \varepsilon^{\mu\nu} \partial_\mu W_\nu$ for a fermion loop in an external vector field W_ν. For the four-dimensional case we refer to Ref. 11. The relevant diagram is:

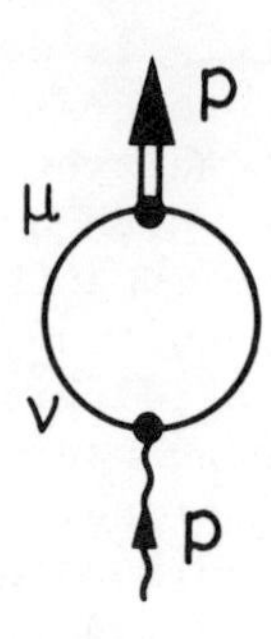

which leads to

$$\left\langle j^5_\mu W_\nu \right\rangle = \int \frac{\mathrm{Tr}\left[\gamma_5 \gamma_\mu \not{\ell} \gamma_\nu (\not{\ell} - \not{p})\right]}{\ell^2 (\ell - p)^2} d^n \ell \ . \tag{2.1}$$

Supersymmetric dimensional regularization tells us that only the momenta p and ℓ are n-dimensional, while all γ-matrix algebra can be done as in two dimensions. Two dimensions is sufficiently simple that we can immediately calculate (2.1), using the usual integrals of 't Hooft and Veltman[10] and the usual two-dimensional γ-matrix algebra

$$\gamma_\mu \gamma_\nu = \delta_{\mu\nu} + i\varepsilon_{\mu\nu}\gamma_5 \ , \quad \gamma_5^2 = 1 \ , \quad \{\gamma_5, \gamma_\mu\} = 0$$

$$\mathrm{Tr}\left(\gamma_\mu \gamma_\nu\right) = 2g_{\mu\nu} \ , \quad \mathrm{Tr}\left(\gamma_5 \gamma_\mu \gamma_\nu\right) = 2i\varepsilon_{\mu\nu} \ . \tag{2.2}$$

Let us first do the integrals. This gives contractions of the γ-matrices with *n-dimensional* $g_{\mu\nu}$'s leading to a factor of (n-2) which cancels the pole in (n-2) of the integral. The final finite result is

$$\left\langle j^5_\mu, W_\nu \right\rangle = 4\pi \left[\varepsilon_{\mu\lambda} \frac{p_\lambda p_\nu}{p^2} - \varepsilon_{\mu\nu} \right] + O(n-2) \ . \tag{2.3}$$

There is nothing dubious about this calculation, nor anything surprising about the result which satisfies for $n \to 2$,

$$p_\nu \left\langle j^5_\mu W_\nu \right\rangle = 0 \ ; \qquad p_\mu \left\langle j^5_\mu W_\nu \right\rangle = -4\pi\varepsilon_{\mu\nu} p_\mu \ . \tag{2.4}$$

This says that the vector current coupled to W_μ is conserved while the axial vector current has an anomalous divergence. What *is* surprising is that had we done the calculation in a slightly different way we would have found a different result! For example, let us first do all the γ-matrix algebra. This produces factors of $g_{\mu\nu}$ which contract the factors of ℓ^μ, p^μ appearing in the integrals.

The $g_{\mu\nu}$ is four-dimensional but only its n-dimensional part is retained when dotted into n-component momenta. Performing the resulting integrals gives

$$\left\langle j_\mu^5 W_\nu \right\rangle = 2\pi \left[\varepsilon_{\nu\lambda} \frac{p_\lambda p_\mu}{p^2} + \varepsilon_{\mu\lambda} \frac{p_\lambda p_\nu}{p^2} \right] + O(n-2) \ . \tag{2.5}$$

This satisfies

$$p_\nu \left\langle j_\mu^5 W_\nu \right\rangle = -2\pi\varepsilon_{\lambda\mu} p_\lambda \ ; \qquad p_\mu \left\langle j_\mu^5 W_\nu \right\rangle = -2\pi\varepsilon_{\lambda\nu} p_\lambda \ , \tag{2.6}$$

which distributes the anomaly equally between the vector and axial vector currents. The result of this calculation is inherently ambiguous because the n-dimensional contraction ($n < 2$) of a two-dimensional expression depends on the precise way in which the two-dimensional expression is written. This ambiguity can be seen to be a consequence of the use of two-dimensional γ-matrix algebra, in particular an anticommuting γ_5, because we could have initially moved the γ_5 in (2.1) from the μ index to the ν index *without* change and then the first of the above calculational routes would have established conservation of the axial current and an anomalous divergence of the vector current. In all cases we get an anomaly, but where it lies is undetermined.

To further examine the consequences of a totally anticommuting γ_5 we shall redo the calculation of $\partial \cdot j^5$ as it is generally done in four dimensions. Accordingly, we multiply (2.1) by $p_\mu = (p-\ell)_\mu + \ell_\mu$ to get

$$p_\mu \left\langle j_\mu^5 W_\nu \right\rangle = - \int \frac{\mathrm{Tr}\left[\gamma_5 (\not\ell - \not p) \not\ell \gamma_\nu (\not\ell - \not p)\right]}{\ell^2 (\ell - p)^2} + \int \frac{\mathrm{Tr}\left[\gamma_5 \gamma_\nu (\not\ell - \not p)\right]}{(\ell - p)^2} \, d^n \ell \ . \tag{2.7}$$

The last integral in (2.7) vanishes on shifting the integration variable, $\ell \to \ell + p$, a manipulation that is allowed in dimensional regularization. Now if we assume in addition that

(1) γ_5 is totally anticommuting;

(2) $\mathrm{Tr}(AB) = \mathrm{Tr}(BA)$; A, B strings of Dirac matrices;

then the first integral in (2.7) can be written in the same form as the second to show that it vanishes also. This would give $\partial \cdot j^5 = 0$. A similar argument shows that also the vector current is conserved, a clear contradiction with the *known* existence of the anomaly. To avoid this we must give up one of the above assumptions. We require assumption (1) for supersymmetry, so we give up assumption (2), the cyclic property of the trace. Before commenting on this unorthodox proposal, let us continue the calculation without using Tr(AB) = Tr(BA). We pull the factor of ($\ell - \not{p}$) in the first integral of (2.7) to the end of the trace using the usual two-dimensional algebra. This gives

$$p_\mu \left\langle j^5_\mu W_\nu \right\rangle = -4i \int \frac{(\ell-p)^\mu \ell^\lambda (\ell-p)^\rho \, d^n\ell}{\ell^2 (\ell-p)^2} A_{\nu\rho\lambda,\mu} \tag{2.8}$$

$$A_{\nu\rho\lambda,\mu} = \left(\varepsilon_{\nu\rho} g_{\lambda\mu} + \varepsilon_{\rho\lambda} g_{\nu\mu} + \varepsilon_{\lambda\nu} g_{\rho\mu}\right) . \tag{2.9}$$

The tensor $A_{\nu\rho\lambda,\mu}$ is antisymmetric in $\nu\rho\lambda$ and so must vanish in two-dimensions. But, as it appears in (2.8), all indices are n-dimensional and $A_{\nu\rho\lambda,\mu}$ need not vanish in n-dimensions. In fact, the n-dimensional trace is

$$A_{\nu\rho\lambda,\mu} g^{\mu\nu} = (n-2)\varepsilon_{\rho\lambda} . \tag{2.10}$$

Doing the integral in (2.8) produces such a trace so that the final result in the limit $n \to 2$ is

$$p_\mu \left\langle j^5_\mu W_\nu \right\rangle = -4\pi \varepsilon_{\mu\nu} p_\mu \tag{2.11}$$

as obtained previously. It is to be expected that this calculation is ambiguous and indeed it is. Different calculational routes lead to additional contributions proportional to tensors antisymmetric in three indices which simply serve to move the anomaly around between the vector and axial vector current. An equivalent ambiguity is already apparent from the fact that we could have started the trace in (2.1) at a different point. This leads to differing results because Tr(AB) ≠ Tr(BA).

Unlike the previous direct calculation this one involves some less obviously allowable manipulations. The result (2.10), for instance, depends on a completeness relation in n-dimensions, i.e. using the notation $\hat{\mu}$ for an index running from 1 to n, that

$$\varepsilon_{\hat{\lambda}\hat{\mu}} \, g_{\hat{\nu}\hat{\rho}} \, g^{\hat{\mu}\hat{\nu}} = \varepsilon_{\hat{\lambda}\hat{\rho}} . \tag{2.12}$$

Our justification for using this relation is really *a posteriori* in that the existence of the anomaly requires (2.10) from which it follows that Tr(AB) ≠ Tr(BA), our initial assumption. One might also wonder whether the unstated assumption that shifting integration variables is allowed may be relaxed. The point here is that without using properties of integrals in n-dimensions one can already derive that the relation $\mathrm{Tr}(\gamma_5\gamma_\mu\gamma_\nu) = 2i\varepsilon_{\mu\nu}$ is inconsistent if the cyclic property of the trace is assumed, so that abandoning this assumption is the minimal relaxation of the usual assumptions that gives the anomaly. We should say here that giving up the *assumption* that Tr(AB) = Tr(BA) does not mean that Tr(AB) ≠ Tr(BA) *always*. This is to be established by calculation. The result is that all traces with an even number of γ_5's *are* cyclic; while those with an odd number of γ_5's are not, except for the simplest case $Tr(\gamma_5\gamma_\mu\gamma_\nu)$. This means that no spurious anomalies will arise, a point particularly emphasized in Ref. 12. Mathematically, the failure of the cyclicity of the trace means that the γ-matrices should not be represented by finite dimensional matrices, but a specific infinite dimensional representation satisfying the desired properties is unknown. This is one fact that makes a genuine proof of the consistency of this regularization scheme difficult.

The fact that supersymmetric dimensional regularization leads to an ambiguous result for the axial current is not necessarily to be regarded as a defect. It is equally a property of every other regularization scheme, and is clearly an unavoidable concomitant of the anomaly. It is plausible that Siegel's recent argument[13] that this regularization scheme is inconsistent is simply a restatement of this ambiguity. It is true that any ambiguity could be regarded as an inconsistency, in which case the only resort is to a regulator independent method of renormalization such as BPHZ. This method has been used successfully in supersymmetric theories by Clark, Piguet and Sibold[14]. However, it is reasonable that we should not have to regard the ambiguity in where to put the anomaly as a real inconsistency. Imposing gauge invariance will allow us to fix anomalous diagrams uniquely.

3. THE SUPERCONFORMAL ANOMALY

The supersymmetry current for super-Yang-Mills theory, now in four-dimensions again, is

$$S_\mu = \sigma_{\alpha\beta}\, G^a_{\alpha\beta}\, \gamma_\mu\, \lambda^a .$$

The first question that must be settled is whether the indices are n or four-dimensional. The natural definition of S_μ within the dimensional reduction scheme is such that the Noether index μ is n-component, while all others are four-dimensional. This follows directly by applying Noether's theorem to the dimensionally reduced Lagrangian. This allows us to get a non-zero result for $\gamma\cdot S$. At first sight it seems obvious that this non-zero result will be the same as obtained using ordinary dimensional regularization, at least at one loop, because if $\gamma_\mu\sigma_{\alpha\beta}\gamma_\mu = (n-4)\sigma_{\alpha\beta}$ produces one factor of (n-4) and ε-scalars another, then it seems that ε-scalars could only contribute at two loops. In fact, this argument is not correct, as pointed out earlier, and the ε-scalars do contribute at one loop. The matrix element of interest is $\langle \bar{S}_\mu(p+q)\lambda^b_\alpha(q)W^c_\rho(p)\rangle$ and the ε-scalar contribution comes from the following graphs

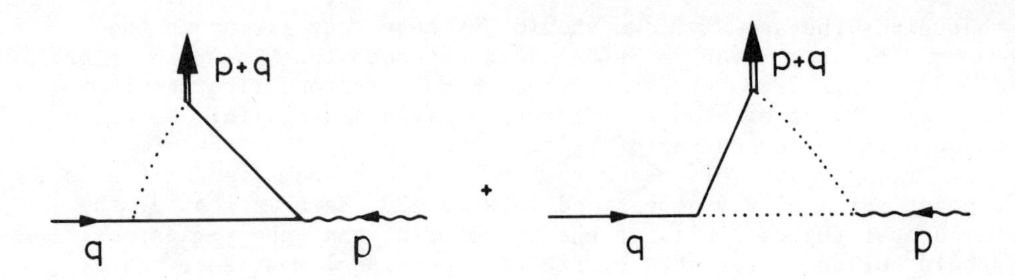

Figure 1.

in which the dotted lines represent ε-scalars. As we have shown in Ref. 11, these contributions give a non-vanishing contribution to the matrix element, but their contributions to $\gamma \cdot S$, $\partial \cdot S$ vanish. Neither do they destroy gauge invariance for the vector W^{ρ}_{μ}. This reduces the calculation to that done by Hagiwara, Pi, and Tsao[15] using ordinary dimensional regularization. These authors find $\gamma \cdot S = 0$ but $\gamma \cdot S = 3N/8\pi^2 \ \sigma \cdot G^a \lambda^a$ for SU(N) (in our notation). They presumably calculated with $n > 4$ rather than $n < 4$, but this should not make a difference. We have redone this calculation and confirm their result.

Had we used a completely four-dimensional S_μ then necessarily $\gamma \cdot S = 0$ so that we should expect to get $\partial \cdot S \neq 0$. This is in fact the result obtained recently by Majumdar, Poggio, and Schnitzer[16]. In rigid supersymmetry this difference is unimportant as no gauge field is coupled to this current. However in local supersymmetry, i.e. supergravity, it is important that the anomaly be in $\gamma \cdot S$ *not* $\partial \cdot S$. Our results are consistent with a recent modification of the dimensional reduction scheme for local supersymmetry proposed by Siegel[17]. The axial superfield $H_\mu(x,\theta)$ that contains the graviton and gravitino fields $h_{\mu\nu}$, ψ_μ, is split under dimensional reduction as $H_\mu \to [H_{\hat{\mu}}, H_{(\mu)}]$. Now $H_{(\mu)}$ must be dropped from the Lagrangian altogether in order not to violate Ward identities of local supersymmetry[17]. This is clearly equivalent to a coupling of supergravity fields to currents whose Noether indices are n-component. Were this not the case ψ_μ would not couple to a conserved current.

We should also mention a recent paper[18] in which the transversality of the gravitino of supergravity was checked at one loop using dimensional regularization. This transversality condition was satisfied only for the supersymmetric form of dimensional regularization.

CONCLUSIONS

We have not yet mentioned the trace anomaly. This is because $\theta^{\alpha}_{\alpha} \propto (n-4)\mathcal{L}$ and ε-scalars provide an extra factor of (n-4), so it is manifest that the one loop result[19] for θ^{α}_{α} will not be affected. Certainly for the axial current and possibly for the others, there is an ambiguity in where to place the anomaly. Imposing physical requirements, such as gauge invariance and $\partial_{\mu}S_{\mu} = 0$ eliminates this ambiguity. We then find the usual results for $\partial\cdot j^5$, θ^{α}_{α} and $\gamma\cdot S$, such that these anomalies form a supersymmetry multiplet at one loop. This result agrees with the superfield results reported at this conference by Grisaru[20]. At two loops the situation is not quite so clear. It seems likely, from our results and from the superfield approach, that the anomalies will continue to form a supersymmetry multiplet. If so, this would be an explicit demonstration that the Adler-Bardeen non-renormalization theorem for the axial anomaly does not apply in this model.

In conclusion we can say that supersymmetric dimensional regularization has been shown to deserve its name.

REFERENCES

1. L.F. Abbott, M.T. Grisaru and H.J. Schnitzer, Phys. Rev. D 16 (1977) 2995.
2. S. Ferrara and B. Zumino, Nucl. Phys. B87 (1975) 207.
3. T. Curtright, Phys. Lett. B71 (1977) 185.
4. J. Lukierski, Phys. Lett. B70 (1977) 183.
5. T. Curtright and G. Ghandour, Ann. of Phys. 106 (1977) 209.
6. P.K. Townsend and P. van Nieuwenhuizen, Phys. Rev. D 20 (1979) 1832;
   E. Sezgin, Nucl. Phys. B, to be published.
7. W. Siegel, Phys. Lett. B84 (1979) 193.
8. D.M. Capper, D.R.T. Jones and P. van Nieuwenhuizen, Stony Brook preprint (1979): Regularization by dimensional reduction of supersymmetric and non-supersymmetric gauge theories.
9. P. Majumdar, E. Poggio and H.J. Schnitzer, Brandeis preprint (1979): The supersymmetry Ward identity for the supersymmetric non-Abelian gauge theory.
10. G. 't Hooft and M. Veltman, Nucl. Phys. B44 (1972) 189.
11. H. Nicolai and P.K. Townsend, CERN preprint TH.2819 (1980): Anomalies and supersymmetric regularization by dimensional reduction.
12. M. Chanowitz, M. Furman and I. Hinchliffe , Nucl. Phys. B159 (1979) 225.
13. W. Siegel, IAS preprint (1980): Inconsistency of supersymmetric dimensional regularization.
14. K. Sibold, in these proceedings, p. 713.
15. T. Hagiwara, S.-Y. Pi and H.-S. Tsao: Regularizations and superconformal anomalies, (unpublished).

16. P. Majumdar, E. Poggio and H.J. Schnitzer, Brandeis preprint (1980): Supersymmetric regulators and supercurrent anomalies.
17. W. Siegel, IAS preprint (1980): Locally supersymmetric dimensional regularization.
18. M.K. Fung, P. van Nieuwenhuizen and D.R.T. Jones, Stony Brook preprint ITP-SB-80-5 (1980): Transversality of the gravitino self-energy, Ward identities, and dimensional regularization.
19. M.J. Duff, Nucl. Phys. B125 (1977) 334.
20. M.T. Grisaru, in these proceedings, p.545.

THE PRODUCTION OF MAGNETIC MONOPOLES IN THE VERY EARLY UNIVERSE

Martin B. Einhorn

NORDITA, Copenhagen, Denmark, and
University of Michigan
Ann Arbor, Michigan 48109

INTRODUCTION

I will talk about the production of magnetic monopoles in the early Universe. Because of time limitations, my talk will not be self-contained. The general properties of magnetic monopoles have been reviewed at this conference by D. Olive[1], and a generalization of the Bogomolny lower bound on their masses has been discussed by D. Scott[2]. In general, a grand unified theory involves a hierarchy of spontaneous symmetry breakdown in which some simple group G is broken down at various energy scales

$$G \to H_n \to H_{n-1} \to \ldots \to H_2 \to H_1 \tag{1}$$

It is presumed that $H_1 = SU(3)_c \otimes U(1)_{em}$ corresponding to QCD $\otimes$ QED, whereas $H_2 = SU(3)_c \otimes SU(2) \otimes U(1)$, where $SU(2) \otimes U(1)$ is the standard model for electroweak unification. Recall that 't Hooft-Polyakov monopoles exist whenever the second homotopy group of the manifold of degenerate vacua is non-trivial[3], i.e., whenever $\pi^2(G/H_j) = \pi_1(H_j) \neq 0$. In particular, this is always true when H_j contains a U(1) factor, so all GUTs give rise to stable magnetic monopoles. Their mass is typically of order[1,2] $(1/\alpha)m_j$, where m_j is the mass of the gauge bosons getting a mass when the remaining unbroken symmetry is reduced to H_j. For example, in the simplest case of SU(5), typically $SU(5) \to H_2$ at a scale for which the vector boson mass $m_X \approx 10^{14}$ GeV. Since $\alpha \sim 1/50$ here, the monopole mass $M \sim 10^{15}$ - - 10^{15} GeV.

Spontaneous symmetry breakdown in GUTs is characterized by a scalar order parameter $\phi = \langle\Phi\rangle$, associated with the vacuum expectation value of some scalar field Φ. Usually, Φ is associated with a fundamental Higgs scalar field appearing in the GUT, although it might be a composite field such as a bilinear product of fermion fields $\bar{\psi}\psi$. At finite temperatures, the vacuum expectation value is replaced by a statistical average[4]

$$\phi_T = \langle\phi\rangle_T = \frac{\mathrm{Tr}\, e^{-\beta H}\phi}{\mathrm{Tr}\, e^{-\beta H}}$$

In principle, ϕ_T is determined by the absolute minimum of the effective potential $V(\phi)$, which is a temperature dependent function. Thus ϕ_T is temperature dependent and so the symmetry can change with temperature. One intuitively expects and typically finds that the symmetry is restored at sufficiently high temperature and that the pattern of symmetry breaking which occurs as the temperature decreases roughly parallels[5] the hierarchy described in Eq. (1). Two such patterns are described in Fig. 1: in Fig. 1a, we illustrate a second order phase transition, characterized by the vanishing of the quadratic term in $V(\phi)$. In mean field theory, near the critical temperature T_c, one has

$$\phi_T = \phi_0 \left[1 - \frac{T^2}{T_c^2}\right]^{1/2}$$

for $T \lesssim T_c$. The scalar Higgs mass is of order $m_H(T) \sim \sqrt{\lambda}\, \phi_T$, where λ is the quartic coupling constant.

The vector bosons which acquire a mass as a result of the symmetry breaking have $m_x(T) \sim g\, \phi_T$, where g is the gauge coupling constant. Monopoles are analogous to magnetic vortices in a type II superconductor. Inside the "core" on the order of the correlation length $\xi_T \sim m_H(T)^{-1}$, the vacuum remains in the symmetric ("normal") phase. It is prevented from collapsing to the broken phase because the magnetic charge ("flux") is conserved. In fact, as pointed out originally by Ginzburg[6,7], because of thermal fluctuations, the Higgs field is effectively not frozen until a temperature T_G somewhat below the critical temperature T_c. The Ginzburg temperature T_G is determined roughly by the requirement that the difference in energy density between the symmetric and broken phases (condensation energy) over a volume on the order of ξ_T^3 be of the order of the thermal energy T. Above this temperature, the Higgs field is so widly fluctuating that magnetic charge conservation is meaningless. One can show[7] that

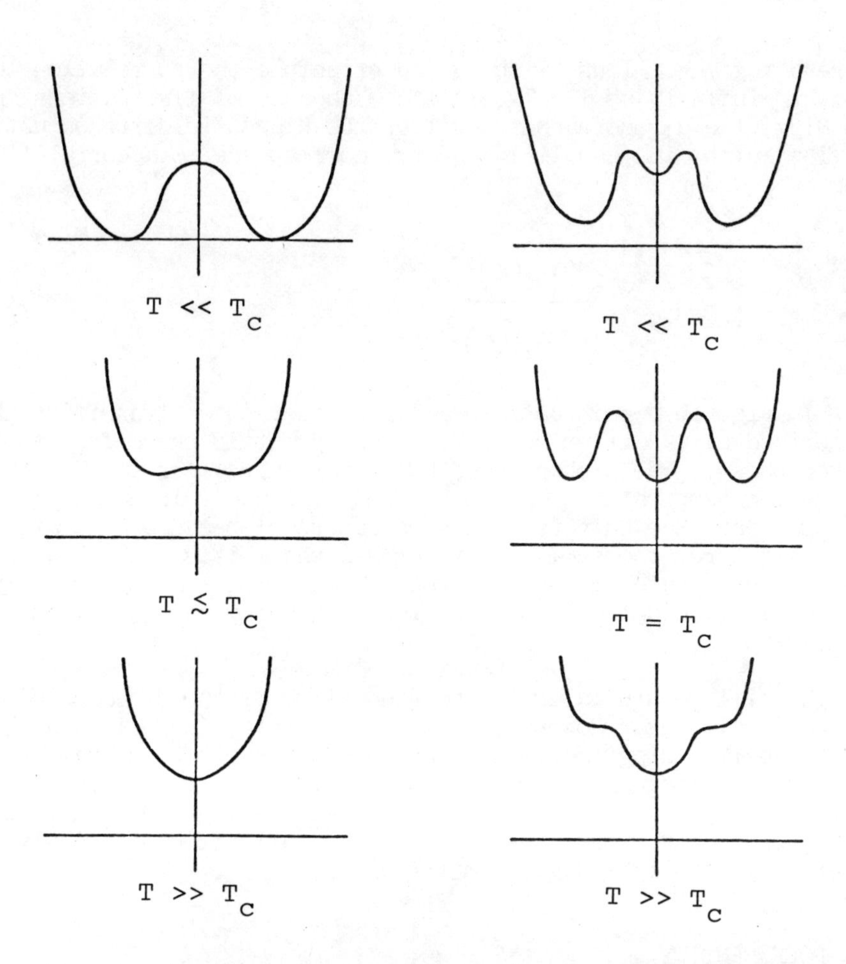

Fig. 1: Temperature dependence of effective potential
(a) Second-order phase transition
(b) First-order phase transtion.

$$1 - \frac{T_G}{T_c} \sim O\left(\alpha\left(\frac{m_H}{m_X}\right)^4\right)$$

where $\alpha \sim g^2/4\pi$. Thus, although in principle the correlation length tends to infinity as $T \to T_C$, in fact, because of fluctuations in the Higgs field, one cannot consider the Higgs field frozen until $T \sim T_G$, which, in mean field theory, corresponds to a correlation length of order

$$\xi_{T_G} \sim \frac{m_X(0)^2}{g\, m_H(0)^3} \tag{2}$$

In Fig. 1b, we illustrate a first order phase transition characterized by a barrier separating the symmetric phase from the broken phase. The broken phase has lower energy for $T \leq T_C$, the transition temperature. As emphasized by Linde[4], in cosmological applications, the Universe tends to supercool because the rate of transition from the symmetric to broken phase is small compared to the expansion rate of the Universe. Transitions can occur by means of thermal fluctuations in ϕ or else by quantum tunnelling through the barrier.

We will return later to the discussion of the dynamics of the phase transitions. However, we would not that, in their analysis of the Abelian Higgs model, Kirzhnits and Linde[8] found that the symmetry breaking phase transition was first order in general. However in fact it could be treated as effectively second order for relatively heavy Higgs masses ($m_H \sim m_X$), while it was strongly first order only for light Higgs ($m_H^2 << 2g^2 m_X$).

MONOPOLE DENSITY

Ellis[10] outlines elsewhere the cosmological limitations on the monopole density and some of the issues involved in estimating their production rate. If the number density were determined by statistical equilibrium, then it would be proportional to

$$\exp(-m(T)/T) \sim \exp(-m_X(T)/\alpha T) \tag{3}$$

Basically, a second-order phase transition gives too many monopoles because $m_X(T_C) = 0$, and so there is no exponential suppression[11,12]. A natural solution would be a first order phase transition occurring effectively at a temperature T_C. In this case, the Higgs field jumps suddenly to the broken phase, so that $m_X(T_C) \approx m_X(0)$. In this case, the exponential can be very small. For $T_C \sim m_X$,

$$\exp(-m(T_C)/T_C) \sim \exp\left(-\frac{1}{\alpha}\frac{m_X}{T_C}\right) \sim \exp(-50) \tag{4}$$

Even in the case of a second order transition, if the Ginzburg temperature T_G is sufficiently below T_C, some suppression may arise. In this case, if the monopole density is given by its value in statistical equilibrium, then it would be suppressed by a Boltzmann factor

$$e^{-\frac{M(T_G)}{T_G}} \sim e^{-C\frac{\sqrt{\lambda}}{g}} \sim e^{-C\frac{m_H}{m_X}} \tag{5}$$

Ellis reported[10] an as yet unpublished calculation by Bais and Rudaz of the constant C in the exponential. It is claimed that C can be rather large, so that for large Higgs masses, considerable sion is possible. Unfortunately, in our opinion, the situation is more complicated than this statistical argument suggests.

When the Higgs field becomes frozen in the broken phase, there is a finite distance d over which the field is correlated with itself. By definition, at distances larger than d, the Higgs field may point in different directions in the manifold of degenerate vacua. Following Kibble[7], we[13] have argued that a certain minimum density of monopoles is to be expected on this account alone. Monopoles (like vortices in a superconductor) can be thought of a measure of the disorder remaining in the system, symmetric (normal) regions trapped by flux quantization in the broken (superconducting) ground state.

To understand how this comes about, consider the Abelian Higgs (Landau-Ginzburg) model in two-space dimensions. The order parameter is a complex scalar field which in the broken phase has a definite magnitude. The vacuum degeneracy is characterized by the freedom of orientation of the phase of the field. Let us partition the space into cells of diameter d. Within each cell, the phase is presumed to be fixed but there is no correlation between the orientation of the phase from one cell to the next. Then consider the situation when three such cells surround a point (Fig. 2a): in each of the three regions, the phase may point in different directions. One can infer that a vortex will be trapped in the centre if the phase rotates by a multiple of 2π as one encircles the centre. Although uncorrelated from the centre of one cell to the next, in fact the phase turns continuously as one passes across cell boundaries. The minimum degree of disorder will result if we assume as little as possible (Fig. 2b). Now the question is, given random assignments of the phase in each cell, how often will the phase rotate by 2π as one follows it around the centre ? A simple calculation shows that this occurs with a probability $p = 1/4$. Therefore the number density of vortices in the plane will be of order pd^{-2}. A similar argument may be applied to an arbitrary gauge theory with vortices (non-trivial first homotopy) group $\pi_1(G/H)$.

It is a simple matter to extend these considerations to monopoles in three space dimensions. Now the cells may be thought of as spheres of diameter d[14], and magnetic flux trapping may be infered by considering four such cells surrounding a central point. The simplest such theory is the Georgi-Glashow SO(3) model[15] in which the manifold of degenerate vacua corresponds to the different orientations of a three-vector of fixed length, i.e., a sphere. Assuming again the vector twists as little as possible in going from one cell to the next, then one finds[7] a non-trivial winding with probability $p = 1/8$. In general, we will obtain a number density of monopoles of order

$$n_M \sim pd^{-3}$$

where p is some group theoretic factor, which we continue to suppose is of order 10^{-1}, and where d is the "phase" correlation length.

At the densities of interest to us, it has been shown[11,12] that the subsequent annihilation of monopole-antimonopole pairs due to magnetic Coulomb attraction is essentially negligible, at least until the time when gravitational interactions cause the formation of galaxies, stars, etc.

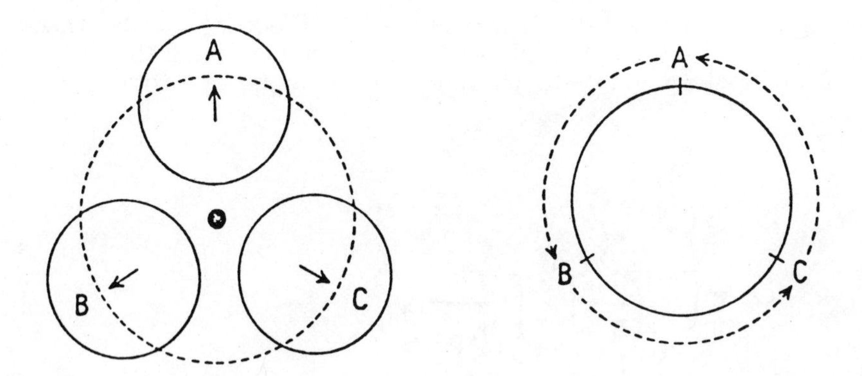

Fig. 2: Two-dimensional U(1) model for vortex trapping
(a) three causal regions A,B,C around a normal point
(b) behaviour of phase of the order parameter in traversing around normal point from A to B to C to A.

So the problem of estimating the production of monopoles in the phase transition has been reduced to a determination of the correlation length d. Now the intrinsic range of coherence of the magnitude of the Higgs field is of order $m_H(T)^{-1}$, while the correlation length of the phase is of order $m_x(T)^{-1}$. In a second order phase transition, these lengths approach infinity at $T = T_C$; however, there are a variety of reasons to suppose that there are some finite limits to the actual length of correlation in a Universe expanding at a finite rate.

I shall review very briefly different estimates of the correlation length. Our first argument[13] avoids complicated dynamics and simply stems from the fact that because the age of the Universe is finite and the speed of light is finite, two points in the Universe which are sufficiently far apart can have no common ancestor, i.e., there is a causal limit d to the distance over which a signal could have passed. One would certainly think this distance, which is closely related to the notion of a particle horizon, provides an upper limit to d. In a radiation-dominated era, one can show this distance as[16]

$$d_0 = 4t = \left[\frac{45}{\pi^3 N}\right]^{1/2} \frac{m_{pl}}{T^2} \qquad (6)$$

where N is the number of spin degrees of freedom[17] in the theory and $m_{Pl} = 10^{19}$ GeV is the Planck mass. Thus we find the ratio r_M of the monopole density to photon density to be of order

$$r_M \sim p(dT)^{-3} \sim p\left[\frac{\pi^3 N}{45}\right]^{3/2}\left[\frac{T}{m_{pl}}\right]^3 \tag{7}$$

Assuming $p \sim 10^{-1}$, $N \sim 10^2$ (as in SU(5)), one finds for temperatures on the order of 10^{14} GeV, $r_M \sim 10^{-13}$. This is to be compared with the upper limit[12] allowed by cosmology of 10^{-19}. Since r_M in Eq. (7) falls like T^3, monopoles could not have been produced above 10^{12} GeV, unless some subsequent phase transition caused them to become unstable or annihilate[17]. It would appear to us that that the usual SU(5) theory is in conflict with cosmology if the symmetry breaking transition at the unification mass scale (> 10^{14} GeV by all estimates) is second-order.

Despite our personal belief that the causal length (Eq. (6)) is a generous overestimate of d, it is possible to dispute its validity. The notion of causal domains in the very early Universe is discomforting, since it renders the cosmological principle (homogeneity and isotropy) implausible. So one school of thought suggests that, due to quantum gravity effects, causal domains must be eliminated.

So we will proceed to give a couple of dynamical estimates of d for a second order phase transition, keeping in mind that, to be compatible with Preskill's upper bound, we would have to increase d_0 by a factor of 100. We have not found any augument which does so. One way to obtain a dynamical estimate of d is to argue, following Kibble[7], that for $T_G < T < T_C$, fluctuations in the magnitude of ϕ are quite likely to occur so that normal regions are being copiously created. Only for temperatures below T_G can spontaneous symmetry breakdown be considered well established. Therefore, the initial size of the cells will be of order $\xi_{T_G} \sim m_H(T_G)^{-1}$, and is given in Eq. (2). Therefore, the monopole-photon ratio will be of order ($g^2 m_X^2 << m_H^2 < m_X^2$)

$$r_M \sim p\left(\frac{g m_H^3}{m_X^2 T_G}\right)^3 \tag{8}$$

For $T_G \sim T_c \sim m_H/g$, we have

$$r_M \sim p\left(\frac{g m_H}{m_X}\right)^6 \tag{9}$$

Since we require $r_M < 10^{-19}$, we have

$$\frac{g m_H}{m_X} < 10^{-3} \tag{10}$$

i.e., the monopole density would be sufficiently suppressed only for sufficiently light Higgs. But, at least in the Abelian Higgs model[8], the phase transition is effectively second order only for $m_H \gg g\, m_X$, so this would be compatible with Eq. (10) only for $g^2 \ll 10^{-3}$, a coupling much smaller than anticipated. This is our first hint that a first-order phase transition may be required.

Another estimate of d may be obtained as follows[13]: for $T \to T_c$ from above, the mean field correlation length ξ_T diverges as $|T^2 - T_c^2|^{-\frac{1}{2}}$. However, the Universe is presumed to be expanding at a certain rate, so d ξ_T/dt also diverges as $T \to T_c$ from above. Thus the mean field formula is expected to break down when this velocity exceeds the ultimate velocity allowed, the speed of light. So if we cut off ξ_T when d $\xi_T/dt = 1$ and estimate its subsequent growth by assuming that it proceeds at the speed of light, then we can estimate its magnitude at a later time. Assuming the rate of expansion of transition, we then find[13]

$$r_M \sim p\left[\frac{N\pi^3}{45}\right]^{1/2} \frac{m_H(0)^2}{m_{pl} T_c} \tag{11}$$

For $m_H(0) \sim T_c \sim 10^{14}$ GeV, $r_M \sim 10^{-6}$.

Another way to see how difficult it is to satisfy the bound at such high temperatures is to note that if we simply identity $d \sim \xi_T$ and require $p(T\, \xi_T)^{-3} < 10^{-19}$, then $T\, \xi_T > 10^6$ for $p \sim 10^{-1}$. If we were to take the mean field formula at face value, then this would mean that

$$1 - \frac{T}{T_c} < \left(10^{-6} \frac{T_c}{M_H(0)}\right)^2$$

which for $M_H > 10^{-1}\, T_c$ (a generous estimate for a second-order phase transition), leads to $1 - (T/T_c) < 10^{-10}$. It is unlikely that the system has even returned to thermal equilibrium for T so close to T_c.

It seems that the only hope of maintaining such a symmetry breaking at $T_c \gtrsim 10^{14}$ GeV is to require a strong first order phase transition, which, in turn, requires[18] a relatively light Higgs mass $m_H \lesssim g\, m_X(0)$. One can imagine many modifications of GUTs which would resolve the monopole problem but first it is interesting to determine whether a scenario like SU(5) is at all compatible.

FIRST-ORDER PHASE TRANSITION

I do not yet know whether the first-order phase transition is sufficient to save SU(5). Let me describe for you what is involved in these considerations[19]. The scenario is as follows: because of the existence of a barrier (see Fig. 1b) between the symmetric and broken phase, the vacuum supercools below the transition temparature T_c [20]. Critical bubbles of the broken vacuum begin to form because of the thermal fluctuations[21] or quantum tunneling[22]. So long as the nucleation rate for bubbles is not too high, the separation between bubbles will usually be large compared to the correlation length ξ_T of the Higgs field. In this case, the orientations of the Higgs field inside different bubbles will be independent of each other. Now the bubbles expand either (A) until they find the entire Universe or (B) until the symmetric minimum disappears at a temperature $T_1 << T_c$. This latter possibility resembles a second-order phase transition in so far as it involves a zero Higgs mass, but resembles a first-order transition in so far as about fluctuations in the Higgs field below T_1, although one does need to be concerned about what happens to this latent heat.

In case (A), the latent heat initially goes into acceleration of the bubble wall and provides the energy which drives the bubble expansion[22]. We have not analyzed how this energy is thermalized through collisions of bubble walls and interactions between bubbles and the light particles running around. The simplest assumption we can make is that no monopoles are created in this process and that we may identify d with the mean size of the bubbles at the time they fill the Universe[23]. This means that the number of monopoles would be approximately the probability p times the number of critical bubbles which were nucleated.

In case (B), it is less clear what happens to the latent heat and what distance to associate with d when the Higgs mass vanishes. Presumably, even though there is no Ginzburg temperature, there is a finite rate at which the Higgs field at a point can grow from zero to its equilibrium value. And, in addition, there will be a certain number of monopoles created because of the nucleation of bubbles as the temperature cooled from T_c or T_1.

The thermalization of the latent heat tends to increase the temperature again, producing more photons and further reducing the monopole/photon ratio[23]. Since one knows how much energy was released and how much energy has gone to monopoles, one can estimate how high temperature this Universe will attain.

THE COSMOLOGICAL CONSTANT

Still to be calculated is the nucleation rate and the effect of all this on the cosmological expansion. Unlike the second-order case, where it was argued that the cosmological constant could be ignored, this is not true in the present case. The point is that, because the cosmological constant is so small today, the vacuum energy density in the symmetric phase has to be exceedingly carefully adjusted so as to be cancelled by the energy released in the symmetry breaking. This fine adjustement is exceedingly unnatural but necessary until a better theory of gravity is available[24].

Fortunately, the effect of the cosmological term on the expansion in a first-order phase transition has been analyzed recently by Sato[25], whose results I will summarize now.

In general, as is well known, the effect of the cosmological term is to cause the cosmic scale factor to increase exponentially after the Universe has expanded beyond the point where the radiation energy density dominates the vacuum energy density. As critical bubbles form, we have the cosmological term inside the bubble much smaller than outside. Thus the Universe would tend to expand much more rapidly in the space outside the bubbles making it that much more difficult for bubbles to fill the Universe. To simplify the situation, Sato takes a coarse grain approach and simply considers the average value of the cosmological term in a large region of space. The equations are still quite complicated, so further approximations are called for. The nucleation rate due to thermal fluctuations[21] involves a barrier factor of $\exp(-E(T)/T)$, where $E(T)$ is the energy of a critical bubble. Although temperature dependent, $E(T)$ is rather slowly varying except right near T_c, so we expect the thermal nucleation rate to peak slightly below T_c, rapidly approaching $\exp(-E(0)/T)$; this is approximate[26] as $p_T\delta(1 - T/T_c)$. The quantum nucleation rate[22] involves the barrier factor $\exp(-S(T))$, where $S(T)$ is the four-dimensional action of a critical bubble. It is a slowly varying function of temperature below T_c, and is approximated as $p_Q\ \theta(T_c - T)$. For the time being, we will assume the symmetric minimum persists all the way down to very low temperatures. Then Sato's results may be simply stated: there is a critical value of p_T above which the phase transition is rapidly completed (but which would lead to far too many monopoles). For values of p_T below this critical value, the phase transition would never finish if it were due to thermal fluctuations ! Because of the exponential increase of the scale factor, the bubbles created near T_c simply cannot catch up with the expansion of the symmetric Universe. However, because of the persistent quantum fluctuations, eventually the phase transition will terminate. Although one would appear to be able to satisfy the upper limit on monopole production, it is not yet clear to me whether this is a believable scenario[27].

Even more interesting is the case when the symmetric minimum disappears at some finite temperature T_1. In this case the phase transition is suddenly and quickly terminated. If T_1 is sufficiently below T_c, it would appear to provide the suppression required[27].

CONCLUSIONS

Since this is an area under active investigation it is too soon to draw firm conclusions. I tend to regard SU(5) as "unnatural" in the sense of Weinberg[28], 't Hooft[29] and Dimopoulos and Susskind[30]. However, if we entertain it as a serious GUT, then unless the Higgs mass is sufficiently small, we expect the phase transition to be effectively second order. In this case, SU(5) seems to be in trouble with the standard cosmology because of the production of too many monopoles. In the case of light Higgs (which is unnatural[29] if it is composite) a strong first order transition is expected. It is not yet clear whether this is compatible with the cosmological upper bound on the monopole density, but it is worth noting that discussions of the production of the baryon excess will require modificatic

It is worth entertaining the possibility that SU(5) or similar GUTs with a "desert" from 100 GeV to 10^{14} GeV will remain in conflict with the standard cosmology. From the point of view of particle theory, the most natural suggestion[29,30,31] is that the desert is replete with states with new forces which provide for the composite scalar order parameters which presently appear as fundamental scalar fields in GUTs. The ultimate GUT will be much larger than SU(5) and, although it will have monopoles, it is not clear at what energies they will be produced. It is also possible that, because of successive phase transitions in the desert region, monopoles produced at 10^{15} GeV will become unstable at lower energies.

Modifications of cosmology can be imagined. Maybe the Universe has a limiting temperature, so it was never in the symmetric state. Perhaps monopoles could be gathered in gravitational clusters where $M\bar{M}$ pairs annihilation (inside galaxies and stars) would reduce the monopole density. The enormous energy so released would presumably cause the Universe to heat up again, so that the galaxies we see today might have formed when the Universe cooled down again. Further thought will have to be given to these possibilities.

ACKNOWLEDGEMENTS

I am much indebted to my collaborators D. Stein and D. Toussaint for filling in some of the gaps in my understanding. I have enjoyed numerous conversations at this conference. I would especially like to thank G. Steigman for discussions about cosmological constraints and to J. Ellis for criticism and for informing me of the work in progress by Bais and Rudaz. I also have benefited from discussions with K. Sato at NORDITA. This work was partially supported by the U.S. Department of Energy.

REFERENCES

1. D. Olive, these proceedings p. 451.
2. D.M. Scott, DAMTP Cambridge preprint 80-2 (1980).
3. See, e.g., S. Coleman, "Classical Lumps and Their Quantum Descendants", in new Phenomena in Subnuclear Physics, ed. A. Zichichi, (Plenum Press, N.Y., 1977).
4. A review of spontaneous symmetry breakdown in finite temperature field theory may be found in A.D. Linde, Rep. Prog. in Phys. 42 (I) (1979) 389.
5. It is not necessary that the pattern precisely follow Eq. (1).
6. V.L. Ginzburg, Yad. Fiz. 2 (1960) 2031[Sov. Phys. Solid State 2 (1960) 1824].
7. This argument has been applied in the present context by T.W.B. Kibble, J. Phys. A: Math. Gen. 9 (1976) 1387.
8. D.A. Kirzhnits and A.D. Linde, Ann. Phys. 101 (1976) 195.
9. It is important to recall also that if the Higgs mass is too small, radiative corrections restore the symmetry even at zero temperature. For the Abelian Higgs model (Ref. 8), it is necessary that $m_H^2 > (3/2)\alpha\ m_X^2$.
10. J. Ellis, M.K. Gaillard and D.V. Nanopoulos, these proceedings p. 461.

11. Y.B. Zel'dovich and M.Y. Khlopov, Phys. Letters 79B (1978) 239.
12. J.P. Preskill, Phys. Rev. Letters 43 (1979) 1365.
13. M.B. Einhorn, D. Stein and D. Toussaint, Phys. Rev. D, to be published.
14. The determination of d is a problem to which we will return shortly.
15. H. Georgi and S. Glashow, Phys. Rev. Letters 32 (1974) 438.
16. Here we are assuming a radiation-dominated epoch and neglecting any cosmological constant. For the second-order phase transition, this is a good approximation. See Linde, Ref. 4, and S.A. Bludman and M.A. Ruderman, Phys. Rev. Letters 18 (1977) 255.
17. G. Steigman (private communication) suggests that the upper limit of 10^{-19} might well be improvable to 10^{-25}, in which case stable monopoles could not have been produced above 10^{10} GeV.
18. This inference seems more general than the Abelian Higgs model. See S. Weinberg, Phys. Rev. D9 (1974) 3357.
19. The suggestion of a first-order phase transition was given in Ref. 13 and also, independently, by A.H. Guth and S.-H.H.Tye, Phys. Rev. Letters 44 (1980) 631, in which the process is described in some detail. The analysis presented here does not agree in some respects with that given by Guth and Tye. In particular, their neglect of the cosmological constant and their approximation of the thermal nucleation rate are questionable.
20. A.D. Linde, Phys. Letters 70B (1977) 306 and Ref. 4.
21. J.S. Langer, Ann. Phys. 41 (1967) 108.
22. S. Coleman, Phys. Rev. D15 (1977) 2929;
C.G. Callan and S. Coleman Phys. Rev. D16 (1977) 1762.
23. These considerations were also described by Guth and Tye, op. cit.
24. If one is considering a GUT like SU(5), other unnatural adjustments are made in the Higgs potential anyway.
25. K. Sato, "First Order Phase Transition of a Vacuum and Expansion of the Universe", NORDITA preprint, January 1980.
26. This is apt to be a much better approximation than the step function assumed by Guth and Tye, op.cit.
27. This and other aspects of the implications of a first-order phase transition on the monopole problem are under consideration by Sato and myself.
28. S. Weinberg, "The Problem of Mass", in Transactions of the New York Academy of Sciences, Series II, Vol. 38 (1977) 185.
29. G. 't Hooft, Cargèse Summer Lecture III (1979).
30. S. Dimopoulos and L. Susskind, Nucl. Phys. B155 (1979) 237.
31. E. Eichten and K. Lane Phys. Letters 90B (1980) 125.

RANDOMNESS AS A SOURCE OF MASSLESS PARTICLES

Giorgio Parisi

INFN

Laboratori Nazionali di Frascati, Italy

It is well known that massless particles arise naturally from symmetry principles, e.g. Goldstone bosons, gauge fields, γ_5 invariance, ... In this talk I speculate that maybe there is a new unforeseen mechanism which could generate massless particles.

Let us construct a very simple method to illustrate this mechanism. We assume that space-time can be divided into flat regions (which will be called points) connected by regions where the structure of space-time is very twisted (bridges or links). We suppose that the points and the links can be organized in such a way as to form a lattice (or an amorphous solid): the links should connect only points which are not too far separated. A macroscopically flat space-time structure can be introduced.

Let us consider a quantum field theory on this space: a self-interacting scalar field ϕ. We assume that the field ϕ will be practically constant inside each point: we denote by ϕ_i the value of the field at the point i; the net effect of the bridges is to couple the fields defined on different bonds. The final Lagrangian is

$$\mathcal{L} = \sum_i \left(\frac{1}{2} m^2 \phi_i^2 + g \phi_i^4 + \sum_k J_{ik} (\phi_i - \phi_k)^2 \right) \tag{1}$$

Equation (1) is the starting point of this analysis; the reader should note that in the derivation of Eq. (1) he (she) is completely free to substitute for the gravitational interaction I have used, any other interaction he (she) may prefer.

It is evident that if all the J_{ik} are positive and m^2 is negative the symmetry (Z_2) $\phi \leftrightarrow -\phi$ is spontaneously broken, but Goldstone bosons are not present as the symmetry group is discrete. Let us consider a more complex case in which the J_{ik} take randomly the values ±1.

Normally we would think that the distribution of the J_{ik} is influenced by their interaction with the field ϕ_i. Let us assume that this effect is very small and can be neglected; we can visualize the J_{ik} as semi-classical macroscopic variables which evolve according to the internal (ergodic) dynamics and are not influenced by the interaction with a quantum field.

We consider now the mean value of the propagators:

$$G^{(1)}(i) = \frac{\sum_k \langle \phi_{i+k} \phi_k \rangle}{\sum_k 1}$$
$$G^{(2)}(i) = \frac{\sum_k \langle \phi_{i+k} \phi_k \rangle^2}{\sum_k 1} \tag{2}$$

The sum over k is needed because the system is not translationally invariant. Although $\langle \phi_{i+k}\phi_k \rangle$ depends on the particular realization of the J_{ik}, it is believed that with probability one in the infinite volume limit the mean propagators will be equal for all the choices of the J_{ik} made according to the same probability law (central limit theorem).

The invariance of the model under the local gauge transformation ($\phi_i \to -\phi_i$, $J_{ik} \to -J_{ik}$) implies that $G^{(1)}(i) = 0$, $i \neq 0$. The only information is contained in the $G^{(2)}(i)$.

It is not clear to me if this model is in contradiction with well-known facts, e.g. energy conservation is only a statistical law; my aim is to point out that when m^2 is negative, one finds in a rather unexpected way that the $G^{(2)}(i)$ is long range, i.e. its Fourier transform $\tilde{G}^{(2)}(K^2)$ contains an infinite number of poles with an accumulation point at zero:

$$\tilde{G}^{(2)}(K^2) \approx \sum_{n=0}^{\infty} C_n / (K^2 + \mu_n^2) \quad : \mu_n \to 0 \text{ as } n \to \infty \tag{3}$$

We have succeeded in generating in a natural way an infinite number of particles from only one field. The reader expert in solid state physics will find this note rather uninteresting: our construction reproduces the definition of a quenched spin glass[1]. Equation (3) has been suggested in Ref. 2 and it is based on the approach of Ref. 3. The arguments leading to Eq. (3) are rather involved and they will not be reproduced here: unfortunately a simple clear-cut physical explanation of the phenomenon is still lacking.

The content of this paper can be summarized as follows: in recent years, solid state physicists have started to study systems with quenched disorder: they have discovered that in particular cases long range correlations may be present (i.e. massless particles in the field theory language). I think that physicists working in the theory of elementary particles should be aware of this effect and study if and how it can be incorporated into a realistic theory.

I am very happy to thank the Academia Sinica and in particular the Institute of Theoretical Physics for their warm hospitality during the writing of this note.

REFERENCES

1. S.F. Edwards and P.W. Anderson, J. Phys. F5 (1975) 965; a good survey of the present theoretical situation can be found in the proceedings of the 1980 Les Houches Winter Advanced Study Institute on "Common Trends in Particle and Condensed Matter Physics", to be published in Phys. Rep. C (1980).
2. D.J. Thouless, J.R.L. de Almeida and J.M. Kosterlitz, Birmingham University preprint "Stability and Susceptibility in Parisi's solution ..." (1980).
3. G. Parisi, Phys. Rev. Letters 43 (1979) 1754 and J. Phys. A13 (1980) 984.

SUPERSYMMETRY, PARTICLE PHYSICS AND GRAVITATION

P. Fayet

Ecole Normale Supérieure, Paris, France and

CERN, 1211 Geneva 23, Switzerland

1. INTRODUCTION

Supersymmetry is a beautiful symmetry between bosons and fermions, although there is no evidence of it in Nature[1]. This does not mean that supersymmetry is not present, but that it must be well hidden. Supersymmetric theories of weak, electromagnetic and strong interactions involve new particles such as heavy bosonic leptons and quarks, and fermionic partners for the photon and gluons[2]. They have a new, and very rich, phenomenology[3-5]. Supersymmetry also provides a natural framework for the introduction of gravity due to its close connection with the geometry of space-time[6], and a possible link between interactions of gravitational and weak types due to a massive spin-3/2 particle, the gravitino[4,7]. (For earlier review articles on supersymmetric theories of particles, see Refs. 8-10.)

Searching for supersymmetry leads one to postulate the existence of new particles, many of them being heavy. This is not unreasonable since, for example, the gauge symmetry of weak and electromagnetic interactions involves mass splittings as large as the mass of the intermediate boson W. A very useful guide in the construction of theories of weak and electromagnetic interactions was the similarity between weak and electromagnetic currents. This could be noticed due to the existence of multiplets of light particles such as (proton, neutron) and (neutrino, electron).

We do not have the same chance for supersymmetry, so that the construction of supersymmetric theories of particles cannot rely on similarities between known currents. Although such a construction could seem arbitrary, this is not the case due to the symmetry itself: supersymmetry is so constraining that, initially, it did not even seem compatible with particle physics. Let us mention some of the problems one has to solve: the identification of multiplets of particles related by supersymmetry; the spontaneous breaking of gauge invariance (the problem is mainly to keep an unbroken $SU(3) \times U(1)$ subgroup when spin-0 fields carry colour and charge) and of supersymmetry; the roles of the fermionic partners of the photon (photino), and of the Goldstone fermion (goldstino). The hardest problem is, in fact, to get a realistic mass spectrum for bosons and fermions, despite the existence of linear relations between the squares of the masses.

In Section 2 we explain why we have to introduce new particles, and present the general features of supersymmetric gauge theories. In Section 3 we discuss spontaneous supersymmetry breaking; we study the couplings of the Goldstone fermion and of its bosonic partner, in connection with the mass spectrum. In Section 4 we examine the phenomenological consequences of these theories. In Section 5 we discuss the interactions of the spin-3/2 gauge particle of supersymmetry, the gravitino; if it is very light it would interact essentially like a massless spin-½ Goldstone fermion. Finally, we comment in Section 6 on the possible relations between supersymmetry and grand unified theories.

2. PARTICLE CONTENT OF THE MULTIPLETS OF SUPERSYMMETRY

2.1 The supersymmetry algebra

Supersymmetry transformations turn bosons into fermions and conversely[1]. They are generated by a (self-conjugate) Majorana spinor Q_α satisfying the algebra

$$[Q_\alpha, P^\mu] = 0 \tag{1}$$

$$\{Q_\alpha, \bar{Q}_\beta\} = -2\, \gamma^\mu{}_{\alpha\beta}\, P_\mu \,. \tag{2}$$

The first relation expresses that the result of a supersymmetry transformation does not depend on the space-time point where it is performed. Since they carry half a unit of spin, the supersymmetry generators satisfy anticommutation relations. Due to the antisymmetry and Lorentz covariance, the anticommutator must be a vector or an antisymmetric tensor. In addition it should commute with the generators of translations P^μ. This leads to formula (2). One can also consider extended supersymmetry algebras with several supersymmetry generators.

It follows from the supersymmetry algebra that all particles in a given representation have the same mass. Therefore supersymmetry has to be broken. A spontaneous breaking of global supersymmetry generates a massless spin-½ Goldstone particle, the goldstino[11]. What happens to it when supersymmetry is realized locally will be discussed in Section 5.

2.2 Why do we need to introduce new fields?

The first task is to find what are the multiplets of supersymmetry. We shall present here what seems to be the simplest solution, although we do not know which one, if any, is chosen by Nature. In any case we have to introduce new fields.

Consider for example the quarks and gluons. Supersymmetry will transform them into boson and fermion fields, respectively. We could try to associate quarks with gluons. This would require the existence of several supersymmetry generators carrying colour and charge, at least six (a colour triplet and an antitriplet), and in fact more. When acting on ordinary fields these would generate many new fields.

Such a feature seems unavoidable if one wants to implement supersymmetry. The simplest and most economic situation corresponds to the smallest supersymmetry algebra with a single gauge-invariant spinorial generator. This one turns leptons and quarks into spin-0 partners, and the photon and gluons into spin-½ partners.

We could also consider the possibility that leptons and quarks have spin-1 partners. If we are in the framework of ordinary gauge theories this would lead to enlarge the gauge group, and also requires the existence of spin-0 fields to make the spin-1 fields massive.

In addition, the world may be invariant under an algebra with several supersymmetry generators. Some may be apparent at relatively low energies, and the others (including, possibly, those carrying charge and colour) only at ultra-high energies. In such a framework leptons and quarks would have both spin-1 and spin-0 partners, and the photon and gluons both spin-3/2 and spin-½ partners.

At the present stage we shall mostly consider the smallest algebra and the lower spin partners of the ordinary particles, although many results on symmetry breaking and phenomenological consequences apply to more general situations. This symmetry may become apparent at relatively low energies of the order of the W mass. It allows for the construction of supersymmetric theories of weak, electromagnetic and strong interactions[2], as we shall see now, leaving for

Section 5 the introduction of gravity.

2.3 Spontaneously broken gauge theories

We now assume that spin-½ leptons and quarks have spin-0 partners. There are two of them for each Dirac fermion, but only one for a two-component fermion. We denote by s_e and t_e the complex spin-0 fields associated with the left-handed and right-handed parts of the electron field, respectively; we use similar notations for other matter fields[2].

The photon and the SU(3) colour octet of gluons have neutral spin-½ partners named the photino and the gluinos, respectively. Their Yukawa couplings between the spin-½ leptons and quarks and their spin-0 partners are fixed by the electrical charge for the photino, and the colour for the gluinos. These interactions are described by Abelian[12] or non-Abelian[13] supersymmetric gauge theories, with U(1) or SU(3) as the gauge group.

In order to describe weak as well as electromagnetic and strong interactions one needs to consider also supersymmetric spontaneously broken gauge theories. In a supersymmetric theory the Higgs mechanism operates as usual. For each generator of the gauge group which is spontaneously broken, a gauge boson acquires a mass while the corresponding would-be Goldstone boson is eliminated. These bosons belong to two originally different representations of supersymmetry, which join to give a single massive representation. It is the massive gauge multiplet describing a vector, a Dirac spinor and a real scalar, which are degenerate in mass as long as supersymmetry is conserved (see, for example, Ref. 14).

The real scalar in this multiplet is the physical Higgs field, so that supersymmetry gives us a relation between intermediate gauge bosons and Higgs bosons. In particular, it leads to the existence of a charged Higgs boson w_- associated with the intermediate gauge boson W_- and a pair of Dirac fermions L_- and ℓ_-.

In order to write Lagrangian densities which are supersymmetric (up to a four-derivative), it is often convenient to use the superfield formalism. We will not need it here, so we only indicate that a superfield is a function of a space-time point and of an anticommuting Grassman co-ordinate θ; the coefficients of its expansion with respect to θ are ordinary fields called components of the superfield[15].

The original theory is formulated in terms of massless gauge superfields (with spins 1 and ½) and chiral superfields (with spins ½ and 0). After the spontaneous breaking of gauge invariance, some of the gauge superfields acquire masses while the corresponding chiral superfields are eliminated. The particle content of such a theory is summarized in Table 1.

Table 1

Particle content of a supersymmetric spontaneously broken gauge theory

Superfields		Spin-1	Spin-$\frac{1}{2}$	Spin-0
Gauge superfields ↗	Massless gauge superfields	Photon Gluons	Photino Gluinos	
Gauge superfields ↘ / Chiral superfields ↗	Massive gauge superfields	Intermediate gauge bosons	Heavy fermions	Higgs bosons
Chiral superfields ↘	Matter superfields		Leptons Quarks	Spin-0 leptons Spin-0 quarks

2.4 R-parity and R-invariance

Before discussing the spontaneous breaking of the supersymmetry we present briefly a useful symmetry denoted by R. In order to distinguish ordinary particles from their partners under supersymmetry it is convenient to define an R-parity character: gauge and Higgs bosons, spin-$\frac{1}{2}$ leptons and quarks are R-even, while their superpartners (photino, gluinos, heavy fermions, spin-0 leptons and quarks) are R-odd. The R-even sector of a supersymmetric gauge theory is like an ordinary gauge theory, but with further constraints due to supersymmetry; for example, relations between gauge and Higgs bosons and constraints on the mass spectrum leading to restrictions on the gauge group, as we shall see later.

It follows from R-parity conservation that ordinary particles cannot exchange R-odd particles at lowest order. Spin-$\frac{1}{2}$ leptons and quarks only exchange gauge or Higgs bosons at the classical level. R-odd particles are produced in pairs[2,8].

Moreover, it is often possible to define R not only as a discrete symmetry but as a continuous one. R-transformations act as phase transformations on spin-0 fields, phase or γ_5 transformations on spin-$\frac{1}{2}$ fields[16,17]. R-invariance can be useful to restrict a Lagrangian density beyond the constraints already imposed by supersymmetry, gauge invariance, and baryon or lepton number conservation. These restrictions are sometimes necessary to get a spontaneous breaking of supersymmetry. In any case, a continuous R-invariance allows for the definition of the spin-½ partners of the gauge bosons as Dirac fermions with their own conserved quantum number. This fact has important consequences, since it may forbid the appearance of masses for the photino and gluinos, as we shall discuss Subsection 4.8.

3. SPONTANEOUS SUPERSYMMETRY BREAKING, THE GOLDSTINO, AND ITS BOSONIC PARTNER

3.1 The spontaneous breaking of the supersymmetry

The spontaneous breaking of the supersymmetry presents particular features due to the presence of the Hamiltonian in the algebra. Let us illustrate them, for global supersymmetry, by the following argument: the Hamiltonian can be expressed in terms of the components of the supersymmetry generator:

$$H = \frac{1}{4} \sum_{\alpha=1}^{4} Q_\alpha^2 \; . \tag{3}$$

Any supersymmetric state has zero energy, certainly the minimum possible one, and therefore is stable.

Despite this fact, spontaneous supersymmetry breaking is possible. One can find models for which no state at all is annihilated by the supersymmetry generator. This is the case, in particular, for the state of minimal energy chosen as the vacuum, so that supersymmetry is spontaneously broken[11].

A massless Goldstone particle appears which has the same quantum numbers as the supersymmetry generator, i.e., a massless, neutral, spin-½, R-odd particle named the goldstino. The masses of bosons and fermions are no longer systematically degenerate, but they turn out to be still related.

3.2 Relations between the goldstino couplings and the boson-fermion mass²-splittings

The boson-fermion pairs which are not directly coupled to the goldstino remain mass-degenerate at lowest order, even if supersymmetry is spontaneously broken. For the others, the couplings of the goldstino are fixed by the mass spectrum, as one sees from the examination of the conserved vector-spinor current of supersymmetry[7].

In a spontaneously broken globally supersymmetric theory this current reads

$$J^{\mu}_{\alpha} = d\, \gamma^{\mu}\gamma_5\, \lambda_g + J_{2\alpha}^{\mu} + \ldots \tag{4}$$

d is a parameter with the dimension of a mass^2; λ_g is the Majorana spinor describing the goldstino, and $J_2{}^{\mu}{}_{\alpha}$, which denotes the bilinear terms in the current, involves all boson-fermion pairs related by supersymmetry.

Let us now express current conservation; it reads:

$$\gamma_5\, \not\partial\, \lambda_g = \frac{1}{d}\, \partial_{\mu}\, J_{2\alpha}^{\mu} + \ldots \tag{5}$$

This gives the equation of motion for the goldstino.

If we are only interested in the trilinear couplings of the goldstino (i.e., to the bilinear terms in its equation of motion) we can use the free equations of motion to compute $\partial_{\mu}\, J_2{}^{\mu}{}_{\alpha}$. This involves only the knowledge of the masses and mixing angles. We rewrite (5) as follows:

$$\gamma_5\, \not\partial\, \lambda_g = \frac{1}{d}\left(\partial_{\mu}\, J_{2\alpha}^{\mu}\right)_{free} + \ldots \tag{6}$$

The ... stand for terms which are, at least, trilinear in the fields. There are no such terms in ordinary supersymmetric gauge theories, in which the goldstino has only gauge and Yukawa couplings.

Mass-degenerate boson-fermion pairs do not contribute to (6) so that they have no direct coupling to the goldstino. The couplings of the goldstino are completely determined by the mass spectrum, up to the parameter d which fixes the magnitude of the spontaneous supersymmetry breaking. We find the following relation between the goldstino Yukawa coupling constant $e_g\sqrt{2}$ to a spin-0 - spin-½ pair, and the mass^2-splitting of the pair[7]:

$$e_g = \mp \frac{\Delta m^2}{d}\,. \tag{7}$$

The sign depends on the chirality of the fermion considered. The same relation is also valid for the coupling of the goldstino to a spin-1 - spin-½ pair, as one finds out easily from the results of Ref. 16; the sign is fixed by the R-quantum number of the fermion. These formulae can be extended to include the effects of mixing angles.

3.3 Are the photino and the goldstino identical?

Which particle could be the goldstino? Low-energy theorems forbid identifying it with the electron neutrino[18]. Identifying the goldstino with one of the neutrinos, or a linear combination of them, could lead to difficulties with the conservation of quantum numbers and universality. But, mostly, the couplings of the goldstino, determined by the mass spectrum, are not like those of a neutrino.

Looking at Table 1 one might think of the photino as the natural candidate for being the goldstino. Identifying the goldstino, coupled to boson-fermion pairs proportionally to their mass^2-splittings, with the photino, which is coupled proportionally to the electrical charges, would lead to a proportionality between the mass^2-splittings Δm^2 and the electrical charges $Q\,e$:

$$\Delta m^2 = \mp Q\, e\, d\,. \tag{8}$$

The sign $\mp$ depends on the chirality, or R-quantum number, of the spin-$\frac{1}{2}$ field considered.

Formula (8), together with the conservation of the electrical charge, leads to relations between mass^2-splittings. To examine whether these are acceptable or not, we shall consider the most simple situation in which spin-$\frac{1}{2}$ leptons and quarks are associated with spin-0 partners.

We denote by s_e and t_e the spin-0 fields associated with the left-handed and right-handed electron fields, respectively, and use similar notations for quarks. Examples of the mass relations we would get are

$$m^2(s_e) + m^2(t_e) = 2m^2(e) \tag{9}$$

$$\left[m^2(s_u) - m^2(u)\right] = -2\left[m^2(s_d) - m^2(d)\right]\,. \tag{10}$$

Such relations are obviously excluded: any attempt to make one of those spin-0 fields heavy would lead to a negative mass^2 for another. Translating the latter would lead to a spontaneous breaking of QED or QCD gauge invariances. Therefore if leptons and quarks have spin-0 partners the photino and the goldstino must be different[19].

It is, precisely, the existence of mass relations such as those written above which makes hard the construction of spontaneously broken supersymmetric theories having a realistic mass spectrum.

3.4 The bosonic partner of the goldstino

Formula (7) relates the boson-fermion mass^2-splitting to the coupling of the goldstino to the pair. Owing to supersymmetry the latter is related to the coupling of the bosonic partner of the goldstino. Therefore the formula

$$\Delta m^2 = \mp e_g\, d \tag{11}$$

also expresses the proportionality of the boson-fermion mass^2-splitting Δm^2 to the coupling e_g of the bosonic partner of the goldstino to the fermion.

Assuming that the electron has spin-0 partners we get the unacceptable mass relation (9) if the goldstino has the photon for superpartner; but, also, if it is any spin-1 particle vectorially coupled to the electron. (If the goldstino is a mixing of the fermionic partner of a spin-1 particle vectorially coupled to the electron with the one of a spin-0 particle, the only effect of the latter could be to mix the two spin-0 fields s_e and t_e and to raise one mass^2 while lowering the other, the average mass^2 remaining equal to the mass^2 of the electron.) The only way to raise the mass^2 of both spin-0 partners of the electron (leptons and quarks) is to make use of a spin-1 particle having an axial coupling with the electron (leptons and quarks).

This leads us to consider the weak neutral currents, which do have axial parts. Could the goldstino be a linear combination of the photon γ and the standard neutral gauge boson Z? Then we would get the mass relations:

$$\Delta m^2 = \pm\, (x\, Q_\gamma + y\, Q_Z)\ . \tag{12}$$

But we know experimentally the electrical and weak charges for the u and d quarks of both chiralities[20], and relations (12) turn out to be unacceptable.

Therefore, if spin-0 leptons and quarks have spin-0 partners getting heavy masses through spontaneous supersymmetry breaking, we need, at least, one neutral spin-1 boson in addition to the bosons of the minimal gauge group SU(3) × SU(2) × U(1) [19].

This is the reason we chose SU(3) × SU(2) × U'(1) × U"(1) as the gauge group for constructing an example of a realistic model[2]. In the simplest case, SU(2) × U'(1) is the standard gauge group of weak and electromagnetic interactions, and there is in addition a new gauge boson axially coupled to leptons and quarks. The goldstino is a linear combination of the spin-½ partners of the new gauge boson and of a spin-0 particle carrying two units of R. The latter does not couple directly to leptons and quarks.

Other gauge groups could be used as well, provided there is one neutral spin-1 boson having couplings of opposite signs with (light) leptons and quarks of opposite chiralities. Its fermionic partner will be the goldstino, up to a mixing angle. Note that the couplings do not need to be exactly axial, but may have vector parts if these are smaller in magnitude than the axial parts.

Note, also, that supersymmetry alone does not require the couplings of the new spin-1 particle to be fixed by a gauge principle. If this is the case the corresponding gauge invariance is spontaneously broken, allowing the new gauge boson, as well as leptons and quarks, to acquire masses.

If the new spin-1 boson is indeed an ordinary gauge particle it is desirable to have a cancellation of anomalies. This requires the introduction of new heavy spin-½ leptons and quarks, with axial couplings of the opposite sign. They would be heavier than their spin-0 partners while ordinary leptons and quarks are lighter. (Consideration of the $N = 2$ extended supersymmetry algebra would also lead to such a new class of leptons and quarks[10].)

However, the interactions of the new boson are also, in a sense which will be specified in Section 5, gravitational interactions (with, possibly, a very small coupling constant), so that the cancellation of anomalies may not be required as in ordinary gauge theories.

3.5 Mass relations involving Higgs bosons and spin-0 leptons and quarks

Whenever we have a symmetry, or a scheme of symmetry breaking, we get mass relations.

The simplest possibility is that supersymmetry is conserved and bosons and fermions are mass-degenerate. This is obviously excluded. The next possibility is that the goldstino is the spin-½ partner of the photon, the boson-fermion mass^2-splittings being proportional to the electrical charges at the classical level. This is also excluded if leptons and quarks have spin-0 partners.

Therefore we considered a more elaborate scheme of spontaneous supersymmetry breaking, in which the goldstino is partly the spin-$\frac{1}{2}$ partner of a new spin-1 boson.

If this new boson is axially coupled (and if the R = 2 spin-0 boson which is also, partly, the partner of the goldstino does not couple directly to leptons and quarks), the masses of the two spin-0 particles s_f and t_f associated with each Dirac lepton or quark f are equal:

$$m^2(s_f) = m^2(t_f) \ . \tag{13}$$

If the couplings of the new spin-1 boson are fixed by a gauge principle, this gauge invariance leads to relations between the various mass^2-splittings. This is the case, in particular, if all the usual leptons and quarks get their masses from couplings with a single pair of Higgs doublets, as in Ref. 2. The simplest set of relations is obtained when the new gauge boson has a universal axial coupling with spin-$\frac{1}{2}$ leptons and quarks. Then we find a universal mass^2-splitting Δm^2 between spin-0 and spin-$\frac{1}{2}$ leptons and quarks

$$m^2(s_f) - m^2(f) = m^2(t_f) - m^2(f) = \Delta m^2 \ . \tag{14}$$

The mass^2-splittings in the lepton and quark multiplets may also be related to the mass^2-splittings between the intermediate boson W_- and the Higgs boson w_-. In a rather general class of models we have:

$$m^2(W_-) - m^2(w_-) = 2\left[m^2(s_f) + m^2(t_f) - 2m^2(f)\right] = 4\Delta m^2 \ . \tag{15}$$

We see, again, that increasing the mass^2 of a spin-0 boson makes another one decrease. This is now acceptable since the latter can be the one of the charged Higgs boson w_-, which is degenerate with the W_- when supersymmetry is conserved. With this scheme of spontaneous supersymmetry breaking the masses of spin-0 leptons and quarks cannot be too large, since they satisfy

$$\left(\frac{m^2(s_f) + m^2(t_f)}{2}\right)^{\frac{1}{2}} \lesssim \tfrac{1}{2}\, m(W_-) \sim 40 \text{ GeV}/c^2 \ . \tag{16}$$

The existence of the mass relation (16) which holds the masses of spin-0 leptons and quarks at reasonable values is encouraging in view of the experimental searches for these particles. However, the absence of spin-0 leptons and quarks below 40 GeV/c^2 would not be a signal against supersymmetry, but only against the general association of spin-½ leptons and quarks with spin-0 partners with the supersymmetry breaking mechanism considered above: (i.e., essentially, the spin-1 partner of the goldstino has its couplings constrained by a gauge principle leading to the mass relation (15)).

4. PHENOMENOLOGICAL CONSEQUENCES OF SUPERSYMMETRIC THEORIES OF PARTICLES

4.1 Foreword

In the supersymmetric theories of particles which have been constructed up to now and appear as conventional gauge theories, leptons and quarks have spin-0 partners while the photon and gluons have spin-½ partners. We shall discuss the experimental consequences of supersymmetry mostly in that framework, but many of the features described can be extended to partners with different spins.

Whether such phenomena can be observed or not depends on how much the supersymmetry is broken, i.e. on quantities which are not yet fixed by the theory:

i) the order of magnitude for the mass^2-splittings Δm^2 inside the multiplets of supersymmetry; is it $\lesssim (m_W/2)^2$, as suggested in Subsection 3.5, or on the contrary much larger, and possibly related to the Planck mass $\sim 10^{19}$ GeV/c^2? In the latter case the chances of finding direct experimental evidence for supersymmetry would be very slim;

ii) the strength e_g of the coupling of the goldstino to boson-fermion pairs; is it of the order of the electromagnetic or weak gauge coupling constants, or on the contrary much smaller, maybe some extremely small number related to the strength of gravitational interactions? (See also Section 5.)

iii) the masses that the fermionic partners of the gluons (and photon) might acquire at the quantum level (see Subsection 4.8).

4.2 The R-even sector

The R-even sector of a supersymmetric gauge theory is similar to any ordinary gauge theory, but with extra constraints due to supersymmetry.

One is the association between intermediate gauge bosons and physical Higgs bosons, together with heavy fermions, in massive gauge multiplets. This implies, in particular, the existence of a charged Higgs boson w_- associated with the intermediate boson W_-. They would be mass-degenerate if supersymmetry were conserved. If the mass relation (15) holds, the Higgs boson w_- is lighter than the W_-. It gets lighter when spin-0 leptons and quarks get heavier; this leads to the upper limit $\sim m_W/2 \sim 40$ GeV/c^2 for the masses of spin-0 leptons and quarks[2,19].

The bosonic partner of the goldstino is an R-even particle which determines the mass^2-splittings between bosons and fermions in the multiplets of supersymmetry. It must be, in part, a new neutral spin-1 boson with, dominantly, axial couplings to leptons and quarks. This particle, related to gravitation, plays a fundamental role in the structure of spontaneously broken supersymmetric theories. It could lead to modifications of the neutral current phenomenology and to many interesting effects which will be discussed elsewhere.

4.3 Bosonic leptons and quarks: indirect constraints

Lower limits on the masses of bosonic leptons and quarks only come from experiments. Let us consider the extra contributions to the muon anomalous magnetic moment due to spin-0 muons s_μ and t_μ.

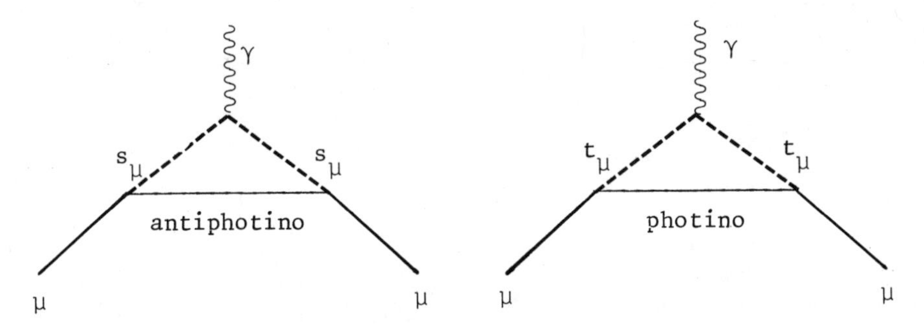

Fig. 1 Contributions of spin-0 muons to the muon anomalous magnetic moment.

The contributions of the diagrams of Fig. 1 to the muon anomaly $a_\mu = (g_\mu - 2)/2$ are:

$$\Delta a_\mu \simeq -\frac{\alpha}{12\pi}\left[\frac{m^2(\mu)}{m^2(s_\mu)} + \frac{m^2(\mu)}{m^2(t_\mu)}\right] \tag{17}$$

For $m(s_\mu) = m(t_\mu) = 15$ GeV/c^2, we find

$$\Delta a_\mu \simeq -2 \times 10^{-8} \ . \tag{18}$$

The sum of all extra contributions to the muon anomaly should not be larger than 2×10^{-8} in magnitude[21]. Unless there is some cancellation mechanism between various contributions, this can be taken as an indirect indication that spin-0 muons cannot be too light:

$$m(s_\mu),\ m(t_\mu) \gtrsim 15 \text{ GeV/c}^2. \tag{19}$$

If,in addition, the leptons and quarks (f) acquire their masses through couplings with a single pair of Higgs doublets, as in Ref. 2, the relations between mass^2-splittings imply that

$$\left[m^2(s_f) - m^2(f)\right]^{\frac{1}{2}}, \ \left[m^2(t_f) - m^2(f)\right]^{\frac{1}{2}} \gtrsim 15 \text{ GeV/c}^2 , \tag{20}$$

i.e., spin-0 leptons and quarks should be heavier than ~ 15 GeV/c^2 (see Ref. 19). However, one should look directly for such particles, irrespectively of this indication on their mass, due to the hypotheses made in the derivation. Similar arguments could be applied if the muons had spin-1 particles for superpartners. Those should not be too light.

4.4 Direct experimental search for bosonic leptons and quarks

Bosonic leptons and quarks are unstable R-odd particles. They decay towards the corresponding spin-½ lepton or quark by emission of a photino or a goldstino. The decay rate for the emission of a photino is $\sim \alpha$ times the mass of the particle (if it is charged). The lifetime is extremely short, at least for charged particles (neutral ones may have longer lifetimes if the interactions of the goldstino are very weak; see Section 5). As a result one should look for their decay products[2,5,8].

R-odd particles can be pair produced, for example in e^+e^- annihilation, or by the Drell-Yan mechanism. (It is also possible to produce a single bosonic lepton or quark in association with its spin-½ partner and a photino or a goldstino.) Figure 2 illustrates the pair production of bosonic leptons or quarks in e^+e^- annihilation.

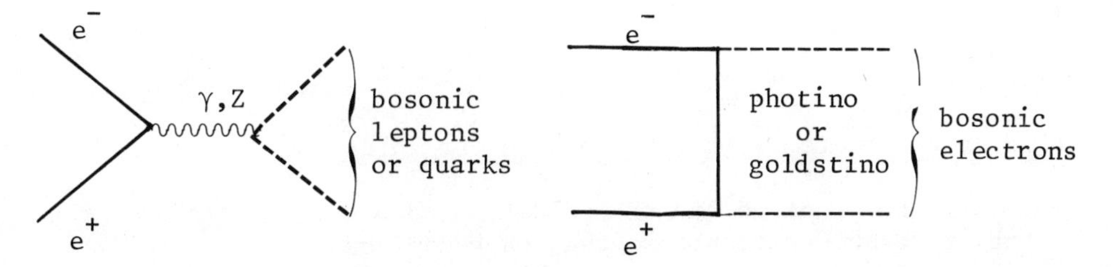

Fig. 2 Pair production of bosonic leptons or quarks in e^+e^- annihilation.

Once produced, the pair decays into the corresponding pair of spin-½ leptons or quarks, by emitting photinos or goldstinos carrying away half of the energy, in average:

$$\begin{aligned} e^+e^- &\rightarrow \text{Pair of bosonic leptons or quarks} \\ &\rightarrow \text{Non coplanar pair } (e^+e^-,\ \mu^+\mu^-,\ \tau^+\tau^-,\ \text{or } q\bar{q}) \\ &\quad + \text{Pair of photinos or goldstinos} . \end{aligned} \tag{21}$$

The pair production of bosonic electrons and muons resemble the pair production of heavy leptons like the τ. But they differ since (at least in the simple case considered here) their decays lead to e^+e^- or $\mu^+\mu^-$ pairs without μ^+e^- or μ^-e^+ pairs. From SPEAR data taken at ~ 7.2 GeV/c^2 [22] one can put a lower limit $\sim 3\frac{1}{2}$ GeV/c^2 on the masses of bosonic electrons and muons, but better limits ($\sim$ 10 to 15 GeV/c^2) are probably already available from PETRA experiments. Bosonic quarks would lead to non-coplanar jets with missing energy.

If the new bosons are spin-0 particles, one expects a β^3-dependence of the production cross-section and an essentially isotropic distribution of their decay products just above threshold. Asymptotically each spin-0 quark would give a contribution to the ratio

$$R = \frac{\sigma(e^+e^- \rightarrow \text{hadrons})}{\sigma(e^+e^- \rightarrow \mu^+\mu^-)}$$

equal to $\frac{1}{4}\, Q_i{}^2$, where Q_i is its electrical charge. Above the threshold for pair production of the spin-½ quarks u, d, c, s, and b and their spin-0 partners, we would get

$$R = \frac{3}{2} \sum_i Q_i^{\,2} = \frac{11}{2} \,. \tag{22}$$

4.5 Interactions of the photino and the goldstino

Both the photino and the goldstino couple leptons and quarks to their bosonic partners so that they can scatter on matter[2,23]. Their couplings are fixed by the electrical charges and the boson-fermion mass^2-splittings, respectively. If the lepton and quark fields, denoted by ψ, have spin-0 partners s and t (for which we assume equal masses for simplicity), the relevant Yukawa couplings are fixed by the following terms in the Lagrangian density:

$$\begin{aligned} \mathcal{L} = {} & i\, Qe \sqrt{2} \left[(\bar{\psi}_L\, s + \bar{\psi}_R\, t)\lambda_\gamma + \bar{\lambda}_\gamma\, (s^* \psi_L + t^* \psi_R) \right] \\ & + i\, e_g \sqrt{2} \left[(-\bar{\psi}_L\, s + \bar{\psi}_R\, t)\, \lambda_g + \bar{\lambda}_g\, (-s^* \psi_L + t^* \psi_R) \right] \end{aligned} \tag{23}$$

λ_γ and λ_g are Majorana spinors representing the photino and the goldstino, Qe is the electrical charge, and

$$e_g = \frac{\Delta m^2}{d} \tag{24}$$

fixes the goldstino coupling constant in terms of the boson-fermion mass^2-splitting Δm^2. In this simple case the spin-1 particle which is, in part, the bosonic partner of the goldstino is axially coupled to the lepton or quark field ψ, and the photino and the goldstino have charge conjugation C = − and C = +, respectively.

We shall study three types of scatterings on matter: photino → photino, photino ↔ goldstino and goldstino → goldstino. They can occur through the exchanges of spin-0 or spin-1 particles as illustrated in Figs 3 and 4, respectively.

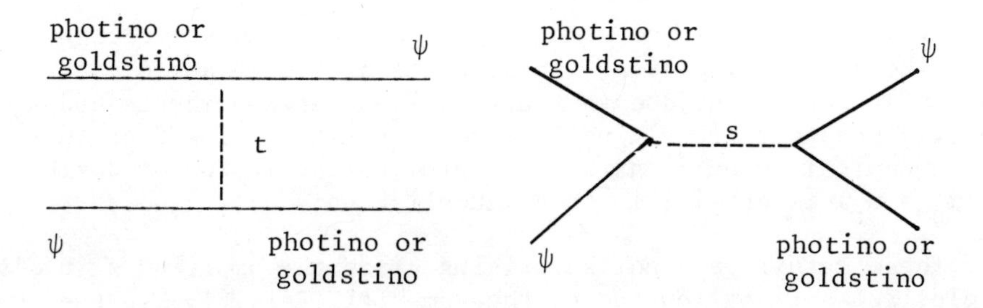

Fig. 3 Scattering of photinos or goldstinos on matter due to the exchange of spin-0 leptons or quarks

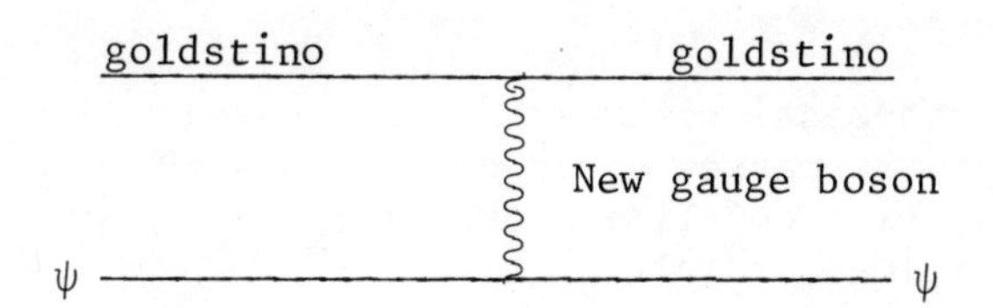

Fig. 4 Scattering of goldstinos on matter due to the exchange of the new spin-1 boson

In the local limit these processes can be described by the effective Lagrangian density

$$\mathcal{L}_{eff} = \frac{Q^2 e^2}{\Delta m^2} (\bar{\lambda}_{\gamma L} \gamma^\mu \lambda_{\gamma L})(\bar{\psi} i\gamma_\mu \gamma_5 \psi)$$

$$+ \frac{Qe\, e_g}{\Delta m^2} (\bar{\lambda}_{\gamma L} \gamma^\mu \lambda_{gL} + \bar{\lambda}_{gL} \gamma^\mu \lambda_{\gamma L})(\bar{\psi} \gamma_\mu \psi) \tag{25}$$

The first term describes the scattering photino → photino. It depends on the electrical charge of the scattered lepton or quark and on the mass of its spin-0 partners.

The second term describes the scatterings photino ↔ goldstino. It involves the electromagnetic current $Qe\, \bar{\psi} \gamma_\mu \psi$. The proportionality coefficient $(e_g/\Delta m^2) = 1/d$ is fixed by the magnitude of the spontaneous supersymmetry breaking (i.e., ultimately, by the mass of the spin-3/2 gauge particle of supersymmetry, the gravitino: see Section 5).

Finally, there is no term responsible for the scattering goldstino → goldstino since the diagrams of Figs 3 and 4 cancel exactly at low energy due to the relations between masses and coupling constants resulting from supersymmetry. The fact that the new spin-1 boson is axially coupled to the lepton or quark (if $m_s = m_t$) is crucial in this cancellation.

These results are not surprising since the amplitudes involving a goldstino satisfy low-energy theorems (cf. Ref.18). In the present situation the photino and the goldstino are orthogonal states. The goldstino → goldstino amplitude must vanish at low energy, since there is no pole term resulting from a boson-fermion mass degeneracy. The goldstino ↔ photino amplitudes are proportional to the matrix elements of the electromagnetic current, due to the photon-photino mass degeneracy.

These cross-sections have been evaluated numerically in Ref. 23 in terms of the masses of spin-0 leptons and quarks, and of the (unknown) goldstino coupling constant e_g. Here we only indicate the most significant result:

$$\sigma(\text{photino} + \text{nucleon} \rightarrow \text{photino} + \text{hadrons}) \simeq$$

$$\simeq \begin{cases} 30 \quad 10^{-38}\ \text{cm}^2\ E(\text{GeV}) \quad \text{for} \quad m_s = m_t = 15\ \text{GeV}/c^2 \\ 0.7 \quad 10^{-38}\ \text{cm}^2\ E(\text{GeV}) \quad \text{for} \quad m_s = m_t = 40\ \text{GeV}/c^2 \end{cases} \qquad (26)$$

The last number is still about three times larger than the usual neutral current cross-section $\sigma^{NC}_{\nu_\mu N}$.

For the goldstino cross-sections we can only give upper limits since the goldstino coupling constant may be very small (in fact these cross-sections are fixed by the ratio

$$\frac{G_{Newton}}{m_{3/2}^2},$$

where $m_{3/2}$ is the gravitino mass: see Section 5).

We expect the photino to have the same effect as a weakly-interacting two-component neutrino on the helium production in the early Universe. This is not necessarily the case for the goldstino: its interaction cross-sections might be very small so that it would drop out of the equilibrium much earlier, and would not affect significantly the evolution of the Universe at the time of helium production[24].

We also mention that both the photino and the goldstino, when scattering on quarks, may produce the fermionic partners of the gluons (gluinos): new hadronic states called R-hadrons can be excited if they are not too heavy[3,23] (see Subsection 4.7).

4.6 The decay $\Psi \to$ goldstino + antiphotino

Lepton-antilepton or quark-antiquark pairs can annihilate into pairs photino-antiphotino, photino-antigoldstino or goldstino-antiphotino. In particular, this can be the case for the Ψ resonance, interpreted as a $c\bar{c}$ bound state[4]. In the simple situation considered earlier, for which the two spin-0 charmed quarks have equal masses, the decay mode photino + antiphotino is forbidden by parity or charge conjugation invariances (see Eq. (25) and Ref. 23).

We only have to consider the decay modes goldstino-antiphotino and photino-antigoldstino. The corresponding diagrams can be obtained easily from those of Fig. 3, and formula (25) gives the effective Lagrangian density responsible for the decay, in the local limit. Comparing with the decay mode $\Psi \to e^+e^-$, we find:

$$r = \frac{\Gamma(\Psi \to \text{goldstino + antiphotino}) + \Gamma(\Psi \to \text{photino + antigoldstino})}{\Gamma(\Psi \to e^+e^-)}$$

$$= \left(\frac{m_\Psi^2}{ed}\right)^2 = \left(\frac{e_g\, m_\Psi^2}{e\ \Delta m^2}\right)^2 \qquad (27)$$

As an example with spin-0 quarks of 15 GeV/c^2 and $e_g = e/2$ we get

$$\Gamma(\text{goldstino + antiphotino}) + \Gamma(\text{photino + antigoldstino}) \simeq \\ \simeq 3 \times 10^{-5} \qquad (28)$$

This number is to be understood as an illustration and not as an estimate. The branching ratio would be completely negligible if the goldstino coupling constant e_g turned out to be very small. It should be compared to the branching ratio expected for a $\nu\bar{\nu}$ decay mode ($\sim 10^{-8}$), and to the present experimental limit: preliminary results from the SLAC-LBL collaboration, obtained from the search for the decay

$$\Psi' \rightarrow \pi^+\pi^-\Psi \qquad (29)$$
$$\hookrightarrow \text{unobserved neutrals}$$

indicate that [25]

$$\Gamma(\Psi \rightarrow \text{unobserved neutrals}) < 7 \times 10^{-3} \ . \qquad (30)$$

As a result we get the lower limit on the parameter d:

$$d = \frac{\Delta m^2}{e_g} > 90\ (\mathrm{GeV}/c^2)^2 \ . \qquad (31)$$

This is not a stringent bound since spin-0 charmed quarks are likely to be heavier than 15 GeV/c^2.

However, this type of experiment is particularly interesting since it gives a direct limit on a fundamental parameter of the theory, d. It measures the magnitude of the spontaneous supersymmetry breaking, and will be related in Section 5 to the mass and effective interaction strength of the spin-3/2 gravitino. It would be useful to be able to improve the limit (30), even if the chances of detecting these decay modes are very slim. The situation would probably be more favourable for the decays of heavier resonances, such as the Υ or the toponium if it is found.

4.7 New hadronic states

Supersymmetry associates to the colour octet of gluons a colour octet of fermions which we call gluinos. They are neutral and R-odd. In the simplest situations they are massless, but they may also acquire a mass by quantum effects, as discussed in Subsection 4.8. The contribution to the β-function of an octet of self-conjugate spin-½ gluinos is the same as for three quark flavours, so that gluinos do not destroy the property of asymptotic freedom of the strong interaction theory.

Gluinos may combine with quarks, antiquarks and gluons to give new R-odd colour-singlet hadronic states called R-hadrons. R-mesons, formed of a $q\bar{q}$ pair and a gluino, would be fermions, while R-baryons, formed of three quarks and a gluino, would be bosons. If R-hadrons are relatively light, one can search for their effects in present experiments[3,8,9].

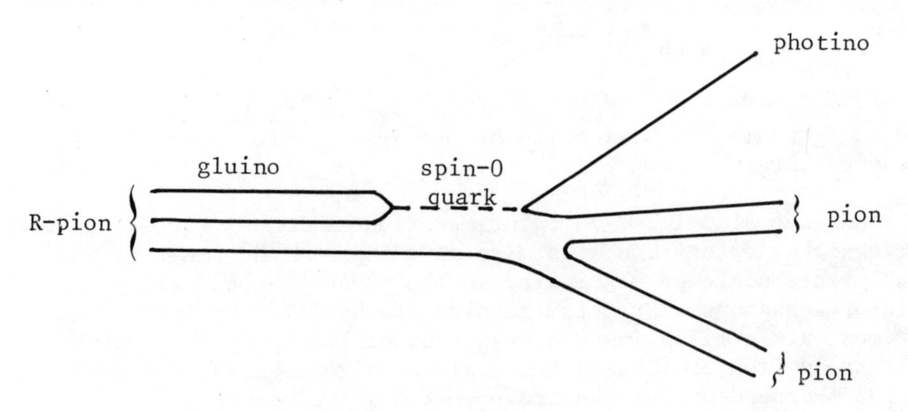

Fig. 5 A diagram responsible for the decay R-pion → ππ photino.

R-hadrons would be unstable and decay into ordinary hadrons by emitting a photino or a goldstino. Figure 5 shows one of the diagrams responsible for the decay R-pion → ππ photino.

The lifetime is comparable to those of non-leptonic weak decays but shorter due to the presence of a strong rather than a weak or electromagnetic coupling constant at the gluino vertex. For an R-pion of mass 1.2 GeV/c^2 the lifetime is estimated to be $\sim 10^{-15}$ to 10^{-12} s.

R-hadrons could be pair-produced, in particular in hadronic reactions, and would decay into ordinary hadrons by emission of a pair of photinos or goldstinos carrying away part of the energy-momentum. Experiments looking for the pair production and semi-leptonic decays of charmed particles can be used to search for R-hadrons. One can look for short tracks, reinteraction of the emitted neutrals, or missing energy in the final state.

From the BEBC beam dump experiment[26] we get the result

$$2\,\sigma_{pN\to R\bar{R}+X} \cdot \sigma_i < 2 \times 10^{-66}\ \text{cm}^4 \tag{32}$$

σ_i is the reinteraction cross-section of the photino or goldstino emitted. If spin-0 quarks are lighter than $m_W/2 \sim 40$ GeV/c^2, the reinteraction cross-section of the photino is larger than 0.7×10^{-38} cm^2 E (GeV) (see Eqs (16) and (26)). Unless the decay modes into the photino are inhibited we get

$$\sigma_{pN\to R\bar{R}+X} \lesssim 6\ \mu b \quad \text{at } \sqrt{s} = 27 \text{ GeV} . \tag{33}$$

The limit would be smaller if the photinos re-excite R-hadrons in the final state[3].

One can also look, in calorimeter experiments, for events without charged leptons and with missing energy in the final state. Such events could be attributed to the production and decay of a pair of R-hadrons. From the results of the Caltech-Stanford experiment at Fermilab[27] one can give upper limits on the R-hadron pair production cross-section. This requires various assumptions on the A-dependence of the cross-section, the masses of R-hadrons and the spectrum of the emitted neutrals. As an example for an R-hadron mass of 2 GeV/c^2 we get[3]

$$\sigma_{pN\to R\bar{R}+X} \lesssim 40\ \mu b \quad \text{at } \sqrt{s} = 27 \text{ GeV} . \tag{34}$$

This limit is less constraining than the one given by formula (33), but it does not depend on estimates of the photino and goldstino reinteraction cross-sections.

From there one can conclude that if R-hadrons have a normal hadronic production they cannot be very light, and their masses should be at least of the order of 1½ or 2 GeV/c^2.

4.8 The question of the mass of gluinos

We do not know whether massless gluinos can lead to R-hadrons somewhat heavier than ordinary hadrons, as experimental results seem to indicate. Therefore it is natural to ask whether gluinos can be massive.

We assume, again, that we are in the usual framework of supersymmetric gauge theories. The Lagrangian density does not contain any direct mass term for gluons, nor for gluinos. None can be generated by the Higgs mechanism since the Yukawa couplings of gluinos involve coloured spin-0 fields which cannot be translated if colour gauge invariance is to be conserved. Therefore, gluinos are massless at the classical level.

Could they get a mass by some other effect? If R-invariance is realized as a continuous symmetry, acting on gluino fields as a γ_5 transformation, it would forbid the appearance of a gluino mass, unless the theory contains a second octet with opposite R-transformation properties. But there is also the possibility that R-invariance is not realized as a continuous symmetry, whether it is broken explicitly or spontaneously (maybe by dynamical effects).

An example shows that if the symmetries of the theory do not prevent the appearance of a gluino mass, this one is, indeed, generated by quantum effects[28]. The value of the mass depends on the detail of the model so that, although it is calculable, it appears in practice as an arbitrary parameter to be determined experimentally. Similar remarks could be applied to the generation of a mass for the photino by quantum effects.

5. SUPERSYMMETRY AND GRAVITATION

5.1 Supergravity: the graviton and the gravitino

There is a close connection between supersymmetry and the geometrical structure of space-time due to the presence of the generators of translations in the anticommutation relations. If the supersymmetry algebra is realized locally, the theory must be invariant under local co-ordinate transformations, so that it includes general relativity. This is the theory of supergravity[6]. The spin-2 graviton is coupled to the conserved energy momentum tensor $T^{\mu\nu}$ with strength $\sim \kappa \sim (G_{Newton})^{\frac{1}{2}}$. Its spin-3/2 partner, the gravitino, is the gauge particle of supersymmetry. It is coupled, also with strength $\sim \kappa$, to the conserved vector-spinor current of supersymmetry, J^{μ}_{α}.

5.2 Spontaneous breaking of local supersymmetry

Spontaneous breaking of global supersymmetry generates a massless Goldstone spin-$\frac{1}{2}$ particle, the goldstino. The magnitude of the spontaneous breaking of the supersymmetry, and therefore the effective strength of the interactions of the goldstino, are characterized by a parameter d with the dimension of a mass^2. It plays the same role as the vacuum expectation value of a spin-0 field when a global internal symmetry is spontaneously broken, generating a massless Goldstone boson.

When supersymmetry is realized locally the spin-$\frac{1}{2}$ goldstino is eliminated while the corresponding gauge particle, the spin-3/2 gravitino, acquires a mass, the total number of degrees of freedom being conserved[29]. The value of this mass, proportional to d and to the gauge coupling constant κ of gravitation, is

$$m_{3/2} = \frac{\kappa d}{\sqrt{6}} \tag{35}$$

with

$$\kappa = (8\pi\, G_{Newton})^{\frac{1}{2}} \simeq 4 \times 10^{-19}\ (\mathrm{GeV}/c^2)^{-1}\ . \tag{36}$$

Formula (35) is quite similar to the formula $m_W \sim gv$ giving the intermediate boson mass in a gauge theory.

We have seen earlier that the boson-fermion mass^2-splitting Δm^2 is proportional to the goldstino coupling constant to the pair, e_g (in the globally supersymmetric theory), which is also the coupling constant of the bosonic partner of the goldstino to the fermion of the pair (formulas (7) and (11)). Disregarding the sign, we can rewrite formula (35) as follows[7]:

$$m_{3/2} = \frac{\kappa\, \Delta m^2}{e_g\, \sqrt{6}} . \qquad (37)$$

5.3 What is the mass of the gravitino?

Even if we knew the mass^2-splitting Δm^2 we could not deduce the gravitino mass unless we also knew the value of the coupling constant e_g. As discussed previously, there are experimental indications that $(\Delta m^2)^{\frac{1}{2}} \gtrsim 15$ GeV/c^2, while it should be $\lesssim 40$ GeV/c^2 if the mass relation (16) holds. But we do not know the value of e_g. It could be of the order of ordinary gauge coupling constants. It could also be much smaller, since it can be viewed as a gravitational effect.

In the first case the gravitino mass may be extremely small. For $(\Delta m^2)^{\frac{1}{2}} = 15$ GeV/c^2 and $e_g = e/2$ we get

$$m_{3/2} \simeq 2 \times 10^{-7} \text{ eV/c}^2 . \qquad (38)$$

This number is not an estimate of the gravitino mass but an illustration of how small it might be. If Δm^2 is larger and/or e_g is smaller, the gravitino mass would not be so tiny. Conceivably it might be as large as the Planck mass $\sim 10^{19}$ GeV/c^2.

Therefore, we should consider the mass of the gravitino as an unknown parameter which, in principle, should be determined from experiment (!), as is the mass of the W in the theory of weak interactions. In fact, we shall see that in both cases the masses of the gauge particles (W or gravitino) determine, together with the gauge coupling constants (g or κ), the effective strength of the corresponding (weak or gravitational) interaction: the exchange of a W boson is fixed by $G_{Fermi} \sim (g^2/m_W{}^2) \sim (1/v^2)$, while the production of a gravitino turns out to be fixed by

$$\frac{G_{Newton}}{m_{3/2}^2} \sim \frac{\kappa^2}{m_{3/2}^2} \sim \frac{1}{d^2} . \tag{39}$$

5.4 A very light gravitino interacts essentially like a goldstino

Since the interactions of the gravitino are, like those of the graviton, fixed by the extremely small coupling constant κ, it could seem, naïvely, that its effects in particle physics would be negligible. However, whether this is indeed true or not depends on the value of the gravitino mass, as we shall see.

It is somewhat surprising that in a spontaneously broken supersymmetric gauge theory, realizing locally the supersymmetry algebra with an extremely small coupling constant κ leads to the disappearance of the goldstino which would, otherwise, have weak-type interactions (with the coupling constant e_g). In fact, the goldstino does disappear while the gravitino acquires new degrees of freedom by getting a mass. But if this mass is very small the four states of the ultra-relativistic massive spin-3/2 gravitino behave as those of massless particles with helicities ±3/2 and ±½. The former have practically negligible gravitational interactions, while the latter do interact as the goldstino of the globally supersymmetric theory.

Let us consider the amplitude for emitting, by gravitational interaction, a very light gravitino of mass $m_{3/2}$, momentum $k^\mu >> m_{3/2}$ and helicity ±½. We first have a factor

$$\frac{\kappa}{2}$$

at the gravitino vertex. Since we select the ±½ helicity states of the massive gravitino we then have an extra factor

$$\frac{k^\mu}{m_{3/2}} \sqrt{\frac{2}{3}} \tag{40}$$

coming from the gravitino wave function. Since $m_{3/2}$ is also proportional to κ, this factor cancels out in the expression

$$\frac{\kappa}{2}\frac{k^\mu}{m_{3/2}}\sqrt{\frac{2}{3}} = \frac{k^\mu}{d} \tag{41}$$

k^μ is to be contracted with the relevant term in the vector-spinor current of supersymmetry.

For example, if a gravitino is produced in the decay of a bosonic partner of a lepton or a quark, the amplitude reads, schematically

$$A \sim \frac{\kappa}{2}\frac{k^\mu}{m_{3/2}}\sqrt{\frac{2}{3}}\, J^\mu_{\frac{1}{2}} \sim \frac{\kappa\ \Delta m^2}{m_{3/2}\ \sqrt{6}} \sim e_g \tag{42}$$

and we recover the coupling constant of the goldstino to the pair (see Ref. 7 for more details).

The amplitude for emitting a gravitino does not vanish in the limit of small κ and $m_{3/2}$ but has a finite limit which is the amplitude for emitting a massless spin-$\frac{1}{2}$ goldstino in a globally supersymmetric theory. The same argument applies to the exchange of these particles:

$$\begin{matrix}\text{Gravitational interactions} \\ \text{of the gravitino}\end{matrix} \xrightarrow[m_{3/2}\,\to\,0]{\kappa\,\to\,0} \begin{matrix}\text{"Weak" interactions} \\ \text{of the goldstino}\end{matrix} \tag{43}$$

This illustrates the continuity of the theory for $\kappa \to 0$, giving an independent determination of the gravitino mass[4,7]. It offers a possibility to look for the effects of the gravitino in particle physics. However, one would not detect its spin-3/2 character, since the only states which could be produced and interact in a non-negligible way would have essentially $\pm\frac{1}{2}$ polarizations.

Consider now a theory which has, in the absence of gravity, a continuous R-invariance, one unit of R being carried by the supersymmetry generator. The introduction of gravity and the generation of a mass for the gravitino breaks R-invariance and reduces it to a discrete symmetry, R-parity. The existence of a mass for the gravitino would lead to $\Delta R = \pm 2$ processes but they would be extremely small (at the classical level): for example, the $\Delta R = \pm 2$ amplitude due to the exchange of a massive gravitino would be

$$\sim \frac{G_{Newton}}{m_{3/2}^2} m_{3/2} \sim \frac{\kappa\, e_g}{\Delta m^2} .$$

5.5 Low energy production of gravitinos

It is not necessary to produce R-odd heavy particles to get gravitinos among their decay products. Gravitinos, if not too heavy, can be produced at low energies since they couple to the photon-photino and gluon-gluino pairs.

As said earlier, a very light gravitino interacts essentially like a goldstino. From formula (41), or equivalently from the low-energy theorems satisfied by a goldstino, we expect the amplitudes to vanish in the low-energy limit, unless there is a pole term contribution associated with a mass-degenerate boson-fermion pair.

Therefore we consider the gravitino-photon-photino vertex[30]. Let q^ρ and k^μ be the momenta of the photon and the gravitino, respectively. The vertex factor is

$$\frac{\kappa}{4} [\gamma^\rho, \gamma^\sigma] \gamma^\mu q_\rho . \tag{44}$$

For a light gravitino of helicity $\pm\frac{1}{2}$ we have the extra factor (40) coming from the wave function. For external gravitinos and photinos we get the effective vertex factor

$$\frac{\kappa}{m_{3/2}\sqrt{6}} \gamma^\sigma q^2 . \tag{45}$$

The factor q^2 cancels the $1/q^2$ coming from the photon propagator. The one-photon exchange contribution to the production of gravitino-antiphotino or photino-antigravitino pairs is exactly given (in the limit of small κ and $m_{3/2}$) by the local four-fermion interaction[4]

$$\mathcal{L}_{eff} = \frac{\kappa}{m_{3/2}\sqrt{6}} (\bar{\lambda}_{gL} \gamma_\mu \lambda_{\gamma L} + \bar{\lambda}_{\gamma L} \gamma_\mu \lambda_{gL}) J^\mu_{electromagnetic} \tag{46}$$

This expression for gravitino production is identical to the second term in expression (25) found for goldstino production[23]. The relation

$$\frac{\kappa}{m_{3/2}\ \sqrt{6}} = \frac{e_g}{\Delta m^2} \tag{47}$$

expresses the continuity of the theory when gravity is neglected. This is illustrated in Fig. 6:

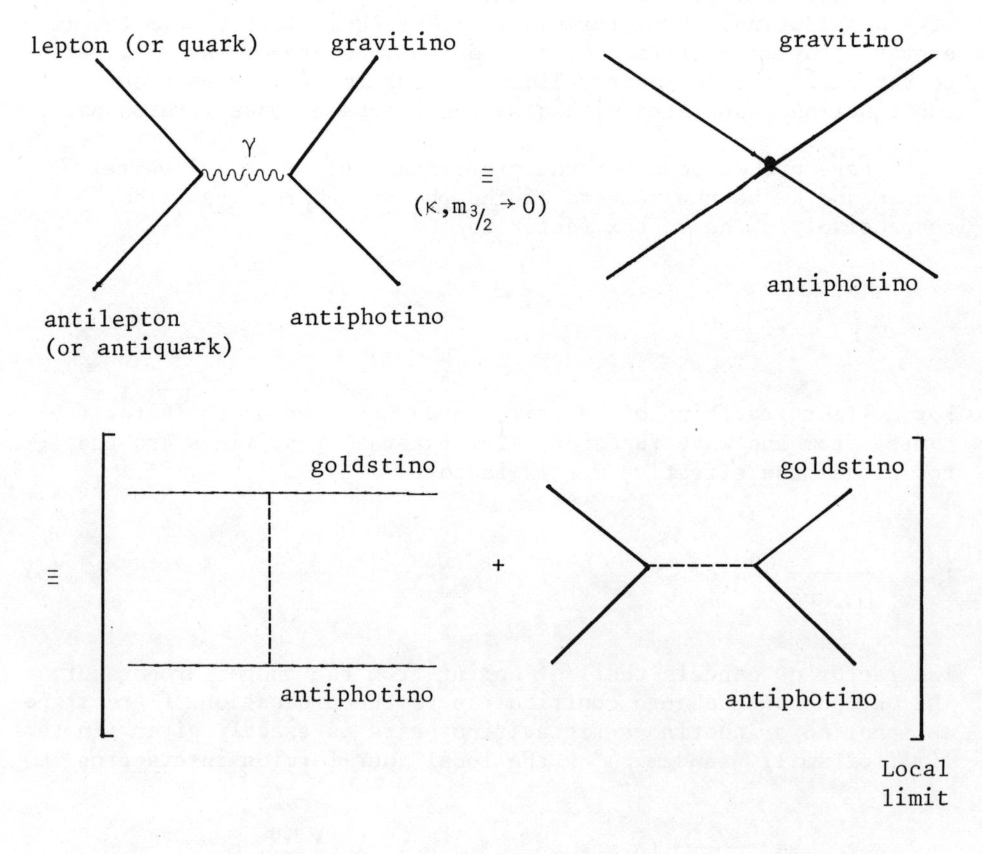

Fig. 6 Equivalence between the production of a gravitino antiphotino and a goldstino-antiphotino pairs.

More precisely a complete study involving other graphs shows that the equivalence between the gravitino and the goldstino formulations holds at any values of the momenta, and not only in the local limit approximation.

5.6 Looking for gravitinos in Ψ decays

It is now easy to re-interpret the results of Subsection 4.6 in terms of gravitinos instead of goldstinos (see Fig. 7).

The experiment gives an upper limit on the effective strength of the gravitino interaction (proportional to $\kappa/m_{3/2} \sim 1/d$), so that we get a lower limit on the gravitino mass[4]:

$$m_{3/2} > 1.5 \times 10^{-8} \text{ eV}/c^2 \ . \tag{48}$$

This is not yet a constraining limit on the theory (although it does exclude the too naïve estimate $m_{3/2} \sim \kappa\, m^2_{proton}$), but it has the merit of being model-independent.

It is the mass of the gravitino which will determine its production rate and its interaction cross-section. Both are proportional to

$$\frac{4\pi}{3}\frac{G_{Newton}}{m^2_{3/2}} = \frac{\kappa^2}{6\, m^2_{3/2}} = \frac{1}{d^2} = \frac{e^2_g}{(\Delta m^2)^2} \ . \tag{49}$$

Only if the gravitino mass is very small do we have a chance to detect directly its effects, i.e., the effects of gravitational interactions, in particle physics. But then it would, also, behave essentially like a spin-½ goldstino.

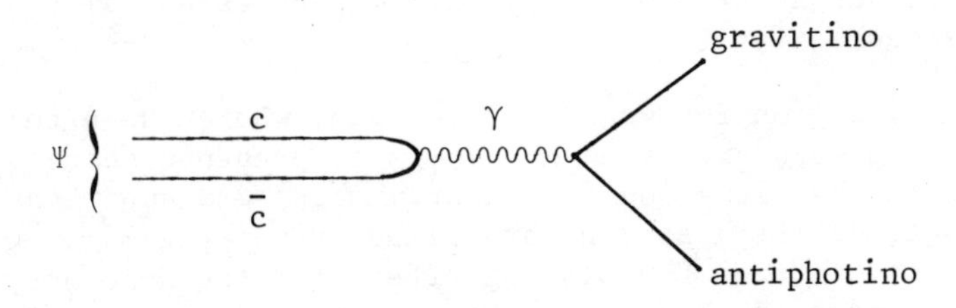

Fig. 7 The main diagram responsible for the decay Ψ → gravitino + antiphotino.

6. SUPERSYMMETRY AND GRAND UNIFICATION

It is believed that weak, electromagnetic and strong interactions are due to the exchanges of SU(2) × U(1) × SU(3) gauge bosons between fermions, leptons and quarks. Among the possible ways to go further are supersymmetry and grand unification[31]. The former relates bosons and fermions, the latter leptons and quarks. We shall discuss briefly the similarities and differences between these two approaches, and whether they can be combined.

6.1 Effective strengths and distinctive features of the interactions

In grand unification schemes SU(2), U(1) and SU(3) are subgroups of a simple group such as SU(5), O(10) or larger groups. The intermediate bosons $W_\pm$ and Z, with masses $\sim 10^2$ GeV/c^2, belong to the same multiplet as the massless photon and gluons together with very heavy gauge bosons with masses estimated to be $\sim 10^{15}$ GeV/c^2. The scale of the breaking of the grand unification group for lepton and quark multiplets is considerably smaller. The existence of these different scales is the so-called hierarchy problem. The effective strength of weak interactions is fixed by the Fermi coupling constant

$$G_F \sim \frac{g^2}{m_W{}^2} \sim \frac{1}{v^2} .$$

The effective strength of the exchanges of the very heavy bosons, collectively denoted by X, is fixed by

$$\frac{g^2_{GUT}}{m_X^2} \sim \frac{1}{v^2_{GUT}} .$$

The observability of their effects, for example in B- or L-violating processes such as proton decay, will depend essentially on the very heavy masses m_X.

The situation for grand unification, where the effects of the very heavy bosons are fixed by $\alpha_{GUT}/m_X{}^2$, presents some formal similarity with the situation for supersymmetry and supergravity, where the effects of the gravitino are fixed by $G_{Newton}/m^2_{3/2}$. One may be able to detect B- or L-violating effects of the X-bosons, e.g., proton decay, even if these gauge bosons are very heavy. On the contrary, because of the smallness of the Newton coupling constant, direct effects of gravitinos may only be observed for those of them, if any, which would be very light. Those would be neutral.

Three facts contribute to make the search for gravitinos harder than the search for proton decay:

i) the Newton coupling constant (with the proton mass as unit) is very small compared to the grand unification fine-structure constant α_{GUT} (or α).

ii) While for grand unification we think we know complete multiplets of leptons and quarks, for supersymmetry we do not have this chance with bosons and fermions.

iii) While the interactions of the very heavy X bosons may violate B- or L-quantum numbers and induce processes such as proton decay, the interactions of a very light gravitino would not have such a striking property. However, a gravitino mass term would violate R-invariance by two units so that it would lead to $\Delta R = \pm 2$ processes, unfortunately both very rare and less remarkable than proton-decay.

6.2 Compatibility of supersymmetry and grand unification

The most naïve approach would be some direct product of these invariances. Then we have of course the consequences of each invariance separately, but we could also look for both of them at the same time by searching for gravitino production in proton decay (see Fig. 8)!

The decay rate is

$$\Gamma \sim \alpha \left[\frac{\alpha_{GUT}}{m_X^2} \right]^2 \frac{G_{Newton}}{m_{3/2}^2} m_p^9 \tag{50}$$

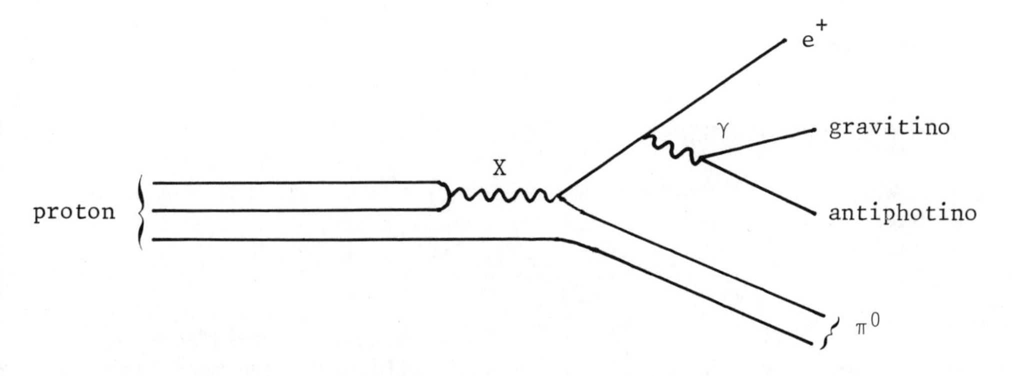

Fig. 8 One of the graphs responsible for the decay proton → π^0 e^+ gravitino antiphotino.

If the gravitino mass lies somewhere between $10^{-(6 \text{ or } 7)}$ eV/c^2 and the proton mass, a very crude estimate of the partial lifetime is:

$$10^{40} \text{ years} \lesssim \tau(\text{proton} \to \text{gravitino} + \ldots) \lesssim 10^{70} \text{ years} . \quad (51)$$

We shall not comment on the feasibility of this experiment, since there are more important consequences. The existence of the superpartners of the ordinary particles (gluinos and photino, bosonic leptons and quarks, heavy fermions) would modify the estimates which have been made for the SU(2) × U(1) mixing angle θ and the proton lifetime in the minimal SU(5) model. A precise evaluation of these effects would require the knowledge of the particle spectrum.

Moreover, the possibility of realizing a spontaneous breaking of supersymmetry implies very strong constraints on a grand unified theory. If we assume that spin-½ leptons and quarks have spin-0 partners, the bosonic partner of the goldstino should be, in part, a new spin-1 boson with axial couplings to leptons and quarks. In the gauge theory framework, the gauge group must be bigger than SU(5). For example, it could be SU(5) × U(1); SU(5) is the grand unified gauge group and the U(1) factor corresponds to a new gauge interaction responsible for the spontaneous supersymmetry breaking and related to gravitation (see Subsection 3.4 for a necessary condition on the gauge group).

If the supersymmetry is such that spin-½ leptons and quarks can have spin-1 partners, the constraints on the gauge group are different. We need either a larger gauge group or the consideration of spin-1 particles which would not have ordinary gauge interactions. In this case, and in particular in the framework of extended supersymmetry which involves large multiplets with many new fields, it seems hard to find out whether supersymmetry is indeed broken or not, and what are the masses and interactions of the boson and fermion states.

7. CONCLUDING REMARKS

Whatever the fundamental theory of Nature is, the framework we have presented may be used as soon as there is a trace of a boson-fermion supersymmetry at energies where field theory is still valid.

The actual observability of this supersymmetry depends on the mass scale at which the boson-fermion multiplet structure becomes apparent. In addition, the direct observability of the effects of the corresponding gauge particle, the gravitino, depends on its mass which fixes the effective strength of its interaction.

There is no incompatibility between the supersymmetry and grand unification approaches, but complementarity. Grand unification aims at a unified description of weak, electromagnetic and strong interactions while supersymmetry seems the natural framework for the introduction of gravity.

We know how to construct spontaneously broken supersymmetric theories of weak, electromagnetic and strong interactions, and how to search experimentally for the effects of supersymmetry. Introducing gravity does not lead to any presently observable consequence, but it is conceptually very important for the connection between gravity and gauge interactions.

REFERENCES

1. Yu.A. Gel'fand and E.P. Likhtman, JETP Letters 13 (1971) 323; J. Wess and B. Zumino, Nuclear Phys. B70 (1974) 39; For a review article, see: P. Fayet and S. Ferrara, Phys. Reports 32C (1977) 249.
2. P. Fayet, Phys. Letters 69B (1977) 489.
3. G.R. Farrar and P. Fayet, Phys. Letters 76B (1978) 575; 79B (1978) 442.
4. P. Fayet, Phys. Letters 84B (1979) 421.
5. G.R. Farrar and P. Fayet 89B (1980) 191.
6. S. Ferrara, D.Z. Freedman and P. Van Nieuwenhuizen, Phys. Rev. D13 (1976) 3214; S. Deser and B. Zumino, Phys. Letters 62B (1976) 335.
7. P. Fayet, Phys. Letters 70B (1977) 461.
8. P. Fayet, Proceedings of the Orbis Scientiae, Coral Gables (Florida, USA) Jan. 1978 (Plenum Publ. Corp., New York) 413.
9. G.R. Farrar, Proceedings of the Int. School of Subnuclear Physics, Erice (Italy), August 1978.
10. P. Fayet, Proceedings of the XVIIth Winter School of Theoretical Physics, Karpacz (Poland), February 1980.
11. P. Fayet and J. Iliopoulos, Phys. Letters 51B (1974) 461.
12. J. Wess and B. Zumino, Nuclear Phys. B78 (1974) 1.
13. S. Ferrara and B. Zumino, Nuclear Phys. B79 (1974) 413; A. Salam and J. Strathdee, Phys. Letters 51B (1974) 353.
14. P. Fayet, Nuovo Cimento 31A (1976) 626.
15. A. Salam and J. Strathdee, Nuclear Phys. B76 (1974) 477. S. Ferrara, J. Wess and B. Zumino, Phys. Letters 51B (1974) 239.
16. P. Fayet, Nuclear Phys. B90 (1975) 104.
17. A. Salam and J. Strathdee, Nuclear Phys. B87 (1975) 85.
18. W.A. Bardeen, unpublished; B. de Wit and D.Z. Freedman, Phys. Rev. Letters 35 (1975) 827.
19. P. Fayet, Phys. Letters B84 (1979) 416.

20. L.F. Abbott and R.M. Barnett, Phys. Rev. Letters 40 (1978) 1303;
F. Dydak, Proc. of the Int. Conf. on High-Energy Physics, Geneva (1979) 25.
21. J. Bailey et al., Nuclear Phys. B150 (1979) 1.
22. F.B. Haile et al., Nuclear Phys. B138 (1978) 189.
23. P. Fayet, Phys. Letters 86B (1979) 272.
24. K. Olive, D. Schramm and G. Steigman, Enrico Fermi Institute preprint No. 80-06 (1980).
25. M. Breidenbach, private communication.
26. P.C. Bosetti et al., Phys. Letters 74B (1978) 143.
27. J.P. Dishaw et al., Phys. Letters 85B (1979) 142.
28. P. Fayet, Phys. Letters 78B (1978) 417.
29. S. Deser and B. Zumino, Phys. Rev. Letters 38 (1977) 1433;
E. Cremmer et al., Phys. Letters 79B (1978) 231.
30. S. Ferrara, F. Gliozzi, J. Scherk and P. van Nieuwenhuizen, Nuclear Phys. B117 (1976) 333.
31. J.C. Pati and A. Salam, Phys. Rev. D8 (1973) 1240;
H. Georgi and S.L. Glashow, Phys. Rev. Letters 32 (1974) 438;
H. Georgi, H.R. Quinn and S. Weinberg, Phys. Rev. Letters 33 (1974) 451;
A.J. Buras, J. Ellis, M.K. Gaillard and D.V. Nanopoulos, Nuclear Phys. B135 (1978) 66.

THE CALCULATION OF THE DECAY RATE OF THE PROTON

D. A. Ross

California Institute of Technology

Pasadena, California

1. INTRODUCTION

All attempts which have been made to construct a model which unifies the strong, weak, and electromagnetic interactions predict the existence of currents which transform quarks into leptons and this generally leads to the prediction that the proton is unstable. Now usually when a theorist predicts the existence of an exotic interaction for which there is no experimental evidence, he keeps one step ahead of the experimentalists by arguing that the mass of the particles which mediate this exotic interaction are arbitrarily massive so that as the experimentalists decrease the upper bound on this interaction the theorist increases these masses. In a certain class of grand unified theories this option is closed since the masses of the particles mediating the interactions which lead to proton decay are exactly calculable. This class of grand unified theories is the class obeying the "desert hypothesis" namely the hypothesis that between the threshold for production of W's and Z's (~ 100 GeV) and grand unification there are no new degrees of freedom which open up. The simplest such model is the SU(5) model of Georgi and Glashow[1), and since the point of unification of the strong, weak and electromagnetic interactions is very insensitive[†] to the exact details of the model provided it obeys the desert hypothesis, I shall work with the SU(5) model.

[†] A possible exception to this is provided by models with accidental cross-overs[2)] in which the strong and weak coupling coincide but do not unify at the point of coincidence. Instead their effective couplings cross over and the 'weak coupling constant' becomes larger than the 'strong coupling constant'.

The 24-gauge bosons of SU(5) may be written as

$$A_\mu = \begin{pmatrix} V+\sqrt{\frac{2}{15}}\,U & V & V & X^r & Y^r \\ V & V+\sqrt{\frac{2}{15}}\,U & V & X^w & Y^w \\ V & V & V+\sqrt{\frac{2}{15}}\,U & X^b & Y^b \\ \bar{X}^r & \bar{X}^w & \bar{X}^b & \frac{W_0}{\sqrt{2}}-\sqrt{\frac{3}{10}}\,U & W^+ \\ \bar{Y}^r & \bar{Y}^w & \bar{Y}^b & W^- & -\frac{W_0}{\sqrt{2}}-\sqrt{\frac{3}{10}}\,U \end{pmatrix} \qquad (1)$$

where the quantities marked V are the eight gluons, the gauge bosons of the $SU(3)_c$ subgroup of SU(5) (there is a tracelessness condition relating the three V's along the diagonal of the above matrix so there are really only eight, not nine as in eq. (1)), the W^+, W^-, W^0 are the gauge bosons of the SU(2) subgroup of SU(5) and U is the gauge boson of the U(1) subgroup. One immediate prediction of this model is that the mixing angle of the SU(2) × U(1) model[3] of weak and electromagnetic interactions is the angle whose tau is the ratio of the two Clebsch-Gordan coefficients in the 4th or 5th diagonal elements of eq. (1). This gives $\sin^2\theta = 3/8$. This rather high value is considerably decreased by the calculable higher order effects (as we shall see later). The gauge bosons marked $X, Y, \bar{X}$ and $\bar{Y}$ (each comes in three colors) are gauge bosons which are not contained in the $SU(3)_c \times SU(2) \times U(1)$ subgroup. It is these gauge bosons that mediate the interactions which lead to proton decay. To see this we consider the way in which the fermions are inserted into the model. Each family of fermions consists of fifteen two-component spinors (left-handed and right-handed quarks each with $I_3 = \pm\frac{1}{2}$ and three colors, a left-handed and right-handed charged lepton and a left-handed neutrino). These are arranged into a 5-plet (fundamental) and 10-plet (rank-2 antisymmetric) representation of SU(5)

$$\psi_5 = \begin{pmatrix} d^r \\ d^w \\ d^b \\ e^+ \\ \bar{\nu} \end{pmatrix}_R \qquad (2)$$

$$\psi_{10} = \begin{pmatrix} 0 & u_c^b & -u_c^w & -u^r & -d^r \\ -u_c^b & 0 & u_c^r & -u^w & -d^w \\ u_c^w & -u_c^r & 0 & -u^b & -d^b \\ u^r & u^w & u^b & 0 & e^- \\ d^r & d^w & d^b & -e^- & 0 \end{pmatrix}_L \qquad (3)$$

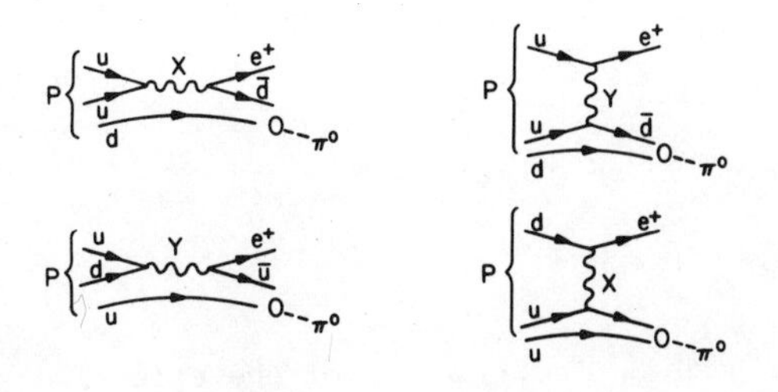

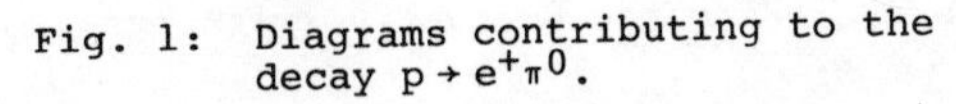
Fig. 1: Diagrams contributing to the decay $p \to e^+\pi^0$.

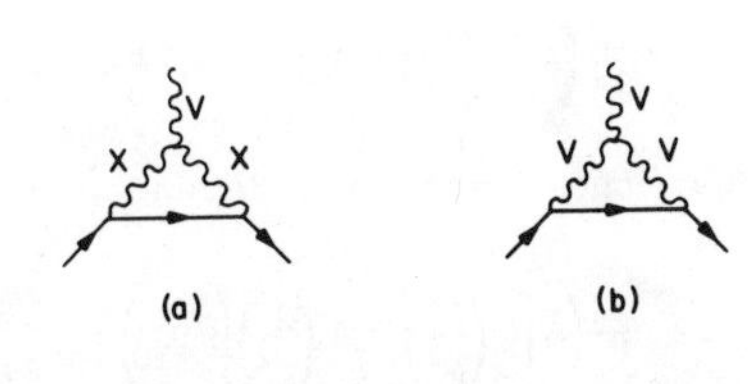

Fig. 2: Some diagrams contributing to the one-loop correction to effective coupling constants.

(the subscript c means charge conjugation). It can immediately be seen that the interaction term $\bar{\psi}_5\gamma^\mu A_\mu\psi_5$ contains interactions in which an X(Y) mediates transitions between d's and e^+ ($\bar{\nu}$). It also turns out that the term $\bar{\psi}_{10}\gamma^\mu A_\mu\psi_{10}$, not only contains more interactions in which X's or Y's couple to a quark and a lepton, but also ones in which X's or Y's can couple to two quarks. It is because of this dual role of the X's or Y's that proton decay is possible. Examples of diagrams contributing to the decay mode $p \to e^+\pi^0$ are shown in Fig. (1). In each diagram two quarks fuse to give a lepton and an antiquark which pairs up with the trivial (spectator) quark inside the proton to make a pion. In each diagram there is an X or Y exchanged and if the mass of these X's or Y's is much larger than the proton mass, the X(Y) propagators may be replaced by $1/M_X^2$ or $1/M_Y^2$, (in the SU(5) model it turns out that $M_X \simeq M_Y$ so from now on I shall use X to mean X or Y) so the proton lifetime is proportional to M_X^4. To calculate the proton lifetime, therefore, we must first calculate M_X.

2. EFFECTIVE COUPLING CONSTANTS

The principle underlying the calculation of M_X is the same as that which solves another problem of grand unified theories. How can a theory with a simple gauge group and therefore only one coupling constant simultaneously describe the strong interactions, whose coupling constant is large, and the weak and electromagnetic interactions whose coupling constant is weak? The answer is that it is not this common renormalized coupling constant that determines the strength of these interactions, but the effective coupling constant for each interaction, defined in some way so that the perturbation expansion for a particular process does not contain powers of large logarithms. These effective couplings are in general different for different interactions as can be seen by looking at the diagrams of Fig. 2 which contribute to the renormalization of a coupling constant in one loop. In Fig. 2a the internal loop contains massive particles of mass M whereas in Fig. 2b we have only massless particles (we neglect for the moment the quark masses). We take the invariant mass of each external leg to be $-Q^2$, with $Q^2 \ll M^2$. Each diagram contains an ultraviolet divergence, which for the moment we will regularize with a cutoff, Λ. In Fig. 2a the finite part may be written as a Taylor series in Q^2/M^2, thus its contribution may be written

$$\frac{\alpha}{4\pi}\left(A\,\ell n\left(\frac{\Lambda^2}{\mu^2}\right) + b_0 + b_1\frac{Q^2}{M^2} + \ldots\right) \tag{4}$$

whereas the functional dependence of Fig. 2b on Q^2 must be

$$\frac{\alpha}{4\pi}\left(A\left[\ell n\left(\frac{\Lambda^2}{\mu^2}\right) - \ell n\left(\frac{Q^2}{\mu^2}\right)\right] + c_0\right) \tag{5}$$

If we now define the contribution to β (the $\ell n\, Q^2$ derivative of the effective coupling constants) from these diagrams by taking the derivative with respect to $\ell n\, Q^2$ we obtain

$$\frac{\alpha}{4\pi}\, b_1 \frac{Q^2}{M^2} + \ldots \tag{6}$$

from Fig. 2a and

$$-\frac{\alpha}{4\pi}\, A \tag{7}$$

from Fig. 2b. This is the usual result that to one-loop β is minus the coefficient of the infinity, whereas for Fig. 2a we obtain a contribution which vanishes as $Q^2 \to 0$. This is the famous decoupling theorem of Appelquist and Carazzone[4]. An approximation which is often made is to assume that if $Q^2 < 4M^2$ (the threshold for the production of the two massive particles in the Minkowskian region) the diagram of Fig. 2a should be completely neglected, and when $Q^2 > 4M^2$ it should be included with its full value as though the masses in the internal loops were zero. We will adopt this approximation temporarily and I shall discuss the errors involved later. Now in Fig. 3 we have all the diagrams contributing to the one-loop renormalization of the SU(5) coupling constant. Fig. 3a contains all the diagrams which involve the heavy X particles and Fig. 3b contains all the other diagrams. When the incoming gauge boson is a gluon we get (for $Q^2 >> M_X^2$) contributions to β of $\frac{\alpha}{4\pi}\cdot\frac{22}{3}$ from Fig. 3a and $\frac{\alpha}{4\pi}(11 - \frac{2}{3}\, n_f)$ from Fig. 3b (where n_f is the number of fermion flavors). When the incoming gauge boson is a W boson we get $\frac{\alpha}{4\pi}\cdot 11$ from Fig. 3a and $\frac{\alpha}{4\pi}(\frac{22}{3} - \frac{2}{3}\, n_f)$ from Fig. 3b. The sum of the contributions is, of course, the same in both cases since for $Q^2 >> M_X^2$ we are in the unified regime where W's and gluons become indistinguishable. For $Q^2 << 4M_X^2$ we cut off the diagrams containing X's and this cuts off a different value for the two subgroups, so we recover the expected result that β for gluons corresponds to that of an SU(3) gauge theory, whereas for W-bosons we have the β corresponding to an SU(2) theory.

The effective coupling constant trajectories are shown in Fig. 4. Above $Q^2 = 4M_X^2$ there is one coupling constant describing the strong, weak and electromagnetic interactions. Below $Q^2 = 4M_X^2$ the three effective couplings split and follow different trajectories (the U(1) coupling constant decreases with decreasing Q^2 since it has a positive β). Below $Q^2 = 4M_W^2$, α_W remains constant since the diagrams renormalizing α_W contain particles whose mass is M_W and they also decouple below their threshold. Thus we have to do two things in order to calculate M_X

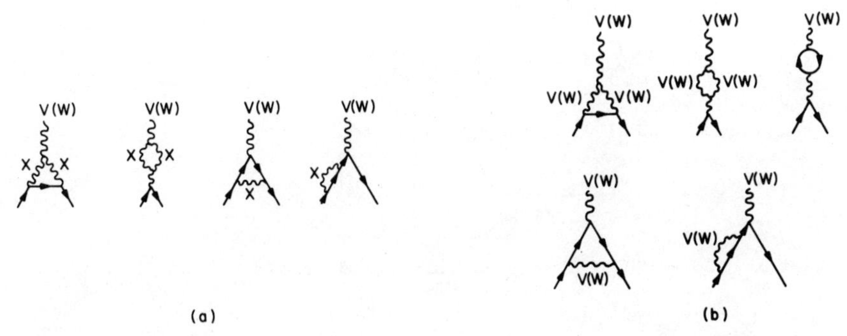

Fig. 3 All diagrams contributing to the strong or weak effective couplings in one loop. Fig. 3a contains all the diagrams with heavy X or Y particles in the internal loops.

i) Project the three trajectories as accurately as possible;

ii) Match the low energy coupling constants with the experimental values.

An approximate calculation can be done very quickly[5]. We have to solve the three differential equations

$$\frac{\partial \bar{\alpha}_i(Q^2)}{\partial \ln Q^2} = \beta_i \, \bar{\alpha}_i(Q^2) \tag{8}$$

for i = S,W or U, to project the three $\bar{\alpha}_i(Q^2)$ from $4M_W^2$ to $4\,M_X^2$ the solutions to eq. (8) are

$$\frac{1}{\alpha_{GUM}} - \frac{1}{\bar{\alpha}_S(4M_W^2)} = \frac{(11 - 2n_f)}{4\pi} \ln \frac{M_X^2}{M_W^2} \tag{9a}$$

$$\frac{1}{\alpha_{GUM}} - \frac{1}{\bar{\alpha}_W(4M_W^2)} = \frac{(\frac{22}{3} - 2n_f)}{4\pi} \ln \frac{M_X^2}{M_W^2} \tag{9b}$$

$$\frac{1}{\alpha_{GUM}} - \frac{1}{\bar{\alpha}_U(4M_W^2)} = - \frac{2n_f}{4\pi} \ln \frac{M_X^2}{M_W^2} \quad , \tag{9c}$$

where α_{GUM} is the common value of the three coupling constants at $Q^2 = 4M_X^2$. We can eliminate two of these equations by using two relations from the Glashow-Weinberg-Salam model

$$\alpha_W = \frac{\alpha_{e.m.}}{\sin^2\theta} \quad , \tag{10}$$

and

$$\alpha_U = \frac{5}{3}\,\alpha'_{G.W.S.} = \frac{5}{3}\,\frac{\alpha_{e.m.}}{\cos^2\theta} \quad , \tag{11}$$

where $\sin^2\theta$ is the mixing angle (whose value is 3/8 if we neglect higher order effects and put $\alpha_W = \alpha_U$). This gives us

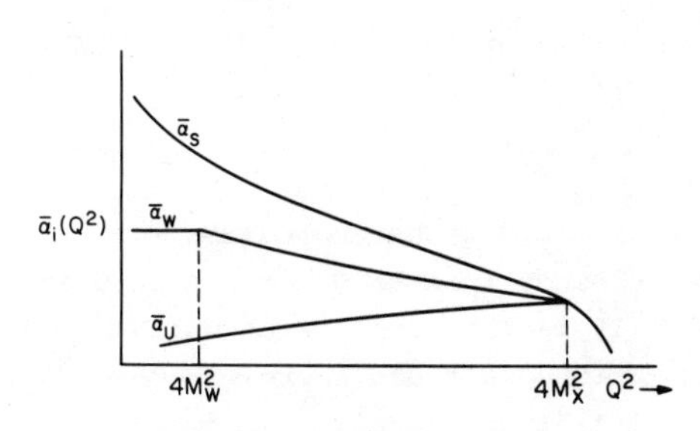

Fig. 4: The trajectory of the three effective couplings in the θ-function approximation.

$$\frac{1}{\alpha_{e.m.}} = \frac{8}{3\alpha_S(4M_W^2)} = \frac{11}{2\pi} \ln\left(\frac{M_X^2}{M_W^2}\right) \tag{12}$$

Now we calculate $\bar{\alpha}_S(4M_W^2)$ from the QCD equation

$$\bar{\alpha}_S(Q^2) = \frac{4\pi}{(11-2f)\ln(Q^2/\Lambda^2)} \tag{13}$$

where we take 400 MeV for Λ and inserting 1/137 for $\alpha_{e.m.}$ we obtain

$$M_X = 2 \times 10^{16} \text{ GeV} \tag{14}$$

We could have solved equations 9), 10) and 11) for $\sin^2\theta$ and obtained

$$\sin^2\theta = .2 \tag{15}$$

The higher order effects[6)] make this quantity .21 which compares favorably with the experimental value of .23 ± .015.

This calculation neglects many things and is clearly only an approximation. Whereas usually in physics approximations are good enough in this case we must be very careful for two reasons

i) The proton lifetime is proportional to the fourth power of M_X.

ii) The calculation gives $\ln M_X$ and not M_X directly so that any error is amplified in the exponentiation.

Thus we see that it is very important to go over the calculation making as many improvements as possible in order to get as accurate an estimate of $\ln M_X$ as possible.

3. FERMION MASS THRESHOLDS

The first such improvement was performed by Buras, Ellis, Gaillard and Nanopoulos[7)]. They observed that the value of Λ was determined from e-p experiments with $Q_2 \leq 10$ GeV^2 (at the time of the writing of their paper) and between 10 GeV^2 and $4M_W^2$ one crosses the threshold for the production of charm, bottom and presumably also top quarks. The exact behavior of β in the threshold region[8)] is given by

$$\beta = \frac{\alpha^2}{4\pi} \sum_i \frac{2}{3}\left(1 + \frac{2m_i^2}{\sqrt{Q^4 + 4m_i^2Q^2}} \ln\left\{\frac{\sqrt{Q^4 + 4m_i^2Q^2} - Q^2}{\sqrt{Q^2 + 4m_i^2Q^2} + Q^2}\right\}\right), \tag{16}$$

where m_i is the mass of the ith fermion threshold. The expression inside the parentheses has the properties that it tends to 1 as $Q^2 \to \infty$, to zero as $Q^2 \to 0$ and has the value .38 for $Q^2 = 4m_i^2$. It is this expression which would be replaced by $\theta(Q^2 - 4m_i^2)$ in the approximation described above. However, Buras et al., integrated the R.H.S. of eq. (16) numerically across the thresholds and found that this increased M_X by a factor of 2.

The next improvement, which turns out to be numerically the most significant, was simultaneously noticed by Marciano[9)] and by Goldman and myself[10)]. The value of $\alpha_{e.m.}$ which was used in the above calculation was 1/137. But this is the value of the Thomson limit, $Q^2 = 0$. We should have used $Q^2 = 4M_W^2$ and thus we must calculate the Q^2 dependence of $\alpha_{e.m.}$ from the diagrams of Fig. 5. We find that $\alpha_{e.m.}$ at the jth threshold is given in terms of $\alpha_{e.m.}$ at the $j\text{-}1^{th}$ threshold by

$$\alpha_{e.m.}(4m_j^2) = \frac{\alpha_{e.m.}(4m_{j-1}^2)}{1 + \frac{\alpha_{e.m.}(4m_{j-1}^2)}{3\pi} \sum_{i=1}^{j-1} q_i^2 \left(\ln \frac{m_j^2}{m_{j-1}^2} + \ln 4 - \frac{5}{3} \right)} \tag{17}$$

where q_i is the charge of the ith quark and m_j is the constituent mass of the jth quark. We find for $\alpha_{e.m.}(4m_W^2)$ the values 1/128.6 and thus reduced M_X by a factor of 10 (4 orders of magnitude in the proton lifetime). We see from eq. (17) that most of the damage is done by the light fermions, since these have large logarithm leverage. Thus the result is rather insensitive to the actual mass of the top quark (which we arbitrarily took to be 25 GeV).

As we know, what actually happens in e^+e^- scattering as one increases the energy is not the production of these quark thresholds, but rather the opening of multipion thresholds and vector meson resonances, etc. Semi-local duality relates the average of this total cross-section to an equation of the form eq. (17) but it is not a precise enough theory to tell us exactly what masses to insert for the thresholds. We took the constituent quark masses (i.e., $m_p/3$ for u and d quarks, $m_s - 2m_p/3$ for the s-quark, etc.). This has recently been checked by Ellis, Gaillard, Nanopoulos, and Rudaz[11)] who, using an idea of Paschos[12)], have "measured" the effective electromagnetic coupling constant by integrating the experimental value of R through the threshold region. They find a value of $\alpha_{e.m.}(4M_W^2)$ which is consistent with ours, but because of the experimental errors in R they have an uncertainty of .5 in $1/\alpha_{e.m.}(4M_W^2)$ (which reflects itself in a 20% uncertainty in M_X).

4. GAUGE BOSON MASS THRESHOLDS

Now let us return to the θ-function approximation for treating vector boson mass thresholds, i.e., take into account the fact that diagrams with

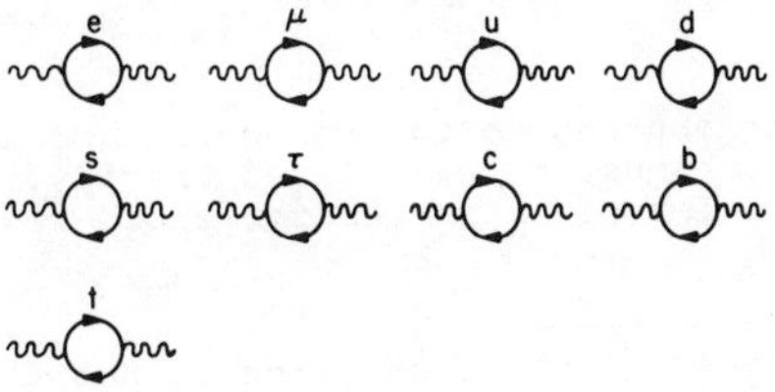

Fig. 5. Diagrams contributing to the correction of $\bar{\alpha}_{e.m.}(4M_W^2)$

particles of mass M in the internal loops give a non-vanishing contribution to β for $Q^2 < 4M^2$ and above $Q^2 = 4M^2$; β only reaches its constant value asymptotically. That means that we have to calculate the contribution to β from the diagrams of Fig. 6 in the threshold region[13]. However, in the threhold region the effective coupling constant is not well-defined. In the tree diagram the coupling of a gauge boson to fermions is given by γ^μ. In higher orders it has the general form†

$$\left[A \ln (Q^2/\mu^2) + C_1(Q^2,M^2)\right]\gamma^\mu + C_2(Q^2,M^2)q^\mu + C_3(Q^2,M^2)q^\nu\sigma^\mu \quad . \tag{18}$$

The coefficient of $\ln (Q^2/\mu^2)$ is unambiguous so that for $Q^2 >> M^2$ (where the functions C_i tend to a constant), β is well-defined. However, the functions C_i depend on the gauge (since we are now looking at off-shell Green's functions) and the exact momentum assignment of the external legs. Furthermore, we would get a different answer if we considered the triple gauge boson coupling instead of the gauge boson coupling to fermions[14]. However this arbitrariness in the definition of the effective coupling constants does not affect the calculation of a physical quantity like M_X. It is only necessary to be consistent in our definitions of the weak, strong, and electromagnetic interaction coupling constants (all sets of consistently defined effective coupling constants will intersect at infinite Q^2 and will have their approach to intersection determined by the same M_X). For convenience we look at the coefficient of γ_μ (defined in Ref. 13) in the Feynman gauge with symmetric momentum (i.e., the invariant mass of all external legs is set equal to $-Q^2$).

Now the strong interaction coupling constant is not usually given in this scheme. Rather it is given as a function of $\Lambda_{\overline{M.S.}}$, i.e., divergent Green's functions are calculated using the dimensional regularization scheme[15] and subtracting the pole at $d = 4$ (d = number of dimensions) together with the accompanying $\ln 4\pi - \gamma_\varepsilon$. In order to compare this coupling constant with the weak and electromagnetic coupling constants whose threshold effects are calculated in the above-mentioned scheme, we must translate the strong coupling constant into that scheme[16]. In other words, we must calculate the diagrams

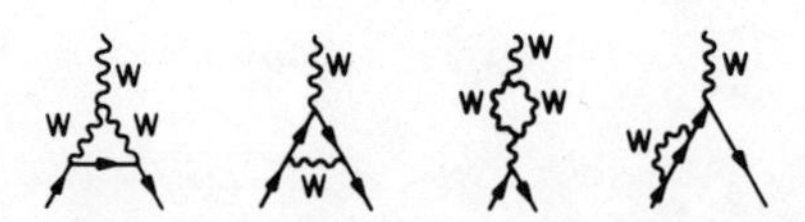

Fig. 6. Diagrams contributing to the connection of the effective weak coupling, which must be calculated exactly, keeping M_W, to determine the behavior of β_W in the threshold region $Q^2 \sim 4M_W^2$.

† This way of defining the effective coupling is not unique. By manipulating the γ-matrices, terms in the coefficient of γ_μ can be expressed as a linear combination of the other two coefficients. Thus there is no unique method for projecting out the coefficient of γ_μ. In Ref. 13 an unambiguous prescription for projecting out the coefficient of γ_μ is described, but many other equally valid prescriptions exist.

of Fig. 6 (with internal gluons), picking out the coefficient of γ_μ, in the Feynman gauge, with symmetric momentum, as a function of $\alpha_{\overline{M.S.}}$. We find for this Green's function

$$\alpha^{SYM}(Q^2) = \alpha_{\overline{M.S.}}(Q^2)\ [1 + 1.7\ \alpha_{\overline{M.S.}}(Q^2)] \quad . \tag{19}$$

Thus the strong interaction coupling constant in this "momentum subtraction" scheme is larger than in the $\overline{M.S.}$ scheme so that the intersection of the weak and strong coupling constant occurs later. The value of M_X is increased by a factor of two from this effect. We must also remember that the value of $\alpha_{e.m.}$, which we discussed above, was not defined in the symmetric momentum subtraction scheme, but rather the scheme in which the fermions are kept on their mass shell and only the photon goes off-shell. When this is translated to the symmetric momentum subtraction scheme we find $\alpha^{SYM}_{e.m.}(4M_W^2) = 1/129.2$.

Fig. 7 shows the β function for $\bar{\alpha}_W(Q^2)$ in the region $Q \sim 4M_W^2$. We see that not only is it different from the value it would have in the θ-function approximation (shown by the broken) line, but that below $Q^2 = 4M_W^2$ the value of β is positive. The reason for this is that the positive contribution to β from fermion loops (Fig. 8) has a suppression factor $Q^2/(Q^2 + M_W^2)$ from the W-boson propagator. If we try to simulate this with a θ-function we find $\theta(Q^2 - 1.1M_W^2)$ so we see that these positive contributions become operative before the negative contributions from diagrams with W-bosons in internal loops (whose thresholds are genuinely around $Q^2 = 4M_W^2$). This enhances the threshold effect, since the effect on the coupling constant above threshold is the integral under the solid line of Fig. 7 minus the integral under the broken line. This is plotted in Fig. 9 where the broken line is the development of $\bar{\alpha}_W(Q^2)$ in the θ-function approximation and the solid line is what really happens. Note that this line rises slightly below $Q^2 = 4M_W^2$ reflecting the positiveness of β in this region. For $Q^2 \gg 4M_W^2$ the 'real' $\bar{\alpha}_W(Q^2)$ is 3% larger than in the θ-function approximation. This has the effect of bringing down the intersection point by 1.7.

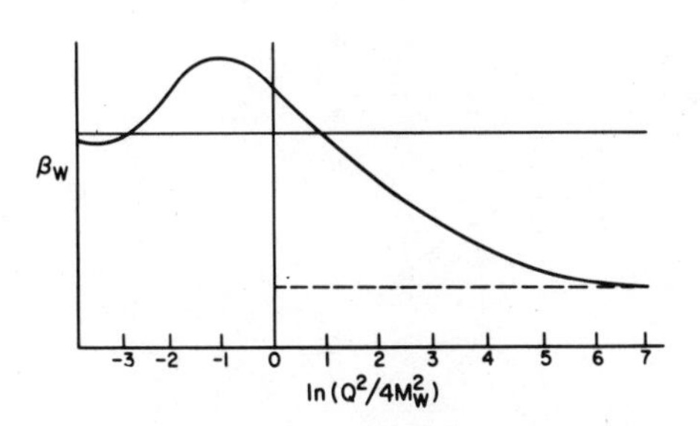

Fig. 7. The behavior of β_W in the threshold region.

Fig. 8. Fermion loop diagram which gives a positive contribution to β_W and has an effective threshold at $Q^2 \sim M_W^2$ from the massive W-propagator.

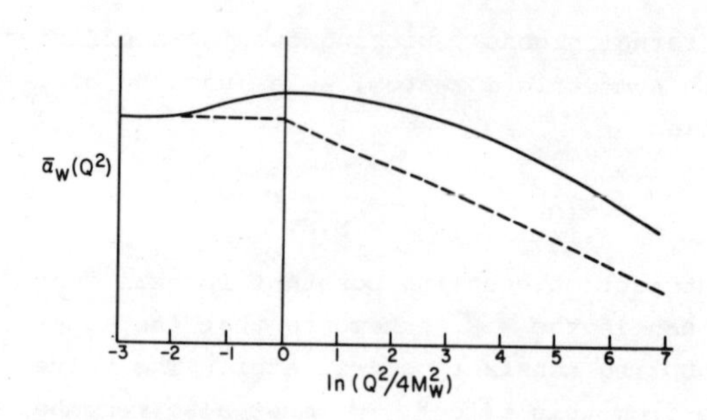

Fig. 9. The trajectory of $\bar{\alpha}_W(Q^2)$ in the threshold region. The broken line is what the trajectory would look like in the θ-function approximation.

In the θ-function approximation the three coupling constants meet at $Q^2 = 4M_X^2$. In fact they only meet at infinite Q^2 (their difference vanishes as M_X^2/Q^2 for $Q^2 >> M_X^2$). This is because for any finite value of Q^2 the effects of masses inside the diagrams with X (or Y) particles in the internal loops can be felt (and these diagrams have different weights for the different coupling constants). If we calculate the effective coupling constants taking these effects into account we obtain the result shown in Fig. 10. We see that at $Q^2 = 4M_X^2$ the coupling constants are visibly separated. This effect is partly cancelled by the fact that below $Q^2 = 4M_X^2$ the trajectories of the coupling constants do not reach their gradients obtained from calculating β in the $SU(3)_c \times SU(2) \times U(1)$ subgroup until we get down to $Q^2 << M_X^2$. The net effect of this threshold is to reduce M_X by a factor of 2.

It has been pointed out[17)] that there is a slight inconsistency in the above threshold analysis. In all gauges except the Landau gauge, the gauge parameter varies with Q^2. Thus if the threshold effects at $Q^2 \sim 4M_X^2$ should be calculated in whatever that gauge transforms into as we move from $Q^2 \sim 10^4$ GeV^2 to $Q^2 \sim 10^{29}$ GeV^2. This will change M_X by about 10%.

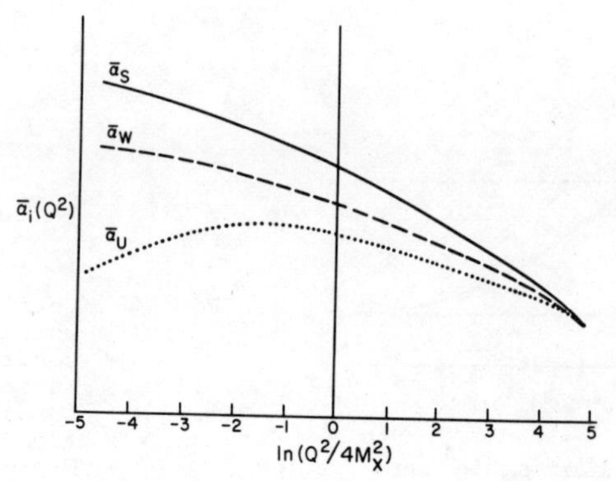

Fig. 10. The trajectories of the three effective coupling constants in the threshold region of the grand unified mass.

Recently a gauge invariant prescription for dealing with thresholds has been given by Binétruy and Schücker[18] (a similar method has also been used by Marciano[19]). The gauge invariant dimensional regularization scheme suffers from the drawback that the poles are independent of the masses, so we get no decoupling at energies below threshold (this means that the perturbation series for a Green's function way below threshold contains powers of large logarithms). In the method of Binétruy and Schücker for energies above threshold the poles are subtracted in the usual way, whereas below the threshold the term $\ln(Q^2/M^2)$ which accompanies the pole is also subtracted. This gives a gauge invariant renormalization prescription with the below threshold decoupling. Furthermore, the effect of passing through a threshold is calculated by calculating a Green's function for $Q^2 \gg M^2$ and $Q^2 \ll M^2$ and subtracting. It is not necessary to perform the complicated calculation in the threshold region†.

5. THE TWO-LOOP BETA

Taking into consideration the threshold effects is looking at the subleading logarithms. To be consistent to this order we must also take into consideration the two-loop contribution to β. In one-loop β_i (the β for the coupling constant α_i) depends only on α_i. In two-loops, diagrams like those shown in Fig. 11 [in fact Fig. 11 contains the only diagrams contributing to the off-diagonal elements of $\beta_{ij}^{(1)}$. All other diagrams cancel by Ward identities.] give a contribution which depends on all the α's. Thus we have

$$\frac{d\bar{\alpha}_i(Q^2)}{d\,\ln(Q^2)} = \beta_i^{(0)}\bar{\alpha}_i^2(Q^2) + \beta_{ij}^{(1)}\bar{\alpha}_j(Q^2) \quad , \tag{20}$$

where

$$\beta_i^{(0)} = \begin{pmatrix} 0 & & \\ & -22/3 & \\ & & -11 \end{pmatrix} + \frac{2}{3}n_f \begin{pmatrix} 1 & & \\ & 1 & \\ & & 1 \end{pmatrix} , \tag{21a}$$

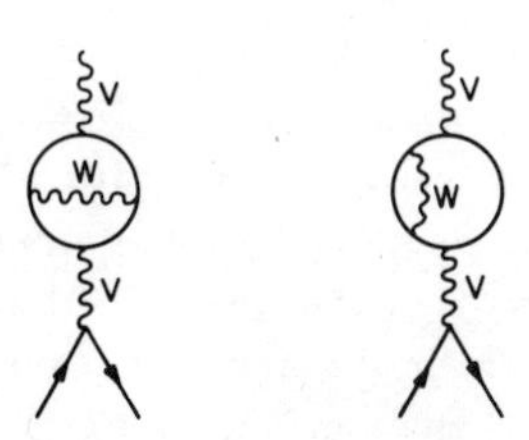

Fig. 11. Diagrams contributing to the off-diagonal elements of $\beta_{ij}^{(1)}$.

† For an alternative method of treating mass thresholds, see Ref. 20.

and

$$\beta_{ij}^{(1)} = \begin{pmatrix} 0 & & \\ & -136/3 & \\ & & -102 \end{pmatrix} + 2n_f \begin{pmatrix} 19/15 & 3/5 & 44/15 \\ 4/5 & 29/3 & 3 \\ 11/5 & 16/3 & 76/3 \end{pmatrix} . \tag{21b}$$

Thus eq. (20) is a set of three coupled differential equations which we solve by iteration. We first solve eq. (20) with $\beta_{ij}^{(1)}$ set equal to zero and plug the answer into the R.H.S. of eq. (20) and solve again. This gives us, for a value of Q^2 between the N-1th and Nth threshold

$$\bar{\alpha}_i(Q^2) = \frac{\bar{\alpha}_i(4m_{N-1}^2)}{1 + \beta_{i,N}^{(0)} \ln (Q^2/4m_{N-1}^2)} \Big\{1 + \beta_{ij,N}^{(1)} \bar{\alpha}_j(4m_{N-1}^2) \times \ln [1 - \beta_{i,N}^{(0)} \ln (Q^2/4m_{N-1}^2)\Big\} , \tag{22}$$

where the subscript N indicates that the value of β must be calculated between the N-1th and Nth threshold. Because the two-loop β has the same sign as the one-loop β, the strong coupling constant fall with Q^2 at a rate which is faster than that calculated in one-loop. This brings forward the 'intersection point' of the coupling constants by a factor of four, of which about a factor of two is due to the fact that $\bar{\alpha}_S(4M_W^2)$ is lower than that calculated in one-loop and another factor of two due to the faster descent of $\bar{\alpha}_S(Q^2)$ between $Q^2 = 4M_W^2$ and $Q^2 = 4M_X^2$.

6. SCALAR PARTICLES

Finally, we consider the effect of the Higgs bosons. Grand unified theories differ qualitatively from QCD in that they require that fundamental scalars couple to the gluons. If all the scalar particles within one multiplet have the same mass there is no effect on M_X, since the Higgs scalars change β by the same amount for each coupling constant. In the SU(5) model the scalars form a 5-plet and 24-plet representation of SU(5). The 5-plet contains the familiar doublet of the SU(2) × U(1) model and its mass is believed to be light[21]. Thus the effect of the Higgs on the weak sector has an effective threshold at $Q^2 \sim M_W^2$ (this comes from the massive W propagator as with the fermion loop diagrams). However, Ellis, Gaillard, and Nanopoulos[22] have shown that the Higgs particles which couple to gluons cannot have a mass lower than $\alpha_{GUM} M_X$ (they acquire such masses in higher orders even if they do not begin with them). Thus between 100 GeV and 10^{12} GeV the β for the weak coupling constant has a positive contribution from Higgs particles but the β for the strong coupling constant does not. This again decreases the effective intersection point by a factor of 1.7. We have assumed that the 24-plet does not affect M_X since all the masses of that multiplet should be approximately the same. However, if one introduces into the model too many scalar multiplets in which the components which couple to the weak group are much lighter than the components which couple to the strong group, the proton becomes unacceptably unstable. Even if we do not have very large mass differences between the components of a multiplet, the inclusion of large multiplets of scalars into the theory can introduce considerable uncertainties in the value of M_X[23].

8. RESULT FOR M_X

Our final result for M_X is

$$M_X = 4.2 \times 10^{14} \text{ GeV} \tag{23}$$

with $\Lambda_{\overline{M.S.}}$ taken equal to 400 MeV. The value of M_X is very sensitive to the value of $\Lambda_{\overline{M.S.}}$ ($M_X \sim (\Lambda_{\overline{M.S.}})^{.98}$) so that our lack of precise knowledge of $\Lambda_{\overline{M.S.}}$ introduces large uncertainties in the proton lifetime. Marciano and Binétruy and Schücker obtain a value of 6.3×10^{14}GeV for the same value of $\Lambda_{\overline{M.S.}}$. This 50% discrepancy seems large but if we compare our values of $\ell n\ (M_X/M_W)$, we agree to within 1%, which is an error of order $\alpha_S^2(4M_W^2)$ and reflects the fact that our different approaches to the threshold problem take into account different pieces of the next order corrections.

Recently Weinberg[24] has suggested a new way of calculating M_X. This method consists of integrating out the heavy X and Y fields in the functional integral and obtain an effective $SU(3)_c \times SU(2) \times U(1)$ theory. L. Hall is currently working out the details of this method in order to calculate M_X.

The value of M_X which we have calculated is the physical mass ($\bar{M}_X(Q^2 = M_X^2)$, the position of the pole in the X-propagator. The mass which is relevant to the calculation of the proton lifetime is $\bar{M}_X(Q^2 \sim m_p^2)$. The diagrams contributing to the renormalization of the mass of M_X are shown in Fig. 12. The diagram in Fig. 12a cuts off for $Q^2 < M_X^2$ by the decoupling theorem so there is no large logarithm in the difference between $\bar{M}_X(Q^2 \sim m_p^2)$ and the physical M_X from this diagram. The diagrams with fermions in the internal loops (Fig. 12b) are transverse only, i.e., they are of the form

$$\alpha_{GUM}(g_{\mu\nu}Q^2 - Q^\mu Q^\nu)\ell n\ (Q^2/M_X^2) \tag{24}$$

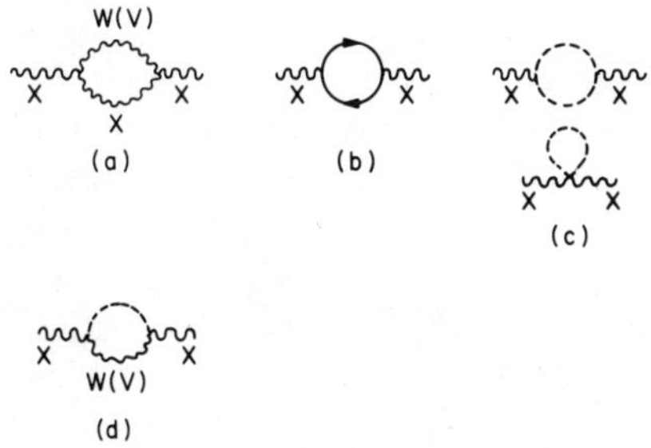

W(V)

X X X (a)

X X (b)

X X X X (c)

X X W(V) (d)

Fig. 12. Diagrams contributing to the renormalization of M_X.

so they give a large logarithm to Z_3 but not to the mass. The same is true for diagrams shown in Fig. 12c. This leaves the diagram of Fig. 12d in which the internal loop consists of one gauge boson and one scalar particle. This gives a contribution proportional to $\alpha_{GUM} \ln (m_H/M_X)$, but as explained above, this cannot be larger than $\alpha_{GUM} \ln \alpha_{GUM}$. The total effect of all these corrections is to reduce M_X by no more than 4%, which is negligible.

9. THE CALCULATION OF THE PROTON LIFETIME

In order to estimate the proton decay rate, having obtained a value for M_X, we must first consider the effect of gluon corrections on the external legs[7]. Diagrams showing this are given in Fig. 13, where the exchange of an X or Y gauge boson has been reduced to a point (these corrections are now nothing to do with the corrections to M_X). With this reduction the diagrams contain an ultraviolet divergence and may be written

$$\frac{\alpha}{2\pi} \ln (\Lambda^2/\mu^2) + \text{finite terms} \tag{25}$$

The coefficient of $\ln \Lambda^2$ is the anomalous dimension of the three-quark operator. The exchange of a massive particle forces the product of the two currents in the proton decay process to be at short distance and one can make a Wilson operator product expansion and write down a renormalization group equation to calculate this enhancement. It is found to be

$$\left(\frac{\bar{\alpha}_S (Q^2 \sim m_p^2)}{\alpha_{GUM}} \right)^{4/(11-2n_f)} = 5.5 \tag{26}$$

One can do the same thing (Refs. 16, 25, 26) for the weak and electromagnetic corrections and we find a further enhancement of a factor of two (the fact that such a large correction may be obtained from weak and electromagnetic interactions may be understood from the fact that if we tried to calculate in ordinary perturbation theory our expansion parameter would be $\alpha_W/\pi \ln (M_X^2/M_W^2) \sim .7$).

All the diagrams contributing to the process $p \to e^+ X$ are shown in Fig. 14. It is believed that the branching ratio into these modes is at least 80%. It can be seen that diagrams 14) and 14b) and diagrams 14d) and 14e) differ only in the channel of the exchanged heavy gauge boson. These pairs of diagrams therefore interfere. Figs. 14c) and 14f) do not interfere with these pairs since the helicity of the outgoing positron is different. We also have SU(6) factors, e.g., the amplitude in Fig. 14a) for finding a

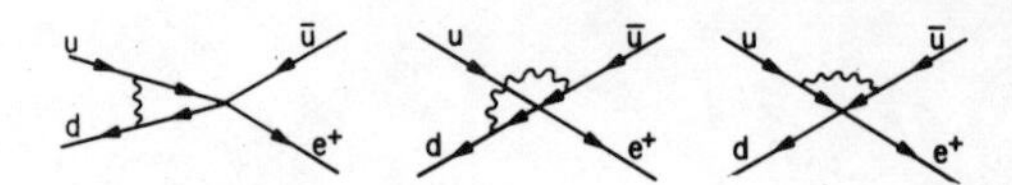

Fig. 13. Diagrams contributing in one-loop to the anomalous dimension of the three-quark operator in proton decay.

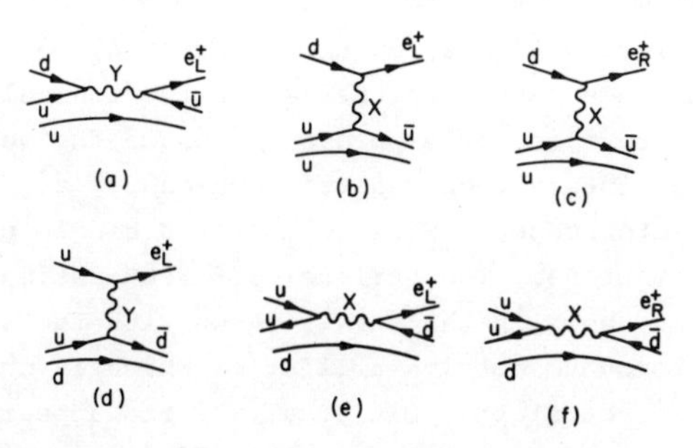

Fig. 14. Diagrams contributing to the decay $p \to e^+ X$.

u-quark and a d-quark in a spin-1 state etc. Taking all these into account we get a combinatoric factor of 171. Then we have a color sum factor of two obtained from contracting the two ϵ_{ijk} color factors in the vertex between an X or Y gauge boson and two quarks. The spin sum and phase space integral gives us a factor $\langle m_{qq}\rangle^2/\pi$, where m_{qq} is the mass of the diquark system. Using the non-relativistic quark model we just insert $2m_p/3$ for this value. Finally there is the question of the probability of finding two quarks on top of each other so that the quark fusion can occur. This is given by the wave-function of the proton at the origin. To obtain this number we compared with an analysis of Λ hyperon decay by Finjord[27] in which the wavefunction at the origin was also needed. His analysis is compatible with ours since he also used SU(6) wavefunction factors and an enhancement factor calculated using the renormalization group. In operator language it is not clear that this is exactly correct since we are actually comparing the matrix elements of different operators, but it seems to us to be a pretty reasonable guess. This gives the rather low value of .001 GeV^3 for $|\psi(0)|^2$. Piecing this all together we have

$$\Gamma = \left(\frac{4\pi\ \alpha_{GUM}}{M_X^2}\right)^2 \frac{\langle m_{qq}\rangle^2}{\pi}\ |\psi(0)|^2 \times 2 \times 171 \times \left[\frac{\alpha_S(Q^2 \sim m_p^2)}{\alpha_{GUM}}\right]^{\gamma_i/\beta_i} . \qquad (27)$$

Putting in the numbers we get a lifetime of

$$\tau = 2 \times 10^{30} \text{ years} \qquad (28)$$

For $\Lambda_{\overline{M.S.}} = 400$ MeV (remember that the proton lifetime is proportional to $(\Lambda_{\overline{M.S.}})^{3.9}$). This is just one order of magnitude above the current experimental limit[28].

A rather cleaner job has been done on the question of the wavefunction at the origin using the bag model by Din, Girardi, and Sorba[29], and by Donoghue[30]. Using the known wavefunctions for quarks inside a proton or pion bag they calculate the overlap integral between the amplitude for the ingoing quarks to be in a proton and the amplitude for the outgoing quarks to be inside a pion. For the mode $p \to e^+\pi^0$, Girardi et al., obtain a lifetime of about 10^{30} years in agreement with our calculation. They also obtain branching ratios which are consistent with those of Machacek[31], who obtained them using SU(6) only. Donoghue, on the other hand, believes that the single pion (and positron) mode cannot be treated in the non-relativistic bag model without including a form factor suppression, since the quark which emerges from the fundamental interaction has too much energy to form a pion and would prefer to form a vector meson. This reduces the single pion branching ratio from about 30% to about 9%. The various different estimates of different branching ratios are shown in the Table. Even with the suppression of the single pion mode, Donoghue obtains a lifetime which is one order of magnitude larger (i.e., $\sim 10^{31}$ years) than Din et al. There appears as yet to be no explanation for this discrepancy.

10. CONCLUSION

I have been considering only the SU(5) model in this calculation. The broad features of the calculation are common to all models with a desert between the W-mass and unification. Slight differences occur from different assignments of Higgs scalars and from threshold effects near unification when there are more X and Y gauge bosons than in the simplest SU(5) model. However, lumping all our uncertainties together and allowing for a reasonable upper bound of 600 MeV for $\Lambda_{\overline{M.S.}}$, we are confident that the proton lifetime in desert-type models must be less than 10^{32} years.

TABLE

Branching Ratios

Mode	Machacek (SU(6))	Donoghue (Bag Model)	Din, et al. (Bag Model)
$p \to e^+ + \pi^0$	34.6%	9%	15-31%
$p \to e^+ + \rho^0$	17.3%	21%	21-32%
$p \to e^+ + \eta$	12.1%	3%	4-5%
$p \to e^+ + \omega$	22.5%	6%	19-29%
$p \to \nu + \pi^+$	8.9%	3%	5-11%
$p \to \nu + e^+$	4.6%	8%	8-12%

* * *

References

1. H. Georgi and S. L. Glashow, Phys. Rev. Letters 32 (1974) 438.

2. T. Goldman and D. A. Ross, Nucl. Phys. B162 (1980) 102.

3. S. L. Glashow, Nucl. Phys. 22 (1961) 579;
S. Weinberg, Phys. Rev. Letters 19 (1967) 1264;
A. Salam, Proceedings of the Eighth Nobel Symposium on Elementary Particle Theory, Relativistic Groups, and Analyticity, edited by N. Svartholm (Wiley, New York, 1969).

4. T. W. Appelquist and J. Carazzone, Phys. Rev. D11 (1975) 2856.

5. H. Georgi, H. Quinn and S. Weinberg, Phys. Rev. Letters 33 (1974) 451.

6. K. T. Mahanthappa and M. Sher, "Precise Determination of $\sin^2\theta_W$ and the Masses of Weak Vector Bosons in SU(5)." Colorado preprint COLO-HEP 13(1979).

7. A. J. Buras, J. Ellis, M. K. Gaillard and D.V. Nanopoulos, Nucl. Phys. B135 (1978) 66.

8. H. Georgi and H. D. Politzer, Phys. Rev. D14 (1976) 1829.

9. W. J. Marciano, Phys. Rev. D20 (1979) 274.

10. T. Goldman and D. A. Ross, Phys. Letters 84B (1979) 208.

11. J. Ellis, M. K. Gaillard, D. V. Nanopoulos and S. Rudaz, LAPP preprint TH-14/CERN preprint TH-2833 (1980).

12. E. A. Paschos, Nucl. Phys. B159 (1979) 285.

13. D. A. Ross, Nucl. Phys. B140 (1978) 1.

14. W. Celmaster and R. J. Gonzalves, Phys. Rev. D20 (1979) 1420.

15. G. 't Hooft and M. J. G. Veltman, Nucl. Phys. B44 (1972) 189; G. Bollini, J. J. Giambiagi, and A. Gonzalez Dominguez, Nuov. Cim. 31 (1964) 550.

16. T. Goldman and D. A. Ross, "How Accurately Can We Estimate the Proton Lifetime in an SU(5) Grand Unified Model?" Caltech Preprint, CALT-68-759 (1980).

17. C. Llewellyn Smith and G. G. Ross, Oxford University preprint in preparation.

18. P. Binétruy and T. Schücker, "Gauge and Renormalization Scheme Dependence in GUTS," CERN preprint TH-2802 (1979).

19. W. Marciano, "Theoretic Aspects of Proton Decay," Rockefeller University preprint, COO-2232B-195 (1980).

20. N. P. Chang, A. Das, and J. Perez-Mercader, "Proton Stability in an Asymptotically Free SU(5) Theory," CCNY-HEP-79-24 (1979) and CCNY-HEP-79-25.

21. J. Ellis, M. K. Gaillard, D. V. Nanopoulos and C. T. Sachrajda, Phys. Letters 83D (1979) 339.

22. J. Ellis, M. K. Gaillard, D. V. Nanopoulos, Phys. Letters 80B (1979) 360.

23. G. R. Cook, K. T. Mahanthappa, and M. A. Sher, Phys. Letters 90B (1980) 398.

24. S. Weinberg, Phys. Letters 91B (1980) 51.

25. F. Wilczek and A. Zee, Phys. Rev. Letters 43 (1979) 1574.

26. J. Ellis, M. K. Gaillard, D. V. Nanopoulos, Phys. Letters 88B (1979) 320.

27. J. Finjord, Phys. Letters 76B (1978) 116.

28. J. Learned, F. Reines and A. Soni, Phys. Rev. Letters 43 (1979) 907.

29. A. Din, G. Girardi and P. Sorba, Phys. Letters 91B (1980) 77.

30. J. Donoghue, "Proton Lifetime and Branching Ratios in SU(5)," MIT preprint, CTP-824 (1979).

31. M. Machacek, Nucl. Phys. B159 (1979) 37.

THE IRVINE-MICHIGAN-BROOKHAVEN† NUCLEON DECAY SEARCH: STATUS REPORT ON AN EXPERIMENT SENSITIVE TO A LIFETIME OF 10^{33} YEARS#

L.R. Sulak*

Randall Laboratory
University of Michigan
Ann Arbor, Michigan 48109

ABSTRACT

We have studied the properties of, and the expected backgrounds in, a totally active 10,000 ton water Čerenkov detector located deep underground and sensitive to many of the conjectured decay modes of the nucleons in it. Identification of (π,μ) and (e,γ) secondaries, good energy resolution, and good angular resolution provide sufficient background rejection in the detector under construction to permit one to obtain significant information about several decay channels, should they be observed. If no events were recorded in the device in one year, a lower limit of $\sim 10^{33}$ years would be placed on the partial lifetime for the most distinct nucleon decay modes. Depending upon the decay channel, this is ~ 3 orders of magnitude longer than previous measurements, and is at or beyond the level suggested by many unifying theories. The sensitivity predicted for this instrument is within an order of magnitude of that achievable in an arbitrarily large detector of this general type, since known background from atmospheric neutrinos imposes an inherent limit.

INTRODUCTION

The Irvine-Michigan-Brookhaven Collaboration is currently building a detector which ultimately will either measure nucleon decays at a lifetime level of 10^{33} years, or establish a lower limit several times longer than that. If no events were observed in it, the limits set would be more than three orders of magnitude greater than existing ones. If the nucleon should decay at a rate which is fortunately close to the present limits (in particular for the modes favored by the unifying theories, $N \rightarrow \ell\pi$) one could record ∿1000 events per year, study branching ratios, and, if muons were copiously produced, measure their polarization.

The detector consists of a 21 metre cube of water with phototubes placed in regular arrays on all six faces. Charged particle tracks (and π^0 mesons through their photon showers) are detected via the cones of Čerenkov light that they emit in water. Angles, energies, and particle types are determined with sufficient precision to identify nucleon decay and reject background at the required level. The detector is particularly sensitive to the decay modes expected in the unification models, e.g., $p \to e^+\pi^0$, $n \to e^+\pi^-$. The signature of these events, two "back-to-back" monoenergetic particles or showers of $\sim\frac{1}{2}$ GeV each, is distinctive relative to all backgrounds. Only atmospheric neutrino-induced $\Delta(3/2,3/2)$ production, the limiting background, might rarely produce a similar pattern. The detector is also capable of observing baryon decays in a variety of other possible decay modes. We have checked the design with two independent Monte Carlo simulations of events and backgrounds. Detector performance and reconstruction of events are discussed.

Secondarily, the detector permits the study of other interesting phenomena taking place in its deep underground environment, chosen to shield out muon-induced background. The search for neutrino oscillations is discussed in a companion article in these proceedings, for example.

Mechanical and electronic construction is well underway. Phototubes and their associated hardware are being produced at the rate of 200 photomultipliers (PM's) per month. The mine site is in preparation. The full sized detector is expected to start operation with half of the PM's around the turn of the year.

This paper is an update on the experiment which has already been described elsewhere.[1]

CERENKOV PATTERNS

The initial configuration that we are constructing employs a 16×16 array of 5" hemispherical phototubes on each face of a 21 m cube of water. Charged particles with $\beta \geq 0.75$ in the detector emit Čerenkov light in cones of half-angle of $\leq 42^\circ$ with respect to their direction. The light travels to the surface where it produces an elliptically shaped annular pattern of triggers on an array of PM tubes, as ideally shown in Fig. 1a. For particles exiting the fiducial volume, the ring becomes a filled in ellipse of light. The energy (range) of a stopping particle is determined by the total number of photoelectrons collected. The distance from the decay vertex to the detector plane is measured by the outer radius of the ring, the range by the thickness of the ring, and the

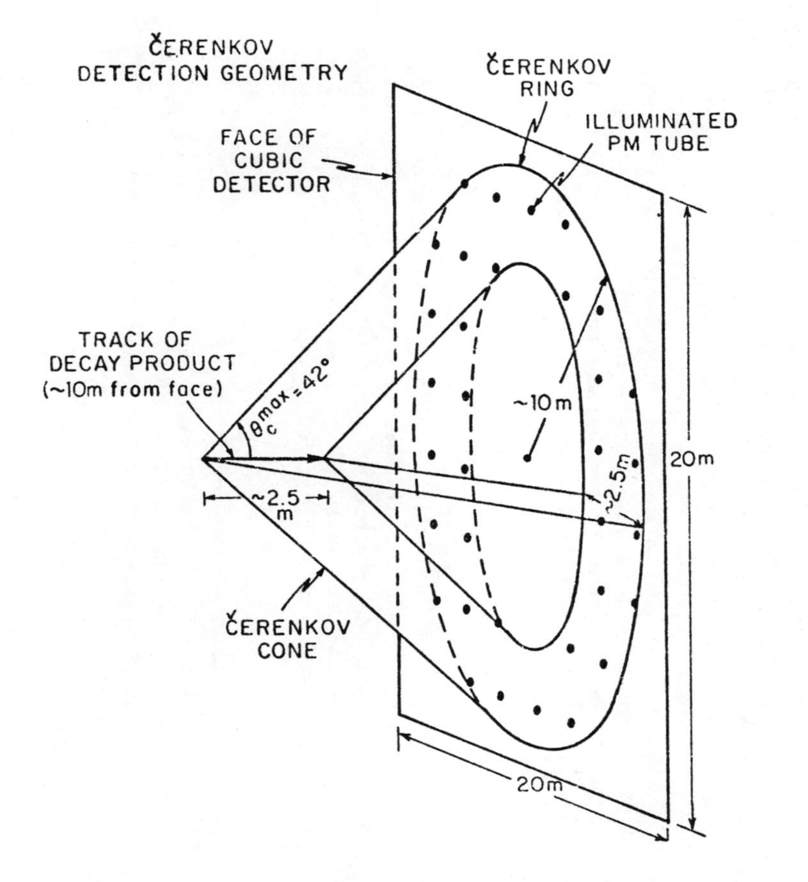

Figure 1a.

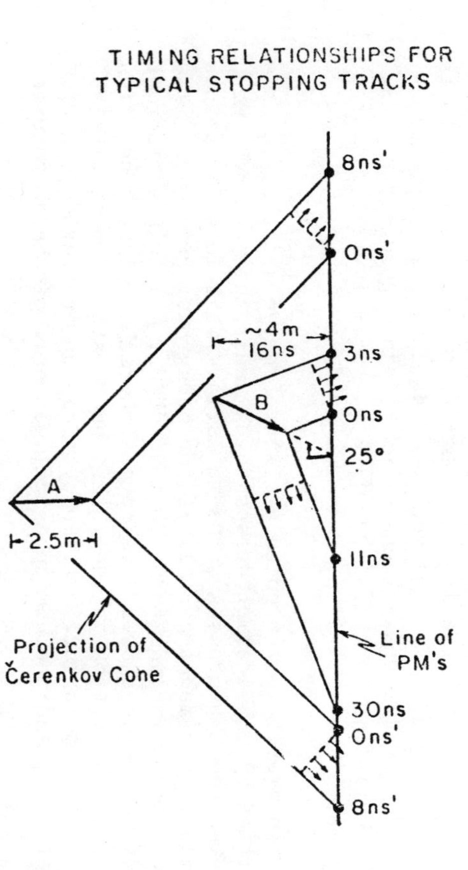

Figure 1b.

direction by the relative time that each PM fired. Due to the ~90° opening angle of the cone and the long lever arm across a ring, time differences (tens of ns) yield a good measurement of angles relative to the plane of PM's, as is ideally illustrated in Fig. 1b.

In practice, the cones of light from decaying nucleons are modified by physical processes which smear the pattern. We have treated these processes in Monte Carlo simulations of various nucleon decay modes as well as for background processes. The calculations include Fermi motion, dE/dx, multiple scattering, electromagnetic shower development, Čerenkov light generation, spectral absorption in the water, photocathode spectral response, photoelectron capture efficiency, Poisson statistics, PM time resolution, and the hemispherical geometry of the phototubes.

From studies of simulated nucleon decays, we find that for $N \rightarrow \pi \ell$ decays, charged pions and muons have typical radiating ranges of 75 to 175 cm, whereas $e^{\pm}$ and π^0 showers have relevant lengths of 100 to 300 cm. The ratio of Čerenkov light emitted, from a 500 MeV $\pi^{\pm}$, $\mu^{\pm}$, and $e^{\pm}$ (or π^0) is about 1:1.3:3. The $e^{\pm}$ or π^0's generally produce a large number of short electron tracks (which are highly efficient Cerenkov radiators) and form a pattern of PM hits that seldom is a perfect ring. Figure 2 shows the phototube amplitude response pattern from a typical nucleon $p \rightarrow e^+\pi^0$ decay in the fiducial volume. Even though the three ring patterns are somewhat filled in, their circular nature is still apparent. The π^0 decay shown is unusually asymmetric. Although we have not exhibited it in this figure, the timing information for each hit PM event supplies a powerful additional constraint on event reconstruction. Our computer codes reconstruct separate tracks from patterns such as these; the reconstructed errors for the event in Figure 2 are indicated there.

EVENT RECONSTRUCTION

The detector is optimally sensitive to nucleon decay signatures of the form $N \rightarrow \pi \ell^{\pm}$, where both the meson or its decay products and the charged lepton are capable of generating a Čerenkov signal. Various instrumental and physical factors limit our ability to recognize these modes and reject backgrounds. The observables to be measured are:

(1) The location of the decay vertex.
(2) The total energy of the event.
(3) The decay opening (collinearity) angle.
(4) The energies of each of the decay products.

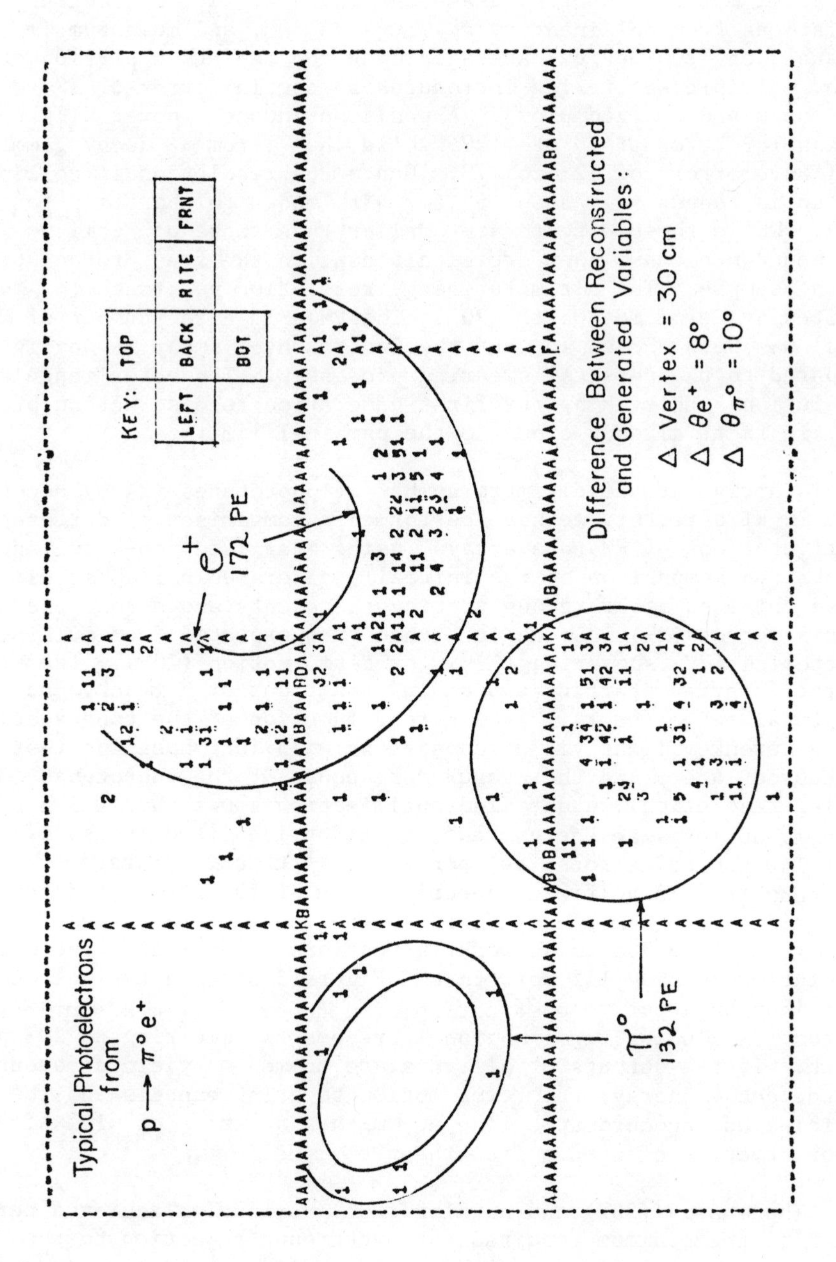
Typical Photoelectrons from
p → π°e+
KEY:
TOP
LEFT
BACK
RITE
FRNT
BOT
e+
172 PE
π°
132 PE
Difference Between Reconstructed and Generated Variables:
Δ Vertex = 30 cm
Δ θe+ = 8°
Δ θπ° = 10°

Figure 2.

The most significant physical effect is the Fermi motion of bound nucleons in oxygen. (Only 20% of the protons in water are free.) The Fermi motion distorts parameters (3) and (4), introducing deviations from collinearity as large as 30°, and momentum imbalances of $\lesssim 200$ MeV/c. Averaged over the radiating portion of the path, multiple scattering introduces an angular error of 5° to 7° for muons and charged pions. A positron-induced shower will have an angular error of 10° to 12°, while those from π^{0} decay gammas will have errors of 12° to 17°. Hence the resolution in collinearity angle ranges from about 5° for $\mu^{+}\pi^{-}$ modes to about 15° for $e^{+}\pi^{0}$. While these effects are smaller than those of Fermi motion for bound nucleons, they are significant in the free proton portion of the sample. The ultimate energy resolution per track is also limited by Fermi motion to $\sim$20%. The PM grid size (number of PM's used) has been chosen so that the instrumental smear is negligible compared to the physical smearing processes. The actual angular resolutions achieved by our first generation reconstruction programs are, in fact, already close to the physical limits.

To arrive at the ultimate number of phototubes (2400) necessary in a 10 kT detector, we have performed a comparison of different configurations of PM tube arrays, using idealized non-showering tracks. A comparison of the initially generated variables with those obtained after reconstructing the events shows that the angular error (2°) is much less than that expected from multiple scattering (7°), showering (12°), or Fermi motion (20°). The error in the inferred starting and ending positions of a track, and in the total range (energy) is a strong function of the tube spacing. Using Čerenkov light yields consistent with long baseline tests discussed later, and the measured response of the phototubes (3 ns timing resolution), a detailed optimization shows that a 1.2 m tube spacing achieves an energy resolution for idealized tracks of $\sim$10% with 140 photoelectrons (pe) per e^{+} or π^{0} track. A larger tube spacing results in significant deterioration of the resolution because the grid size becomes comparable to the range of the tracks. When energy smearing due to shower fluctuations is included, the energy resolution becomes 15% per track. Figure 3 shows a typical energy distribution after reconstruction of a decay e^{+} from a stationary proton. Including Fermi motion increases the smearing to 20% per track. If the outputs of all PM's are summed to yield a measure of the decay energy, the Fermi motion smearing cancels and the anticipated monochromatic line at $\sim m_{p}$ has an error of $\sim$10% after reconstruction of events from the $\pi^{0}e^{+}$ mode.

The energy (20%) and angular (20%) resolution achieved per track is the minimum required for background rejection to the 10^{33} year level. Since the physical smearing processes (and not the detector configuration) limit the resolution, collecting more photoelectrons by more PM's, addition of waveshifter, use of mirrors, etc., provides no gain in resolution.

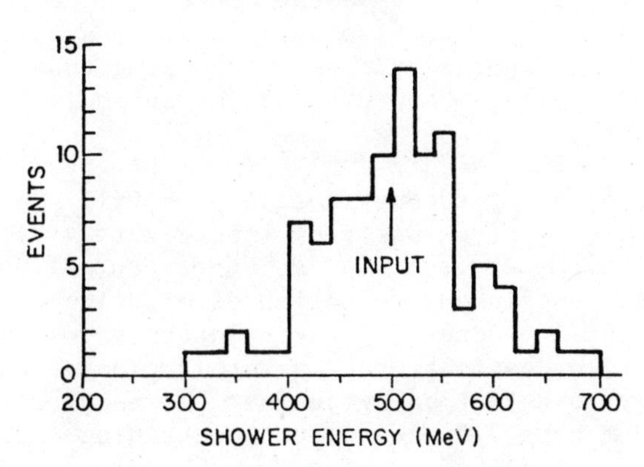

Figure 3.

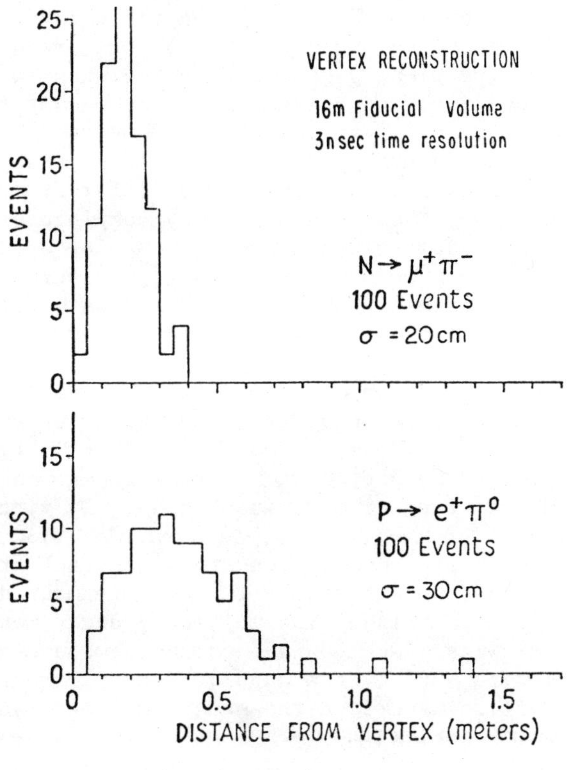

Figure 4.

For non-showering tracks ($\mu^{\pm}$, $\pi^{\pm}$) an average number of $\sim$50 photoelectrons per track appears necessary for reliable reconstruction. Fewer photoelectrons result in a significant loss of background rejection. We find that we can achieve 50 pe for a π^- track (from the $n \to e^+\pi^-$ mode) also using a 1.2 m spacing. Thus, an additional line of approach leads to the same tube spacing and the same number of tubes, $\sim$2400 5" PM's, or an equivalent mix of 5" and 8" tubes.

The results of event reconstruction with full smearing show that the error on the vertex position (see Figure 4) is $\sim$20(30) cm for the $\mu^+\pi^-$ ($e^+\mu^o$) mode, essentially independent of position in the detector. For sufficient rejection of entering cosmic ray muons and the neutral products of their interactions in the rock near the detector, we estimate that a veto region of thickness $\sim$2 m is necessary to reject background at the required level. This leaves a 17 m cube fiducial volume containing 4000 tons of water with 2.5×10^{33} nucleons.

The separation of particles into the categories of "showering" (π^o, $e^{\pm}$) and "non-showering" ($\pi^{\pm}$,$\mu^{\pm}$) is relatively simple. In Fig. 5 we show the distribution of total photoelectron yields from various tracks from nucleon decays. When the decays are from nucleons at rest (hydrogen nuclei, Fig. 5a), the resulting monoenergetic tracks give clearly separable peaks whose widths reflect the combined geometrical and Poisson fluctuations of the events. Although decays from oxygen nuclei (Fermi momentum = 235 MeV/c, Fig. 5b) have broadened peaks, they are still separable at the 95% level. The pattern of hits is also quite different for showering and non-showering tracks so that an essentially complete separation of the two categories should be possible (except for a correction due to conversion of $\pi^- \to \pi^o$ via nuclear charge exchange). However, it appears that the majority of π^o induced showers will be difficult to distinguish from $e^{\pm}$ showers of the same primary energy.

Positive identification of most π^+'s and μ^+'s,and $\sim$70% of μ^-'s will be achieved by detection of the light produced by the electrons from delayed μ decay. Typically $\sim$10 PM's fire in coincidence over a 50 ns period. Figure 6 shows the photoelectron yield from 40 MeV $e^{\pm}$ typical of a muon decay. By virtue of their temporal and spatial distributions, delayed hits are identifiable with the information produced by the stopping track from the event that triggered the detector. Two or more muon decays can be recognized as well, thus labelling multibody decay modes. It also appears that with $\sim$300 muon events, one could measure the muon polarization to $\sim$30% by the angular decay correlation. This would be important to ascertain the scalar vs vector nature of the superheavy boson mediating the decay.

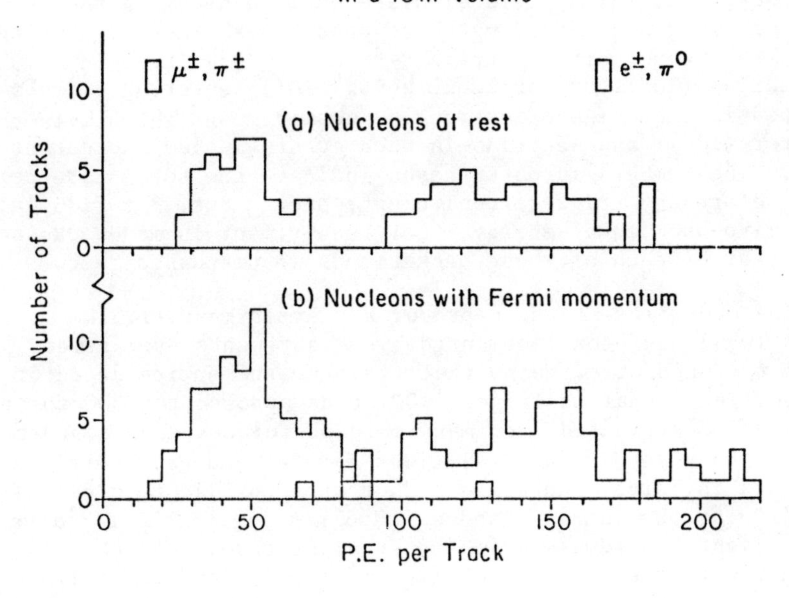

Figure 5

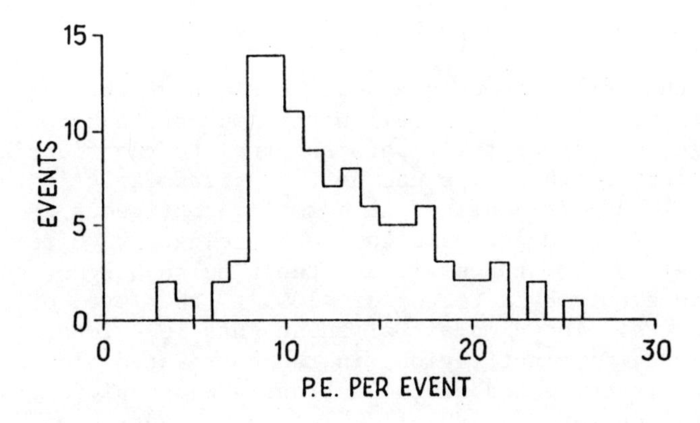

Figure 6

BACKGROUND REJECTION

Background which could simulate nucleon decay in the detector could be induced by three possible sources: entering cosmic ray muons, either straight-through (10^8), stopping (10^6), interacting (10^5), or corner clipping (10^7); entering hadrons produced in the neighboring rock by cosmic muons which miss the detector (10^3); and neutrino-induced events inside the detector (10^3). The numbers in parentheses indicate the total rate per year, before any trigger requirement, energy cut, 2 m fiducial cut, or back-to-back,equal-energy,1 GeV requirement is made. Reduction to $\lesssim$1/year of each of these backgrounds is discussed below.

The muon flux at the depth of the experiment (1.5 km water equivalent (we)) has been documented by measurements over several years at the proposed site. Muons that traverse the entire detector pose no problem. The signal is large (1100 pe as opposed to $\sim$300 for proton decay) and distributed over many more phototubes than a proton decay event. The spread of arrival times for far and near portions of the track is large, and a local "hot spot" of light is deposited in PM's near the entrance and exit points due to the 1/r divergence of the light. Hardware selected straight-through tracks will be recorded with the data for on-line monitoring and for calibration purposes, as well as other physics studies. Similarly, tagged entering muons which stop in the detector provide an important check of the ability of the apparatus to respond to low light level decays within it.

Those muons that interact in the detector produce other tracks, but in general they are at small angles to the initial track and the total pulse-height will be even greater. An upper energy threshold can remove them from the trigger. However, the characteristics of these events give valuable information on the nature of similar interactions in the neighboring rock where the muon goes undetected but the resulting hadrons penetrate the detector.

Muons that traverse only a small portion of the detector ("corner-clippers") and thus fall within our pulse-height cuts have a large portion of their path external to the fiducial volume. Events of this type can be recognized early in the analysis, since their signal is confined to a small, contiguous group of phototubes. A single software cut (or a trigger requirement) based on the number of phototubes hit and their pulse-heights can reduce these events by a factor of $\sim$1000 to the level of one per several minutes. The simplest of these cuts is evident from Fig. 7. There, distinct regions in the correlated plot of tube multiplicity vs collected photoelectrons are occupied by nucleon decay events generated in the fiducial volume for three different decay modes. Cosmic ray muons entering the detector are clearly differentiated from the decay events by having fewer tubes fired

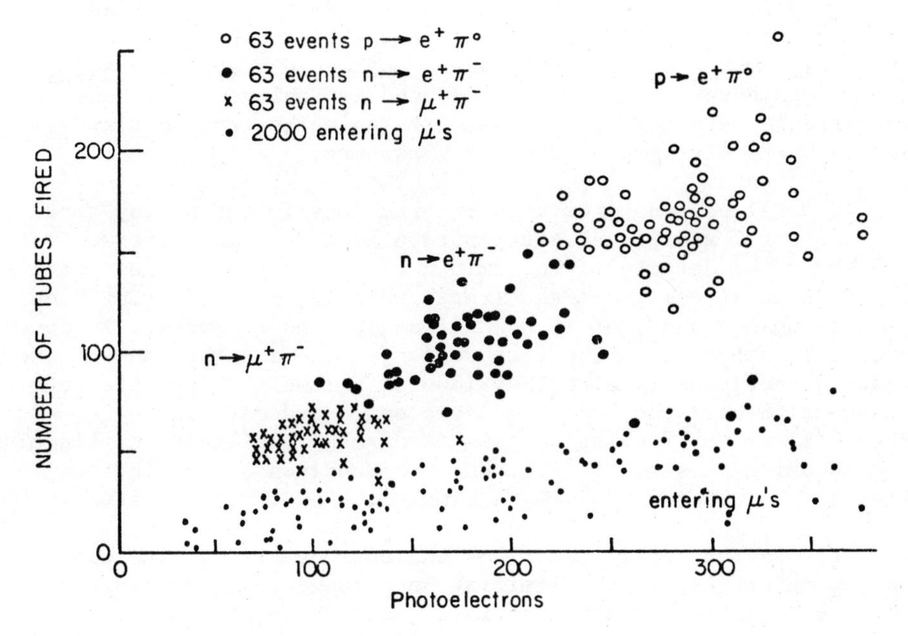

Figure 7.

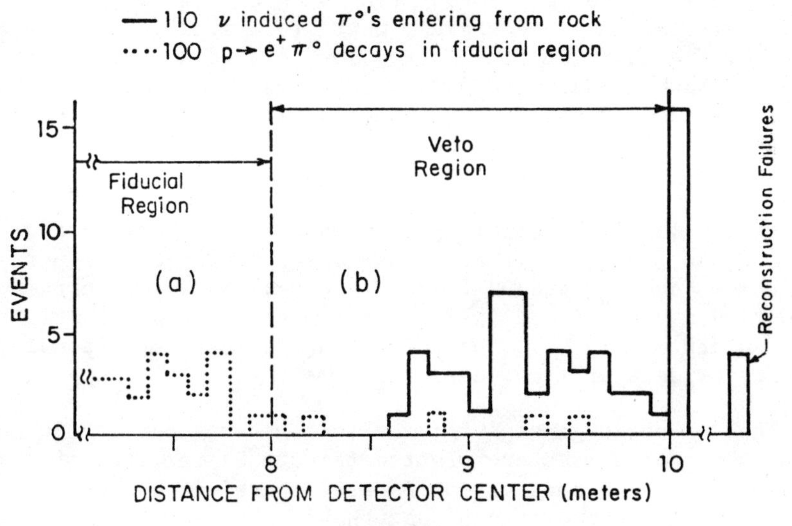

Figure 8.

per photoelectron due to their path near PM's at the surface of the detector. Cutting the events in the region highly populated by muons removes a negligible number of the nucleon decays ($\lesssim 5\%$). A further reduction in muon background is achieved by reconstructing the track's path through the veto region. The remainder are rejected by requiring a back-to-back pattern.

Neutral hadrons from muon interactions in the nearby rock (with a small additional contribution from neutrino interactions) can enter the detector. To mimic nucleon decay, they must either be produced at a large angle (rare) relative to the parent muon passing near a side, or at a small angle (and traverse a large amount of material in the veto region). The interactions of the hadrons that penetrate to the fiducial volume ($\sim$20/yr) are not expected to have the characteristic nucleon decay signature. Our level (i) vertex finding program is already capable of distinguishing incoming π^0 showers (Fig. 8b) from nucleon decays in the fiducial volume (Fig. 8a) by a simple cut on vertex position.

Events induced by neutrinos and anti-neutrinos with two radiating tracks are the limiting background to the experiment. Using estimates of the cosmic ray ν_μ and ν_e spectra[2] (with energies between 0.3 and 2 GeV) and known neutrino interaction cross-sections and q^2 distributions,[3] we have undertaken two independent Monte Carlo simulations of events of this type, including the effects of nuclear Fermi momentum and detector resolution.[4]

In general, recoil protons from such events are below Cerenkov threshold. (At higher energies, where relativistic protons are possible, the neutrino flux is too low to give a substantial background.) Thus elastic charged-current events do not contribute to the background. Double pion production in neutral current interactions is appreciable only at energies that give pulse-heights and opening angles outside our range of acceptance.

We are left with single-pion production by charged current interactions, which occur at a rate of about 200 events per year, 140 from ν_μ and $\bar{\nu}_\mu$, and 60 from ν_e and $\bar{\nu}_e$. We simulate these Δ production events with a Breit-Wigner mass spectrum at 1232 MeV with a width of 110 MeV. We generated a sample corresponding to 25 years of data from our detector.

The ν_μ events that survive a total energy cut of 300 MeV around the proton mass are plotted (Figure 9) in terms of two configuration-dependent variables used to reject background, the angle between the two tracks, and the fraction of the total energy carried by one of them. The region inside the broken line is that defined by the Fermi motion from bound nucleon decays. Three events fall on the boundary of that region, with 7 or 8 more within

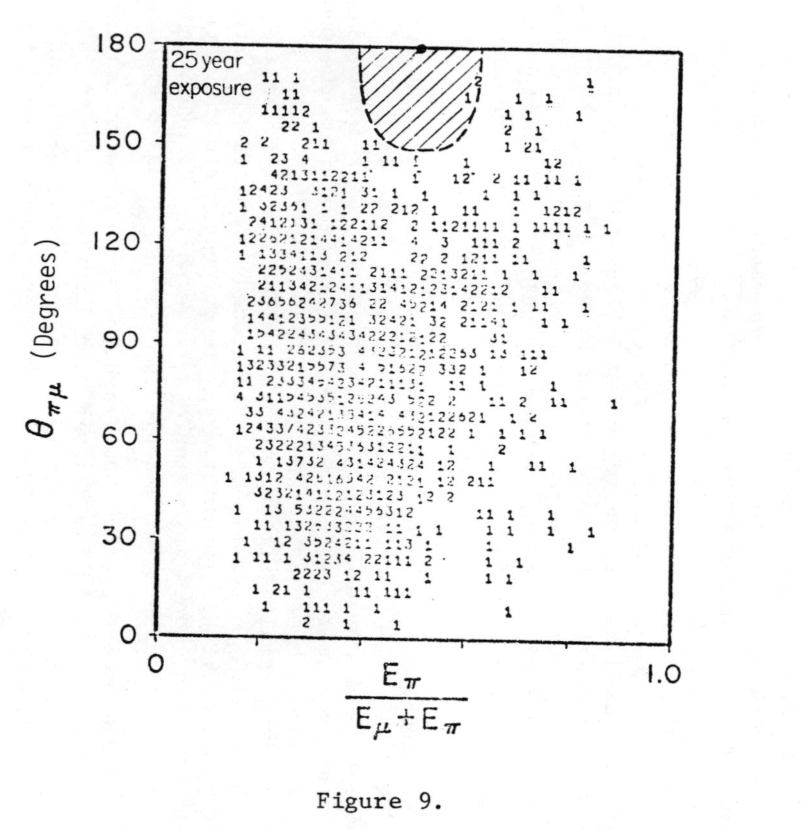

Figure 9.

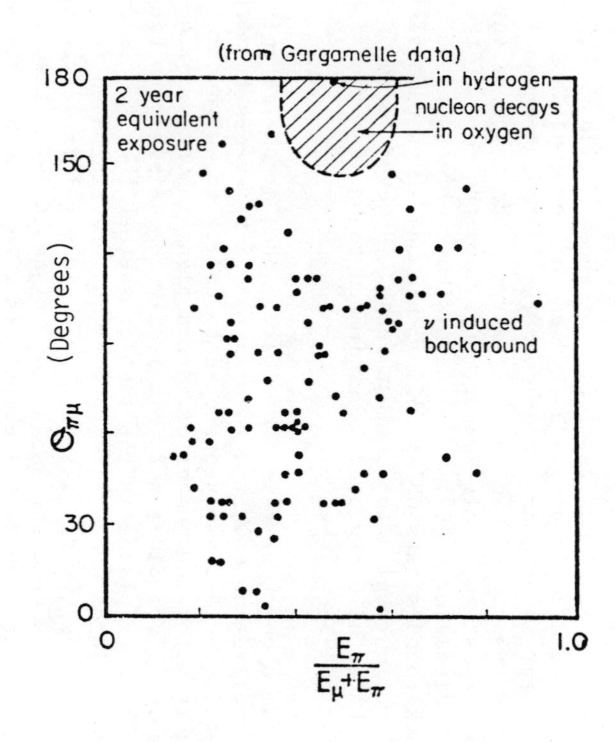

Figure 10.

Kinematic Characteristics of Neutrino Background

one standard deviation in our resolution on these variables. On this basis, we would anticipate a background of somewhat less than one event per year in or near the allowed region for $\mu\pi$ channels and a factor of two fewer for $e\pi$ channels. Varying the shape of the Δ mass spectrum (Gaussian, Breit-Wigner, etc.) and the form of the q^2 distribution (dipole or e^{-aq^2}) has a significant effect on the number of background events which fall within one standard deviation of the Fermi motion region for nucleon decay. However, even in the worst case, the total background is estimated to be less than 1 event/year for all $N \to \pi\ell$ channels.

To provide a check of this Monte Carlo, the Gargamelle collaboration[5] applied our criteria to a sample of ν_μ events from their data tapes equivalent to two years of data collection with our detector (in terms of the number of interactions in the allowed energy range). All candidates topologically compatible with single-pion production were placed in the sample, whether they successfully fit this hypothesis or not. This retains any unanticipated reaction channels, as well as events distorted by nuclear reinteraction of pions exiting from a decaying nucleus. The sample is plotted in Fig. 10. Again, the results are compatible with an event rate of less than one per year, and the background only approaches the boundary of the nucleon decay region of the plot. The nuclear scattering effects in freon are worse than in water, so the background indicated is probably an overestimate.

The foregoing represents what we believe to be a realistic assessment of background expected in the detector. However, in the event that increased background rejection is required, we can increase the thickness of the veto region by suitable adjustment of the selection criteria. An increase in the effective veto thickness from the assumed 2 to as much as 4 metres provides for a further background reduction by a factor of 13 (with an attendant decrease in fiducial volume from 4000 to 1700 tons).

In the actual analysis the precise kinematic cuts and the surviving background will be decay-mode dependent. In addition to shower-non-shower discrimination, ν_μ charged current background can be eliminated from searches for $e^\pm$ modes by the signal from the muon decay. This should be reliable enough to eliminate 70% to 80% of μ^+, and a somewhat smaller fraction of μ^-, since 22% of these are captured on oxygen. In addition, π^+ backgrounds in π^- final states can be eliminated by this techique. We are currently studying our sensitivity to decay modes involving other mesons (ρ, K, etc.) as well as multi-body modes. For example, $p \to e^+\omega^o$, $\omega^o \to \pi^+\pi^-\pi^o$ has a signature comprised of one monoenergetic shower and two close showers consistent with a π^o, as well as one delayed muon decay pointing to the vertex. Although this is a fairly restrictive signature, it can be mimicked by $\nu_e N \to eN^*$, $N^* \to \pi^o\pi^+X$ or $nN \to \pi^o\pi^o\pi^+X$. For the modes expected from various selection rules in the decay, Table I summarizes our best estimates of the sensitivity of the detector. The second column of the table shows that for each selection rule at least one characteristic proton decay exists that is unique to that selection rule. Thus, this detector has the sensitivity to unfold the physics of possible baryon decays, should they occur.

It is appropriate to mention two ancillary questions relative to the water Čerenkov scheme. First, are the nuclear rescattering effects of the pions born in oxygen nuclei important? Table II, which has been tabulated by R. Barloutaud, addresses this issue. Note that at least ∿60% of the pions escape unscattered. Approximately 15% each are either elastically scattered or charge exchange scattered. Although these processes confuse the information relating to the branching ratio, they do not necessarily remove the events from the search for nucleon decay. In effect, pion charge exchange just "exchanges" decaying protons for decaying neutrons, and elastic scattering just spoils the back-to-back signature: the 1 GeV, equal-energy signature is still preserved. Typically we only expect ∿10% of the nucleon decays to be removed from the search by absorption.

A second question relates to the detection technique. Why a surface rather than a volume array? Although a volume array always has some tubes close to the track and these usually capture a large number of photoelectrons (however, with large fluctuations), we found in designing the detector that the <u>total number</u> of tubes fired by an event to be far more important to reconstruction than the total number of photoelectrons. For each additional tube hit, 3 space coordinates, 1 time, and 1 pulse-height are added to the number of constraints available to the reconstruction routine. Further, since most of the fiducial volume is near the faces of the cube (as well as most of interactions due to the entering, non-neutrino background), a surface detector has many more tubes lit up per trigger than a volume array outfitted with an equal number of PM's. Figure 11 shows the number of tubes hit by a 2m-long track going towards a wall as a function of its distance from the wall. Results for both surface and volume arrays of PM's are shown. The surface array typically has a factor of two or more PM's hit than a volume array.

PROGRESS IN DETECTOR CONSTRUCTION

The underground site that we have chosen is located in the Morton Salt mine at Headlands Beach State Park, Fairport Harbor, Ohio. The depth is 1500m we (∿550 m at density 2.7 g cm^{-3}). A tailor built cavity is currently being excavated to house our detector. Simultaneously, manufacture and assembly of detector components is proceeding apace. The complete electronic readout system -- from vertical strings of 16 underwater PM tubes to mini-

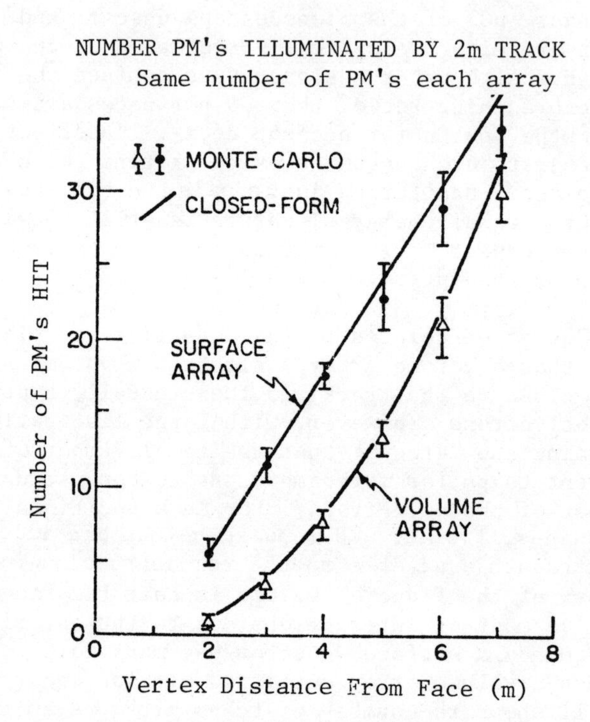
NUMBER PM's ILLUMINATED BY 2m TRACK
Same number of PM's each array
MONTE CARLO
CLOSED-FORM
SURFACE ARRAY
VOLUME ARRAY
Number of PM's HIT
0
10
20
30
0
2
4
6
8
Vertex Distance From Face (m)

Figure 11.

computer recording of data -- is currently under systems development at the IMB 21 m vertical test tank at Ann Arbor. For each tube the electronics digitizes one pulse-height and records time on two scales. One time scale has sufficient resolution for the initial event reconstruction while the other records delayed pulses from π, μ decays. Both pulse-height and timing requirements are more modest than in typical accelerator experiments, and are particularly easy to meet in our constant 75°F thermal environment. A modest 3 ns timing resolution, 10% energy resolution, and a low factor of 15 dynamic range in energy per tube, and the low ($\sim$4 sec^{-1}) trigger rate permit an electronics design that is simple, economical, and highly reliable for operation in the unusual experimental site.

The housing and support structure for the photomultiplier tubes has been perfected in both the vertical and in a 10 m horizontal test tank as well as in other pools. Each tube can be positioned to a few centimeters in space and a few degrees in angle. Features include easy PM retrieval for above-water servicing, construction with corrosion-free materials, maintenance of high purity water, and a minimum of structure so as to permit easy redeployment of the detector.

The support structure is illustrated in Fig. 12. The tubes are positioned and accessed by a "conveyor belt" made of a parallel pair of nylon monofilaments. Any one of the 16 tubes on each string can be brought to the surface for maintenance by rotating the string. The support structure is stable because the tube modules are designed to be neutrally buoyant and to have no net torque when suspended in the water. Polyethylene sealed anchors maintain the positions of the strings.

After manufacture and assembly, the PM support structure, the PM housings, the cables, etc. are tested in the high purity water (>40m attenuation length at 450 nm) of the vertical and horizontal test tanks for at least two days. In addition, a large number will have been operated for months in this simulated detector environment before installation at the site.

CONCLUSION

The two-body nucleon decay signature is remarkably distinct in a straightforward, though massive, Čerenkov detector. The backgrounds are tractable up to a sensitivity of 10^{33} years. Civil construction of the underground laboratory is proceeding on schedule. Occupancy of the laboratory is expected in the fall of 1980 when 1200- 5" PM tubes, one-half the anticipated final detector, will be installed. Manufacture and long term underwater proofing of PM's, housings and electronics is progressing at a production rate of 200 PM detector modules per month.

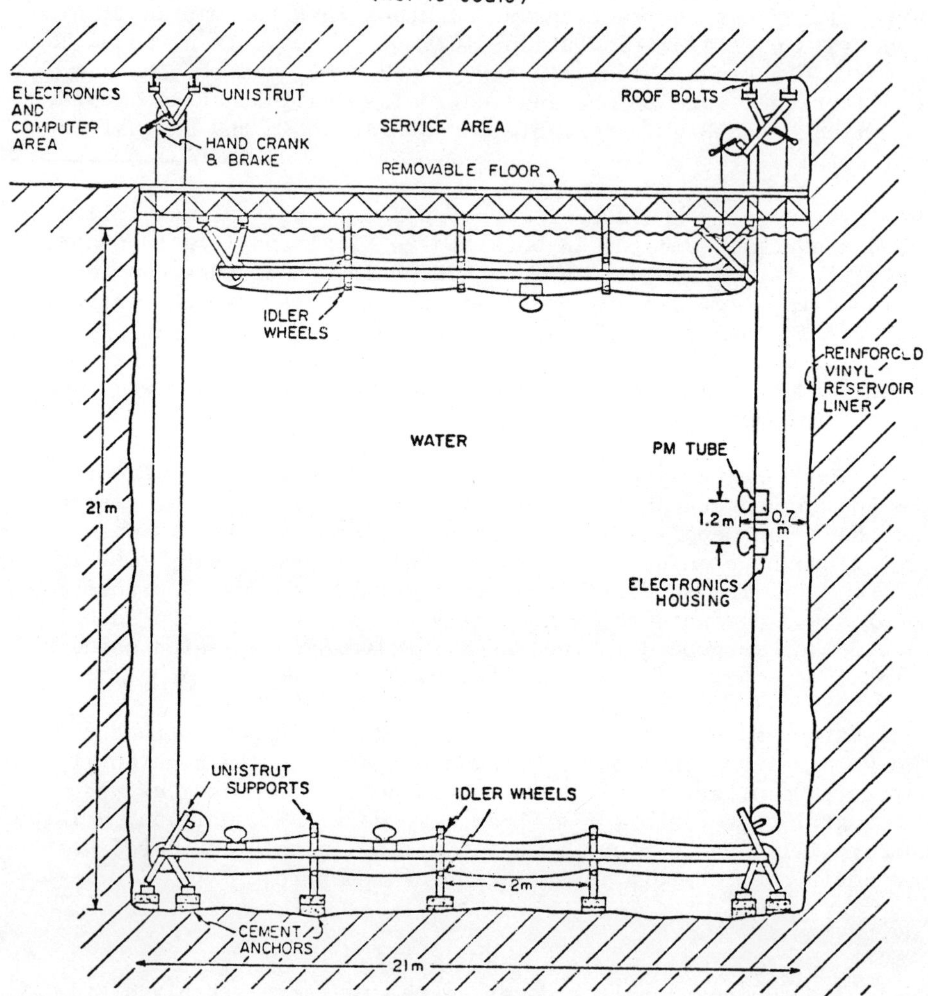
Schematic Cross-Section of Detector
(not to scale)
ELECTRONICS AND COMPUTER AREA
UNISTRUT
HAND CRANK & BRAKE
SERVICE AREA
ROOF BOLTS
REMOVABLE FLOOR
IDLER WHEELS
REINFORCED VINYL RESERVOIR LINER
WATER
PM TUBE
1.2 m
0.7 m
ELECTRONICS HOUSING
21 m
UNISTRUT SUPPORTS
IDLER WHEELS
~2m
CEMENT ANCHORS
21m

Figure 12.

TABLE I
DETECTOR SENSITIVITY TO VARIOUS SELECTION RULES

Δ(B–L)	Proton Mode Signature	Decay Mode		Approximate Sensitivity
		Proton	Neutron	
-4	No line 1 shower 2 μ decays	$\nu\nu\nu\pi^+$	$\nu\nu\nu\pi^o$	$\gtrsim 10^{30}$yr
		$\nu\nu e^-\pi^+\pi^+$	$\nu\nu e^-\pi^+$	$\gtrsim 10^{30}$yr
-2	line at 1 GeV 2 μ decays	$\ell^-\pi^+\pi^+$	$\ell^-\pi^+$	$\sim 10^{33}$yr
		$\nu\pi^+$	$\nu\pi^o$	$\sim 3\times10^{31}$yr
		$\nu_1\nu_2 e^+$	$\nu_1\nu_2 e^+\pi^-$	$\gtrsim 10^{30}$yr
0	line at 1 GeV 2 body	$\ell^+\pi^o$	$\ell^+\pi^-$	$\sim 10^{33}$yr
		$\bar{\nu}\pi^+$	$\bar{\nu}\pi^o$	$\sim 3\times10^{31}$yr
		ℓ^+K^o	ℓ^+K^-	$\sim 10^{32}$yr
+2	event deep in detector	$\overline{\nu\nu\nu}\pi^+$	$\overline{\nu\nu\nu}\pi^o$	$\gtrsim 10^{30}$yr
		$\overline{\nu\nu}e^+$	$\overline{\nu\nu}e^+\pi^-$	$\gtrsim 10^{30}$yr

TABLE II
PION INTERACTION IN NUCLEUS

	No Internal Interaction	Elastic Scattering	Charge Exchange	Absorption	Reference
^{12}C	0.64 ± 0.06	π^o 0.14±.04 π^- 0.19±.04	0.11 ± .03 0.06 ± .02	0.11 ± .03	6
^{16}O	0.58				7
^{27}Al	0.51				7
	0.68	π^o 0.11 π^- 0.18	0.14 0.07	0.06	8

ACKNOWLEDGEMENTS

During the last year our theoretical colleagues have continued to find more ways for the nucleon to decay. We take the opportunity to thank them here for many enlightening comments: J. Ellis, M. K. Gaillard, M. Gell-Mann, H. Georgi, S. Glashow, T. J. Goldman, C. Jarlskog, P. Langacker, M. Machacek, R. Mohapatra, D. V. Nanopoulos, J. C. Pati, H. R. Quinn, P. Ramond, D. A. Ross, A. Salam, R. Slansky, S. B. Treiman, S. Weinberg, F. Wilczek, L. Wolfenstein, F. J. Yndurain, and A. Zee. We apologize to them and their friends for neglecting to cite the voluminous literature they have generated on the subject of this report. In addition, conversations with C. Baltay, C. Bennett, J. Cronin, and C. Rubbia, are gratefully acknowledged.

REFERENCES

[†]The members of the collaboration are the following:
M. Goldhaber, Brookhaven National Laboratory
B. Cortez, G. Foster, L. Sulak, Harvard University and University of Michigan
C. Bratton, W. Kropp, J. Learned, F. Reines, J. Schultz, D. Smith, H. Sobel, C. Wuest, University of California at Irvine
J. LoSecco, E. Shumard, D. Sinclair, J. Stone and J. Vander Velde, University of Michigan

[#]Supported in part by the U.S. Department of Energy.
*Visiting Scholar, Harvard University, Cambridge, MA 02138.

1. M. Goldhaber et al., Proceedings of the International Conference on Neutrino Physics, Bergen, June 1979 (C. Jarlskog, ed.) p.121.
2. A. W. Wolfendale and E. C. M. Young, CERN 69-28, p.95, 10 Nov. 1969; E. C. M. Young in Cosmic Rays at Ground Level (A. W. Wolfendale, ed.) Institute of Physics Press, London and Bristol, 1973, p.105; and E. C. M. Young, private communication, 1977.
3. R. T. Ross, Proceedings of the 1978 Conference on Neutrino Physics (E. C. Fowler, ed.) p.929; M. Derrick, Proceedings of the 1977 Conference on Neutrino Physics (H. Faissner et al., ed.) p.374; and the references cited in the two works.
4. Experimental results with which the Monte Carlo has been compared include: E. C. M. Young, CERN Yellow report 67-12, "Neutrino Charged Current Single Pion Production in Freon"; S. J. Barish, et al., Phys. Rev. Lett. 36, 179 (1976); and J. Campbell, et al., Phys. Rev. Lett. 30, 335 (1973).
5. We would like to thank Drs. J. Morfin and M. Pohl for providing the data from the Gargamelle experiment.

6. Experimental results from Gargamelle for 250 MeV/c pions. B. Degrange, private communication via R. Barloutaud.
7. Monte Carlo results of T. W. Jones, private communication.
8. Monte Carlo Study of C. Longuemare, private communication via R. Barloutaud.

A LONG BASELINE NEUTRINO OSCILLATION EXPERIMENT SENSITIVE TO MASS DIFFERENCES OF HUNDREDTHS OF AN ELECTRON VOLT[#]

B. Cortez[†] and L.R. Sulak[*]

Harvard University
Cambridge, MA 02138

University of Michigan
Ann Arbor, MI 48109

ABSTRACT

We detail the capabilities of a search for neutrino oscillations in a massive underground neutrino detector sensitive to low energy atmospheric ν_e's and ν_μ's using as a baseline the diameter of earth. A flux independent asymmetry in the up/down ratio of the two neutrino species is the primary signal. For concreteness, we study the 10,000T water Čerenkov detector of the Irvine-Michigan-Brookhaven Collaboration[§] and find that the full 10,000T is necessary; smaller detectors would have insufficient statistical power. For a two year exposure, this detector provides a several standard deviation signal for maximal mixing of either species over the mass difference range of 10^{-3} to 10^{-1} eV. The upper end of this range coincides with the lower end of conceivable accelerator-based searches for neutrino oscillations.

INTRODUCTION

Although dedicated to the search for nucleon decay, an apparatus now under construction and scheduled for completion within the year -- the Irvine-Michigan-Brookhaven Underground Detector Facility[1] -- comprises the largest mass cosmic neutrino detector (10,000T) yet proposed. In fact, the limiting backgrounds to the proton decay experiment are induced by atmospheric neutrino events. As such, these events must be understood in sufficient detail so as not to be confused with potential nucleon decay candidates. In an attempt to use the information from these background events creatively, we have undertaken a feasibility study of a neutrino oscillation experiment using them.

The search for oscillations[2] of atmospherically produced neutrinos is straightforward in the detector. Unlike neutrino beam lines at accelerators, the muons produced in the upper atmosphere have flight paths long enough for them to decay. A typical production chain for 1 GeV neutrinos is

$$\pi^+ \rightarrow \mu^+ \nu_\mu$$
$$\quad \hookrightarrow e^+ \nu_e \bar{\nu}_\mu .$$

Superficially, this suggests that the expected neutrino ratio should be $\nu_e:\nu_\mu = 1:2$. In fact, an integration of cross sections and calculated fluxes shows that (in a detector that does not measure charge, as the one we consider) the ratio of electron to muon events is expected to be $\sim 1:2$. In principle, one would measure this ratio as a function of the neutrino flight path ℓ from the neutrino production point (directly as a function of zenith angle θ_Z) and as a function of neutrino momentum p. As depicted in Figure 1 and plotted in Figure 2, the dynamic range in ℓ is 3×10^3, from downward coming neutrinos produced $\sim$ 5 km above the detector (which is located 0.5 km below the earth's surface) to upward coming neutrinos that have traversed the 13,000 km diameter of the earth. The dynamic range in p is insignificant relative to that in ℓ. On the lower end it is bounded by the onset of the cross-section for elastic neutrino scattering (see Figure 3) and on the upper by the rapidly falling ($\propto E^{-2.5}$) energy spectra of neutrinos (see Figures 4 and 5). Although one would like to measure $\nu_e:\nu_\mu$ as a function of ℓ/p to map out explicitly any oscillatory behavior, the statistics in even this massive detector will be a limiting factor. However, we will show that a comparison of the ratios for the upper and the lower hemispheres (equal solid angles) will be sufficient to search for neutrino oscillations out of either channel for large mixing (Pontecorvo) angles[3] ξ and eigenmass differences $\Delta m = \sqrt{m_1^2 - m_2^2}$ between 10^{-3} and 10^{-1} eV.

The evolution of an electron neutrino beam of intensity I_e that is mixing with another neutrino beam of intensity I_x is described by

$$I_e = I_e(0)[1-\sin^2 2\xi/2\ (1-\cos 2\pi\ell/L)]$$
$$+ I_x(0)\ \sin^2 2\xi/2\ (1-\cos 2\pi\ell/L)\ ,$$

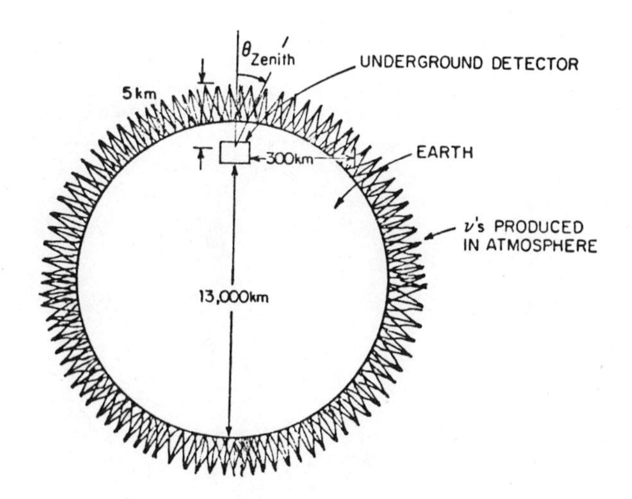

Figure 1.

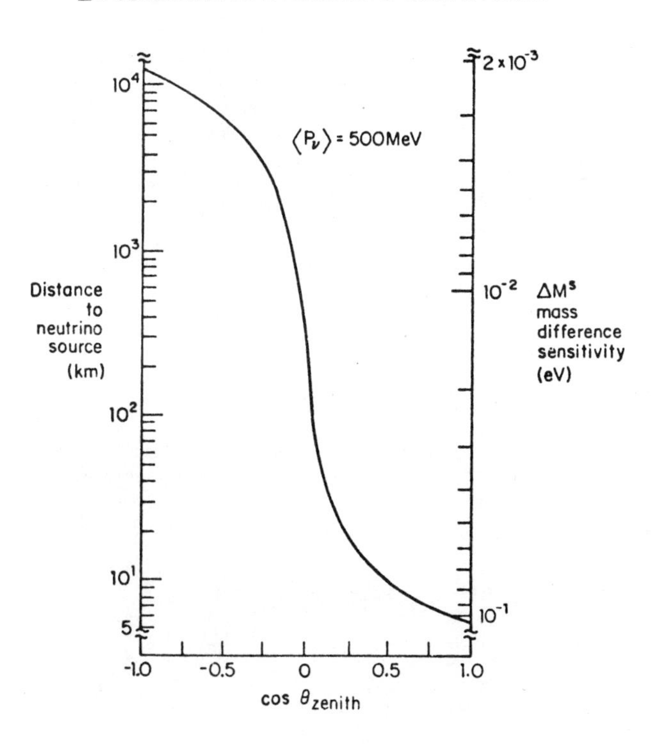

Figure 2.

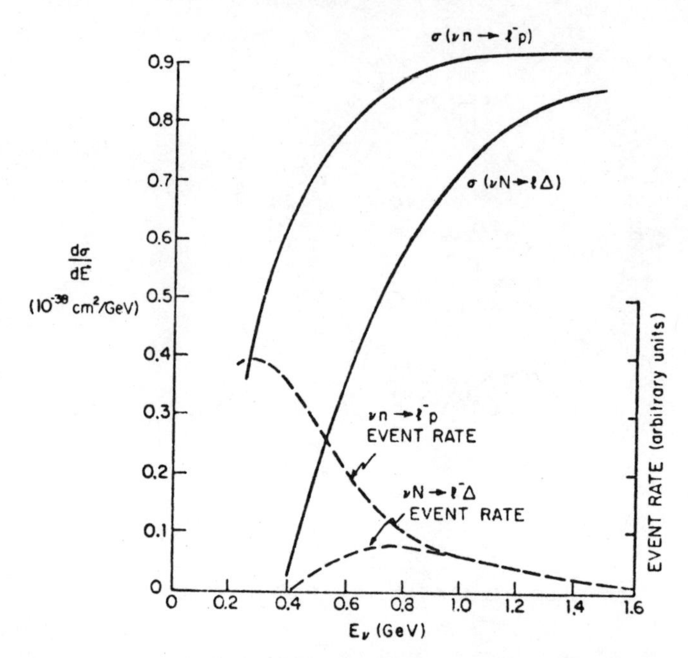

Figure 3.

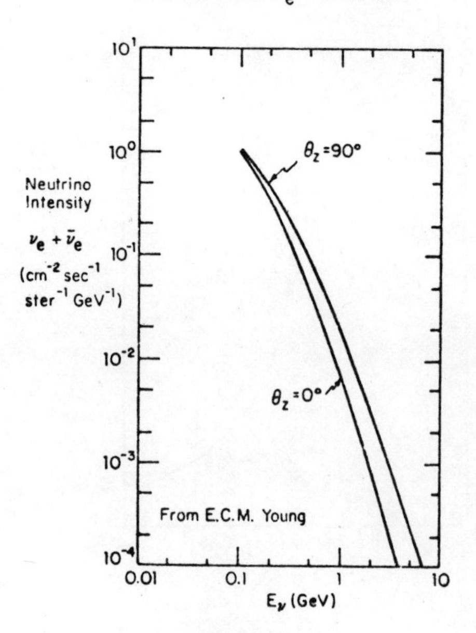

Figure 4.

which is familiar from the two component evolution of the $K°$-$\overline{K}°$ system. The oscillation length L is $L(m) = 2.5\ p(\text{MeV})/\Delta m\ (\text{eV}^2)$, and the characteristic measure of the proper evolution time in the rest frame of the neutrino system is proportional to ℓ/p. The sensitivity in Δm of any experiment is characterized both by ℓ/p and by statistics. The achievable value Δm^S is typically

$$\Delta m^S \cong 2(N_e/M_x)^{1/4}\ (p/\ell \sin 2\xi)^{1/2}$$

for $N_e(N_x)$ observed events induced by ν_e and ν_x's. For maximal mixing ($\xi = 45°$), $p \sim 500$ MeV, and $\ell = 1.3\times10^7$m, the ultimate sensitivity of an atmospheric neutrino experiment is 2×10^{-3} eV. Figure 2 shows the sensitivity as a function of both ℓ and of $\cos\theta_Z$. The $\cos\theta_Z$ dependence is experimentally relevant since equal statistics come in equal intervals of solid angle. Note that the dynamic range in Δm^S spans the region from 10^{-3} eV to 10^{-1} eV. The upper limit of this range coincides with the lower limit of the best sensitivity of accelerator and reactor based experiments.[4,5]

For communication between the three neutrino states, the phenomenology is generalized to two eigenmass differences, three Pontecorvo angles, and one CP violation phase.

Recent reports of possible evidence for neutrino oscillations[6] and a non-zero neutrino mass[7] are consistent with the previous experimentally allowed regions of these variables, as analyzed in Reference 5. In particular, a small mass difference ($\sim10^{-2}$ eV) may exist between the ν_e and ν_μ, and a large one ($\sim$ 10 eV) between these two states and the ν_τ. The experiment we outline is sensitive both to 1) ν_e-ν_τ mixing for $\Delta m > 0.1$ eV by an unexpected loss of ν_e's relative to ν_μ's in the downward coming neutrino flux, and to 2) ν_e-ν_μ mixing of $\sim10^{-2}$ eV by comparison of downward and upward ratios of e's to μ's.

SIMULATION OF NEUTRINO EVENTS IN THE DETECTOR

We have investigated the characteristics of neutrino-induced events in the I.M.B. detector to determine the event rates and the resolution in the direction of the incident neutrinos. To simulate events in the detector, several ingredients are necessary, including the neutrino flux and the charged-current cross sections. The atmospheric neutrino spectra for ν_e and ν_μ have been calculated some years ago by E.C.M. Young (References 8 and 9, resp.) and are reproduced in Figures 4 and 5, resp. The curves are similar in shape; on average, the ν_μ's dominate the ν_e's by a factor of two. Also the relative composition of neutrinos vs. antineutrinos can be inferred from the μ^+/μ^- charge ratio (1.2) of cosmic ray muons at the earth's surface[10]. The zenith angle distribution[8,9] of the fluxes

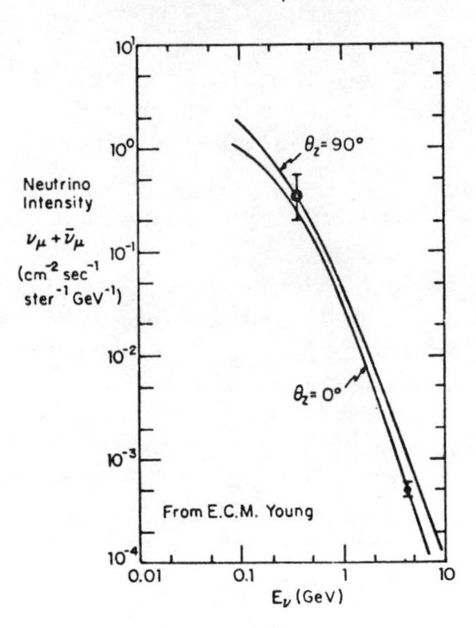

Figure 5.

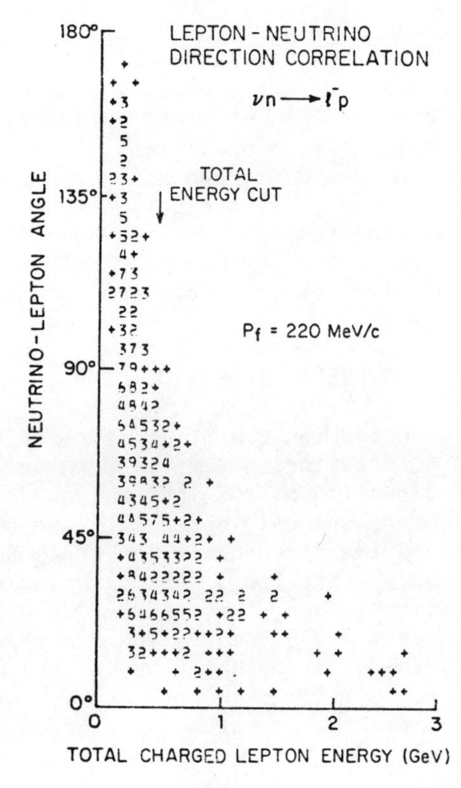

Figure 6.

is also taken from Young. The flux decreases by 25% from $\theta_Z=90°$ to $\theta_Z=0°$. Although no observations of cosmic ray ν_e's have been reported, and only a handful of ν_μ events have been recorded, the errors in these flux calculations are only expected to be $\sim$30% around 1 GeV since they are inferred from the observed muon flux at sea level. (With the vastly better knowledge of the π/K ratios etc. today (relative to 1969 when the calculations were made) a reevaluation of the spectra is strongly encouraged.) The ν_μ charged current cross sections near 1 GeV are well known from Argonne and Gargamelle data.[11,12] The energy dependence for elastic scattering and $\Delta(3/2, 3/2)$ production are shown in Figure 3. The products of flux and cross section, also shown in Fig. 3, demonstrate that only elastic charged-current scattering is statistically important. This is fortuitous, since otherwise charged pions from resonance production could confuse the μ/e event identification, which relies on stopping muon decay signatures.

Thus, the dominant signal in the detector for the oscillation experiment is

$$\nu_\mu n \rightarrow \mu^- p$$

and

$$\nu_e n \rightarrow e^- p.$$

However, the recoil proton initiated by 1 GeV neutrinos is below Cerenkov threshold ($\beta=0.75$) and therefore is invisible in a water Cerenkov detector. The correlation between the direction of the charged lepton and the incident neutrino will be used to measure the incoming neutrino direction. This correlation is shown as a function of the lepton total energy in Fig. 6. In this plot, the Fermi motion of the nucleons is assumed to be that of water, with p_f = 220 MeV/c. To insure that the angular correlation between outgoing charged lepton and the incoming neutrino is sufficient to sense an up/down asymmetry, a cut at $\sim$ 0.3 GeV in total lepton energy is necessary. Events with this cut have the angular distribution of Figure 7. The mean correlation angle is 45°. Although the energy cut eliminates 1/2 of the events from the data sample, the rejected events predominantly have low energies and therefore are more susceptible to background and to misidentification. After the charged-lepton energy cut, the mean neutrino energy of the accepted events is $\sim$ 0.7 GeV. Neutrinos of this energy produce μ's and e's with a radiating range of typically a factor of three shorter than that encountered in two body, 1 GeV decays of the proton. Thus, the available containment volume in the detector for this experiment could be somewhat larger than that for the proton decay experiment. Rather than the 2 m veto region there, we make a 1.2m cut on all faces but the top, giving an anticipated containment volume of $\sim 4\times10^{33}$ nucleons.

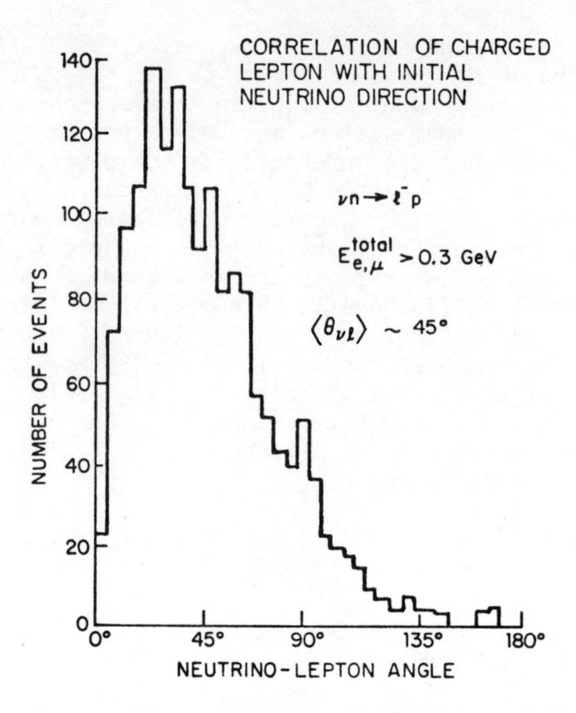

Figure 7.

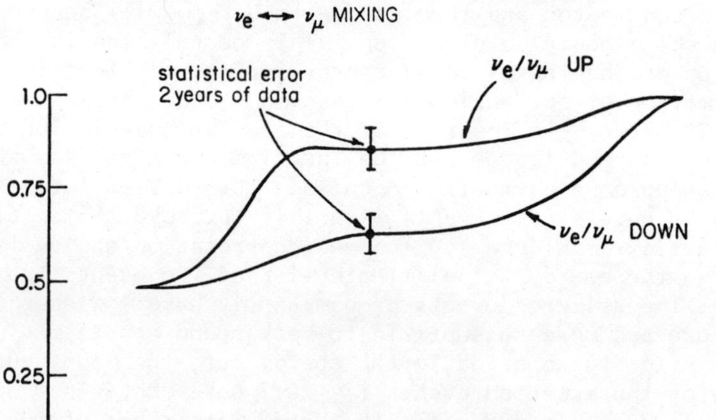

Figure 8.

Conservatism suggests the added veto region on the top as insurance against the entrance of cosmic ray muons, which are inherently angular asymmetric toward the vertical.

The apparatus resolution in angle for the charged lepton ($\sim$20°) is sharp enough to preserve any up/down asymmetry. Also, the energy cut is not significantly smeared by the energy resolution of the detector, which is $\sim 15\%/\sqrt{E(\mathrm{GeV})}$.

Muon events are distinguished from electron events by the delayed pulse produced by stopping, decaying muons. In the $\sim 10\mu$sec live period after each muon trigger, $\sim$90% of μ^- decay events are expected to register in $>$ 3 tubes at coincident times. These times must be consistent with the stopping point of the muon registered during the initial trigger. When compounded with the nuclear absorption probability of 20% in oxygen, the detection efficiency for stopping μ^-'s should be $\sim$ 70%.

The sensitivity of the experiment can be characterized by mean event rates expected under various oscillation hypotheses, as is shown in Table I. The total number of fully contained events per year is expected to be 500 (1000) induced by ν_e (ν_μ). This rate is halved by the 0.3 GeV cut on charged lepton energy. Thus, for no oscillations, the e/μ event ratio should be 0.5 ± 0.07, whether initiated by down or up going neutrinos. Both down and up going e/μ ratios will be unity if ν_e,ν_μ oscillations exist with L $\lesssim$ 300 km. The e/μ ratio will be unity only for upcoming neutrinos if L $\sim$ radius of the earth. And if muon neutrinos do not oscillate, but ν_e,ν_τ oscillations exist with a large mass differences, the e/μ event ratio will be 0.25 ± 0.04 for both up and down going events.

For e,μ oscillations, the sensitivity of the detector is characterized as a function of the mass difference in Fig. 8. The e/μ ratio is shown, both for upward coming and downward going neutrinos. For two years of data, the statistical power is $\sim$ four standard deviations for each ratio in the region between 0.006 eV and 0.15 eV. Thus for e,μ mixing the experiment is optimized in the low mass range, but the full 10 KT size of the detector is necessary to have sufficient statistical power.

We should also consider the effect on vacuum neutrino oscillations induced by the different forward scattering amplitudes in the matter through which the upward coming neutrinos have passed. ν_e's can scatter from electrons by both neutral and charged currents where as ν_μ's and ν_τ's only have charge-current interactions. Wolfenstein has shown that both the vacuum Pontecorvo angles and oscillation lengths are modified by the matter at oscillation lengths comparable to the earth's radius.[13] Since this results in exploring a somewhat different range of variables than in vacuum over the same distance, we ignore these effects in the current paper.

In conclusion, the massive detector under construction for the baryon decay experiment may provide the most sensitive detector for oscillation searches between ν_μ and ν_e species, and may yield a signal for ν_e,ν_τ oscillations soon after turn on if the expected 0.5 ratio of ν_e/ν_μ is not observed.

ACKNOWLEDGMENTS

We acknowledge innumerable enlightening discussions with R. Barbieri, A. De Rújula, J. Ellis, P. Frampton, M.K. Gaillard, H. Georgi, S. Glashow, D. Nanopoulos, D. Ross, A. Yildiz on the topic of neutrino oscillations. We also are deeply thankful to L. Wolfenstein for continued clarification of these matters. Conversations with R. Barloutaud revealed that he has independently done some of the calculations presented here. See Reference 14.

Table 1. Typical Neutrino Event Rates Under Various Oscillation Hypotheses

Errors noted are purely statistical. Entries are events/year in the fiducial volume above a total energy cut of 0.3 GeV on the charged lepton.

Neutrinos mixing	Oscillation length	Event Rates ν's down	Event Rates ν's up
ν_e,ν_μ	L>13,000km	125±12e 250μ	125e 250μ
ν_e,ν_μ	13,000>L>300km	125e 250μ	187e 187μ
ν_e,ν_μ	300km >> L	187e 187μ	187e 187μ
ν_e,ν_τ	13,000>L>300km (ν_μ, L>13,000km)	125e 250μ	60±8e 250μ

REFERENCES

#Supported in part by the U.S. Department of Energy.
†Also, Research Assistant, University of Michigan, Ann Arbor, MI 48109.
* Also, Visiting Scholar, Harvard University, Cambridge, MA 02138.
§The members of the collaboration are the following:
M. Goldhaber, Brookhaven National Laboratory; B. Cortez, G. Foster, L. Sulak, Harvard University and the University of Michigan; C. Bratton, W. Kropp, J. Learned, F. Reines, J. Schultz, D. Smith, H. Sobel, C. Wuest, University of California at Irvine; J. LoSecco, E. Schumard, D. Sinclair, J. Stone, and J. Vander Velde, University of Michigan.
1. For a detailed description of the I.M.B. proton decay detector see the companion article by L. Sulak which appears in these proceedings and M. Goldhaber, et al. Proceedings of International Neutrino Conference, p. 121, C. Jarlskog, ed. Bergen (1979).
2. For a review of the theory of neutrino oscillations see S.M. Bilenky and B. Pontecorvo, Phys. Reports 41:225 (1978).
3. We choose to refer to the neutrino mixing angle in this way since B. Pontecorvo has continually emphasized the open nature of this question.
4. The status of the most sensitive accelerator based oscillations search is described by one of the authors (LS) in the Proc. of the BNL Workshop on the Fixed Target AGS Program, BNL 50947, Nov. 8-9, 1978, p.212.
5. A. DeRújula, et al., "A Fresh Look at Neutrino Oscillations", Ref. TH. 2788-CERN; V. Barger et al., "Mass and Mixing Scales of Neutrino Oscillations", COO-881-135 (U. Wisconsin preprint).
6. F. Reines, H. Sobel and E. Pasierb, "Evidence for Neutrino Instability", (U. Irvine Preprint) submitted to PRL (1980).
7. Private communication; E.T. Tretyakov, et al., Proceedings of the International Neutrino Conference, p. 663 Aachen (1976), has a description of the experiment and a preliminary analysis of the data.
8. E.C.M. Young, Cosmic Rays at Ground Level, Institute of Physics Press p. 105, A.W. Wolfendale, ed., London and Bristol (1973); and E.C.M. Young, private communication (1977).
9. A.W. Wolfendale and E.C.M. Young, CERN 69-28, p. 95, (1969).
10. M.G. Thompson in Cosmic Rays at Ground Level, Institute of Physics Press, A.W. Wolfendale, ed., London and Bristol (1973).
11. R.T. Ross, Proceedings of the 1978 Conference on Neutrino Physics p. 929, E.C. Flower, ed.; M. Derrick, Proceedings of the 1977 Conference on Neutrino Physics, p. 374, H. Faissner et al., ed; and the references cited in the two works.

12. Experimental results with which the Monte Carlo has been compared include: E.C.M. Young, "Neutrino Charged Current Single Pion Production in Freon", CERN Yellow report 67-12; S.J. Barish, et al., Phys. Rev. Lett. 36:179 (1976); J. Campbell, et al., Phys. Rev. Lett. 30:335 (1973).
13. L. Wolfenstein, Proc. of the Clyde L. Cowan Memorial Symposium, A. Saenz ed. Catholic University of Washington 1978; Phys. Rev. D. 17:2369 (1978).
14. R. Barloutaud, Test of Neutrino Oscillations in a Large Underground Detector, 12 March 1980, Draft.

INVESTIGATIONS ON NUCLEON STABILITY IN EUROPE

E. Bellotti

Istituto di Fisica dell'Università and Sezione INFN

Via Celoria 16, 20133 Milano - Italy

This communication will be mainly devoted to the description of the so called NUSEX (Nucleon Stability Experiment) project(1) which, at present, has been funded and is in the preparation stage. The NUSEX apparatus will be installed in the Mont Blanc Tunnel and the data taking is expected to start in the middle of 1981. Finally, some considerations on possible developments of this kind of experiment in Europe will be reported.

The search for possible nucleon instability violating baryon number conservation is an old experimental problem(2), but in recent years the interest on this subject has rapidly grown because it is a crucial test of the grand unification models.

The existing limit on nucleon lifetime (with $\Delta B \neq 0$) is $\sim 10^{30}$ y's(3) and most of the theoretical estimates range between 10^{31}-10^{33} yrs; according to the SU(5) model, the favoured decay channels are $p(n) \rightarrow e^+\pi^0(\pi^-)$; $e^+\rho^0(\rho^-)\ldots$, but other models prefer final states with μ's(4).

The task of detecting such a small decay rate is a very hard one and requires very massive apparatuses operating in extremely low background conditions. In order to tackle this problem, the groups of Frascati and Milano begun ($\sim$ Spring 1979) to think of a small prototype detector of $\sim 1\ m^3$ of volume, corresponding to a mass of $\sim$ 3 tons, to be run in an underground laboratory. But in a short time it was realized that such a detector was too small for an accurate study of the background competing with nucleon decay. About at the same time, the Torino group, which was designing an experiment using the Cerenkov Technique(5), agreed to undertake a common effort with the above mentioned groups. Then it was possible to design a much bigger detector and ask successfully for funds. Finally, the collaboration has been joined by D.C. Cundy and M. Price from CERN.

General remarks on apparatus for nucleon decay search

In any new apparatus designed to search for nucleon decay looking directly at the decay products, the useful mass of the source of nucleons (which is identical to the detector mass) has a lower bound due to the fact that a limit of $\sim 10^{30}$ yrs has already been reached. Assuming 1 year of running time and 100% efficiency, the mass must be $\gtrsim$ 3 tons.

The upper limit on the mass depends on the capability of the detector in rejecting background interactions, on the space available in the laboratory and on the cost of the entire apparatus.

As will be seen later, the known background which certainly cannot be avoided is due to atmospheric neutrino interactions. Assuming an energy resolution of ± 20% at 1 GeV (the total energy of nucleon decay products), 100% efficiency and no rejection of background based on the topology of the events, the number of background events due to ν's is $\sim 3 \cdot 10^{-2}$ events/(year x ton). The upper limit on the fiducial mass is then $\sim$30 tons and the attainable limit on $\tau_{nucleon} \sim 2 \cdot 10^{31}$ yrs.

It appears that a detector of total mass in the range 100-300 tons [6], capable of detecting many final states, simple and apt for reliable long runs is an appropriate choice for a first generation experiment. Furthermore, a detector of such a mass and with mean density $\sim$3 fits well in the Mont Blanc Laboratory, and its cost is not excessive.

The detector

The detector will consist of a uniform sandwich of iron plates 1 cm thick (0.5 rad. length), interleaved with layers of limited streamer resistive tubes (Fig.1). The total volume is 3.5 x 3.5 x x 3.5 m^3 corresponding to a total mass of $\sim$150 tons ($\sim$ 134 tons of iron and $\sim$ 16 tons of PVC). Iron and PVC are the sources of nucleon decays and the iron plates act as passive elements of the detector. Resistive tubes operated in the limited streamer mode are the basic elements of the sensitive part of the detector. These tubes, developed in Frascati by Battistoni et al.[7], show a large high voltage plateau (Fig.2), large output signals and a good efficiency.

The use of resistive cathodes allows a very simple method for bidimensional read-out of multitube layers. Because the resistive cathodes are transparent to the pulsed e.m. field associated with any pulsed current generated around the sense wire, pick-up electrodes, shaped like strips, can be placed outside the tubes and the pulses collected by these electrodes. Both x and y coordinate of the streamer will be read by means of strips (Fig. 3) in order to avoid the high voltage capacitive connections to the h.v. wires, which would represent the weaker part of the entire apparatus

The behaviour of resistive tubes, filled with standard gas

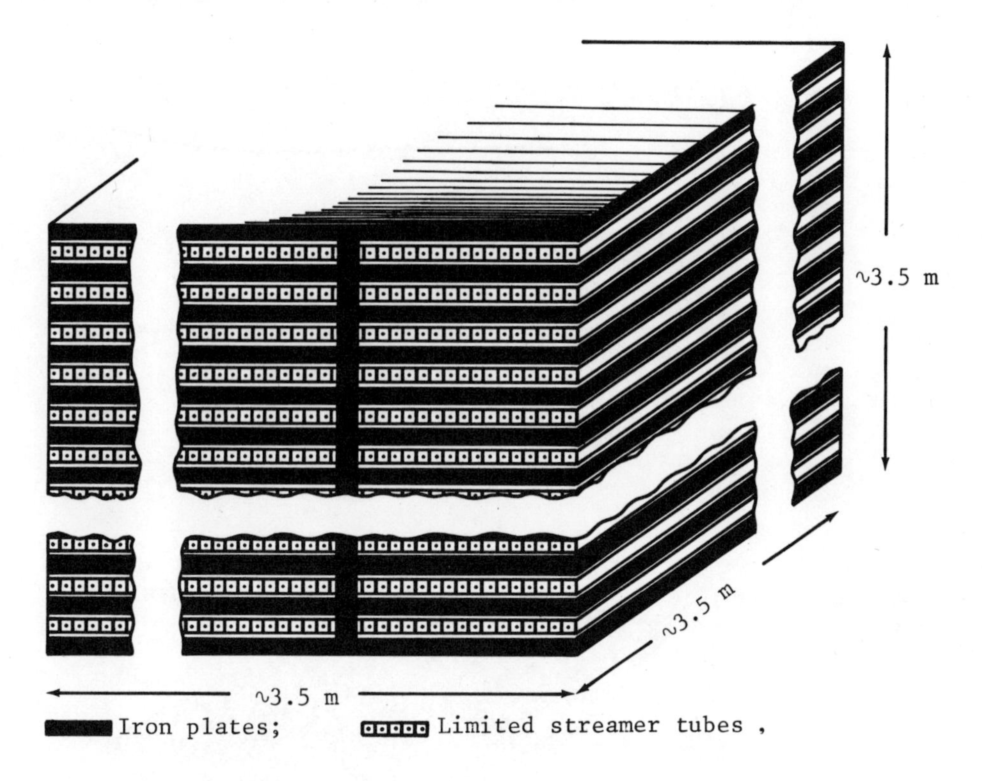

Fig.1 - Schematic view of the NUSEX detector.

mixture similar to those we will use in the detector has been tested and found satisfactory. Tests with tubes filled with different gas mixtures and with sealed tubes are in progress.

Tubes, strips, iron plates and spacers will be assembled as shown in Fig.4; details on the assembling procedure are given in(1).

The detector will consist of 134 iron plates, 47,000 tubes and 94,000 read-out channels. Some space will be left free around the detector to add, if necessary, some anticoincidence counters or shielding.

The basic block of the read-out system is a card containing the circuitry for detecting, storing and reading out 32 tube signals in a compact package for easy mounting directly on the detector. A prompt OR signal will be available from each card and will be the basic element for trigger and monitor. Cards and controller will be provided by Le Croy Research Systems S.A.

Accepted events will be read out, enclosed and stored by controllers (one each 128 cards) and finally transmitted to a PDP 11/60 or similar computer.

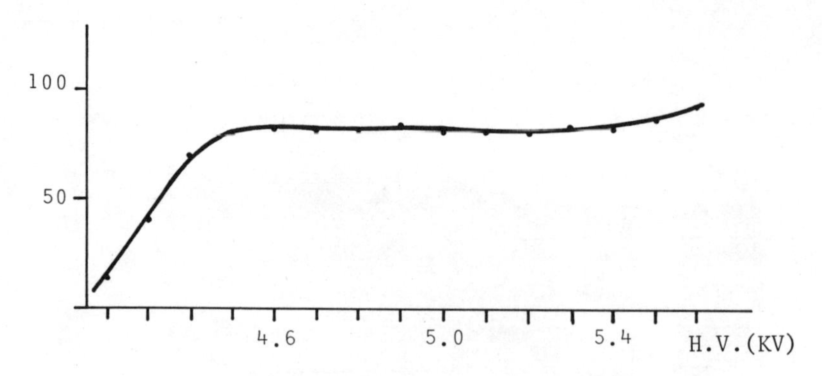

Fig.2 - Single rate (^{90}Sr) vs high voltage.

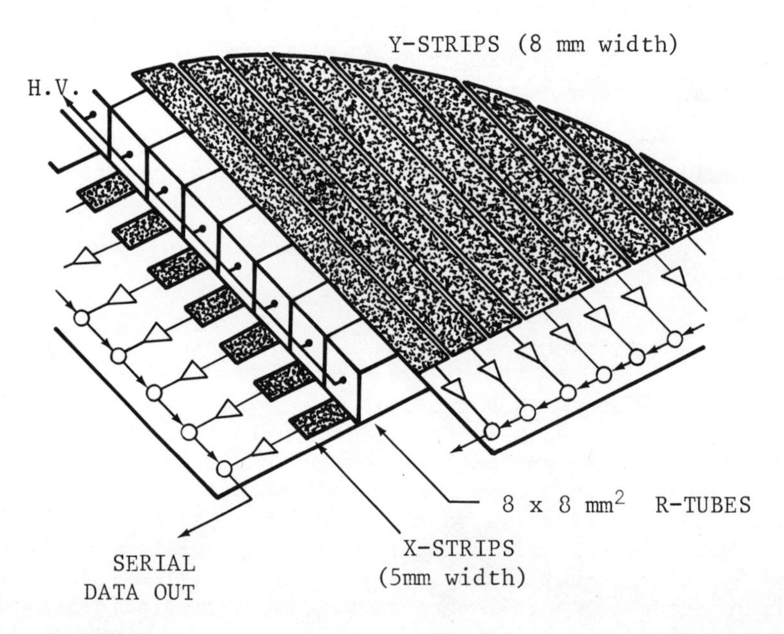

Fig.3 - Proposed structure of the resistive tube planes with x-, y-strip read-out.

Tests and Monte Carlo calculations

The behaviour of resistive tubes in an electromagnetic calorimeter has been investigated in Frascati some time ago. A small

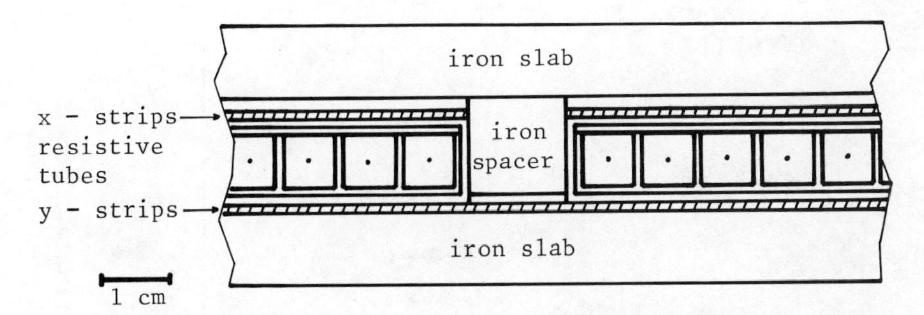

Fig.4 - Details of the mechanical assembling of the set-up.

calorimeter of resistive tubes and lead plates (0.4 rad. lengths) was exposed to an e^+ beam of low energy. The energy and angular resolutions were in agreement with the expected ones[1].

Events simulating proton and neutron decay were generated in the present detector by a Monte Carlo programme. Fermi motion of the decaying nucleon was included. Charged pions are propagated taking into account elastic, inelastic and absorption reactions; electromagnetic showers have been computed with the EGS3[8] code developed at SLAC.

The simulated events are used to investigate the possibility of identifying the topology of the events, to evaluate the energy resolution and as input for a simple reconstruction programme. Preliminary results show that electromagnetic showers originated by the $e^+\pi^0$ decay are identifiable by the transverse dimensions of the pattern and the vertex of the event is reconstructed with the accuracy of few centimetres.

For these events, the number of signals on strips N_s is shown in Fig.5.

The resulting energy resolution is ∿20% for event normal to the iron plates and changes slowly with the shower direction.

The identification of μ's does not pose any problem if their energy is not too low (>150 MeV) because they cross the iron plates without strong scattering.

Major problems arise from pions. In fact, π's of ∿500 MeV have a large probability (80%) of interacting before the end of their range. Then the energy estimate is very difficult and, in some cases (∿ 40%) the identification of the topology becomes very doubt.

Despite these difficulties, the detector has a very low energy threshold; e.g., an electron or γ ray of 100 MeV can fire tubes in

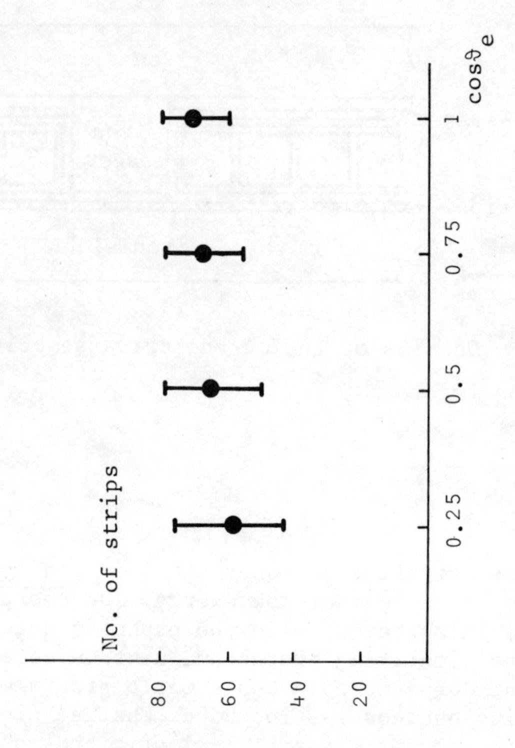

Fig.5 - No. of hits on strips as function of the cosine with the normal to the plates

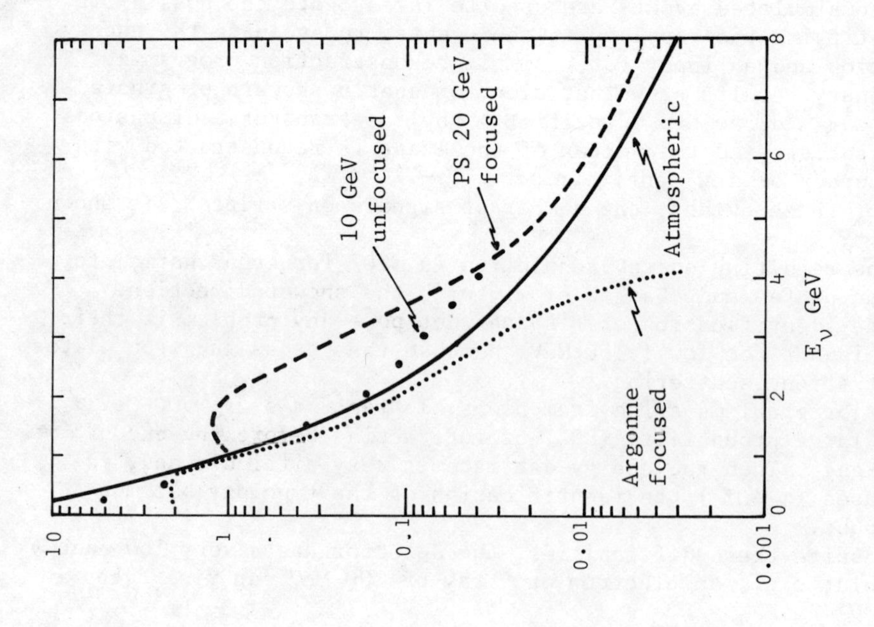

Fig.6 - Relative flux shapes

3 or more different layers, and that will be enough for triggering the data collection. It will be possible to study the background without important limitations on the energy deposited.

In the second half of the year 1980, a detector of (3.5 x 1 x x 1)m will be exposed to electrons and hadrons test beams at CERN.

This small detector will be also tested in a ν beam. In fact the shape of the neutrino flux from a primary proton beam of $\sim$10 GeV impinging on a bare target (no horn) is similar to the shape of the atmospheric ν flux(9) (Fig.6).

The most important difference concerns the composition, the accelerator beam consisting of ν_μ and $\bar{\nu}_\mu$, while $\sim$1/3 of atmospheric ν's are ν_e and $\bar{\nu}_e$. The problem can partly be solved combining data from ν_μ and e induced events.

The laboratory

The apparatus will be located in the Mont Blanc Tunnel, in a "garage" between kilometers 5 and 6 from the French entrance. The site can be easily reached by car in few hours from CERN, Torino and Milano.

The rock thickness covering the laboratory, expressed in metres of water equivalent (m.w.e.) is shown in Fig. 7 as a function of the azimuthal and zenithal angles. In every direction the covering is greater than $5 \cdot 10^3$ m.w.e. and provides an excellent shielding against cosmic μ's.

Background

The known sources of background are: local radioactivity atmospheric muons, associated neutrons and atmospheric neutrinos.

The γ radioactivity has been measured(10) with Ge(Li) detectors and its level is quite high (e.g., 3 times that in Milano). This activity should produce a counting rate of $\sim$30 Hz in the external tubes and <1 Hz in the inner part of the detector.

This background would not cause any trouble because of the limited energy of γ rays and, on the contrary, will be used to monitor the state of the tubes.

The muon flux has been recently measured by the Torino group(11) in different positions of the Tunnel. About 10^4 muons/ /year are expected to enter the detector. We believe that the high energy μ's will be easily identified. The small fraction of μ's stopping in the detector ($\sim$ 3%)(4) will be recognized by their topology; if it proves necessary, a suitable veto could be installed.

Low energy neutrons can be produced by spontaneous fission in the rock and neutrons of any energy by cosmic muons.

To my knowledge a direct measurement of high energy neutrons deep underground does not exist(12). An estimate has been done in(1), by D.H. Perkins(13) and by A.L. Grant(14). The expected number of neutron interactions of energy in the range of interest ($\sim$ 1 GeV), without any other hadron, is less than one per year in the present detector.

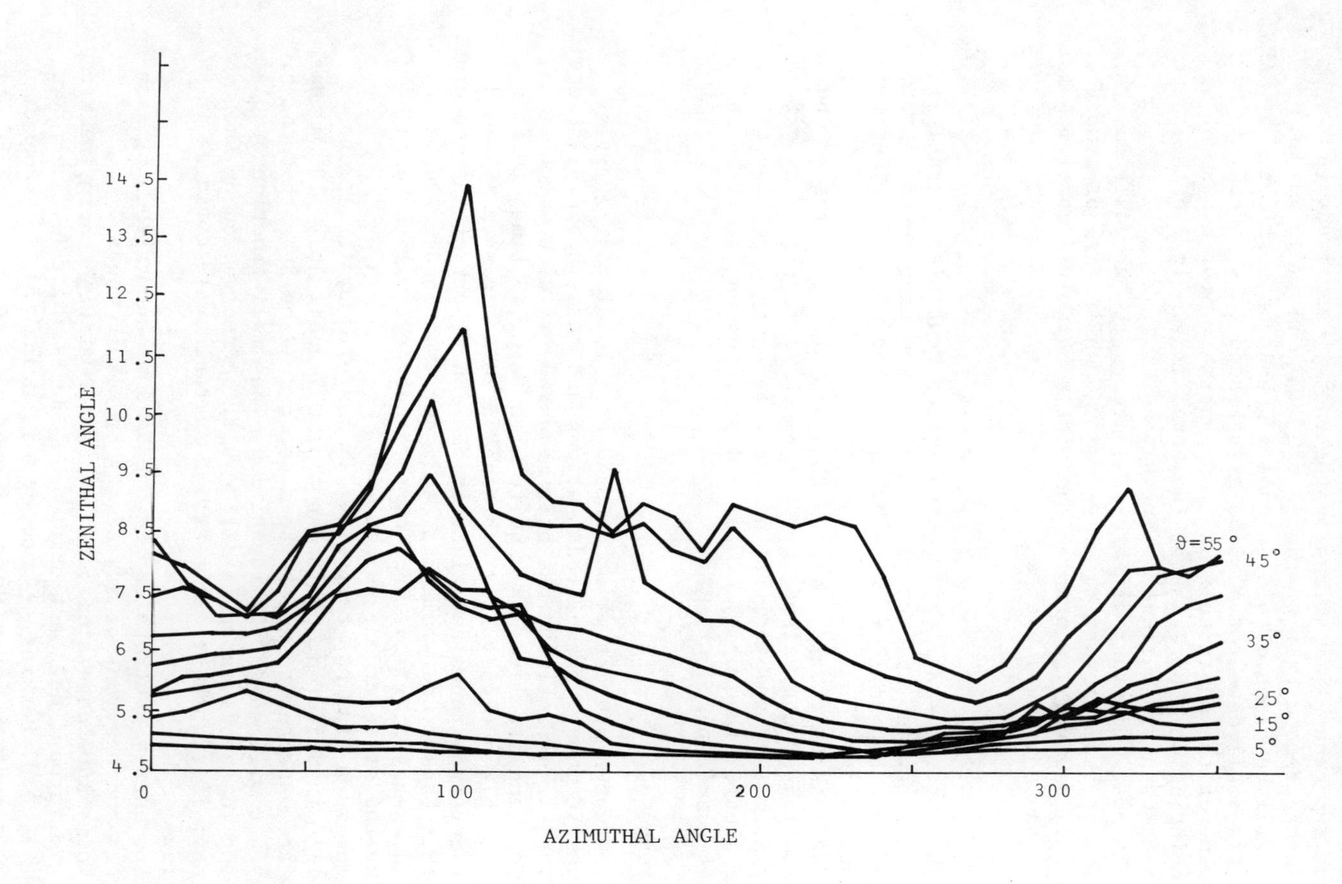
ZENITHAL ANGLE
14.5
13.5
12.5
11.5
10.5
9.5
8.5
7.5
6.5
5.5
4.5
AZIMUTHAL ANGLE
0
100
200
300
ϑ=55°
45°
35°
25°
15°
5°

Figure 7.

Probably the only serious source of background is due to atmospheric neutrinos. The ν flux has been computed many years ago by Osborne et al.(15) and by Wolfendale et al.(16) and it is shown in Fig.8. The result obtained by Crouch et al.(17) on the μ flux induced by ν's deep underground, is in agreement with the computed flux:

$$\frac{(\mu \text{ from } \nu)_{\text{computed}}}{(\mu \text{ from } \nu)_{\text{observed}}} \sim 1.6 \pm 0.4$$

The expected number of ν events/(year x 100 tons) is summarized in the following table.

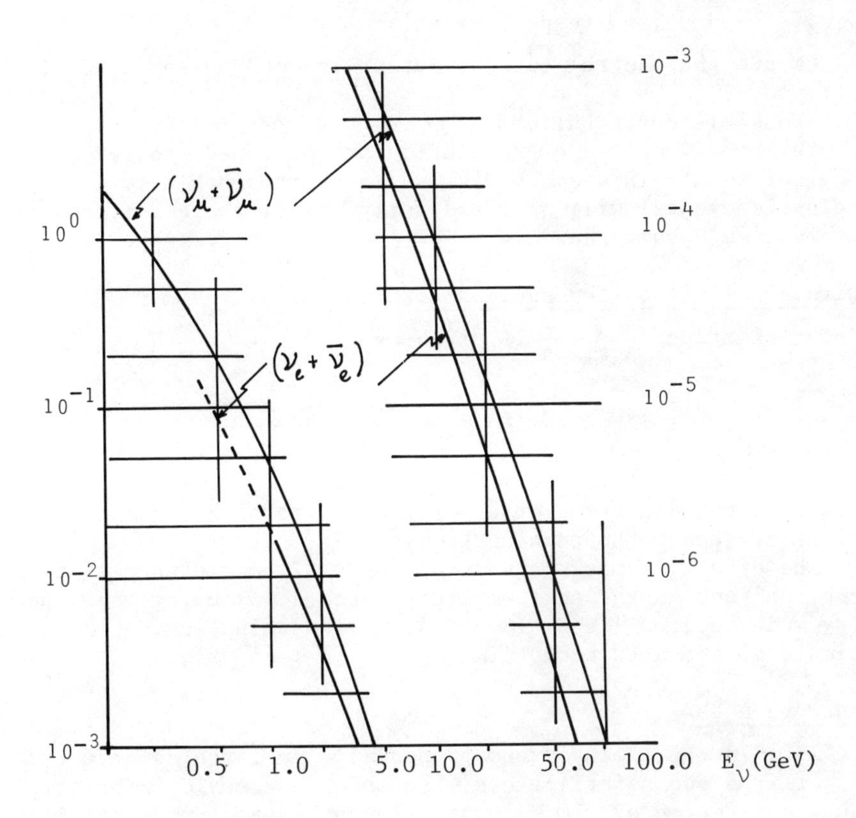

Fig.8 - ν flux (horizontal component) /(cm^2 x s x sr x x GeV).

Visible Energy (GeV)	$CC_{\nu_\mu \bar{\nu}_\mu}$	$CC_{\nu_e \bar{\nu}_e}$	CN	Tot
0.4 - 0.6	2.06	0.936	1.02	3.96
0.6 - 0.8	1.34	0.637	0.511	2.49
0.8 - 1.0	1.13	0.469	0.314	1.91
1.0 - 1.2	0.89	0.379	0.227	1.51
1.2 - 1.4	0.74	0.310	0.105	1.22

Visible energy is the energy of the incident ν for charged current interaction and the energy of the hadronic shower produced in neutral current interaction.

[$CC_{\nu_\mu,\bar{\nu}_\mu}$ and $CC_{\nu_e,\bar{\nu}_e}$ are the charged current events produced by $\nu_\mu,\bar{\nu}_\mu$ and $\nu_e,\bar{\nu}_e$ respectively.]

CN are the neutral current events induced by any kind of neutrinos.

(Possible contributions from ν_τ have been neglected.)

Taking 20% of energy resolution around 1 GeV, the expected number of events induced by ν's is $\sim$ 4. Part of them (e.g., ν_μ and $\bar{\nu}_\mu$ elastic events) will show an easily identifiable pattern. Thus the expected ν background is of the order of 1 event/year.

Attainable limit on the nucleon lifetime

The limit on $\tau_{nucleon}$ is given by

$$\tau_{nucleon} = \frac{6 \cdot 10^{29} \times m \times t}{N_{events}}$$

where m is the effective mass = (fiducial mass) x (detection efficiency) and t the running time.

The effective mass depends on the real consistency of the background and on the nucleon decay channel. Assuming that the decay mode nucleon $\rightarrow e^+X$ $(X = \pi, \rho ...)$ is dominating, a rough estimate of the attainable limit is $\tau \sim 4 \cdot 10^{31}$ yrs.

Other projects

In 1979 the Saclay group proposed[18] a large calorimeter made of iron and scintillators with which one could in principle reach a lifetime of $\sim 10^{32}$ years (in one year of running time) for the partial lifetime $p \rightarrow e^+\pi^0$.

Presently, a French Collaboration (Orsay - Ecole Polytechnique and Saclay[19]) is working on a test detector to investigate possible solutions for a second generation experiment. The groups

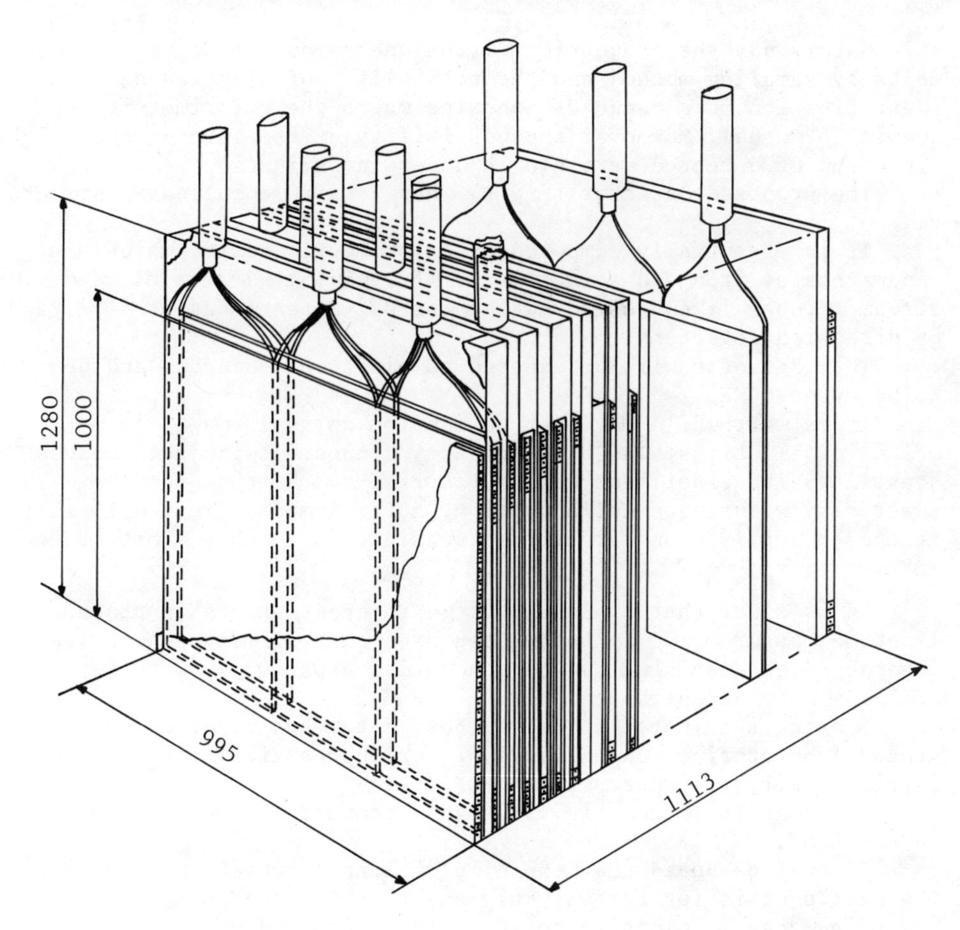

Fig.9 - Schematic view of the French prototype detector. (Lengths in mm).

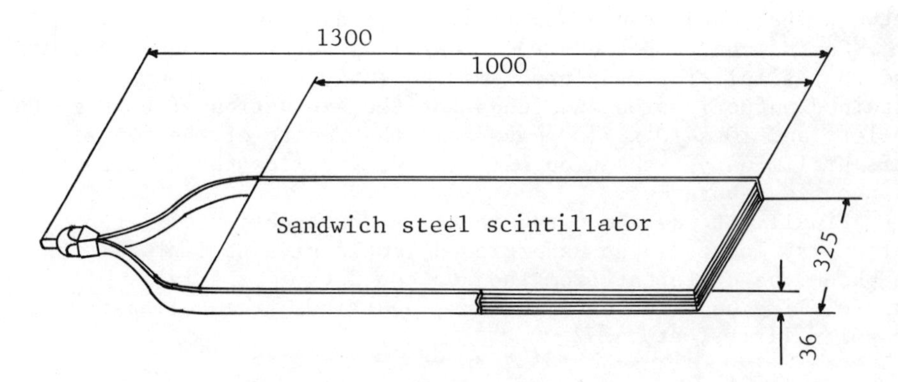

Fig.10 - Calorimetric cell.

intend to study the properties of the energy and direction measurements by sampling method and the possibility of electron and hadron identification in a range of energies where the calorimetric technique is not usual. Moreover the possibility of recognizing the flight direction of detected particle is under investigation.

The proposed detector (Fig.9) has a quite complicated structure.

It is essentially composed by calorimetric cells, which are sandwiches of iron and scintillator (Fig.10) with layers of limited streamer tubes like those used for the DM2 detector at LAL[20], and by direction detectors.

This detector will be exposed to electron, hadron and neutrino beams at CERN.

It is worthwhile moreover to mention an experiment[21] on $n \leftrightarrows \bar{n}$ oscillations, which are clearly connected with the nucleon stability. The experiment will be carried out at the Grenoble reactor by a European Collaboration. A first measurement will start at the end of 1980 and it will be possible to reach a limit of few 10^6 sec on $\tau_{n \to \bar{n}}$.

It is clear that in Europe large interest exists in proton lifetime measurements and connected problems and many groups are interested in designing second generation experiments with detectors of large mass and high accuracy.

Because the installation of these detectors requires underground laboratories, an analysis of various possibilities has been carried out by H. Laporte from CERN[22].

The next table and Fig. 11 show a comparison among 4 tunnels in the Alps.

It must be noted that the Simplon Tunnel, which appears to be the best one, is for railway only and this fact poses many problems regarding access, electric noise, ventilation and so on.

Moreover it seems very difficult, if not impossible, to enlarge the present laboratory in the Mont Blanc Tunnel.

In fact, this tunnel is one of the most important connections between Italy and France and it is crossed by heavy traffic.

On the contrary the Frejus Tunnel is not yet opened to traffic and that simplifies many problems. Recently French groups have obtained authorization and funds for the excavation of a 40 m long gallery and 10 x 10 x 7 m^3 cave near the centre of the tunnel. This preliminary excavation will in future allow the construction of a large laboratory.

Finally it must be noted that the construction of large detectors, located deep underground, could give new information on phenomena different from the nucleon decay, especially of cosmic origin, like neutrino oscillations, neutrinos from collapsing stars or solar flares, etc.[23].

Tunnel \ Parameters	H (m)	ℓ_{max}	ℓ_{min}	$<\ell>$	V 10^9 m^3
Mont Blanc	1747	3500	1700	2260	8.9
Frejus	1687	2500	1450	1677	3.0
Simplon	2198	3510	1770	2331	8.8
Gotthard	1446	2270	1210	1466	2.0

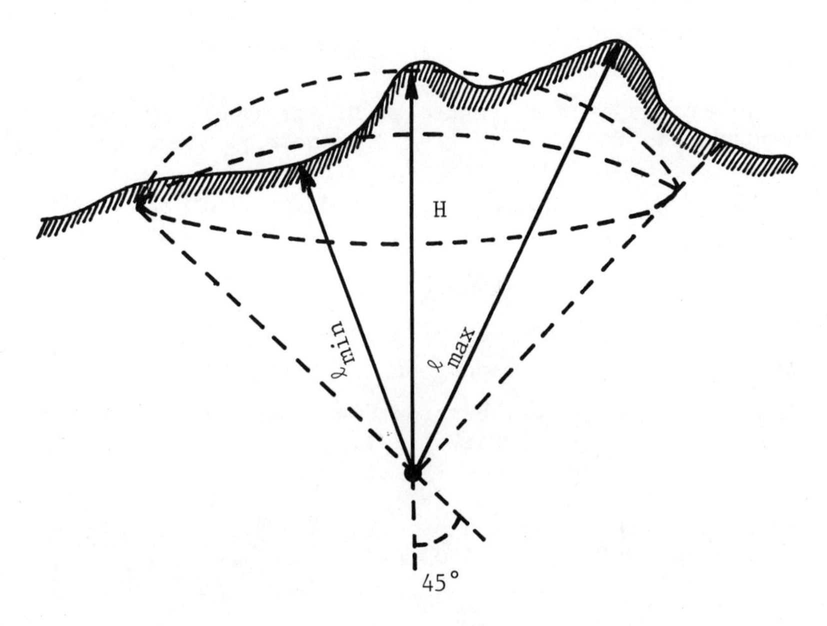

Fig.11 - V is the volume in a cone of 45°.

Acknowledgments

I would like to acknowledge all the participants to the NUSEX experiment.

References

1) G. Battistoni(+), E. Bellotti(x), G. Bologna(°), C. Castagnoli(°) V. Chiarella(+), D.C. Cundy(=), B. D'Ettorre Piazzoli(°), E. Fiorini(x), E. Iarocci(+), G. Mannocchi(°), G.P. Murtas(+), P. Negri(x), L. Periale(°), P. Picchi(°), M. Price(=), A. Pullia(x), S. Ragazzi(x), M. Rollier(x), O. Saavedra(°), L. Trasatti(+), and L. Zanotti(x),

(+) L.N.F. - Frascati; (x) Ist. di Fisica and INFN - Milano; (°) L.C.G.F.-C.N.R. - Torino; (=) CERN

Proposal for an experiment on nucleon stability with a fine grain detector, INFN - Internal report (1979).

2) H.S. Gurr et al. Phys. Rev. 158:1321 (1968), also for previous references.
I. Bergamasco et al., Lett. al Nuovo Cim. 11:493 (1974).
F. Reines et al., Phys. Rev. Lett. 32:636 (1974).

3) J. Learned, F. Reines and A. Soni, Phys. Rev. Lett. 43:907 (1979).

4) See, e.g., M. Machacek, Nuclear Phys. B159 (1979) 37.

5) C. Castagnoli, Report given at the "Conferenza Nazionale dell'INFN", Frascati (1979).

6) This solution was also suggested by D. Perkins - Referee's report to the Feasibility test for an experiment on nucleon stability, CERN/EP/D.H. Perkins/mk (1979).

7) G. Battistoni et al., Nucl. Instr. and Meth. 152:423 (1978).
G. Battistoni et al., Nucl. Instr. and Meth. 164:57 (1979).
G. Battistoni et al., Resistive cathode detectors with bidimensional strip read out: tubes and drift chamber, L.N.F. preprint (1979).

8) R. Ford and W.R. Nelson, Report SLAC/210 (1978).

9) D.C. Cundy, Background calibration for proton lifetime detectors CERN/EP/D.C. Cundy/mk (Oct. 1979).

10) E. Fiorini et al., Nuovo Cim. 13:747 (1974).

11) B. Baschiera et al., Absolute muon intensity angular distribution underground between 4000 and 6000 hg/cm^2 s.c. at the Mont Blanc Station, Proc. 16th Int. Cosmic Ray Conf., Kyoto (1979).

12) See ref. 1 for a more detailed discussion.

13) D.H. Perkins, Muon induced backgrounds in the Mont Blanc and Frejus Tunnels, CERN/EP/DP/mk, (Sept. 1979).

14) A.L. Grant, Isolated neutron background in the Mont Blanc and Frejus Tunnels, CERN/EF/ALG/ed (Dec.1979).

15) J.L. Osborne et al., Proc. Phys. Soc. 86:93 (1965).

16) A.W. Wolfendale et al., Cosmic ray muon neutrino intensities below 1 GeV, Neutrino Meeting CERN 69-28 (1969).

17) M.F. Crouch et al., Phys. Rev. D18:2239 (1978).

18) R. Barloutaud et al., Search for proton decay using a large calorimeter made of iron and polystyrene scintillator, Saclay Int. Report (1979).

19) R. Barloutaud et al., (Orsay-Ecole Polytechnique-Saclay) Etude de performances d'un dispositif calorimetrique en vue de la mesure de la vie moyenne du nucleon, Int. Report(1979).

20) See previous reference.

21) M. Baldo Ceolin, Search for $n \leftrightarrows \bar{n}$ oscillation, presented at Astrophysics and elementary particle common problems, Rome (21-22 Feb. 1980).

22) H. Laporte, CERN SB/A/142 (1979).

23) P. Galeotti, NUSEX as a detector of neutrinos from point cosmic sources, L.C.G.F. Int. Rep. 42 (1980).

SUPERSYMMETRY WITHOUT ANTICOMMUTING VARIABLES

H. Nicolai

CERN

Geneva, Switzerland

1. INTRODUCTION

We all know about the importance of anticommuting variables in the context of supersymmetry[1]. They play an essential role in the formulation of supersymmetric theories and they greatly facilitate complex calculations, for example in perturbation theory. Nonetheless, as mathematical objects, they are not always as convenient and wieldy as ordinary numbers. For instance, they do not have any positivity properties, and we do not know how to attribute an intrinsic meaning to a functional integral over superfields beyond perturbation theory (of course, Gaussian integrals can always be defined) without going back to the component fields. Thus, their main advantage lies in algebraic applications such as proving divergence cancellations, whereas they appear to be unsuitable for analytical applications such as proving correlation inequalities for theories containing fermions in interaction with bosons.

It is therefore of interest that there does exist a characterization of supersymmetric theories directly in terms of the functional measure by means of which expectation values of a supersymmetric theory are defined[2]. This result, which enables us to reconstruct supersymmetric models without recourse to anticommuting variables, is essentially a consequence of the fact that in supersymmetric theories the vacuum energy vanishes identically[3]. In the absence of interactions, this just means that the bosonic and fermionic degrees of freedom add up to zero (fermions being counted as negative), a basic principle known to all those working in supersymmetry. This counting rule is sufficient to determine the supermultiplets, but not to deduce the possible non-trivial supersymmetric interactions.

Our theorem may be viewed as a generalization of this rule to encompass the interacting case as well.

Let us suppose that the multiplet contains some bosonic fields A_i -- where the index i may be either vectorial, internal, or both -- and some Majorana spinors ψ_i; auxiliary fields are assumed to have been eliminated. The number of space-time dimensions is arbitrary, but the multiplets will, of course, depend on the dimension. A Euclidean metric is assumed throughout the rest of this paper. We will furthermore make repeated use of the fact that the fermions may be "integrated out"[4]. If

$$\frac{1}{2}\bar{\psi}M(A)\psi \equiv \frac{1}{2}\int \bar{\psi}_{i\alpha}(x)M_{i\alpha,j\beta}(x,y,\lambda;A_k)\psi_{j\beta}(y)\ dx\ dy \tag{1.1}$$

denotes the fermionic part of the action, which we henceforth assume to be quadratic in the fermions, we have, for instance*,

$$\int d\psi\ \exp\left[-\frac{1}{2}\bar{\psi}M(A)\psi\right] = \det M(\lambda;A)^{1/2} \equiv \left[\det M(\lambda=0;A)\right]^{1/2} D(\lambda;A)\ , \tag{1.2}$$

since the fermions are Majorana ($\bar{\psi} = \psi^T\mathbb{C}$). $D(\lambda;A)$ is the Matthews-Salam-Seiler (MSS) determinant[5] of the model. We can now state our main theorem.

<u>Theorem</u>: Supersymmetric theories are characterized by the existence of a generally non-linear and non-local transformation T_λ of the bosonic fields

$$T_\lambda\ :\ A_i(x) \to A_i'(x,\lambda;A) \tag{1.3}$$

with the following properties

i) T_λ is invertible in the sense of formal power series.

ii) $S(\lambda;A) = S_0(A'(\lambda;A))$, where S denotes the full bosonic part of the action and S_0 its free part.

iii) The Jacobi determinant of the transformation T_λ equals the MSS determinant in the case of "scalar" supersymmetry and the product of the MSS determinant and the Faddeev-Popov determinant[6] in the presence of an additional gauge symmetry.

---

*λ (or g) stands for the various coupling parameters.

We will not give the details of the proof in this talk, but rather concentrate on some explicit examples which illustrate and, we hope, clarify the content of the theorem.

2. SOME "SCALAR" SUPERSYMMETRIC EXAMPLES

As a prelude, let us begin with an extremely simplified example in zero space-time dimensions, where no non-local (kinetic) couplings exist and where the functional integral becomes an ordinary integral. In a nutshell, this example neatly displays all the relevant features. For a "multiplet" A, F real, ψ_1, ψ_2 anticommuting, consider the "Lagrangian"

$$\mathcal{L} = \frac{1}{2}F^2 + iFp(A) - \frac{1}{2}p'(A)\psi_\alpha \varepsilon^{\alpha\beta}\psi_\beta \ , \tag{2.1}$$

where p(A) is a globally invertible but otherwise arbitrary C^1 function. This "Lagrangian" is invariant under the supersymmetry transformations

$$\begin{aligned} &\delta A = \zeta_\alpha \varepsilon^{\alpha\beta}\psi_\beta \ , \quad \delta\psi_\alpha = i\zeta_\alpha F \ , \quad \delta F = 0 \\ &\zeta_\alpha \text{ anticommuting} \ , \quad \varepsilon_{\alpha\beta} = \begin{pmatrix} 0 & -1 \\ 1 & 0 \end{pmatrix} \end{aligned} \tag{2.2}$$

"Vacuum expectation values" are given by

$$\frac{1}{2\pi}\int R(A)e^{-\mathcal{L}(A,F,\psi_\alpha)} \, dA \, dF \, d\psi_1 \, d\psi_2 \ . \tag{2.3}$$

Integrating out F (the "auxiliary field") and ψ_1, ψ_2, we find

$$\langle R(A)\rangle = \frac{1}{\sqrt{2\pi}} \ R(A)e^{-1/2[p(A)]^2}p'(A) \, dA \ , \tag{2.4}$$

where $p'(A)$ is the MSS determinant. Indeed, the transformation $A \rightarrow A' = p(A)$ reduces (2.4) to a Gaussian integral and has Jacobi determinant $p'(A)$! Moreover, we can derive "super Ward identities" without using ζ's, so, for instance, the identity $\langle\psi_\alpha\psi_\beta\rangle = -i\varepsilon_{\alpha\beta}\langle FA\rangle$ is nothing but

$$\int_{-\infty}^{+\infty} \frac{\partial}{\partial A}\left[Ae^{-1/2 \ p(A)^2}\right] dA = 0 \ , \tag{2.5}$$

and all other "Ward identities" can be obtained in a similar fashion. Conversely, if we had postulated the existence of a transformation which reduces the measure to a Gaussian, we would have been immediately led to (2.1).

In two space-time dimensions, the minimal supermultiplet contains a scalar A and a two-component Majorana spinor ψ_α [7]. The most general invariant action is[7]

$$S(A,\psi) = \frac{1}{2}\int dx\Big[(\partial_\mu A)^2 + p(A)^2\Big] + \frac{1}{2}\int dx\; \overline{\psi}_\alpha\Big[\gamma^\mu_{\alpha\beta}\partial^\mu + \delta_{\alpha\beta}p'(A)\Big]\psi_\beta \tag{2.6}$$

and, for definiteness, we will take $p(A) = mA + \lambda A^3$ which $m, \lambda > 0$. From (2.6), we may now compute the MSS determinant*

$$\Big[\det\Big\{\gamma^\mu_{\alpha\beta}\partial^\mu + \delta_{\alpha\beta}(m+3\lambda A^2(x))\Big\}\delta(x-y)\Big]^{1/2} =$$

$$= \det\,(-\Delta+m^2)^{1/2}\,\exp\Big[3m\lambda C(0)\int A^2(x)\,dx + O(\lambda^2)\Big]\,, \tag{2.7}$$

where $C(x) = (-\Delta+m^2)^{-1}(x)$ is the usual propagator. The transformation of A that we are looking for reads up to first order

$$A'(x,\lambda;A) = A(x) + m\lambda\int C(x-y)A^3(y)\,dy + O(\lambda^2) \tag{2.8}$$

(it has been written out up to third order in Ref. 2). It is not difficult to verify that indeed

$$\det\frac{\delta A'(x,\lambda;A)}{\delta A(y)} = \text{MSS determinant} + O(\lambda^2) \tag{2.9}$$

and

$$\frac{1}{2}\int dx\Big[(\partial_\mu A')^2 + m^2A'^2\Big] = \frac{1}{2}\int dx\Big[(\partial_\mu A)^2 + m^2A^2 + 2m\lambda A^4\Big] + O(\lambda^2)\,. \tag{2.10}$$

Had we chosen, say, a Yukawa coupling constant other than that prescribed by supersymmetry, either (2.9) or (2.10) would cease to hold. For $\lambda \to 0$, (2.8) becomes the identity transformation, whereas in the ultralocal limit we obtain $mA' = mA + \lambda A^3$, and the map is invertible only if $m, \lambda > 0$ (or both < 0). The question of global invertibility

*Remember, it is the determinant of a matrix with space-time indices (= x,y) and spinor indices (= α,β).

of the transformation is thus related to the question of vacuum degeneracy, since the condition $m, \lambda > 0$ just ensures that the scalar field potential has only one absolute minimum.

Let us now turn to the case of four dimensions. Here, the minimal multiplet consists of one scalar A, one pseudoscalar B, and one four-component Majorana spinor[8] ψ_α and the simplest non-trivial action is given by[8]

$$S(A,B,\psi) = \frac{1}{2}\int dx\Big[(\partial_\mu A)^2 + (\partial_\mu B)^2 + m^2(A^2+B^2) + 2mgA(A^2+B^2) + g^2(A^2+B^2)^2\Big] + \frac{1}{2}\int dx\ \bar{\psi}_\alpha\Big[\gamma^\mu_{\alpha\beta}\partial^\mu + \delta_{\alpha\beta}(m+2gA) - 2g\gamma^5_{\alpha\beta}B\Big]\psi_\beta \ . \tag{2.11}$$

As before, we calculate the MSS determinant from this action

$$\Big[\det\Big\{\gamma^\mu_{\alpha\beta}\partial^\mu + \delta_{\alpha\beta}\big[m+2gA(x)\big] - 2g\gamma^5_{\alpha\beta}B(x)\Big\}\delta(x-y)\Big]^{1/2} = \det(-\Delta+m^2)\exp\Big[4mgC(0)\int A(x)\,dx + O(g^2)\Big] \tag{2.12}$$

and, as before, we verify that the transformation

$$\begin{aligned} A'(x,g;A,B) &= A(x) + mg\int C(x-y)(A^2(y)-B^2(y))\,dy + O(g^2) \\ B'(x,g;A,B) &= B(x) + 2mg\int C(x-y)A(y)B(y)\,dy + O(g^2) \end{aligned} \tag{2.13}$$

has all the desired properties up to first order in g. The novel feature is that now, in the ultralocal limit, we get

$$mA' = mA + g(A^2-B^2)\ , \quad mB' = mB + 2gAB\ , \tag{2.14}$$

so, although being locally invertible almost everywhere, the transformation is no longer globally invertible. In general, if there are at least two scalar fields, the winding number of the transformation equals the number of absolute minima of the potential. As a technical remark, we mention that, in quantum field theory, the bosonic fields are distributions in general and therefore the transformations (2.8) and (2.13) are not really well defined as they stand. However, it is not difficult to find a supersymmetry-preserving UV cut-off which remedies this defect.

3. SUPERSYMMETRIC GAUGE THEORIES

Supersymmetric gauge theories[9] are physically more interesting. Because of the additional gauge invariance, a gauge-fixing procedure is needed[6] which either explicitly violates supersymmetry[10] or, through additional ghost multiplets, renders the theory considerably more complicated[11]. We will disregard the second possibility, since the statement loses much of its transparency in that case. In a Euclidean space-time, the model is given by the Lagrangian[9]

$$\mathcal{L} = \frac{1}{4} F^a_{\mu\nu} F^a_{\mu\nu} + \frac{1}{2}\bar{\psi}^a \not{D}\psi^a + \frac{1}{2} D^{a2} , \qquad (3.1)$$

where, obviously,

$$F^a_{\mu\nu}(A) = \partial_\mu A^a_\nu - \partial_\nu A^a_\mu + g f^{abc} A^b_\mu A^c_\nu \qquad (3.2)$$

and

$$\not{D}\psi^a = \gamma^\mu \partial^\mu \psi^a + g f^{abc} A^b_\mu \gamma_\mu \psi^c . \qquad (3.3)$$

It is by now standard folklore that (3.1) is invariant (up to a total divergence) with respect to the transformations

$$\delta A^a_\mu = -\bar{\varepsilon}\gamma_\mu \psi^a , \quad \delta D^a = \bar{\varepsilon}\gamma_5 \not{D}\psi^a , \quad \delta\psi^a = (\sigma_{\alpha\beta} F^a_{\alpha\beta} - \gamma^5 D^a)\varepsilon . \qquad (3.4)$$

We now fix a gauge by adding a term $1/2(\partial_\mu A^a_\mu)^2$ to the Lagrangian[12], so the purely bosonic part of the action becomes (dropping auxiliary fields)

$$\frac{1}{4}\int F^a_{\mu\nu}(A)^2 \, dx + \frac{1}{2}\int (\partial_\mu A^a_\mu)^2 \, dx =$$

$$= \frac{1}{2}\int A^a_\mu (-\delta_{\mu\nu}\delta^{ab}\Delta) A^b_\nu \, dx + O(g) . \qquad (3.5)$$

It is only for convenience that we have set the conventional gauge parameter $\alpha = 1$, for we could have even chosen a non-linear gauge-fixing term without altering our final result. To compensate for the explicit breaking of gauge invariance, the functional measure has to be weighted with the Faddeev-Popov determinant[6] which, in our case, reads

$$(\det \delta^{ac}\Delta) \cdot \det \left\{ \delta^{ac}\delta(x-y) - g f^{abc} \partial_\mu C(x-y) A^b_\mu(y) \right\} , \qquad (3.6)$$

where, now, $C(x) = -\Delta^{-1}(x)$. The MSS determinant is obtained from

$$\left[\det\left\{\delta^{ac}\gamma^{\mu}_{\alpha\beta}\partial^{\mu} + gf^{abc}\gamma^{\mu}_{\alpha\beta}A^{b\mu}(x)\right\}\delta(x-y)\right]^{1/2} . \tag{3.7}$$

After a slightly more tedious calculation, we get for the product of the two determinants

$$\exp\Big[ng^2\int dx\, dy\,\Big\{\frac{3}{2}\,\partial_\mu C(x-y)A^a_\mu(y)\partial_\nu C(y-x)A^a_\nu(x) - \\ - \partial_\mu C(x-y)A^a_\nu(y)\partial_\mu C(y-x)A^a_\nu(x) + \\ + \partial_\mu C(x-y)A^a_\nu(y)\partial_\nu C(y-x)A^a_\mu(x)\Big\} + O(g^3)\Big] , \tag{3.8}$$

where we made use of $f^{abc}f^{a'bc} = n\delta^{aa'}$ and omitted a trivial factor $(\det \delta^{ac}\Delta)^2$ (it is not entirely trivial, because it is exactly the factor needed for the elementary counting rule cited in the Introduction, as the reader may easily check!). The required field transformation is found to be

$$A'^a_\mu(x,g;A) = A^a_\mu(x) + gf^{abc}\int dy\ \partial_\lambda C(x-y)A^b_\mu(y)A^c_\lambda(y) + \frac{g^2}{2}f^{abc}f^{bde} \\ \int dy\, dz\,\Big\{\partial_\nu C(x-y)A^c_\lambda(y)\partial_\nu C(y-z)A^d_\mu(z)A^e_\lambda(z) - \\ - \partial_\nu C(x-y)A^c_\lambda(y)\partial_\lambda C(y-z)A^d_\mu(z)A^e_\nu(z) + \\ + \partial_\nu C(x-y)A^c_\lambda(y)\partial_\mu C(y-z)A^d_\lambda(z)A^e_\nu(z)\Big\} \tag{3.9}$$

up to and including second order in g. It satisfies both

$$\det \frac{\delta A'^a_\mu(x,g;A)}{\delta A^b_\nu(y)} = (3.8) + O(g^3) \tag{3.10}$$

and

$$\frac{1}{2}\int A'^a_\mu(-\delta_{\mu\nu}\delta^{ab}\Delta)A'^b_\nu\ dx = \frac{1}{4}\int F^a_{\mu\nu}(A)^2\ dx + \frac{1}{2}\int(\partial_\mu A^a_\mu)^2\ dx + O(g^3). \tag{3.11}$$

The general proof to all orders in g in the case of supersymmetric gauge theories will be given in a forthcoming publication[13]. We note that, in contradistinction to the "scalar" supersymmetric case, the problem of making (3.9) mathematically acceptable for distribution-valued $A_\mu^a(x)$ has not been solved: there exists up to date no non-perturbative regularization prescription which respects both supersymmetry and gauge invariance.

REFERENCES

1. P. Fayet and S. Ferrara, Phys. Rep. C32:249 (1977).
2. H. Nicolai, Phys. Lett. 89B:341 (1980).
3. B. Zumino, Nucl. Phys. B89:535 (1975).
4. F.A. Berezin, "The Method of Second Quantization", Academic Press, New York (1966).
5. P.T. Matthews and A. Salam, Nuovo Cimento 12:563 (1954): 2:120 (1955).
   E. Seiler, Commun. Math. Phys. 42:163 (1975).
6. L.D. Faddeev and V.N. Popov, Phys. Lett. 25B:29 (1967).
7. S. Ferrara, Lett. Nuovo Cimento 13:629 (1975).
   B. Zumino, *in* "Renormalization and Invariance in Quantum Field Theory", Plenum Press, New York (1973).
8. J. Wess and B. Zumino, Nucl. Phys. B70:39 (1974; Phys. Lett. 49B:52 (1974).
9. J. Wess and B. Zumino, Nucl. Phys. B78:1 (1974).
   A. Salam and J. Strathdee, Phys. Lett. 51B:353 (1974).
   R. Delbourgo, A. Salam and J. Strathdee, Phys. Lett. 51B:475 (1974).
   S. Ferrara and B. Zumino, Nucl. Phys. B79:413 (1974).
10. B. de Wit and D.Z. Freedman, Phys. Rev. D12:2286 (1975).
11. B. de Wit, Phys. Rev. D12:1628 (1975).
   S. Ferrara and O. Piguet, Nucl. Phys. B93:361 (1975).
   F. Honerkamp, M. Schlindwein, F. Krause and M. Scheunert, Nucl. Phys. B95:397 (1975).
12. G.'t Hooft, Nucl. Phys. B33:173 (1971).
13. H. Nicolai, "Supersymmetry and functional integration measures", submitted for publication.

WHAT IS A NON-TRIVIAL SOLUTION IN SUPERGRAVITY?

P.C. Aichelburg*

Institut für Theoretische Physik

Universität Wien, Austria

When Einstein formulated general relativity more than 60 years ago, one of the main tools used to understand the physical implications of his theory was finding and analyzing exact solutions to the field equations. Supergravity is the locally supersymmetric extension of general relativity and at the same time a promising way of introducing the fundamental concept of anticommuting fermion fields into gravity theory. So far, however, very little is known about the physical implications of this theory. A step towards a better physical understanding (at the semi-classical level) would be to find and interpret exact solutions of the field equations.

Written in differential forms, the equations read[1]:

$$\Theta^{bc} \wedge \eta_{abc} = {}^*T_a = \frac{i}{2} \bar{\Psi} \wedge \gamma_5 \gamma_a D\Psi \tag{1a}$$

$$i\gamma_5\gamma_a \wedge D\Psi = 0 \quad , \quad S^a = \frac{i}{4} \bar{\Psi} \wedge \gamma^a \Psi \tag{1b}$$

(In our notation $\psi \equiv \psi_a e^a$ is an anticommuting spinor valued 1-form, e^a are the Cartan basis forms, $D = d + \frac{1}{2} \omega^{ab} \sigma_{ab} \wedge$ is the exterior covariant-derivative and the connection forms ω^{ab} include the torsion S in terms of ψ; θ^{ab} is the curvature 2-form, $\gamma = \gamma_a e^a$, $\sigma_{ab} = \frac{1}{4} [\gamma_a, \gamma_b]$ and η^{abc} the dual $e^a \wedge e^b \wedge e^c$.)

---

* Work supported in part by the Einstein Memorial Foundation.

For $\psi \equiv 0$ the system (1a,b) reduces to Einstein's vacuum equations. Any solution of the system which is related by a supersymmetry transformation to an Einstein vacuum solution is called trivial. Only non-trivial solutions can give new physical insight into the theory and one would like to have a criterion by which trivial solutions can be identified.

The question that I would like to consider here is: when does there exist, for a given solution (e,ψ) of Eqs (1a,b), a supersymmetry transformation which brings ψ to zero? In answering this question I restrict myself to infinitesimal supersymmetry transformations, i.e., to solutions lying in the neighbourhood of $\psi = 0$.

Two infinitesimal neighbouring solutions (e,ψ) and (e',ψ') are related by a supersymmetry transformation if an infinitesimal spinor field $\alpha(x)$ exists, such that

$$e^a - e^{a'} = i\bar{\alpha}\gamma^a\psi \tag{2a}$$

$$\psi - \psi' = D\alpha \tag{2b}$$

If $\psi' = 0$, then

$$\psi = D\alpha \tag{3}$$

and e is invariant because α and ψ are infinitesimal, which implies that the metric formed from the vierbein e is (already) a solution to Einstein's vacuum equations. This is in agreement with Eqs (1a,b), since neglecting quadratic terms in ψ the right-hand side of (1a) vanishes, as well as the torsion contributions to the connection.

So the question is reduced to the problem of when, for a given (e,ψ), a solution to Eq. (3) for α exists. From Eq. (3) one derives, by differentiating, the integrability conditions

$$D\psi = \frac{1}{2}\theta^{ab}\sigma_{ab}\alpha \tag{4}$$

Given (e,ψ), Eq. (4) is a system of 24 algebraic equations for the spinor field α. The existence of a solution to Eq. (4) is necessary for (e,ψ) to be trivial.

Because the curvature tensor enters on the right-hand side of Eq. (4) and the Ricci tensor from e vanishes (Einstein vacuum equation), it is natural to analyze the system in terms of the algebraic Petrov types of space-time[2]. Requiring that Eq. (4) be solvable for α leads to different conditions on the antisymmetric derivatives of ψ corresponding to the different algebraic types. It turns out that, if these necessary conditions are satisfied, one can obtain α uniquely for all Petrov types except type N.

Now one may ask if the necessary conditions following from Eq. (4) are also sufficient. Differentiating Eq. (4) once more yields

$$\theta^{ab}\sigma_{ab}(\psi - D\alpha) = 0 \tag{5}$$

If this system of homogeneous equations for ψ-$D\alpha$ admits only the trivial solution, then $\psi = D\alpha$ and therefore the conditions on ψ following from Eq. (4) are necessary and sufficient. Analyzing Eq. (5) with the help of the two-component spinor formalism shows that for general Petrov type I (not all principal null directions lying in one hyperplane), type D and type II only the trivial solution is possible.

Summing up, our results can be stated as follows: the necessary conditions for a supergravity solution to be trivial are given by Eq (4) and can explicitly be written down for the different algebraic Petrov types. These conditions are also <u>sufficient</u> for the physically interesting cases: type I (general), type D and type II.

APPLICATIONS

i) Exact solutions to the supergravity equations representing plane-fronted waves for the metric and the ψ-field were given by T. Dereli and the author[3]. It can be shown that these solutions do not satisfy the conditions of Eq. (4) and therefore are non-trivial.

ii) P. Cordero and C. Teitelboim[4] have conjectured a no-hair theorem for supersymmetric black holes by showing that a spherical symmetric ψ-field on a Schwarzschild background is necessarily a gauge field. One may show that their ansatz satisfies Eq. (4) for type D, which is then necessary and sufficient.

I finally remark that restricting the problem to infinitesimal supersymmetry transformations means that, in principle, one can identify all trivial solutions that can be continuously transformed to $\psi = 0$. There might however exist finite isolated trivial solutions.

Details of the calculations and the explicit conditions on the ψ-field for different Petrov types will be published elsewhere.

REFERENCES

1. D.Z. Freedman, P. Van Nieuwenhuizen and S. Ferrara, Phys. Rev. D13 (1976) 3214.
   S. Deser and B. Zumino, Phys. Letters 62B (1976) 335.
2. See, for example, P. Jordan, J. Ehlers and W. Kundt, Akad. Wiss. Lit. Mainz. Abh. Math.-Nat. Nr 2 (1960).
3. P.C. Aichelburg and T. Dereli, Phys. Rev. D18 (1978) 1754.
4. P. Cordero and C. Teitelboim, Phys. Letters 78B (1978) 80.

SUPERSYMMETRIC PRE-QCD DYNAMICS

Jerzy Lukierski

Institute for Theoretical Physics

University of Wroclaw, Wroclaw, Poland

1. INTRODUCTION

It is known from the efforts to describe consistently spin 3/2-field interactions that supersymmetry can be useful in deriving new couplings of fermions and bosons. Because confinement as a consequence of quark dynamics remains rather a mystery in QCD, it is interesting to look for supersymmetric dynamics with quark fields in fundamental and gluons in adjoint representation of the colour group as a possible alternative to the conventional QCD scheme. It is known that the simplest solution - the N = 1 supersymmetric Yang-Mills theory[1,2] with quarks and gluons in one gauge supermultiplet - does not work, because quarks and gluons are both in adjoint colour representations. In order to supersymmetrize the interaction of quarks and gluons we have to introduce two N = 1 supermultiplets:

- a chiral scalar supermultiplet - with quark fields ψ_α^{Ji} (i = 1 ... n is a colour index; J = 1 ... m describes flavour) supplemented by n × m scalar "quarkinos" ϕ^{Ji}, both in fundamental SU(m) × SU(n) representations;

- a gauge supermultiplet, describing gluon fields V_μ^{ij} together with spin 1/2 "gluinos", both in an adjoint SU(n) representation (we choose the Wess-Zumino gauge).

In the conventional approach to the supersymmetrization of QCD both the supermultiplets listed above are elementary and one gets supersymmetry through a considerable increase in the number of elementary fields. Here we shall present another approach with the chiral supermultiplet elementary and the gauge supermultiplet composite. More explicitly, we assume that gluons and gluinos are composite, and can be expressed as local functions of elementary spin 1/2 quark and scalar "quarkino" fields. The compositeness condition can be simply formulated if we use the superfield notation. We shall show that such a construction is described by a supersymmetric generalization of the four-dimensional generalized (quadrilinear) $G_{m,n}(C)$ σ-model.

Bosonic generalized σ-models with quadrilinear Lagrangians have been considered recently by several authors as describing "pure" Yang-Mills theory with composite gauge fields[3-8]. The supersymmetric extension presented here has been obtained independently by Bars and Günaydin[9] and by Milewski with the present author[10]. We interpret such a model as supersymmetric quark dynamics with elementary subcanonical quark fields.

2. QUADRILINEAR LAGRANGIANS

It is known that σ-models with bilinear free parts are geometric because they describe the mappings of two Riemannian manifolds. One can construct the following sequence of action integrals:

a) Free point particle in R^N in proper time parametrization (we put c = 1):

$$S = m \int d\tau \left(\frac{dY_\mu}{d\tau}\right)^2 \qquad \mu = 1 \ldots N \tag{1}$$

b) One-dimensional σ-model which describes geodesic motions on the Riemannian manifold $\mathcal{M}$ with the metric tensor g_{ij}

$$\left(\frac{dY^\mu}{d\tau}\right)^2 \rightarrow g_{ik}(\xi) \frac{d\xi_i}{d\tau} \frac{d\xi_j}{d\tau} \tag{2}$$

c) k-dimensional σ-model, describing the harmonic mappings of the co-ordinate space $\mathcal{S}$ (for simplicity $\mathcal{S}$ is assumed to be flat, Minkowski or Euclidean) into $\mathcal{M}$ (k = dim $\mathcal{M}$)

$$S = m^{2-k} \int d^k x \, g_{ik}(\xi) \frac{\partial \xi_i}{\partial x_\mu} \frac{\partial \xi_j}{\partial x^\mu} \tag{3}$$

The prototype for quadrilinear σ-models, analogous to (1) for bilinear ones, is the action describing the motion of a free string in R^N, parametrized invariantly by two parameters τ_1, τ_2 [11,12]. We get in place of (1)

$$S = M^2 \int d\tau_1 d\tau_2 \left(\frac{\partial(Y_\mu, Y_\nu)}{\partial(\tau_1, \tau_2)} \right)^2 = M^2 \int dS \tag{4}$$

where the Jacobian

$$\frac{dS_{\mu\nu}}{dS} = \frac{\partial(Y_\mu, Y_\nu)}{\partial(\tau_1, \tau_2)} = \frac{\partial Y_\mu}{\partial \tau_1}\frac{\partial Y_\nu}{\partial \tau_2} - \frac{\partial Y_\nu}{\partial \tau_1}\frac{\partial Y_\mu}{\partial \tau_2} \tag{4a}$$

describes the two-dimensional "area" derivative (see, e.g., Ref. 13) and the orthonormal co-ordinate system (τ_1, τ_2) parametrizes the surface swept by the motion of string in such a way that

$$dS = d\tau_1 d\tau_2 = dS_{\mu\nu} dS^{\mu\nu}$$

with $dS_{\mu\nu} = dY_\mu \wedge dY_\nu$ denoting the projections of the infinitesimal area on $\frac{1}{2}N(N-1)$ co-ordinates planes (μ, ν). Formula (4) describes the simplest quadrilinear Lagrangian, and we obtain two generalizations, corresponding to (2) and (3) as follows:

a) One-dimensional generalized σ-model
We replace in (4)

$$dY_\mu \rightarrow E_i^a(\xi) d\xi_i \tag{5a}$$

where E_i^a denotes the "generalized vierbein" for the manifold $\mathcal{M}$, i.e.,

$$g_{ik}(\xi) = E_i^a(\xi) E_{ka}(\xi) \tag{5b}$$

b) k-dimensional generalized σ-model

$$S = M^{4-k} \int d^k x \, g_{ik}(\xi) g_{jl}(\xi) \frac{\partial(\xi_i, \xi_j)}{\partial(x_\mu, x_\nu)} \frac{\partial(\xi_k, \xi_l)}{\partial(x^\mu, x^\nu)} = \tag{6}$$

$$= M^{4-k} \int d^k x \, F^{ab}{}_{\mu\nu} F_{ab}{}^{\mu\nu}$$

where

$$F^{ab}_{\mu\nu} = E^a_i(\xi) E^b_j(\xi) \frac{\partial(\xi_i, \xi_j)}{\partial(x_\mu, x_\nu)} \tag{7}$$

describes a composite Yang-Mills strength with the "flat" indices a,b in the tangent bundle of the manifold $\mathcal{M}$.

The formula (6) is valid for <u>any Riemannian manifold</u> with σ-fields parametrized by an atlas of independent local co-ordinate maps $(\xi_1 \ldots \xi_N)$. One can also use global parametrization, e.g., describe $\mathcal{M}$ as a linear manifold with constraints and equivalence relations. In particular the complex Grassmann manifold $G_{m,n}(C)$ is conveniently described by rectangular complex matrices, satisfying the constraints

$$\bar{\varphi}^{iJ} \varphi^{Jj} = \delta^{ij} \qquad i,j = 1 \ldots n, \; J = 1 \ldots m \; (m > n) \tag{8}$$

and the equivalence relation $\phi^{Ji} = \phi^{Jj} U^{ji}$, $U \in U(n)$. Formula (7) for the composite U(n) Yang-Mills field strength has the form

$$F^{ij}_{\mu\nu} = \nabla^{ik}_{[\mu} \bar{\varphi}^{kJ} \nabla^{jl}_{\nu]} \varphi^{Jl} \tag{9}$$

where

$$\nabla^{ik}_\mu = \delta^{ik} - i A^{ik}_\mu$$

and

$$A^{ik}_\mu = i \bar{\varphi}^{iJ} \partial_\mu \varphi^{Jk} \tag{10}$$

describe the composite U(n) gauge potentials [3-8, 14, 15].

Let us observe that the "conventional" σ-models (3) are conformal invariant if k = 2, and the generalized σ-models (6) do not depend on conformal frames if k = 4. Because the σ-fields always have canonical dimensionality zero, the quadrilinearity for d = 4 follows from the property that the action depends only on the σ-field and its first derivatives. One can argue that bilinear two-dimensional σ-models and quadrilinear generalized σ-models for k = 4 have many common features: instanton solutions[5-8], spontaneous mass generation[17] and probably renormalizability.

3. COMPOSITE SUPERSYMMETRIC U(n) YANG-MILLS THEORY

Let us supplement $\phi^{iJ} \in G_{m,n}(C)$ with additional components of the chiral Wess-Zumino superfield

$$\varphi^{Ji} \longrightarrow (\varphi^{Ji}, \psi_\alpha^{Ji}, F^{Ji}) \tag{11}$$

Denoting $\phi_- = \phi$, $\phi_+ = (\phi)^+$ (Hermitean conjugate) and $F_- = F$, $F_+ = (F^+)$ one can introduce m × n chiral Minkowski superfields Φ_+^{iJ}, Φ_-^{iJ} as follows

$$\begin{aligned}\Phi_\pm = \varphi_\pm + \bar\theta\psi_\pm + \tfrac{1}{4}\bar\theta\theta F_\pm \pm \tfrac{i}{4}\bar\theta\gamma_5\theta F_\pm \mp \\ \mp \tfrac{1}{4}\bar\theta\gamma_\mu\gamma_5\theta\,\partial^\mu\varphi_\pm - \tfrac{i}{4}(\bar\theta\theta)\,\bar\theta\,\partial\!\!\!/\psi_\pm - \tfrac{1}{32}(\bar\theta\theta)^2\Box\varphi_\pm\end{aligned} \tag{12}$$

We supplement the subsidiary condition (8) by

$$\varphi_+^{iJ}\psi_\alpha^{J\gamma} = 0 \qquad\qquad \varphi_+^{iJ}F^{J\gamma} = 0 \tag{13}$$

and we define the composite SU(n) gauge superfield $V^{ij}(x;\theta)$ by generalizing (10) in the following way:

$$e^{-2V} = \Phi_+\Phi_- \tag{14}$$

It appears that due to the subsidiary conditions (8) and (13) the relation (14) can be written in terms of the component fields

$$\begin{aligned} V_\mu &= i\varphi^+\partial_\mu\varphi + \tfrac{1}{2}\psi_+^T C\gamma_\mu\psi_- = A_\mu + j_\mu \\ \chi_+ &= -i\partial^\mu\varphi^+\gamma_\mu\psi_- - \psi_+ F \\ \chi_- &= -i\gamma^\mu\psi_+\partial_\mu\varphi - F^+\psi_- \\ D &= -2(\nabla_\mu\varphi)^+\nabla^\mu\varphi + i\psi_+^T C(\overleftarrow{\nabla\!\!\!/} - \overrightarrow{\nabla\!\!\!/})\psi_- + 2j_\mu j^\mu - 2F^+F \end{aligned} \tag{15}$$

where the composite vector superfield V is in the Wess-Zumino gauge, i.e.,

$$V = \frac{1}{4}\bar{\theta} i \gamma_\mu \gamma_5 V^\mu + \frac{1}{4}\bar{\theta}\theta\bar{\theta}\chi + \frac{1}{32}(\bar{\theta}\theta)^2 D \tag{16}$$

The formula (16) for the composite U(n) Yang-Mills Lagrangian with $F^{ij}_{\mu\nu}$ given by (9) is generalized as follows (we do not write explicitly the subsidiary conditions which should be added using Lagrange multipliers):

$$\mathcal{L} = -\frac{1}{4}F_{\mu\nu}F^{\mu\nu} \longrightarrow -\frac{1}{4}\left(F_{\mu\nu} + \nabla_{[\mu}j_{\nu]} - i\,[j_\mu, j_\nu]\right)^2 - \\ -\frac{i}{2}\chi_+^T C\left(\not{\partial}\chi_- - i\gamma^\mu [j_\mu, \chi_-]\right) + \frac{1}{2}D^2 \tag{17}$$

Setting $\psi_+ = \psi_- = 0$ in the formula (17) one gets the purely bosonic quadrilinear model

$$\mathcal{L} = -\frac{1}{4}F_{\mu\nu}F^{\mu\nu} + 2\left[(\nabla_\mu\varphi)^+\nabla^\mu\varphi + F^+F\right]^2 \tag{18}$$

Using the field equations for F one obtains the result that the composite gauge fields A^{ij}_μ (see (10)) satisfy the SU(n) Yang-Mills equations.

The Lagrangian (17) is invariant under

a) U(n) local gauge transformations

$$\delta\varphi = -i\varphi a \qquad \delta\psi_- = -i\psi_- a \qquad \delta F = -iFa \tag{19}$$

b) Supersymmetry transformations:

$$\begin{aligned} \delta\varphi &= \bar{\epsilon}_-\psi_+ \qquad\qquad \delta\varphi^+ = \bar{\epsilon}_+\psi_- \\ \delta\psi_+ &= F\epsilon_+ - i\not{\partial}\varphi\epsilon_- - V_\mu\gamma^\mu\epsilon_-\varphi^+ \\ \delta\psi_- &= F^+\epsilon_- - i\not{\partial}\varphi^+\epsilon_+ + \varphi V_\mu\gamma^\mu\epsilon_+ \\ \delta F &= -i(\bar{\epsilon}_+\not{\partial} - \tfrac{1}{2}\partial_\mu\bar{\epsilon}_+\gamma^\mu)\psi_+ - V_\mu\epsilon_-^T C\gamma^\mu\psi_+ + \bar{\epsilon}_+\chi_-\varphi \\ \delta F^+ &= -i(\bar{\epsilon}_-\not{\partial} - \tfrac{1}{2}\partial_\mu\epsilon_-\gamma^\mu)\psi_- - \psi_-^T C\gamma^\mu\epsilon_- V_\mu + \bar{\epsilon}_-\varphi\chi_+ \end{aligned} \tag{20}$$

where $\varepsilon_{\pm} = \varepsilon^1_{\pm} + \gamma_\mu X^\mu \varepsilon^2_{\pm}$, and $\varepsilon^{1,2}_{\pm}$ are the chiral projections of two 4-component Majorana spinors. These transformations describe the algebra of the superconformal group supplemented by the field-dependent generalized U(n) gauge transformations.

The compositeness conditions (10) and (14) can be generalized in a way which permits the introduction of elementary gauge fields for the flavour group U(m). One gets instead of (10)

$$A^{ij}_\mu = i\,\bar{\varphi}^{iJ}\,\tilde{\nabla}^{JK}_\mu\,\varphi^{Kj} \qquad \tilde{\nabla}^{JK}_\mu = \partial_\mu \delta^{JK} - i\,a^{JK}_\mu \tag{21}$$

and in the supersymmetric case

$$e^{-2V} = \Phi_+\, e^{-2W}\, \Phi_- \tag{22}$$

where W^{JK} is the matrix of elementary gauge supermultiplets. The formula (22) in its component form is given in Ref. 17.

4. PRE-QCD WITH SUBCANONICAL ELEMENTARY FIELDS

Let us summarize briefly the properties of the supersymmetric composite U(n) gauge theory:

a) gluon fields V^{ij}_μ have the canonical dimension d = 1 and are composite, together with their supersymmetric spin 1/2 partners (gluinos) with d = 3/2.

b) quark fields are <u>subcanonical</u> (dimension d = 1/2), as are their supersymmetric scalar partners (d = 0). In the model these fields are fundamental and elementary.

c) one can formulate <u>local</u> dynamics on two levels:
- for canonical composite fields, with bilinear free parts in the Lagrangians
- for subcanonical elementary fields, with the action at least quadrilinear in these fundamental field variables.

The idea that quark fields are subcanonical is an old one and can be traced to the concept of Heisenberg's "Urfeld"[18]. There are, however, two problems in Heisenberg's unified theory:

- subcanonical dimensions are obtained by assuming that the bilinear part of the Lagrangian contains higher order wave operators. In such a framework asymptotic states with negative norms (ghosts) appear on-shell which leads to serious difficulties with the unitarity relation.

- one lacks any clear principle for constructing a local dynamics of composite canonical operators.

In the model presented here the first difficulty caused by negative metric states also occurs, but possibly in a less dangerous form. The quadrilinear Lagrangians do not introduce any mass-shell, and it is not clear how the asymptotic ghost states are defined. The second problem, however, is solved: the relation between the dynamical at the elementary and composite levels are fixed by the supercovariant compositeness conditions (see (14)). It is the notion of superspace and superfield which simplifies the considerations. One can say that supersymmetry indicates how to formulate the local dynamics with bosons and fermions on both the canonical and subcanonical levels.

The composite operators describing gluons and gluinos are not the ones introducing colourless hadronic degrees of freedom. The compositeness condition describing mesons should be obtained by multiplying the fundamental chiral superfields in transposed order, with a summation over the colour index i = 1 ... n. One can choose

$$\text{mesons} \quad \left(e^{-2M}\right)^{JK} = \Phi_-^{Ji}\,\Phi_+^{iK} \tag{23a}$$

$$\begin{matrix}\text{baryons} \\ \text{(we put } n = 3)\end{matrix} \quad \left(B_\pm\right)^{JKL} = \epsilon_{ijk}\,\Phi_\pm^{iJ}\,\Phi_\pm^{jK}\,\Phi_\pm^{kL} \tag{23b}$$

It is easy to see that the subcanonical dimension d = 1/2 of the elementary quark field matches perfectly with canonical dimensionalities for the composite meson and baryon states:

$$d(\psi) = \tfrac{1}{2} \quad \begin{matrix} \nearrow \; d(\bar{\psi}\psi) = 1 \quad \text{(mesons)} \\ \searrow \; d(\psi\psi\psi) = \tfrac{3}{2} \quad \text{(baryons)} \end{matrix} \tag{24}$$

The formulae (23) describe the supersymmetric generalization of the relations well known from the "naïve" quark model. We obtain

- additional contributions from "quarkino" fields:

- mesons and baryons occurring together with their supersymmetric partners.

It remains to be seen how the equations for the fundamental pre-QCD variables determine the dynamics of superfields M and $B_{\pm}$, as a first step towards a comparison with the phenomenology of hadronic interactions.

5. FINAL REMARKS

At present the main problem is to learn how to deal with quadrilinear Lagrangians even in the purely bosonic case, in particular how to define the quantum theory via the functional integral. The quantization using functional methods of the quadrilinear composite U(n) Yang-Mills theory can be formulated in two different ways:

a) the conventional one which involves considering the Hamiltonian formalism and writing down the functional integral in phase space à la Faddeev[19]. In such a way one obtains the measure in the space of field values with a non-covariant determinant, obtained by integrating over field momenta which are the canonical conjugates of the ϕ^{Ji}.

b) using the property that quadrilinear Lagrangians (6) are obtained by the replacement of the "line element" by an "area element" in Riemannian space (see Section 2) it appears promising to consider the functional integral as a limit of a discrete approximation, with the notion of discretized virtual paths replaced by suitably discretized virtual surfaces. Such an approach was already proposed for the description of glue strings in Yang-Mills theory[20], and should provide the meaning of the quadrilinear Lagrangians obtaind on the lattice in the local limit (for difficulties see Ref. 3).

In order to calculate explicitly functional integrals we should be able to generate in the action some bilinear terms. For the quadrilinear Lagrangians such terms can be obtained by introducing background fields (e.g., describing instanton solutions) or using the 1/N method, where N describes the number of quark colour multiplets. Both methods are now under consideration.

REFERENCES

1. S. Ferrara and B. Zumino, Nucl. Phys. B79 (1974) 413.
2. A. Salam and J. Strathdee, Phys. Letters 51B (1974) 353.
3. J. Fröhlich, "A new look at generalized non-linear σ-models and Yang-Mills theory", IHES preprint (1979).
4. I. Bars, Proc. VIIIth Intern. Colloquium on Group-Theor. Methods in Physics, March 1979, Kiriat Anavim, Israel.

5. J. Lukierski, CERN preprint TH.2678 (1979); improved version publ. in Proc. of Summer Institute on Field Theoretic Methods in Particle Physics, Kaiserslautern, August 1979, (Springer Verlag, Germany 1979).
6. F. Gürsey and H.C. Tze, Yale preprint (1979), to be published in Annals of Physics.
7. D. Maison, Max Planck Inst. preprint (1979).
8. J. Lukierski, Proc. of the Symposium on Differential Geometric Methods in Physics, Aix-en-Provence, September 1979, (Springer Verlag, Germany 1979).
9. I. Bars and M. Günaydin, "Theory of Ternons", Yale preprint YTP-79-05 (1979).
10. J. Lukierski and B. Milewski, Wroclaw University preprint No 494 (1980); Phys. Letters B, in press.
11. A. Schild, Phys. Rev. D16 (1977) 1722.
12. T. Eguchi, Phys. Rev. Letters 44 (1980) 126.
13. T. Rado, "Length and Area", (AMS publications, vol. XXX, New York, 1948).
14. M.S. Narasimhan and S. Ramanan, Am. J. Math. 83 (1961) 563.
15. M. Dubois-Violette and Y. Georgelin, Phys. Letters 82B (1979) 251.
16. G. Jona-Lasinio, S. Pierini and A. Vulpiani, Lett. al Nuovo Cim. 23 (1978) 353.
17. J. Lukierski, ICTP preprint IC/80/48 (1980); to be published in the Proc. of XVII Winter School, Karpacz, Poland, February 1980, (to be published by Plenum Press, 1980).
18. W. Heisenberg, "Introduction to the Unified Theory of Elementary Particles", (Interscience, London, 1966).
19. L.D. Faddeev, "Methods in Field Theory", Les Houches Summer School, 1975, Ed. R. Balian and J. Zinn-Justin (North-Hooland, Amsterdam, 1976).
20. A.M. Polyakov, Nucl. Phys. B164 (1980) 171.

THE GAUGE INVARIANT SUPERCURRENT IN SQED

K. Sibold

Institut für Theoretische Physik
Universität Karlsruhe
D-75 Karlsruhe, FRG

Supersymmetric QED (SQED)[1] is the minimal model where supersymmetry is combined with Abelian gauge invariance including a charged Dirac spinor. In fact it is just scalar QED amended by some Yukawa couplings added to spinor QED, all interactions being governed by one coupling constant. As far as supersymmetry is concerned its most economical presentation is in terms of superfields ϕ (vector), $\phi_\pm$ (chiral) with transformation laws

$$i[Q_\alpha, \varphi] = \delta_\alpha \varphi \quad \delta_\alpha \varphi = \left(\frac{\partial}{\partial \theta^\alpha} + i\sigma^\mu_{\alpha\dot\alpha}\bar\theta^{\dot\alpha}\partial_\mu\right)\varphi \tag{1}$$

$$\begin{aligned} \delta_g \phi &= i(\Lambda - \bar\Lambda) \\ \delta_g \phi_\pm &= \mp i g \Lambda \phi_\pm \\ \delta_g \bar\phi_\pm &= \pm i g \bar\Lambda \bar\phi_\pm \end{aligned} \tag{2}$$

where (1) are the supersymmetry and (2) the gauge transformations. If we require in addition parity and charge conjugation invariance, the most general invariant action is seen to be

$$\begin{aligned} -i\Gamma = \int dV \Big(&-2\phi D\bar D\bar D D\phi + \frac{2}{\alpha} DD\phi\bar D\bar D\phi \\ &-16M^2\phi^2 + \phi_+\bar\phi_+ e^{g\phi} + \phi_-\bar\phi_- e^{-g\phi} + 4m\left(\int dS\, \phi_+\phi_- + h.c.\right) \end{aligned} \tag{3}$$

Here a gauge fixing and vector mass term have already been added. Up to the matter mass term this action is in fact invariant under a much larger set of symmetries, namely those of the superconformal group[2]. In particular there is a γ_5 symmetry R

$$[R, Q_\alpha] = Q_\alpha \qquad \delta_R \varphi = (n + \theta\partial_\theta + \bar\theta\partial_{\bar\theta})\varphi \tag{4}$$

where n is the R-weight, with $n(\phi) = 0$, $n(\phi_+)$ = a real number, which yields a one-parameter family of axial currents, conserved in the tree approximation up to contributions from the masses of matter fields. Of course these additional invariances are expected to be spoiled by anomalies in higher orders of perturbation theory, to which we now turn.

The first aim is to prove the Ward identities

$$W_\alpha \Gamma = 0 \qquad \bar W_{\dot\alpha} \Gamma = 0 \tag{5}$$

$$W_g \Gamma = \frac{-32i}{\alpha}(\Box + \alpha M^2)\bar D\bar D\phi, \quad \bar W_g \Gamma = -\frac{32i}{\alpha}(\Box + \alpha M^2) DD\phi \tag{6}$$

$$W_R \Gamma = 8i(n+1)m\left(\int dS\, \phi_+\phi_- - \int d\bar S\, \bar\phi_+\bar\phi_-\right) \tag{7}$$

expressing supersymmetry, gauge invariance and softly broken (global) R-invariance. This can in fact be done[3] employing the BPHZ normal product algorithm in the version of Zimmermann and Lowenstein as extended to superspace in Ref. 4. The breaking term in (7), e.g., has to be interpreted as an N_3 normal product. This subtraction scheme has the virtue of being regulator-free and manifestly supersymmetric, but has the drawback that gauge invariant operators need special treatment in order to avoid possible α-dependence[5]. We will come back to this question once we have constructed the supercurrent - the second aim and the object of principal interest in the theory since it contains all information about the symmetries of the model. The supercurrent is a (real, axial-vector) superfield[6] which contains amongst its components the axial current R_μ, the supersymmetry current $Q_{\mu\alpha}$ and the energy-momentum-tensor $T_{\mu\nu}$. All the remaining currents associated with the superconformal group can be constructed[6,7] as x-moments of $T_{\mu\nu}$.

From the renormalized field equations one can derive the following fundamental relations (to be used between physical states, $M = 0$) [3]

$$\bar{D}^{\dot{\alpha}} V_{\alpha\dot{\alpha}} = -2(n+1) D_\alpha S - 6\left(n+\tfrac{2}{3}\right) \bar{D}\bar{D} D_\alpha I_5 \tag{8}$$

$$\partial^\mu V_\mu = i(n+1)(DDS - \bar{D}\bar{D}\bar{S}) \tag{9}$$

where

$$V_{\alpha\dot{\alpha}} = N_3\left[-4n V^1_{\alpha\dot{\alpha}} - 8 V^2_{\alpha\dot{\alpha}} - 16 V^3_{\alpha\dot{\alpha}} + V^g_{\alpha\dot{\alpha}}\right] + \text{rad. corr.} \tag{10}$$

$$V^1_{\alpha\dot{\alpha}} = [D_\alpha, \bar{D}_{\dot{\alpha}}] I_5$$

$$V^2_{\alpha\dot{\alpha}} = D_\alpha(\phi_+ e^{g\phi}) \bar{D}_{\dot{\alpha}}(\bar{\phi}_+ e^{g\phi}) e^{-g\phi} + D_\alpha(\phi_- e^{-g\phi}) \bar{D}_{\dot{\alpha}}(\bar{\phi}_- e^{-g\phi}) e^{g\phi}$$

$$V^3_{\alpha\dot{\alpha}} = \bar{D}\bar{D} D_\alpha \phi \, DD\bar{D}_{\dot{\alpha}} \phi$$

$$I_5 = \phi_+ \bar{\phi}_+ e^{g\phi} + \phi_- \bar{\phi}_- e^{-g\phi}$$

$$S = (m+a) L_m - r_K L_{kin} - \tfrac{1}{2} r_g \bar{D}\bar{D} I_5 \tag{11}$$

$$L_m = 4 N_2 [\phi_+ \phi_-], \quad L_{kin} = L_1 + L_2 - L_0$$

$$L_1 = N_3 [\bar{D}\bar{D} D\phi \, \bar{D}\bar{D} D\phi], \quad L_2 = N_3 [\bar{D}\bar{D} DD\phi \, \bar{D}\bar{D}\phi]$$

$$L_0 = \bar{D}\bar{D} DD\phi \, \bar{D}\bar{D}\phi$$

(Here $V^g_{\alpha\dot{\alpha}}$, L_2 and L_0 ensure gauge invariance and infra-red-finiteness and are terms of technical interest only.) Note that S is a chiral superfield, $S = A + \theta\psi + \theta^2 F$.

As far as the parameter n is concerned there are three cases[8] of interest:

$n = -2/3$ R closes with Q_α and P_μ onto the superconformal algebra and one finds

$$\partial^\mu R^{conf}_{\mu 5} = -2i(F - \bar{F}), \quad \partial^\mu Q_{\mu\alpha} = 0, \quad \partial^\mu T_{\mu\nu} = 0 \tag{12}$$

$$Q^\alpha_\mu \sigma^\mu_{\alpha\dot{\alpha}} = 12 i \bar{\psi}_{\dot{\alpha}}, \quad T_\lambda{}^\lambda = 3(F + \bar{F}), \quad T_{\mu\nu} = T_{\nu\mu}$$

i.e., the breaking terms in $\partial^\mu R_{\mu 5}^{(conf)}$ and the traces of $Q_{\mu\alpha}$, $T_{\mu\nu}$ transform as part of a chiral multiplet; the coefficients are renormalized and given by the Callan-Symanzik functions β,γ:

$$\begin{aligned} \beta &= -\frac{1}{2}(1+\sigma)r_K \\ \gamma &= -\frac{1}{2}(1+\sigma)r_g \end{aligned} \qquad (1+\sigma)^{-1} = -r_g + \frac{1}{2} r_K g \partial_g \ln(m+a) \tag{13}$$

<u>$n = \infty$</u> R commutes with the superconformal algebra.

$$V_{\infty\alpha\dot\alpha} = \lim_{n\to\infty} \frac{1}{n} V_{\alpha\dot\alpha} = -4[D_\alpha, \bar D_{\dot\alpha}] I_5 + \text{rad.corr.} \tag{14}$$

It is thus compelling to absorb into the current the r_g piece of S and thus find

$$\begin{aligned} \partial^\mu V_{\infty\mu} &= i DD\left(-2(m+a)L_m - r_K L_{kin}\right) \\ &\quad - i\bar D\bar D\left(-2(m+a)\bar L_m - r_K \bar L_{kin}\right) \end{aligned} \tag{15}$$

Here $r = -\frac{1}{8}(g^2/(16\pi)^2)$ is the Adler-Bardeen anomaly and $V_{\infty\alpha\dot\alpha}$ no longer contains the supersymmetry current or the energy-momentum-tensor but only the $R_{\infty\mu}$ current (as is suggested from the transformation law generated by $\partial^\mu V_{\infty\mu}$). In terms of the superfield components:

$$\partial^\mu R_{\infty\mu} = 2i(m+a)N_3\left[\bar\Psi_D \gamma_5 \Psi_D + \ldots\right] + 64 r_K N_4\left[\varepsilon_{\mu\nu\rho\sigma} \partial^\mu A^\nu \partial^\rho A^\sigma + m\right] \tag{16}$$

<u>$n = -1$</u> $R_{n=-1}$ is an automorphism of the algebra of $M_{\mu\nu}$, P_μ, Q_α, $\bar Q_{\dot\alpha}$ and no longer closes within the superconformal algebra, and

$$\partial^\mu V_\mu^{(n=-1)} = 0 \qquad \text{(on shell)} \tag{17}$$

so that R_μ, $Q_{\mu\alpha}$, $T_{\mu\nu}$ are conserved, $T_{\mu\nu} = T_{\nu\mu}$.

Having a conserved axial current $R_\mu^{(n=-1)}$ is no surprise: $\delta_R\psi_\pm = i(n+1)\psi_\pm = 0$, i.e., the charged spinor does not transform as $\delta_R\lambda = \gamma_5\lambda$ but λ is uncharged, and does not couple to the vector field A_μ.

We now study the gauge invariance, i.e., the α-independence of the theory, in particular of the (formally gauge invariant) supercurrent[9]. To this end we consider first the massive theory ($M \neq 0$) and define $\mathcal{H}_p$ as that subspace of the indefinite metric Fock space $\mathcal{H}$ for which

$$(DD\phi)^- |\Phi\rangle = 0 \qquad (\bar{D}\bar{D}\phi)^- |\Phi\rangle = 0 \tag{18}$$

We call an operator Q gauge invariant if

$$\begin{aligned} &[Q, DD\phi] = 0 = [Q, \bar{D}\bar{D}\phi] \\ &\frac{\partial}{\partial\alpha}\langle\Psi|Q|\Phi\rangle = 0 \qquad |\Psi\rangle, |\Phi\rangle \in \mathcal{H}_p \end{aligned} \tag{19}$$

First of all, one has to ensure the α-independence of the S-matrix since unitarity follows already from the Ward identity: $(\Box + \alpha M^2)DD\phi = 0$ on shell, i.e., the ghosts $DD\phi$, $\bar{D}\bar{D}\phi$ are free. The requirement of α-independence fixes the coefficients a and z which denote the mass of the matter and the interaction counter-terms respectively.

Next one constructs a basis of gauge invariant operators of dimension three enabling one to express the supercurrent in terms of them. Finally one has to study the limit $M \to 0$. The results are as follows:

1) For $M \neq 0$: a gauge invariant supercurrent is obtained, i.e., in particular, a gauge invariant energy-momentum tensor.

2) For $M = 0$:
   (a) transversal Green's functions are indeed α-independent.
   (b) the formally gauge invariant supercurrent is α-independent.
   (c) β, γ are α-independent.

REFERENCES

1. J. Wess and B. Zumino, Nucl. Phys. B70 (1974) 207.
2. S. Ferrara, Nucl. Phys. B77 (1974) 73.
3. T.E. Clark, O. Piguet and K. Sibold, The Renormalized Supercurrents in SQED, Los Alamos preprint LA-UR 79-3439 (1979).
4. T.E. Clark, O. Piguet and K. Sibold, Ann. of Phys. 109 (1977) 418.

5. J.H. Lowenstein and B. Schroer, Phys. Rev. D7 (1973) 1929.
6. S. Ferrara and B. Zumino, Nucl. Phys. B87 (1975) 207.
7. T.E. Clark, O. Piguet and K. Sibold, Nucl. Phys. B143 (1978) 445.
8. T.E. Clark, O. Piguet and K. Sibold, Nucl. Phys. B159 (1979) 1.
9. T.E. Clark, O. Piguet and K. Sibold, The Gauge Invariance of the Supercurrent in SQED, Los Alamos preprint LA-UR 80 228 (1980).

LIST OF AUTHORS

PARTICIPANTS

Yoav ACHIMAN	Physics Department University of Wuppertal D-56 WUPPERTAL, Germany
Peter C. AICHELBURG	Institut für Theoretische Physik Boltzmanngasse 5 A-1090 WIEN, Austria
Richard ARNOWITT	Northeastern University BOSTON, MA 02115, U.S.A.
Stephen AVIS	DAMTP Silver Street CAMBRIDGE CB3 9EW, U.K.
Riccardo BARBIERI	Theory Division CERN CH-1211 GENEVA 23, Switzerland
Stephen BEDDING	Max Planck Institute für Physik Fohringer Ring 6, D-8000 MUNCHEN 40, Germany
Enrico BELLOTTI	Istituto di Fisica Via Celoria, 16 I-20133 MILANO, Italy
Haluk BERKMEN	Middle East Tech. University ANKARA, Turkey
Peter BREITENLOHNER	Max Planck Institut für Physik Fohringer Ring 6 D-8000 MUNCHEN 40, Germany
Lars BRINK	Institute of Theoretical Physics Chalmers Tekniska Hogskola S-41296, GOTEBORG, Sweden

Franco BUCCELLA	Istituto di Fisica Guglielmo Marconi Piazzale Aldo Moro 5 I-00185 ROMA, Italy
Ali CHAMSEDDINE	Theory Division CERN CH-1211 GENEVA, Switzerland
Eugene CREMMER	Laboratoire de Physique Théorique Ecole Normale Supérieure 24 rue Lhomond F-75231 PARIS, France
Riccardo D'AURIA	Istituto di Fisica Corso Massimo D'Azeglio 46 I-10125 TORINO, Italy
Stanley DESER	Brandeis University WALTHAM, MA 02254, U.S.A.
Bernard DE WIT	NIKHEF-H NL-1018 AMSTERDAM, The Netherlands
B. DOWNES-MARTIN	Niels Bohr Institute Blegdamsvej 17 DK-2100 COPENHAGEN, Denmark
Michael DUFF	Imperial College Physics Department LONDON SW7, U.K.
Martin B. EINHORN	Nordita Blegdamsvej 17 DK-2100 COPENHAGEN, Denmark
John ELLIS	Theory Division CERN CH-1211 GENEVA 23, Switzerland
Robert ELLIS	Melbourne University PARKVILLE, Victoria, Australia
Pierre FAYET	Laboratoire de Physique Théorique Ecole Normale Supérieure 24 rue Lhomond F-75231 PARIS, France
Sergio FERRARA	Laboratoire de Physique Théorique Ecole Normale Supérieure 24 rue Lhomond F-75231 PARIS, France

Jan FINJORD — Institut für Theoretische Physik
Sidlerstrasse 5
CH-3012 BERN, Switzerland

Pietro FRE — Istituto di Fisica
Corso Massimo D'Azeglio 46
I-10125 TORINO, Italy

Daniel FREEDMAN — State University at New York
STONY BROOK, NY 11194 , U.S.A.

Mary K. GAILLARD — LAPP
Chemin de Bellevue
F-74019 ANNECY-LE-VIEUX, France

Raoul GATTO — Dept. de Physique Théorique
Univ. de Genève
32 Bd d'Yvoy
CH-1211 GENEVA, Switzerland

Michel GOURDIN — LPTHE
Université Pierre et Marie Curie
4 Place Jussieu
F-75230 PARIS, France

Richard GRIMM — Institut für Theoretische Physik
Universität Karlsruhe
D-75 KARLSRUHE, Germany

Marc GRISARU — Brandeis University
WALTHAM, MA 02254, U.S.A.

P. HUT — Astronomical Institute
Roetersstraat 15
NL-1018 AMSTERDAM, The Netherlands

Zoltan KUNSZT — DESY
Notkestr. 85
D-2000 HAMBURG, Germany

D.R. Tim JONES — Rutherford Laboratory
CHILTON, DIDCOT, Oxon, U.K.

Ulf LINDSTROM — Inst. of Theoretical Physics
University of Stockholm
S-11346 STOCKHOLM, Sweden

Jerzy A. LUKIERSKI — Institute of Theoretical Physics
Cybulskiego 36
WROCLAW, Poland

Demetres NANOPOULOS	Theory Division CERN CH-1211 GENEVA 23, Switzerland
Yuval NE'ÈMAN	Tel Aviv University TEL AVIV, Israel
Hermann NICOLAI	Theory Division CERN CH-1211 GENEVA 23, Switzerland
David OLIVE	Imperial College LONDON SW7, U.K.
Fabrizio PALUMBO	Laboratori Nazionale Via E. Fermi I-00044 FRASCATI, Italy
Giorgio PARISI	INFN I-00044 FRASCATI, Italy
Jogesh PATI	International Centre for Theoretical Physics Miramare I-34100 TRIESTE, Italy
Pran NATH	Northeastern University BOSTON, MA 02215, U.S.A.
Subhash RAJPOOT	Imperial College Prince Consort Road LONDON SW7, U.K.
Finn RAVNDAL	Department of Physics University of Oslo OSLO 3, Norway
Martin ROCEK	DAMTP Silver Street CAMBRIDGE CB3 9EW, U.K.
Douglas ROSS	California Institute of Technology PASADENA, CA 91125, U.S.A.
Christopher SACHRAJDA	University of Southampton SOUTHAMPTON SO9 5NH, U.K.
Abdus SALAM	International Centre for Theoretical Physics Miramare I-34100 TRIESTE, Italy

Gianfranco SARTORI — Dept de Physique Théorique
Univ. de Genève
Bd. d'Yvoy 32
CH-1211 GENEVA 4, Switzerland

Carlos SAVOY — Dept de Physique Théorique
Univ. de Genève
Bd. d'Yvoy 32
CH-1211 GENEVA 4, Switzerland

Thomas SCHUCKER — Dept de Physique Théorique
Univ. de Genève
Bd. d'Yvoy 32
CH-1211 GENEVA 4, Switzerland

Antonino SCIARRINO — Università di Napoli
Istituto de Fisica Teorica
Mostra d'Oltremare - Pad 19
I-80125 NAPOLI, Italy

David M. SCOTT — DAMTP
Silver Street
CAMBRIDGE CB3 9EW, U.K.

Marc SHER — University of Colorado
Department of Physics
BOULDER, CO 80309, U.S.A.

Klaus SIBOLD — Inst. Theor. Physics
University of Karlsruhe
D-75 KARLSRUHE, Germany

Joel SCHERK — Laboratoire de Physique Théorique
Ecole Normale Supérieure
24 rue Lhomond
F-75231 PARIS, France

Martin SOHNIUS — Imperial College
Prince Consort Road
LONDON SW7, U.K.

Berthold STECH — University of Heidelberg
Theoretical Institut
D-69 HEIDELBERG, Germany

Gary STEIGMAN — Bartol Research Foundation
University Delaware
NEWARK, DE 19711, U.S.A.

Kellogg S. STELLE
Imperial College
Department of Physics
LONDON SW7, U.K.

Larry R. SULAK
Physics Department
University of Michigan
ANN ARBOR, MI 48104, U.S.A.

Leonard SUSSKIND
Stanford University
Physics Department
STANFORD, CA 94305, U.S.A.

Paul K. TOWNSEND
Theory Division
CERN
CH-1211 GENEVA 23, Switzerland

Peter VAN NIEUWENHUIZEN
State University of New York
STONY BROOK, NY 11794, U.S.A.

A. VAN PROEYEN
K.U. Leuven
Celestijnenlann 200 D
B-3030 LEUVEN, Belgium

Peter C. WEST
King's College
Mathematics Department
LONDON WC2, U.K.

Christof WETTERICH
Univ. Freiburg
Herm-Herder-Str. 3
D-78 FREIBURG, Germany

Bruno ZUMINO
Theory Division
CERN
CH-1211 GENEVA 23, Switzerland

INDEX